AUTOMATIC SEQUENCES

Uniting dozens of disparate results from different fields, this book combines concepts from mathematics and computer science to present the first integrated treatment of sequences generated by the simple model of computation called the finite automaton.

The authors develop the theory of automatic sequences and their generalizations, such as Sturmian words and k-regular sequences. Further, they discuss applications to number theory (particularly formal power series and transcendence in finite characteristic), physics, computer graphics, and music.

Results are presented from first principles wherever feasible, and the book is supplemented by a collection of 460 exercises, 85 open problems, and over 1600 citations to the literature. Thus this book is suitable for graduate students or advanced undergraduates, as well as for mature researchers wishing to know more about this fascinating subject.

Jean-Paul Allouche is Directeur de Recherche at CNRS, LRI, Orsay. He has written some 90 papers in number theory and combinatorics on words. He is on the editorial board of *Advances in Applied Mathematics* and on the scientific committee of the *Journal de Théorie des Nombres de Bordeaux*.

Jeffrey Shallit is Professor of Computer Science at the University of Waterloo. He has written 80 articles on number theory, algorithms, formal languages, combinatorics on words, computer graphics, history of mathematics, algebra and automata theory. He is the editor-in-chief of the *Journal of Integer Sequences* and coauthor of *Algorithmic Number Theory*.

AUTOMATIC SEQUENCES

Theory, Applications, Generalizations

JEAN-PAUL ALLOUCHE

CNRS, LSI, Orsay

JEFFREY SHALLIT

University of Waterloo

CAMBRIDGE
UNIVERSITY PRESS

CAMBRIDGE
UNIVERSITY PRESS

32 Avenue of the Americas, New York NY 10013-2473, USA

Cambridge University Press is part of the University of Cambridge.

It furthers the University's mission by disseminating knowledge in the pursuit of
education, learning and research at the highest international levels of excellence.

www.cambridge.org
Information on this title: www.cambridge.org/9780521823326

First published 2003

A catalogue record for this publication is available from the British Library

Library of Congress Cataloguing in Publication data
Allouche, Jean-Paul, 1953-
Automatic sequences : theory, applications, generalizations / Jean-Paul Allouche, Jeffrey Shallit.
p. cm.
Includes bibliographical references and index.
ISBN 0-521-82332-3
1. Sequential machine theory. 2. Sequences (Mathematics) I. Shallit, Jeffrey Outlaw. II. Title.
QA267.5.S4 A55 2003
515'.24 – dc21 2002041262

ISBN 978-0-521-82332-6 Hardback

Nous dédions ce livre à Michel Mendès France
en signe de gratitude et d'amitié

Contents

Preface

Goals of This Book

Sequences, both finite and infinite, are ubiquitous in mathematics and theoretical computer science. Sloane and Plouffe's book, *The Encyclopedia of Integer Sequences*, lists over 5,000 interesting sequences from the mathematical literature. Sloane's web site,

http://www.research.att.com/~njas/sequences/index.html

gives access to more than 69,000 sequences. There is a web-based scholarly journal, the *Journal of Integer Sequences*, devoted to sequence-related topics, and even a periodic international conference, SETA (Sequences and Their Applications), devoted to the study of sequences.

Sequences come in all flavors. Some, such as periodic sequences, are highly ordered and very easy to describe, while others, such as random sequences, are unordered and have no simple description.

The subject of this book is *automatic sequences* and their generalizations. Automatic sequences form a class of sequences somewhere between simple order and chaotic disorder. This class contains such celebrated sequences as the Thue–Morse sequence (see Chapters 1 and 6) and the Rudin–Shapiro sequence (see Chapter 3), which play important roles in many different areas of mathematics.

Automatic sequences are generated by finite automata, one of the most basic models of computation. Finite automata and other computational models are introduced in Chapter 5. Automatic sequences are also generated by iterating a simple kind of map, called a uniform morphism; see Chapter 6. By generalizing this to arbitrary morphisms, we obtain another interesting class of sequences called morphic sequences, which are discussed in Chapter 7.

Other generalizations of automatic sequences discussed in this book include multidimensional sequences (Chapter 14), sequences over infinite alphabets (Chapter 16), and sequences that are generated by slowly growing automata (Chapter 15).

One of the main reasons to study automatic sequences and their generalizations is the large number of interesting connections with number theory. To cite the most prominent example, methods of automata theory have recently been applied to prove new results in transcendence theory in positive characteristic; see Chapter 12.

While hundreds of papers discussing the relationship between automata theory and number theory have appeared in the literature, up to now there has been little attempt to bring these results together in any sort of consistent framework, using a unified notation. Books on automata theory rarely discuss results with a number-theoretic flavor, and when they do, these applications are often relegated to footnotes. On the other hand, the techniques of theoretical computer science are rarely incorporated in books on number theory, since they require unfamiliar language and notation.

Because our subject incorporates results from both mathematics and computer science, papers are scattered widely in the literature and often use inconsistent notation. Sometimes important results have appeared in obscure journals or remained unpublished because they did not find a home in more mainstream journals devoted to pure mathematics or theoretical computer science. Furthermore, since many researchers in the area are French, some important results have appeared only in the French language, making them less accessible to non-Francophones. Many of these results appear in this book in English for the first time.

We have attempted to present the material in as self-contained a way as feasible. Unfortunately, some results, such as Roth's theorem and Ridout's theorem, require rather detailed and complicated proofs, and we have chosen to omit the proofs.

Since this book is intended as an introduction, we do not always present results in the most general possible formulation. For example, in Chapter 9 we focus on characteristic words, and do not prove many theorems on the more general case of Sturmian words. Sometimes results are presented largely for their illustrative and pedagogical value. In particular, material in this book intersects with symbolic dynamics and ergodic theory, as well as other fields, but this book is not intended to be an introduction to those fields.

Each chapter ends with sections entitled Exercises and Notes. Some exercises are very easy, while the solution of others is a significant accomplishment. (Indeed, researchers should not be insulted if they find their own favorite results listed as exercises.) Exercises are arranged more or less randomly, with order having no implication for difficulty. Hints and solutions to selected exercises, as well as references, can be found in the Appendix. The Notes sections provide the reader with a detailed set of over 1600 references to pursue further work. Finally, dozens of unsolved research problems are listed in a section of each chapter entitled Open Problems.

Prerequisites

We hope the material in this book will be useful to readers at many levels, from advanced undergraduates to experts in the area. Experts may want to turn immediately to the new results in Chapters 12 and 13, for example, while novices may first need the background material in Chapters 1–5.

The main prerequisite is a degree of mathematical sophistication. Familiarity with the basic concepts of formal languages and number theory will be useful, but not absolutely essential. We have attempted to make the material self-contained wherever feasible.

A typical graduate or advanced undergraduate course on formal languages and number theory might cover the material in Chapters 1–6, 9, 12, and 13. Readers familiar with number theory and algebra might plan to skip Chapter 2, while those familiar with theoretical computer science would skip Chapter 4.

Algorithm Descriptions

Algorithms in this book are described in a pseudocode notation similar to Pascal or C, which should be familiar to most readers. We do not provide a formal definition of this notation.

Acknowledgments

We learned much of this material from Michel Mendès France, our good friend and colleague. We also benefited greatly from conversations with Jean Berstel, James Currie, David Damanik, Will Gilbert, and Luca Zamboni.

Drew Vandeth typed a very early version of the manuscript, suggested many improvements and changes, drew some of the diagrams, and wrote solutions to some of the exercises. Jennifer Keir provided expert TEX advice. Andy Poon verified many of the references, and found many typographical errors. Eric Bach, Céline Barberye, Jean Berstel, Valérie Berthé, Richard Crandall, Larry Cummings, Michael Domaratzki, Anna Frid, Marton Kósa, Bryan Krawetz, Scott Lindhurst, Andrew Martinez, Michel Mendès France, Simon Plouffe, Joe Rideout, Patrice Séébold, Gentcho Skordev, Troy Vasiga, Ming-wei Wang, David Wilson, and David Yeung were kind enough to read drafts of this book; they pointed out many errors and suggested many improvements. We are grateful to all of them. We also owe a huge debt to Jia-Yan Yao, who performed a Herculean task by reading every chapter with great care and sending us 67 pages of corrections. Of course, any errors that remain are our responsibility. Doug Bowman generously shared with us some of his unpublished results for use as exercises. We also thank Christophe Genolini for his assistance in creating Figures 17.1 and 17.2. Finally, we thank Mikael Kristiansen and Thor Bak for sharing their knowledge about the work of composer Per Nørgård with us.

The authors would appreciate hearing about any errors, no matter how trivial. Current errata for the book can be found in Shallit's home page;

http://www.math.uwaterloo.ca/~shallit/

is the URL.

This work was supported in part by grants from the Natural Sciences and Engineering Research Council (Canada). Part of this book was prepared during several visits of the first author to the University of Waterloo; he thanks that institution for its assistance. Part of this book was prepared while the second author enjoyed the kind hospitality of the Departments of Mathematics and Computer Science at the University of Arizona; he thanks John Brillhart and Peter Downey for their assistance.

Jean-Paul Allouche
Paris, France; September 2002
Jeffrey Shallit
Waterloo, Ontario; September 2002

1

Stringology

In this chapter we introduce the basic objects of interest to this book: finite and infinite words. A set of words forms a language, a concept introduced in Section 1.3. Morphisms, discussed in Section 1.4, provide a way to transform words. Two of the basic theorems on words – the theorems of Lyndon and Schützenberger – are discussed in Section 1.5.

Repetitions in words are introduced in Section 1.6. Section 1.7 discusses the particular case of binary words avoiding a certain type of repetition called an overlap; this section is rather technical and can be omitted on a first reading. Finally, Section 1.8 briefly introduces some additional topics about repetitions.

1.1 Words

One of the fundamental mathematical objects we will study in this book is the *word*. A word is made up of *symbols* (or *letters*), which are usually displayed in a typewriter-like font like `this`. (We treat the notion of symbol as primitive and do not define it further.) Let Σ denote a nonempty set of symbols, or *alphabet*; in this book, Σ will almost always be finite. One alphabet is so important that we give it a special symbol: if k is an integer ≥ 2, then we define

$$\Sigma_k = \{0, 1, \ldots, k - 1\}.$$

Note that we sometimes identify symbols with the integers they represent, so that, for example,

$$\Sigma_2 = \{0, 1\} = \{0, 1\}.$$

We typically denote variables whose domain is Σ using the lowercase italic letters a, b, c, d. A *word* or *string* (we use the terms interchangeably) is a finite or infinite list of symbols chosen from Σ. Although we usually denote words by simply juxtaposing their symbols, such as 3245, for clarity (particularly when negative integers are involved) we sometimes write them using an explicit concatenation operator, e.g., Concat(3, 2, 4, 5). If unspecified, a word is assumed to be finite.

1

We typically use the lowercase italic letters s, t, u, v, w, x, y, z to represent finite words.

More precisely, let $[m..n]$ denote the set of integers $\{m, m + 1, \ldots, n\}$. Then a finite word is a map from either $[0..n - 1]$ or $[1..n]$ to Σ. (The choice of the initial index gives us a little flexibility in defining words.) If $n = 0$, we get the *empty word*, which we denote by ϵ. The set of all finite words made up of letters chosen from Σ is denoted by Σ^*. For example, if $\Sigma = \{a, b\}$, then $\Sigma^* = \{\epsilon, a, b, aa, ab, ba, bb, aaa, \ldots\}$. We let Σ^+ denote the set of all nonempty words over Σ.

If w is a finite word, then its *length* (the number of symbols it contains) is denoted by $|w|$. For example, if $w = \texttt{five}$, then $|w| = 4$. Note that $|\epsilon| = 0$. We can also count the occurrences of a particular letter in a word. If $a \in \Sigma$ and $w \in \Sigma^*$, then $|w|_a$ denotes the number of occurrences of a in w. Thus, for example, if $w = \texttt{abbab}$, then $|w|_a = 2$ and $|w|_b = 3$.

One of the fundamental operations on words is *concatenation*. We concatenate two finite words w and x by juxtaposing their symbols, and we denote this by wx. For example, if $w = \texttt{book}$ and $x = \texttt{case}$, then $wx = \texttt{bookcase}$. Concatenation of words is, in general, not commutative; for example, we have $xw = \texttt{casebook}$. However, concatenation is associative: we have $w(xy) = (wx)y$ for all words w, x, y. Notationally, concatenation is treated like multiplication, so that w^n denotes the word $www \cdots w$ (n times). Note that the set Σ^* together with concatenation becomes an algebraic structure called a *monoid*, with the empty word ϵ playing the part of the identity element.

We say a word y is a *subword* or *factor* of a word w if there exist words x, z such that $w = xyz$. We say x is a *prefix* of w if there exists y such that $w = xy$. We say x is a *proper prefix* of w if $y \neq \epsilon$. We say that z is a *suffix* of w if there exists y such that $w = yz$. If $w = a_1 a_2 \cdots a_n$, then for $1 \le i \le n$, we define $w[i] = a_i$. If $1 \le i \le n$ and $i - 1 \le j \le n$, we define $w[i..j] = a_i a_{i+1} \cdots a_j$. Note that $w[i..i] = a_i$ and $w[i..i - 1] = \epsilon$.

A *language* over Σ is a (finite or infinite) set of words – that is, a subset of Σ^*.

Example 1.1.1 The following are examples of languages:

$$\texttt{PRIMES2} = \{\texttt{10}, \texttt{11}, \texttt{101}, \texttt{111}, \ldots\}$$

(the primes expressed in base 2),

$$\texttt{EQ} = \{x \in \{0, 1\}^* : |x|_0 = |x|_1\}$$

(words containing an equal number of each symbol).

We now define infinite words (or infinite sequences – we use the terms interchangeably). We let $\mathbb{Z}$ denote the integers, $\mathbb{Z}^+$ denote the positive integers, $\mathbb{Z}^-$ denote the negative integers, and $\mathbb{N}$ denote the non-negative integers. Then we will usually take a *one-sided* (or *unidirectional*) right-infinite word $\mathbf{a} = a_0 a_1 a_2 \cdots$ to be a map from $\mathbb{N}$ to Σ. We can form an infinite word by concatenating infinitely many finite words; for example,

$$\prod_{i \geq 1} w_i$$

denotes a word $w_1 w_2 w_3 \cdots$, which is infinite if and only if $w_i \neq \epsilon$ infinitely often.

Example 1.1.2 The following is an example of a right-infinite word:

$$\mathbf{q} = (q_n)_{n \geq 0} = 11001000010000001 \cdots,$$

where $q_n = 1$ if n is a square and 0 otherwise. The sequence $\mathbf{q}$ is called the *characteristic sequence* of the perfect squares.

Sometimes, if subscripts become too cumbersome, we write $\mathbf{a} = a(0)a(1)$ $a(2) \cdots$ instead. Also, instead of beginning indices at 0, occasionally we will use a map from $\mathbb{Z}^+$ to Σ, beginning our indices at 1, as the following example shows.

Example 1.1.3 Define

$$\mathbf{p} = (p_n)_{n \geq 1} = 0110101000101 \cdots,$$

the characteristic sequence of the prime numbers.

The set of all one-sided right-infinite words over Σ is denoted by Σ^{ω}. We define $\Sigma^{\infty} = \Sigma^* \cup \Sigma^{\omega}$.

A *left-infinite word* $\cdots a_{-3} a_{-2} a_{-1}$ is a map from $\mathbb{Z}^-$ to Σ. The set of all left-infinite words is denoted by $^{\omega}\Sigma$.

A *two-sided* (or *bidirectional*) *infinite word* is a map from $\mathbb{Z}$ to Σ. Such a word is of the form $\cdots c_{-2} c_{-1} c_0 . c_1 c_2 c_3 \cdots$; the decimal point is a notational convention and not part of the word itself. We denote the set of all two-sided infinite words over Σ by $\Sigma^{\mathbb{Z}}$. In this book, infinite words are typically denoted in boldface. Unless otherwise indicated, infinite words are assumed to be one-sided and right-infinite.

We can produce one-sided infinite words from two-sided infinite words by ignoring the portion to the right or left of the decimal point. Suppose $\mathbf{w} = \cdots c_{-2} c_{-1} c_0 . c_1 c_2 c_3 \cdots$. We define

$$L(\mathbf{w}) = \cdots c_{-2} c_{-1} c_0,$$

a left-infinite word, and

$$R(\mathbf{w}) = c_1 c_2 c_3 \cdots,$$

a right-infinite word.

The notions of subword, prefix, and suffix for finite words have evident analogues for infinite words. Let $\mathbf{w} = a_0 a_1 a_2 \cdots$ be an infinite word. For $i \geq 0$ we define $\mathbf{w}[i] = a_i$. For $i \geq 0$ and $j \geq i - 1$, we define $\mathbf{w}[i..j] = a_i a_{i+1} \cdots a_j$. We also define $\mathbf{w}[i..\infty] = a_i a_{i+1} \cdots$. If

$$\lim_{n \to \infty} \frac{|\mathbf{w}[0..n-1]|_b}{n}$$

exists and equals r, then the *frequency* of the symbol b in $\mathbf{w}$ is defined to be r. We denote this frequency as $\mathrm{Freq}_b(\mathbf{w})$.

Example 1.1.4 Consider the word $\mathbf{q}$ from Example 1.1.2. Then $\mathrm{Freq}_0(\mathbf{q}) = 1$ and $\mathrm{Freq}_1(\mathbf{q}) = 0$.

Infinite words may be specified by the limit of a sequence of finite words. If $w_1, w_2, w_3, \ldots$ form a sequence of words with w_i a proper prefix of w_j for $i < j$, then $\lim_{n \to \infty} w_n$ is the unique infinite word of which $w_1, w_2, \ldots$ are all prefixes.

Let k be an integer ≥ 2. A *k-aligned subword* of an infinite word $\mathbf{x} = a_0 a_1 a_2 \cdots$ is a subword of the form $a_{ki} a_{ki+1} \cdots a_{ki+k-1}$ for some integer $i \geq 0$.

We can also concatenate a finite word on the left with an infinite word on the right, but not vice versa. Clearly we cannot concatenate two right-infinite or two left-infinite words, but it is possible to concatenate a left-infinite word with a right-infinite word; see below. If x is a nonempty finite word, then x^ω is the right-infinite word $xxx \cdots$. Such a word is called *purely periodic*. An infinite word $\mathbf{w}$ of the form $x\, y^\omega$ for $y \neq \epsilon$ is called *ultimately periodic*. If $\mathbf{w}$ is ultimately periodic, then we can write it in the form $x\, y^\omega$ for finite words x, y with $y \neq \epsilon$. Then x is called a *preperiod* of $\mathbf{w}$, and y is called a *period*. (Sometimes we abuse terminology by calling the length $|x|$ the preperiod and $|y|$ the period.) If $|x|$, $|y|$ are chosen as small as possible, then x is called the *least preperiod*, and y is called the *least period*.

If L is a language, we define

$$L^\omega = \{w_1 w_2 w_3 \cdots : w_i \in L \setminus \{\epsilon\} \text{ for all } i \geq 1\}.$$

Thus, for example, Σ_2^ω is the set of all right-infinite words over $\{0, 1\}$. Similarly, we define

$$^\omega L = \{\cdots w_{-2} w_{-1} w_0 : w_i \in L \setminus \{\epsilon\} \text{ for all } i \leq 0\}.$$

If w is a nonempty finite word, then by $w^{\mathbb{Z}}$ we mean the two-sided infinite word $\cdots www.www \cdots$. Using concatenation, we can join a left-infinite word $\mathbf{w} = \cdots c_{-2} c_{-1} c_0$ with a right-infinite word $\mathbf{x} = d_0 d_1 d_2 \cdots$ to form a new two-sided infinite word, as follows:

$$\mathbf{w}.\mathbf{x} := \cdots c_{-2} c_{-1} c_0 . d_0 d_1 d_2 \cdots .$$

If L is a language, we define

$$L^{\mathbb{Z}} := \{\cdots w_{-2} w_{-1} w_0 . w_1 w_2 \cdots : w_i \in L \setminus \{\epsilon\} \text{ for all } i \in \mathbb{Z}\}.$$

If $w = a_1a_2 \cdots a_n$ and $x = b_1b_2 \cdots b_n$ are finite words of the same length, then by $w \amalg x$ we mean the word $a_1b_1a_2b_2 \cdots a_nb_n$, the *perfect shuffle* of w and x. For example, clip $\amalg$ aloe = calliope. A similar definition can be given for infinite words.

If $w = a_1a_2 \cdots a_n$ is a finite word, then by w^R we mean the *reversal* of the word w, that is, $w^R = a_na_{n-1} \cdots a_2a_1$. For example, $(\text{drawer})^R = \text{reward}$. Note that $(wx)^R = x^Rw^R$. A word w is a *palindrome* if $w = w^R$. Examples of palindromes in English include deified, rotator, repaper, and redivider.

If $\mathbf{w} = a_0a_1a_2 \cdots$ is a one-sided right-infinite word, then we define the *shift map* $\mathcal{S}(\mathbf{w})$ to be the word $a_1a_2a_3 \cdots$. Similarly, for $k \geq 0$, we have $\mathcal{S}^k(\mathbf{w}) = a_ka_{k+1}a_{k+2} \cdots$. For $k < 0$, we define $\mathcal{S}^k(\mathbf{w}) = u\mathbf{w}$ for an arbitrarily chosen word u of length k. For two-sided infinite words and $k \in \mathbb{Z}$, we define

$$\mathcal{S}^k(\cdots a_{-2}a_{-1}a_0.a_1a_2a_3 \cdots) = a_{k-2}a_{k-1}a_k.a_{k+1}a_{k+2}a_{k+3} \cdots .$$

This notation is also extended to finite words, where for $k \geq 0$ we define

$$\mathcal{S}^k(a_0a_1 \cdots a_j) = \begin{cases} a_ka_{k+1} \cdots a_j & \text{if } 0 \leq k \leq j, \\ \epsilon, & \text{otherwise.} \end{cases}$$

If $\mathbf{w} = a_0a_1a_2 \cdots$ is an infinite word over Σ and $\mathbf{x} = b_0b_1b_2 \cdots$ is an infinite word over Δ, then by $\mathbf{w} \times \mathbf{x}$ we mean the infinite word $c_0c_1c_2 \cdots$ over $\Sigma \times \Delta$ defined by $c_i = (a_i, b_i)$. We also extend the notation $\times$ to apply to finite words of the same length.

If Σ is an ordered set, we can define an ordering on words of Σ^ω. We define a lexicographic order Σ^ω as follows: let $\mathbf{w} = a_1a_2a_3 \cdots$ and $\mathbf{x} = b_1b_2b_3 \cdots$. Define $\mathbf{w} < \mathbf{x}$ if there exists an index $i \geq 0$ such that $a_j = b_j$ for $j \leq i$ and $a_{i+1} < b_{i+1}$. A similar definition can be given for finite words of the same length.

Let $\mathbf{w} = a_0a_1a_2 \cdots$ be an infinite word over Σ, and let k be an integer ≥ 1. The *k-block compression* of $\mathbf{w}$, which we write as $\text{comp}(\mathbf{w}, k)$, is an infinite word $b_0b_1b_2 \cdots$ over the alphabet Σ^k defined by $b_i = (a_{ki}, a_{ki+1}, \ldots, a_{ki+k-1})$.

If L_1, L_2 are languages over Σ, we define $L_1L_2 = \{xy : x \in L_1, y \in L_2\}$.

1.2 Topology and Measure

Let A be a set, and let $\mathcal{T}$ be a collection of subsets of A. Recall that we say $(A, \mathcal{T})$ is a *topological space*, or just a *topology*, if

(i) $\emptyset$ and A are members of $\mathcal{T}$;
(ii) if $(X_i)_{i \in I}$ are members of $\mathcal{T}$, then so is $\bigcup_{i \in I} X_i$;
(iii) if $(X_i)_{1 \leq i \leq n}$ are members of $\mathcal{T}$ for some integer $n \geq 1$, then so is $\bigcap_{1 \leq i \leq n} X_i$.

The elements of $\mathcal{T}$ are called *open sets*. A subset $F \subseteq A$ is called *closed* if its complement $A \setminus F$ is open. A topology may be specified by providing a *base* B; this is a collection of open sets such that each element of $\mathcal{T}$ may be expressed as a union of elements of B. A topology may be also specified by providing a *sub-base*

D of B; this is a collection of open sets such that each element of B can be expressed as a nonempty finite intersection of elements of D.

Example 1.2.1 Let $A = \mathbb{R}$, and let the topology T be specified by letting B, a base, consist of the open intervals of the form (a, b) with $a, b \in \mathbb{R}$ and $a < b$.

Let Σ be a finite alphabet. We can specify a natural topology on Σ^ω, the set of one-sided right-infinite words over Σ, by specifying a sub-base D as follows:

$$D = \bigcup_{\substack{j \geq 0 \\ a \in \Sigma}} D_{j,a},$$

where $D_{j,a}$ consists of those words $\mathbf{w}$ such that $\mathbf{w}[j] = a$. Base elements, which are nonempty finite intersections of the $D_{j,a}$, are of the form $\Sigma^{i_1} a_1 \Sigma^{i_2} a_2 \cdots \Sigma^{i_j} a_j \Sigma^\omega$, where $j, i_1, i_2, \ldots, i_j \geq 0$ are integers and $a_1, a_2, \ldots, a_j \in \Sigma$. Such a set is called a *cylinder*.

Theorem 1.2.2 *The open sets in Σ^ω are precisely those sets of the form $L\Sigma^\omega$, with $L \subseteq \Sigma^*$.*

Proof. Since by definition the $D_{j,a}$ form a sub-base for the topology, every base element is of the form $\Sigma^{i_1} a_1 \Sigma^{i_2} a_2 \cdots \Sigma^{i_j} a_j \Sigma^\omega$, where $j, i_1, i_2, \ldots, i_j \geq 0$ are integers and $a_1, a_2, \ldots, a_j \in \Sigma$. Thus every base element is of the form $L\Sigma^\omega$, where $L = \Sigma^{i_1} a_1 \Sigma^{i_2} a_2 \cdots \Sigma^{i_j} a_j$. Now by definition each open set is a union of sets of the form $L_i \Sigma^\omega$. But $\bigcup_{i \in I} L_i \Sigma^\omega = (\bigcup_{i \in I} L_i) \Sigma^\omega$.

For the converse, we need to show that $L\Sigma^\omega$ is open. But $L\Sigma^\omega = \bigcup_{x \in L} x\Sigma^\omega$, and each element of the form $x\Sigma^\omega$ for $x \in \Sigma^*$ is clearly a base element. ∎

Let (X, T) be a topological space, and $A \subseteq X$. We say that $x \in X$ is a *limit point* of A if every open set containing x intersects $A \setminus \{x\}$. The set of all limit points of A is called the *derived set* and is sometimes written as A'. If $A = A'$, we say A is *perfect*.

Recall that a *metric* on a set A is a map $d : A \to \mathbb{R}^{\geq 0}$ such that

(i) for $x, y \in A$ we have $d(x, y) = 0$ if and only if $x = y$;
(ii) for $x, y \in A$ we have $d(x, y) = d(y, x)$;
(iii) for $x, y, z \in A$ we have the *triangle inequality* $d(x, z) \leq d(x, y) + d(y, z)$.

Here by $\mathbb{R}^{\geq 0}$ we mean the non-negative real numbers. The pair (A, d) is called a *metric space*.

A metric d induces a topology as follows: we take as a base the family of all open balls of the form $\{x \in A : d(x, y) < r\}$ for $y \in A$ and $r > 0$.

We can make Σ^ω into a metric space by defining

$$d(\mathbf{x}, \mathbf{y}) = \begin{cases} 0 & \text{if } \mathbf{x} = \mathbf{y}, \\ 2^{-n} & \text{otherwise,} \end{cases}$$

where $n = \min\{i : \mathbf{x}[i] \neq \mathbf{y}[i]\}$. Intuitively, two infinite sequences are "close together" if they agree on a long prefix. Note that in addition to the triangle inequality, d satisfies the stronger *ultrametric inequality*

$$d(\mathbf{x}, \mathbf{y}) \leq \max(d(\mathbf{x}, \mathbf{z}), d(\mathbf{y}, \mathbf{z}))$$

for all $\mathbf{x}, \mathbf{y}, \mathbf{z} \in \Sigma^\omega$. It is not difficult to see that the topology induced by d is the same as the topology mentioned above.

The *closure* of a set $X \subseteq \Sigma^\omega$ is defined to be the intersection of all closed subsets of Σ^ω containing X; it is denoted by $\mathrm{Cl}(X)$. Alternatively, $\mathbf{w} \in \mathrm{Cl}(X)$ if for all real $\delta > 0$ there exists $\mathbf{x} \in X$ such that $d(\mathbf{w}, \mathbf{x}) < \delta$.

Theorem 1.2.3 *Let $X \subseteq \Sigma^\omega$, and let $\mathbf{w} \in \Sigma^\omega$. Then $\mathbf{w} \in \mathrm{Cl}(X)$ if and only if every prefix of $\mathbf{w}$ is the prefix of some word in X.*

Proof. We have $\mathbf{w} \in \mathrm{Cl}(X)$ if and only if for all $k \geq 0$ there exists $\mathbf{x} \in X$ with $d(\mathbf{w}, \mathbf{x}) \leq 2^{-k}$, if and only if for all $k \geq 0$ there exists $\mathbf{x} \in X$ which agrees with $\mathbf{w}$ on the first k terms. ∎

We can extend the metric d to Σ^∞ by introducing a new symbol b, not in Σ, and identifying each finite word w with the right-infinite word $wb^\omega \in (\Sigma \cup \{b\})^\omega$.

Finally, we can put a measure m on Σ^ω by defining the measure of the cylinders

$$m(\Sigma^{i_1} a_1 \Sigma^{i_2} a_2 \cdots \Sigma^{i_j} a_j \Sigma^\omega) = k^{-j},$$

where $k = \mathrm{Card}\,\Sigma$.

1.3 Languages and Regular Expressions

As we have seen above, a language over Σ is a subset of Σ^*. Languages may be of finite or infinite cardinality. We start by defining some common operations on languages.

Let $L, L_1, L_2 \subseteq \Sigma^*$ be languages. Recall that we define the *product of languages* by

$$L_1 L_2 = \{wx : w \in L_1, x \in L_2\}.$$

We define $L^0 = \{\epsilon\}$ and define L^i as LL^{i-1} for $i \geq 1$. We define

$$L^{\leq i} = L^0 \cup L^1 \cup \cdots \cup L^i.$$

We define L^* as $\bigcup_{i \geq 0} L^i$; the operation L^* is sometimes called *Kleene closure*. We define $L^+ = LL^*$; the operation $+$ in the superscript is sometimes called *positive* *closure.* If L is a language, then the *reversed* language is defined as follows: $L^R = \{x^R : x \in L\}$. Finally, we define the *quotient* of languages as follows:

$$L_1/L_2 = \{x \in \Sigma^* : \exists\, y \in L_2 \text{ such that } xy \in L_1\}.$$

We now turn to a common notation for representing some kinds of languages. A *regular expression* over the alphabet Σ is a well-formed word over the alphabet

$$\Sigma \cup \{\epsilon,\ \emptyset,\ (,\ \ ),\ \ +,\ \ *\}.$$

(Exercise 64 makes this more precise.) In evaluating such an expression, $*$ represents Kleene closure and has highest precedence. Concatenation is represented by juxtaposition, and has next highest precedence. Finally, $+$ represents union and has lowest precedence. Parentheses are used for grouping.

If the word u is a regular expression, then $L(u)$ represents the language that u specifies. For example, consider the regular expression $u = (0+10)*(1+\epsilon)$. Then $L(u)$ represents all finite words of 0's and 1's that do not contain two consecutive 1's. Frequently we will abuse the notation by referring to the language as the naked regular expression without the surrounding $L()$. A language L is said to be *regular* if $L = L(u)$ for some regular expression u.

Theorem 1.3.1 *Every finite language is regular.*

Proof. If $L = \{w_1, w_2, \ldots, w_i\}$, then a regular expression for L is just $w_1 + w_2 + \cdots + w_i$. ∎

1.4 Morphisms

In this section we introduce a fundamental tool of formal languages, the *homomorphism*, or just *morphism* for short. Let Σ and Δ be alphabets. A morphism is a map h from Σ^* to Δ^* that obeys the identity $h(xy) = h(x)h(y)$ for all words $x, y \in \Sigma^*$. Typically, we use the Latin letters f, g, h and Greek letters $\varphi, \theta, \mu, \sigma, \rho$ to denote morphisms.

Clearly if h is a morphism, then we must have $h(\epsilon) = \epsilon$. Furthermore, once h is defined for all elements of Σ, it can be uniquely extended to a map from Σ^* to Δ^*. Henceforth, when we define a morphism, we will always give it by specifying its action on Σ.

Example 1.4.1 Let $\Sigma = \{e, m, o, s\}$, let $\Delta = \{a, e, l, n, r, s, t\}$, and define

$$h(m) = \text{ant},$$
$$h(o) = \epsilon,$$
$$h(s) = \text{ler},$$
$$h(e) = \text{s}.$$

Then $h(\text{moose}) = \text{antlers}$.

If $\Sigma = \Delta$, then we can iterate the application of h. We define $h^0(a) = a$ and $h^i(a) = h(h^{i-1}(a))$ for all $a \in \Sigma$.

Example 1.4.2 Let $\Sigma = \Delta = \{0, 1\}$. Define the *Thue–Morse morphism* $\mu(0) = $ 01 and $\mu(1) = $ 10. Then $\mu^2(0) = $ 0110 and $\mu^3(0) = $ 01101001.

There are various parameters associated with a morphism $h : \Sigma^* \to \Delta^*$. We define Width $h = \max_{a \in \Sigma} |h(a)|$, Depth $h = $ Card Σ, and Size $h = \sum_{a \in \Sigma} |h(a)|$.

We can classify morphisms into different groups, as follows: if there is a constant k such that $|h(a)| = k$ for all $a \in \Sigma$, then we say that h is k-*uniform* (or just *uniform*, if k is clear from the context). A 1-uniform morphism is called a *coding*. We typically use the Greek letters τ and ρ to denote codings. A morphism is said to be *expanding* if $|h(a)| \geq 2$ for all $a \in \Sigma$.

If $h(a) \neq \epsilon$ for all $a \in \Sigma$, then h is *nonerasing*. If $h(a) = \epsilon$ for all $a \in \Sigma$, then we say h is *trivial*. If there exists an integer $j \geq 1$ such that $h^j(a) = \epsilon$, then the letter a is said to be *mortal*. The set of mortal letters associated with a morphism h is denoted by M_h. The *mortality exponent* of a morphism h is defined to be the least integer $t \geq 0$ such that $h^t(a) = \epsilon$ for all $a \in M_h$. (If $M_h = \emptyset$, we take $t = 0$.) We write the mortality exponent as $\exp(h) = t$. It is easy to prove that $\exp(h) \leq $ Card M_h; see Exercise 3.

We also define the notion of *inverse homomorphism* of languages. Given $h : \Sigma^* \to \Delta^*$ and a language L, we define

$$h^{-1}(L) = \{x \in \Sigma^* : h(x) \in L\}.$$

We can also apply morphisms to infinite words. If $\mathbf{w} = c_0 c_1 c_2 \cdots$ is a right-infinite word, then we define

$$h(\mathbf{w}) = h(c_0)h(c_1)h(c_2) \cdots.$$

If $\mathbf{w} = \cdots c_{-2} c_{-1} c_0 . c_1 c_2 \cdots$ is a two-sided infinite word, and h is a morphism, then we define

$$h(\mathbf{w}) := \cdots h(c_{-2})h(c_{-1})h(c_0).h(c_1)h(c_2) \cdots. \tag{1.1}$$

We now introduce the notion of a *primitive morphism*. A morphism $h : \Sigma^* \to \Sigma^*$ is said to be *primitive* if there exists an integer $n \geq 1$ such that for all $a, b \in \Sigma$, a occurs in $h^n(b)$.

One reason why primitive morphisms are of interest is that if h is primitive, then the growth rate of $|h^r(a)|$ is essentially independent of a. We have the following

Theorem 1.4.3 *Let $h : \Sigma^* \to \Sigma^*$ be a primitive morphism. Then there exists a constant C (which does not depend on n but may depend on* Width h *and* Depth h) *such that $|h^n(b)| \leq C|h^n(c)|$ for all $b, c \in \Sigma$ and all $n \geq 0$.*

Proof. Let $W = \text{Width } h$. Since h is primitive, there exists an integer $e \geq 1$ such that for all $b, c \in \Sigma$ we have $h^e(c) \in \Sigma^* b \Sigma^*$. Thus for $r \geq 1$ we have

$$
\begin{aligned}
|h^{er}(c)| &= |h^{e(r-1)}(h^e(c))| \\
&= |h^{e(r-1)}(xby)| \qquad \text{for some } x, y \in \Sigma^* \\
&\geq |h^{e(r-1)}(b)|.
\end{aligned}
$$

Also $|h^{er}(b)| = |h^e(h^{e(r-1)}(b))| \leq W^e|h^{e(r-1)}(b)|$. Putting these bounds together, we get $|h^{er}(b)| \leq W^e|h^{er}(c)|$.

Now write $n = er + i$ for some $r \geq 0$ and $0 \leq i < e$. If $r = 0$, we have $|h^i(b)| \leq W^i \leq W^i|h^i(c)|$. If $r \geq 1$, then

$$
\begin{aligned}
|h^{er+i}(b)| &\leq W^i|h^{er}(b)| \\
&\leq W^{i+e}|h^{er}(c)| \\
&\leq W^{i+e}|h^{er+i}(c)|.
\end{aligned}
$$

It follows that $|h^n(b)| \leq W^{2e-1}|h^n(c)|$, so we may take $C = W^{2e-1}$. ∎

Exercise 8.8 explores how big e can be. Note: Theorem 1.4.3 is made more precise in Proposition 8.4.1.

Let $h : \Sigma^* \to \Sigma^*$ be a morphism. A finite or infinite word w such that $h(w) = w$ is said to be a *fixed point* of h. If there exists a letter $a \in \Sigma$ such that $h(a) = ax$, and $x \notin M_h^*$, we say h is *prolongable* on a. In this case, the sequence of words $a, h(a), h^2(a), \ldots$ converges, in the limit, to the infinite word

$$
h^\omega(a) := a\, x\, h(x)\, h^2(x) \cdots,
$$

which is a fixed point of h, that is, $h(h^\omega(a)) = h^\omega(a)$. Furthermore, it is easy to see that $h^\omega(a)$ is the unique fixed point of h which starts with a. If $\mathbf{w} = h^\omega(a)$, then we call $\mathbf{w}$ a *pure morphic sequence*. If there is a coding $\tau : \Sigma \to \Delta$ and $\mathbf{w} = \tau(h^\omega(a))$, then we call $\mathbf{w}$ a *morphic sequence*.

1.5 The Theorems of Lyndon and Schützenberger

In this section, we prove two beautiful and fundamental theorems due to Lyndon and Schützenberger. We start with one of the simplest and most basic results on words, sometimes known as Levi's lemma:

Theorem 1.5.1 *Let $u, v, x, y \in \Sigma^*$, and suppose that $uv = xy$. If $|u| \geq |x|$, there exists $t \in \Sigma^*$ such that $u = xt$ and $y = tv$. If $|u| < |x|$, there exists $t \in \Sigma^+$ such that $x = ut$ and $v = ty$.*

Proof. Left to the reader. ∎

Now we can state the first theorem of Lyndon and Schützenberger:

Theorem 1.5.2 *Let $y \in \Sigma^*$ and $x, z \in \Sigma^+$. Then $xy = yz$ if and only if there exist $u, v \in \Sigma^*$ and an integer $e \geq 0$ such that $x = uv$, $z = vu$, and $y = (uv)^e u = u(vu)^e$.*

Proof. $\Longrightarrow$: The proof is by induction on $|y|$. If $|y| = 0$, then we can take $v = x = z$, $u = \epsilon$, and $e = 0$. Thus suppose that $|y| \geq 1$. There are two cases.

Case I: If $|x| \geq |y|$, then we have a situation like the following:

By Levi's lemma there exists $w \in \Sigma^*$ such that $x = yw$ and $z = wy$. Now take $u = y$, $v = w$, $e = 0$, and we are done.

Case II: Now suppose that $|x| < |y|$. Then we have a situation like the following:

By Levi's lemma there exists $w \in \Sigma^+$ such that $y = wz = xw$. By induction (since $|w| = |y| - |z| < |y|$), we know there exist u, v, e such that

$$x = uv,$$
$$z = vu,$$
$$w = (uv)^e u = u(vu)^e,$$

so it follows that $y = u(vu)^{e+1} = (uv)^{e+1} u$.

$\Longleftarrow$: We have

$$xy = uv(uv)^e u = (uv)^{e+1} u,$$
$$yz = u(vu)^e vu = u(vu)^{e+1},$$

and these words are identical. ∎

We now state and prove the second theorem of Lyndon and Schützenberger.

Theorem 1.5.3 *Let* $x, y \in \Sigma^+$. *Then the following three conditions are equivalent:*

(1) $xy = yx$.
(2) There exist integers $i, j > 0$ *such that* $x^i = y^j$.
(3) There exist $z \in \Sigma^+$ *and integers* $k, l > 0$ *such that* $x = z^k$ *and* $y = z^l$.

Proof. We show that (1) $\Longrightarrow$ (3), (3) $\Longrightarrow$ (2), and (2) $\Longrightarrow$ (1).

(1) $\Longrightarrow$ (3): By induction on $|xy|$. If $|xy| = 2$, then $|x| = |y| = 1$, so $x = y$ and we may take $z = x = y$, $k = l = 1$.

Now assume the implication is true for all x, y with $|xy| < n$. We prove it for $|xy| = n$. Without loss of generality, assume $|x| \geq |y|$. Then we have a situation like the following:

x	y

w

y	x

Hence there exists $w \in \Sigma^*$ such that $x = wy = yw$. If $|w| = 0$ then $x = y$ and we can take $z = x = y$, $k = l = 1$.

Otherwise $|w| \geq 1$. We have $|wy| = |x| < |xy| = n$, so the induction hypothesis applies, and there exists $z \in \Sigma^+$ and integers $k, l > 0$ such that $w = z^k$, $y = z^l$. It follows that $x = wy = z^{k+l}$.

(3) $\Longrightarrow$ (2): By (3) there exist $z \in \Sigma^+$ and integers $k, l > 0$ such that $x = z^k$ and $y = z^l$. Hence, taking $i = l$, $j = k$, we get

$$x^i = (z^k)^i = z^{kl} = (z^l)^k = (z^l)^j = y^j,$$

as desired.

(2) $\Longrightarrow$ (1): We have $x^i = y^j$. If $|x| = |y|$ then we must have $i = j$ and so $x = y$. Otherwise, without loss of generality assume $|x| > |y|$. Then we have a situation like the following:

x	x	x	x

w

y	y	y	y	y	y

That is, there exists $w \in \Sigma^+$ such that $x = yw$. Hence $x^i = (yw)^i = y^j$, and so $y(wy)^{i-1}w = y^j$. Therefore $(wy)^{i-1}w = y^{j-1}$, and so, by multiplying by y on the right, we get $(wy)^i = y^j$. Hence $(yw)^i = (wy)^i$, and hence $yw = wy$. It follows that $x = yw = wy$ and $xy = (yw)y = y(wy) = yx$. ∎

We now make the following definition: a word $w \in \Sigma^+$ is said to be *primitive* if there is no solution to $w = z^k$ for k an integer ≥ 2 and $z \in \Sigma^+$. For example, abcb is primitive, but abab $=$ ab^2 is not.

Theorem 1.5.4 *Every nonempty word w can be expressed uniquely in the form $w = x^n$, where $n \geq 1$ and x is primitive.*

Proof. Choose n as large as possible so that $w = x^n$ has a solution; clearly $1 \leq n \leq |w|$. We claim that the resulting x is primitive. For if not, we could write $x = y^k$ for some $k \geq 2$ and then $w = y^{kn}$, where $kn > n$.

To prove uniqueness, suppose w has two representations $w = x^n = y^m$, where both x, y are primitive and $n, m \geq 1$. Then by Theorem 1.5.3, there exists z with $|z| \geq 1$ such that $x = z^k$ and $y = z^\ell$. Since x, y are primitive, however, we must have $k = \ell = 1$. But then $x = y = z$, and hence $n = m$, and the two representations are actually the same. ■

We now recall the definition of the Möbius function,

$$\mu(n) = \begin{cases} 0 & \text{if } n \text{ is divisible by a square} > 1, \\ (-1)^j & \text{if } n = p_1 p_2 \cdots p_j \text{ where the } p_i \text{ are distinct primes.} \end{cases} \quad (1.2)$$

(Hopefully there will be no confusion with the morphism μ defined in Section 1.4.)

Theorem 1.5.5 *There are $\sum_{d \mid n} \mu(d) k^{n/d}$ distinct primitive words of length n over a k-letter alphabet.*

Proof. Let $\psi_k(n)$ denote the number of primitive words of length n over a k-letter alphabet. There are k^n distinct words of length n over a k-letter alphabet, and each such word w can, by the previous theorem, be represented uniquely in the form $w = x^d$, where $d \geq 1$. Clearly we must have $d \mid n$. Hence

$$k^n = \sum_{d \mid n} \psi_k(d),$$

and by Möbius inversion (Exercise 13) we have

$$\psi_k(n) = \sum_{d \mid n} \mu(d) k^{n/d}. \quad ■$$

The following theorem, due to Fine and Wilf, can be considered a generalization of Theorem 1.5.3:

Theorem 1.5.6 (Fine–Wilf) *Let $\mathbf{a} = a_0 a_1 a_2 \cdots$ (respectively $\mathbf{b} = b_0 b_1 b_2 \cdots$) be a purely periodic infinite word of period $m \geq 1$ (respectively $n \geq 1$), and suppose $\mathbf{a}$ and $\mathbf{b}$ agree on a prefix of length $m + n - \gcd(m, n)$. Then $\mathbf{a} = \mathbf{b}$.*

Proof. We assume that $\gcd(m, n) = 1$. We show how to remove this assumption at the end of the proof.

Since **a** and **b** are purely periodic, we have $a_i = a_{i+m}$ and $b_i = b_{i+n}$ for all $i \geq 0$. By hypothesis we also have $a_i = b_i$ for $0 \leq i < m + n - 1$. If $m = n = 1$ the result is trivial, so we may assume without loss of generality that $m > n$. Then for $0 \leq i < m - 1$ we have

$$a_i = b_i = b_{i+n} = a_{i+n}, \tag{1.3}$$

where the indices on a are taken mod m and the indices on b are taken mod n.

Starting with a_{n-1} and applying Eq. (1.3) $m - 1$ times, it follows that

$$a_{n-1} = a_{2n-1} = a_{3n-1} = \cdots = a_{(m-1)n-1} = a_{mn-1}, \tag{1.4}$$

where the indices are taken mod m. Since $\gcd(m, n) = 1$, it follows that all m indices (mod m) are represented in Eq. (1.4). Hence all the symbols of **a** equal a_0, and the same result holds for **b**.

Now we explain how to remove the assumption that $\gcd(m, n) = 1$. Let $g = \gcd(m, n)$. We assume $\mathbf{a}, \mathbf{b} \in \Sigma^\omega$. Now consider grouping the symbols of **a** and **b** into blocks of length g, i.e., we form the g-block compression $\mathbf{a}' = \mathrm{comp}(\mathbf{a}, g)$ and $\mathbf{b}' = \mathrm{comp}(\mathbf{b}, g)$. Then $\mathbf{a}'$ and $\mathbf{b}'$ are purely periodic infinite words of period m/g and n/g, respectively, over the new alphabet Σ^g. Now $\gcd(m/g, n/g) = 1$, so by the proof above we have $\mathbf{a}', \mathbf{b}'$ are both purely periodic of period 1. Hence $\mathbf{a}, \mathbf{b}$ are purely periodic of period g, and these words are identical. ∎

1.6 Repetitions in Words

A *square* is a word of the form xx, such as the English word `hotshots` $= (\text{hots})^2$. If w is a (finite or infinite) word containing no nonempty subword of this form, then it is said to be *squarefree*. Note that the word `square` is squarefree, while the word `squarefree` is not.

It is easy to verify (see Exercise 14) that there are no squarefree words of length > 3 over a two-letter alphabet. However, there are infinite squarefree words over a three-letter alphabet. We construct one below in Theorem 1.6.2.

Similarly, a *cube* is a word of the form xxx, such as the English sort-of-word `shshsh` (an admonition to be quiet). If w contains no nonempty cube, it is said to be *cubefree*. The word `cubefree` is not squarefree, since it contains two consecutive occurrences of the word `e`, but it is cubefree.

An *overlap* is a word of the form $cxcxc$, where x is a word, possibly empty, and c is a single letter. (The term "overlap" refers to the fact that such a word can be viewed as two overlapping occurrences of the word cxc.) The English word `alfalfa`, for example, is an overlap with $c = \text{a}$ and $x = \text{lf}$. If w contains no overlap, it is said to be *overlap-free*.

In this section, we prove some simple results in the theory of repetitions in words. We start by constructing an infinite squarefree word over a finite alphabet. Define

$$t_n = \begin{cases} 0 & \text{if the number of 1's in the base-2 expansion of } n \text{ is even,} \\ 1 & \text{if the number of 1's in the base-2 expansion of } n \text{ is odd.} \end{cases}$$

We define $\mathbf{t} = t_0 t_1 t_2 \cdots = 01101001 \cdots$. The infinite word $\mathbf{t}$ is usually called the *Thue–Morse* word, named after two of the first mathematicians to study its properties. We have the following theorem:

Theorem 1.6.1 *The Thue–Morse infinite word $\mathbf{t}$ is overlap-free.*

Proof. Observe that $t_{2n} = t_n$ and $t_{2n+1} = 1 - t_n$ for $n \geq 0$.

Assume, contrary to what we want to prove, that $\mathbf{t}$ contains an overlap. Then we would be able to write $\mathbf{t} = uaxaxav$ for some finite words u, x, an infinite word $\mathbf{v}$, and a letter a. In other words, we would have $t_{k+j} = t_{k+j+m}$ for $0 \leq j \leq m$, where $m = |ax|$ and $k = |u|$. Assume $m \geq 1$ is as small as possible. Then there are two cases: (i) m is even, and (ii) m is odd.

(i) If m is even, then let $m = 2m'$. Again there are two cases: (a) k is even, and (b) k is odd.

    (a) If k is even, then let $k = 2k'$. Then we know $t_{k+j} = t_{k+j+m}$ for $0 \leq j \leq m$, so it is certainly true that $t_{k+2j'} = t_{k+2j'+m}$ for $0 \leq j' \leq m/2$. Hence $t_{2k'+2j'} = t_{2k'+2j'+2m'}$ for $0 \leq j' \leq m'$, and so $t_{k'+j'} = t_{k'+j'+m'}$ for $0 \leq j' \leq m'$. But this contradicts the minimality of m.

    (b) If k is odd, then let $k = 2k' + 1$. Then as before we have $t_{k+2j'} = t_{k+2j'+m}$ for $0 \leq j' \leq m/2$. Hence $t_{2k'+2j'+1} = t_{2k'+2j'+2m'+1}$ for $0 \leq j' \leq m'$, and so $t_{k'+j'} = t_{k'+j'+m'}$ for $0 \leq j' \leq m'$, again contradicting the minimality of m.

(ii) If m is odd, then there are three cases: (a) $m \geq 5$, (b) $m = 3$, and (c) $m = 1$. For $n \geq 1$, we define $b_n = (t_n + t_{n-1}) \bmod 2$. (Here by $x \bmod y$ we mean the least non-negative remainder when x is divided by y.) Note that $b_{4n+2} = (t_{4n+2} + t_{4n+1}) \bmod 2$. Since the base-2 representations of $4n + 2$ and $4n + 1$ are identical, except that the last two bits are switched, we have $t_{4n+2} = t_{4n+1}$, and so $b_{4n+2} = 0$. On the other hand, $b_{2n+1} = (t_{2n+1} + t_{2n}) \bmod 2$, and the base-2 representations of $2n + 1$ and $2n$ are identical except for the last bit; hence $b_{2n+1} = 1$.

    (a) m odd, ≥ 5. We have $b_{k+j} = b_{k+j+m}$ for $1 \leq j \leq m$. Since $m \geq 5$, we can choose j such that $k + j \equiv 2 \pmod 4$. Then for this value of $k + j$, we have from above that $b_{k+j} = 0$, but $k + j + m$ is odd, so $b_{k+j+m} = 1$, a contradiction.

    (b) $m = 3$. Again, $b_{k+j} = b_{k+j+3}$ for $1 \leq j \leq 3$. Choose j such that $k + j \equiv 2$ or $3 \pmod 4$. If $k + j \equiv 2 \pmod 4$, then the reasoning of the previous case applies. Otherwise $k + j \equiv 3 \pmod 4$, and then $b_{k+j} = 1$, while $b_{k+j+3} = 0$.

    (c) $m = 1$. Then $t_k = t_{k+1} = t_{k+2}$. Hence $t_{2n} = t_{2n+1}$ for $n = \lceil k/2 \rceil$, a contradiction.

This completes the proof. ∎

Using the fact that **t** is overlap-free, we may now construct a squarefree infinite word over the alphabet $\Sigma_3 = \{0, 1, 2\}$.

Theorem 1.6.2 *For $n \geq 1$, define c_n to be the number of 1's between the nth and $(n+1)$st occurrence of 0 in the word* **t**. *Set* $\mathbf{c} = c_1 c_2 c_3 \cdots$. *Then* $\mathbf{c} = 210201 \cdots$ *is an infinite squarefree word over the alphabet Σ_3.*

Proof. First, observe that **c** is over the alphabet $\{0, 1, 2\}$. For if there were three or more 1's between two consecutive occurrences of 0 in **t**, then **t** would not be overlap-free, a contradiction.

Next, assume that **c** is not squarefree. Then it contains a square of the form xx, with $x = x_1 x_2 \cdots x_n$ and $n \geq 1$. Then, from the definition of **c**, the word **t** would contain a subword of the form

$$01^{x_1} 01^{x_2} 0 \cdots 01^{x_n} 01^{x_1} 01^{x_2} 0 \cdots 01^{x_n} 0,$$

which constitutes an overlap, a contradiction.   ∎

For alternate definitions of **c**, see Exercise 33.

1.7 Overlap-Free Binary Words

In this section we prove the remarkable result that an infinite binary word which is a fixed point of a morphism either has an overlap or is equal to **t**, the Thue–Morse infinite word, or its complement $\overline{\mathbf{t}}$. (Here the overbar is shorthand for the morphism that maps 0 to 1 and 1 to 0.) Recall that $\Sigma_2 = \{0, 1\}$, and recall from Example 1.4.2 the Thue–Morse morphism $\mu : \Sigma_2^* \to \Sigma_2^*$ by $\mu(0) = 01$, $\mu(1) = 10$. We first make the following easy remarks.

Let $u \in \Sigma_2^*$. There exists $v \in \Sigma_2^*$ such that $u = \mu(v)$ if and only if $u \in \{01, 10\}^*$.
Let $u, v \in \Sigma_2^*$. If $uv \in \{01, 10\}^*$ and $|u|$ is even, then $v \in \{01, 10\}^*$.

We start with some technical lemmas.

Lemma 1.7.1 *Suppose $y \in \Sigma_2^*$ and $a \in \Sigma_2$. If the word $a\,\overline{a}\,\overline{a}\,y$ is overlap-free, then at least one of the following holds:*

(a) y begins with aa;
(b) $|y| \leq 3$;
(c) y begins with $a\,\overline{a}\,a\,a$.

Proof. If y begins with aa or $|y| \leq 3$, then we are done. Hence assume y does not begin with aa and $|y| \geq 4$.

Now y cannot begin with $\overline{a}$, since $a\,\overline{a}\,\overline{a}\,y$ is assumed to be overlap-free. Hence y begins with a. By assumption y does not begin with aa, so it begins with $a\,\overline{a}$. Let $y = a\,\overline{a}\,z$, where $|z| \geq 2$. If z begins with $\overline{a}$, then $z = \overline{a}\,w$ for a nonempty word w, which implies $a\,\overline{a}\,\overline{a}\,y = a\,\overline{a}\,\overline{a}\,a\,\overline{a}\,\overline{a}\,w$. Now any choice for the first letter of w gives an overlap. Hence z must begin with a, say $z = a\,w$ for a nonempty word w. But then $a\,\overline{a}\,\overline{a}\,y = a\,\overline{a}\,\overline{a}\,a\,\overline{a}\,a\,w$, and the first letter of w must be a. ∎

Lemma 1.7.2 *If $y, y' \in \Sigma_2^*$, and if there exist $c, d \in \Sigma_2$ such that $u = c\mu(y) = \mu(y')d$, then $u = c\,(\overline{c}\,c)^{|y|}$.*

Proof. Clearly $|y| = |y'|$. Let $y = a_1 a_2 \cdots a_t$ and $y' = b_1 b_2 \cdots b_t$. Then we have

$$c\,a_1\,\overline{a_1}\,a_2\,\overline{a_2}\, \cdots \,a_{t-1}\,\overline{a_{t-1}}\,a_t\,\overline{a_t} = b_1\,\overline{b_1}\,b_2\,\overline{b_2}\,\cdots\,b_{t-1}\,\overline{b_{t-1}}\,b_t\,\overline{b_t}\,d.$$

Hence

$$
\begin{aligned}
b_1 &= c, \\
a_1 &= \overline{b_1} = \overline{c}, \\
b_2 &= \overline{a_1} = c, \\
a_2 &= \overline{b_2} = \overline{c}, \\
&\;\;\vdots \\
b_t &= \overline{a_{t-1}} = c, \\
a_t &= \overline{b_t} = \overline{c}, \\
\overline{a_t} &= d.
\end{aligned}
$$

Hence $c = d$ and so $u = c(\overline{c}\,c)^t$. ∎

Lemma 1.7.3 *Suppose $y, z \in \Sigma_2^*$ and $\mu(y) = zz$. Then there exists $x \in \Sigma_2^*$ such that $z = \mu(x)$.*

Proof. Suppose that $\mu(y) = zz$.

If $|z|$ is even, the result is clear, since then $|\mu(y)| \equiv 0 \pmod 4$, and hence $|y|$ is even. Hence $z = \mu(w)$, where w is the prefix of y of length $|y|/2$.

Let us show that $|z|$ cannot be odd. If it were, let $z = au = vb$, where $a, b \in \Sigma_2$ and u and v are words of even length. Then $\mu(y) = zz = vbau$; hence there exist $r, s \in \Sigma_2^*$ such that $u = \mu(r)$ and $v = \mu(s)$. Hence $zz = \mu(s)bau\mu(r)$, and $b = \overline{a}$. But $z = a\mu(r) = \mu(s)b$, hence by Lemma 1.7.2 we have $z = a\,(\overline{a}\,a)^{|r|}$. Then the last letter of z equals a and $\overline{a}$, a contradiction. ∎

We say a morphism h is overlap-free if the image under h of every finite overlap-free word is overlap-free. Our next lemma states that μ, the Thue–Morse morphism, is overlap-free.

Lemma 1.7.4 *Let $w \in \Sigma_2^*$ and $\mathbf{x} \in \Sigma_2^\omega$. Then*

(a) *w contains an overlap if and only if $\mu(w)$ contains an overlap;*
(b) *$\mathbf{x}$ contains an overlap if and only if $\mu(\mathbf{x})$ contains an overlap.*

Proof. (a) $\Longrightarrow$: Suppose w contains an overlap, say $w = x\,c\,y\,c\,y\,c\,z$ for $x, y, z \in \Sigma_2^*$ and $c \in \Sigma_2$. Then

$$\mu(w) = \mu(x)\,\mu(c)\,\mu(y)\,\mu(c)\,\mu(y)\,\mu(c)\,\mu(z) = \mu(x)\,c\,\bar{c}\,\mu(y)\,c\,\bar{c}\,\mu(y)\,c\,\bar{c}\,\mu(z),$$

and so $\mu(w)$ contains the overlap $c\,v\,c\,v\,c$, where $v = \bar{c}\,\mu(y)$.

$\Longleftarrow$: Now suppose that $\mu(w)$ contains an overlap, say $\mu(w) = x\,c\,v\,c\,v\,c\,y$, with $c \in \Sigma_2$ and $x, v, y \in \Sigma_2^*$. Since $|\mu(w)|$ is even, the number $|x| + |y|$ has to be odd. There are now two cases: (1) $|x|$ is even and (2) $|x|$ is odd.

Case 1: $|x|$ is even. Hence $|y|$ is odd. Then $\mu(w) = (x)(cvcv)(cy) \in \{01, 10\}^*$ implies that x, $cvcv$, and cy are all images under μ of binary words. By Lemma 1.7.3 above, this implies that cv is the image by μ of a binary word. Let r, s, t be words in Σ_2^* such that $\mu(r) = x$, $\mu(s) = cv$, and $\mu(t) = cy$. Then $\mu(w) = \mu(r)\mu(s)\mu(s)\mu(t) = \mu(rsst)$. Hence $w = rsst$. But $\mu(s)$ and $\mu(t)$ both begin with c; hence s and t both begin with c. This implies that sst (and hence w) contains an overlap.

Case 2: $|x|$ is odd. Hence $|y|$ is even. Then $\mu(w) = (xc)(vcvc)y$, so by the same reasoning as in case 1, there exist $r, s, t \in \Sigma_2^*$ such that $\mu(r) = xc$, $\mu(s) = vc$, and $\mu(t) = y$. Then $\mu(w) = \mu(r)\mu(s)\mu(s)\mu(t) = \mu(rsst)$; hence $w = rsst$. But $\mu(r)$ and $\mu(s)$ both end in c. This implies that r and s both end in $\bar{c}$. Hence rss contains an overlap and so does w.

(b): The argument in part (a) easily extends to infinite words. ∎

This lemma will now allow us to prove the following "factorization" result, as well as a first consequence of it.

Proposition 1.7.5

(a) *If $x \in \Sigma_2^*$ is overlap-free, then there exist u, v, y with $u, v \in \{\epsilon, 0, 1, 00, 11\}$ and $y \in \Sigma_2^*$ an overlap-free word, such that $x = u\mu(y)v$. Furthermore this factorization is unique if $|x| \geq 7$, and u (v) is completely determined by the prefix (suffix) of length 7 of x. The bound 7 is best possible.*
(b) *If $\mathbf{x} \in \Sigma_2^\omega$ is an infinite overlap-free word, then there exist $u \in \{\epsilon, 0, 1, 00, 11\}$ and an infinite overlap-free word $\mathbf{y} \in \Sigma_2^\omega$ such that $\mathbf{x} = u\mu(\mathbf{y})$. The prefix u is completely determined by the prefix of $\mathbf{x}$ of length 4, except if $\mathbf{x}$ begins with 0010 or 1101, in which case the word u is completely determined by the prefix of $\mathbf{x}$ of length 5.*

Proof. (a): First, note that if the decomposition exists, then y must be overlap-free by Lemma 1.7.4.

First, we show the existence of a factorization $x = u\mu(y)v$. We prove this by induction on $|x|$. The assertion is easily proved for $|x| = k \leq 2$. Now suppose it is true for all x with $|x| < k$; we prove it for $|x| = k$. Let x be an overlap-free word of length ≥ 3. Write $x = az$, where $a \in \Sigma_2$ and $z \in \Sigma_2^*$. The word z is overlap-free. Since $|z| < |x|$, the induction hypothesis shows that $z = u\mu(y)v$, where $u, v \in \{\epsilon, 0, 1, 00, 11\}$.

If $u = \epsilon$ or if $u = a$, then $x = (au)\mu(y)v$ gives the desired factorization.

If $u = \overline{a}$, then $x = (a\overline{a})\mu(y)v$ and we get the factorization $x = \mu(ay)v$.

If $u = a\,a$, then x begins with aaa, which is impossible.

If $u = \overline{a}\,\overline{a}$, then $x = a\,\overline{a}\,\overline{a}\,\mu(y)v$.

  If $|y| = 0$, then $x = a\,\overline{a}\,\overline{a}\,v$. Hence $v \in \{\epsilon, a, aa\}$. The three corresponding desired factorizations for x are respectively $x = \mu(a)\overline{a}$, $x = \mu(a\overline{a})$, and $x = \mu(a\overline{a})a$.

  If $|y| = 1$, then $y = a$, hence $x = a\,\overline{a}\,\overline{a}\,a\,\overline{a}\,v$. Thus, if $v = \epsilon$, then $x = \mu(a\overline{a})\overline{a}$. If $v = a$, then $x = \mu(a\,\overline{a}\,\overline{a})$. If $v = \overline{a}$, then $x = \mu(a\overline{a})\,\overline{a}\,\overline{a}$. If $v = aa$, then $x = \mu(a\,\overline{a}\,\overline{a})a$. Finally, v cannot be equal to $\overline{a}\,\overline{a}$, because then x would contain the overlap $\overline{a}\,\overline{a}\,\overline{a}$.

  If $|y| \geq 2$, then $|\mu(y)| \geq 4$. But $\mu(y)$ cannot begin with aa; hence by Lemma 1.7.1, $\mu(y)$ must begin with $a\overline{a}aa$. This is not possible for the image of a word under μ; hence this case cannot occur.

We now show that the factorization above is unique provided $|x| \geq 7$. Suppose x is an overlap-free word of length ≥ 7, and suppose

$$x = u\mu(y)v = u'\mu(y')v', \qquad (1.5)$$

where y and y' are overlap-free words in Σ_2^*, and $u, v, u', v' \in \{\epsilon, 0, 1, 00, 11\}$. We note that the assumption $|x| \geq 7$ implies that $|y|, |y'| \geq 2$, since $|\mu(y)|, |\mu(y')| \geq 3$.

If $|u| = |u'|$, then $u = u'$; hence $\mu(y)v = \mu(y')v'$. Hence $|v| \equiv |v'| \pmod 2$.

  If $|v| = |v'|$, then $v = v'$; hence $\mu(y) = \mu(y')$, which gives $y = y'$, and the two factorizations of x are identical.

  If $|v| \neq |v'|$, then $|v| = 2$ and $|v'| = 0$ (or vice versa); say, $\mu(y)v = \mu(y')$. The word v is equal to 00 or 11, and must be a suffix of the word $\mu(y')$, and hence equal to $\mu(0)$ or $\mu(1)$, a contradiction.

Suppose $|u| \neq |u'|$. Without loss of generality, assume $|u'| < |u|$. Then u' is a prefix of u. Let $u = u'w$, with $w \neq \epsilon$. Then, by canceling u' from both sides of Eq. (1.5), we get $w\mu(y)v = \mu(y')v'$. Since y' is not empty, the word w is a prefix of $\mu(y')$, and hence cannot be equal to 00 or 11. This implies that $w = a \in \Sigma_2$. Then $a\mu(y)v = \mu(y')v'$. So we have $|v| \not\equiv |v'| \pmod 2$; hence $|v| \neq |v'|$.

  If $|v| < |v'|$, then v is a suffix of v', say $v' = zv$, with $z \neq \epsilon$. Hence $a\mu(y) = \mu(y')z$. Now z has odd length; hence it must be a letter $b \in \Sigma_2$ (remember that $|v'| \leq 2$). But then $a\mu(y) = \mu(y')b$, which is not possible from Lemma 1.7.2, since then $a\mu(y)$ would be equal to $a\,(\overline{a}\,a)^{|y|}$ and hence would contain an overlap, as $|y| \geq 2$.

  If $|v'| < |v|$, then v' is a suffix of v, say $v = zv'$, with z nonempty. Hence $a\mu(y)z = \mu(y')$. Now z has odd length, hence it must be a letter $b \in \Sigma_2$. But then $a\mu(y)b = \mu(y')$. Hence

Table 1.1.

x	u	x	u	x	u	x	u
$0010011\cdots$	00	$0100110\cdots$	0	$1001011\cdots$	ϵ	$1011010\cdots$	1
$0010110\cdots$	0	$0101100\cdots$	ϵ	$1001100\cdots$	ϵ	$1100100\cdots$	1
$0011001\cdots$	0	$0101101\cdots$	ϵ	$1001101\cdots$	ϵ	$1100101\cdots$	1
$0011010\cdots$	0	$0110010\cdots$	ϵ	$1010010\cdots$	ϵ	$1100110\cdots$	1
$0011011\cdots$	0	$0110011\cdots$	ϵ	$1010011\cdots$	ϵ	$1101001\cdots$	1
$0100101\cdots$	0	$0110100\cdots$	ϵ	$1011001\cdots$	1	$1101100\cdots$	11

x	v	x	v	x	v	x	v
$\cdots0010011$	1	$\cdots0100110$	ϵ	$\cdots1001011$	1	$\cdots1011010$	ϵ
$\cdots0010110$	ϵ	$\cdots0101100$	0	$\cdots1001100$	0	$\cdots1100100$	00
$\cdots0011001$	ϵ	$\cdots0101101$	1	$\cdots1001101$	1	$\cdots1100101$	ϵ
$\cdots0011010$	ϵ	$\cdots0110010$	0	$\cdots1010010$	0	$\cdots1100110$	ϵ
$\cdots0011011$	11	$\cdots0110011$	1	$\cdots1010011$	1	$\cdots1101001$	ϵ
$\cdots0100101$	ϵ	$\cdots0110100$	0	$\cdots1011001$	ϵ	$\cdots1101100$	0

y' must begin with a, say $y' = at$, with $t \neq \epsilon$, since $|y| \geq 2$. Hence $\mu(y)b = \overline{a}\,\mu(t)$. This is again not possible, using Lemma 1.7.2 and the fact that $|y| \geq 2$.

Finally in this factorization of x, the word u (v) depends only on the prefix (suffix) of x of length 7. This is shown in Table 1.1, which gives all possible prefixes (suffixes) of length ≥ 7: by inspection, we see that the word u (v) is uniquely determined, assuming the factorization does indeed exist. (Note, however, that some of the words in the table, e.g., 0011011, might not be extendable to longer overlap-free words.)

The bound 7 is best possible, as shown by the example $x = 001011$, which has the two different factorizations $x = 00\mu(1)11 = 0\mu(00)1$.

(b): For any prefix x_n of $\mathbf{x}$ such that, say, $|x_n| = n$, part (a) gives the existence of $u_n, v_n \in \{\epsilon, 0, 1, 00, 11\}$ and an overlap-free word $y_n \in \Sigma_2^*$ such that $x_n = u_n\mu(y_n)v_n$. Furthermore, u_n does not depend on n for $n \geq 7$. Define $u = u_7$. Hence $\mathbf{x} = \lim_{n\to\infty} x_n = \lim_{n\to\infty} u\mu(y_n)v_n$. Since $|\mu(y_n)|$ goes to infinity, this implies $\mathbf{x} = \lim_{n\to\infty} u\mu(y_n)$. Hence $\lim_{n\to\infty} \mu(y_n)$ exists, which gives the existence of $\mathbf{y} = \lim_{n\to\infty} y_n$. This sequence is overlap-free, as it is a limit of overlap-free words; and we have $\mathbf{x} = u\mu(\mathbf{y})$.

Write $x_n = u\mu(y_n)v_n$ for $n \geq 7$. Now $|v_n| \leq 2$; hence $\mu(y_n)v_n$ is a prefix of $\mu(y_k)$ for k sufficiently large. Hence v_n is a prefix of the image by μ of some word. It follows that in the factorization of x_n, the word v_n cannot be equal to 00 or 11, since these two words are not images of a word by μ.

Hence v_n is either empty or is equal to 0 or 1. Finally, inspecting Table 1.1 shows that the prefix of length 4 of a 7-letter word determines the word u in all cases but the two cases where the word begins with 0010 or 1101, when we need to look at the prefix of length 5. ∎

Lemma 1.7.6 *Let $w \in \Sigma_2^*$ be an overlap-free word with $|w| \geq 52$. Then w contains $\mu^3(0) = 01101001$ and $\mu^3(1) = 10010110$ as subwords.*

Proof. Since w is overlap-free, we know by Proposition 1.7.5(a) that we can write

$$w = \alpha \, \mu(w') \, \eta, \qquad (1.6)$$

with $|w'| \geq 24$. Further, by the same result, we know w' is overlap-free.

Similarly, we can write

$$w' = \beta \mu(w'') \, \theta, \qquad (1.7)$$

with w'' overlap-free and $|w''| \geq 10$.

Finally, we can write

$$w'' = \gamma \mu(y) \zeta \qquad (1.8)$$

with y overlap-free and $|y| \geq 3$.

Now, putting together Eqs. (1.6)–(1.8), we get

$$w = \alpha \, \mu(\beta) \, \mu^2(\gamma) \, \mu^3(y) \, \mu^2(\zeta) \, \mu(\theta) \, \eta$$

where y is overlap-free, $|y| \geq 3$, and $\alpha, \beta, \gamma, \zeta, \eta, \theta \in \{\epsilon, 0, 1, 00, 11\}$. Since y is overlap-free, it contains both 0 and 1, and thus $\mu^3(y)$ contains both words $\mu^3(0) = 01101001$ and $\mu^3(1) = 10010110$. So does w. ∎

We can now prove the following theorem on overlap-free morphisms.

Theorem 1.7.7 *Let μ be the morphism defined by $\mu(0) = 01$ and $\mu(1) = 10$. Let E be the morphism defined by $E(0) = 1$ and $E(1) = 0$. Let $h : \Sigma_2^* \to \Sigma_2^*$ be a nonerasing morphism. If the image by h of any overlap-free binary word of length 3 is overlap-free, then there exists an integer $k \geq 0$ such that either $h = \mu^k$ or $h = E \circ \mu^k$.*

Proof. Let $h(0) = u$ and $h(1) = v$ with $|u|, |v| \geq 1$. We prove the result by induction on $|u| + |v|$.

By hypothesis $uuv = h(001)$ is overlap-free; hence u and v cannot begin with the same letter. In the same way, $uvv = h(011)$ is overlap-free; hence u and v cannot end with the same letter. Let $u = a \cdots = \cdots b$ and $v = \overline{a} \cdots = \cdots \overline{b}$, with $a, b \in \Sigma_2$.

Now neither u nor v can begin or end in 00 or 11: if $u = aa \cdots$, then since $u = \cdots b$ and $v = \cdots \overline{b}$, we have $uu = \cdots baa \cdots$ and $vu = \cdots \overline{b}aa \cdots$. Then one of these words contains the cube, and uu is a subword of $h(001)$ and vu is a subword of $h(100)$. In the same way, u cannot end in the square of a letter in Σ_2, and v cannot begin or end in the square of a letter in Σ_2.

The base case of the induction is $|u| + |v| = 2$. If $|u| = |v| = 1$, then we are done, and either $h = \mu^0$ or $h = E \circ \mu^0$.

Now suppose the result is true for all u, v with $|u| + |v| < j$. We prove it for $|u| + |v| = j$. Suppose that $|u| \geq 2$. We claim that $|v| \geq 2$. For if $|v| = 1$, say $v = b$, then, from what precedes, u begins with $\overline{b}b$ and ends with $b\overline{b}$. Hence, $h(010) = uvu$ contains the overlap $b\overline{b}bb\overline{b}$.

Similarly, if $|v| \geq 2$, then $|u| \geq 2$. Thus we can suppose $|u|, |v| \geq 2$.

Let us write $u = a\overline{a} \cdots = \cdots \overline{b}b$ and $v = \overline{a}a \cdots = \cdots b\overline{b}$. We will prove that $u = \mu(w)$ for some $w \in \Sigma_2^*$.

If $|u| = 2$, then $u = a\overline{a} = \overline{b}b = \mu(a)$.

If $|u| = 3$, then $\overline{a} = \overline{b}$, whence $a = b$, and we have $uP = a\overline{a}a$ and $v = \cdots a\overline{a}$. Then $vu = \cdots a\overline{a}a\overline{a}a$ would contain an overlap although it is a subword of $vuu = h(100)$.

If $|u| = 4$, then $u = a\overline{a}\overline{b}b = \mu(a\overline{b})$.

If $|u| \geq 5$, then $u = a\overline{a}z\overline{b}b$ for some word $z \in \Sigma_2^*$ of length ≥ 1. Hence $vuv = \cdots \overline{b}b\overline{a}az\overline{b}b\overline{a}a \cdots$. There are now two cases to consider.

Case 1: $b = a$. Then $vuv = \cdots a\overline{a}a\overline{a}z\overline{a}a\overline{a}a \cdots$. We know that $|z| \geq 1$, and clearly z cannot begin in a; hence let $z = \overline{a}t$, with $t \in \Sigma_2^*$. Then $vuv = \cdots a\overline{a}a\overline{a}a\overline{a}t\overline{a}a\overline{a}a \cdots$. Now t cannot be the empty word, for otherwise vuv would contain the cube $\overline{a}\,\overline{a}\,\overline{a}$. Furthermore t cannot end in a. Let $t = x\overline{a}$ for a word $x \in \Sigma_2^*$. Hence $vuv = \cdots a\overline{a}a\overline{a}a\overline{a}x\overline{a}a\overline{a}a \cdots$. The word $a\overline{a}\overline{a}x\overline{a}a\overline{a}a$ is a subword of $vuv = h(101)$ and thus overlap-free. Hence we can apply the existence and uniqueness of the factorization of this word given by Proposition 1.7.5 to obtain that $a\overline{a}\overline{a}x\overline{a}a\overline{a}a = \epsilon\mu(y_1)\epsilon$, i.e., $u = \mu(y_1)$.

Case 2: $b = \overline{a}$. Then $vuv = \cdots \overline{a}a a\overline{a}z a\overline{a}\,\overline{a}a \cdots$. By again applying the existence and uniqueness of the factorization given by Proposition 1.7.5, we obtain that $\overline{a}a a\overline{a}z a\overline{a}\,\overline{a}a = \epsilon\mu(y_2)\epsilon$; hence y_2 begins and ends in $\overline{a}$, say $y_2 = \overline{a}\,y_3\overline{a}$. Hence, $u = a\overline{a}z a\overline{a} = \mu(y_3)$.

The same reasoning shows that v is also the image by μ of a word in Σ_2^*.

Hence there exist two words u' and v' such that $u = \mu(u')$ and $v = \mu(v')$. Looking at the morphism h' defined by $h'(0) = u'$ and $h'(1) = v'$, we see that $h = \mu \circ h'$; hence the image by h' of any overlap-free word of length 3 is overlap-free by Lemma 1.7.4. Furthermore $|u'| < |u|$ and $|v'| < |v|$. Hence the induction hypothesis applies, and $h' = \mu^k$ or $h' = E \circ \mu^k$. (Note that $E \circ \mu = \mu \circ E$.) The result now follows. ∎

The following corollary is easily obtained.

Corollary 1.7.8 *With the same notation as above, let h be a morphism on the alphabet Σ_2 such that $h(01) \neq \epsilon$. Then the following conditions are equivalent.*

(a) *The morphism h is nonerasing and maps any overlap-free word of length 3 to an overlap-free word.*

(b) There exists $k \geq 0$ such that $h = \mu^k$ or $h = E \circ \mu^k$.

(c) The morphism h maps any infinite overlap-free word to an infinite overlap-free word.

(d) There exists an infinite overlap-free word whose image under h is overlap-free.

(e) The morphism h maps the word 01101001 to an overlap-free word.

Proof. (a) $\Longrightarrow$ (b) was proved in Theorem 1.7.7 above.

(b) $\Longrightarrow$ (c) is an easy consequence of Lemma 1.7.4.

(c) $\Longrightarrow$ (d): It suffices to know that there exists an infinite overlap-free word: the Thue–Morse word is an example, by Theorem 1.6.1.

(d) $\Longrightarrow$ (e): The hypothesis asserts the existence of an infinite overlap-free infinite word $\mathbf{x}$ whose image under h is overlap-free. Using Lemma 1.7.6 above, we see that $\mathbf{x}$ contains the subword 01101001, hence the image $h(01101001)$ is certainly overlap-free.

(e) $\Longrightarrow$ (a): If $h(0) = \epsilon$, then since $h(01) \neq \epsilon$, we know $h(1) \neq \epsilon$. Then $h(01101001) = h(1)^4$ contains an overlap, a contradiction. Similarly, $h(1) \neq \epsilon$. Thus h is nonerasing. But every overlap-free word of length 3 on Σ_2 is a subword of 01101001, and we are done. ∎

We say a morphism $h : \Sigma_2^* \to \Sigma_2^*$ is the identity morphism if $h(0) = 0$ and $h(1) = 1$. The following corollary completely characterizes the binary sequences that are fixed points of non-identity morphisms.

Corollary 1.7.9 *An infinite overlap-free binary word is a fixed point of a non-identity morphism if and only if it is equal to* $\mathbf{t}$, *the Thue–Morse word, or its complement* $\overline{\mathbf{t}}$.

Proof. Let h be a non-identity morphism on the alphabet Σ_2. Let $\mathbf{x}$ be a fixed point of h that is overlap-free. Using Corollary 1.7.8 above, we see that the morphism h, mapping an overlap-free infinite word to an overlap-free infinite word, must be of the form μ^k or $E \circ \mu^k$, for some $k \geq 0$. Since h has a fixed point, it cannot be of the form $E \circ \mu^k$. Since h is non-identity, $h = \mu^k$ for some $k \geq 1$. But $\mathbf{t}$ and $\overline{\mathbf{t}}$ are the only fixed points of μ^k, and the corollary is proved. ∎

1.8 Additional Topics on Repetitions

There are many other topics dealing with repetitions in words. We content ourselves with a brief survey.

We can define the concept of *fractional power*. We say a (finite or infinite) word w contains an α-*power* (real $\alpha > 1$) if w has a subword of the form $x^{\lfloor \alpha \rfloor} x'$ where x' is a prefix of x and $|x^{\lfloor \alpha \rfloor} x'| \geq \alpha |x|$. For example, the word

$$2301 \overbrace{01234567\ 01234567\ 0123}\ 310$$

has a $\frac{5}{2}$-power. A word is α-*power-free* if it contains no α-power.

Given an infinite word **w** it is an interesting and challenging task to determine its *critical exponent* e, such that **w** contains α-powers for all $\alpha < e$, but has no α-powers for $\alpha > e$. (It may or may not have e-powers.)

Theorem 1.8.1 *The critical exponent of the Thue–Morse word* **t** *is* 2.

Proof. The word **t** begins $011\cdots$ and hence contains a square. If **t** contained a $(2 + \epsilon)$-power for any $\epsilon > 0$, then it would contain an overlap. But **t** is overlap-free, by Theorem 1.6.1. ∎

There also exist various generalizations of squarefreeness. We say a word is an *abelian square* if it is of the form $w\,w'$ where w' is a permutation of w. A word is *abelian squarefree* if it contains no abelian squares. It is possible to prove that there exists an infinite abelian squarefree sequence on 4 symbols; see the Notes (Section 1.11) for more information.

Another generalization is to study more general pattern avoidance problems. Let w, p be words. We say w *avoids the pattern* p if there does not exist a nonerasing morphism h such that $h(p)$ is a subword of w. For example, w avoids the pattern xx if and only if w is squarefree, and w avoids the pattern xxx if and only if w is cubefree. We say that pattern p is *avoidable on a k-letter alphabet* Σ_k if there exists $\mathbf{x} \in \Sigma_k^\omega$ such that $\mathbf{x}$ avoids p. Clearly x and xyx are not avoidable on finite alphabets. However, xx is avoidable on a 3-letter alphabet, by Theorem 1.6.2, and $xyxyx$ is avoidable on a 2-letter alphabet, by Theorem 1.6.1.

1.9 Exercises

1. What are the shortest overlap-free words which are not subwords of **t**, the Thue–Morse infinite word?

2. Show that a set S of infinite words is perfect (see Section 1.2) if, for all $\mathbf{w} \in S$ and all integers $n \geq 0$, there exists a word $\mathbf{v} \in S \setminus \{\mathbf{w}\}$ such that $\mathbf{v}$ and $\mathbf{w}$ have a common prefix of length $\geq n$.

3. Prove that $\exp(h) \leq \operatorname{Card} M_h$, where exp is the mortality exponent. (See Section 1.4.)

4. Suppose **w** is a purely periodic right-infinite word, of period j and k. Show that **w** is periodic of period $\gcd(j, k)$.

5. In this exercise we explore a property of two-sided infinite words. We say a two-sided infinite word $\mathbf{w} = \cdots c_{-2}c_{-1}c_0c_1c_2 \cdots$ is *periodic* if there exists an integer $p \geq 1$ such that $c_k = c_{k+p}$ for all integers $k \in \mathbb{Z}$. Show that **w** is periodic if and only if there exist a right-infinite word **x** and an infinite sequence of negative indices $0 > i_1 > i_2 > \cdots$ such that $\mathbf{x} = c_{i_j}c_{i_j+1}c_{i_j+2} \cdots$ for all $j \geq 1$.

6. Suppose **w** is an ultimately periodic infinite word. Show that the frequency of each letter in **w** exists and is rational.

7. Suppose **w** is an infinite word. We define its sequence of *run lengths* to be a word over the alphabet $\mathbb{N} \cup \{\infty\}$ giving the number of adjacent identical elements. For example, the sequence of run lengths of $0110001111000001\cdots$ is $12345\cdots$. Show that
   (a) if **w** is ultimately periodic, then the sequence of run lengths of elements of **w** is finite or ultimately periodic;
   (b) the converse is true if **w** is over an alphabet with ≤ 2 letters, but is false for alphabets of size ≥ 3.

8. Suppose $\mathbf{w}_1, \mathbf{w}_2, \ldots, \mathbf{w}_j$ are ultimately periodic infinite words over $\Sigma_2 = \{0, 1\}$. Show that any integer linear combination of the $\mathbf{w}_i$ is ultimately periodic.

9. Let $x, y \in \Sigma^*$ be words. Show that there exists $z \in \Sigma^*$ such that $x^2 y^2 = z^2$ if and only if $xy = yx$.

10. In the Lyndon–Schützenberger theorems, we proved a necessary and sufficient condition for $xy = yx$ and $xy = yz$. Find similar necessary and sufficient conditions for the following to hold:
    (a) $xy = y^R x$;
    (b) $xy = y^R z$.

11. Let x, y, z, w be words. Find necessary and sufficient conditions for the following two equations to hold simultaneously: $xy = zw$ and $yx = wz$.

12. Find all solutions to the system of equations $x_1 x_2 x_3 = x_2 x_3 x_1 = x_3 x_1 x_2$ in words.

13. Recall the definition of the Möbius function μ from Section 1.5. Prove the Möbius inversion formula: if $g(n) = \sum_{d \mid n} f(d)$, then $f(n) = \sum_{d \mid n} \mu(n/d) g(d)$.

14. Show that there are no squarefree words of length > 3 over a 2-letter alphabet. Show that there are no abelian squarefree words of length > 8 over a 3-letter alphabet.

15. Show that the binary word

    $$0010010110010010110010011011\cdots,$$

    obtained by iterating the morphism h mapping $0 \to 001$ and $1 \to 011$, is cubefree.

16. Show that Jacobs's "Mephisto waltz" infinite word

    $$0010011100010011101101101001\cdots,$$

    obtained by iterating the morphism which maps $0 \to 001$ and $1 \to 110$, is fourth-power-free.

17. (Berstel) A *Langford string* is a nonempty word over the infinite alphabet $\{2, 3, 4, \ldots\}$ such that two consecutive occurrences of the letter a are separated by exactly $a - 1$ letters. For example, 24262425262 is a Langford string. Prove that every Langford string is squarefree.

18. (Euwe and Morse) According to official rule (10.12) of the game of chess, a player can claim a draw if "at least 50 consecutive moves have been made by each side without the capture of any piece and without the movement of any pawn". Actually, this is not enough to accommodate checkmate for certain positions, such as king + rook + bishop versus king + 2 knights, so the rule also stipulates that "This number of 50 moves can be increased for certain positions, provided that this increase in number and these positions have been clearly announced by the organisers before the event starts." Another rule (10.10) allows a draw to be claimed if the same position occurs for the third time. By "same position" we mean that the pieces are in the same position, including the rights to castle or capture a pawn *en passant*. Without these two rules, infinite games are clearly possible. However, can rule (10.10) be weakened and still disallow infinite chess games? Consider the following alternative rule: a draw occurs if the same sequence of moves occurs twice in succession and is immediately followed by the first move of a third repetition. Using an infinite overlap-free sequence, show that this formulation would permit an infinite game of chess.

19. Find some examples of words in languages other than English that are squares, cubes, fourth powers, overlaps, or palindromes.

20. Find an English word containing two distinct squares, each of length ≥ 4.

21. Is the decimal expansion of π squarefree? Cubefree? How about the decimal expansion of $\sqrt{2}$, or e?

22. Let $u(n)$ denote the number of overlap-free binary words of length n. Show that $u(n)$ is bounded by a polynomial in n.

23. Let $\mathbf{w} = c_0 c_1 c_2 \cdots = 00100110100010110011 \cdots$ be the infinite word defined by $c_n =$ the number of 0's (mod 2) in the binary expansion of n. Show that $\mathbf{w}$ is overlap-free.

24. Show how to extend the Thue–Morse infinite word $\mathbf{t}$ on the left to a two-sided infinite word that is still overlap-free. Is there only one way to do this?

25. Show that the Thue–Morse word avoids the pattern $\alpha^2 \beta \alpha^2 \beta$.

26. Give another proof of the Fine–Wilf theorem (Theorem 1.5.6) as follows: without loss of generality assume the symbols of $\mathbf{a}$ and $\mathbf{b}$ are chosen from $\mathbb{Q}$. Let X be an indeterminate, and let $A(X) := \sum_{i \geq 0} a_i X^i$ and $B(X) := \sum_{i \geq 0} b_i X^i$. Since $\mathbf{a}$ (respectively $\mathbf{b}$) is purely periodic, we have $A(X) = (1 - X^m)^{-1} P(X)$ and $B(X) = (1 - X^n)^{-1} Q(X)$ for a polynomial P of degree $< m$ (respectively Q of degree $< n$). Then

$$H(X) := A(X) - B(X) = (1 - X^{\gcd(m,n)})(1 - X^m)^{-1}(1 - X^n)^{-1} R(X)$$

where R is a polynomial of degree $< m + n - \gcd(m, n)$. Now the first $m + n - \gcd(m, n)$ coefficients of H are 0, so $R = 0$, and hence $H = 0$.

27. Prove that the bound of $m + n - \gcd(m, n)$ in the Fine–Wilf theorem (Theorem 1.5.6) is best possible.

28. Prove the following theorem: for all integers m, n with $1 \leq m \leq n$, there exist finite words $x, y, z \in \{0, 1\}^*$ with $|x| = m$, $|yz| = n$ such that the first $m + n - 2$ symbols of x^ω coincide with yz^ω, but there is a mismatch at the $(m + n - 1)$th symbol. Also show that this bound of $m + n - 1$ is best possible.

29. Let m, n be distinct integers ≥ 0. What is the length of the longest word over $\mathbb{Z}$ with the property that every subword of length m has a positive sum, but every subword of length n has a negative sum?

30. Let h be the morphism defined by $h(a) = \text{abcab}$; $h(b) = \text{acabcb}$; and $h(c) = \text{acbcacb}$. Show that $\lim_{n \to \infty} h^n(a)$ is squarefree.

31. Let $\Sigma = \{0, 1\}$, and for $w \in \Sigma^\omega$ let $\overline{w}$ denote the image of w under the coding that maps $0 \to 1$ and $1 \to 0$. Let S be the shift map, as defined in Section 1.1. Show that if w is the least non-ultimately-periodic infinite word such that $\overline{w} \leq S^k(w) \leq w$ for all $k \geq 0$, then $w = S(\mathbf{t})$, where $\mathbf{t}$ is the Thue–Morse word.

32. Let h be the morphism defined over $\{a, b, c, d\}$ by $h(a) = \text{abcd}$, $h(b) = \text{bacd}$, $h(c) = \text{cabd}$, and $h(d) = \text{cbad}$. Let $(w_n)_{n \geq 1}$ be defined by $w_n = h^n(a)$. Show that w_n is squarefree for $n \geq 1$.

33. In this exercise we explore some alternative constructions of $\mathbf{c}$, the infinite squarefree word introduced in Section 1.6.

    (a) Let $t_0 t_1 t_2 \ldots$ be the Thue–Morse word. Define $b_n = \tau(t_n, t_{n+1})$, where

    $$\tau(0, 0) = 1, \qquad \tau(0, 1) = 2;$$
    $$\tau(1, 0) = 0, \qquad \tau(1, 1) = 1.$$

    Show that $\mathbf{c} = b_0 b_1 b_2 \ldots$.

    (b) Let f be the morphism mapping $2 \to 210$, $1 \to 20$, and $0 \to 1$. Show that $f(\mathbf{c}) = \mathbf{c}$.

    (c) Let g be the morphism mapping $a \to ab$, $b \to ca$, $c \to cd$, $d \to ac$, and let τ be the coding mapping $a \to 2, b \to 1, c \to 0, d \to 1$. Show that $\mathbf{c} = \tau(g^\omega(a))$.

34. Develop efficient algorithms to check if a finite word is squarefree, or overlap-free.

35. Show that there are an uncountable number of infinite squarefree words over a three-letter alphabet.

36. Define $\nu_2(n)$ to be the *ruler function*, i.e., the exponent of the largest power of 2 that divides n. Show that the infinite word (over the infinite alphabet $\mathbb{N}$)

$$\nu_2(1)\nu_2(2)\nu_2(3) \cdots = 0102010301020104 \cdots$$

    is squarefree.

37. For every real number $\Omega \geq 2$, do there exist infinite words over a finite alphabet that have Ω as a critical exponent?

38. (Allouche) Let $t_0 t_1 t_2 \cdots$ be the Thue–Morse sequence. Let $n \geq 1$ be an integer, and let $0 \leq j < 2^n$. Let $\text{sgn } x = -1$ if $x < 0$; 0 if $x = 0$; and 1 if $x > 0$.

    (a) Show that $\text{sgn}\left(\prod_{0 \leq i < n} \sin(2^i x)\right) = (-1)^{t_j}$ for all $x \in (j\pi/2^n, (j+1)\pi/2^n)$.

(b) Let x be a real number, and let n be an integer ≥ 1. Show that

$$\prod_{0 \leq l < 2n} \sin(2^l x) = \frac{(-1)^{n+1}}{2^{2n-1}} \sum_{0 \leq j < 2^{2n-1}} (-1)^{t_j} \cos((2j+1)x).$$

39. Give an $O(kn^2)$ algorithm for determining whether a word s of length n over an alphabet Σ of size k contains an abelian square. Can this running time be improved?

40. Let Σ be a finite alphabet, and let A be an infinite subset of Σ^*. Show by means of an example that there need not be an infinite word $\mathbf{w} \in \Sigma^\omega$ such that infinitely many members of A are prefixes of $\mathbf{w}$.

41. Prove the following version of the König infinity lemma: let Σ be a finite alphabet, and let A be an infinite subset of Σ^*. Prove that there exists an infinite word $\mathbf{w} \in \Sigma^\omega$ such that any prefix of $\mathbf{w}$ is a prefix of at least one word in A.

42. (Currie) Let $w \in \{0, 1, 2\}^+$. We say w is a *minimal square* if w is a square (i.e., $w = xx$ for some $x \in \{0, 1, 2\}^+$) and no proper nonempty subword of w is a square. Show there are infinitely many distinct minimal squares.

43. A *border* of a finite word w is a word $x \notin \{\epsilon, w\}$ that is both a prefix and a suffix of w. A word w is said to be *unbordered* if it has no borders. Define $b_0 = 1$, and for $n \geq 1$ define $b_{2n+1} = kb_{2n}$, $b_{2n} = kb_{2n-1} - b_n$.

    (a) Show that there are b_n unbordered words of length n over a k-letter alphabet. Note: Unbordered words are sometimes called "bifix-free" or "primary" in the literature.

    (b) Prove that $b_{2n}/k^{2n} + \sum_{1 \leq i \leq n} b_i k^{-2i} = 1$.

44. Show that all prefixes of the Thue–Morse infinite word $\mathbf{t}$ are primitive. Conclude that a primitive word of arbitrary length can be generated in linear time.

45. Show how to determine if a word w is primitive in linear time. Furthermore, show how to determine the largest k such that $w = x^k$ in the same time bound.

46. Show how to generate a random primitive word in expected linear time.

47. Let $k \geq 2$ be an integer. Show that the set of all subwords of the primes expressed in base k is $\{0, 1, \ldots, k-1\}^*$.

48. Let $\mu : 0 \rightarrow 01, 1 \rightarrow 10$ be the Thue–Morse morphism.

    (a) Show that the lexicographically least infinite overlap-free word over $\Sigma_2 = \{0, 1\}$ starting with 1 is $\mu^\omega(1)$.

    (b) Show that the lexicographically least infinite overlap-free word starting with 0 is $001001\varphi^\omega(1)$.

    (c) What is the lexicographically least infinite overlap-free word over Σ_2 that can be extended to a two-sided infinite overlap-free word?

49. Define $w_1 = 01$, and $w_{n+1} = w_n w_n w_n^R$ for $n \geq 1$. Prove that all the words w_i, $i \geq 1$, are cubefree. Can $\lim_{i \rightarrow \infty} w_i$ be generated as the image of the fixed point of a uniform morphism?

50. Suppose w is a primitive word, and suppose $w = uv$, where u and v are both nonempty palindromes. Show that this factorization into palindromes is in fact unique.

51. Let $x, y \in \Sigma^+$. We say x is a *conjugate* of y, and we write $x \sim y$, if there exist $u, v \in \Sigma^*$ such that $x = uv$ and $y = vu$. In other words, x is a cyclic shift of the symbols of y.
    (a) Show that $\sim$ is an equivalence relation.
    (b) Suppose $x \in \Sigma^+$, $j \geq 1$, is an integer, and there exist $u, v \in \Sigma^*$ such that $x^j = uv$. Then there exist $r, s \in \Sigma^*$ such that $x = rs$, and $vu = y^j$ for $y = sr$.
    (c) Suppose w is primitive and $w \sim x$. Then x is primitive.
    (d) Suppose w, x are primitive words and there exist $i, j \geq 1$ such that $w^i \sim x^j$. Then $w \sim x$.

52. Let $i, j, k \geq 2$ be integers. Consider the equation in words $x^i y^j = z^k$. Show that the only solutions x, y, z to this equation are of the form $x = w^a, y = w^b$, and $z = w^{(ai+bj)/k}$, where $k \mid ai + bj$.

53. Let $\mathbf{t} = t_0 t_1 t_2 \cdots$ be the Thue–Morse sequence. Show that $t_n = 0$ for infinitely many integers n that are either prime or the product of two primes.

54. Let $k \geq 2$ be an integer, and let $\Sigma_k = \{0, 1, \ldots, k - 1\}$. Show that if $k = 5$, then it is possible to construct an infinite word over Σ that simultaneously avoids xx and xx', where x' denotes $(x + 1) \bmod k$, while if $k \leq 4$, this is not possible.

55. Let $k \geq 2$ be an integer, and let $\Sigma_k = \{0, 1, \ldots, k - 1\}$. Show that if $k = 3$, it is possible to construct an infinite word over Σ that avoids xx', where $|x| = |x'| \geq 2$ and $x' \equiv x + a \pmod{k}$ for all a, while if $k = 2$, this is not possible.

56. Is the Thue–Morse sequence *mirror-invariant*, that is, if w is a finite subword of $\mathbf{t}$, need w^R also be a subword of $\mathbf{t}$?

57. Let $M(n)$ be the maximum number of distinct squares occurring in any binary word of length n. Show that $\lfloor n/2 \rfloor \leq M(n) < 2n$ for all $n > 0$.

58. (Friedman) Let $n(k)$ be the length of the longest sequence $w = (a(i))_{1 \leq i \leq n}$ over the alphabet $\{1, 2, \ldots, k\}$ with the following property: there do not exist two subwords $w[i..2i], w[j..2j]$ with $1 \leq i < j \leq n/2$ such that the first is a subsequence of the second. (We say x is a *subsequence* of y if we can obtain x by striking out 0 or more letters of y.)
    (a) Show that $n(k)$ is well defined (i.e., there really is a longest sequence) for all $k \geq 1$.
    (b) Compute $n(k)$ for $k = 1, 2, 3, 4$.

59. In analogy with Exercise 58, does there exist an infinite word $\mathbf{x}$ over the alphabet $\Sigma_3 = \{0, 1, 2\}$ such that for all i, j with $1 \leq i < j$ we have that $\mathbf{x}[i..2i]$ is not a subword of $\mathbf{x}[j..2j]$?

60. Let $f(n, k)$ denote the number of primitive words over $\{0, 1\}$ of length n that contain exactly k ones. Show that

$$f(n, k) = \sum_{d \mid \gcd(k, n)} \mu(d) \binom{n/d}{k/d}.$$

61. Show that no infinite squarefree word over $\Sigma_3 = \{0, 1, 2\}$ can contain all possible finite squarefree words.

62. Show that if $x, y \in \Sigma^*$ with $xy \neq yx$, then $xyxxy$ is primitive.

63. What is the expected number of consecutive letters of an infinite word over $\{0, 1\}$ that must be examined until an overlap is found? Express your answer in terms of $(g(n))_{n \geq 0}$, the number of distinct overlaps of length n, and estimate it numerically.

64. Give an unambiguous context-free grammar that generates the well-formed regular expressions over $\{0, 1\}$.

65. Prove that there exists a cubefree infinite word over $\{0, 1\}$ without arbitrarily large squares.

66. What name in the index to this book contains a cube?

1.10 Open Problems

1. Show that for $k \geq 5$ every sufficiently long word over a k-letter alphabet contains a $k/(k-1)$ power, and this bound is best possible. (Remark: This is *Dejean's conjecture*; see the notes to Section 1.8 in Section 1.11 below for further information.)

2. Characterize the lexicographically least squarefree word over $\{0, 1, 2\}$.

3. Is there a pattern w which is avoidable on 5 letters but not on 4 letters? (Remark: See Baker, McNulty, and Taylor [1989]; Currie [1993].)

4. (Currie) Give a simple closed form for the number $c(n)$ of squarefree words of length n over the 3-letter alphabet $\{0, 1, 2\}$. (Remarks: See the notes to Section 1.6 in Section 1.11 below, or Currie [1993].)

5. (Currie) Recall the definition of perfect set from Exercise 2.
   (a) Is the set of cubefree infinite words over $\{0, 1\}$ perfect?
   (b) Prove or disprove: if a pattern p is avoidable on Σ, then the set of infinite words on Σ avoiding p is a perfect set.

6. Is there an algorithm which, given a pattern p and a natural number k, decides whether p is avoidable on k letters? (Remark: See Bean, Ehrenfeucht, and McNulty [1979]; Currie [1993].)

7. Suppose two morphisms φ, ρ commute on all elements of Σ. What can be said about them?

8. Suppose $w, y \in \{0, 1, 2\}^*$ with the property that wx is squarefree for infinitely many words $x \in \{0, 1, 2\}^*$ and xy is squarefree for infinitely many words $x \in \{0, 1, 2\}^*$. Prove or disprove: there exists a word $z \in \{0, 1, 2\}^*$ such that wzy is squarefree. (Remark: See, e.g., Restivo and Salemi [1985b].)

9. Let $w \in \{0, 1, 2\}^+$ be a squarefree word. If there is an integer $n \geq 1$ such that
   (a) there exists a word y with $|y| = n - 1$ such that wy is squarefree, and
   (b) wx is not squarefree for all x with $|x| \geq n$,
   then we say w is of *s-index* n. For example, 0102010 is of s-index 1, and

$$0210120102012021012010200$$

   is of s-index 2. Prove or disprove that there exist words of arbitrarily large s-index.

10. Let $\mathbf{t} = t_0 t_1 t_2 \cdots$ be the Thue–Morse sequence. Prove or disprove that there exist infinitely many primes p such that $t_p = 0$. (Remarks: This problem is apparently due to Gelfond. Olivier [1971a] claimed a proof, but it is apparently incomplete. See Montgomery [1994, Item 67, p. 208]. Also see Dartyge and Mauduit [2000].)

11. Let $\mathbf{t} = t_0 t_1 t_2 \cdots$ be the Thue–Morse sequence, and define $\mathbf{u} = (t_{n^2})_{n \geq 0}$. Does $\mathbf{u}$ contain arbitrarily long squares?

12. Minimal squares were defined in Exercise 42. Let s_n be the number of length-n minimal squares over $\{0, 1, 2\}$.

   (a) Given an efficient algorithm to compute s_n.
   (b) Estimate s_n asymptotically.
   (c) Are there infinitely many indices n for which $s_n = 0$?

1.11 Notes on Chapter 1

1.1 The books of Lothaire [1983, 2002] are good introductions to combinatorics on words, as is Choffrut and Karhumäki [1997].

   Europeans tend to use the term "factor" for what we call "subword", and they use "subword" to mean a word obtained by deleting some letters, not necessarily adjacent, from a given word – what we call "subsequence".

   The notation $\tilde{w}$ is sometimes used in the literature to denote the reversal w^R.

   In the literature, an infinite word is sometimes called an ω-word.

   Two-sided infinite words are also called "bi-infinite" or "doubly infinite" in the literature. The term "bisequence" is also sometimes used. Our notation comes from Beauquier [1985] and Nivat and Perrin [1982, 1986]. The set of two-sided infinite words is sometimes denoted $^\omega \Sigma^\omega$.

   For algorithms to detect palindromes in words, see, for example, Apostolico, Breslauer, and Galil [1992, 1995].

1.2 Kelley [1955] is a good reference for topology. For topology of infinite words, see, for example, Perrin and Pin [1993].

   For measure theory, see, for example, Halmos [1950].

1.3 Hopcroft and Ullman [1979] is a good introduction to the theory of languages. Regular expressions were invented by Kleene [1956]. For properties of regular expressions, see Brzozowski [1962] and S. Yu [1997].

1.4 For more information about morphisms, see Hopcroft and Ullman [1979]; Harju and Karhumäki [1997].

   In the literature, codings (1-uniform morphisms) are sometimes called *length-preserving* morphisms. Nonerasing morphisms are sometimes called ϵ-*free*, or *propagating*.

   Primitive morphisms derive their name from the fact that their associated incidence matrices M (see Section 8.2) are primitive. This means there exists an integer e such that M^e has all positive entries. For a good discussion of primitive matrices, see, for example, Minc [1988].

1.5 Levi's lemma is due to Levi [1944].

For Theorems 1.5.2 and 1.5.3, see Lyndon and Schützenberger [1962]. Theorem 1.5.6 is due to Fine and Wilf [1965]. For other proofs of these facts, see, e.g., Lothaire [1983]. For another simple proof of the Fine–Wilf theorem, see Halava, Harju, and Ilie [2000].

For generalizations of the Fine–Wilf theorem, see Giancarlo and Mignosi [1994]; Castelli, Mignosi, and Restivo [1999]; Berstel and Boasson [1999]; Justin [2000]; Mignosi, Shallit, and Wang [2001]; Blanchet-Sadri and Hegstrom [2002].

There is a large literature on primitive words and their properties; for an introduction, see Lothaire [1983] and Shyr [1991]. The term "primitive" seems to be due to Lyndon and Schützenberger [1962], and Theorem 1.5.4 appears there. For other papers, see Shyr and Thierrin [1977]; M. Ito, Katsura, Shyr, and Yu [1988]; M. Ito and Katsura [1991]; Shyr and Tu [1991]; Dömösi, Horváth, and Ito [1993]; Grounds and Silberger [1993]; Dömösi, Horváth, Ito, Kászonyi, and Katsura [1993, 1994a, 1994b]; Shyr and Yu [1994a, 1994b, 1994c]; Petersen [1994, 1996]; Horváth [1995]; Chang and Shyr [1995]; Shyr [1996]; Păun and Thierrin [1997]; Mitrana [1997a, 1997b]; Kászonyi and Katsura [1997]; Kari and Thierrin [1998]; Horváth and Ito [1999]; Dömösi, Hauschildt, Horváth, and Kudlek [1999]; Păun, Santean, Thierrin, and Yu [2002].

1.6 The literature on repetitions in words is huge. Allouche [1984] and particularly Berstel [1984a] are surveys of the area. A reasonably complete bibliography can be found in Guy [1994, §E21].

In the literature, the terms *nonrepetitive*, *repeat-free*, *aperiodic*, and *strongly primitive* have sometimes been used in place of squarefree.

The Thue–Morse sequence **t** was introduced in Thue [1912], although it was hinted at sixty years earlier by Prouhet [1851]. In an earlier paper, Thue [1906] constructed an infinite squarefree word on three symbols. For a discussion, in English, of Thue's work, see Hedlund [1967] and especially Berstel [1992, 1995]. Morse [1921] introduced the sequence **t** independently, as did Mahler [1927] and the Dutch chess master Max Euwe [1929]; see Exercise 18. (Euwe's work was the inspiration for some English cryptography during World War II; see Good [1993].) Arshon [1937] introduced the Thue–Morse sequence in a slightly disguised form (as a fixed point of the map $1 \to 12$, $2 \to 21$) and observed that it was cubefree. Gardner [1961a, 1961b] popularized the problem of finding an infinite squarefree string in his *Mathematical Games* column; the columns were reprinted in Gardner [1967a, pp. 32–33, 90–95]. Noland [1962] proposed the problem of finding an infinite squarefree word over three symbols, and a solution was given by Braunholtz [1963]. Istrail [1977] rediscovered previous results. Berstel [1978] showed that constructions of Thue, Braunholtz, and Istrail all generate the same word, up to renaming. Also see Berstel [1980a]. Pansiot [1981a] showed that **t** is generated by iterating a morphism only if that morphism is a power of the Thue–Morse morphism μ. Restivo and Salemi [2002] discussed patterns that occur in **t**.

The Danish composer Per Nørgård (1932–) independently rediscovered the Thue–Morse sequence, and used it in some of his compositions, such as the first movement of his *Symphony No. 3*. Nørgård [1999] gives percussion compositions that involve the Thue–Morse sequence. For more about the Thue–Morse sequence in music, see the beginning of the Notes to Chapter 7.

Christol, Kamae, Mendès France, and Rauzy [1980] generalized Thue's construction by considering the parity of the number of occurrences of an arbitrary pattern in the binary expansion of n. Černý [1986] studied the repetitions in such sequences, as did Séébold [1985a, 1986].

In the literature, overlap-free words have been called both *strongly cubefree* and *weakly cubefree*; another term used is *irreducible*. The French term for overlap is *chevauchement*.

Gottschalk and Hedlund [1964] described all overlap-free binary words. Fife [1980] gave a description of all infinite overlap-free words. Carpi [1993a] found an alternative description, using a finite-state machine. Berstel [1994] found an alternative presentation of Fife's results. Also see Shur [1996b].

Karhumäki [1981] showed it is decidable whether a binary morphism generates an overlap-free word. Restivo and Salemi [1983, 1985a] showed that the number $g(n)$ of overlap-free words of length n over a 2-letter alphabet is $O(n^{\log_2 15})$. This was improved to $g(n) = O(n^{1.7})$ by Kfoury [1988a], and to $g(n) = O(n^{1.587})$ by Kobayashi [1988]. In this last paper it is also proved that $g(n) = \Omega(n^{1.155})$. Cassaigne [1993b] proved the surprising fact that if $\beta = \inf\{r : g(n) = O(n^r)\}$ and $\alpha = \sup\{r : n^r = O(g(n))\}$, then $\alpha \neq \beta$.

Kfoury [1988a, 1988b] gave an $O(n)$ algorithm to test whether a word contains an overlap.

Many writers have constructed squarefree words through iterating a morphism. For example, Hawkins and Mientka [1956] considered the word which is the fixed point of the (nonuniform) morphism

$$\text{a} \rightarrow \text{bacbcacbabcbaca},$$
$$\text{b} \rightarrow \text{bacbabcbacbcacbaca},$$
$$\text{c} \rightarrow \text{bacbcabacabcbaca}$$

and proved it is squarefree. Leech [1957] proved a similar result for a fixed point of the 13-uniform morphism

$$0 \rightarrow 0121021201210,$$
$$1 \rightarrow 1202102012021,$$
$$2 \rightarrow 2010210120102.$$

His results were rediscovered by Elser [1983]. Zech [1958] used the 12-uniform morphism

$$a \to \text{cacbcabacbab},$$
$$b \to \text{cabacbcacbab},$$
$$c \to \text{cbacbcabcbab}.$$

Pleasants [1970] proved a similar result for the fixed point of the morphism $a \to \text{abcab}$, $b \to \text{acabcb}$, and $c \to \text{acbcacb}$; see also Harrison [1978, §1.8]. Allouche, Astoorian, Randall, and Shallit [1994] showed that a sequence derived from the tower of Hanoi is squarefree; see Exercise 6.3. Also see Hinz [1996].

The topological properties of the set of squarefree words were investigated by Shelton [1983]. Fife [1983] investigated similar questions for the set of overlap-free words; also see Kobayashi [1986]. Currie and Shelton [1996a, 1996b] showed that the set of infinite kth-power-free words ($1 < k < 2$) is a perfect set.

Karhumäki [1983] showed it is decidable whether a binary morphism generates an infinite cubefree word. Mossé [1992] showed for an infinite word $\mathbf{w}$ which is the fixed point of a primitive morphism that either $\mathbf{w}$ is ultimately periodic or there exists an integer n such that $\mathbf{w}$ is nth-power-free.

One of the most interesting applications of squarefree words is the Burnside problem for groups. The *exponent* of a multiplicative group G is the smallest positive integer e such that $a^e = 1$ for all $a \in G$, if it exists; otherwise we say G is of infinite exponent. Burnside [1902] asked whether every finitely generated group of finite exponent must necessarily be finite. Suppose G has m generators and exponent n. Then it is easy to see that G must be finite if $m = 1$ or $n = 2$, so it suffices to consider the case $m > 1$. Burnside [1902] proved that G must be finite for $n = 3$. The case $n = 4$ was resolved positively by Sanov [1940], and the case $n = 6$ by Hall [1957, 1958]. Finally, Novikov and Adian [1968] resolved Burnside's question in the negative, showing that G could indeed be infinite when $m > 1$ and $n \geq 4381$ is odd. The proof is extremely long and difficult, and makes use of squarefree words in an essential way. The bound 4381 was improved to 665 by Adian [1979]. Britton [1973] attempted to provide a simpler proof of Novikov and Adian's results, but later acknowledged (Britton [1980]) that his proof was flawed. Later, Ol'shanskii [1982] gave a simpler geometric proof that G could be infinite for the case of odd $n > 10^{10}$, and Ivanov [1992, 1994] and Lysënok [1992, 1996] independently answered Burnside's question negatively for almost all n.

For more on the Burnside problem, see Hall [1964]; Magnus, Karrass, and Solitar [1976, §5.12, 5.13]; Adian [1980]; and Gupta [1989].

One can generalize Burnside's question to monoids and semigroups. For example, if S is a finitely generated semigroup for which there exist integers

$m > n \geq 0$ such that $x^m = x^n$ for all $x \in S$, must S necessarily be finite? The answer is yes for $(m, n) = (2, 1)$, as proved by Green and Rees [1952]. However, for $(m, n) = (3, 2)$ the answer is no, and Morse and Hedlund [1944] observed that a counterexample can easily be constructed using the infinite squarefree word of Theorem 1.6.2; also see Brzozowski, Culik, and Gabrielian [1971]. For more on the Burnside question in semigroups, see McNaughton and Zalcstein [1975]; Lallement [1979, Chapter 10]; and de Luca [1983, 1990].

If a morphism h has the property that $h(w)$ is squarefree if w is, then it is called squarefree. Several writers, such as Crochemore [1982a] and Ehrenfeucht and Rozenberg [1982b], have studied the properties of such morphisms. Carpi [1983a, 1983b] showed that any squarefree morphism on a 3-letter alphabet must be of size at least 18; this bound is best possible. Similarly, De Felice [1983] showed that the minimum width of a uniform squarefree morphism on a 3-letter alphabet is 11. Keränen [1985, 1986] and Leconte [1985] studied the properties of kth-power-free morphisms. Richomme and Wlazinksi [2000, 2002] proved that a morphism h is cubefree over $\{a, b\}$ if and only if $h(aabbababbabbaabaababaabb)$ is cubefree. Wlazinski [2001] discussed test sets for k-power-free morphisms.

A naive algorithm running in $O(n^3)$ steps can detect squares in a word of length n, but this can be considerably improved. Crochemore [1981] gave an $O(n \log n)$ algorithm for the problem, as did Apostolico and Preparata [1983]. Crochemore [1983a, 1984, 1986] improved the result to $O(n)$ provided the alphabet is of fixed size. Apostolico [1984] and Main and Lorentz [1984, 1985] obtained similar results. Rabin [1985] gave a simple efficient randomized algorithm for the square detection problem. Kosaraju [1994] gave an $O(n)$ algorithm to determine the shortest square beginning at every position of a word. Crochemore and Rytter [1991], Apostolico [1992], Apostolico, Breslauer, and Galil [1992], and Apostolico and Breslauer [1996] gave efficient parallel algorithms for the square detection problem. Karhumäki, Plandowski, and Rytter [2000] developed an $O(n \log n)$ algorithm to find the subword with the largest (fractional) power in a given string. For other papers on algorithmic aspects of repetitions, see Main [1989] and Kolpakov and Kucherov [1999b]. For a survey on algorithmic problems in repetitions, see Smyth [2000].

One may ask how a finite squarefree or overlap-free word can be extended to get a longer word with the same property. For results in this area, see Cummings [1978, 1981]; Shelton [1981a, 1981b]; Shelton and Soni [1982]; and Currie [1995a, 1996].

Read [1979] proposed the problem of enumerating the squarefree words over a 3-letter alphabet. Brinkhuis [1983] proved that if $s(n)$ denotes the number of such words of length n and $\lambda = \inf_{n \geq 1} s(n)^{1/n}$, then $1.029 < \lambda < 1.316$. Elser [1983] independently found weaker bounds. The upper bound has been improved to 1.302128 by Noonan and Zeilberger [1999]. Brandenburg [1983] improved the lower bound to 1.032. (He also gave a similar result for the density of cubefree words over a 2-letter alphabet, as did Edlin [1999].) Ekhad and Zeilberger [1998] showed $\lambda > 1.0416$. Currently, the best lower bound is due to Grimm [2001], who showed that $\lambda > 1.109999$. For more on enumeration of ternary squarefree words, see Baake, Elser, and Grimm [1997].

Berstel [1979] showed it was decidable that an infinite word on three letters obtained as the fixed point of a morphism is squarefree. De Luca [1984] investigated the question of when the product of squarefree words is squarefree.

Although infinite binary words cannot be squarefree, it is possible to avoid arbitrarily long squares. Such a word is sometimes said to be of *bounded repetition*. This line was pursued by Entringer, Jackson, and Schatz [1974]; Dekking [1976]; Prodinger and Urbanek [1979]; Prodinger [1983]; Allouche [1984]; P. Roth [1991]; and Allouche and Bousquet-Mélou [1994b].

The construction of squarefree and cubefree infinite words has been used in other areas of mathematics. For example, Burris and Nelson [1971] used **t** in a theorem on lattices of equational theories satisfying $x^2 = x^3$; also see Ježek [1976]. Brzozowski, Culik, and Gabrielian [1971]; Goldstine [1976]; Reutenauer [1981]; Gabarró [1985]; and Main, Bucher, and Haussler [1987] used this and similar constructions in formal language theory. Autebert, Beauquier, Boasson, and Nivat [1979] conjectured that the set of all words containing a square subword is not a context-free language. A proof was found by Ross and Winklmann [1982] and Ehrenfeucht and Rozenberg [1983d], independently. For applications to dynamic logic, see Stolboushkin and Taitslin [1983]; Urzyczyn [1983]; Stolboushkin [1983]; and Kfoury [1985].

Fraenkel and Simpson [1995] investigated the longest binary string containing at most n distinct squares.

The notions of squarefree and cubefree have been extended to two-dimensional arrays (Siromoney and Subramanian [1985]; Siromoney [1987]).

For other papers dealing with repetitions in words, see Lyndon [1951]; Dean [1965]; Restivo [1977]; Berstel [1982]; Carpi [1984]; Carpi and de Luca [1986]; Crochemore [1982b, 1983b]; Cummings [1983]; Shelton and Soni [1985]; Keränen [1987]; Carpi and de Luca [1990]; de Luca and Varricchio [1990]; Crochemore and Goralcik [1991]; Currie and Bendor-Samuel [1992]; X. Yu [1995]; Currie [1995b].

Our proof of Theorem 1.6.1 is based on a proof of a weaker result by D. Bernstein (personal communication); also see Currie [1984] and Klosinski, Alexanderson, and Larson [1993].

1.7 The results in this section are due to Séébold [1982, 1985c, 1985d]. Also see Harju [1986]; Kfoury [1988b]; Berstel and Séébold [1993a]; and Richomme and Séébold [1999].

1.8 Dejean [1972] introduced the notion of fractional powers. She proved that every sufficiently long word over a three-letter alphabet contains a $\frac{7}{4}$ power, and this bound is best possible. She conjectured that this repetition threshold was equal to $k/(k-1)$ on alphabets of size $k \geq 5$; this conjecture is still open. Pansiot [1984c] showed that the repetition threshold for an alphabet of size 4 is $\frac{7}{5}$. Moulin-Ollagnier [1992] showed that Dejean's conjecture is true for $5 \leq k \leq 11$.

Mignosi and Pirillo [1992] proved that the critical exponent for the Fibonacci word (discussed in Section 7.1) is $\frac{\sqrt{5}+5}{2} \doteq 3.618$. For other results on critical exponents, see Klepinin and Sukhanov [1999], Vandeth [2000], and Damanik and Lenz [2002].

Erdős [1961, p. 240] first raised the problem of the existence of infinite abelian squarefree words. (These words are also called *strongly squarefree*, *strongly nonrepetitive*, and *permutation-free* in the literature.) Evdokimov [1968] constructed such a sequence on 25 symbols. Pleasants [1970] improved this to 5 symbols. T. Brown [1971] gave a survey on constructing strongly nonrepetitive sequences. Entringer, Jackson, and Schatz [1974] proved that every infinite word over a 2-letter alphabet contains arbitrarily long abelian squares. Keränen [1992] solved Erdős's problem by exhibiting a strongly nonrepetitive sequence over a 4-letter alphabet. Carpi [1998] showed that there are uncountably many abelian squarefree words over a 4-letter alphabet, and that the number of abelian squarefree words of each length grows exponentially. Cummings [1993] found an interesting connection between Gray codes and abelian squarefree strings. Cummings and Smyth [1997] gave a simple quadratic algorithm to compute all abelian squares in a string of length n. Carpi [1999] studied morphisms preserving abelian squarefreeness.

Cori and Formisano [1990, 1991] introduced the notion of *partially abelian squarefree words*, which are defined with respect to a commutation relation. Also see Justin [1972].

One can also study abelian nth-power-free words. Justin [1972] proved the existence of infinite abelian 5th-power-free words over a 2-letter alphabet. Dekking [1979b] proved the existence of infinite abelian cubefree words over a 3-letter alphabet, and infinite abelian 4th-power-free words over a 2-letter alphabet. Also see Carpi [1993b].

More general pattern avoidance problems were studied by Bean, Ehrenfeucht, and McNulty [1979] and Zimin [1982], independently. U. Schmidt [1987a, 1989] showed that all binary patterns of length at least 13 are avoidable on a 2-letter alphabet. This was improved to length 6 by P. Roth [1992], who classified all unavoidable binary patterns modulo the status of the pattern $xxyyx$. Finally, Cassaigne [1993a] completed the classification of avoidable binary patterns by proving that $xxyyx$ is avoidable on a 2-letter alphabet; also see Goralcik and Vanicek [1991]. For other papers on pattern avoidance, see U. Schmidt [1987b]; Cummings and Mays [2001]; and particularly the influential paper of Baker, McNulty, and Taylor [1989]. Shur [1996a] determined the words avoided by the Thue–Morse sequence. A summary of some open problems in the area can be found in Currie [1993].

It is also possible to study words avoiding specific subwords. See, for example, Schützenberger [1964]; Crochemore, Le Rest, and Wender [1983]; and Rosaz [1995, 1998].

2

Number Theory and Algebra

This chapter is a smorgasbord of some of the basic results from algebra and number theory that are used in the rest of this book.

2.1 Divisibility and Valuations

Let k, n be integers with $k \geq 2$. We define $v_k(n)$ to be the exponent of the highest power of k that divides n. More precisely, $v_k(n) = a$ if $k^a \mid n$ but $k^{a+1} \nmid n$. We define $v_k(0) = +\infty$.

Lemma 2.1.1 *For all integers k, n, n' with $k \geq 2$ we have*

$$v_k(n + n') \begin{cases} = \min(v_k(n), v_k(n')) & \text{if } v_k(n) \neq v_k(n'), \\ \geq v_k(n) & \text{if } v_k(n) = v_k(n'). \end{cases}$$

We also have

$$v_k(nn') \geq v_k(n) + v_k(n'),$$

with equality if $k = p$, a prime.

Proof. Left to the reader. ∎

If p is a prime, the function v_p is sometimes called the *p-adic valuation*. In this case we sometimes write $p^a \mid\mid n$ if $a = v_p(n)$.

2.2 Rational and Irrational Numbers

Let α be a real number. If $\alpha = a/b$ for some integers a, b, then α is said to be rational; otherwise it is irrational.

The valuation v_p can be extended to rational numbers as follows. If $r = a/b$ for integers a, b with $b \neq 0$, then we define $v_p(r) = v_p(a) - v_p(b)$. It is easy to see that this definition is independent of the particular representation chosen for r.

Theorem 2.2.1 *The number $\sqrt{2}$ is irrational.*

Proof. Suppose $\sqrt{2}$ is rational. Then $\sqrt{2} = a/b$ for integers a, b, with $b \neq 0$. Hence $a^2 = 2b^2$. But $v_2(a^2)$ is even, while $v_2(2b^2)$ is odd, a contradiction. ∎

Theorem 2.2.2 *The number e is irrational.*

Proof. Suppose e is rational. Then $e = a/b$ for integers $a, b \geq 1$. But then, since

$$e = \frac{1}{0!} + \frac{1}{1!} + \frac{1}{2!} + \cdots,$$

we have

$$b!e = b! \left(\frac{1}{0!} + \frac{1}{1!} + \frac{1}{2!} + \cdots + \frac{1}{b!} \right) + \frac{1}{b+1} + \frac{1}{(b+1)(b+2)} + \cdots.$$

Hence, using the fact that

$$\frac{1}{b+1} + \frac{1}{(b+1)(b+2)} + \cdots < \frac{1}{b+1} + \frac{1}{(b+1)^2} + \cdots = \frac{1}{b}$$

we find

$$b! \sum_{0 \leq i \leq b} \frac{1}{i!} < b!e < b! \sum_{0 \leq i \leq b} \frac{1}{i!} + \frac{1}{b},$$

so that $b!e$ is not an integer, a contradiction. ∎

Theorem 2.2.3 *The number π is irrational.*

Proof. For integers $n \geq 0$, and real a, define

$$I_n = \int_{-1}^{1} (1 - x^2)^n \cos ax \, dx.$$

Integrate by parts to obtain

$$I_n = \frac{2n}{a} \int_{-1}^{1} x(1 - x^2)^{n-1} \sin ax \, dx$$

$$= \frac{2n}{a} H_{n-1}$$

for $n \geq 1$, where $H_n := \int_{-1}^{1} x(1 - x^2)^n \sin(ax) \, dx$. Now integrate H_{n-1} by parts to obtain

$$H_{n-1} = \frac{1}{a} \int_{-1}^{1} (1 - x^2)^{n-1} \cos ax \, dx - \frac{2(n-1)}{a} \int_{-1}^{1} x^2 (1 - x^2)^{n-2} \cos ax \, dx$$

$$= \frac{1}{a} I_{n-1} - \frac{2(n-1)}{a} \int_{-1}^{1} (1 - (1 - x^2))(1 - x^2)^{n-2} \cos ax \, dx$$

$$= \frac{1}{a} I_{n-1} - \frac{2(n-1)}{a} I_{n-2} + \frac{2(n-1)}{a} I_{n-1}$$

for $n \geq 2$. It follows that

$$I_n = \frac{(2n)(2n-1)}{a^2} I_{n-1} - \frac{(4n)(n-1)}{a^2} I_{n-2} \tag{2.1}$$

for $n \geq 2$. Now define $J_n = a^{2n+1} I_n$ for integers $n \geq 0$. By direct calculation we find $J_0 = 2 \sin a$ and $J_1 = 4 \sin a - 4a \cos a$. The recurrence relation (2.1) gives

$$J_n = 2n(2n-1)J_{n-1} - 4a^2 n(n-1) J_{n-2}.$$

It now easily follows by induction that

$$J_n = n!(P_n \sin a + Q_n \cos a),$$

where P_n, Q_n are polynomials in a with integer coefficients and degree $\leq 2n$.

Now take $a = \pi/2$, and suppose $a = c/d$ for integers c, d. Then we have

$$\frac{c^{2n+1}}{n!} I_n = d^{2n+1} P_n.$$

The right side is an integer for all $n \geq 0$. However, $0 < I_n < 2$, and so

$$0 < \frac{c^{2n+1}}{n!} I_n < 1$$

for large n. This is a contradiction, and so $\pi/2$, and hence π, cannot be rational. ∎

2.3 Algebraic and Transcendental Numbers

We say a complex number is *algebraic* if it is the root of an equation of the form

$$a_n x^n + a_{n-1} x^{n-1} + \cdots + a_1 x + a_0 = 0$$

where $a_0, a_1, \ldots, a_n \in \mathbb{Z}$ and $a_n \neq 0$. Examples of algebraic numbers include the rational numbers, and other numbers such as $i = \sqrt{-1}$, $\sqrt{2}$, $\sqrt[3]{2}$, etc. If a complex number is not algebraic, we say it is *transcendental*.

Theorem 2.3.1 *The set of algebraic numbers is countable.*

Proof. A number is algebraic if and only if it is the root of

$$a_n x^n + a_{n-1} x^{n-1} + \cdots + a_0 = 0 \qquad (2.2)$$

with $a_0, a_1, \ldots, a_n \in \mathbb{Z}$ and $a_n \neq 0$. Define the *rank* of Eq. (2.2) as $n + |a_0| + |a_1| + \cdots + |a_n|$. For any given rank, there are clearly only a finite number of equations, and each equation has finitely many roots. By listing the roots in order of increasing rank, we can arrange a 1–1 correspondence between the algebraic numbers and $1, 2, 3, \ldots$. ∎

Corollary 2.3.2 *Almost all real numbers are transcendental.*

Proof. The real algebraic numbers, being countable, form a set of measure 0. ∎

Let θ be a real number. We say that θ is *approximable by rationals to order n* if there exists a constant $c(\theta)$ such that the inequality

$$\left| \frac{p}{q} - \theta \right| < \frac{c(\theta)}{q^n}$$

has infinitely many distinct rational solutions p/q.

Theorem 2.3.3 *A real algebraic number of degree n is not approximable to any order greater than n.*

Proof. Suppose that θ is a real number satisfying

$$f(\theta) = a_n \theta^n + \cdots + a_1 \theta + a_0 = 0$$

with $a_0, a_1, \ldots, a_n \in \mathbb{Z}$, $a_n \neq 0$. Then there exists a number $M(\theta)$ such that

$$|f'(x)| < M(\theta) \qquad \text{for } \theta - 1 < x < \theta + 1.$$

Suppose that p/q is an approximation to θ. Without loss of generality we may assume that $\theta - 1 < p/q < \theta + 1$, and that p/q is closer to θ than any other root of f, so $f(p/q) \neq 0$. Then

$$\left| f\left(\frac{p}{q} \right) \right| = \frac{|a_n p^n + \cdots + a_1 p + a_0|}{q^n} \geq \frac{1}{q^n},$$

and, by the mean value theorem, we have

$$f\left(\frac{p}{q} \right) = f\left(\frac{p}{q} \right) - f(\theta) = \left(\frac{p}{q} - \theta \right) f'(x)$$

for some x lying between p/q and θ. Thus

$$\left| \frac{p}{q} - \theta \right| = \frac{|f(p/q)|}{|f'(x)|} > \frac{1}{M(\theta) q^n}.$$

So θ is not approximable to any order higher than n. ∎

Corollary 2.3.4 (Liouville, 1844) *The number* $\theta = \sum_{k \geq 1} 10^{-k!} =$ $0.110001000 \cdots$ *is transcendental.*

Proof. Define $\theta_n = \sum_{1 \leq k \leq n} 10^{-k!} = p/10^{n!} = p/q$. Now

$$0 < \theta - \frac{p}{q} = \theta - \theta_n = \sum_{k \geq n+1} 10^{-k!} < 2 \cdot 10^{-(n+1)!} \leq 2q^{-n}.$$

Thus θ is approximable to order n for any n. Hence by Theorem 2.3.3, θ cannot be algebraic. ∎

Note that $\sum_{k \geq 1} 10^{-k!}$ is an example of a *Liouville number*, i.e., one for which the inequality

$$\left| \frac{p}{q} - \theta \right| < \frac{1}{q^n}$$

has a solution for every $n \geq 1$. Not all transcendental numbers are Liouville, but every Liouville number is transcendental.

We give the following theorem without proof.

Theorem 2.3.5 *The numbers π and e are transcendental.*

Here is another useful lemma.

Lemma 2.3.6 *Let β be an algebraic number of degree g. For each $N \geq 1$, there exists a constant $C > 0$ that depends only on β and N, such that, for every polynomial Q of degree N with integer coefficients, we have*

$$Q(\beta) = 0 \quad \text{or} \quad |Q(\beta)| \geq \frac{C}{\|Q\|^{g-1}},$$

where $\| \sum_{0 \leq j \leq N} a_j X^j \| = \max_{0 \leq j \leq N} |a_j|$.

Proof. Let P be the minimal polynomial of β having integer coefficients and leading coefficient $a_g \geq 1$. Let $\beta_1 = \beta, \beta_2, \ldots, \beta_g$ be the conjugates of β, i.e., the roots of P, taken with multiplicity. Let Q be a polynomial of degree N, with leading coefficient c_N, such that $Q(\beta) \neq 0$, and let $\gamma_1, \gamma_2, \ldots, \gamma_N$, be its roots. We note that the β_i's and the γ_j's are all different. We have

$$0 < \left| \frac{1}{a_g^N} \prod_{1 \leq j \leq N} P(\gamma_j) \right| = \left| \prod_{1 \leq j \leq N} \prod_{1 \leq i \leq g} (\gamma_j - \beta_i) \right|$$

$$= \left| \prod_{1 \leq i \leq g} \prod_{1 \leq j \leq N} (\gamma_j - \beta_i) \right| = \left| \frac{1}{c_N^g} \prod_{1 \leq i \leq g} Q(\beta_i) \right|.$$

Hence

$$0 < |Q(\beta)| = \frac{\left| c_N^g \prod_{1 \leq j \leq N} P(\gamma_j) \right|}{a_g^N \left| \prod_{2 \leq i \leq g} Q(\beta_i) \right|}. \tag{2.3}$$

Now the expression $\prod_{1 \leq j \leq N} P(\gamma_j)$ is a symmetric polynomial in the γ_j's with integer coefficients, so the quantity $\left| c_N^g \prod_{1 \leq j \leq N} P(\gamma_j) \right|$ is a positive integer. Hence we deduce from Eq. (2.3) that

$$Q(\beta) \geq \frac{1}{a_g^N \left(\sup_{2 \leq i \leq g} |Q(\beta_i)| \right)^{g-1}}.$$

Note that

$$\sup_{2 \leq i \leq g} |Q(\beta_i)|^{g-1} \leq \|Q\|^{g-1} \sup_{1 \leq i \leq g} (1 + |\beta_i| + \cdots + |\beta_i|^N).$$

Hence

$$Q(\beta) \geq \frac{C}{\|Q\|^{g-1}},$$

where C depends only on β and N. ∎

Many more advanced techniques exist for proving transcendence. One of the most useful is Roth's theorem, which we state without proof:

Theorem 2.3.7 *Let θ be a real number, and suppose there exists a real number $\epsilon > 0$ such that*

$$\left| \theta - \frac{p}{q} \right| < \frac{1}{q^{2+\epsilon}}$$

for infinitely many distinct rational numbers p/q. Then θ is transcendental.

2.4 Continued Fractions

A *finite continued fraction* is an expression of the form

$$a_0 + \cfrac{1}{a_1 + \cfrac{1}{a_2 + \cfrac{1}{a_3 + \cdots + \cfrac{1}{a_n}}}}, \tag{2.4}$$

which we abbreviate as $[a_0, a_1, \ldots, a_n]$. If the a_i are integers such that $a_i \geq 1$ for $1 \leq i \leq n$, then the continued fraction is called *simple*.

The following result is fundamental to the theory of continued fractions.

Theorem 2.4.1 *Let $a_0, a_1, \ldots, a_n$ be real numbers with $a_i > 0$ for $i \geq 1$. Define p_n and q_n as follows:*

$$p_{-2} = 0, \quad q_{-2} = 1; \qquad p_{-1} = 1, \quad q_{-1} = 0,$$

and

$$p_k = a_k p_{k-1} + p_{k-2}, \qquad q_k = a_k q_{k-1} + q_{k-2} \tag{2.5}$$

for $0 \leq k \leq n$. If $\alpha > 0$ is a real number, then

$$[a_0, a_1, \ldots, a_n, \alpha] = \frac{\alpha p_n + p_{n-1}}{\alpha q_n + q_{n-1}}.$$

Proof. The proof is by induction. The result clearly holds for $n = -1, 0$. Now assume the result holds for n. We find

$$[a_0, a_1, \ldots, a_{n+1}, \alpha] = \left[a_0, a_1, \ldots, a_n, a_{n+1} + \frac{1}{\alpha} \right]$$

$$= \frac{\left(a_{n+1} + \frac{1}{\alpha} \right) p_n + p_{n-1}}{\left(a_{n+1} + \frac{1}{\alpha} \right) q_n + q_{n-1}} \qquad \text{(by the induction hypothesis)}$$

$$= \frac{\alpha p_{n+1} + p_n}{\alpha q_{n+1} + q_n} \qquad \text{by (2.5)},$$

and so the result follows. ∎

From (2.5) we get

$$\begin{bmatrix} p_n & p_{n-1} \\ q_n & q_{n-1} \end{bmatrix} \begin{bmatrix} a_{n+1} & 1 \\ 1 & 0 \end{bmatrix} = \begin{bmatrix} p_{n+1} & p_n \\ q_{n+1} & q_n \end{bmatrix},$$

and hence an easy induction gives

$$\begin{bmatrix} p_n & p_{n-1} \\ q_n & q_{n-1} \end{bmatrix} = \begin{bmatrix} a_0 & 1 \\ 1 & 0 \end{bmatrix} \begin{bmatrix} a_1 & 1 \\ 1 & 0 \end{bmatrix} \cdots \begin{bmatrix} a_n & 1 \\ 1 & 0 \end{bmatrix}. \tag{2.6}$$

By taking the determinants of both sides, we get

Theorem 2.4.2

$$p_n q_{n-1} - p_{n-1} q_n = (-1)^{n+1}. \tag{2.7}$$

We now define infinite simple continued fractions. Provided the limit exists, we may define the symbol $[a_0, a_1, \ldots, a_n, \ldots]$ as $\lim_{n \to \infty}[a_0, a_1, \ldots, a_n]$. If the continued fraction is simple, that is, if the a_i are integers and $a_i \geq 1$ for $i \geq 1$, then it is easy to see that the limit exists. For by Eq. (2.7) we see that

$$\left| \frac{p_n}{q_n} - \frac{p_{n-1}}{q_{n-1}} \right| = \frac{|p_n q_{n-1} - p_{n-1} q_n|}{q_n q_{n-1}} = \frac{1}{q_n q_{n-1}}.$$

Since $q_n \geq n$ by (2.5), we see that p_n / q_n is a Cauchy sequence, and hence converges to a limit.

Henceforth, unless otherwise stated, we assume all continued fractions are simple.

We state the following theorem without proof.

Theorem 2.4.3 *The value of a simple continued fraction is rational if and only if the continued fraction is finite.*

For example, we have

$$\pi = [3, 7, 15, 1, 292, 1, 1, 1, 2, 1, 3, \ldots],$$
$$e = [2, 1, 2, 1, 1, 4, 1, 1, 6, 1, 1, 8, \ldots].$$

There is no known simple pattern in the partial quotients for π, but the evident pattern in the expansion for e can be proved.

We now introduce some notation for continued fractions with an ultimately periodic sequence of partial quotients. If $a_{n+t} = a_n$ for all $n > N$, we write

$$[a_0, a_1, a_2, \ldots] = [a_0, a_1, \ldots, a_N, \overline{a_{N+1}, a_{N+2}, \ldots, a_{N+t}}].$$

The following is a classical theorem of Lagrange, which we state without proof.

Theorem 2.4.4 *The partial quotients in the continued fraction expansion for real α are ultimately periodic if and only if α is an irrational number that satisfies a quadratic equation with integer coefficients.*

For example, we have

$$\sqrt{2} = [1, 2, 2, 2, 2, \ldots],$$
$$\frac{1 + \sqrt{5}}{2} = [1, 1, 1, 1, 1, \ldots].$$

We now prove some basic facts about approximation. Let α be an irrational real number with continued fraction $[a_0, a_1, a_2, \ldots]$, and define $\eta_k = (-1)^k (q_k \alpha - p_k)$ for $k \geq -1$.

Theorem 2.4.5 *We have*

(a) $\eta_k > 0$ *for* $k \geq 0$;
(b) $\eta_{k+1} < \eta_k$ *for* $k \geq 1$;
(c) $\eta_{k+1} = -a_{k+1}\eta_k + \eta_{k-1}$ *for* $k \geq 0$.

Proof. Define $a'_k = [a_k, a_{k+1}, \dots]$. We have

$$\alpha = [a_0, a_1, \dots, a_k, a_{k+1}, \dots]$$
$$= [a_0, a_1, \dots, a_k, [a_{k+1}, a_{k+2}, \dots]]$$
$$= [a_0, a_1, \dots, a_k, a'_{k+1}]$$
$$= \frac{a'_{k+1} p_k + p_{k-1}}{a'_{k+1} q_k + q_{k-1}}.$$

It follows that

$$\alpha - \frac{p_k}{q_k} = \frac{a'_{k+1} p_k + p_{k-1}}{a'_{k+1} q_k + q_{k-1}} - \frac{p_k}{q_k}$$

$$= \frac{p_{k-1} q_k - p_k q_{k-1}}{q_k(a'_{k+1} q_k + q_{k-1})}$$

$$= \frac{(-1)^k}{q_k(a'_{k+1} q_k + q_{k-1})}.$$

Hence

$$\eta_k = (-1)^k(q_k \alpha - p_k) = \frac{1}{a'_{k+1} q_k + q_{k-1}} > 0. \tag{2.8}$$

This proves part (a).

To prove part (b), note that $a_{k+1} < a'_{k+1} < a_{k+1} + 1$. Hence, by (2.8), we get

$$\frac{1}{q_{k+2}} \leq \frac{1}{q_{k+1} + q_k} = \frac{1}{(a_{k+1} + 1)q_k + q_{k-1}} < \eta_k < \frac{1}{a_{k+1} q_k + q_{k-1}} = \frac{1}{q_{k+1}},$$

i.e.,

$$\frac{1}{q_{k+2}} < \eta_k < \frac{1}{q_{k+1}}.$$

Since $q_{k+1} < q_{k+2}$ for $k \geq 1$, this proves part (b).

Finally, from (2.5), we get

$$p_{k+1} = a_{k+1} p_k + p_{k-1},$$
$$q_{k+1} = a_{k+1} q_k + q_{k-1}.$$

Subtract the first equation from α times the second; we get

$$\alpha q_{k+1} - p_{k+1} = a_{k+1}(\alpha q_k - p_k) + (\alpha q_{k-1} - p_{k-1}),$$

or, in other words, $\eta_{k+1} = -a_{k+1}\eta_k + \eta_{k-1}$. ∎

Corollary 2.4.6 *For $k \geq 1$ we have*

$$a_{k+1} < \frac{\eta_{k-1}}{\eta_k} < a_{k+1} + 1.$$

Proof. By part (c) of the preceding theorem, we have $\eta_{k-1} = \eta_{k+1} + a_{k+1}\eta_k$. Dividing by η_k, we get

$$\frac{\eta_{k-1}}{\eta_k} = a_{k+1} + \frac{\eta_{k+1}}{\eta_k}.$$

But by parts (a) and (b), we know

$$0 < \frac{\eta_{k+1}}{\eta_k} < 1,$$

so the desired result follows.   ∎

2.5 Basics of Diophantine Approximation

Let $x \bmod y$ denote the least non-negative remainder when x is divided by y. Let $\{x\}$ denote $x \bmod 1 = x - \lfloor x \rfloor$, the fractional part of the real number x.

Theorem 2.5.1 (Dirichlet) *For all real numbers θ and positive integers N, there exists a positive integer $n \leq N$ and an integer r such that*

$$|n\theta - r| < \frac{1}{N}.$$

Proof. Consider the $N + 1$ numbers

$$\{0\}, \{\theta\}, \{2\theta\}, \ldots, \{N\theta\},$$

and consider the N intervals $\left[\frac{m}{N}, \frac{m+1}{N}\right)$ for $0 \leq m < N$. There are $N + 1$ numbers in N intervals, so by the pigeonhole principle, at least two numbers, say $\{i\theta\}$ and $\{j\theta\}, 0 \leq i < j \leq N$, fall into some interval, say $\left[\frac{m}{N}, \frac{m+1}{N}\right)$. Then define $n = j - i$. Clearly $0 < n \leq N$. Also write $i\theta = s + \{i\theta\}$, $j\theta = t + \{j\theta\}$. Then $(j - i)\theta = t - s + \{j\theta\} - \{i\theta\}$; define $r = t - s$, and then we get

$$|n\theta - r| < \frac{1}{N}.$$ ∎

Example 2.5.2 Find integers n, r such that $|n\pi - r| < \frac{1}{100}$. We arrange the fractional parts $\{k\pi\}, 0 \leq k \leq 100$, into intervals of size $1/100$ and look for duplicates; we get for example

$$50\pi \doteq 157.07963$$
$$57\pi \doteq 179.07078$$

Thus $7\pi \doteq 21.99115$ and $|7\pi - 22| < \frac{1}{100}$.

Dirichlet's theorem states that for any real number θ, some multiple of θ must be an integer or arbitrarily close to an integer. The following theorem, due to Kronecker, says that if θ is irrational, then the fractional part of some multiple of θ must be arbitrarily close to any given number.

Theorem 2.5.3 (Kronecker) *Let θ be an irrational number. For all real α and all $\epsilon > 0$, there exist integers a, c with $|a\theta - \alpha - c| < \epsilon$.*

We say a set S is *dense* in an interval I if every subinterval of I contains an element of S. Similarly, we say a sequence $(a_n)_{n\geq 0}$ is dense in I if the associated set $\{a_0, a_1, \dots\}$ is dense in I. Kronecker's theorem can be expressed by saying that the sequence $(\{n\theta\})_{n\geq 0}$ is dense in the interval $[0, 1)$.

Proof. By Dirichlet's theorem (Theorem 2.5.1), for all $\epsilon > 0$ there exist integers a, b with $|a\theta - b| < \epsilon$. Furthermore, since θ is irrational, $0 < |a\theta - b|$. Then the series of points $0, \{a\theta\}, \{2a\theta\}, \dots$ form a *chain* across the interval $[0, 1)$ whose *mesh* (i.e., the distance between links in the chain) is less than ϵ. (Note that the chain goes left to right if $a\theta - b > 0$, and right to left if $a\theta - b < 0$.) Thus $\{\alpha\}$ must fall between links, and so there exist integers a, c with

$$|a\theta - \alpha - c| < \epsilon. \qquad \blacksquare$$

Example 2.5.4 Find a, c such that

$$|a\pi - \sqrt{2} - c| < \frac{1}{100}.$$

By the previous calculation we have that $7\pi \doteq 21.99115$. Thus $\{7a\pi\}$ forms a mesh from right to left of the interval $[0, 1)$, with width $\epsilon \doteq 0.00885$, and thus we get within ϵ of $\{\sqrt{2}\}$ when

$$a \doteq \frac{2 - \sqrt{2}}{22 - 7\pi} \doteq 66.1799.$$

Thus

$$|462\pi - \sqrt{2} - 1450| < \frac{1}{100}.$$

Kronecker's Theorem 2.5.3 is quite important, but does not tell the full story. Not only is $\{n\theta\}$ dense in $[0, 1)$, it is also uniformly distributed. By *uniformly distributed*, we mean that the fraction of n for which $\{n\theta\}$ falls in interval I is equal to the measure of I. More precisely, let $m_I(n)$ be the number of points $\{k\theta\}_{1\leq k\leq n}$

that fall in the interval I. If

$$\lim_{n \to \infty} \frac{m_I(n)}{n} = |I|$$

for all $I \subseteq (0, 1)$, we say $\{n\theta\}$ is uniformly distributed.

Theorem 2.5.5 (Weyl) *If θ is an irrational number then $\{n\theta\}$ is uniformly distributed.*

Proof. Let $0 < \epsilon \leq \frac{1}{3}$. By Theorem 2.5.3 we can find a j such that $0 < \{j\theta\} = \delta < \epsilon$. Write $k = \lfloor \frac{1}{\delta} \rfloor$. For $0 \leq h \leq k$, define I_h to be the interval

$$(\{hj\theta\}, \{(h + 1)j\theta\}].$$

(When $h = k$, this "interval" actually wraps around 1 to 0.) Define, as above, $m_h(n) := m_{I_h}(n)$ to be the number of points $\{\theta\}, \{2\theta\}, \ldots, \{n\theta\}$ that lie in I_h. Now if $\{t\theta\} \in I_0$ then $\{(t + hj)\theta\} \in I_h$, and conversely. Hence

$$m_h(n) - m_h(hj) = m_0(n - hj),$$

because for $n \geq hj$, the points counted by $m_0(n - hj)$ and $m_h(n) - m_h(hj)$ are in 1–1 correspondence. But $m_h(hj) \leq hj$, and $m_0(n - hj) \geq m_0(n) - hj$. Hence

$$m_0(n) - hj \leq m_0(n - hj) \leq m_h(n) = m_0(n - hj) + m_h(hj)$$
$$\leq m_0(n) + m_h(hj)$$
$$\leq m_0(n) + hj.$$

It follows that

$$1 - \frac{hj}{m_0(n)} \leq \frac{m_h(n)}{m_0(n)} \leq 1 + \frac{hj}{m_0(n)},$$

as $n \to \infty$, $m_0(n) \to \infty$; and we have

$$\lim_{n \to \infty} \frac{m_h(n)}{m_0(n)} = 1, \qquad 0 \leq h \leq k.$$

Now

$$\sum_{0 \leq h \leq k-1} m_h(n) \leq n \leq \sum_{0 \leq h \leq k} m_h(n),$$

so we have

$$\frac{m_0(n) + \cdots + m_{k-1}(n)}{m_0(n)} \leq \frac{n}{m_0(n)},$$

so

$$\liminf_{n \to \infty} \frac{m_0(n) + \cdots + m_{k-1}(n)}{m_0(n)} \leq \liminf_{n \to \infty} \frac{n}{m_0(n)}.$$

But

$$\liminf_{n\to\infty} \frac{m_0(n) + \cdots + m_{k-1}(n)}{m_0(n)} = \lim_{n\to\infty} \frac{m_0(n) + \cdots + m_{k-1}(n)}{m_0(n)}$$

$$= \lim_{n\to\infty} \frac{m_0(n)}{m_0(n)} + \cdots + \frac{m_{k-1}(n)}{m_0(n)}$$

$$= k.$$

Hence

$$k \leq \liminf_{n\to\infty} \frac{n}{m_0(n)},$$

and so

$$\frac{1}{k} \geq \limsup_{n\to\infty} \frac{m_0(n)}{n}.$$

Similarly,

$$\frac{1}{k+1} \leq \liminf_{n\to\infty} \frac{m_0(n)}{n},$$

and of course

$$\liminf_{n\to\infty} \frac{m_0(n)}{n} \leq \limsup_{n\to\infty} \frac{m_0(n)}{n}.$$

Hence

$$\frac{1}{k+1} \leq \liminf_{n\to\infty} \frac{m_0(n)}{n} \leq \limsup_{n\to\infty} \frac{m_0(n)}{n} \leq \frac{1}{k}. \qquad (2.9)$$

Now let $I = (\alpha, \beta)$ with $|I| = \beta - \alpha \geq \epsilon$. Then there are integers u, v such that

$$0 \leq \{uj\theta\} \leq \alpha \leq \{(u+1)j\theta\} \leq \{(u+v)j\theta\} \leq \beta < \{(u+v+1)j\theta\}$$

(again using the circular representation of intervals), so that

$$\sum_{u+1\leq h\leq u+v-1} m_h(n) \leq m_I(n) \leq \sum_{u\leq h\leq u+v} m_h(n).$$

Hence, dividing by $m_0(n)$ and applying the reasoning that gave us inequality (2.9), we get

$$v - 1 \leq \liminf_{n\to\infty} \frac{m_I(n)}{m_0(n)} \leq \limsup_{n\to\infty} \frac{m_I(n)}{m_0(n)} \leq v + 1,$$

and so, multiplying by inequality (2.9),

$$\frac{v-1}{k+1} \leq \liminf_{n\to\infty} \frac{m_I(n)}{n} \leq \limsup_{n\to\infty} \frac{m_I(n)}{n} \leq \frac{v+1}{k}.$$

Now $k\delta \leq 1 \leq (k+1)\delta$ and $(v-1)\delta < |I| < (v+1)\delta$. Hence $1+\delta \geq (k+1)\delta$; $|I| - 2\delta < (v-1)\delta$; and so

$$\frac{|I| - 2\delta}{\delta + 1} \leq \frac{(v-1)\delta}{(k+1)\delta} = \frac{v-1}{k+1}.$$

Similarly, $|I| + 2\delta > (v+1)\delta$, $1 - \delta \leq k\delta$, and so

$$\frac{|I| + 2\delta}{1 - \delta} \geq \frac{(v+1)\delta}{k\delta} = \frac{v+1}{k}.$$

Therefore

$$\frac{|I| - 2\delta}{\delta + 1} \leq \liminf_{n\to\infty} \frac{m_I(n)}{n} \leq \limsup_{n\to\infty} \frac{m_I(n)}{m_0(n)} \leq \frac{|I| + 2\delta}{1 - \delta}.$$

But ϵ can be arbitrarily small, hence δ is also arbitrarily small, and thus

$$\lim_{n\to\infty} \frac{m_I(n)}{n} = |I|.$$

The proof is complete.   ∎

Weyl also gave the following necessary and sufficient condition for uniform distribution, which we state without proof.

Theorem 2.5.6 *The sequence of real numbers $(a_n)_{n\geq 1}$ is uniformly distributed if and only if for all integers $h \neq 0$ we have*

$$\lim_{N\to\infty} \frac{1}{N} \sum_{1\leq n\leq N} e^{2\pi i h a_n} = 0.$$

We conclude this section by proving results on multiplicative independence. We say two integers $k, l \geq 1$ are *multiplicatively independent* if $\log k$ and $\log l$ are linearly independent over $\mathbb{Q}$. Otherwise k and l are *multiplicatively dependent*.

Theorem 2.5.7 *Let k, l be integers ≥ 2. The following are equivalent:*

(a) k and l are multiplicatively dependent.
(b) $\log_k l$ is rational.
(c) $\log_l k$ is rational.
(d) There exist integers $r, s \geq 1$ such that $k^r = l^s$.
(e) There exist an integer $n \geq 2$ and two integers $b, c \geq 1$ such that $k = n^b$ and $l = n^c$.

Proof. (a) $\Longrightarrow$ (b): If k and l are multiplicatively dependent, then there exist rational numbers r, s, not both 0, such that $r \log k + s \log l = 0$. Since $k, l \geq 2$, we have

$\log k > 0$ and $\log l > 0$. Hence $r \neq 0$ and $s \neq 0$. Thus $\log_k l = (\log l)/(\log k) = -r/s$, which is rational.

(b) $\Longrightarrow$ (c): We have $\log_k l = (\log_l k)^{-1}$.

(c) $\Longrightarrow$ (d): By hypothesis $\log_l k = s/r$ for positive integers r, s. Then $k = l^{s/r}$ and hence $k^r = l^s$.

(d) $\Longrightarrow$ (e): Suppose $k^r = l^s$. Let $g = \gcd(r, s)$. Without loss of generality we may assume $g = 1$, for otherwise $k^{r/g} = l^{s/g}$. Let p be a prime dividing k. Then $p^{a_p} \,\|\, k$ for some integer $a_p \geq 1$, and $p^{b_p} \,\|\, l$ for some integer $b_p \geq 1$. Then $a_p r = b_p s$. Since $\gcd(r, s) = 1$, we have $r \mid b_p$ and $s \mid a_p$. Let $n = \prod_{p \mid k} p^{a_p/s} = \prod_{p \mid l} p^{b_p/r}$. Then $n^s = k$ and $n^r = l$, as desired.

(e) $\Longrightarrow$ (a): If $k = n^b$ and $l = n^c$, then $\log k - \frac{b}{c} \log l = 0$, so k and l are multiplicatively dependent. $\blacksquare$

Example 2.5.8 36 and 216 are multiplicatively dependent, since $36^3 = 216^2$.

We now prove

Lemma 2.5.9 *If k and l are multiplicatively independent, then the set of quotients*

$$\{k^p / l^q : p, q \geq 0\}$$

is dense in the positive reals.

Proof. We show how to get arbitrarily close to any $x > 0$. Let $\theta = \frac{\log k}{\log l} = \log_l k$; then θ is irrational by Proposition 2.5.7. Let $\alpha = \frac{\log x}{\log l}$. By Kronecker's theorem (Theorem 2.5.3), for all $\epsilon > 0$ there exist integers $a, c \geq 0$ with $|a\theta - \alpha - c| < \epsilon$. Hence $|a \log k - \log x - c \log l| < \epsilon \log l$, or $a \log k - c \log l \in (\log x - \epsilon \log l, \log x + \epsilon \log l)$. Thus $k^a / l^c \in (xl^{-\epsilon}, xl^{\epsilon})$, which can fit inside any open interval around x by taking ϵ sufficiently small. $\blacksquare$

2.6 The Three-Distance Theorem

In this section we sketch the proof of a theorem that was originally conjectured by Steinhaus. It is variously called the three-distance theorem, the three-gap theorem, or Steinhaus's conjecture.

Let α be an irrational real number with $0 < \alpha < 1$, and let its continued fraction expansion be $\alpha = [0, a_1, a_2, a_3, \dots]$. As in Section 2.4, let $p_k/q_k = [0, a_1, a_2, \dots, a_k]$ and $\eta_k = (-1)^k (q_k \alpha - p_k)$. Let n be a positive integer. Consider the $n + 2$ numbers

$$0, \{\alpha\}, \{2\alpha\}, \dots, \{n\alpha\}, 1.$$

Arrange these numbers in increasing order $0 = c_0(n) < c_1(n) < \cdots < c_n(n) < c_{n+1}(n) = 1$, and for $0 \le i \le n$, define the half-open interval $L_i(n) = [c_i(n), c_{i+1}(n))$. The three-distance theorem says that for fixed n, there are at most three distinct interval lengths.

If we consider the numbers $\{i\alpha\}$ inserted consecutively, then intervals of various lengths appear and disappear through time. With each interval length, we can associate numbers indicating at what point in time they are introduced.

Theorem 2.6.1 *Every integer $n \ge 1$ can be written uniquely in the form $n = mq_k + q_{k-1} + r$, where $1 \le m \le a_{k+1}$ and $0 \le r < q_k$. Then there are only three distinct interval lengths among the $L_i(n)$, $0 \le i \le n$, and they are distributed as follows. There are*

$r + 1$ intervals of length $\eta_{k-1} - m\eta_k$ (called type I), numbered $0, \ldots, r$;
$n + 1 - q_k$ intervals of length of length η_k (called type II), numbered $0, \ldots, n + 1 - q_k$;
$q_k - r - 1$ intervals of length $\eta_{k-1} - (m - 1)\eta_k$ (called type III), numbered $r + 1, \ldots, q_k - 1$.

Proof. First, some notation. As the points $\{\alpha\}, \{2\alpha\}, \ldots$ are inserted into the interval $[0, 1)$, let the segments of a given length be numbered $0, 1, \ldots$ in order of their appearance. By $\{x\}^+$ we mean $x + 1 - \lceil x \rceil$.

Now we outline the proof of the theorem by induction on n. For our induction hypothesis, we take the above, plus the following:

Interval number s of length $\{t\alpha\}$, where $t = mq_k + q_{k-1}$ and $0 \le m < a_{k+1}$ and k is even and $0 \le s < q_k$, has left endpoint $\{s\alpha\}$ and right endpoint $\{(s + t)\alpha\}^+$.
Interval number s of length $1 - \{t\alpha\}$, where $t = mq_k + q_{k-1}$ and $0 \le m < a_{k+1}$ and k is odd and $0 \le s < q_k$, has left endpoint $\{(s + t)\alpha\}$ and right endpoint $\{s\alpha\}^+$.
The operation of inserting $\{n\alpha\}$ destroys interval number r of type III, and creates new intervals number r of type I, and number $n - q_k$ of type II.

The details are left to the reader. ∎

Theorem 2.6.2 *Let $q_k \le n < q_{k+1}$. Then*

$$\min_{0 \le i \le n} |L_i(n)| = \eta_k = (-1)^k (q_k \alpha - p_k).$$

(Note that for $n = 1$ and $a_1 = 1$ one must choose $k = 1$.)

Proof. Since the length of an interval of type III is the sum of the lengths of the intervals of types I and II, the smallest interval must be of either type I or type II. There are two cases to consider.

First, if $q_k \le n < q_k + q_{k-1}$, then in the representation for n mentioned in Theorem 2.6.1, we have $n = a_k q_{k-1} + q_{k-2} + r$, where $0 \le r < q_{k-1}$. In this case, the smallest intervals have length η_{k-1} and $\eta_{k-2} - a_k\eta_{k-1}$. Now by Corollary 2.4.6, we have $a_k + 1 > \eta_{k-2}/\eta_{k-1}$. From this we have $\eta_{k-1} > \eta_{k-2} - a_k\eta_{k-1}$, so the smallest interval is of type II and has length $\eta_{k-2} - a_k\eta_{k-1}$. But by Theorem 2.4.5(c), this equals η_k.

Second, if $q_k + q_{k-1} \leq n < q_{k+1}$, then in the representation for n we have $n = mq_k + q_{k-1} + r$, and hence $m < a_{k+1}$. Thus $m \leq a_{k+1} - 1$, and so, using Corollary 2.4.6, we have $m + 1 < \eta_{k-1}/\eta_k$. Thus $\eta_k < \eta_{k-1} - m\eta_k$, and so the smallest interval is of type I and of length η_k. ∎

Theorem 2.6.3 *Suppose $n \geq q_k - 1$. Then*

$$\max_{0 \leq i \leq n} |L_i(n)| \leq \eta_k + \eta_{k-1}.$$

Proof. It is easily seen that $\max_{0 \leq i \leq n} |L_i(n)|$ is a nonincreasing function of n. Hence it suffices to prove the result for $n = q_k - 1$. There are two cases to consider: $a_k > 1$ and $a_k = 1$.

If $a_k > 1$, then the representation for n from Theorem 2.6.1 is $(a_k - 1)q_{k-1} + q_{k-2} + (q_{k-1} - 1)$. Also, there are no intervals of type III, so the longest interval must be of length η_{k-1} or $\eta_{k-2} - (a_k - 1)\eta_{k-1}$. But by Theorem 2.4.5(c), this last quantity equals $\eta_k + \eta_{k-1}$.

If $a_k = 1$, then the representation for n from Theorem 2.6.1 is $a_{k-1}q_{k-2} + q_{k-3} + (q_{k-2} - 1)$. Again, there are no intervals of type III, so the longest interval must be of length η_{k-2} or $\eta_{k-3} - a_{k-1}\eta_{k-2}$. By Theorem 2.4.5(c) and the fact that $a_k = 1$, the first quantity equals $\eta_k + \eta_{k-1}$, while the second quantity equals η_{k-1}.

Thus in all cases, the largest interval is bounded by $\eta_k + \eta_{k-1}$. ∎

Corollary 2.6.4 *Suppose $q_k \leq n < q_{k+1}$. Then*

$$\frac{\max_{0 \leq i \leq n} |L_i(n)|}{\min_{0 \leq i \leq n} |L_i(n)|} \leq a_{k+1} + 2.$$

Proof. By Theorem 2.6.2 and 2.6.3, we have

$$\frac{\max_{0 \leq i \leq n} |L_i(n)|}{\min_{0 \leq i \leq n} |L_i(n)|} \leq \frac{\eta_k + \eta_{k-1}}{\eta_k} = 1 + \frac{\eta_{k-1}}{\eta_k} < a_{k+1} + 2,$$

where we have used Corollary 2.4.6. ∎

2.7 Algebraic Structures

A *semigroup* is a nonempty set S together with a binary operation defined on $S \times S$. Further, this operation is associative: for all $a, b, c \in S$ we have $(ab)c = a(bc)$.

A *monoid* is a semigroup M with a special element $e \in M$, called the identity, such that $ea = ae = a$ for all $a \in M$. A *free monoid* over a finite set Σ is what we have previously called Σ^*, the set of all finite strings over Σ, together with the operation of concatenation and the identity ϵ.

A *group* is a monoid G such that, for all $a \in G$, there exists an element $a^{-1} \in G$ such that $aa^{-1} = a^{-1}a = e$. A group G is *abelian* if it is commutative, that is, if $ab = ba$ for all $a, b \in G$.

A *semiring* is a set S containing elements 0 and 1, together with binary operations $+$ and $\cdot$ such that $(S, +, 0)$ is a commutative monoid, $(S, \cdot, 1)$ is a monoid, the additive identity 0 satisfies $a \cdot 0 = 0 \cdot a = 0$, and $\cdot$ distributes over $+$, that is, $a \cdot (b + c) = a \cdot b + a \cdot c$ and $(a + b) \cdot c = a \cdot c + b \cdot c$. If $n \in \mathbb{N}$, then by na we mean the element $\overbrace{a + a + \cdots + a}^{n}$.

A *ring* is a semiring R such that $(R, +, 0)$ is an abelian group.

A *field* is a ring R such that $(R \setminus \{0\}, \cdot, 1)$ also forms an abelian group.

2.8 Vector Spaces

In this section we recall the basic notions about vector spaces.

A nonempty set V is called a *vector space* over a field F if V is an abelian group under an operation $+$, and for every $a \in F$, $v \in V$, there exists an element of V called av such that, for all $a, b \in F$, $v, w \in V$ we have

(1) $a(v + w) = av + aw$,
(2) $(a + b)v = av + bv$,
(3) $a(bv) = (ab)v$, and
(4) $1(v) = v$.

The members of V are called *vectors*. If $W \subseteq V$ itself forms a vector space (under the same operations as for V), then W is a *subspace* of V.

If V is a vector space over F, and if $v_1, v_2, \ldots, v_n$ are elements of V, then by a *linear combination* over F we mean the quantity

$$a_1 v_1 + a_2 v_2 + \cdots + a_n v_n,$$

where $a_1, a_2, \ldots, a_n$ are elements of F. The vectors $v_1, v_2, \ldots, v_n$ are *linearly dependent* if 0 can be written as a nonzero linear combination of $v_1, v_2, \ldots, v_n$. Otherwise they are *linearly independent*.

2.9 Fields

In this section we recall the basic facts about fields.

One typical way to obtain a field is by starting with an integral domain, i.e., a commutative ring R with the property that $a, b \in R$ and $ab = 0 \implies a = 0$ or $b = 0$. In this case we can form the *field of fractions* by considering all quotients

a/b for $a, b \in R$ and $b \neq 0$, and identifying a/b with c/d if $ad = bc$. For example, the field of fractions of $\mathbb{Z}$ is $\mathbb{Q}$.

Suppose K and F are fields. Then K is called an *extension* of F if $F \subseteq K$. In this case, it is easy to see that K forms a vector space over F. The dimension of K as a vector space over F is called the *degree* of the extension, and is denoted by $[K : F]$.

Suppose $a \in K$. Then a is said to be *algebraic* over F if there exist elements $c_0, c_1, \ldots, c_n \in F$, not all 0, such that $c_n a^n + \cdots + c_1 a + c_0 = 0$. This is a generalization of the term "algebraic", which was introduced in Section 2.3. More precisely, a complex number is said to be algebraic if it is algebraic over the field $\mathbb{Q}$.

Theorem 2.9.1 *Let K, F be fields with K an extension of F. Let α, β be elements of K that are algebraic over F. Then the following are all algebraic over F:*

(a) $-\alpha$;
(b) α^{-1} if $\alpha \neq 0$;
(c) $\alpha + \beta$;
(d) $\alpha\beta$.

Proof. Let α be algebraic of degree m, i.e., there exist $a_0, a_1, \ldots, a_{m-1}, a_m$ with $a_m \neq 0$ such that

$$a_0 + a_1\alpha + \cdots + a_m\alpha^m = 0$$

and β be algebraic of degree n, i.e., there exist $b_0, b_1, \ldots, b_{n-1}, b_n$ with $b_n \neq 0$ such that

$$b_0 + b_1\beta + \cdots + b_n\beta^n = 0.$$

By dividing by a_m and b_n respectively, we may assume that $a_m = b_n = 1$.

(a): $-\alpha$ satisfies the polynomial

$$a_0 - a_1 X + a_2 X^2 - a_3 X^3 + \cdots + (-1)^m a_m X^m.$$

(b): α^{-1} satisfies the polynomial

$$a_0 X^m + a_1 X^{m-1} + \cdots + a_{m-1} X + 1.$$

(c): Let v be the vector

$$(1, \alpha, \ldots, \alpha^{m-1}, \beta, \beta\alpha, \ldots, \beta\alpha^{m-1}, \ldots, \beta^{n-1}, \beta^{n-1}\alpha, \ldots, \beta^{n-1}\alpha^{m-1}).$$

Then $(\alpha + \beta)v$ is a linear combination of elements of v, i.e., there exists a matrix M such that $(\alpha + \beta)v = Mv$. Hence, letting $x = \alpha + \beta$, we have $(M - Ix)v = 0$, where I is the identity matrix. Since $v \neq 0$, it follows that $\det(M - Ix) = 0$. This gives a polynomial for which $X = \alpha + \beta$ is a root.

(d): The same proof as in (c) works. ∎

Example 2.9.2 Show that $\sqrt{2} + \sqrt{3}$ is algebraic over $\mathbb{Q}$.

Let $v = (1, \sqrt{2}, \sqrt{3}, \sqrt{6})$. Then if $\alpha = \sqrt{2} + \sqrt{3}$ we have

$$\alpha v = \begin{pmatrix} 0 & 1 & 1 & 0 \\ 2 & 0 & 0 & 1 \\ 3 & 0 & 0 & 1 \\ 0 & 2 & 2 & 0 \end{pmatrix} \begin{pmatrix} 1 \\ \sqrt{2} \\ \sqrt{3} \\ \sqrt{6} \end{pmatrix}$$

and

$$\det \begin{pmatrix} -X & 1 & 1 & 0 \\ 2 & -X & 0 & 1 \\ 3 & 0 & -X & 1 \\ 0 & 3 & 2 & -X \end{pmatrix} = X^4 - 10X^2 + 1,$$

so α is a root of $X^4 - 10X^2 + 1$.

We say a field F is of *positive characteristic* if there exists an integer $n \geq 1$ such that $na = 0$ for all $a \in F$. It is easy to see that if F is of positive characteristic n, then n must be a prime.

Theorem 2.9.3 *In a field F of characteristic p, we have $(a + b)^p = a^p + b^p$ for all elements $a, b \in F$.*

Proof. By the binomial theorem,

$$(a + b)^p = \sum_{0 \leq i \leq p} \binom{p}{i} a^i b^{p-i}.$$

But it is easy to see that $p \mid \binom{p}{i}$, except when $i = 0, p$. The result follows. ∎

The following theorem is stated without proof.

Theorem 2.9.4 *A finite field with n elements exists if and only if $n = p^k$, where p is a prime and k is a positive integer. In this case, there is only one such field, up to isomorphism.*

2.10 Polynomials, Rational Functions, and Formal Power Series

Let R be a commutative ring with unit element. Then $R[X]$, the set of polynomials in the indeterminate X over R, is the set of all expressions of the form

$$a_0 + a_1 X + \cdots + a_n X^n$$

together with the operations of addition and multiplication as usually defined. It is easy to see that $R[X]$ is a commutative ring with unit element 1.

Now suppose $R = K$, a field. Then it is easy to see that $K[X]$ is an integral domain. Hence we may consider the field of fractions of $K[X]$, denoted $K(X)$, which consists of all fractions f/g where $f, g \in K[X]$ and $g \neq 0$. The field $K(X)$ is called the field of *rational functions* over K.

There are similar definitions for $R[X_1, X_2, \ldots, X_n]$ and $K(X_1, X_2, \ldots, X_n)$.

Next, we discuss formal power series in a single indeterminate X. A *formal power series* is an expression of the form

$$A(X) = \sum_{0 \leq i < \infty} a_i X^i$$

where the a_i take their values in some commutative ring with unit, R. (By formal we mean that we do not think of $A(X)$ as a "function" of the "variable" X, and we are not concerned with the "convergence" or "divergence" of the series.) We define addition and multiplication as follows: if $A(X) = \sum_{i \geq 0} a_i X^i$ and $B(X) = \sum_{i \geq 0} b_i X^i$, then

$$A(X) + B(X) := \sum_{i \geq 0} (a_i + b_i) X^i$$

and

$$A(X)B(X) := \sum_{n \geq 0} X^n \sum_{i+j=n} a_i b_j.$$

It is not hard to verify that, with these operations, the set of all formal power series in one variable over R is a commutative ring with unit element. We denote this ring by $R[[X]]$.

Theorem 2.10.1 *If R is an integral domain, then so is $R[[X]]$.*

Proof. Let $A(X)$ and $B(X)$ be formal power series, neither of them zero, with coefficients in R. Let $a_i X^i$ be the least index term in $A(X)$ with $a_i \neq 0$, and let $b_j X^j$ be the term of least index in $B(X)$ with $b_j \neq 0$. Then the coefficient of the term X^{i+j} of $A(X)B(X)$ is

$$\sum_{\substack{i'+j'=i+j \\ i' \geq 0 \\ j' \geq 0}} a_{i'} b_{j'} = a_i b_j \neq 0,$$

since R is an integral domain, and $a_{i'} = 0$ for $i' < i$, $b_{j'} = 0$ for $j' < j$. ■

We will be mainly interested in the case where $R = K$, a field. Let us first determine the *units*, or invertible elements, in $K[[X]]$.

Theorem 2.10.2 *Let K be a field. Then the units in $K[[X]]$ are those formal power series $A(X) = \sum_{i \geq 0} a_i X^i$ with $a_0 \neq 0$.*

Proof. There exists a $B = \sum_{j \geq 0} b_j X^j \in K[[X]]$ such that $A(X)B(X) = 1$ if and only if

$$\sum_{n \geq 0} X^n \sum_{i+j=n} a_i b_j = 1.$$

Hence we find that

$$a_0 b_0 = 1,$$
$$a_0 b_1 + a_1 b_0 = 0,$$
$$a_0 b_2 + a_1 b_1 + a_2 b_0 = 0,$$
$$\vdots$$

The first equation is satisfied if and only if $a_0 \neq 0$. If $a_0 \neq 0$, then b_0 is uniquely determined. Then in the next equation, b_1 is uniquely determined. In general we have that

$$b_n = -a_0^{-1} \sum_{1 \leq i \leq n} a_i b_{n-i}. \qquad \blacksquare$$

Example 2.10.3 Compute the inverse of $1 - X - X^2$ (considered as a formal power series) over $\mathbb{Q}[[X]]$.

We must solve $A(X)B(X) = 1$ for $A(X) = 1 - X - X^2$. We find that

$$b_n = -a_0^{-1} \sum_{1 \leq i \leq n} a_i b_{n-i} = -(-b_{n-1} - b_{n-2}) = b_{n-1} + b_{n-2}.$$

Thus we have $b_n = F_{n+1}$, the $(n+1)$th Fibonacci number. Hence

$$(1 - X - X^2)(1 + X + 2X^2 + 3X^3 + 5X^5 + \cdots) = 1.$$

Formal power series are particularly pleasant to work with when the underlying field K equals $GF(q)$, where $q = p^n$, p is prime, and $n \geq 1$. In this case we have

Theorem 2.10.4 *Let* $A(X) = \sum_{n \geq 0} a_n X^n$. *Then* $A(X^q) = A(X)^q$.

Proof. Left to the reader as Exercise 16. $\blacksquare$

Let K be a field. As we have seen, $K[[X]]$ is an integral domain. Hence we may consider the field of fractions of $K[[X]]$, which is denoted by $K((X))$. This field coincides with the field of *formal Laurent series*, which are expressions of the form

$$\sum_{i \geq a} a_i X^i$$

for some integer $a \in \mathbb{Z}$. Notice that such an expression may have infinitely many nonzero positive powers of X, but only finitely many negative powers.

Indeed, if we are given two formal power series, say

$$A(X) = a_0 + a_1 X + a_2 X^2 + \cdots$$

and

$$B(X) = b_0 + b_1 X + b_2 X^2 + \cdots,$$

then we may compute the Laurent series expansion of $A(X)/B(X)$ by long division.

Example 2.10.5 Let $A(X) = 1 + 2X + 3X^2 + 4X^3 + \cdots$ and $B(X) = X + X^2 + X^4 + X^8 + \cdots$. Then in $\mathbb{Q}((X))$ we have

$$A(X)/B(X) = X^{-1} + 1 + 2X + X^2 + 3X^3 + X^4 + 5X^5 - X^6 + \cdots.$$

When working with formal Laurent series, some writers prefer setting $X = 1/T$, where T is an indeterminate. In this case, the analogy with the base-k expansion of a real number becomes more obvious, since if we set $T = k$, a real number has an expansion with only finitely many positive powers of k, but potentially infinitely many negative powers. The distinction between $K((X))$ and $K((1/T))$ is similar to the distinction between the Taylor (or Laurent) expansion of a function around $x = 0$ and that around $x = \infty$. For example, in $K((X))$ we have

$$\frac{1}{1 - X} = 1 + X + X^2 + X^3 + \cdots,$$

while in $K((1/T))$ we have

$$\frac{1}{1 - T} = -\frac{1}{T} - \frac{1}{T^2} - \frac{1}{T^3} - \cdots.$$

In this book, we will sometimes use $K((X))$ and sometimes $K((1/T))$, depending on which is more customary in the literature.

As we have seen in Section 2.4, every real number can be expanded as a continued fraction in an essentially unique way. This fact also holds for formal Laurent series. We state the following two theorems, leaving the proof to the reader as Exercise 31.

Theorem 2.10.6 *Let $t(X) = \sum_{k \geq -k_0} a_k X^{-k}$ be a formal Laurent series in $K((X^{-1}))$, where K is a field. Then $t(X)$ has a unique expansion as a finite or infinite continued fraction $[a_0, a_1, a_2, \ldots]$ where $a_i \in K[X]$ for $i \geq 0$ and $\deg a_i > 0$ for $i > 0$.*

As usual for continued fractions, we define the numerators p_n and denominators q_n of the convergents as follows: $p_{-2} := 0$, $p_{-1} := 1$, $q_{-2} := 1$, $q_{-1} := 0$, and

$$p_n = a_n p_{n-1} + p_{n-2},$$
$$q_n = a_n q_{n-1} + q_{n-2}$$

for $n \geq 0$.

Example 2.10.7 Here is an example in $\mathbb{Q}((X^{-1}))$. Define

$$f(X) := \frac{1}{\sqrt{X^2 - 1}} = \sum_{k \geq 0} 2^{-2k} \binom{2k}{k} X^{-2k-1}$$

$$= X^{-1} + \frac{1}{2} X^{-3} + \frac{3}{8} X^{-5} + \frac{5}{16} X^{-7} + \frac{35}{128} X^{-9} + \cdots.$$

Then it is not hard to see that

$$f(X) = [0, \ X, \ -2X, \ 2X, \ -2X, \ 2X, \ -2X, \ldots].$$

Here are the first few convergents to the continued fraction for f:

n	0	1	2	3	4
a_n	0	X	$-2X$	$2X$	$-2X$
p_n	0	1	$-2X$	$-4X^2+1$	$8X^3-4X$
q_n	1	X	$-2X^2+1$	$-4X^3+3X$	$8X^4-8X^2+1$

In fact, it can be proved that

$$p_n(X) = (-1)^{\lfloor n/2 \rfloor} U_{n-1}(X),$$
$$q_n(X) = (-1)^{\lfloor n/2 \rfloor} T_n(X),$$

where T and U are the Chebyshev polynomials of the first and second kinds, respectively.

The following theorem is about Diophantine approximation in $K((X^{-1}))$.

Theorem 2.10.8 *Let $t(X)$ be a formal Laurent series with continued fraction expansion $[a_0, a_1, a_2, \ldots]$. Let p_n/q_n be the nth convergent, and let p, q be polynomials. Then*

(a) $\deg(q_n t - p_n) = -\deg q_{n+1} < -\deg q_n$;
(b) *if* $\deg(qt - p) < -\deg q$, *then* p/q *is a convergent to* t.

2.11 *p*-adic Numbers

Let p be a prime number. The field of p-adic numbers, denoted by $\mathbb{Q}_p$, is the set of all formal expressions of the form

$$a_{-m} p^{-m} + \cdots + a_{-1} p^{-1} + a_0 + a_1 p + a_2 p^2 + \cdots$$

where $0 \leq a_i < p$ for $i \geq 0$. In this field, the four operations of addition, subtraction, multiplication, and division are defined as in base p, with the notational difference that carries go to the right instead of the left.

Alternatively, for $x \in \mathbb{Q}$ we first define the map $|x|_p$ as follows:

$$|x|_p = \begin{cases} p^{-\nu_p(x)} & \text{if } x \neq 0, \\ 0 & \text{otherwise.} \end{cases} \tag{2.10}$$

Then we consider the set of Cauchy sequences of rational numbers for $|\ |_p$, i.e., the set of sequences $(a_i)_{i \geq 0}$ such that for all $\epsilon > 0$ there exists an N such that $|a_i - a_j|_p < \epsilon$ for all $i, j \geq N$. We now define an equivalence relation on Cauchy sequences as follows: $(a_i)_{i \geq 0}$ and $(b_i)_{i \geq 0}$ are equivalent if $|a_i - b_i|_p \to 0$ as $i \to \infty$. The field $\mathbb{Q}_p$ is now defined to be the set of equivalence classes of Cauchy sequences. The norm $|a|_p$ of an equivalence class a is defined to be $\lim_{i \to \infty} |a_i|_p$, where $(a_i)_{i \geq 0}$ is any representative of a.

We define the p-adic integers as follows:

$$\mathbb{Z}_p = \{x \in \mathbb{Q}_p : |x|_p \leq 1\}.$$

Another way to say this is that $\mathbb{Z}_p$ is the set of all p-adic numbers for which the p-adic expansion involves only non-negative powers of p. It is easy to see that $\mathbb{Z}_p$ is a subring of $\mathbb{Q}_p$.

Example 2.11.1 Let $p = 3$. We have

$$-1 = 2 + 2 \cdot 3 + 2 \cdot 3^2 + 2 \cdot 3^3 + 2 \cdot 3^4 + 2 \cdot 3^5 + 2 \cdot 3^6 + \cdots,$$
$$\frac{1}{2} = 2 + 1 \cdot 3 + 1 \cdot 3^2 + 1 \cdot 3^3 + 1 \cdot 3^4 + 1 \cdot 3^5 + 1 \cdot 3^6 + \cdots,$$
$$\sqrt{7} = \pm(1 + 1 \cdot 3 + 1 \cdot 3^2 + 0 \cdot 3^3 + 2 \cdot 3^4 + 0 \cdot 3^5 + 0 \cdot 3^5 + \cdots).$$

2.12 Asymptotic Notation

We reminder the reader of the standard notation for asymptotic estimates.

We write $f = O(g)$ if there exist constants $c > 0$, $n_0 \geq 0$ such that $f(n) \leq cg(n)$ for all $n \geq n_0$.

We write $f = \Omega(g)$ if there exist constants $c > 0$, $n_0 > 0$ such that $f(n) \geq cg(n)$ for all $n \geq n_0$.

We write $f = o(g)$ if $\lim_{n \to \infty} f(n)/g(n) = 0$.

Finally, we write $f = \Theta(g)$ if $f = O(g)$ and $g = O(f)$.

2.13 Some Useful Estimates

In this section we state, without proof, some estimates from analytic number theory. These will prove useful for Chapter 15.

Theorem 2.13.1 *Let $\varphi(n)$ be the Euler φ-function. Then $\varphi(n) \geq \frac{n}{5 \log \log n}$ for $n \geq 3$.*

Theorem 2.13.2 *Let* $\pi(x)$ *denote the number of primes* $\leq x$. *Then* $\pi(x) < 1.25506 \frac{x}{\log x}$ *for* $x > 1$.

Theorem 2.13.3 *Let* a, b *be integers with* $1 \leq a < b$ *and* $\gcd(a, b) = 1$. *Then there exists a prime* $p \equiv a \pmod{b}$ *with* $p = O(b^{11/2})$.

Theorem 2.13.4 *Let* $\vartheta(x) = \sum_{p \leq x} \log p$, *where the sum is over primes only. Then* $\vartheta(x) < 1.000081x$ *for* $x > 0$ *and* $\vartheta(x) \geq 0.84x$ *for* $x \geq 101$.

Theorem 2.13.5 *Let* $x > 1$. *Then*

$$\frac{e^{-\gamma}}{\log x}\left(1 - \frac{1}{(\log x)^2}\right) < \prod_{p \leq x}\left(1 - \frac{1}{p}\right) < \frac{e^{-\gamma}}{\log x}\left(1 + \frac{1}{2(\log x)^2}\right). \quad (2.11)$$

2.14 Exercises

1. Prove the division theorem: given integers a and b with $b > 0$, there exists a unique pair of integers q and r such that

$$a = qb + r \quad \text{with } 0 \leq r < b.$$

   Moreover, $r = 0$ if and only if $b \mid a$.

2. Prove Kummer's theorem: if p is a prime and $p^e \parallel \binom{m}{n}$, then e is the number of carries that occur when adding $m - n$ to n in base p.

3. Show that

$$\{x\} - \{y\} = \begin{cases} \{x - y\} & \text{if } \{x\} \geq \{y\}, \\ \{x - y\} - 1 & \text{if } \{x\} < \{y\}. \end{cases}$$

4. Show that Dirichlet's theorem (Theorem 2.5.1) is in fact true for all real $N \geq 1$.

5. A *partial order* is a relation $\leq$ that is reflexive, transitive, and antisymmetric. Consider a relation on $\mathbb{N}^k$ defined as follows: $a \leq b$ if and only if $a_i \leq b_i$ for $1 \leq i \leq k$.

   (a) Prove that $\leq$ is a partial order.

   (b) An *antichain* in a set with a partial order is a set of mutually incomparable items. Show that $\mathbb{N}^k$ possesses no infinite antichains under the $\leq$ ordering.

6. Suppose $(a_n)_{n \geq 0}$ is an integer-valued sequence that satisfies a linear recurrence over $\mathbb{Q}$. Show that it satisfies a linear recurrence over $\mathbb{Z}$.

7. Fill in the details of the following proof of the irrationality of $\sqrt{2}$: if $\sqrt{2}$ were rational, say $\sqrt{2} = m/n$ with n minimal, then $\sqrt{2} = (2n - m)/(m - n)$, contradicting the minimality of the denominator.

8. Generalize the proof in the preceding exercise to apply to the irrationality of all $\sqrt{k}$ where k is not a perfect square.

9. Find an infinite set of positive integers with the property that no finite subset sums to a nontrivial integer power.

10. Show that $v_2(1 + 1/2 + 1/3 + \cdots + 1/n) = -\lfloor \log_2 n \rfloor$.

11. Compute $v_2(1 + 1/3 + 1/5 + \cdots + 1/(2n - 1))$.

12. We can generalize simple continued fractions to the case where the numerators are not necessarily 1. Suppose

$$\begin{bmatrix} p_n & p_{n-1} \\ q_n & q_{n-1} \end{bmatrix} = \begin{bmatrix} X_0 & 1 \\ 1 & 0 \end{bmatrix} \begin{bmatrix} X_1 & 1 \\ Y_1 & 0 \end{bmatrix} \cdots \begin{bmatrix} X_n & 1 \\ Y_n & 0 \end{bmatrix}.$$

Then show that

$$\frac{p_n}{q_n} = X_0 + \cfrac{Y_1}{X_1 + \cfrac{Y_2}{X_2 + \cfrac{\ddots}{+\cfrac{Y_n}{X_n}}}}.$$

(This form of continued fraction is sometimes written as

$$X_0 + \frac{Y_1}{X_1} + \frac{Y_2}{X_2} + \cdots + \frac{Y_n}{X_n}.)$$

13. Show that

$$\prod_{i \geq 1}(X_i + 1) = 1$$

$$+ \frac{X_1}{1} - \frac{X_2(X_1 + 1)}{X_1 + X_2 + X_1 X_2} - \frac{X_1 X_3(X_2 + 1)}{X_2 + X_3 + X_2 X_3} - \frac{X_2 X_4(X_3 + 1)}{X_3 + X_4 + X_3 X_4} - \cdots.$$

14. Define

$$A_n = \sum_{0 \leq k \leq n} (F_{2^k})^{-1},$$

where F_j is the jth Fibonacci number. Show that for $n \geq 3$ the continued fraction expansion of A_n is given by

$$[2, 2, \overbrace{1, \ldots, 1}^{2^n - 5}, 2].$$

Conclude that $\lim_{n \to \infty} A_n = \frac{7 - \sqrt{5}}{2}$.

15. Let K be an algebraically closed field of characteristic 0. What is the algebraic closure of $K((X))$?

16. Prove that if $A(X)$ is a formal power series over the finite field $GF(q)$, then $A(X^q) = A(X)^q$.

17. Give an example of two formal power series $A, B \in K[[X_1, X_2, \ldots, X_n]]$ for $n \geq 2$ such that the quotient A/B is not expressible as a single formal power series, even allowing negative exponents.

18. Prove that, for any field K, $X^{1/2}$ cannot be expressed as a formal Laurent series in $K((X))$.

19. (Erdős) Suppose $n_1 < n_2 < \cdots$ is a strictly increasing sequence of positive integers with

$$\lim_{k \to \infty} \frac{n_k}{n_1 n_2 \cdots n_{k-1}} = \infty.$$

Show that $\sum_{i \geq 1} 1/n_i$ is irrational.

20. Let $f_q(n)$ be the probability that an $n \times n$ matrix with entries chosen uniformly at random from $GF(q)$ is invertible.
    (a) Show that $\lim_{n \to \infty} f_q(n) = \prod_{k \geq 1}(1 - q^{-k})$.
    (b) Show that the quantity in (a) is irrational for all integers $q \geq 2$.

21. An integer n is said to be *squarefree* if it is not divisible by a square > 1. Let $r_2(n)$ be the number of squarefree integers j with $1 \leq j \leq n$. Show that $\lim_{n \to \infty} r_2(n)/n = \frac{6}{\pi^2}$.

22. Let q be an integer ≥ 2, and let k, n be integers ≥ 1. Show that $q^k - 1 \mid q^n - 1$ if and only if $k \mid n$.

23. Using Theorem 2.13.4, show that if $0 \leq i < j \leq n$, and $n \geq 2$, then there exists a prime $p \leq 4.4 \log n$ such that $i \not\equiv j \pmod{p}$.

24. An integral domain R is called a *Euclidean ring* if for all nonzero $a \in R$ there exists an integer $d(a) \geq 0$ such that (i) for all $a, b \in R \setminus \{0\}$ we have $d(a) \leq d(ab)$; (ii) for all $a, b \in R \setminus \{0\}$, there exist $t, r \in R$ such that $a = tb + r$, where $r = 0$ or $d(r) < d(b)$. Show that if F is a field, then $F[X]$ is a Euclidean ring, where X is an indeterminate.

25. Let R be an integral domain. An element $a \in R$ is called *irreducible* if $a = bc$ with $b, c \in R$ implies either b or c is a unit. Two elements a, b are called *associates* if $a = cb$ for some unit c. Then R is called a *unique factorization domain* if any nonzero element of R is either a unit or can be written as the product of a finite number of irreducible elements of R, and the decomposition is unique up to the order and associates of the irreducible elements. Show that if R is a unique factorization domain, then so is $R[X]$, where X is an indeterminate.

26. Show that every finitely generated group of exponent 2 with m generators is of cardinality at most 2^m.

27. Give an example of an infinite semigroup S on three generators such that $x^3 = x^2$ for all $x \in S$.

28. Prove that

$$\sum_{t \geq 0} \binom{2t}{t} X^t = (1 - 4X)^{-\frac{1}{2}}.$$

29. Let p be a prime number. Show that $(1 + X)^{p^\ell} \equiv (1 + X^p)^{p^{\ell-1}} \pmod{p^\ell}$.

30. Let α, β be transcendental real numbers. Show that either $\alpha + \beta$ or $\alpha\beta$ is transcendental.

31. Prove Theorems 2.10.6 and 2.10.8.

32. Let $(u_n)_{n\geq 1}$ be a sequence with values in $\mathbb{C}$. Its *Cesáro mean* is the sequence $(v_n)_{n\geq 1}$ defined by

$$v_n := \frac{u_1 + u_2 + \cdots + u_n}{n}.$$

Show that if $(u_n)_{n\geq 1}$ tends to a limit ℓ, then its Cesáro mean converges and its limit is ℓ. Is the converse true?

33. Let $a, c_0, c_1, c_2, \ldots$ be positive integers. Show that

$$\frac{1}{a}[ac_0, c_1, ac_2, c_3, ac_4, c_5, \ldots] = [c_0, ac_1, c_2, ac_3, c_4, ac_5, \ldots].$$

2.15 Open Problems

1. Is $\log \log 2$ irrational? Is it transcendental? (Remark: See Baker [1990, Theorem 12.2, p. 119].)

2. Let α be a real algebraic number of degree > 2. Are the partial quotients in the continued fraction expansion of α unbounded? (Remarks: This is one of the outstanding open problems in the theory of continued fractions. See Shallit [1992] for more information.)

2.16 Notes on Chapter 2

2.1 In the literature, especially the p-adic literature, $v_p(x)$ is often written as $|x|_p$. For more on valuations and p-adic numbers, see Koblitz [1984] or Gouvêa [1993].

2.2 Niven [1963] is a delightful introduction to the theory of irrational numbers. Also see Perron [1960].

For other proofs of the irrationality of $\sqrt{2}$ and similar numbers, see Bloom [1995].

Lambert [1761] proved the irrationality of π and e. For other proofs of the irrationality of π, see, for example, Niven [1947] or Breusch [1954]. Our proof for π is due to M. Cartwright, and is taken from Jeffreys [1973, p. 268].

2.3 For an introduction to the theory of transcendental numbers, see Mahler [1976a]; Baker [1990]. Theorem 2.3.3 and Corollary 2.3.4 are due to Liouville [1844]. Our presentation is based on that in Hardy and Wright [1985].

The transcendence of e is due to Hermite [1873]. The transcendence of π is due to Lindemann [1882].

Lemma 2.3.6 is from LeVeque [1956, pp. 167–168]. Theorem 2.3.7 is due to K. Roth [1955].

2.4 For an introduction to the theory of continued fractions, see, for example, Hardy and Wright [1985] or Rockett and Szüsz [1992]. Our presentation, based on 2×2 matrices, is due to Hurwitz; see Hurwitz and Kritikos [1986]. It was rediscovered by Frame [1949] and Kolden [1949], independently. For a more recent treatment, see, for example, van der Poorten [1990].

2.5 For more on Diophantine approximation, see Koksma [1936], Cassels [1957], Mahler [1961], W. Schmidt [1980], and Lang [1995].

Our proof of Kronecker's theorem (Theorem 2.5.3), including the "chain" and "mesh" terminology, is based on that in Hardy and Wright [1985, §23.2].

Our proof of Weyl's theorem (Theorem 2.5.5) is taken nearly verbatim from Hardy and Wright [1985, §23.10].

For the proof of Theorem 2.5.6, see, for example, Kuipers and Niederreiter [1974].

2.6 The three-distance theorem was apparently first noticed by H. Steinhaus. The first proofs were published by Sós [1957, 1958], Świerczkowski [1958], and Surányi [1958]. Our presentation is taken nearly verbatim from Knuth [1973, §6.4, Exercise 8].

For other papers on this and related theorems, see Slater [1950]; Florek [1951]; Slater [1964]; Halton [1965]; Slater [1967]; Graham and van Lint [1968]; van Ravenstein [1985, 1988, 1989]; van Ravenstein, Winley, and Tognetti [1990]; Langevin [1991]; Fraenkel and Holzman [1995]. Berthé [1996] and Alessandri and Berthé [1998] found interesting connections between the three-distance theorem and formal language theory.

Graham conjectured a generalization of the three-distance theorem; see Knuth [1973, §6.4, Exercise 10]. This generalization was proved by Chung and Graham [1976]. Later, Liang [1979] found an extremely simple proof of this theorem.

A 2-dimensional version of Steinhaus's theorem was obtained by Geelen and Simpson [1993]. Fried and Sós [1992] generalized the theorem to groups.

Corollary 2.6.4 is due to Mukherjee and Karner [1998].

2.7 For an introduction to semigroups, with emphasis on languages and machines, see Arbib [1968]. For an introduction to groups and rings, see, for example, Herstein [1975]. For semirings, see Kuich and Salomaa [1986].

2.8 For an introduction to vector spaces, see, for example, Herstein [1975].

2.9 For an introduction to fields, see, for example, Herstein [1975] or Jacobson [1974].

2.10 For a brief introduction to formal power series, see, for example, Hungerford [1974] or Niven [1969]. A much more detailed treatment is given in Zariski and Samuel [1960].

2.11 Gouvêa [1993] and Koblitz [1984] are good introductions to p-adic numbers.

2.13   Theorem 2.13.1 is due to Rosser and Schoenfeld [1962, Theorem 15], and
       Theorem 2.13.2 is from Rosser and Schoenfeld [1962, Corollary 1]. Theo-
       rem 2.11 is also from Rosser and Schoenfeld [1962].

       Theorem 2.13.3, a version of Linnik's theorem, is due to Heath-Brown
       [1992].

       The upper bound in Theorem 2.13.4 is due to Schoenfeld [1976, p. 360],
       and the lower bound is due to Rosser and Schoenfeld [1962, Theorem 10].

3

Numeration Systems

In this chapter, we discuss how numbers can be represented by strings over a finite alphabet. Our emphasis is on the representation of integers, although we briefly discuss representations for real numbers in Section 3.4.

We start with the classical base-k representation, and then discuss less familiar representations such as representation in base $-k$, Fibonacci representation, and representation in complex bases.

3.1 Numeration Systems

A *numeration system* is a way of expressing an integer n (or, more generally, an element of a given semiring S) as a finite linear combination $n = \sum_{0 \leq i \leq r} a_i u_i$ of base elements u_i. The a_i are called the *generalized digits*, or just *digits*. The finite string of digits $a_r a_{r-1} \cdots a_1 a_0$ is then said to be a *representation* of the number n. Note that our convention is to write representations starting with the *most* significant digit, although admittedly this choice is somewhat arbitrary.

For example, in ordinary decimal notation the base elements are the powers of 10. As is certainly familiar to most readers, every non-negative integer can be expressed as a non-negative integer linear combination $\sum_{0 \leq i \leq r} a_i 10^i$ with $0 \leq a_i < 10$.

The *leading-zeros problem* is a minor annoyance we must deal with. For example, each of the strings $101, 0101, 00101, \ldots$ is a different way to express the number 5 in base 2. In this book, we will assume unless stated otherwise that the leading digit of a representation, if it exists, is nonzero. In particular, the empty string ϵ is the usual representation for 0 in every numeration system.

More formally, we define a numeration system $\mathcal{N}$ for a semiring S to be a triple $\mathcal{N} = (U, D, R)$, where $U = \{u_0, u_1, u_2, \ldots\}$ is an infinite sequence of elements of S called the *base sequence*, D is a (usually finite) subset of S, called the *digit set*, and $R \subseteq D^*$ is the set of *valid representations*. (For an example of a numeration system with unbounded digits, see Exercise 37.) We define a mapping $[w]_U$ from D^* to S as follows: if $w = a_t a_{t-1} \cdots a_1 a_0$, then $[w]_U = \sum_{0 \leq i \leq t} a_i u_i$.

Two desirable properties of a numeration system $\mathcal{N}$ are

(a) that there be at least one valid representation for every element of the underlying semiring – in which case we say $\mathcal{N}$ is *complete*; and
(b) that every element have no more than one valid representation – in which case we say $\mathcal{N}$ is *unambiguous*.

If $\mathcal{N}$ is both complete and unambiguous, we say it is *perfect*. In this case, the mapping $[w]_U$ is invertible. Given U, we are particularly interested in specifying D and R in such a way that the resulting numeration system (U, D, R) is perfect.

Numeration systems can be specified in a number of ways. One way is to fix D and then agree that the set R is some simple subset of D^*. For example, a common choice for R is $\{\epsilon\} \cup (D \setminus \{0\})D^*$; that is to say, any sequence of digits chosen from D is permissible provided the leading digit is nonzero. Another way to specify a numeration system is to choose U and then specify R with a method called the *greedy algorithm*; it produces *greedy representations*. For purposes of concreteness assume that $S = \mathbb{N}$, and let

$$u_0 < u_1 < u_2 < \cdots$$

be an infinite increasing sequence of positive integers. We also assume that $u_0 = 1$. The following algorithm expresses any integer $N \geq 1$ as a linear combination $\sum_{0 \leq i \leq t} a_i u_i$.

GREEDY(N)

$t := 0$
while $u_{t+1} \leq N$ do
$\quad t := t + 1$
for $i = t$ downto 0 do
$\quad a_i := \lfloor N/u_i \rfloor$
$\quad N := N - a_i u_i$
$\quad$output(a_i)

Since we have assumed that $u_0 = 1$, the algorithm GREEDY always terminates. If GREEDY(N) = $a_t a_{t-1} \cdots a_1 a_0$ then we clearly have

$$N = \sum_{0 \leq i \leq t} a_i u_i. \tag{3.1}$$

We define the set of valid representations by

$$R = \{\text{GREEDY}(N) : N \geq 1\} \cup \{\epsilon\}.$$

In this case, the digit set D is defined *implicitly*, through the output of the algorithm. It is easy to see that $a_i < u_{i+1}/u_i$. If $c = \sup_{i \geq 0} u_{i+1}/u_i$ is finite, then D is finite, and we may take $D = \Sigma_{\lfloor c \rfloor + 1} = \{0, 1, \dots, \lfloor c \rfloor\}$.

One nice property of greedy representations is that they are order-preserving. By *order-preserving*, we mean the following: let w (respectively, x) be the greedy representation of m (n). Then pad the shorter of w and x with leading zeros so the representations are of the same length, obtaining w' (x'). Then w' precedes x' in lexicographic order if and only if $m < n$.

The greedy algorithm clearly produces a perfect numeration system. However, it is often useful instead to find some easily computable condition that exactly characterizes R. The following theorem accomplishes this when the underlying semiring is $\mathbb{N}$.

Theorem 3.1.1 *Let $u_0 < u_1 < u_2 < \cdots$ be an increasing sequence of integers with $u_0 = 1$. Every non-negative integer N has exactly one representation of the form $\sum_{0 \le i \le s} a_i u_i$ where $a_s \ne 0$, and for $i \ge 0$, the digits a_i are non-negative integers satisfying the inequality*

$$a_0 u_0 + a_1 u_1 + \cdots + a_i u_i < u_{i+1}. \tag{3.2}$$

Proof. For $N = 0$, the result is true, since then N is represented by the empty sum over 0 terms.

First, we show that the greedy algorithm produces a representation satisfying (3.2). The computations of the greedy algorithm may be represented as follows:

$$
\begin{aligned}
N &= a_t u_t + r_t & (0 \le r_t < u_t), \\
r_t &= a_{t-1} u_{t-1} + r_{t-1} & (0 \le r_{t-1} < u_{t-1}), \\
r_{t-1} &= a_{t-2} u_{t-2} + r_{t-2} & (0 \le r_{t-2} < u_{t-2}), \\
&\ \ \vdots \\
r_2 &= a_1 u_1 + r_1 & (0 \le r_1 < u_1), \\
r_1 &= a_0 u_0.
\end{aligned}
$$

Now one easily sees by induction that

$$r_{i+1} = a_i u_i + a_{i-1} u_{i-1} + \cdots + a_0 u_0,$$

and furthermore we have $r_{i+1} < u_{i+1}$, so (3.2) is satisfied.

Next, we prove uniqueness of the representation. Suppose N has two distinct representations

$$
\begin{aligned}
N &= a_s u_s + \cdots + a_0 u_0 \\
&= b_s u_s + \cdots + b_0 u_0,
\end{aligned}
$$

where the a_i and b_i are non-negative integers and the shorter representation has been padded with zeros, if necessary, to make the representations have the same length. Note that s can be chosen such that at least one of a_s, b_s is nonzero. Let i be the largest integer such that $a_{i+1} \neq b_{i+1}$. Without loss of generality, assume $a_{i+1} > b_{i+1}$. Then

$$
\begin{aligned}
u_{i+1} &\leq (a_{i+1} - b_{i+1})u_{i+1} \\
&= (b_i - a_i)u_i + \cdots + (b_0 - a_0)u_0 \\
&\leq b_i u_i + \cdots + b_0 u_0,
\end{aligned}
$$

contradicting condition (3.2). ∎

As a corollary to Theorem 3.1.1, we obtain the fundamental theorem of base-k representation. Let $\Sigma_k = \{0, 1, \ldots, k - 1\}$, and define C_k to be the regular language

$$
C_k := \{\epsilon\} \cup (\Sigma_k \setminus \{0\}) \Sigma_k^*.
$$

Then the next theorem states that every non-negative integer has a unique representation by a word in C_k.

Corollary 3.1.2 *Let $k \geq 2$ be an integer. Then every non-negative integer has a unique representation of the form*

$$
N = \sum_{0 \leq i \leq t} a_i k^i,
$$

where $a_t \neq 0$ and $0 \leq a_i < k$ for $0 \leq i \leq t$.

Proof. In fact, this representation is obtained from the greedy algorithm. By (3.2) with $u_i = k^i$ we have $a_i k^i < k^{i+1}$, so $a_i < k$.

On the other hand, every string $a_r a_{r-1} \ldots a_0$ with $a_r \neq 0$ and $0 \leq a_i < k$ for $0 \leq i \leq r$ represents a valid output of the greedy algorithm, since clearly

$$
a_0 + a_1 k + \cdots + a_r k^r \leq (k - 1)(1 + k + \cdots + k^r) = k^{r+1} - 1 < k^{r+1}. \quad \blacksquare
$$

We now define the *canonical base-k representation*. We have proved that for each non-negative integer N there is a unique representation of the form $N = \sum_{0 \leq i \leq t} a_i k^i$ with $a_t \neq 0$ and $0 \leq a_i < k$. In this case, we define $(N)_k = a_t a_{t-1} \cdots a_1 a_0 \in C_k$. Thus, for example, $(19)_2 = 10011$.

We also define an inverse operation. For $w = b_1 b_2 \cdots b_r$, define $[w]_k = \sum_{1 \leq i \leq r} b_i k^{r-i}$. Clearly we have $[(N)_k]_k = N$.

Although we have defined base-k expansion for integers $k \geq 2$, it is possible to define it for $k = 1$. In this case we get the so-called *unary representation* in which an integer n is represented by the string

$$
1^n = \overbrace{11 \cdots 1}^{n}.
$$

3.2 Sums of Digits

Let k be an integer ≥ 2. We now define two natural functions based on $(n)_k$, the canonical base-k representation of a non-negative integer. If $(n)_k = b_r b_{r-1} \cdots b_0$, we define $\ell_k(n) = r + 1$, the base-k *length function*. Clearly we have

$$\ell_k(n) = \begin{cases} 1 + \lfloor \log_k n \rfloor & \text{if } n \geq 1, \\ 0 & \text{if } n = 0. \end{cases}$$

We also define $s_k(n)$, the base-k *sum-of-digits function*, as follows:

$$s_k(n) := \sum_{0 \leq i \leq r} b_i.$$

The functions $\ell_k(n)$ and $s_k(n)$ appear in many different areas of mathematics, particularly number theory. In this section we focus on some properties of $s_k(n)$.

We start with the following theorem. Recall that $v_k(r)$ denotes the exponent of the highest power of k dividing r.

Theorem 3.2.1 *Let k, n be integers with $k \geq 2$, $n \geq 0$. Then*

$$\sum_{1 \leq m \leq n} v_k(m) = \frac{n - s_k(n)}{k - 1}.$$

Proof. The result is clearly true for $n = 0$. Let $m \geq 1$ be an integer, and suppose the base-k representation of m is $w\, c\, 0^a$ for some $w \in \Sigma_k^*$, $c \in \Sigma_k \setminus \{0\}$, and integer $a \geq 0$. Note $a = v_k(m)$. Then the base-k representation of $m - 1$ is Concat(w, $(c - 1)$, $(k - 1)^a$) where the exponent denotes repetition. Thus we have

$$s_k(m) - s_k(m - 1) = 1 - (k - 1)v_k(m). \tag{3.3}$$

Now, summing both sides from $m = 1$ to $m = n$, we get

$$s_k(n) = n - (k - 1) \sum_{1 \leq m \leq n} v_k(m),$$

from which the desired result follows for $n \geq 1$. ∎

As a corollary we get the following classic result of Legendre:

Corollary 3.2.2 *Let p be a prime number. Then for all $n \geq 0$ we have*

$$v_p(n!) = \frac{n - s_p(n)}{p - 1}.$$

Proof. The result follows from Theorem 3.2.1 and the fact that $\sum_{1 \leq m \leq n} v_p(m) = v_p(n!)$. ∎

The function $s_k(n)$ is not terribly well behaved; infinitely often it can be as large as $(k-1)\log_k(n+1)$, or as small as 1. In order to discuss the "average" behavior of $s_k(n)$ as $n \to \infty$, we introduce the summatory function $S_k(x)$, defined as follows:

$$S_k(x) := \sum_{0 \leq i < x} s_k(i).$$

It is easy to see that

$$S_k(k^n) = \frac{k-1}{2} k^n n, \tag{3.4}$$

which suggests that $S_k(x) \approx \frac{k-1}{2\log k} x \log x$. In fact, this is true, as the following theorem shows:

Theorem 3.2.3 *Let k be an integer ≥ 2. Then*

$$S_k(x) = \frac{k-1}{2\log k} x \log x + O(x),$$

where the constant in the big-O term depends on k but not on x.

Proof. Suppose that $0 \leq t < k$ and $j \geq 0$. Define $\varepsilon_j^{(k)}(n)$ to be the coefficient of k^j in the base-k representation of n. Then $\varepsilon_j^{(k)}(n) = t$ if and only if there exist $m, u \geq 0$ such that

$$n = mk^{j+1} + u,$$

where $m \geq 0$ and $tk^j \leq u < (t+1)k^j$. Now define $f(x, j, t)$ to be the number of positive integers $n < x$ such that $\varepsilon_j^{(k)}(n) = t$. Then

$$f(x, j, t) = \sum_{\substack{m, u \geq 0 \\ mk^{j+1}+u<x \\ tk^j \leq u < (t+1)k^j}} 1$$

$$= \sum_{tk^j \leq u < (t+1)k^j} \left(\frac{x}{k^{j+1}} + O(1) \right)$$

$$= \frac{x}{k} + O(k^j).$$

Then

$$S_k(x) = \sum_{\substack{0 \le j < \frac{\log x}{\log k} \\ 0 \le t < k}} t f(x, j, t)$$

$$= \sum_{\substack{0 \le j < \frac{\log x}{\log k} \\ 0 \le t < k}} t \left(\frac{x}{k} + O(k^j) \right)$$

$$= \frac{k(k-1)}{2} \left(\frac{\log x}{\log k} + O(1) \right) \frac{x}{k} + \frac{k(k-1)}{2} \sum_{0 \le j < \frac{\log x}{\log k}} O(k^j)$$

$$= \frac{k-1}{2 \log k} x \log x + O(x) + \frac{k(k-1)}{2} O \left(\frac{kx-1}{k-1} \right)$$

$$= \frac{k-1}{2 \log k} x \log x + O(x).$$ ∎

The next theorem shows that the result in Theorem 3.2.3 is, in some sense, best possible.

Theorem 3.2.4 *There exists an infinite increasing sequence of positive integers* $(x_n)_{n \ge 1}$ *and a real constant* $C > 0$ *such that*

$$\left| S_k(x_n) - \frac{k-1}{2 \log k} x_n \log x_n \right| \ge C x_n$$

for all $n \ge 1$.

Proof. Let $x_n = (k+1)k^{n-1}$. Then

$$S_k((k+1)k^{n-1}) - S_k(k^n) = k^{n-1} + S_k(k^{n-1}),$$

and from (3.4) we know that

$$S_k(k^n) = \frac{k-1}{2} \cdot k^n \cdot n.$$

Hence

$$S_k((k+1)k^{n-1}) = k^{n-1} + \frac{k-1}{2} k^n n + \frac{k-1}{2} k^{n-1} (n-1).$$

Now define

$$C(k) = \frac{\frac{k-1}{2 \log k} x_n \log x_n - S_k(x_n)}{x_n};$$

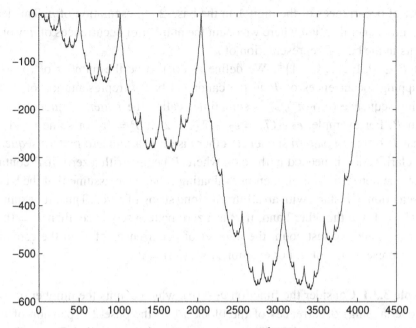

Figure 3.1: Graph of the Function $S_2(x) - \frac{x \log x}{2 \log 2}$ for $1 \leq x \leq 4096$.

we will see below that $C(k)$ does not depend on n. In fact, we have

$$
C(k) = \frac{k-1}{2 \log k} \log(k^{n-1}(k+1)) - \left(\frac{k^{n-1} + \frac{k-1}{2}k^n n + \frac{k-1}{2}k^{n-1}(n-1)}{(k+1)k^{n-1}} \right)
$$

$$
= \frac{k-1}{2} \log_k(k+1) + \frac{k-1}{2}(n-1) - \frac{nk^2 - n + 3 - k}{2(k+1)}
$$

$$
= \frac{k-1}{2} \log_k(k+1) + \frac{(k^2-1)(n-1) - nk^2 + n - 3 + k}{2(k+1)}
$$

$$
= \frac{1}{2} \left((k-1) \log_k(k+1) - \frac{k^2 - k + 2}{k+1} \right).
$$

It is not hard to see that $C(k) \geq \frac{1}{8}$ for $k \geq 2$ (see Exercise 17), so we may take $C = \frac{1}{8}$, and the theorem is proved. ∎

Figure 3.1 shows a plot of the difference $S_2(x) - \frac{x \log x}{2 \log 2}$ for $1 \leq x \leq 4096$. In Section 3.5, we study this difference in more detail.

3.3 Block Counting and Digital Sequences

In the previous section, we examined the properties of the sum-of-digits function, $s_k(n)$. In the case $k = 2$, the function $s_2(n)$ can also be thought of as counting the number of occurrences of the digit 1 in the base-2 representation of n. This leads to a natural generalization where we count the number of occurrences of *any* block of digits in the base-k representation of n.

Let $P \in \{0, 1, \ldots, k-1\}^+$. We define $e_{k;P}(n)$ to be the number of (possibly overlapping) occurrences of P in the canonical base-k representation of n. The resulting sequence $(e_{k;P}(n))_{n \geq 0}$ is sometimes called the *pattern sequence* for the pattern P. For example, $e_{2;11}(7) = e_{2;11}(27) = 2$. If $P = 0^i$ for some $i \geq 1$, we say that P is a *zero pattern sequence*; otherwise it is a *nonzero pattern sequence*. Some clarification is needed in the case where P begins with a zero. If P contains at least one nonzero symbol, then in evaluating $e_{k;P}(n)$ we assume that the base-k representation of n starts with an arbitrarily long string of zeros. Thus, for example, $e_{2;010}(5) = 1$. On the other hand, if P is a zero pattern sequence, then $P = 0^i$ for some $i \geq 1$, and we just count the number of occurrences of P in the canonical base-k representation of n. For example, $e_{2;00}(36) = 2$.

Example 3.3.1 Consider the function $e_{2;11}(n)$, which counts the number of (possibly overlapping) occurrences of the block 11 in the base-2 expansion of n. If for $n \geq 0$ we let $r_n = (-1)^{e_{2;11}(n)}$, then we get $\mathbf{r}$, the celebrated *Rudin–Shapiro sequence*, whose first few terms are as follows:

$$n = 0\ 1\ 2\ \ \ 3\ 4\ 5\ \ \ 6\ 7\ 8\ 9\ 10\ \ \ 11\ \ \ 12\ \ \ 13\ 14\ \ \ 15 \cdots$$
$$r_n = 1\ 1\ 1\ -1\ 1\ 1\ -1\ 1\ 1\ 1\ \ \ 1\ -1\ -1\ -1\ \ \ 1\ -1 \cdots$$

The Rudin–Shapiro sequence was first studied because of an interesting relationship with the maxima of certain functions. Define the L^2 norm of a function as follows:

$$\|f\|_2 = \left(\frac{1}{2\pi} \int_0^{2\pi} |f(t)|^2 dt \right)^{\frac{1}{2}}.$$

Then it is not hard to show that, for any sequence $(a_n)_{n \geq 0}$ over $\{-1, +1\}$, we have

$$\sup_{\theta \in \mathbb{R}} \left| \sum_{0 \leq n < N} a_n e^{in\theta} \right| \geq \left\| \sum_{0 \leq n < N} a_n e^{in\theta} \right\|_2 = \sqrt{N};$$

see Exercise 18.

On the other hand, it can be shown that for "almost all" sequences $(a_n)_{n \geq 0}$ over $\{-1, +1\}$, we have

$$\sup_{\theta \in \mathbb{R}} \left| \sum_{0 \leq n < N} a_n e^{in\theta} \right| = O(\sqrt{N \log N}). \tag{3.5}$$

For the Rudin–Shapiro sequence, however, we have the following result.

Theorem 3.3.2 *The Rudin–Shapiro sequence* **r** *has the square-root property, i.e., there exists a positive constant* C *such that for all* $N \geq 0$

$$\sup_{\theta \in \mathbb{R}} \left| \sum_{0 \leq n < N} r_n e^{in\theta} \right| \leq C\sqrt{N}. \tag{3.6}$$

Furthermore we can take $C = 2 + \sqrt{2}$.

Proof. Define the 2-dimensional vector V_n and the 2×2 matrices A_0 and A_1 by

$$V_n := \begin{pmatrix} r_n \\ r_{2n+1} \end{pmatrix}, \qquad A_0 := \begin{pmatrix} 1 & 0 \\ 1 & 0 \end{pmatrix}, \qquad A_1 := \begin{pmatrix} 0 & 1 \\ 0 & -1 \end{pmatrix}.$$

Using the relations $r_{2n} = r_n$, $r_{4n+1} = r_n$, and $r_{4n+3} = -r_{2n+1}$, we have for all $n \geq 0$

$$V_{2n} = A_0 V_n,$$
$$V_{2n+1} = A_1 V_n.$$

If we now define

$$R(N, \theta) := \sum_{0 \leq n < 2^N} V_n e^{in\theta},$$
$$M(\theta) := A_0 + e^{i\theta} A_1,$$

we easily obtain the following: for all $N \geq 1$

$$R(N, \theta) = \sum_{0 \leq n < 2^N} V_n e^{in\theta}$$

$$= \sum_{0 \leq n < 2^{N-1}} V_{2n} e^{i(2n)\theta} + \sum_{0 \leq n < 2^{N-1}} V_{2n+1} e^{i(2n+1)\theta}$$

$$= \sum_{0 \leq n < 2^{N-1}} A_0 V_n e^{i(2n)\theta} + \sum_{0 \leq n < 2^{N-1}} A_1 V_n e^{i(2n+1)\theta}$$

$$= A_0 \left(\sum_{0 \leq n < 2^{N-1}} V_n e^{i(2n)\theta} \right) + A_1 \left(\sum_{0 \leq n < 2^{N-1}} V_n e^{i(2n+1)\theta} \right).$$

Hence

$$R(N, \theta) = A_0 R(N - 1, 2\theta) + e^{i\theta} A_1 R(N - 1, 2\theta) = M(\theta) R(N - 1, 2\theta).$$

Iterating, this gives

$$R(N, \theta) = M(\theta)\, R(N - 1, \, 2\theta)$$
$$= M(\theta)\, M(2\theta)\, R(N - 2, \, 4\theta)$$
$$= \cdots$$
$$= M(\theta)\, M(2\theta) \cdots M(2^{N-1}\theta)\, R(0, \, 2^N \theta)$$
$$= M(\theta)\, M(2\theta) \cdots M(2^{N-1}\theta)\, V_0. \tag{3.7}$$

We now make a brief digression to discuss norms of vectors and matrices. Let us start with some notation. If $v = (v_1, v_2, \ldots, v_r)$ is a vector of real or complex numbers, then by $\|v\|$ we mean the Euclidean norm $(\sum_{1 \le i \le r} |v_i|^2)^{1/2}$. Let M be a square matrix of dimension d with entries in $\mathbb{R}$ or $\mathbb{C}$. By $\|M\|$ we mean the L^2 norm, which is the matrix norm associated with the usual Euclidean norm on $\mathbb{R}^d$ or $\mathbb{C}^d$ by the formula $\|M\| = \sup_{\|x\|=1} \|Mx\|$. If M is a matrix, then its *spectral radius* $\rho(M)$ is defined to be the largest absolute value of its eigenvalues. The following theorem is well known:

Theorem 3.3.3 *The square of the L^2 norm of M equals $\rho(MM^*)$, where M^* is the conjugate transpose of M.*

We can now return to the proof of Theorem 3.3.2. Taking the L^2 norm of both sides of (3.7), we obtain

$$\left| \sum_{0 \le n < 2^N} r_n e^{in\theta} \right| \le \|R(N, \theta)\| = \|M(\theta)M(2\theta) \cdots M(2^{N-1}\theta)V_0\|$$

$$\le \|M(\theta)M(2\theta) \cdots M(2^{N-1}\theta)\| \, \|V_0\|$$

$$\le \left(\prod_{0 \le j < N} \|M(2^j\theta)\| \right) \sqrt{2} = \sqrt{2} \prod_{0 \le j < N} \sqrt{\rho(M(2^j\theta)M^*(2^j\theta))}.$$

An easy calculation shows that for any real number α,

$$\rho\left(\begin{pmatrix} 1 & e^{i\alpha} \\ 1 & -e^{i\alpha} \end{pmatrix} \begin{pmatrix} 1 & e^{i\alpha} \\ 1 & -e^{i\alpha} \end{pmatrix}^* \right) = \rho\left(\begin{pmatrix} 2 & 0 \\ 0 & 2 \end{pmatrix} \right) = 2.$$

Thus

$$\left| \sum_{0 \le n < 2^N} r_n e^{in\theta} \right| \le \sqrt{2}(2^{N/2}).$$

This gives the square-root property when the range of summation for the sum in Theorem 3.3.2 is $[0, 2^N - 1]$. In order to bound the full sum

$$S(N, \theta) = \sum_{0 \le n < N} r_n e^{in\theta}$$

we let the binary expansion of N be $2^{N_0} + 2^{N_1} + \cdots$, where $N_0 > N_1 > \cdots \ge 0$. Hence, abbreviating $\sum_{a \le n < b} r_n e^{in\theta}$ by $\sum_{a \le n < b}$, we have

$$\sum_{0 \le n < N} = \sum_{0 \le n < 2^{N_0}} + \sum_{2^{N_0} \le n < 2^{N_0} + 2^{N_1}} + \sum_{2^{N_0} + 2^{N_1} \le n < 2^{N_0} + 2^{N_1} + 2^{N_2}} + \cdots.$$

Each of these sums has the form $\sum_{K \le n < K+2^J}$ where 2^{J+1} divides K. But it is easy to see that if n belongs to $[0, 2^J - 1]$ and if 2^{J+1} divides K, then $r_{n+K} = r_n r_K$.

Thus

$$\sum_{K \le n < K+2^J} r_n e^{in\theta} = \sum_{0 \le n < 2^J} r_{n+K} e^{i(n+K)\theta} = \sum_{0 \le n < 2^J} r_n r_K e^{in\theta} e^{iK\theta}.$$

Hence

$$\left| \sum_{K \le n < K+2^J} r_n e^{in\theta} \right| = \left| \sum_{0 \le n < 2^J} r_n e^{in\theta} \right| \le \sqrt{2}(2^{J/2}).$$

Going back to the sum $S(N, \theta)$ we thus have

$$|S(N, \theta)| \le \left| \sum_{0 \le n < 2^{N_0}} \right| + \left| \sum_{2^{N_0} \le n < 2^{N_0}+2^{N_1}} \right| + \cdots \le \sqrt{2}(2^{N_0/2} + 2^{N_1/2} + \cdots).$$

In order to finish this proof we only need to show that

$$(2^{N_0/2} + 2^{N_1/2} + \cdots) \le (1 + \sqrt{2})(2^{N_0} + 2^{N_1} + \cdots)^{1/2}.$$

Define the functions h_j by $h_j(x) := (x + 2^{N_j})^{1/2} - x^{1/2}$. These functions are clearly decreasing for $x \ge 0$. Hence, remembering that $N_0 > N_1 > \cdots$, we have for all j that

$$h_j(2^{N_j}) \le h_j(2^{N_{j+1}} + 2^{N_{j+2}} + \cdots),$$

i.e.,

$$(2^{N_j} + 2^{N_j})^{1/2} - (2^{N_j})^{1/2} \le (2^{N_j} + 2^{N_{j+1}} + \cdots)^{1/2} - (2^{N_{j+1}} + 2^{N_{j+2}} + \cdots)^{1/2}.$$

Thus

$$(\sqrt{2} - 1)2^{N_j/2} \le (2^{N_j} + 2^{N_{j+1}} + \cdots)^{1/2} - (2^{N_{j+1}} + 2^{N_{j+2}} + \cdots)^{1/2}.$$

Summing over j, we obtain

$$(\sqrt{2} - 1)(2^{N_0/2} + 2^{N_1/2} + \cdots) \le (2^{N_0} + 2^{N_1} + \cdots)^{1/2},$$

i.e.,

$$(2^{N_0/2} + 2^{N_1/2} + \cdots) \le (\sqrt{2} + 1)(2^{N_0} + 2^{N_1} + \cdots)^{1/2}. \qquad \blacksquare$$

We now turn to a different topic, namely, sums of pattern sequences.

Theorem 3.3.4 *Let $(S(n))_{n \ge 0}$ be a sequence of real numbers with $S(0) = 0$, and let $k \ge 2$ be an integer. Then we can express S uniquely as a sum of pattern sequences as follows:*

$$S(n) = \sum_{i \ge 1} \hat{S}(i) e_{k;(i)_k}(n), \qquad (3.8)$$

where the $\hat{S}(i)$ are real numbers. Furthermore, if S is integer-valued, then so is the sequence $\hat{S}$.

The map $(S(n)) \to (\hat{S}(n))$ is called the *pattern transform* of the sequence S.

Proof. Note that the sum (3.8) is well defined, since for any particular n there are only a finite number of terms.

To see that the coefficients $\hat{S}(i)$ exist, we note that we may take $\hat{S}(1) = S(1)$, since $e_{k;1}(1) = 1$. Now consider the function $S(n) - \hat{S}(1)e_{k;1}(n)$. This function is 0 at $n = 0, 1$. Hence we may choose $\hat{S}(2)$ to be $S(2) - \hat{S}(1)e_{k;1}(2)$. Now consider $S(n) - \hat{S}(1)e_{k;1}(n) - \hat{S}(2)e_{k;(2)_k}(n)$, etc.

To see that the expansion (3.8) is unique, note that $\hat{S}(1)$ is completely determined by setting $n = 1$. Once $\hat{S}(1)$ is determined, $\hat{S}(2)$ is completely determined by setting $n = 2$, etc.

The coefficients $\hat{S}(i)$ are integer-valued, if S is integer-valued, since at each step we only perform subtractions, never divisions.   ■

We now give a closed form for the sequence $\hat{S}$. Define

$$\lambda_k(n) = \begin{cases} n - a \cdot k^j & \text{if } a \cdot k^j \le n < (a+1) \cdot k^j \text{ for some } j \ge 0, 1 \le a < k, \\ 0 & \text{if } n = 0. \end{cases}$$

In other words, $\lambda_k(n)$ is the number that results from removing the most significant digit in the base-k representation of n. Define the function $x_{k;w}(n)$ to be 1 if the base-k expansion of n ends in the string w, and 0 otherwise. Finally, define the relation $m \triangleright n$ to be true if $(m)_k$ is a suffix of $(n)_k$, and false otherwise.

Theorem 3.3.5 *For all integers $r \ge 1$ we have*

$$\hat{S}(r) = S(r) - S(\lfloor r/k \rfloor) - S(\lambda_k(r)) + S(\lfloor \lambda_k(r)/k \rfloor).$$

Proof. For all $n \ge 0$ and $0 \le a < k$ we have

$$S(kn + a) = S(0) + \sum_{j \ge 1} \hat{S}(j)e_{k;(j)_k}(kn + a)$$

$$= S(0) + \sum_{j \ge 1} \hat{S}(j)(e_{k;(j)_k}(n) + x_{k;(j)_k}(kn + a))$$

$$= S(n) + \sum_{j \ge 1} \hat{S}(j)x_{k;(j)_k}(kn + a)$$

$$= S(n) + \sum_{\substack{m \ge 1 \\ m \triangleright kn + a}} \hat{S}(m)$$

$$= S(n) + \hat{S}(a) + \sum_{\substack{m \ge 1 \\ m \triangleright n}} \hat{S}(km + a).$$

It follows that

$$S(kn + a) - S(n) = \hat{S}(a) + \sum_{\substack{m \geq 1 \\ m \rhd n}} \hat{S}(km + a)$$

$$= \hat{S}(a) + \hat{S}(kn + a) + \sum_{\substack{m \geq 1 \\ m \rhd \lambda_k(n)}} \hat{S}(km + a).$$

But we also have

$$S(k\lambda_k(n) + a) - S(\lambda_k(n)) = \hat{S}(a) + \sum_{\substack{m \geq 1 \\ m \rhd \lambda_k(n)}} \hat{S}(km + a).$$

Hence it follows that

$$S(kn + a) - S(n) = \hat{S}(kn + a) + S(k\lambda_k(n) + a) - S(\lambda_k(n)),$$

and therefore, setting $r = kn + a$, we get

$$\hat{S}(r) = S(r) - S(\lfloor r/k \rfloor) - S(\lambda_k(r)) + S(\lfloor \lambda_k(r)/k \rfloor).$$ ∎

Example 3.3.6 Let $k = 2$. Consider the function $s_2(3n)$, or, in other words, $e_{2;1}(3n)$. Letting $f_w(n) = e_{2;w}(n)$, we find

$$f_1(3n) = 2f_1(n) - 2f_{11}(n) + f_{111}(n) - 2f_{1011}(n)$$

$$+ f_{11011}(n) - 2f_{101011}(n) + f_{1101011}(n) - \cdots$$

$$= 2f_1(n) - 2\sum_{i \geq 0} f_{(10)^i 11}(n) + \sum_{i \geq 0} f_{11(01)^i 1}(n).$$

For a proof, see Exercise 21.

We now turn to the concept of digital sequence. Informally, a sequence is said to be *digital* if it can be defined as the sum, over all length-r *windows* in the base-k representation of n, of some function of those windows.

More formally, if the base-k representation of n is $n = \sum_{0 \leq i \leq t} a_i k^i$, then, as in the proof of Theorem 3.2.3 above, define $\varepsilon_i^{(k)}(n) = a_i$. (If the base is clear from the context, we will often omit the superscript.) Note that

$$\varepsilon_i^{(k)}(n) = \lfloor n/k^i \rfloor - k\lfloor n/k^{i+1} \rfloor.$$

Then a sequence $(b(n))_{n \geq 0}$ is said to be a *digital sequence* if there exist an integer $r \geq 1$ and a function F from $\mathbb{Z}^r \to \mathbb{C}$ such that $F(0, 0, \ldots, 0) = 0$ and

$$b(n) = \sum_{i \geq 0} F(\varepsilon_i(n), \ldots, \varepsilon_{i+r-1}(n)).$$

Note that the sum is well defined because $F(0, 0, \ldots, 0) = 0$.

Example 3.3.7 The sequence $s_k(n)$ is digital. It corresponds to the case $r = 1$ and $F(x) = x$.

Example 3.3.8 Let $B_k(n)$ denote the number of blocks of adjacent identical digits in the base-k representation of n. For example, the number $B_{10}(3331000022) = 4$. Then $B_k(n)$ is digital, since we may take $r = 2$ and $F(x, y) = 1$ if $x \neq y$, and 0 otherwise.

Theorem 3.3.9 *A sequence $(b(n))_{n \geq 0}$ is digital if and only if it can be expressed as a finite linear combination of nonzero pattern sequences.*

Proof. Let $\Sigma_k = \{0, 1, \ldots, k - 1\}$. Suppose $(b(n))_{n \geq 0}$ is digital. Then

$$b(n) = \sum_{i \geq 0} F(\varepsilon_i(n), \ldots, \varepsilon_{i+r-1}(n)).$$

It is now easy to see that

$$b(n) = \sum_{(t_0, \ldots, t_{r-1}) \in \Sigma_k^r} F(t_0, t_1, \ldots, t_{r-1}) e_{k; t_{r-1} t_{r-2} \cdots t_0}(n).$$

On the other hand, suppose $(b(n))_{n \geq 0}$ is a finite linear combination of nonzero pattern sequences, say

$$b(n) = \sum_{P \in \mathcal{P}} a_P e_{k; P}(n).$$

If r is the length of the longest string in $\mathcal{P}$, we can write

$$b(n) = \sum_{i \geq 0} F(\varepsilon_i(n), \ldots, \varepsilon_{i+r-1}(n)),$$

where

$$F(t_0, t_1, \ldots, t_{r-1}) = \sum_{\substack{P \in \mathcal{P} \\ P \triangleright t_{r-1} \cdots t_1 t_0}} a_P. \qquad \blacksquare$$

3.4 Representation of Real Numbers

In previous sections, we considered the representation of integers in base k. In this section, we briefly discuss representations for real numbers, focusing on representation in base k.

Theorem 3.4.1 *Let k be an integer ≥ 2. Every real number x can be represented in the form*

$$\lfloor x \rfloor + \sum_{i \geq 1} a_i k^{-i},$$

where $0 \leq a_i < k$. *If x is not of the form b/k^r for some integers b, r with $r \geq 0$, then the representation is unique. If x is of the form b/k^r with $r \geq 0$, then there are two different representations, one where $a_i = 0$ for $i > r$, and another where $a_i = k - 1$ for $i > r$.*

Proof. The following algorithm provides one base-k representation for x_0:

REALREP(k, x_0)
$a_0 := \lfloor x_0 \rfloor$
$i := 0$
while $a_i \neq x_i$ do
    $x_{i+1} := k(x_i - a_i)$
    $i := i + 1$
    $a_i := \lfloor x_i \rfloor$
    output(a_i)

If the algorithm terminates on input (k, x), then it is clear that $x = a_0 + \sum_{1 \leq i \leq r} a_i k^{-i}$ for some $r \geq 0$. On the other hand, if the algorithm does not terminate, then it is easy to see that the sequence $(a_0 + \sum_{1 \leq i \leq r} a_i k^{-i})_{r \geq 1}$ tends to x from below. Hence every number has at least one representation.

Suppose there are integers b, r such that $x = b/k^r$. Then we can write $x = \lfloor x \rfloor + \{x\}$, where $\{x\} = c/k^r$ for some integer $c \geq 0$. Let the base-k representation of the integer c be $w = (c)_k$, and let $w' = 0^{r - |w|} w$. Then if $w' = d_1 d_2 \cdots d_r$, we have

$$x = a_0 + \sum_{i \geq 1} a_i k^{-i} = a_0' + \sum_{i \geq 1} a_i' k^{-i},$$

where $a_0 = \lfloor x \rfloor$, and $a_i = d_i$ for $1 \leq i \leq r$, and $a_i = 0$ for $i > r$, and $a_i' = d_i$ for $1 \leq i < r$; $a_r' = d_r - 1$, and $a_i' = k - 1$ for $i > r$. Finally, $a_0' = a_0$ unless $r = 0$, in which case $a_0' = a_0 - 1$. We leave it to the reader to verify that these two representations are the only ones possible.

Now suppose that there exist no integers b, r such that $x = b/k^r$, and assume that x has at least two different representations, say

$$x = a_0 + \sum_{i \geq 1} a_i k^{-i}$$

and

$$x' = a_0' + \sum_{i \geq 1} a_i' k^{-i},$$

where $x = x'$. Since these representations differ, there must exist a smallest index $j \geq 0$ such that $a_j \neq a_j'$. Without loss of generality assume $a_j < a_j'$. Then there exists an index $l > j$ such that $a_l < k - 1$; for if not, we would have $x = b/k^r$ for some integers b, r. Then $x' - x > k^{-l}$, contradicting the assumption that $x' = x$. ∎

Theorem 3.4.2 *Let k be an integer ≥ 2, and let $\{x\} = 0.a_1a_2a_3 \cdots$ be the base-k representation of the fractional part of x. Then x is a rational number if and only if the infinite word*

$$\mathbf{a} = a_1a_2a_3 \cdots$$

is ultimately periodic.

Proof. Suppose $\mathbf{a}$ is ultimately periodic. Then we can write

$$\{x\} = .a_1a_2 \cdots a_r(a_{r+1} \cdots a_{r+s})^\omega$$

for some integers r, s with $r \geq 0$ and $s > 0$. Then it is easy to verify that

$$\{x\} = k^{-r}\left([a_1a_2 \cdots a_r]_k + \frac{[a_{r+1}a_{r+2} \cdots a_{r+s}]_k}{k^s - 1}\right),$$

so x is rational.

On the other hand, if x is rational, then $\{x\} = b/c$ for some integers b, c with $b \geq 0, c > 0$. Each step of the algorithm REALREP produces a new digit a_i and an x_i of the form b_i/c, with $0 \leq b_i < c$. If $b_i = 0$, the algorithm terminates, which corresponds to an ultimately periodic representation with period equal to the single digit 0. If the algorithm does not terminate, there are at most c different possibilities for b_i; when one occurs for the second time, the output of the algorithm becomes ultimately periodic.   ∎

Let $k \geq 2$ be an integer. Let $w, x \in \Sigma_k^*$ be finite words with $w = a_1a_2 \ldots a_i$, $x = b_1b_2 \cdots b_j$. We define $[w.x]_k = [w]_k + k^{-j}[x]_k$. Similarly, if $\mathbf{x} \in \Sigma_k^\omega$ and $\mathbf{x} = b_1b_2 \cdots$, then we define $[w.\mathbf{x}]_k = \lim_{n \to \infty}([w]_k + k^{-j}[x_n]_k)$, where $x_n = b_1b_2 \cdots b_n$.

3.5 Sums of Sums of Digits

In Section 3.2 we showed that $S_k(x) = ((k - 1)/(2 \log k))x \log x + O(x)$, where S_k is the summatory function of the sum-of-digits function. In this section, we prove even more; namely, that the error term is a continuous, nowhere differentiable periodic function of $(\log x)/(\log k)$. Actually, we prove a more general result below in Theorems 3.5.1 and 3.5.3 for certain well-behaved sums of the form

$$\sum_{0 \leq n < x} f(n),$$

where f is some function of the base-k digits of n.

We define the exponential of a matrix as follows:

$$\exp M = \sum_{i \geq 0} \frac{M^i}{i!};$$

it is easy to see that this sum always converges. Let I be the identity matrix. We define the logarithm as follows:

$$\log(I + M) = \sum_{i \geq 1} \frac{(-1)^{i+1}}{i} M^{i+1};$$

this is well defined if $\lim_{i \to \infty} \|M^i\| = 0$. Finally, assuming that $\log M$ is well defined, we define M^x for a real number x as $\exp(x \log M)$.

Theorem 3.5.1 *Let $k \geq 2$ be an integer. Suppose there exist an integer $d \geq 1$, a sequence of vectors $(V_n)_{n \geq 0}$, $V_n \in \mathbb{C}^d$, defined by*

$$V_n = \begin{pmatrix} V_n^{(1)} \\ V_n^{(2)} \\ \vdots \\ V_n^{(d)} \end{pmatrix},$$

and k square matrices $\Gamma_0, \Gamma_1, \ldots, \Gamma_{k-1}$ of dimension d such that

(a) $V_{kn+r} = \Gamma_r V_n$ *for all $n \geq 0$ and all r, $0 \leq r < k$;*
(b) $\|V_n\| = O(\log n)$; *and*
(c) *there exist a $d \times d$ matrix Λ and a constant $c > 0$ such that either $\|\Lambda\| < c$ or Λ is nilpotent, i.e., $\Lambda^\tau = 0$ for some integer τ, such that $\Gamma := \Gamma_0 + \Gamma_1 + \cdots + \Gamma_{k-1} = cI + \Lambda$. Furthermore, Γ being clearly invertible, we assume that $\|\Gamma^{-1}\| < 1$.*

Then there exists a continuous function $G : \mathbb{R} \to \mathbb{C}^d$ of period 1 such that if $A(N) := \sum_{0 \leq n < N} V_n$, then

$$A(N) = N^{\frac{\log c}{\log k}} (I + c^{-1} \Lambda)^{\frac{\log N}{\log k}} G\left(\frac{\log N}{\log k}\right). \tag{3.9}$$

Proof. Here is an outline of the proof. First, we prove the existence of a function G such that (3.9) holds. Next, we show that G is of period 1. Finally, we show it is continuous.

Note that the conditions on Λ imply that Γ is invertible and

$$\Gamma^{-1} = c^{-1}(I + c^{-1}\Lambda)^{-1} = \sum_{i \geq 0} (-1)^i c^{-i-1} \Lambda^i.$$

We turn to the computation of $A(kN + r)$ for $0 \leq r < k$. We first compute $A(kN)$ for any integer N. We have

$$A(kN) = \sum_{0 \leq n < kN} V_n = \sum_{0 \leq r < k} \sum_{kn+r < kN} V_{kn+r}$$

$$= \sum_{0 \leq r < k} \sum_{0 \leq n < N} V_{kn+r} = \sum_{0 \leq r < k} \sum_{0 \leq n < N} \Gamma_r V_n$$

$$= \sum_{0 \leq r < k} \Gamma_r \sum_{0 \leq n < N} V_n = \left(\sum_{0 \leq r < k} \Gamma_r \right) \left(\sum_{0 \leq n < N} V_n \right)$$

$$= \Gamma A(N).$$

Now we note that, for every r with $0 \leq r < k$, we have

$$A(kN + r) = A(kN) + \sum_{0 \leq l < r} V_{kN+l}.$$

Then for $0 \leq r < k$ we have

$$A(kN + r) = \Gamma A(N) + \left(\sum_{0 \leq l < r} \Gamma_l \right) V_N. \tag{3.10}$$

Let x be a positive real number. We now define three sequences associated with x. Let the base-k representation of x be

$$x = \varepsilon_0 + \sum_{r \geq 1} \varepsilon_r k^{-r},$$

where ε_0 is a non-negative integer, $0 \leq \varepsilon_r < k$, for all $r \geq 1$, and the ε_r's are not all eventually equal to $k - 1$. Let

$$x_i := \lfloor k^i x \rfloor = \sum_{0 \leq r \leq i} \varepsilon_r k^{i-r},$$

and, finally, let

$$T_i(x) := \Gamma^{-i} A(x_i).$$

Since $x_i = k x_{i-1} + \varepsilon_i$, we have, by Eq. (3.10), that

$$T_i(x) - T_{i-1}(x) = \Gamma^{-i}(A(x_i) - \Gamma A(x_{i-1})) = \Gamma^{-i} \left(\sum_{0 \leq l < \varepsilon_i} \Gamma_l \right) V_{x_{i-1}}. \tag{3.11}$$

Hence

$$\|T_i(x) - T_{i-1}(x)\| \leq \|\Gamma^{-1}\|^i \left(\sup_{0 \leq r \leq k-1} \left\| \sum_{0 \leq l < r} \Gamma_l \right\| \right) \|V_{x_{i-1}}\|.$$

But $\|V_{x_{i-1}}\| = O(\log x_{i-1}) = O(i \log x)$. Hence for any $x > 0$, the series

$$\sum_{i \geq 0} \|T_i(x) - T_{i-1}(x)\|$$

converges for any $x > 0$. Then, by telescopic cancellation, the sequence $(T_i(x))_{i \geq 0}$ converges.

Let

$$\Phi(x) := \lim_{i \to \infty} T_i(x). \tag{3.12}$$

Summing the equalities (3.11) and letting i go to infinity, we obtain

$$\Phi(x) = T_0(x) + \sum_{i \geq 1} (T_i(x) - T_{i-1}(x)).$$

Hence

$$\Phi(x) = A(\lfloor x \rfloor) + \sum_{i \geq 1} \Gamma^{-i} \left(\sum_{0 \leq l < \varepsilon_i} \Gamma_l \right) V_{x_{i-1}}. \tag{3.13}$$

Furthermore, if $x = N$ is an integer, then $\varepsilon_i = 0$ for all $i \geq 1$, Hence Eq. (3.13) reduces to

$$\Phi(N) = A(N). \tag{3.14}$$

Now if $x' = kx$, then $x'_i := \lfloor k^i x' \rfloor = \lfloor k^{i+1} x \rfloor = x_{i+1}$. Hence

$$T_i(kx) = T_i(x') = \Gamma^{-i} A(x'_i) = \Gamma^{-i} A(x_{i+1}) = \Gamma T_{i+1}(x).$$

Letting i go to infinity, we obtain

$$\Phi(kx) = \Gamma \Phi(x). \tag{3.15}$$

We now define the function $G : \mathbb{R} \longrightarrow \mathbb{C}^d$ by

$$G(y) := c^{-y} (I + c^{-1} \Lambda)^{-y} \Phi(k^y). \tag{3.16}$$

Property (3.15) implies immediately that G is periodic, of period 1, and

$$\Phi(x) = c^{\frac{\log x}{\log k}} (I + c^{-1} \Lambda)^{\frac{\log x}{\log k}} G\left(\frac{\log x}{\log k} \right) = x^{\frac{\log c}{\log k}} (I + c^{-1} \Lambda)^{\frac{\log x}{\log k}} G\left(\frac{\log x}{\log k} \right). \tag{3.17}$$

In particular, if $x = N$ is an integer, then, using (3.14), we find

$$A(N) = \Phi(N) = N^{\frac{\log c}{\log k}} (I + c^{-1} \Lambda)^{\frac{\log N}{\log k}} G\left(\frac{\log N}{\log k} \right). \tag{3.18}$$

It remains to show that G is continuous. Of course, in view of (3.16), it suffices to prove that Φ is continuous. We first need a lemma on the digits of a converging sequence.

Lemma 3.5.2 *Let x be a number in $\mathbb{R}^{>0}$, and let the base-k representation of x be $x = \varepsilon_0 + \sum_{r \geq 1} \varepsilon_r k^{-r}$, with ε_0 a non-negative integer, $0 \leq \varepsilon_r < k$ for all $r \geq 1$, and the ε_r's not all eventually equal to $k - 1$. Let y tend to x^-, (i.e., y tends to x and $y \leq x$), and $y = \alpha_0 + \sum_{r \geq 1} \alpha_r k^{-r}$, with $\alpha_0 \in \mathbb{N}$, the digits $\alpha_r \in [0, k-1]$ for all $r \geq 1$, and the α_r's not all eventually equal to $k - 1$. Define as before $x_i := \lfloor k^i x \rfloor$ and $y_i := \lfloor k^i y \rfloor$. Then*

Case 1. If x is not of the form $\frac{a}{k^b}$, then the sequence $(\alpha_r)_{r \geq 0}$ converges to the sequence $(\varepsilon_r)_{r \geq 0}$ and y_i tends to x_i for every i.

Case 2. If $x = a/k^b$, where a and b are integers ≥ 0 and b is minimal, i.e., $x = \varepsilon_0 + \sum_{1 \leq r \leq b} \varepsilon_r k^{-r}$, and $\varepsilon_b \neq 0$, then the sequence $(\alpha_r)_{r \geq 0}$ converges to the sequence $(\widetilde{\varepsilon}_r)_{r \geq 0}$ defined by

$$\widetilde{\varepsilon}_r = \begin{cases} \varepsilon_r & \text{if } r < b, \\ \varepsilon_r - 1 & \text{if } r = b, \\ k - 1 & \text{if } r > b. \end{cases}$$

Furthermore we have

$$y_i = \begin{cases} x_i & \text{if } i < b, \\ x_i - 1 = k^{i-b} x_b - 1 & \text{if } i \geq b \end{cases}$$

for all y sufficiently close to x^-.

Proof. The proof is left to the reader. ∎

In order to prove the continuity of our function Φ at the point x, it suffices to prove left continuity and right continuity. The right continuity is straightforward. For the left continuity, we distinguish three cases.

Case 1. If x is not of the form a/k^b with a and b integers, and if y tends to x^-, then, from Lemma 3.5.2, the sequence of digits of y in base k converges to the sequence of digits of x in base k. On the other hand, x is not an integer; hence $A(\lfloor y \rfloor)$ tends to $A(\lfloor x \rfloor)$, and Eq. (3.12) shows that $\Phi(y)$ tends to $\Phi(x)$.

Case 2. If $x = a/k^b$, let $(\varepsilon_r)_{r \geq 0}$ and $(\widetilde{\varepsilon})_{r \geq 0}$ be the sequences defined in Lemma 3.5.2. If y tends to x^+, then $\Phi(y)$ tends to $\Phi(x)$. Let y tend to x^-, and let $y = \alpha_0 + \sum_{r \geq 1} \alpha_r k^{-r}$ with $\alpha_0 \in \mathbb{N}$ and, for $r \geq 1$, the digits α_r in $[0, k-1]$ not all eventually equal to $k - 1$. Also let $y_i = \lfloor k^i y \rfloor$. We have from Eq. (3.13)

$$\Phi(y) = A(\lfloor y \rfloor) + \sum_{i \geq 1} \Gamma^{-i} \left(\sum_{0 \leq l < \alpha_i} \Gamma_l \right) V_{y_{i-1}}.$$

Using Lemma 3.5.2, and supposing that x is not an integer (hence $b \neq 0$), we obtain that

$$\sum_{i\geq 1} \Gamma^{-i} \left(\sum_{0\leq l<\alpha_i} \Gamma_l \right) V_{y_{i-1}}$$

$$\longrightarrow \sum_{0\leq i<b} \Gamma^{-i} \left(\sum_{0\leq l<\varepsilon_i} \Gamma_l \right) V_{x_{i-1}} + \Gamma^{-b} \left(\sum_{0\leq l\leq\varepsilon_b-2} \Gamma_l \right) V_{x_b-1}$$

$$+ \sum_{i\geq b+1} \Gamma^{-i} \left(\sum_{0\leq l\leq k-2} \Gamma_l \right) V_{x_{i-1}-1}. \tag{3.19}$$

But $x_{i-1} - 1 = k^{i-1-b} x_b - 1$, since $x = a/k^b$. Then

$$V_{x_{i-1}-1} = V_{k^{i-1-b}x_b-1} = \Gamma_{k-1}^{i-1-b} V_{x_b-1}.$$

Hence

$$\sum_{i\geq b+1} \Gamma^{-i} \left(\sum_{0\leq l\leq k-2} \Gamma_l \right) V_{x_{i-1}-1} = \sum_{i\geq b+1} \Gamma^{-i} (\Gamma - \Gamma_{k-1}) \Gamma_{k-1}^{i-1-b} V_{x_b-1}.$$

By telescoping cancellation, this last quantity equals

$$\Gamma^{-b} V_{x_b-1} = \Gamma^{-b} V_{kx_b-1+\varepsilon_b-1} = \Gamma^{-b} \Gamma_{\varepsilon_b-1} V_{x_b-1},$$

which finally shows, using (3.19),

$$\sum_{i\geq 1} \Gamma^{-i} \left(\sum_{0\leq l<\alpha_i} \Gamma_l \right) V_{y_{i-1}} \longrightarrow \sum_{0\leq i\leq b} \Gamma^{-i} \left(\sum_{0\leq l<\varepsilon_i} \Gamma_l \right) V_{x_{i-1}}.$$

This last quantity equals

$$\sum_{i\geq 1} \Gamma^{-i} \left(\sum_{0\leq l<\varepsilon_i} \Gamma_l \right) V_{x_{i-1}},$$

since the ε_i are zero for $i \geq b+1$. On the other hand, since x is not integer, $A(\lfloor y \rfloor) \longrightarrow A(\lfloor x \rfloor)$ and the continuity of Φ at x is proven.

Case 3. It remains to handle the case where x is an integer. This is done by looking at the argument above in case 2, when $b = 0$. We have

$$\sum_{i\geq 1} \Gamma^{-i} \left(\sum_{0\leq l<\alpha_i} \Gamma_l \right) V_{y_{i-1}} \longrightarrow \sum_{i\geq 1} \Gamma^{-i} \left(\sum_{0\leq l\leq k-2} \Gamma_l \right) V_{x_{i-1}-1},$$

since previously $x_{i-1} - 1 = k^{i-1-b} x_b - 1 = k^{i-1} x - 1$ and $V_{x_{i-1}-1} = \Gamma_{k-1}^{i-1-b} V_{x_b-1} = \Gamma_{k-1}^{i-1} V_{x-1}$. Hence

$$\sum_{i\geq 1} \Gamma^{-i} \left(\sum_{0\leq l\leq k-2} \Gamma_l \right) V_{x_{i-1}-1} = \sum_{i\geq 1} \Gamma^{-i}(\Gamma - \Gamma_{k-1}) \Gamma_{k-1}^{i-1} V_{x-1} = V_{x-1}.$$

Now, since x is an integer, $A(\lfloor y \rfloor) \longrightarrow A(x-1)$, and hence

$$\Phi(y) \longrightarrow A(x-1) + V_{x-1} = A(x) = \Phi(x)$$

using (3.14).   ∎

We now state a proposition about the nondifferentiability of these sums.

Theorem 3.5.3 *Let* $(a_n)_{n \geq 0}$ *be a sequence of complex numbers such that there exist a function* L, *a continuous function* F *with period* 1, *and a positive real number* $\alpha \leq 1$ *with*

$$\sum_{0 \leq n < N} a_n = L(N) + N^\alpha F\left(\frac{\log N}{\log k}\right). \tag{3.20}$$

Let λ *and* e *be two integers* ≥ 1. *Let* x *be a real number in* $(0, 1)$. *Define*

$$k^x = \sum_{r \geq 0} \varepsilon_r k^{-r}, \quad \text{with } \varepsilon_r \in [0, k-1],$$
$$\qquad \qquad \text{and the } \varepsilon_r \text{ not all eventually equal to } k-1,$$

$$N_i := k^\lambda \sum_{0 \leq r \leq i} \varepsilon_r k^{i-r} + e,$$

$$x_i := \frac{\log N_i}{\log k} - i - \lambda, \tag{3.21}$$

$$y_i := \frac{\log(N_i+1)}{\log k} - i - \lambda.$$

Then

$$a_{N_i} = L(N_i + 1) - L(N_i) + \frac{F(y_i) - F(x_i)}{((y_i - x_i)\log k)^\alpha} + \alpha N_i^{\alpha-1} F(x) + o(N_i^{\alpha-1}). \tag{3.22}$$

In particular:

(i) *If* $\alpha < 1$, $L = 0$ *and if there exists a constant* γ *such that for all* n, $a_n \geq \gamma > 0$, *then the function* F *is nowhere differentiable. More precisely, we have*

$$F(x+h) - F(x) = \Omega(|h|^\alpha) \qquad \text{when } h \to 0.$$

(ii) *If* $\alpha = 1$, $L(x) = \delta x \log x$, *and there exist an integer* $\lambda_1 \geq 1$ *and an integer* e_1 *such that for all* n, $a_{k^{\lambda_1}n+e_1} - a_{k^{\lambda_1}n} = a_{e_1} - a_0 \neq 0$, *then* F *is nowhere differentiable.*

Proof. Using (3.20) and the fact that F has period 1 and is bounded, we have

$$a_{N_i} = \sum_{0 \leq n < N_i+1} a_n - \sum_{0 \leq n < N_i} a_n$$
$$= L(N_i+1) - L(N_i) + (N_i+1)^\alpha F(\tfrac{\log(N_i+1)}{\log k}) - N_i^\alpha F(\tfrac{\log N_i}{\log k})$$
$$= L(N_i+1) - L(N_i) + N_i^\alpha(F(y_i) - F(x_i)) + ((N_i+1)^\alpha - N_i^\alpha)F(y_i)$$
$$= L(N_i+1) - L(N_i) + N_i^\alpha(F(y_i) - F(x_i)) + \alpha N_i^{\alpha-1} F(x) + o(N_i^{\alpha-1}),$$

since

$$(N_i + 1)^\alpha - N_i^\alpha = \alpha N_i^{\alpha-1} + o(N_i^{\alpha-1})$$

and F is continuous. But

$$y_i - x_i = \frac{\log(N_i + 1)}{\log k} - \frac{\log N_i}{\log k} = \frac{1}{N_i \log k}(1 + O(1/N_i)).$$

Hence, since F is continuous,

$$a_{N_i} = L(N_i + 1) - L(N_i) + \frac{F(y_i) - F(x_i)}{((y_i - x_i) \log k)^\alpha} + \alpha N_i^{\alpha-1} F(x) + o(N_i^{\alpha-1}).$$

(3.23)

Let us prove assertion (i). Suppose that $\alpha < 1$ and $L = 0$. Then we deduce from Eq. (3.22)

$$a_{N_i} = \frac{F(y_i) - F(x_i)}{((y_i - x_i) \log k)^\alpha} + o(1).$$

Hence, if furthermore $a_n \geq \gamma > 0$, we have

$$\frac{F(y_i) - F(x_i)}{((y_i - x_i) \log k)^\alpha} \gg 1.$$

Since

$$|k^x - k^{x_i}| = \left| k^x - \frac{N_i}{k^{i+\lambda}} \right| = \left| \sum_{r \geq i+1} \varepsilon_r k^{-r} - e k^{-i-\lambda} \right| \leq (1 + e k^{-\lambda}) k^{-i},$$

we have $|x - x_i| = O(k^{-i})$, and similarly $|x - y_i| = O(k^{-i})$. Hence the property $|F(x + h) - F(x)| = o(|h|^\alpha)$ when $h \to 0$ does not hold.

Let us now prove (ii). Suppose that $\alpha = 1$, $L(x) = \delta x \log x$, and there exist two integers $\lambda_1, e_1 \geq 1$ such that $a_{k^{\lambda_1} n + e_1} - a_{k^{\lambda_1} n} = a_{e_1} - a_0 \neq 0$ for all n. Then, from Eq. (3.22), we have

$$a_{N_i} = \delta(1 + \log N_i) + \frac{F(y_i) - F(x_i)}{(y_i - x_i) \log k} + F(x) + o(1).$$

Hence, if F is differentiable at x, we have

$$a_{N_i} = \delta(1 + \log N_i) + \frac{F'(x)}{\log k} + F(x) + o(1)$$

$$= \delta(1 + (i + x_i + \lambda) \log k) + \frac{F'(x)}{\log k} + F(x) + o(1)$$

$$= \delta(1 + (i + x + \lambda) \log k) + \frac{F'(x)}{\log k} + F(x) + o(1). \qquad (3.24)$$

Recall that

$$N_i = N_i(\lambda, e) = k^\lambda \sum_{0 \leq r \leq i} \varepsilon_r k^{i-r} + e,$$

and observe that the dominant term in the right-hand side of Eq. (3.24) does not depend on e. Hence, writing this formula for $e \neq 0$ and $e = 0$,

$$a_{N_i(\lambda,e)} - a_{N_i(\lambda,0)} = o(1).$$

Now take $\lambda = \lambda_1$, $e = e_1$, and note that $N_i(\lambda_1, e_1) = k^{\lambda_1} n(i) + e_1$ and $N_i(\lambda_1, 0) = k^{\lambda_1} n(i)$, with $n(i) = \sum_{0 \leq r \leq i} \varepsilon_r k^{i-r}$. Using the property of λ_1 and e_1, we obtain

$$a_{e_1} - a_0 = a_{k^{\lambda_1} n(i) + e_1} - a_{k^{\lambda_1} n(i)} = a_{N_i(\lambda_1, e_1)} - a_{N_i(\lambda_1, 0)} = o(1).$$

Hence

$$a_{e_1} - a_0 = 0,$$

which contradicts the hypothesis.   ∎

We now apply our results to three celebrated sums. First, we handle the case $S_k(N) = \sum_{0 \leq n < N} s_k(n)$:

Theorem 3.5.4 *Let k be an integer ≥ 2. Then there exists a continuous nowhere differentiable function $F = F_k$ of period 1 such that for all $N \geq 0$ we have*

$$\sum_{0 \leq n < N} s_k(n) = \frac{k-1}{2} N \frac{\log N}{\log k} + N F_k \left(\frac{\log N}{\log k} \right). \qquad (3.25)$$

Proof. Define the sequence of vectors $(V_n)_{n \geq 0}$ in $(\mathbb{R}^k)^{\mathbb{N}}$ by

$$V_n := \begin{pmatrix} s_k(n) \\ 1 \\ 2 \\ \vdots \\ k-1 \end{pmatrix}.$$

Let us define k square matrices of dimension k by

$$\Gamma_0 := I := \begin{pmatrix} 1 & 0 & 0 & \cdots & 0 \\ 0 & 1 & 0 & \cdots & 0 \\ 0 & 0 & 1 & \cdots & 0 \\ \vdots & \vdots & \vdots & & \vdots \\ 0 & 0 & 0 & \cdots & 1 \end{pmatrix}$$

and, for $1 \leq r \leq k-1$,

$$\Gamma_r := I + \begin{pmatrix} 0 & \cdots & 0\ 1\ 0 & \cdots & 0 \\ 0 & \cdots & \cdots & \cdots & 0 \\ \vdots & & & & \vdots \\ 0 & \cdots & \cdots & \cdots & 0 \end{pmatrix},$$

the only 1 in the above matrix being in row 1 and in column $r+1$. Then, for $1 \leq r \leq k-1$,

$$V_{kn+r} = \Gamma_r V_n,$$

and

$$\Gamma = \sum_{0 \leq r < k} \Gamma_r = kI + \begin{pmatrix} 0 & 1 & \cdots & 1 & \cdots & 1 \\ 0 & 0 & \cdots & \cdots & \cdots & 0 \\ \vdots & \vdots & & & & \vdots \\ 0 & 0 & \cdots & \cdots & \cdots & 0 \end{pmatrix}.$$

Hence we can apply Theorem 3.5.1, with $d = k$, $c = k$, and

$$\Lambda = \begin{pmatrix} 0 & 1 & \cdots & 1 & \cdots & 1 \\ 0 & 0 & \cdots & \cdots & \cdots & 0 \\ \vdots & \vdots & & & & \vdots \\ 0 & 0 & \cdots & \cdots & \cdots & 0 \end{pmatrix}.$$

Now $\|\Lambda\| = \sqrt{(k-1)} < k$ and $\Lambda^2 = 0$. Hence

$$\log(I + k^{-1}\Lambda) = k^{-1}\Lambda$$

and, for any real number x,

$$(I + k^{-1}\Lambda)^x = I + xk^{-1}\Lambda.$$

Hence, writing Eq. (3.9), we find that there exists a continuous (vector) function G of period 1 such that

$$A(N) = \sum_{0 \leq n < N} V_n = N \left(I + \frac{\log N}{k \log k} \Lambda \right) G \left(\frac{\log N}{\log k} \right). \tag{3.26}$$

But, inverting Eq. (3.26), we have

$$G\left(\frac{\log N}{\log k}\right) = \frac{1}{N}\left(I - \frac{\log N}{k \log k}\Lambda\right)A(N)$$

$$= \frac{A(N)}{N} - \frac{\log N}{kN \log k}\begin{pmatrix} 0 & 1 & \cdots & 1 & \cdots & 1 \\ 0 & 0 & \cdots & \cdots & \cdots & 0 \\ \vdots & \vdots & & & & \vdots \\ 0 & 0 & \cdots & \cdots & \cdots & 0 \end{pmatrix} A(N)$$

$$= \frac{A(N)}{N} - \frac{\log N}{kN \log k}\begin{pmatrix} 0 & 1 & \cdots & 1 & \cdots & 1 \\ 0 & 0 & \cdots & \cdots & \cdots & 0 \\ \vdots & \vdots & & & & \vdots \\ 0 & 0 & \cdots & \cdots & \cdots & 0 \end{pmatrix}\begin{pmatrix} \displaystyle\sum_{0 \le n < N} s_k(n) \\ N \\ 2N \\ \vdots \\ (k-1)N \end{pmatrix}$$

Hence

$$G\left(\frac{\log N}{\log k}\right) = \frac{1}{N}\begin{pmatrix} \displaystyle\sum_{0 \le n < N} s_k(n) - \frac{(k-1)N \log N}{2 \log k} \\ N \\ 2N \\ \vdots \\ (k-1)N \end{pmatrix},$$

which implies

$$\Lambda G\left(\frac{\log N}{\log k}\right) = \frac{k(k-1)}{2}\begin{pmatrix} 1 \\ 0 \\ \vdots \\ 0 \end{pmatrix}.$$

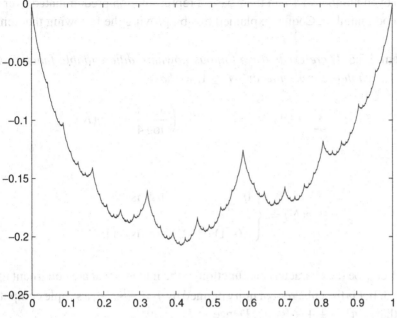

Figure 3.2: Graph of the Function $F_2(x)$ for $x \in [0, 1]$.

Now, going back to Eq. (3.26), we obtain

$$A(N) = \sum_{0 \le n < N} V_n = N \left(I + k^{-1} \frac{\log N}{\log k} \Lambda \right) G \left(\frac{\log N}{\log k} \right)$$

$$= NG \left(\frac{\log N}{\log k} \right) + \frac{k-1}{2 \log k} N \log N \begin{pmatrix} 1 \\ 0 \\ \vdots \\ 0 \end{pmatrix}.$$

Hence, if F is the first component of G, this is a continuous function with period 1, and

$$\sum_{0 \le n < N} s_k(n) = \frac{k-1}{2 \log k} N \log N + N F \left(\frac{\log N}{\log k} \right).$$

To show the non-differentiability of the function F, we apply Theorem 3.5.3, with $\lambda_1 = e_1 = 1$. $\blacksquare$

Figure 3.2 is a graph of the function F_2.

Next, we consider the sum $\sum_{0 \le n < N} (-1)^{s_2(3n)}$. This sum originally arose following an observation of Moser. He noted that among the first few multiples of 3, those with an even number of 1's in their base-2 representation predominated over those with an odd number. Coquet explained this by proving the following theorem.

Theorem 3.5.5 *There exists a continuous nowhere differentiable function G_0 of period 1 such that, for all integers $N \geq 1$, we have*

$$\sum_{0 \leq n < N} (-1)^{s_2(3n)} = N^{\frac{\log 3}{\log 4}} G_0 \left(\frac{\log N}{\log 4} \right) + \eta(N), \qquad (3.27)$$

where

$$\eta(N) = \begin{cases} 0 & \text{if } N \text{ is even,} \\ \frac{1}{3}(-1)^{s_2(3N-1)} & \text{if } N \text{ is odd.} \end{cases}$$

Proof. Let χ be the characteristic function of the integers that are congruent to zero modulo 3. If j is the cube root of unity defined by $j = \frac{-1+i\sqrt{3}}{2}$, a simple computation shows that $\chi(n) = \frac{1}{3} + \frac{2}{3} \operatorname{Re} j^n$. Hence

$$\sum_{0 \leq n < N} (-1)^{s_2(3n)} = \sum_{0 \leq n < 3N} (-1)^{s_2(n)} \chi(n) = \frac{1}{3} \sum_{0 \leq n < 3N} (-1)^{s_2(n)} (1 + 2 \operatorname{Re} j^n)$$

$$= \frac{1}{3} \sum_{0 \leq n < 3N} (-1)^{s_2(n)} + \frac{2}{3} \operatorname{Re} \sum_{0 \leq n < 3N} (-1)^{s_2(n)} j^n.$$

Defining the sequence $(a(n))_{n \geq 0}$ by $a_n = (-1)^{s_2(n)} j^n$, we easily obtain

$$\begin{aligned} a_{4n} &= a_n, \\ a_{4n+1} &= -j a_n, \\ a_{4n+2} &= -j^2 a_n, \\ a_{4n+3} &= a_n. \end{aligned}$$

Hence we can apply Theorem 3.5.1 with $k = 4$, $d = 1$, $\Gamma_0 = 1$, $\Gamma_1 = -j$, $\Gamma_2 = -j^2$, and $\Gamma_3 = 1$, so that $\Gamma = \Gamma_0 + \Gamma_1 + \Gamma_2 + \Gamma_3 = 3$. This gives $c = 3$ and $\Lambda = 0$. Hence there exists a continuous function G with period 1 such that

$$A(N) = \sum_{0 \leq n < N} a_n = \sum_{0 \leq n < N} (-1)^{s_2(n)} j^n = N^{\frac{\log 3}{\log 4}} G \left(\frac{\log N}{\log 4} \right). \qquad (3.28)$$

Hence, if we define

$$G_0(x) = \frac{2}{3^{1 - \frac{\log 3}{\log 4}}} \operatorname{Re} G \left(x + \frac{\log 3}{\log 4} \right), \qquad (3.29)$$

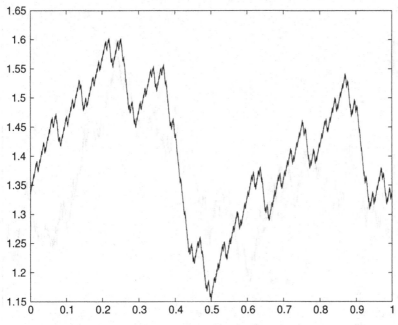

Figure 3.3: Graph of the Function $G_0(x)$ for $x \in [0, 1]$.

then G_0 is continuous, has period 1, and satisfies the equation

$$\sum_{0 \leq n < N} (-1)^{s_2(3n)} = \frac{1}{3} \sum_{0 \leq n < 3N} (-1)^{s_2(n)} + N^{\frac{\log 3}{\log 4}} G_0 \left(\frac{\log N}{\log 4} \right). \qquad (3.30)$$

Finally, since $(-1)^{s_2(n)}$ and $(-1)^{s_2(2n+1)}$ always have opposite signs, we get

$$\sum_{0 \leq n < 3N} (-1)^{s_2(n)} = \begin{cases} 0 & \text{if } N \text{ is even,} \\ (-1)^{s_2(3N-1)} & \text{if } N \text{ is odd.} \end{cases} \qquad (3.31)$$

Now, to show that G_0 is nowhere differentiable, it suffices to show that G is nowhere differentiable. This is a straightforward consequence of Theorem 3.5.3(i) with $\alpha = \frac{\log 3}{\log 4}$. ∎

Figure 3.3 is a graph of the function G_0.

Finally, we apply the technique to the Rudin–Shapiro sequence.

Theorem 3.5.6 *Let $e_{2;11}(n)$ be the number of (possibly overlapping) 11's in the base-2 representation of the integer n, and define $r_n = (-1)^{e}_{2;11}(n)$. Then there exists a continuous nowhere differentiable function G_1 of period 1 such that, for*

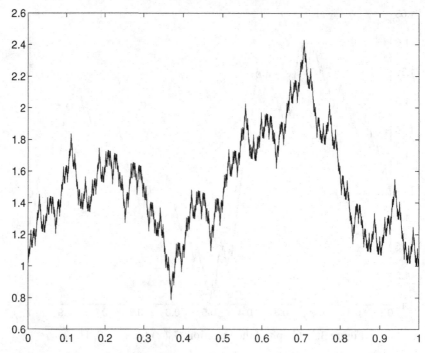

Figure 3.4: Graph of the Function $G_1(x)$ for $x \in [0, 1]$.

all integers $N \geq 1$ we have

$$\sum_{0 \leq n < N} r_n = \sqrt{N} G_1 \left(\frac{\log N}{\log 4} \right). \qquad (3.32)$$

Proof. Define the sequence of vectors $(V_n)_{n \geq 0}$ in $(\mathbb{R}^2)^{\mathbb{N}}$ by

$$V_n = \begin{pmatrix} r_n \\ r_{2n+1} \end{pmatrix}.$$

Then for $0 \leq i \leq 3$ we have $V_{4n+i} = \Gamma_i V_n$, where

$$\Gamma_0 = \begin{pmatrix} 1 & 0 \\ 1 & 0 \end{pmatrix}, \qquad \Gamma_1 = \begin{pmatrix} 1 & 0 \\ -1 & 0 \end{pmatrix},$$

$$\Gamma_2 = \begin{pmatrix} 0 & 1 \\ 0 & 1 \end{pmatrix}, \qquad \Gamma_3 = \begin{pmatrix} 0 & -1 \\ 0 & 1 \end{pmatrix}.$$

Note that $\Gamma = \Gamma_0 + \Gamma_1 + \Gamma_2 + \Gamma_3 = 2I$. Hence, applying Theorem 3.5.1 with $k = 4$, the Γ's above, and $\Lambda = 0$, there exists a continuous (vector) function G of period 1 such that

$$A(N) = \sum_{0 \leq n < N} V_n = \sqrt{N} G \left(\frac{\log N}{\log 4} \right).$$

Hence, taking the first projection, there exists a continuous function G_1 of period 1 such that

$$\sum_{0 \le n < N} r_n = \sqrt{N} G_1 \left(\frac{\log N}{\log 4} \right).$$

To show that G_1 is nowhere differentiable, we apply Theorem 3.5.3(i) with $\alpha = \frac{1}{2}$. ∎

Figure 3.4 is a graph of the function G_1.

3.6 Base-*k* Representation with Alternate Digit Sets

The previous two sections discussed representation in base k using the digit set $\Sigma_k = \{0, 1, \ldots, k - 1\}$. This gives rise to a numeration system for the non-negative integers where the valid representations are given by the set $C_k = \{\epsilon\} \cup (\Sigma_k \setminus \{0\})\Sigma_k^*$. In this section, we examine the use of alternate digit sets.

We begin by examining the digit set $E_k = \{1, 2, \ldots, k\}$. This digit set has the pleasant property of avoiding the leading-zeros problem. As the following theorem shows, this gives rise to a perfect numeration system for the non-negative integers, where the set of valid representations is simply E_k^*.

Theorem 3.6.1 *Let k be an integer ≥ 2. Every integer $n \ge 0$ can be represented uniquely in the form*

$$n = \sum_{0 \le i \le r} a_i k^i,$$

where the a_i are integers chosen from $\{1, 2, \ldots, k\}$.

Proof. Left to the reader as Exercise 28. ∎

This representation is called the *bijective representation*.

We next turn to digit sets that include both positive and negative digits. The inclusion of negative digits means there is the potential to represent all integers, not just the non-negative ones. Probably the most famous example of this type of numeration system is the so-called *balanced ternary* system, which represents numbers in base 3 using the digit set $F = \{-1, 0, 1\}$. This is a perfect numeration system for $\mathbb{Z}$ whose corresponding set of valid representations is given by $\{\epsilon\} \cup (F \setminus \{0\})F^*$.

Call a finite set D of integer digits *basic* for base $k \ge 2$ if $0 \in D$ and

$$((1, k, k^2, \ldots), D, \{\epsilon\} \cup (D \setminus \{0\})D^*)$$

is a perfect numeration system. Then the following natural question arises: given a base k, which digit sets are basic? The following theorem provides an answer.

We say a set S is a *complete residue system* (mod k) if Card $S = k$ and for all integers n there exists $m \in S$ such that $m \equiv n$ (mod k).

Theorem 3.6.2 *Let $k \geq 2$ be an integer. Then the digit set D containing 0 is basic for k if and only if the following two conditions hold:*

(a) *D forms a complete residue system (mod k);*
(b) *for all $n \geq 1$ and all $w \in D^n$, $[w]_k$ is never equal to any nonzero multiple of $k^n - 1$.*

Before we prove the theorem, we state and prove some lemmas.

Lemma 3.6.3 *If D is basic for k, then $|D| \leq k$.*

Proof. Suppose $|D| > k$. Let $m = \max_{d \in D} |d|$; clearly $m \geq 1$. Consider the strings in $S = (D \setminus \{0\})D^n$. There are at least $k(k+1)^n$ such strings, and if $w, x \in S$, we must have $[w]_k \neq [x]_k$, since D is basic. But for $w \in S$ we have

$$|[w]_k| \leq mk^n + \cdots + mk + m \leq m(k^{n+1} - 1).$$

Hence there are at most $2m(k^{n+1} - 1) + 1 \leq 2mk^{n+1}$ possible different values for representations of length exactly $n + 1$. But there are $k(k+1)^n$ different representations. Since $k(k+1)^n > 2mk^{n+1}$ for n sufficiently large, some number must have at least two different representations. Hence D cannot be basic. ∎

Lemma 3.6.4 *If D is basic for k then D must be a complete residue system mod k.*

Proof. Since D is basic for k, for all i with $0 \leq i < k$ there must be a representation w such that $[w]_k = i$. Suppose $w = a_r a_{r-1} \cdots a_1 a_0$. Then $i = a_r k^r + \cdots + a_1 k + a_0$, so taking both sides modulo k, we find $i \equiv a_0$ (mod k). Since by Lemma 3.6.3 D contains at most k members, the result follows. ∎

Lemma 3.6.5 *If D is basic for k, then $j(k-1) \notin D$ for all $j \neq 0$.*

Proof. Assume D is basic, and assume there exists a $j \neq 0$ such that $j(k-1) \in D$. Since D is basic, there exists a string w such that $[w]_k = -j$. Let $w = a_r a_{r-1} \cdots a_1 a_0$. Then since $a_0 \equiv -j$ (mod k), we have, by Lemma 3.6.4, that $a_0 = j(k-1)$. Hence

$$[a_r a_{r-1} \cdots a_1]_k = \frac{(-j) - a_0}{k} = \frac{-j - jk + j}{k} = -j,$$

so $-j$ has two different representations, a contradiction. ∎

Lemma 3.6.6 *For any integer $t \geq 1$ the digit set D is basic for k if and only if the digit set $\{[w]_k : w \in D^t\}$ is basic for k^t.*

Proof. Left to the reader. ∎

We are now ready to prove Theorem 3.6.2.

Proof. If D is a basic digit set, then claim (a) follows by Lemma 3.6.4, and claim (b) follows by combining Lemmas 3.6.5 and 3.6.6.

Now suppose that conditions (a) and (b) hold. We wish to conclude that D is basic for k. Define $f(j) = (j - \overline{j})/k$, where $\overline{j}$ is the unique member of D congruent to $j \pmod{k}$. Define $f^0(j) = j$, and $f^s(j) = f(f^{s-1}(j))$ for $s \geq 1$. Clearly, if $f^{s+1}(j) = 0$ for some s, then

$$j = \overline{j} + \overline{f(j)}k + \cdots + \overline{f^s(j)}k^s.$$

Now for $|j|$ sufficiently large we have $f(j) < j$. Hence either there exists an s with $f^s(j) = 0$, or there exist $s \geq 0$, $t \geq 1$ with $f^s(j) = f^{s+t}(j) = a$ for some $a \neq 0$. In the former case, we get a representation for j that is necessarily unique. In the latter case, we have $f^t(a) = a$. Hence

$$a = \overline{a} + \cdots + \overline{f^{t-1}(a)}k^{t-1} + ak^t,$$

and so

$$-a(k^t - 1) = \overline{a} + \cdots + \overline{f^{t-1}(a)}k^{t-1}.$$

It follows that $-a(k^t - 1) \in D^t$, which contradicts (b). ∎

Example 3.6.7 Let k, l be positive integers. The (k, l) numeration system is defined to be a form of base-$(k + l + 1)$ representation using the digit set $\{-k, 1 - k, 2 - k, \ldots, -1, 0, 1, 2, \ldots, l - 1, l\}$. (Balanced ternary is the $(1, 1)$ numeration system.) By Theorem 3.6.2 these numeration systems are perfect.

3.7 Representations in Negative Bases

In some of the previous sections we examined the properties of various representations of non-negative integers in base k, $k \geq 2$. In this section, we briefly examine representations for *all* integers in base $-k$. First, we prove a general result about representations for rings.

Let S be a commutative ring with unit element, let $k \in S$, and let $U = (1, k, k^2, \ldots)$. We usually specify a digit set D containing 0 and then define the set of valid representations to be

$$R = \{\epsilon\} \cup (D \setminus \{0\})D^*.$$

We now generalize concept of complete residue systems, introduced in the previous section, to arbitrary commutative rings. We say that D constitutes a *complete residue system* for S (mod k) if D satisfies the following two conditions.

(a) For each $s \in S$, there exists $d \in D$ such that $s \equiv d$ (mod k).
(b) If $d, d' \in D$ and $d \neq d'$, then $d \not\equiv d'$ (mod k).

We then have the following useful lemma.

Lemma 3.7.1 *Let S be a ring with $D \subseteq S$ and $k \in S$. Let $\mathcal{N} = (U, D, R)$ be a numeration system with $U = (1, k, k^2, \ldots)$ and $R = \{\epsilon\} \cup (D \setminus \{0\})D^*$.*

(a) *If $\mathcal{N}$ is complete then D contains a complete residue system for S (mod k).*
(b) *If D is a complete residue system with $0 \in D$, and if $kx = ky$ in S implies $x = y$ and $\mathcal{N}$ is complete, then $\mathcal{N}$ is unambiguous.*

Proof. (a): Since D is complete, any $s \in S$ has a representation $s = \sum_{0 \leq i \leq r} a_i k^i$. Then the set of all possible values for a_0 contains a complete residue system, since $s \equiv a_0$ (mod k).

(b): Suppose D is a complete residue system for S (mod k). Let $\sum_{0 \leq i \leq r} a_i k^i$ and $\sum_{0 \leq j \leq r'} a'_j k^j$ be two different representations for some $s \in S$, with $a_i, a'_j \in D$ for $0 \leq i \leq r, 0 \leq j \leq r'$. Without loss of generality, we may assume that $r \leq r'$ and r is the smallest among all elements with two different representations. Then we have $s \equiv a_0$ (mod k), and $s \equiv a'_0$ (mod k), so $a_0 \equiv a'_0$ (mod k), and hence $a_0 = a'_0$. Now consider $x := \sum_{1 \leq i \leq r} a_i k^{i-1}$ and $y := \sum_{1 \leq j \leq r'} a'_j k^{j-1}$. We see that $kx = ky = s - a_0$. By the cancellation property, $x = y$. But then x has two different representations and one is of shorter length than before, a contradiction. ∎

As an application, we prove the following:

Theorem 3.7.2 *Let $k \geq 2$ be an integer. Then every integer $N \in \mathbb{Z}$ has a unique base-$(-k)$ representation of the form*

$$N = \sum_{0 \leq i \leq t} a_i (-k)^i,$$

where $a_t \neq 0$ and $0 \leq a_i < k$ for $0 \leq i \leq t$.

Proof. First, we prove that every integer can be represented in base $(-k)$ using the digits $\Sigma_k = \{0, 1, \ldots, k - 1\}$. We do so with the following algorithm:

NEGBASEREP(N, k)
$i := 0$
while $N \neq 0$ do
(1) $a_i := N \bmod k$
(2) $N := (N - a_i)/(-k)$
(3) $i := i + 1$

It is easy to see that if the algorithm terminates, then there exists r such that

$$N = \sum_{0 \le i \le r} a_i(-k)^i.$$

It remains to see that NEGBASEREP terminates. We do this by induction on $|N|$. It is easy to see that the algorithm terminates if $|N| \le 1$. Otherwise, in step (2) we replace N with $N' = -(N - a_i)/k$. Now if $N \notin \{0, -1\}$, we have $|N'| < |N|$, and the proof is complete.

To see that the representation is unique, we just apply Lemma 3.7.1. ∎

If $N = \sum_{0 \le i \le r} a_i(-k)^i$ and $a_r \ne 0$, we define $(N)_{-k}$ to be the string $a_r a_{r-1} \cdots a_1 a_0$. Also, if $w = c_1 c_2 \cdots c_t$, we write $[w]_{-k}$ for the integer $\sum_{1 \le i \le t} c_i(-k)^{t-i}$.

3.8 Fibonacci Representation

We now turn to examine another exotic numeration system, based on the Fibonacci numbers. Recall that the Fibonacci numbers are defined by $F_0 = 0$, $F_1 = 1$, and $F_n = F_{n-1} + F_{n-2}$ for $n \ge 2$. We have the following:

Theorem 3.8.1 *Every integer* $n \ge 0$ *can be uniquely expressed as* $n = \sum_{2 \le i \le r} a_i F_i$ *with* $a_i \in \Sigma_2 = \{0, 1\}$, $a_r \ne 0$, *and* $a_i a_{i+1} = 0$ *for* $2 \le i < r$.

This representation is called the *Fibonacci representation*.

Proof. We use Theorem 3.1.1. Suppose $a_i \in \{0, 1\}$. It suffices to see that $a_t F_t + a_{t-1} F_{t-1} + \cdots + a_2 F_2 < F_{t+1}$ for all t if and only if $a_i a_{i+1} = 0$ for $2 \le i < r$.

Suppose that there is an index i with $a_i a_{i+1} = 1$, i.e., $a_i = a_{i+1} = 1$. Then $a_{i+1} F_{i+1} + a_i F_i + \cdots + a_2 F_2 \ge F_i + F_{i+1} = F_{i+2}$, a contradiction for $t = i + 1$.

On the other hand, suppose $a_i a_{i+1} = 0$ for $2 \le i < r$. Then the maximum possible value of $a_t F_t + \cdots + a_2 F_2$ clearly occurs when $a_t = 1$, $a_{t-1} = 0$, $a_{t-2} = 1$, $a_{t-3} = 0$, etc. But it is easy to prove by induction that $F_t + F_{t-2} + \cdots + F_2 = F_{t+1} - 1$ if t is even, and $F_t + F_{t-2} + \cdots + F_3 = F_{t+1} - 1$ if t is odd. ∎

It follows from Theorem 3.8.1 that there is a one–one correspondence between the non-negative integers and representations of the form $\sum_{2 \le i \le r} a_i F_i$ with $a_i a_{i+1} = 0$ for $2 \le i < r$. We write $(n)_F$ for $a_r a_{r-1} \cdots a_2$, the Fibonacci representation of n, and if $w = a_1 a_2 \cdots a_j \in \{0, 1\}^*$, we write $[w]_F = \sum_{1 \le i \le j} a_i F_{j-i+2}$.

The set of all valid Fibonacci representations is given by the regular expression

$$\epsilon + 1(0 + 01)^*.$$

3.9 Ostrowski's α-Numeration System

In this section we introduce a numeration system, originally due to Ostrowski, which is based on continued fractions. It can be viewed as a generalization of the Fibonacci numeration system discussed in the previous section.

As we have seen in Section 2.4, every irrational real number α can be expressed uniquely as an infinite simple continued fraction $\alpha = [a_0, a_1, a_2, \ldots]$. Further, if we define $p_{-2} = 0$, $p_{-1} = 1$, $q_{-2} = 1$, $q_{-1} = 0$, and $p_n = a_n p_{n-1} + p_{n-2}$, $q_n = a_n q_{n-1} + q_{n-2}$ for $n \geq 0$, then $p_n/q_n = [a_0, a_1, \ldots, a_n]$. The sequence $(q_n)_{n \geq 0}$ of the denominators of the convergents then forms the basis for a numeration system based on α.

We have the following theorem:

Theorem 3.9.1 *Let α be an irrational real number, and let $(q_n)_{n \geq 0}$ be the sequence of the denominators of the convergents to the continued fraction for α. Then every non-negative integer N can be represented uniquely in the form*

$$N = \sum_{0 \leq i \leq j} b_i q_i,$$

where the b_i are integers satisfying the following three conditions:

1. $0 \leq b_0 < a_1$.
2. $0 \leq b_i \leq a_{i+1}$ for $i \geq 1$.
3. For $i \geq 1$, if $b_i = a_{i+1}$ then $b_{i-1} = 0$.

Proof. We use Theorem 3.1.1. It suffices to show that the inequality

$$b_0 q_0 + b_1 q_1 + \cdots + b_i q_i < q_{i+1} \tag{3.33}$$

is equivalent to the three stated conditions.

Assume that (3.33) holds. Then, since $q_0 = 1$ and $q_1 = a_1$, we see that the inequality $b_0 q_0 < q_1$ implies condition 1.

Now we prove condition 2. Inequality (3.33) implies that $b_i q_i < q_{i+1} = a_{i+1} q_i + q_{i-1}$. Dividing by q_i, we get $b_i < a_{i+1} + q_{i-1}/q_i$. Since $q_{i-1} \leq q_i$, we get $b_i \leq a_{i+1}$.

To prove condition 3, note that if $b_i = a_{i+1}$ and $b_{i-1} \geq 1$, then

$$b_0 q_0 + \cdots + b_{i-1} q_{i-1} + b_i q_i \geq a_{i+1} q_i + q_{i-1} = q_{i+1},$$

contradicting inequality (3.33).

It remains to see that conditions 1–3 imply (3.33). We prove this by induction on i. For $i = 0$, we know by condition 1 that $b_0 < a_1$. Hence $b_0 q_0 < q_1$. For $i = 1$, we know by conditions 2 and 3 that $b_1 \leq a_2$ and $b_0 = 0$ if $b_1 = a_2$; hence $b_0 q_0 + b_1 q_1 = b_0 + b_1 a_1 \leq a_1 a_2 < a_1 a_2 + 1 = q_2$.

Now assume the implication for $i < r$; we prove it for $i = r$. We know $b_r \leq a_{r+1}$ by condition 2; if $b_r \leq a_{r+1} - 1$, then $b_r q_r \leq (a_{r+1} - 1)q_r \leq q_{r+1} - q_r$. Together with the hypothesis $b_0 q_0 + \cdots + b_{r-1} q_{r-1} < q_r$, this yields $b_0 q_0 + \cdots + b_r q_r < q_{r+1}$.

On the other hand, if $b_r = a_{r+1}$ then by condition 3 we know $b_{r-1} = 0$. It follows that $b_{r-1} q_{r-1} + b_r q_r = a_{r+1} q_r = q_{r+1} - q_{r-1}$. Together with the hypothesis $b_0 q_0 + \cdots + b_{r-2} q_{r-2} < q_{r-1}$, this gives $b_0 q_0 + \cdots + b_r q_r < q_{r+1}$, as desired. ■

3.10 Representations in Complex Bases

In previous sections we have considered representations for integers or positive integers. In this section, we turn to representation of the Gaussian integers,

$$\mathbb{Z}[i] = \{a + bi : a, b \in \mathbb{Z}\},$$

where $i = \sqrt{-1}$. We confine our discussion to base sets consisting of the powers of a single Gaussian integer, θ. As in previous sections, we seek representations that are natural and that represent each integer exactly once.

As we have seen in Section 3.1, in order to get a complete numeration system, we need a digit set that is a complete residue system. For $\theta = d + ei$, we define the norm $N(\theta) = d^2 + e^2$.

The next lemma tells us how to compute the representative of a particular element.

Lemma 3.10.1 *Let d, e be integers with $\gcd(d, e) = 1$. Define $\theta = d + ei$. Let $\overline{d}$ be an integer such that $d\overline{d} \equiv 1 \pmod{N(\theta)}$. Then for all integers a, b there exists a unique integer c, $0 \leq c < N(\theta)$, such that $a + bi \equiv c \pmod{d + ei}$. Furthermore, we may take $c = (a + be\overline{d}) \bmod N(\theta)$.*

Proof. It is easy to verify that, for all $c \in \mathbb{Z}$, we have

$$(a + bi - c)(d - ei) = (a - c)d + be + (bd + (c - a)e)i.$$

It follows that

$$a + bi - c = (f + gi)(d + ei),$$

where

$$f = \frac{(a - c)d + be}{d^2 + e^2}$$

and

$$g = \frac{bd + (c - a)e}{d^2 + e^2}.$$

Now let $c \equiv a + be\overline{d} \pmod{d^2 + e^2}$. For this choice of c, it is clear that $f \in \mathbb{Z}$. To see that $g \in \mathbb{Z}$, note that

$$bd + (c - a)e \equiv bd + be^2\overline{d} \pmod{d^2 + e^2}$$
$$\equiv b\overline{d}(d^2 + e^2) \pmod{d^2 + e^2}$$
$$\equiv 0 \pmod{d^2 + e^2}.$$

Hence $a + bi \equiv c \pmod{d + ei}$.

To see that c is unique, assume that for two distinct integers c, c' with $0 \leq c, c' < N(\theta)$ we have $c \equiv c' \pmod{d + ei}$. Then if $h = c - c'$, we have $h \equiv 0 \pmod{d + ei}$, and furthermore

$$1 \leq h < N(\theta). \tag{3.34}$$

Hence there exist integers r, s such that $h = (r + si)(d + ei) = rd - es + (ds + er)i$. Since h is a real integer, it follows that $ds + er = 0$. Since $\gcd(d, e) = 1$, there exists an integer k such that $s = -ek$ and $r = dk$. Hence $h = rd - es = (d^2 + e^2)k = N(\theta)k$. But this contradicts (3.34). ∎

Example 3.10.2 Find an integer c such that $17 + 23i \equiv c \pmod{\theta}$, where $\theta = 31 + 45i$. In this case we have $a = 17, b = 23, d = 31, e = 45$, and $N(\theta) = 2986$. Then $\overline{d} = 289$. Thus we can take $c = (a + be\overline{d}) \bmod N(\theta) = 532$. Indeed, we have $a + bi - c = (-5 + 8i)(d + ei)$.

Lemma 3.10.3 *Let d, e be integers with $\gcd(d, e) = \lambda > 0$. Given integers a, b, there exists a unique pair of integers (x, y) with $0 \leq x < (d^2 + e^2)/\lambda$ and $0 \leq y < \lambda$ such that $a + bi \equiv x + yi \pmod{d + ei}$.*

Proof. Let A, B be integers with $0 \leq A, B < \lambda$ such that $A \equiv a \pmod{\lambda}$ and $B \equiv b \pmod{\lambda}$. By Lemma 3.10.1 there is a unique c with $0 \leq c < N(\theta)/\lambda^2$ such that

$$\frac{a - A}{\lambda} + \frac{b - B}{\lambda}i \equiv c \pmod{\frac{d + ei}{\lambda}} \tag{3.35}$$

Hence, multiplying by λ, we get

$$a - A + (b - B)i \equiv \lambda c \pmod{d + ei}.$$

It follows that

$$a + bi \equiv \lambda c + A + Bi \pmod{d + ei},$$

so we may take $x = \lambda c + A$ and $y = B$. Since $0 \leq c \leq N(\theta)/\lambda^2 - 1$, it follows that $0 \leq x < N(\theta)/\lambda$.

To prove uniqueness, assume there are integers x, x', y, y' with $0 \leq x, x' < N(\theta)/\lambda, 0 \leq y, y' < \lambda$, and $(x, y) \neq (x', y')$ such that

$$x + yi \equiv x' + y'i \pmod{d + ei}.$$

Then, letting $m = x - x'$ and $n = y - y'$, there must exist integers r, s such that $m + ni = (d + ei)(r + si)$. It follows that $m = dr - es$ and $n = ds + er$, and $0 \leq |m| < N(\theta)/\lambda$, $0 \leq |n| < \lambda$. Since $\lambda = \gcd(d, e)$, we have $\lambda \mid n$, so it follows that $n = 0$. Hence $ds = -er$, and there exists an integer k such that $r = kd/\lambda$ and $s = -ke/\lambda$. Hence $m = k(d^2 + e^2)/\lambda$, and hence $k = 0$. But this means $x = x'$ and $y = y'$, a contradiction. ∎

We have now proved the following result:

Theorem 3.10.4 *Let d, e be integers with $\gcd(d, e) = \lambda > 0$, and define $\theta = d + ei$. Then the set*

$$\left\{ x + yi : 0 \leq x < \frac{N(\theta)}{\lambda},\ 0 \leq y < \lambda \right\}$$

forms a complete residue system (mod θ).

Let $U = (1, \theta, \theta^2, \dots)$ and $D = \{0, 1, \dots, N(\theta) - 1\}$. We now investigate the following natural question: for which Gaussian integers θ is the numeration system $(U, D, \{\epsilon\} \cup (D \setminus \{0\})D^*)$ perfect?

Theorem 3.10.5 *Let $\theta \in \mathbb{Z}[i]$. The numeration system $\mathcal{N} = (U, D, \{\epsilon\} \cup (D \setminus \{0\})D^*)$, where $U = \{1, \theta, \theta^2, \dots\}$ and $D = \{0, 1, \dots, N(\theta) - 1\}$, is perfect if and only if $\theta = -A \pm i$ for some integer $A \geq 1$.*

Proof. First, we show that if θ is not of the form $-A \pm i$, then some element of $\mathbb{Z}[i]$ has no representation.

Let $\theta = a + bi$. Suppose $a \geq 1$. Then we prove that the number $\alpha := (1 - a) + bi$ is not represented by $\mathcal{N}$. Suppose it were. Then we would have

$$\alpha = c_0 + c_1\theta + c_2\theta^2 + \cdots + c_k\theta^k.$$

Define $\rho := \alpha(1 - \theta) = a^2 + b^2 - 2a + 1$. We now have

$$\rho = c_0 + (c_1 - c_0)\theta + \cdots + (c_k - c_{k-1})\theta^k - c_k\theta^{k+1}. \tag{3.36}$$

It follows that $\rho \equiv c_0 \pmod{\theta}$. Since $0 \leq \rho < N(\theta)$, it follows that $\rho = c_0$. Now, from (3.36), we get

$$(c_1 - c_0)\theta + \cdots + (c_k - c_{k-1})\theta^k - c_k\theta^{k+1} = 0.$$

Dividing by θ, we get

$$(c_1 - c_0) + (c_2 - c_1)\theta + \cdots + (c_k - c_{k-1})\theta^{k-1} - c_k\theta^k = 0.$$

Taking both sides mod θ, we see that $c_1 \equiv c_0 \pmod{\theta}$. Hence $c_1 = c_0$. Similarly $c_2 = c_1, \dots, c_k = c_{k-1}$, and finally $c_k = 0$. Hence $c_0 = c_1 = c_2 = \cdots = c_k = 0$ and $\rho = 0$. It follows that $\theta = 1$, which is clearly not the base of a perfect numeration system.

Next we show that $b = \pm 1$. From the binomial theorem, we easily see that b is a divisor of Im $\theta^j = $ Im $(a + bi)^j$ for each $j \geq 0$. It follows that if

$$\alpha = c_0 + c_1\theta + \cdots + c_k\theta^k,$$

then

$$\text{Im } \alpha = c_1\text{Im } \theta + \cdots + c_k\text{Im } \theta^k.$$

Hence $b \mid \text{Im } \alpha$. Since α was arbitrary, we must have $b = \pm 1$, a contradiction.

The remaining possibility, $\theta = \pm i$, is easily seen not to be the base of a perfect numeration system.

It now remains to see that if $\theta = -A \pm i$, then the associated numeration system $\mathcal{N}$ is perfect. By Lemma 3.7.1 and Theorem 3.10.4, it suffices to show that each Gaussian integer possesses at least one representation. We first consider the case where $\theta = -A + i$.

The idea behind the proof is the following: we first show how to express any $\alpha = \mathbb{Z}[i]$ in the form $\sum_{0 \leq i \leq r} c_i\theta^i$ where the c_i are non-negative integers that are not necessarily less than $N(\theta)$. We call such a representation *unnormalized*. Next, we show how to convert this representation to a normalized one where the digits c_i satisfy $0 \leq c_i < N(\theta)$ by a series of steps called the *clearing algorithm*.

We start by obtaining an unnormalized representation. Let $\alpha = e + fi$. Let $c = e + Af$. Then a simple computation shows that $c + f\theta = e + fi = \alpha$. Now it is easy to see that $\theta^2 + 2A\theta + A^2 = -1$, so if $c < 0$ we may replace c with $(-c)(\theta^2 + 2A\theta + A^2)$. Similarly, if $f < 0$ we may replace f with $(-f)(\theta^2 + 2A\theta + A^2)$. Hence, we get a representation

$$\alpha = d_0 + d_1\theta + d_2\theta^2 + d_3\theta^3,$$

where d_0, d_1, d_2, d_3 are non-negative integers.

This representation may not be normalized, so we now show how to convert it to a normalized one by a series of steps. Each step will preserve the invariant that the digits are all non-negative, while not increasing the sum of those digits.

More generally, consider a representation

$$\alpha = d_0 + d_1\theta + \cdots + d_k\theta^k \tag{3.37}$$

where $k \geq 3$ and $d_i \geq 0$ for $0 \leq i \leq k$. Define $t_0 = \sum_{0 \leq j \leq k} d_j$. Then t_0 is a non-negative integer, and $t_0 = 0$ if and only if $\alpha = 0$. By the division theorem (Exercise 2.1), we may write

$$d_0 = q(A^2 + 1) + u_0, \tag{3.38}$$

where $0 \leq u_0 \leq A^2$ and $q \geq 0$. An easy computation shows that

$$A^2 + 1 = (A - 1)^2\theta + (2A - 1)\theta^2 + \theta^3. \tag{3.39}$$

Now we substitute (3.39) in (3.38) and get

$$d_0 = u_0 + q((A - 1)^2\theta + (2A - 1)\theta^2 + \theta^3). \tag{3.40}$$

We substitute this expression for d_0 in (3.37) to get

$$\begin{aligned}
\alpha &= u_0 + (d_1 + q(A - 1)^2)\theta + (d_2 + q(2A - 1))\theta^2 \\
&\quad + (d_3 + q)\theta^3 + d_4\theta^4 + \cdots + d_k\theta^k \\
&= d_0' + d_1'\theta + \cdots + d_k'\theta^k.
\end{aligned}$$

If we set $t_0' = \sum_{0 \le j \le k} d_j'$, then we find

$$\begin{aligned}
t_0' - t_0 &= u_0 - d_0 + q(A - 1)^2 + q(2A - 1) + q \\
&= q(-(A^2 + 1) + (A - 1)^2 + (2A - 1) + 1) \\
&= 0,
\end{aligned}$$

so $t_0' = t_0$.

Now write $\alpha = d_0' + \theta\alpha_1$. We see that

$$\alpha_1 = d_1' + d_2'\theta + \cdots + d_k'\theta^{k-1}.$$

If we define $t_1 = d_1' + d_2' + \cdots + d_k'$, then clearly $t_1 \le t_0' = t_0$ with equality if and only if $u_0 = 0$. Now continue this process. We thus obtain

$$\alpha = \alpha_1\theta + u_0, \qquad \alpha_1 = \alpha_2\theta + u_1, \cdots$$

and the corresponding sums of digits $t_0, t_1, t_2, \ldots$. Since this latter sequence is non-increasing and non-negative, eventually we find $t_q = t_{q+1} = t_{q+2} \cdots$ for some $q \ge 0$. It follows that $0 = u_q = u_{q+1} = u_{q+2} = \cdots$. Hence $\alpha_q = \alpha_{q+1}\theta$, $\alpha_{q+1} = \alpha_{q+2}\theta, \ldots$, which implies that $\theta^k \mid \alpha_q$ for all $k \ge 0$. Hence $\alpha_q = 0$. Thus we obtain the normalized representation

$$\alpha = u_0 + u_1\theta + \cdots + u_{q-1}\theta^q,$$

which completes the proof for $-A + i$.

To handle the case of representation in base $(-A - i)$, simply expand the complex conjugate $\overline{\alpha}$ in base $(-A + i)$ as before. With $\theta = -A + i$, we obtain

$$\overline{\alpha} = u_0 + u_1\theta + \cdots + u_{q-1}\theta^q.$$

Now take the complex conjugate of both sides to obtain the representation of α in base $-A - i$. ∎

3.11 Exercises

1. Show that for integers $n, m \geq 0$:

   (i) $v_p(n!) = \sum_{k=1}^{\infty} \left\lfloor \dfrac{n}{p^k} \right\rfloor$;

   (ii) the binomial coefficients $\binom{m}{n} := \dfrac{m!}{n!(m-n)!}$ are integers for $m \geq n$.

2. Show that if p is a prime number and $n \geq 1$ is an integer, then

$$\frac{n}{p-1} - \frac{\log(n+1)}{\log p} \leq v_p(n!) \leq \frac{n-1}{p-1}.$$

3. Show that for a prime p and real numbers $x > 1$ we have

$$S_p(x) \leq \frac{p-1}{\log p} \left(\sum_{\substack{p' \leq x+1 \\ p' \text{ prime}}} \frac{\log p'}{p'} + \sum_{j \leq x+1} \frac{\log j}{j(j-1)} \right) (x+1).$$

4. Let $b \geq 2$ be an integer, and let p be a prime such that $p \nmid b - 1$. Let a and c be any residues modulo p. Show that if $N(x)$ is the number of positive integers $n < x$ for which $n \equiv a \pmod{p}$ and $s_b(n) \equiv c \pmod{p}$, then

$$\lim_{x \to \infty} \frac{N(x)}{x} = \frac{1}{p^2}.$$

5. Let k be an integer ≥ 2, and let N be a non-negative integer.
   (a) Show that $S_k(kN) = kS_k(N) + k(k-1)N/2$.
   (b) Let $0 < a < k$. Show that $S_k(kN + a) = S_k(kN) + as_k(N) + a(a-1)/2$.
   (c) Conclude that $S_k(N)$ can be computed in time polynomial in $\log k$ and $\log N$.

6. Show that $s_k(n) \equiv n \pmod{k-1}$ for all integers $k \geq 2, n \geq 0$.

7. Let k be an integer ≥ 2, and let m, n be non-negative integers. Prove that $s_k(m+n) \leq s_k(m) + s_k(n)$, with equality if and only if there are no carries if m is added to n in base k.

8. Let $k \geq 2$ be an integer. Show that $(s_k(n))_{n \geq 0}$ does not satisfy a linear recurrence with constant rational coefficients.

9. Let k, m be integers with $k \geq 2, m \geq 1$.
   (a) Show that $(s_k(n) \bmod m)_{n \geq 0}$ is ultimately periodic if and only if $m \mid k-1$.
   (b) Prove the following generalization of Thue's result on overlap-free words: the sequence $(s_k(n) \bmod m)_{n \geq 0}$ is overlap-free if and only if $m \geq k$.

10. Does there exist an integer $n \geq 0$ such that $s_{10}(2^n) = s_{10}(2^{n+1})$?

11. Let $a_k(n)$ denote the alternating sum of digits, base k. More precisely, let $(n)_k = b_t b_{t-1} \cdots b_0$. Define $a_k(n) = \sum_{0 \leq i \leq t} (-1)^i b_i$. Show that $a_k(n) \equiv n \pmod{k+1}$ for all integers $k \geq 2, n \geq 0$.

12. Let k be an integer ≥ 2, and for $n \geq 0$ define $f_k(n) = \sum_{1 \leq i \leq n} i \cdot v_k(i)$. Show that

$$f_k(n) = \frac{n(n+1)/2 - n s_k(n) + S_k(n)}{k-1}.$$

13. (D. Wilson) Let $a > b > 0$ be integers, and consider expansion in "base a/b". (Note that $\gcd(a, b)$ need not be 1.) In this base, normalization is defined by subtracting a from a digit position until the result lies in $[0, a)$, and adding b for each subtraction to the digit immediately to the left.
    (a) Prove that every positive integer has exactly one representation in this system.
    (b) For which pairs (a, b) is the set of all valid representations for all positive integers a regular language?

14. Define $G(1) = 0$, and $G(n) = \max_{1 \leq i \leq n/2}(G(i) + G(n - i) + i)$ for $n > 1$. Show that $G(n) = S_2(n)$.

15. (Clements and Lindström) If m, n are integers, let the base-2 representations of m and n (possibly with leading zeros) be $m = \sum_{0 \leq i \leq r} a_i 2^i$ and $n = \sum_{0 \leq i \leq r} b_i 2^i$. Then define $\alpha(m, n) = \sum_{0 \leq i \leq r} a_i b_i$. Finally, define the $n \times n$ matrix $M(n) = [(-1)^{\alpha(i,j)}]_{0 \leq i,j < n}$. Show that $|\det M(n)| = 2^{S_2(n)}$.

16. (Roberts) Let $\varepsilon_j(n)$ be the j th least significant digit in the base-2 representation of n (as in Section 3.3). Show that, for all $n \geq 0$,

$$\sum_{0 \leq i \leq n} \left(\binom{n}{i} \bmod 2 \right) 2^i = \prod_{\substack{j \\ \varepsilon_j(n)=1}} (2^{2^j} + 1).$$

17. Consider the function $C(k)$ in the proof of Theorem 3.2.4 as a function of k. Show that $C(k) \geq \frac{1}{8}$ for all $k \geq 2$. Also show that C is minimized when $k \doteq 85.34646$.

18. Show that for any sequence $(a_n)_{n \geq 0}$ over $\{-1, +1\}$, we have

$$\sup_\theta \left| \sum_{0 \leq n < N} a_n e^{in\theta} \right| \geq \left\| \sum_{0 \leq n < N} a_n e^{in\theta} \right\|_2 = \sqrt{N}.$$

19. Let k be an odd integer ≥ 3. Show that every integer can be represented uniquely in the form $\sum_{0 \leq i \leq r} a_i k^i$, where $a_i \in \{k, 1, 2, 3, \ldots, (k-1)/2, -1, -2, -3, \ldots, -(k-1)/2\}$. This is a representation for all integers that does not use the digit 0.

20. Let the representation of $n \geq 0$ in base -2 be

$$n = \sum_{k \geq 0} e_k(n)(-2)^k.$$

Prove that

$$e_k(n) \equiv \left\lfloor \frac{3n + 2(4^{\lceil \frac{1}{2}(k-1) \rceil} - 1)}{3 \cdot 2^k} \right\rfloor \pmod 2.$$

21. Let $f_w(n) = e_{2;w}(n)$. Show that

$$f_1(3n) = 2f_1(n) - 2f_{11}(n) + f_{111}(n) - 2f_{1011}(n)$$
$$+ f_{11011}(n) - 2f_{101011}(n) + f_{1101011}(n) - \cdots$$
$$= 2f_1(n) - 2\sum_{i \geq 0} f_{(10)^i 11}(n) + \sum_{i \geq 0} f_{11(01)^i 1}(n).$$

22. Define two sequences of polynomials $P_n(x)$ and $Q_n(x)$ as follows: $P_0(x) = Q_0(x) = 1$, and for $n \geq 1$,

$$P_{n+1}(x) = P_n(x) + x^{2^n} Q_n(x),$$
$$Q_{n+1}(x) = P_n(x) - x^{2^n} Q_n(x).$$

Prove the following:
(a) $Q_n(x) = (-1)^n x^{2^n-1} P_n(-1/x)$ for $n \geq 0$.
(b) $P_{n+1}(x) = P_n(x^2) + x P_n(-x^2)$ for $n \geq 0$.
(c) $Q_{n+1}(x) = Q_n(x^2) + x Q_n(-x^2)$ for $n \geq 1$.
(d) $|P_n(e^{i\theta})|^2 + |Q_n(e^{i\theta})|^2 = 2^{n+1}$ for $n \geq 0$, θ real.
(e) $P_{m+n+1}(x) = P_m(x)P_n(x^{2^{m+1}}) + x^{2^m} Q_m(x)P_n(-x^{2^{m+1}})$ for $m, n \geq 0$.
(f) $Q_{m+n+1}(x) = P_m(x)Q_n(x^{2^{m+1}}) + x^{2^m} Q_m(x)Q_n(-x^{2^{m+1}})$ for $m \geq 0, n \geq 1$.
(g) $P_{n-1}(x)P_{2n-1}(-x) - P_{n-1}(-x)P_{2n-1}(x) = (-2)^n x^{2^n-1} P_{n-1}(-x^{2^n})$ for $n \geq 1$.
(h) $P_{n+2}(x) = (1 - x^{2^{n+1}})P_{n+1}(x) + 2x^{2^{n+1}} P_n(x)$ for $n \geq 0$.
(i) $Q_{n+2}(x) = (1 - x^{2^n})Q_{n+1}(x) + 2x^{2^n} Q_n(x)$ for $n \geq 0$.
(j) $P_n(x)P_n(1/x) + Q_n(x)Q_n(1/x) = 2^{n+1}$ for $n \geq 0$.

23. (Brillhart and Morton) Let $r_n = (-1)^{e_{2;11}(n)}$ be the Rudin–Shapiro sequence. Consider the sequence of partial sums defined by $p_n = \sum_{0 \leq i \leq n} r_n$. Then in the sequence $(p_n)_{n \geq 0}$ prove that 0 occurs 0 times, 1 occurs once, 2 occurs twice, and in general n occurs n times.

24. Prove the existence of a bijection $f : \mathbb{N} \to \mathbb{Q}$ such that f and f^{-1} are both polynomial-time computable. (Note that f is polynomial-time computable if there exists a constant c such that $f(n)$ is computable using $O((\log n)^c)$ steps, and f^{-1} is polynomial-time computable if there exists a constant d such that $f^{-1}(p/q)$ is computable using $O((\log p + \log q)^d)$ steps.)

25. Let k be an integer ≥ 2. Show that

$$\sum_{1 \leq i \leq n} \lfloor \log_k i \rfloor = (n+1)\lfloor \log_k n \rfloor - k(k^{\lfloor \log_k n \rfloor} - 1)/(k-1).$$

26. Consider the Barbier infinite word

$$\mathbf{B} = b_1 b_2 b_3 \cdots = 1234567891011121314151617181920212223 \cdots$$

over the alphabet Σ_{10} defined by concatenating the base-10 representation of the natural numbers together. Compute $b_{10^{1000}}$.

27. The Cantor ternary set C is defined to be the set of all real numbers in $[0, 1]$ that can be written using only the digits 0 and 2 in base 3. Show that $C + C = [0, 2]$, where by $C + C$ we mean the set $\{c + c' : c, c' \in C\}$.

28. Prove Theorem 3.6.1: if $k \geq 2$ is an integer, then every non-negative integer can be represented uniquely in the form $\sum_{0 \leq i \leq r} b_i k^i$, where $b_i \in \{1, 2, \ldots, k\}$.

29. (Lenard) Let S be the set of non-negative real numbers whose representation in base 4 uses only the digits 0 and 1.
    (a) Prove that if $x \in S$ and $y \in S$, and $x \neq y$, then $(x + y)/2 \notin S$.
    (b) Suppose that T is a set with $S \subseteq T$, $S \neq T$. Show that there exist $x, y \in T$ with $x \neq y$ such that $(x + y)/2 \in T$.

30. Show that the integers $\{0, 1, \ldots, a^k - 1\}$ can be partitioned into a disjoint sets $S_1, S_2, \ldots, S_a$ such that for each i $(0 \leq i < k)$ the sum $\sum_{s \in S_j} s^i$ is independent of j.

31. Consider representing a positive integer n as the sum of squares using the greedy algorithm.
    (a) Let r_n be the least positive integer whose representation requires n squares. Show that $r_1 = 1, r_2 = 2, r_3 = 3$, and
    $$r_{n+1} = \left(\frac{r_n + 1}{2}\right)^2 + r_n \qquad \text{for } n \geq 3.$$
    (b) Conclude that the greedy algorithm provides a representation using $O(\log \log n)$ squares.

32. (Trigg) Let k be an integer ≥ 2. Do there exist any words $w \in C_k = \{\epsilon\} \cup (\Sigma_k \setminus \{0\})\Sigma_k^*$ such that $[w]_k^j = [w^j]_k$ for some $j > 1$?

33. (Samborski) Let $\sum_{k \geq 1} 2^{-n_k}$ be the binary representation of $(\sqrt{5} - 1)/2$. Show that $n_k \leq 2^{k-1} - 2$ for $k \geq 4$.

34. Show that every power of 5 has a multiple whose representation in base 10 contains no zeros.

35. (Knuth) Let m and n be non-negative integers, with Fibonacci representations $\sum_{2 \leq i \leq r} a_i F_i$ and $\sum_{2 \leq j \leq s} b_j F_j$, respectively. Show that the operation $\circ$ defined by
$$m \circ n = \sum_{2 \leq i \leq r} \sum_{2 \leq j \leq s} a_i b_j F_{i+j}$$
is associative.

36. (Glaisher) Show that the number of odd binomial coefficients $\binom{n}{j}$ with $0 \leq j \leq n$ is $2^{s_2(n)}$.

37. Show that every positive integer can be represented uniquely in the form $\sum_{1 \leq i \leq r} a_i \cdot i!$, where $0 \leq a_i < i$.

38. Find a function $h : \mathbb{N} \to \mathbb{N}$ such that
    (a) $h(1) = 2$;
    (b) $h(h(n)) = h(n) + n$;
    (c) $h(n + 1) > h(n)$.

39. (Bateman and Bradley) Let k be an integer ≥ 1, and let X be an indeterminate. Prove that
$$\sum_{0 \leq j < 2^k} (-1)^{s_2(j)}(X + j)^{k+1} = (-1)^k \cdot (k + 1)! \cdot 2^{k(k-1)/2} \left(X + \frac{2^k - 1}{2}\right).$$

40. Let k be an integer ≥ 1, and let X be an indeterminate. Show that

$$\sum_{0 \leq j < 2^k} (-1)^{s_2(j)}(X+j)^k = (-1)^k \cdot k! \cdot 2^{k(k-1)/2}.$$

41. (a) Give a simpler proof of Theorem 3.10.5 in the case of representation in base $-1+i$, following the sketch below: First, prove that a representation

$$g = \sum_{0 \leq k \leq r} a_k(-1+i)^k \tag{3.41}$$

exists by showing that a_0 and a_1 are determined by the following rule:

$$a_0 \equiv \text{Re } g + \text{Im } g \pmod{2},$$
$$a_1 \equiv \text{Im } g \pmod{2}.$$

Now define

$$g' = \frac{g - (a_0 + (-1+i)a_1)}{-2i}.$$

Then g' is a Gaussian integer, and $|g'| < |g|$ if $|g| > \sqrt{2}$. Thus we can recursively expand g' as $b_r b_{r-1} \cdots b_0$; the expansion for g is then $(b_r b_{r-1} \cdots b_0)(a_1 a_0)$. It remains to see expansions exist for $|g| \leq \sqrt{2}$.

   (b) Using a similar technique, prove that there exist constants c, c' such that

$$2\log_2(|z|) + c < \ell_{-1+i}(z) < 2\log_2(|z|) + c',$$

for all $z \neq 0$, where $\ell_{-1+i}(z)$ denotes the length of the canonical representation of z in base $-1+i$.

42. Let $k \geq 2$ be an integer. Show that

$$\sum_{n \geq 1} \frac{1}{n(\ell_k(n))^2} \sim \frac{\pi^2}{6} \log k.$$

Here ℓ_k is the base-k length function, defined in Section 3.2.

43. Let $g(n)$ denote the number of digits greater than 4 in the base-10 representation of n. Show that $\sum_{n \geq 1} g(n)/2^n = \frac{2}{9}$.

44. (Segal and Lepp) Let $f(n) = \sum 1/k$, where the sum is over all integers $k \geq 1$ having no 0 in their base-n representation. Show that $f(n) = n \log n + O(1/n)$.

45. (Graham and Pollak) Let $u_1 = 1$, and for $n \geq 1$ define $u_{n+1} = \lfloor \sqrt{2}(u_n + \frac{1}{2}) \rfloor$. Prove that if the base-2 representation of $\sqrt{2}$ is $1.a_1 a_2 a_3 \cdots$, then $a_n = u_{2n+1} - 2u_{2n-1}$.

46. (Knuth) Let m, n be positive integers. Simplify the expression

$$\sum_{0 \leq k \leq 2^{mn}-1} \binom{k^m}{n} (-1)^{s_2(k)}.$$

47. (M. Golomb) Show that

$$\sum_{n\geq 1} \frac{1}{4^{\lfloor \log_2 n \rfloor} s_2(n)} = 4 \log \tfrac{3}{2}.$$

48. Let T be a set with n elements. Call a collection C of subsets of T a *separating collection* if for all pairs of elements $(x, y) \in T \times T$ with $x \neq y$ there exists $S \in C$ such that $|S \cap \{x, y\}| = 1$. Show that, for all $n \geq 1$, there exists a separating collection of size $\lceil \log_2 n \rceil$, and that this bound is best possible.

49. Suppose S is a collection of subsets of $G_n = \{1, 2, \ldots, n\}$ such that every subset of G_n can be obtained by taking some number of intersections and unions of sets of S. How small can S be, as a function of n?

50. Consider a numeration system with $U = \{2, 3, 5, 9, \ldots, 2^k + 1, \ldots\}$ and $D = \{0, 1\}$. Show that there exists a constant $c > 0$ such that infinitely many positive integers n have $\geq c \log^* n$ different representations. Here $\log^* x$ is defined recursively as follows:

$$\log^* x = \begin{cases} 0 & \text{if } x \leq 1, \\ 1 + \log^*(\log x) & \text{otherwise.} \end{cases}$$

51. (Olivier) Show that the density of the integers n such that $\gcd(n, s_2(n)) = 1$ is $6/\pi^2$.

52. Let $k \geq 2$ be an integer. Show that

$$\sum_{n\geq 1} \frac{s_k(n)}{n(n+1)} = \frac{k}{k-1} \log k.$$

Generalize.

53. Let j be a fixed integer ≥ 2. Show that $\limsup_{n\to\infty} s_2(n^j)/(\log_2 n) = j$.

54. (D. Bowman) Suppose $(n)_k = a_r a_{r-1} \cdots a_0$. As usual, let $s_k(n) = \sum_{0 \leq i \leq r} a_i$. Define $h_k(n) = \sum_{0 \leq i \leq r} i a_i$. Show that

$$\sum_{n\geq 1} \frac{(-1)^{s_2(n)}}{h_2(2n)} = 6 - \frac{4\pi\sqrt{3}}{3}.$$

55. Show that every positive integer can be written as the sum of terms of the form $2^a 3^b$ for some $a, b \geq 0$, such that no summand divides another.

56. Consider representations of integers in base 2, using only the digits $D = \{-1, 0, 1\}$.
    (a) Show that every integer n has such a representation without leading zeros, subject to the restriction that no two consecutive digits are nonzero.
    (b) Show this representation (called the Reitwiesner representation) is unique.
    (c) Show that the Reitwiesner representation has the smallest number of nonzero digits over all representations in base 2 using the digit set D.

    (d) Give a finite-state transducer that converts a number represented in (ordinary) base-2 representation to the Reitwiesner representation.

57. Show that the Rudin–Shapiro sequence contains subwords of length $1, 2, 3, 4, 5, 6, 7, 8, 10, 12, 14$ that are palindromes, but palindromes of no other lengths.

58. Let $a \geq b \geq 0$ be integers. Show that the solution to the recurrence

$$S_n = \min_{1 \leq k \leq n/2} a S_{n-k} + b S_k$$

    for $n \geq 2$ is

$$S_n = S_1 + (a + b - 1) S_1 \sum_{1 \leq i \leq n-1} a^{e_{2;0}(i)} b^{e_{2;1}(i)-1}.$$

59. (Hickerson) Let a, x be integers with $a \geq 2$, $x \geq 1$. Prove that

$$\frac{a+1}{a} \sum_{1 \leq n \leq x} a^{-v_a(n)} = x + \frac{[(x)_a^R]_a}{a^{|(x)_a|}}.$$

60. Estimate the minimum of the function $F_2(x)$ defined in Section 3.5.

61. Are there integers $a, b, c \geq 0$ such that $s_{10}(a + b) < 5$, $s_{10}(a + c) < 5$, $s_{10}(b + c) < 5$, but $s_{10}(a + b + c) > 50$?

3.12 Open Problems

1. Find a closed form for or efficient way to compute $\sum_{0 \leq n < N} s_k(n^2)$. (Remark: See, e.g., Agronomof [1926].)

2. Is there an algorithm to compute $s_k(a^n)$ in time polynomial in $\log k$, $\log a$, and $\log n$?

3. (Kamae) Let $\alpha(n) = (-1)^{t_n}$ where $t_0 t_1 t_2 \cdots$ is the Thue–Morse word. Prove or disprove that

$$\lim_{n \to \infty} \frac{1}{n} \sum_{i=1}^{n-1} \alpha(3^i) = 0.$$

    (Remark: See Kamae [1990].)

4. Is the sequence $(\{3^n/2^n\})_{n \geq 1}$ uniformly distributed? (Remark: See, for example, Mahler [1968]; Choquet [1980]; Kamae [1990].)

5. (Demaine) Is there a simple closed form for $s_2\left(\binom{2n}{n}/(n+1)\right)$?

6. (Sloane) Let $a_r a_{r-1} \cdots a_1 a_0$ be the base-k representation of n. Define $p_k(n) = \prod_{0 \leq i \leq r} a_i$ and p_k^j to be the j-fold iterate of p_k. Define $P_k(n)$, the base-k persistence of n to be the least j such that $p_k^j(n) \leq k$.

    (a) Is there a number with base-k persistence > 11?

    (b) Is the following true? For all bases $k \geq 2$ there exists a number $d(k)$ such that $P_k(n) \leq d(k)$ for all n.

    (Remarks: See Gottlieb [1969]; Sloane [1973].)

7. (Selfridge and Lacampagne) Can every $k \equiv \pm 1 \pmod 3$ be written as a/b, where a, b have representations in base 3 using digits $1, -1$, but not 0? (Remark: See Guy [1994, p. 267].)

8. Let $\mathbf{s} = (s(n))_{n \geq 0}$ be an integer sequence over $\Sigma_2 = \{0, 1\}$ with $s(0) = 0$. Define $T_0 = \mathbf{s}$, and for $i \geq 1$, define $T_i = \hat{T}_{i-1}$, where $\hat{S}$ denotes the pattern transform of a sequence S. Prove or disprove: T_{2^i} agrees with $\mathbf{s}$, modulo 2, on the first $2^{2^i} - 1$ terms. (Remark: Here is an alternative form of the problem. For $n \geq 1$, create a matrix $M = (m_{i,j})_{1 \leq i, j \leq 2^{2^n}}$ where $m_{i,j} = e_{2;(j)_2}(i)$. Then the conjecture is that $M^{2^n} \equiv I + E_{2^{2^n},1} \pmod 2$, where I is the $2^{2^n} \times 2^{2^n}$ identity matrix, and $E_{i,j}$ is a $2^{2^n} \times 2^{2^n}$ matrix of all zeros except for a lone 1 in row i and column j.)

3.13 Notes on Chapter 3

3.1 Theorem 3.1.1 is a "folk theorem" whose precise origins are unknown to us. It can be found, for example, in Yaglom and Yaglom [1967, pp. 13–14]. Also see Fraenkel [1985, 1989].

Base-k representation apparently originated with the Babylonians, who preferred $k = 60$. The Maya Indians used $k = 20$. Modern decimal representation (i.e., $k = 10$) apparently originated with the Hindus c. 600 C.E. Binary representation (i.e., $k = 2$) was used by Francis Bacon in 1605; see Heath [1972]. For a good survey of the origins of base-k representation, see Knuth [1981, §4.1].

For more on greedy numeration systems, see Frougny [1986, 1988, 1989a, 1989b, 1992a]; Shallit [1994]; Loraud [1995]; and Hollander [1998].

3.2 In a famous passage, Hardy [1967, p. 105] denigrated the mathematics of digital problems such as *find all positive integers equal to the sum of the cubes of their decimal digits*:

These are odd facts, very suitable for puzzle columns and likely to amuse amateurs, but there is nothing in them which appeals much to a mathematician. The proofs are neither difficult nor interesting – merely a little tiresome. The theorems are not serious; and it is plain that one reason (though perhaps not the most important) is the extreme speciality of both the enunciations and the proofs, which are not capable of significant generalization.

Hardy's criticism has merit, but there are ways of answering his objections. We can, for example, try to make our theorems more general, working in an arbitrary base k whenever possible.

Corollary 3.2.2 is due to Legendre [1830, Vol. I, p. 10]. The function $s_2(n)$ is sometimes called *Hamming weight*, after Hamming [1950].

Bush [1940] seems to have been the first to examine $S_k(x) = \sum_{0 \leq n < x} s_k(n)$. He proved that $S_k(x) \sim ((k-1)/(2 \log k)) x \log x$. Bellman and Shapiro [1948] improved this for $k = 2$ by showing that $S_2(x) = (x \log x)/(2 \log 2) + O(x \log \log x)$, and stated in a footnote that they could improve the error term to $O(x)$. Theorem 3.2.3 is due to Mirsky [1949]; we have followed his proof.

Theorem 3.2.3 was also discovered by Cheo and Yien [1955]. The paper of Tang [1963] (in English) appears to be virtually word for word the same as the paper in Chinese of Cheo and Yien [1955].

Drazin and Griffith [1952] found bounds for the error term $S_k(x) - ((k - 1)/(2 \log k))x \log x$. They also discussed sums of *powers* of digits, as did Porges [1945] and B. Stewart [1960]. Clements and Lindström [1965], Trollope [1968], McIlroy [1974], and Shiokawa [1974b] found bounds for the error term $S_2(x) - (x \log x)/(2 \log x)$. Foster [1987] found bounds for

$$\frac{S_k(n)}{n} - \frac{k-1}{2} \left\lfloor \frac{\log n}{\log k} \right\rfloor,$$

for $k = 2, 3$. See also Foster [1991, 1992] and the Notes to Section 3.5.

Trollope [1967] generalized Bush's result to the more general numeration systems discussed in Section 3.1.

Fine [1965] examined the distribution of $s_k(n)$ in different residue classes, as did Gelfond [1968]. Also see Bésineau [1972], Solinas [1989], and Mauduit and Sárközy [1996]. Thuswaldner [2000] discussed this problem for complex bases, and Hoit [1999] for other numeration systems.

Bellman and Shapiro [1948] and Gelfond [1968] introduced the notion of *k-additive function* (one for which $f(ak^r + b) = f(ak^r) + f(b)$ for $1 \leq a < k$ and $0 \leq b < k^r$). It is also possible to study *k-multiplicative functions* (as before, but with multiplication replacing addition). These were first studied by Mendès France [1967, 1970, 1973b, 1973c] and later by by Delange [1972]. A sequence is *strongly k-additive* (or *strongly k-multiplicative*, respectively) if in addition we have $f(ak) = f(a)$ for $1 \leq a < k$. For other papers on the topics of k-additive and k-multiplicative functions, see Shiokawa [1974], Coquet and Mendès France [1977], Coquet, Kamae, and Mendès France [1977], Coquet [1979], Kawai [1984], Mauclaire [1987, 1997], Murata and Mauclaire [1988], Grabner [1993a], Bassily and Kátai [1995], Toshimitsu [1997, 1998], and Uchida [1999].

D. Newman and Slater [1975] examined the distribution of $s_k(a(n))$ (mod 2) for some "naturally defined" sequences $a(n)$. Coquet and Toffin [1981] studied the statistical independence of the sum of digits in base k and Fibonacci representation.

It follows from a theorem of Senge and Straus [1973] that if $a > 1$ is an odd integer, then $s_2(a^j) \to \infty$ as $j \to \infty$. See also C. Stewart [1980]. Stolarsky [1978] studied the ratio $s_2(a^j)/s_2(a)$.

Kátai [1967] discussed the analogue of the sum $S_k(x)$ where the sum is taken over prime terms only. His bound depended on the assumption of the Riemann hypothesis. Also see Kátai and Mogyoródi [1968]. Shiokawa [1974a] improved the bound and removed the dependence on the Riemann hypothesis. Also see Heppner [1976]; Kátai [1977a].

Èminyan [1991] studied the quantity $\sum_{\substack{n \le x \\ s_2(n) \equiv 0 \ (\text{mod } 2)}} \tau(n)$, where τ is the number-of-divisors function. Èminyan [1994] later studied the quantity $\sum_{\substack{n \le x \\ s_2(n) \equiv 0 \ (\text{mod } 2)}} r_2(n)$, where r_2 counts the number of representations as a sum of two squares.

Bellman and Shapiro [1948] studied the summatory function of iterates of $s_2(n)$. Also see Stein and Stux [1978]; Stein [1982].

Several authors have studied the distribution of $\{\alpha s_k(n)\}$ where α is an irrational number. See, for example, Mendès France [1967]; Coquet [1980]; Tichy and Turnwald [1986]; Drmota and Larcher [2001].

McIlroy [1974] discussed the relationship of $S_2(n)$ to several merging algorithms; see Exercise 14. Also see Li and Reingold [1989].

Problems associated with sums of digits have attracted much attention in the recreational mathematics literature. For example, the notion of "self-number" – a positive integer that cannot be represented in the form $n + s_{10}(n)$ – was popularized by Kaprekar [1956], although some earlier references exist, e.g., Goormaghtigh [1949]. Similar notions can be defined for bases other than 10. For papers on self-numbers and generalizations, see Makowski [1966], A. Rao [1966], Vaidya [1969], Joshi [1971], Recamán [1973], Gardner [1975a, 1975b], Zannier [1982], Patel [1990, 1991a, 1991b], Troi and Zannier [1995], and Cai [1996b].

Another notion is that of the *Niven number* – a positive integer n such that $s_{10}(n) \mid n$. There is an evident generalization to bases other than 10. For papers on this topic, see Kennedy, Goodman, and Best [1980], Kennedy [1982], Kennedy and Cooper [1984, 1989a], Cooper and Kennedy [1985, 1988, 1989, 1993], Grundman [1994], and Cai [1996a].

Finally, Wilansky [1982] introduced the concept of a *Smith number* – a positive integer n such that $s_{10}(n) = \sum_{1 \le i \le r} e_i s_{10}(p_i)$, where the prime factorization of n is $p_1^{e_1} \cdots p_j^{e_j}$. See, for example, Oltikar and Wayland [1983]; McDaniel and Yates [1989]. There is an evident generalization to bases other than 10. For a critical appraisal of the Smith-number literature, see Dudley [1994].

Terr [1996] discussed those indices k for which $s_{10}(F_k) = k$, where F_k is the k th Fibonacci number.

3.3 The Rudin–Shapiro sequence was studied by Shapiro [1952] and later Rudin [1959]; they obtained the square-root property. However, the sequence appeared previously (in a somewhat disguised form) in papers of Golay [1949, 1951]. The Rudin–Shapiro polynomials appear in Problem 19 in Littlewood's delightful book of unsolved problems [1968]. Edwards and Price [1970], Figà-Talamanca and Price [1973], McMullen and Price [1976], Allouche and Liardet [1991], and Mauclaire [1994] studied various generalizations of the Rudin–Shapiro construction to compact groups and to locally compact groups. Körner [1979] studied a generalization to finite abelian groups.

The connection between the Rudin–Shapiro sequence and paperfolding was first observed by Mendès France and Tenenbaum [1981]; also see Mendès France [1990b]. Another family of sequences with the Rudin–Shapiro property was described in Allouche and Liardet [1991]: following a suggestion of Mendès France, they studied a generalization of the Rudin–Shapiro sequence, where instead of counting the number of 11's in the binary expansion of n, one counts the total number of occurrences of patterns of the form $1w1$, where w is any string of fixed length.

Coifman, Geshwind, and Meyer [2001] studied noiselets, which are certain complex-valued functions and distributions closely related to the Rudin–Shapiro sequence.

Brillhart and Carlitz [1970] studied discriminants and resultants involving the Rudin–Shapiro polynomials, as well as some of the formulas in Exercise 22. (Additional identities of interest can be found in Brillhart [1973]; Brillhart, Lomont, and Morton [1976]; and Morton [1981].)

Brillhart and Morton [1978] and Brillhart, Erdős, and Morton [1983] studied the summatory function of the Rudin–Shapiro sequence. An exposition of their work is Brillhart and Morton [1996]. Also see Blecksmith and Laud [1995].

The exponent $\frac{1}{2}$ of N in (3.6) is optimal; see Exercise 18. The constant $C = 2 + \sqrt{2}$ in (3.6) was improved to $C = (2 + \sqrt{2})\sqrt{\frac{3}{5}}$ by Saffari [1986]; also see Saffari [1987]. Brillhart and Morton [1978] proved that $C \geq \sqrt{6}$. Probably $\sqrt{6}$ is the optimal value for C; in fact, Saffari announced a proof of this fact c. 1995, but it has not yet been published.

The estimate (3.6) was generalized to sequences taking values in $1, e^{2\pi i/r}, \cdots, e^{2\pi i(r-1)/r}$ by Rider [1966]. We may also consider sequences of modulus-1 complex numbers satisfying (3.6). For this, see Littlewood [1966], Byrnes [1977], Körner [1980], Kahane [1980], and Beck [1991]. Allouche and Mendès France [1985b] generalized (3.6) to all unimodular 2-multiplicative sequences. Also see Alzer [1995].

The last part of our proof of Theorem 3.3.2, which gives a bound for the sum $2^{N_0/2} + 2^{N_1/2} + \cdots$, is adapted from a proof due to Balazard. This inequality was generalized by Allouche, Mendès France, and Tenenbaum [1988].

It can be asked whether similar inequalities hold for other automatic sequences with values ± 1. The following bound is due to Gelfond [1968]: Let $\mathbf{t} = (t_n)_{n \geq 0}$ be the Thue–Morse sequence. Then

$$\left| \sup_{\theta \in \mathbb{R}} \sum_{0 \leq n < N} (-1)^{t_n} e^{i\theta n} \right| \leq C N^{(\log 3)/(\log 4)},$$

and the exponent $(\log 3)/(\log 4)$ is optimal.

Kervaire, Saffari, and Vaillancourt [1986] found non-Rudin–Shapiro polynomials satisfying the Rudin–Shapiro identity $|P(e^{i\theta})|^2 + |Q(e^{i\theta})|^2 =$ constant.

Queffélec [1987a] studied the spectral properties of the dynamical system associated with the Rudin–Shapiro sequence.

Newman and Byrnes [1990] studied the L^4 norm of the Rudin–Shapiro polynomials; also see Borwein and Mossinghoff [2000].

Allouche and Mendès France [1985a] studied a physical interpretation of the Rudin–Shapiro sequence in connection with the Ising model; also see Mendès France [1990b] and Chapter 17.

The Rudin–Shapiro sequence seems to be particularly useful in constructing certain kinds of counterexamples. Rider [1969] used it to construct a counterexample in Banach algebras. (He also gave a characterization of the Rudin–Shapiro sequence in terms of the number of occurrences of the pattern 11, but only for odd n. This was generalized to all n by Brillhart and Carlitz [1970].) Bazinet and Siddiqi [1972] used the Rudin–Shapiro sequence to construct an almost periodic matrix that does not satisfy a Borel property. Gupta, Madan, and Tewari [1994] used it to construct a counterexample in Fourier analysis.

Shepherd, Van Eetvelt, Wyatt-Millington, and Barton [1995] used the Rudin–Shapiro polynomials in a coding scheme.

The estimate (3.5) is due to Salem and Zygmund [1954].

The notion of expansion as a sum of pattern sequences was introduced by Morton and Mourant [1989]. Theorem 3.3.5 is from Allouche and Shallit [1992]. Also see Allouche, Morton, and Shallit [1992].

Digital sequences were studied by Cateland [1992], where Theorem 3.3.9 can be found.

Blecksmith, Filaseta, and Nicol [1993] studied the number of blocks of adjacent identical digits in the base-k expansion of n. Also see Blecksmith and Laud [1995].

Prodinger [1982] studied the summatory function of $e_{2;11\cdots1}(n)$, and Kirschenhofer [1983] studied the summatory functions for any binary block not starting with a zero. Also see Flajolet, Grabner, Kirschenhofer, Prodinger, and Tichy [1994].

3.4 We have covered only a tiny part of a very large area. For example, there is a natural generalization to expansion in base β, where β is a real number > 1, not necessarily an integer. The interested reader will want to explore the vast literature on β-expansions. To list just a few references, see Rényi [1957]; Parry [1960]; Galambos [1973]; Bertrand-Mathis [1989]; Blanchard [1989]; Frougny [1989b, 1992b, 1992d]; Solomyak [1992]; Frougny and Solomyak [1992]; Berend and Frougny [1994]; Akiyama [1999, 2000]; Burdík, Frougny, Gazeau, and Krejcar [2000]; and S. Ito and Sano [2001]. Frougny [2000] and Lothaire [2002, §7] give surveys of β-expansions.

Ribenboim [1985] discussed representations for positive real numbers in terms of series involving powers of Fibonacci numbers.

3.5 For Theorem 3.3.3, see, for example, Atkinson [1978].

The first person to recognize the oscillatory behavior of the error term in $S_k(n)$ (Theorem 3.5.4) was apparently Trollope [1968], for the case $k = 2$. Delange [1975] made this connection more precise, and also obtained the Fourier expansion of the function $F_k(x)$. Delange's work was generalized to the k-additive functions by Murata and Mauclaire [1988].

Stolarsky [1977] and Coquet [1986] studied $\sum_{0 \le k < x} (s_2(x))^d$. Also see Kennedy and Cooper [1991, 1993]; T. Brown [1994]; Okada, Sekiguchi, and Shiota [1995].

Kirschenhofer [1990] studied the variance of $S_2(x)$. The higher moments were studied by Grabner, Kirschenhofer, Prodinger, and Tichy [1993]. Flajolet, Grabner, Kirschenhofer, Prodinger, and Tichy [1994] showed how to obtain these estimates, and many others, using the powerful technique of the Mellin transform.

Moser observed empirically that if n is an integer divisible by 3, then $s_2(n)$ is more likely to be odd than even. This was explained by D. Newman [1969], who showed that if $R(n) = \sum_{0 \le i < n} (-1)^{s_2(3i)}$, then $R(n) > 0$ and is approximately $n^{\log_4 3}$. Coquet [1983] obtained more precise results. Also see Cateland [1992]; Flajolet, Grabner, Kirschenhofer, Prodinger, and Tichy [1994]. Grabner [1993b] examined similar questions where the number 3 is replaced by 5. For related papers, see Dumont [1983]; Goldstein, Kelly, and Speer [1992]; Drmota and Skalba [1995, 2000b]; and Leinfellner [1999].

Osbaldestin [1991] is a survey of digital summations with oscillatory behavior.

Kátai [1977b] discussed the difference $s_2(3n) - s_2(n)$, and Stolarsky [1980] examined the numbers n for which $s_2(jn) \ge s_2(n)$ for all $j \ge 1$. Also see Dringó and Kátai [1981]. Stolarsky [1979] discussed $s_2(\prod_{1 \le i \le r} (2^{a_i} - 1))$.

Dumont and Thomas [1993] obtained estimates similar to those in our Section 3.5 for more general numeration systems. Also see Dumont and Thomas [1997]. Grabner, Kirschenhofer, and Prodinger [1998] found similar estimates for bases $-A + i$, and Thuswaldner [1998a, 1999a] obtained results on sums of digits for number fields.

Our exposition has been based heavily on the article of Tenenbaum [1997].

3.6 Colson [1726] suggested the use of both positive and negative digits in base-10 representation, and Cauchy [1840] did the same for arbitrary bases. Lalanne [1840] followed up on Cauchy's discussion to develop what is now called the balanced ternary system, and stated without proof that the resulting system is perfect. De Morgan [1840] reported on a calculating machine invented by Thomas Fowler that worked with balanced ternary.

Shannon [1950] discussed numeration systems using both negative digits and positive digits. Tompkins and Wakelin [1950, pp. 287–289] discussed addition circuits based on balanced ternary.

Theorem 3.6.1 is a "folk theorem" whose precise origins are unknown to us. Knuth [1969, Solution to Exercise 4.1-24, p. 495] mentioned the result for base 10. Salomaa [1973, Note 9.1, pp. 90–91] discussed the result for all bases $k \geq 2$. It has been rediscovered many times; for example, see Davis and Weyuker [1983, pp. 70–76], Forslund [1995], and Boute [2000].

Theorem 3.6.2 is due to Matula [1976, 1978, 1982]. Matula [1982] also obtained results for representation of real numbers, as did Odlyzko [1978].

3.7 Grünwald [1885] appears to have been the first to discuss representations in negative bases. Kempner [1936] mentioned the possibility in a footnote. The idea again became popular with the advent of electronic digital computers; see, for example, Songster [1956, 1962, 1963]; Pawlak and Wakulicz [1957]; Balasiński and Mrówka [1957]; Wadel [1957, 1961]; Pawlak [1959, 1960]; Lazarkiewicz and Balasiński [1961]; Wells [1963]; Dietmeyer [1963]; Twaddle [1963]; Penney [1964]; A. Nelson [1967]; Zohar [1970]; Gardner [1973]; Sankar, Chakrabarti, and Krishnamurthy [1973a, 1973b]; Kanani and O'Keefe [1973]; Houselander [1974]; G. Rao, Rao, and Krishnamurthy [1974]; Agrawal [1974a, 1974b, 1975a, 1975b, 1975c, 1977, 1978]; Yuen [1975]; Murugesan [1977]; Gilbert and Green [1979]. The term "negabinary" for base -2 was introduced by de Regt [1967]. Knuth [1981, §4.1] provides valuable technical and historical remarks. For sums of digits in negative bases, see Grabner and Thuswaldner [2000].

3.8 Theorem 3.8.1 first appeared in print in the article of Lekkerkerker [1952], although the result is due to Zeckendorf in 1939; see Zeckendorf [1972]. For this reason Fibonacci representation is sometimes called Zeckendorf representation. For a biography of Zeckendorf, see Kimberling [1998b].

For applications of Fibonacci representation and its generalizations, see Daykin [1960]; J. Brown [1961, 1963, 1964, 1965]; Erdős [1962]; Ferns [1965]; Carlitz [1968]; Hoggatt [1972]; Keller [1972]; Carlitz, Scoville, and Hoggatt [1972a, 1972b]; Filipponi [1986].

Freitag and Phillips [1996] found the Fibonacci representation for differences and quotients of Fibonacci numbers.

For applications to coding theory, see Kautz [1965]; Apostolico and Fraenkel [1987]; Capocelli [1989].

Frougny [1986] discussed normalization of Fibonacci representations. Frougny [1991] discussed addition in the Fibonacci numeration system. For generalizations, see Frougny [1988, 1992a, 1992b]. Frougny and Sakarovitch [1999] found a beautiful connection between expansion in base $(1 + \sqrt{5})/2$ and Fibonacci representation.

In analogy with the sum-of-digits function defined in Section 3.2, if $n = \sum_{2 \le i \le r} a_i F_i$, one can define $s_F(n) = \sum_{2 \le i \le r} a_i$, the *sum of the Fibonacci digits*. Similarly, one can define $S_F(n) = \sum_{0 \le i < n} s_F(i)$ and study its asymptotic properties. Coquet and van den Bosch [1986] proved the following theorem: there exists a continuous, nowhere differentiable real-valued function G of period 1 such that

$$S_F(n) = \frac{3 - \alpha}{5 \log \alpha} n \log n + nG \left(\frac{\log n}{\log \alpha} \right) + O(\log n),$$

where $\alpha = (1 + \sqrt{5})/2$. The parity of the sum of the Fibonacci digits was studied by Drmota and Skalba [2000].

3.9 Theorem 3.9.1 is due to Ostrowski [1922]. Also see Fraenkel [1985, Theorem 3]; Berthé [2001]. There are many applications. For example, Dupain [1979] and Ramshaw [1981] used Ostrowski representations to compute the discrepancy of the sequence $(k\theta)_{k \ge 0}$, where θ is irrational. Also see Dupain and Sós [1980].

3.10 Theorem 3.10.4 is essentially due to Gauss [1832]. Other complete residue systems are discussed in Davio, Deschamps, and Gossart [1978].

Knuth [1960] discussed representation in base $2i$, using the digits $0, 1, 2, 3$. In this system, every complex number has a unique representation, but Gaussian integers with odd imaginary part cannot be represented without the use of negative powers, i.e., digits to the right of the radix point. For example, the representation of i is 10.2. Nadler [1961] developed algorithms for division and square root in this system. See also Slekys and Avizienis [1978].

Theorem 3.10.5 is due to Kátai and Szabó [1975]. Earlier, Penney [1965] stated without proof that $-1 + i$ was a suitable base for a perfect numeration system for $\mathbb{Z}[i]$. Theorem 3.10.5 was generalized by Kátai and Kovács [1980], who proved an analogous result for real quadratic fields, and Kátai and Kovács [1981], who proved a result for imaginary quadratic fields. These results were also obtained independently by Gilbert [1981a]. Kovács [1981a], Kovács and Pethö [1991], and Akiyama and Pethö [2002] obtained some results for more general algebraic number fields. Grossman [1985] studied the lengths of representations in quadratic number fields. Körmendi [1986b] studied representations in a particular cubic number field.

Davio, Deschamps, and Gossart [1978] explored digit sets other than $\{0, 1, \ldots, N(\theta) - 1\}$ for base $\theta \in \mathbb{Z}[i]$. Among other things, they proved that every θ with $N(\theta) \ge 5$ has a digit set leading to a perfect numeration system.

We can also consider representations of complex numbers (not just Gaussian integers) using positive and negative powers of a complex base. Kátai and Szabó [1975] proved that every complex number can be so represented in base $-A + i$, where $A \ge 2$ is an integer. However, the expansions need not be unique; for example, $(1 - 2i)/5$ has three different representations in base $-1 + i$. Gilbert [1982b] described all numbers with multiple representations in base $-A + i$.

For general complex bases (not necessarily of the form $-A + i$) the set of representable complex numbers often has a fractal structure, and in some cases it is possible to explicitly compute the fractal dimension. For papers along these lines, see Gilbert [1981b, 1982a, 1986, 1987], S. Ito [1989], and Thuswaldner [1998b]. For topological properties, see Akiyama and Thuswaldner [2000].

Robert [1994] studied a numeration system for $\mathbb{Q}(\sqrt{-3})$ with base $\sqrt{-3}$ and digits $\{0, 1, (1 + \sqrt{-3})/2\}$.

4

Finite Automata and Other Models of Computation

In this chapter, we introduce some simple models of computation, focusing particularly on finite automata and their variants.

4.1 Finite Automata

A *deterministic finite automaton,* or DFA, is one of the simplest possible models of computation. It is an *acceptor*; that is, strings are given as input and are either accepted or rejected.

A DFA starts in an *initial state* and after reading the input can be in one of a finite number of states. The DFA takes as input a string w and – based on the symbols of w, read in order from left to right – moves from state to state. If after reading all the symbols of w the DFA is in a distinguished state called an *accepting state* (or *final state*), then the string is accepted; otherwise, it is rejected. The language accepted by the DFA is the set of all accepted strings.

A DFA can be represented by a directed graph called a *transition diagram*. A directed edge labeled with a letter indicates the new state of the machine if the given letter is read. By convention, the initial state is drawn with an unlabeled arrow entering the state, and accepting states are drawn with double circles.

For example, Figure 4.1 shows the transition diagram of a finite automaton that accepts all strings over $\{0, 1\}$ that do not contain two consecutive 1's.

More formally, a DFA M is defined to be a 5-tuple

$$M = (Q, \Sigma, \delta, q_0, F)$$

where

Q is a finite set of states,
Σ is the finite input alphabet,
$\delta : Q \times \Sigma \rightarrow Q$ is the *transition function*,
$q_0 \in Q$ is the *initial state*, and
$F \subseteq Q$ is the set of accepting states.

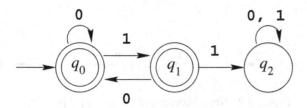

Figure 4.1: DFA Accepting Strings with No Two Consecutive 1's.

Note that a DFA is assumed to be *complete*, i.e., δ is defined for all pairs in its range. In order to formally define acceptance by a DFA, we need to extend the domain of δ to $Q \times \Sigma^*$. We do this as follows: first, we define $\delta(q, \epsilon) = q$ for all $q \in Q$, and define $\delta(q, xa) = \delta(\delta(q, x), a)$ for all $q \in Q$, $x \in \Sigma^*$, and $a \in \Sigma$. Then $L(M)$, the language accepted by M, is defined to be

$$L(M) = \{w \in \Sigma^* : \delta(q_0, w) \in F\}.$$

We call a state q of a DFA *reachable* if there exists $x \in \Sigma^*$ such that $\delta(q_0, x) = q$, and *unreachable* otherwise. Clearly, unreachable states may be deleted without changing the language accepted by a DFA.

In Section 1.3, we defined the class of regular languages. Below we will prove Kleene's theorem, which shows that every regular language is accepted by some DFA, and vice versa. Hence DFAs accept exactly the regular languages.

First, however, we prove some closure properties of the languages accepted by DFAs:

Theorem 4.1.1 *Let L be accepted by a DFA with m states. Then $\overline{L} = \Sigma^* \setminus L$ is accepted by a DFA with m states.*

Proof. Suppose $L = L(M)$ for $M = (Q, \Sigma, \delta, q_0, F)$. Define a new DFA M' that is identical to M, except that accepting and nonaccepting states have had their roles interchanged. More formally, let $M' = (Q, \Sigma, \delta, q_0, Q \setminus F)$. Then it is easy to see that $L(M') = \overline{L}$. ∎

Theorem 4.1.2 *Let $L_1, L_2 \subseteq \Sigma^*$ be languages accepted by DFAs with m states and n states, respectively. Then $L_1 \cap L_2$ and $L_1 \cup L_2$ can be accepted by DFAs with mn states.*

Proof. Let $L_1 = L(M_1)$ and $L_2 = L(M_2)$, where $M_1 = (Q_1, \Sigma, \delta_1, q_1, F_1)$ and $M_2 = (Q_2, \Sigma, \delta_2, q_2, F_2)$. We construct a new DFA M whose state set is the direct product of the state sets of M_1 and M_2. Hence, states of M are ordered pairs of the form $[p, q]$. In the first component, we simulate the computation of M_1, and in the second component, we simulate the computation of M_2. More formally, let

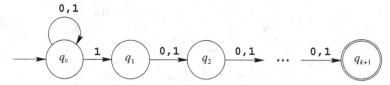

Figure 4.2: NFA Accepting $L_k = \{0, 1\}^*1\{0, 1\}^k$.

$M = (Q_1 \times Q_2, \Sigma, \delta, [q_1, q_2], F)$, where $\delta([p, q], a) = [\delta_1(p, a), \delta_2(q, a)]$ and $F = F_1 \times F_2$. It is now easy to see that $L(M) = L_1 \cap L_2$.

The proof for union is similar, except now we define $F = (F_1 \times Q_2) \cup (Q_1 \times F_2)$. ∎

We claim that the concept of finite automaton is very natural. As evidence, we argue that the definition is *robust*: small changes to the model do not change the class of languages accepted. For example, we now generalize the DFA concept to allow *nondeterminism,* and show that nondeterministic finite automata accept the same class of languages as deterministic finite automata.

A *nondeterministic finite automaton,* or NFA, is like a DFA, but we also allow the possibility of zero or two or more distinct transitions on a state-input symbol pair. We accept an input string if there exists *any* choice of transitions that leads to an accepting state. More formally, an NFA is a 5-tuple $M = (Q, \Sigma, \delta, q_0, F)$, where $\delta : Q \times \Sigma \to 2^Q$. (By 2^Q we mean the set of all subsets of Q.)

Nondeterminism is useful in part because it allows more succinct representation of some languages. As an example, consider the NFA represented in Figure 4.2. This NFA has $k + 2$ states and accepts the language $L_k = \{0, 1\}^*1\{0, 1\}^k$. Exercise 6 asks you to prove the fact that no DFA with fewer than 2^{k+1} states can accept L_k.

Theorem 4.1.3 *If L is accepted by an NFA with n states, then L is accepted by a DFA with at most 2^n states.*

Proof. Let L be accepted by the NFA $M = (Q, \Sigma, \delta, q_0, F)$. We construct a simulating DFA whose states are subsets of the state set for M; this is often called the "subset construction". More precisely, let $M' = (2^Q, \Sigma, \delta', q_0', F')$ with

$$\delta'(S, a) = \bigcup_{s \in S} \delta(s, a),$$

and

$$q_0' = \{q_0\}, \qquad F' = \left\{S \in 2^Q : S \cap F \neq \emptyset\right\}.$$

It is now easy to prove by induction w is accepted by M if and only if it is accepted by M'. ∎

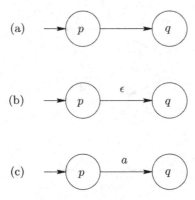

Figure 4.3: Base Case for Theorem 4.1.4.

A generalization of the NFA commonly found in the literature is the NFA-ϵ; this is an NFA in which we allow the machine to spontaneously move from one state to another without consuming any input. Such a move is called an ϵ-transition, and is denoted by $\delta(p, \epsilon) = q$. A further generalization of NFAs allows transitions labeled with *any* regular expression. A *generalized finite automaton*, or GFA, is a 5-tuple $M = (Q, \Sigma, f, q_0, F)$, where $f : Q \times Q \to 2^{\Sigma^*}$ is a mapping such that $f(p, q)$ is a regular language for all $p, q \in Q$. In this more general model, an input x is accepted if and only if there exists a sequence of states $q_0, q_1, \ldots, q_i$ where $q_i \in F$ and $x \in f(q_0, q_1)f(q_1, q_2) \cdots f(q_{i-1}, q_i)$.

Theorem 4.1.4 *If L is accepted by a GFA, then L is accepted by an NFA.*

Proof. First, we show how to replace a transition from state p to state q, labeled with a regular expression r, with a subautomaton where the labels are either single symbols or ϵ. This is done inductively, where the induction is on the number of operator symbols in r.

The base case is where r contains no operator symbols. In this case, either (a) $r = \emptyset$, or (b) $r = \epsilon$, or (c) $r = a$ for some $a \in \Sigma$. We use the appropriate subautomaton in Figure 4.3.

In the general case, we can write either (a) $r = s + t$, (b) $r = st$, or (c) $r = s^*$, where s, t are regular expressions containing fewer operators. In this case, we use the appropriate subautomaton in Figure 4.4.

Hence, we may assume without loss of generality that every transition is labeled with a subset of $\Sigma \cup \{\epsilon\}$. Suppose $M = (Q, \Sigma, f, q_0, F)$ is a GFA with $f(p, q) \in \Sigma \cup \{\epsilon\}$ for all $p, q \in Q$. Then we define an NFA $M' = (Q, \Sigma, \delta, q_0, F')$ as follows. For $q \in Q$ and $a \in \Sigma$, let $\delta(q, a)$ be the set of all states q_i such that there exist $q_1, q_2, \ldots, q_{i-1}$ with $a \in f(q, q_1)f(q_1, q_2) \cdots f(q_{i-1}, q_i)$. Also define F' to be F unless there exist $q_1, q_2, \ldots, q_i$ such that $\epsilon \in f(q_0, q_1)f(q_1, q_2) \cdots f(q_{i-1}, q_i)$ and $q_i \in F$, in which case we let $F' = F \cup \{q_0\}$.

Clearly M' is an NFA. We leave it to the reader to show that $L(M') = L(M)$.

∎

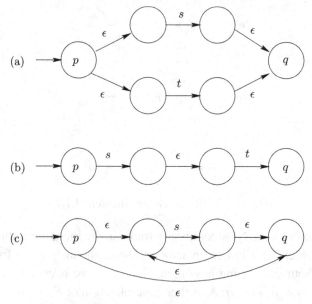

Figure 4.4: Induction Step for Theorem 4.1.4.

We are now ready to prove the fundamental theorem linking regular expressions and automata.

Theorem 4.1.5 (Kleene's Theorem) *A language is accepted by a DFA if and only if it can be specified by a regular expression.*

Proof. Suppose L is specified by a regular expression r. Then we can create a GFA M accepting $L(r)$ with exactly two states, q_0 and q_1, where $f(q_0, q_1) = L(r)$ (and $f(p, q) = \emptyset$ otherwise), and $F = \{q_1\}$. By Theorem 4.1.4, L can be accepted by an NFA, and by Theorem 4.1.3, L can be accepted by a DFA.

On the other hand, suppose L is accepted by a DFA M. Then we can find a regular expression for L as follows. Treat M as a GFA $M' = (Q, \Sigma, f, q_0, F)$. First, we ensure that all transitions into the initial state are labeled $\emptyset$, by creating a new initial state q_0', if necessary, and adding a transition from q_0' to q_0 labeled ϵ. Next, we ensure that there is exactly one final state, and that all transitions out of this final state are labeled $\emptyset$, by creating a new final state q_f, and adding transitions labeled ϵ from all $q \in F$ to q_f. We make our new set of final states $\{q_f\}$.

Now eliminate states successively, one by one, as follows. Choose a state q other than the initial or final state to eliminate. For all other states p, r, if $f(p, q) = r_1$, $f(q, q) = r_2$, and $f(q, r) = r_3$, then update $f(p, r)$ by setting $f(p, r) \leftarrow f(p, r) + r_1 r_2^* r_3$. Now eliminate state q and all transitions associated with it.

We leave it to the reader to verify that this process eventually produces a GFA with exactly two states, one initial, and one final, with a transition connected them labeled with a regular expression specifying L. ■

Recall the definition of quotient of languages from Section 1.3:

$$L_1/L_2 = \{x \in \Sigma^* : \exists y \in L_2 \text{ such that } xy \in L_1\}.$$

Then we have the following theorem:

Theorem 4.1.6 *If L_1 is regular and L_2 is arbitrary, then L_1/L_2 is regular.*

Proof. Let $M = (Q, \Sigma, \delta, q_0, F)$ be a DFA accepting L_1. Define $M' = (Q, \Sigma, \delta, q_0, F')$ as follows:

$$F' = \{q \in Q : \exists y \in L_2 \text{ such that } \delta(q, y) \in F\}.$$

Then $\delta(q_0, x) \in F'$ if and only if there exists $y \in L_2$ such that $xy \in L_1$. ∎

Recall the definition of inverse morphism of languages from Section 1.4:

$$h^{-1}(L) = \{x \in \Sigma^* : h(x) \in L\}.$$

Then we have the following theorem:

Theorem 4.1.7 *If L is regular and h is a morphism, then $h^{-1}(L)$ is regular.*

Proof. Let $M = (Q, \Delta, \delta, q_0, F)$ be a DFA accepting $L \subseteq \Delta^*$. Then define $M' = (Q, \Sigma, \delta', q_0, F)$ as follows: $\delta'(q, a) = \delta(q, h(a))$. It is easy to prove by induction on x that $\delta(q_0, h(x)) = \delta'(q_0, x)$, and hence M' accepts x if and only if M accepts $h(x)$. ∎

We now turn to the Myhill–Nerode theorem, an interesting characterization of regular languages based on equivalence relations. A *relation* R on a set S is a subset of $S \times S$. If $(x, y) \in R$, we write $x \, R \, y$. A relation is said to be *reflexive* if $x \, R \, x$ for all $x \in S$. If $x \, R \, y$ implies $y \, R \, x$ for all $x, y \in S$, then R is *symmetric*. Finally, if $x \, R \, y$ and $y \, R \, z$ implies $x \, R \, z$ for all $x, y, z \in S$, then R is said to be *transitive*. If a relation is reflexive, symmetric, and transitive, then it is called an *equivalence relation*.

An equivalence relation R partitions a set S into a number of disjoint *equivalence classes*. If $x \in S$, then by $[x]$ we mean the equivalence class containing x. If there are only a finite number of equivalence classes, then R is said to be of *finite index*.

An equivalence relation on Σ^* is said to be *right-invariant* if $x \, R \, y$ implies $xz \, R \, yz$ for all $z \in \Sigma^*$. A particularly important right-invariant equivalence relation, called the *Myhill–Nerode equivalence relation*, is the relation R_L defined as follows: $x \, R_L \, y$ if and only if for all $z \in \Sigma^*$ one has $xz \in L$ if and only if $yz \in L$. It is easy to see R_L is right-invariant: Suppose $x \, R_L \, y$, and let $v \in \Sigma^*$. Then, for all $w \in \Sigma^*$, $xw \in L$ if and only if $yw \in L$. Now let $w = vz$. We see $xvz \in L$ if and only if $yvz \in L$. It follows that $xv \, R_L \, yv$.

Theorem 4.1.8 (Myhill–Nerode) *The following three statements are equivalent:*

(a) L is a regular language;

(b) There exists a right-invariant equivalence relation E of finite index such that L is the union of some of E's equivalence classes;

(c) The Myhill–Nerode equivalence relation R_L is of finite index.

Proof. (a) $\Longrightarrow$ (b): Let L be accepted by a DFA M, where $M = (Q, \Sigma, \delta, q_0, F)$. Let E be the equivalence relation defined by $x \, E \, y$ if and only if $\delta(q_0, x) = \delta(q_0, y)$. Then E is right-invariant, since

$$\delta(q_0, xz) = \delta(\delta(q_0, x), z) = \delta(\delta(q_0, y), z) = \delta(q_0, yz)$$

for all $z \in \Sigma^*$. The relation E is of finite index, since it is bounded above by Card Q. Also, we have $L = \bigcup_{\delta(q_0, x) \in F} [x]$, where $[x]$ is the equivalence class containing x.

(b) $\Longrightarrow$ (c): We show that E is a refinement of R_L; that is, each equivalence class of E is contained in some equivalence class of R_L. Thus the index of R_L is less than or equal to the index of E, and hence R_L is finite.

Let $x \, E \, y$. Since E is right-invariant, we have $xz \, E \, yz$. Since L is the union of some of E's equivalence classes, we have $xz \in L$ if and only if $yz \in L$. Hence $x \, R_L \, y$. Hence each equivalence class of E is contained in an equivalence class of R_L.

(c) $\Longrightarrow$ (a): We build a DFA $M = (Q, \Sigma, \delta, q_0, F)$ accepting L. Let Q be the finite set of equivalence classes of R_L, and define $\delta([x], a) = [xa]$ for $x \in \Sigma^*, a \in \Sigma$. Let $q_0 = [\epsilon]$ and $F = \{[x] : x \in L\}$. Since R_L is right-invariant, the definition is consistent, and it is easy to see that M accepts L. ∎

Corollary 4.1.9 *The DFA with the minimum number of states accepting a regular language L is given by M in the last paragraph of the previous proof, and is unique up to renaming of the states.*

Proof. Left to the reader as Exercise 4. ∎

We finish this section with a technical lemma that will prove useful in Chapter 11. We say a DFA M is *minimal* if it has the smallest number of states among all DFAs M' with $L(M') = L(M)$.

Lemma 4.1.10 *Let L be a regular language, and $A = (Q, \Sigma, \delta, q_0, F)$ a minimal DFA accepting L. Define $A_q = \{x \in \Sigma^* : \delta(q_0, x) = q\}$. Then each A_q can be expressed as a finite boolean combination of sets of the form $L/\{w\}$ for $w \in \Sigma^*$. Moreover, $L = \bigcup_{q \in F} A_q$.*

Proof. Since A is minimal, for each pair of distinct states (q, q') there exists a string $w = w_{qq'}$ such that $\delta(q, w) \in F$ and $\delta(q', w) \notin F$, or $\delta(q, w) \notin F$ and $\delta(q', w) \in F$.

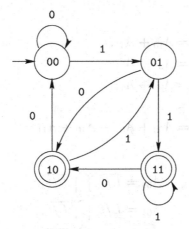

Figure 4.5: Minimal DFA for $L = \{0, 1\}^*1\{0, 1\}$.

Otherwise, we could identify q and q' and obtain a smaller DFA. Then we claim that

$$A_q = \bigcap_{q' \neq q} B_{qq'},$$

where

$$B_{qq'} = \begin{cases} L/\{w_{qq'}\} & \text{if } \delta(q, w_{qq'}) \in F, \\ \overline{(L/\{w_{qq'}\})} & \text{if } \delta(q, w_{qq'}) \notin F. \end{cases}$$

To see this, let us prove that $A_q \subseteq \bigcap_{q' \neq q} B_{qq'}$. Then let $x \in A_q$, i.e., $\delta(q_0, x) = q$. Choose a state $q' \neq q$. Then if $\delta(q, w_{qq'}) \in F$, we must show that $x \in L/\{w_{qq'}\}$. But if $\delta(q_0, xw_{qq'}) = \delta(\delta(q_0, x), w_{qq'}) \in F$, so indeed $x \in L/\{w_{qq'}\}$.

If $\delta(q, w_{qq'}) \notin F$, then we must show that $x \notin L/\{w_{qq'}\}$. But $\delta(q_0, xw_{qq'}) \notin F$, so indeed $x \notin L/\{w_{qq'}\}$.

Now let us show that

$$\overline{A_q} \subseteq \bigcup_{q' \neq q} B_{qq'}.$$

Let $x \in \overline{A_q}$. Then $\delta(q_0, x) = p \neq q$. Consider B_{pq}. If $\delta(p, w_{pq}) \in F$, then $B_{pq} = L/\{w_{pq}\}$. Then $\delta(q_0, xw_{pq}) \in F$, so indeed $x \in B_{pq}$. Similarly if $\delta(p, w_{pq}) \notin F$, then $B_{pq} = \overline{L/\{w_{pq}\}}$. But $\delta(q_0, xw_{pq}) \notin F$, so indeed $xw_{pq} \notin L$, and $x \notin L/\{w_{pq}\}$. ∎

Example 4.1.11 Let $L = \{0, 1\}^*1\{0, 1\}$. Then the minimal DFA for L is given in Figure 4.5.

It is easy to verify that

$$L/\epsilon = A_{10} + A_{11},$$
$$L/0 = A_{01} + A_{11},$$
$$L/1 = A_{01} + A_{11},$$
$$L/00 = \emptyset,$$
$$L/10 = A_{00} + A_{01} + A_{10} + A_{11} = \{0, 1\}^*.$$

Then we have

$$A_{00} = \overline{L/0} \cap \overline{L/\epsilon},$$
$$A_{01} = L/0 \cap \overline{L/\epsilon},$$
$$A_{10} = L/\epsilon \cap \overline{L/1},$$
$$A_{11} = L/\epsilon \cap L/0.$$

4.2 Proving Languages Nonregular

The following property of regular languages, known as the *pumping lemma*, is often an effective tool for proving languages nonregular.

Lemma 4.2.1 *Let $L \subseteq \Sigma^*$ be a regular language. Then there exists a constant $n \geq 1$ such that for all strings $z \in L$ with $|z| \geq n$, there exists a decomposition $z = uvw$, where $u, v, w \in \Sigma^*$ and $|uv| \leq n$ and $|v| \geq 1$, such that $uv^i w \in L$ for all $i \geq 0$. Furthermore, the constant n can be taken to be the number of states in the minimal DFA for L.*

Proof. If L is regular, then there exists a DFA with r states accepting L. Now choose $n = r$. Let $z \in L$ with $|z| \geq n$, and write $z = z_1 z_2 \ldots z_m$ with each $z_j \in \Sigma$. Consider the $n + 1$ states

$$q_0, \delta(q_0, z_1), \ldots, \delta(q_0, z_1 z_2 \cdots z_n).$$

By the pigeonhole principle, some state must occur at least twice in this list. Let q be the first state that occurs at least twice on the list, and let $\delta(q_0, z_1 z_2 \cdots z_s)$ and $\delta(q_0, z_1 z_2 \cdots z_t)$ be the first two occurrences. Then if we take $u = z_1 z_2 \cdots z_s$, $v = z_{s+1} \cdots z_t$, and $w = z_{t+1} \cdots z_m$, the result easily follows.　　∎

Example 4.2.2 Let us prove that the language

$$L = \{1^j 0^j : j \geq 0\}$$

is not regular. Suppose it were, and let n be the pumping lemma constant. Then let $z = 1^n 0^n$. Write $z = uvw$ with $|uv| \leq n$ and $|v| \geq 1$; then $u = 1^r$, $v = 1^s$, and $w = 1^{n-(r+s)} 0^n$. Then $uv^0 w = uw = 1^{n-s} 0^n \notin L$, since $s \geq 1$.

Another effective tool for proving languages nonregular is the Myhill–Nerode theorem (Theorem 4.1.8). Let us reprise the previous example, but this time with an alternate proof.

Example 4.2.3 Let us prove that

$$L = \{1^j 0^j : j \geq 0\}$$

is not regular, using the Myhill–Nerode theorem. Then we claim that, for $j \geq 0$, the equivalence classes $\{[1^j] : j \geq 0\}$ of the relation R_L are pairwise distinct. To see this, let $k \neq j$. Then $1^j 0^j \in L$, but $1^k 0^j \notin L$. Since there are infinitely many equivalence classes, L cannot be regular.

Finally, it may sometimes be useful to apply the fact that regular languages are closed under intersection and inverse morphism. Consider the following example.

Example 4.2.4 Let us prove that

$$L = \{x\,x^R\,w : x, w \in \{a, b\}^+\}$$

is not regular. Suppose it were. Then

$$L' = L \cap (ab)^+ (ba)^+$$

would also be regular. But it is not hard to see that

$$L' = \{(ab)^m\,(ba)^n : 1 \leq m < n\}.$$

Define $\varphi(c) = ab$ and $\varphi(d) = ba$. Then by Theorem 4.1.7,

$$\varphi^{-1}(L') = \{c^m\,d^n : 1 \leq m < n\}$$

would be regular. But a simple argument using the pumping lemma shows it is not.

4.3 Finite Automata with Output

In Section 4.1 we introduced the basic finite automaton model, which either accepts or rejects any given input string. Another way to look at this model is that a given DFA computes a function $f : \Sigma^* \to \{0, 1\}$, where 1 represents acceptance and 0 rejection. We are now interested in more general models of function computation by finite automata.

Informally, this is how an automaton computes a function: the string w is input, and the automaton moves from state to state according to its transition function δ, while reading the symbols of w. When the end of the string is reached, the automaton halts in a state q. At this point the automaton outputs the symbol $\tau(q)$, where τ is the output mapping.

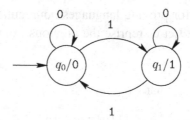

Figure 4.6: DFAO Computing the mod-2 Sum of its Input Bits.

More formally, a *deterministic finite automaton with output* (DFAO) is defined to be a 6-tuple

$$M = (Q, \Sigma, \delta, q_0, \Delta, \tau)$$

where Q, Σ, δ, q_0 are as in the definition of DFA in Section 4.1, Δ is the output alphabet, and $\tau : Q \to \Delta$ is the output function. Such a machine M defines a function from Σ^* to Δ, which we denote as $f_M(w)$, as follows:

$$f_M(w) = \tau(\delta(q_0, w)).$$

If a function $f : \Sigma^* \to \Delta$ can be computed in this fashion, it is called a *finite-state function*.

We can represent a DFAO with a transition diagram, much the same way we did for DFAs; the only difference is that a state labeled q/a indicates that the output associated with the state q is the symbol a.

As an example, consider the DFAO represented in Figure 4.6. It computes the sum, modulo 2, of the bits of the input string $w \in \{0, 1\}^*$.

DFAOs clearly are strongly related to DFAs. The following two theorems make this relationship more precise.

First, we prove that if M is a DFAO, then the words that result in a given output form a regular language.

Theorem 4.3.1 *Let* $M = (Q, \Sigma, \delta, q_0, \Delta, \tau)$ *be a DFAO. Then for all* $d \in \Delta$, *the set*

$$I_d(M) = \{w \in \Sigma^* : \tau(\delta(q_0, w)) = d\}$$

is a regular language.

Proof. Let L_q be the language accepted by the DFA $M_q = (Q, \Sigma, \delta, q_0, \{q\})$. Then it is easy to see that

$$I_d(M) = \bigcup_{\substack{q \in Q \\ \tau(q)=d}} L_q.$$

Since regular languages are closed under finite unions, $I_d(M)$ is regular. ∎

Next we prove the converse: a DFAO can be constructed by "gluing together" the DFAs for different regular languages.

Theorem 4.3.2 *Let* Σ, Δ *be finite alphabets. Let* r *be an integer* ≥ 1, *and let* $L_1, L_2, \ldots, L_r$ *be* r *regular languages that partition* Σ^*, *i.e.,* $\bigcup_{1 \leq i \leq r} L_i = \Sigma^*$ *and* $L_i \cap L_j = \emptyset$ *for* $i \neq j$. *Let* $\Delta = \{a_1, a_2, \ldots, a_r\}$. *Then there exists a DFAO* $M = (Q, \Sigma, \delta, q_0, \Delta, \tau)$ *such that* $\tau(\delta(q_0, w)) = a_i$ *if and only if* $w \in L_i$.

Proof. Since L_i is regular for $1 \leq i \leq r$, there exists a DFA $M_i = (Q_i, \Sigma, \delta_i, q_{0i}, F_i)$ accepting L_i for $1 \leq i \leq r$.

Define $Q = Q_1 \times Q_2 \times \cdots \times Q_r, q_0 = [q_{01}, q_{02}, \ldots, q_{0r}]$, and

$$\delta([q_1, q_2, \ldots, q_r], a) = [\delta_1(q_1, a), \delta_2(q_2, a), \ldots, \delta_r(q_r, a)]$$

for all $q_1 \in Q_1, q_2 \in Q_2, \ldots, q_r \in Q_r, a \in \Sigma$. Also define $\tau([q_1, q_2, \ldots, q_r]) = a_j$ if there exists a unique index j such that $q_j \in F_j$ and (say) a_1 otherwise. The map τ is well defined, since the L_i partition Σ^*. It is now easy to see that M is a DFAO with the desired property. ∎

We now prove that a finite-state function remains finite-state when its input is read in reverse order.

Theorem 4.3.3 *Let* $f : \Sigma^* \to \Delta$ *be a finite-state function. Then so is the function* f^R *defined by* $f^R(w) = f(w^R)$.

Proof. If f is a finite-state function, then there is a DFAO $M = (Q, \Sigma, \delta, q_0, \Delta, \tau)$ such that $f = f_M$. Define a new DFAO $M' = (S, \Sigma, \delta', q_0', \Delta, \tau')$, where $S = \Delta^Q$ (so that elements of S are functions g mapping Q to Δ). We let q_0' be the function that maps q to $\tau(q)$, $\tau'(g) = g(q_0)$ for all $g \in S$, and for $g \in S, a \in \Sigma$, we have $\delta'(g, a) = h$, where $h(q) = g(\delta(q, a))$.

We now prove by induction on $|w|$ that $\delta'(q_0', w) = h$, where h is the function mapping q to $\tau(\delta(q, w^R))$. This is clearly true for $|w| = 0$. Now suppose the result is true for $|w| = n$; we prove it for $|w| = n + 1$. Let $w = xa$, where $|x| = n$. Then $\delta'(q_0', xa) = \delta'(\delta'(q_0', x), a) = \delta'(g, a) = h$, where, by induction, g is the function that maps q to $\tau(\delta(q, x^R))$. Then $h(q) = g(\delta(q, a)) = \tau(\delta(\delta(q, a), x^R)) = \tau(\delta(q, ax^R)) = \tau(\delta(q, (xa)^R)) = \tau(\delta(q, w^R))$.

Since $\tau'(h) = h(q_0)$, it now follows that M' computes the function f^R. ∎

Corollary 4.3.4 *If* $f : \Sigma^* \to \Delta$ *is a finite-state function computable by a DFAO with* n *states, then* f^R *is computable with a DFAO with at most* $|\Delta|^n$ *states.*

Corollary 4.3.5 *If* L *is a regular language, then so is* L^R.

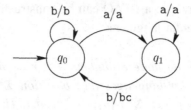

Figure 4.7: A Finite-State Transducer.

Proof. Regular languages correspond to an output alphabet $\Delta = \{0, 1\}$. ∎

Next, we examine another model of automata with output, the *finite-state transducer*. A *finite-state transducer* is a machine

$$T = (Q, \Sigma, \delta, q_0, \Delta, \lambda)$$

where Q is a finite set of states, Σ is is the input alphabet, δ is the transition function, q_0 is the initial state, Δ is the output alphabet, and λ is the output function, mapping $Q \times \Sigma$ to Δ^*. If there exists an integer $k \geq 1$ such that $|\lambda(q, a)| = k$ for all $q \in Q$, $a \in \Sigma$, then the transducer is said to be *k-uniform*. Transducers can be represented graphically in a form similar to finite automata; an arrow labeled a/y from state p to state q corresponds to $\delta(p, a) = q$ and $\lambda(p, a) = y$.

A transducer can be viewed as a means to define functions: on input $w = w_1 w_2 \cdots w_r$ the transducer enters states $\delta(q_0, \epsilon), \delta(q_0, w_1), \ldots, \delta(q_0, w_1 \cdots w_r)$ and produces the outputs

$$\lambda(q_0, w_1), \lambda(\delta(q_0, w_1), w_2), \ldots, \lambda(\delta(q_0, w_1 w_2 \cdots w_{r-1}), w_r).$$

The function $T(w)$ is then defined to be

$$T(w) = \prod_{0 \leq j \leq r-1} \lambda(\delta(q_0, w_1 w_2 \cdots w_j), w_{j+1}).$$

If $L \subseteq \Sigma^*$ and $L' \subseteq \Delta^*$ are languages, we define

$$T(L) := \{T(w) : w \in L\},$$
$$T^{-1}(L') := \{w \in \Sigma^* : T(w) \in L'\}.$$

As an example, consider the transducer defined by $Q = \{q_0, q_1\}$, $\Sigma = \{a, b\}$, $\delta(q_0, a) = \delta(q_1, a) = q_1$, $\delta(q_0, b) = \delta(q_1, b) = q_0$, $\Delta = \{a, b, c\}$, $\lambda(q_0, a) = \lambda(q_1, a) = a$, $\lambda(q_0, b) = b$, $\lambda(q_1, b) = bc$. This transducer inserts a c after every occurrence of ab in the input string, and is represented graphically in Figure 4.7.

Theorem 4.3.6 *If L is a regular language, and T is a finite-state transducer, then $T(L)$ is regular.*

Proof. Let $M = (Q, \Sigma, \delta, q_0, F)$ be a DFA accepting L. Let $T = (Q', \Sigma, \delta', q_0', \Delta, \lambda)$ be a finite-state transducer, with output alphabet Δ, and $\lambda : Q \times \Sigma \to \Delta^*$ the output function. We want to show that $T(L)$ is regular.

We do this by constructing an NFA-ϵ for $T(L)$. Define $M'' = (Q'', \Delta, \delta'', q_0'', F'')$ as follows:

$$Q'' = Q \times Q' \times (\Sigma \cup \{\epsilon\}) \times \{0, 1, \ldots, k\},$$
$$\text{where } k = \max_{a \in \Sigma, \, q \in Q'} |\lambda(q, a)|,$$
$$q_0'' = [q_0, q_0', \epsilon, 0],$$
$$\delta''([q, q', \epsilon, 0], \epsilon) = \{[q, q', b, 0] : b \in \Sigma\},$$
$$\delta''([q, q', a, i - 1], y_i) = \{[q, q', a, i]\} \quad \text{for } 1 \leq i \leq j$$
$$\text{if } \lambda(q', a) = y_1 y_2 \cdots y_j,$$
$$\delta''([q, q', a, j], \epsilon) = \{[\delta(q, a), \delta'(q', a), \epsilon, 0]\} \quad \text{if } \lambda(q', a) = y_1 y_2 \cdots y_j,$$
$$F'' = \{[q, q', \epsilon, 0] : q \in F\}.$$

The idea behind this construction is as follows: if $w \in T(L)$, then there is a word $x \in L$ such that $w = T(x)$. Let $x = x_1 x_2 \cdots x_r$, with $x_i \in \Sigma$, and let $w_i \in \Delta^*$ be the output of T on x_i, so that $w = w_1 w_2 \cdots w_r$. On input w_i, M'' simulates M on the (guessed) word x; this is done in the first component of states of M''. M'' also simulates the computation of T on x; this is done in the second component of states of M''. The third component is used to keep track of what letters of x have been guessed; it is initially ϵ. When a letter a is guessed, it changes to a. The fourth component keeps track of how many letters of w_i have been seen so far; it is initially 0, and is incremented each time the next letter of w_i is seen. Eventually the fourth component becomes $|w_i|$; at this point, there is an ϵ-transition that resets the third component to ϵ and the fourth component to 0, advancing the states in the first and second components at the same time.

The formal proof is a bit messy. Here is just a sketch: we claim that $T(L) = L(M'')$. First, let us prove that $T(L) \subseteq L(M'')$. If $w \in T(L)$, then there exists an $x \in L$ such that $w = T(x)$. Let $k_i = |w_i|$. Now on input x, M successively enters states $p_0 = q_0, p_1, \ldots, p_r$, and $p_r \in F$. Also, T enters states $p_0' = q_0', p_1', \ldots, p_r'$ and emits outputs $w_1, w_2, \ldots, w_r$, where $w_i = \lambda(p_{i-1}', x_i)$. Then, on input w, M'' enters states

$$q_0'' = [p_0, p_0', \epsilon, 0], [p_0, p_0', x_1, 0], \ldots, [p_0, p_0', x_1, k_1],$$
$$[p_1, p_1', \epsilon, 0], [p_1, p_1', x_2, 0], \ldots, [p_1, p_1', x_2, k_2],$$
$$\vdots$$
$$[p_{r-1}, p_{r-1}', \epsilon, 0], [p_{r-1}, p_{r-1}', x_r, 0], \ldots, [p_{r-1}, p_{r-1}', x_r, k_r],$$
$$[p_r, p_r', \epsilon, 0],$$

and this last state is an accepting state.

Similarly, let us prove that $L(M'') \subseteq T(L)$. Let $w \in L(M'')$. We claim that if there is an accepting computation for w, then it is of the form mentioned in the previous paragraph, and then, using this computation, we can trace out the value of $x = x_1 x_2 \cdots x_r$ for which $w = T(x)$, by examining the third components of the states encountered. ∎

Corollary 4.3.7 *Let $L \subseteq \Sigma^*$ be a regular language, and let $\varphi : \Sigma^* \to \Delta^*$ be a morphism. Then $\varphi(L) = \{\varphi(w) : w \in L\}$ is regular.*

Proof. This follows immediately from Theorem 4.3.6, as a morphism can be implemented as a finite-state transducer with exactly one state. ∎

Theorem 4.3.8 *If T is a finite-state transducer and L is a regular language, then $T^{-1}(L)$ is regular.*

Proof. Let $L \subseteq \Delta^*$ be accepted by a DFA $M = (Q, \Delta, \delta, q_0, F)$, and $T = (Q', \Sigma, \delta', q_0', \Delta, \lambda)$ be a finite-state transducer. We define a DFA $M'' = (Q'', \Sigma, \delta'', q_0'', F'')$ accepting $T^{-1}(L)$. The idea is simple: on input w we simulate T on w, getting string x, and we simulate M on x, accepting if M accepts.

More precisely, let $Q'' = Q' \times Q$, $q_0'' = [q_0', q_0]$, $F'' = Q' \times F$, and $\delta''([q', q], a) = [\delta'(q', a), \delta(q, \lambda(q', a))]$. It is now easy to prove by induction that the construction works. ∎

Here are some applications.

Lemma 4.3.9 *Let $S \subseteq \mathbb{N}$. If $\{(n)_k : n \in S\}$ is a regular language, then so is $\{(n + 1)_k : n \in S\}$.*

Proof. Let $L = \{(n)_k : n \in S\}$. Since L is regular, so is $L^R = \{(n)_k^R : n \in S\}$. We design a finite-state transducer T to "add one" to the k-ary expansion of $(n)_k^R$. One small problem is that $(n + 1)_k^R$ may have one more digit than $(n)_k^R$. To get around this problem we feed the transducer with $L^R 0$, and then adjust the results.

Our transducer has two states. Let $Q = \{0, 1\}$. Then let $\Sigma = \{0, 1, \ldots, k - 1\}$, $q_0 = 1$, and $\Delta = \{0, 1, \ldots, k - 1\}$. We have

$$\delta(c, a) = \begin{cases} 0 & \text{if } c = 0 \text{ or } (c = 1 \text{ and } a \leq k - 2), \\ 1 & \text{if } c = 1 \text{ and } a = k - 1, \end{cases}$$

and

$$\lambda(c, a) \equiv a + c \pmod{k}.$$

Then if $T = (Q, \Sigma, \delta, q_0, \Delta, \lambda)$, we leave it to the reader to verify that

$$T((n)_k^R 0) = w,$$

where $[w^R]_k = n + 1$. Let $L' = T(L^R 0)$. By Theorems 4.3.5 and 4.3.6 we know that L' is regular. Then so is $\left(L'/0^*\right)^R \cap C_k = \{(n+1)_k : n \in S\}$, where C_k is defined as in Section 3.1. ∎

Corollary 4.3.10 *For all integers $c \geq 0$, if $\{(n)_k : n \in S\}$ is a regular language, then so is $\{(n + c)_k : n \in S\}$.*

Lemma 4.3.11 *Let $S \subseteq \mathbb{N}$. Suppose that $\{(n)_k : n \in S\}$ is a regular language. Then so is $\{(bn)_k : n \in S\}$ for any integer $b \geq 0$.*

Proof. As in the previous lemma, we construct a finite-state transducer that "multiplies by b". The construction is similar to that in the previous lemma, except now we allow carries as much as $b - 1$. Let $M = (Q, \Sigma, \delta, q_0, \Delta, \lambda)$, where

$$\Sigma = \Delta = \{0, 1, 2, \ldots, k - 1\},$$
$$Q = \{0, 1, 2, \ldots, b - 1\},$$
$$q_0 = 0;$$
$$\delta(c, a) = \left\lfloor \frac{c + ba}{k} \right\rfloor,$$
$$\lambda(c, a) \equiv c + ba \bmod k.$$

As before, we feed the transducer with $(n)_k^R 0^{\lfloor \log_k b \rfloor}$, and then strip off the extra zeros. ∎

4.4 Context-Free Grammars and Languages

In this section, we briefly introduce the concept of context-free grammars (CFGs) and context-free languages (CFLs).

A context-free grammar is a type of rewriting system on two kinds of symbols: variables and terminals. Whenever a variable appears, it can be rewritten using one of the rewriting rules. Terminals, however, are never rewritten, and when the rewriting rules produce a string of all terminals, the process stops.

More formally, a *context-free grammar* is a quadruple $G = (V, \Sigma, P, S)$ where:

V is a finite set of variables.
Σ is a finite set of terminals.
P is a set of rewriting rules called *productions*; it consists of finitely many pairs of the form (A, α) where $A \in V$ and $\alpha \in (V \cup \Sigma)^*$. Productions are typically written in the form $A \to \alpha$.
S is a distinguished variable called the *start symbol*.

Rewriting takes place as follows: if $A \rightarrow \alpha$ is a production of G, and $\beta A \gamma$ is a member of $(V \cup \Sigma)^*$, then we may replace A with α to obtain $\beta \alpha \gamma$. In this case we write $\beta A \gamma \Longrightarrow \beta \alpha \gamma$. If the grammar G is clear from the context, we may simply write $\beta A \gamma \overset{G}{\Longrightarrow} \beta \alpha \gamma$. If $\alpha_1 \Longrightarrow \alpha_2 \Longrightarrow \cdots \Longrightarrow \alpha_n$, then we write $\alpha_1 \overset{n}{\Longrightarrow} \alpha_n$. If $\alpha \overset{n}{\Longrightarrow} \beta$ for some $n \geq 0$, we write $\alpha \overset{*}{\Longrightarrow} \beta$.

The language generated by a context-free grammar is defined to be

$$L(G) = \{x \in \Sigma^* : S \overset{*}{\underset{G}{\Longrightarrow}} x\}.$$

Example 4.4.1 The following grammar generates the language $\{a^n b^n : n \geq 1\}$: $G = (\{S\}, \{a, b\}, P, S)$, where P is the set of productions

$$S \rightarrow aSb,$$
$$S \rightarrow ab.$$

We typically combine productions that have the same left-hand side using the following notation:

$$S \rightarrow aSb \mid ab.$$

A language is said to be *context-free* if it is generated by a context-free grammar. A *derivation* of a word $w \in L(G)$ is a series of steps

$$S \Longrightarrow \alpha_1 \Longrightarrow \alpha_2 \Longrightarrow \cdots \Longrightarrow \alpha_n \Longrightarrow w.$$

If at each step, the variable in α_i that gets rewritten is the rightmost one, then the derivation is said to be *rightmost*. A grammar G is said to be *unambiguous* if, for each word $w \in L(G)$, there is exactly one rightmost derivation of w. Otherwise G is said to be *ambiguous*. A context-free language with no unambiguous grammar is said to be *inherently ambiguous*.

With any language $L \subseteq \Sigma^*$ we can associate a power series $e_L(X) = \sum_{n \geq 0}$ Card $(L \cap \Sigma^n) X^n$ that counts the number of words of each length in L. We state the following classical theorem without proof.

Theorem 4.4.2 (Chomsky–Schützenberger) *If L is a context-free language, and there is at least one unambiguous context-free grammar G with $L = L(G)$, then its associated formal power series $e_L(X)$ is algebraic over $\mathbb{Q}(X)$.*

Example 4.4.3 Consider the unambiguous grammar

$$S \rightarrow M \mid U,$$
$$M \rightarrow 0M1M \mid \epsilon,$$
$$U \rightarrow 0S \mid 0M1U,$$

which represents strings of "if–then–else" clauses. This grammar has the following commutative image:

$$S = M + U,$$
$$M = x^2M^2 + 1,$$
$$U = Sx + x^2MU.$$

This system of equations has the following power series solutions:

$$M = 1 + x^2 + 2x^4 + 5x^6 + 14x^8 + 42x^{10} + \cdots,$$
$$U = x + x^2 + 3x^3 + 4x^4 + 10x^5 + 15x^6 + 35x^7 + \cdots,$$
$$S = 1 + x + 2x^2 + 3x^3 + 6x^4 + 10x^5 + 20x^6 + \cdots.$$

By Theorem 4.4.2, each variable satisfies an algebraic equation over $\mathbb{Q}(x)$. We have

$$x^2M^2 - M + 1 = 0,$$
$$x^2(2x - 1)U^2 + (x + 1)(2x - 1)U + x = 0,$$
$$x(2x - 1)S^2 + (2x - 1)S + 1 = 0.$$

As in the case of regular languages, which we defined first in terms of regular expressions, there is a machine model that accepts precisely the context-free languages. This model is called the *pushdown automaton*, or PDA. Roughly speaking, a pushdown automaton is an NFA equipped with a potentially infinite stack.

Formally, a PDA M is a 7-tuple $(Q, \Sigma, \Gamma, \delta, q_0, Z_0, F)$ where

Q is a finite set of states;
Σ is the finite input alphabet;
Γ is the finite stack alphabet;
δ, which maps $Q \times (\Sigma \cup \{\epsilon\}) \times \Gamma$ into finite subsets of $Q \times \Gamma^*$ is the transition function;
q_0 is the initial state;
Z_0 is the initial stack symbol;
$F \subseteq Q$ is the set of final states.

To formally describe how a PDA functions, we must introduce the notion of configuration. A *configuration* is a triple (q, w, α) where $q \in Q$ is the current state, w is the *unexpended* input (the input not yet read), and $\alpha \in \Gamma^*$ is the current stack contents. For $a \in \Sigma \cup \{\epsilon\}$ and $X \in \Gamma$, we say $(q, aw, X\alpha) \vdash (p, w, \beta\alpha)$ if $(p, \beta) \in \delta(q, a, X)$. We write $\overset{*}{\vdash}$ for the reflexive, transitive closure of $\vdash$. Then the language accepted by M is

$$L(M) = \{x \in \Sigma^* : (q_0, x, Z_0) \overset{*}{\vdash} (q, \epsilon, \alpha) \text{ for some } q \in F \text{ and } \alpha \in \Gamma^*\}.$$

The following classical theorem relates PDAs and CFGs. We state it without proof.

Theorem 4.4.4 *If L is generated by a CFG, then it is accepted by some PDA, and vice versa.*

4.5 Context-Sensitive Grammars and Languages

In this section we introduce a new type of grammar, the *context-sensitive* grammar (CSG). In such a grammar, the left-hand side of a production no longer needs to be a single variable; now it can be any nonempty string of variables and terminals. However, we insist in any production of the form $\alpha \rightarrow \beta$ that $|\beta| \geq |\alpha|$. Such a production is, for historical reasons, called context-sensitive, and a grammar is context-sensitive if all of its productions are of this form.

A language is said to be a *context-sensitive language* (CSL) if it is generated by a CSG.

Example 4.5.1 The following context-sensitive grammar generates the language $L = \{a^n b^n c^n : n \geq 1\}$.

$$S \rightarrow abc \mid aYbc,$$
$$Yb \rightarrow bY,$$
$$Yc \rightarrow Zbcc,$$
$$bZ \rightarrow Zb,$$
$$aZ \rightarrow aaY,$$
$$aZ \rightarrow aa.$$

To see why it works, consider a derivation starting from the sentential form $a^n Y b^n c^n$. The only possibility is to apply the production $Yb \rightarrow bY$ n times, getting $a^n b^n Y c^n$. Now the production $Yc \rightarrow Zbcc$ must be used, getting $a^n b^n Z b c^{n+1}$. Now the production $bZ \rightarrow Zb$ must be used n times, getting $a^n Z b^{n+1} c^{n+1}$. Now we have a choice. If we use $aZ \rightarrow aa$, we generate the string $a^{n+1} b^{n+1} c^{n+1}$. If we use $aZ \rightarrow aaY$, then we get $a^{n+1} Y b^{n+1} c^{n+1}$, which is the same as the starting sentential form with the exception that the numbers of a's, b's, and c's have each increased by 1.

4.6 Turing Machines

A *Turing machine* is a very general model of computation. Informally, a Turing machine consists of a tape divided into cells, and a finite control with a tape head that can read and write the tape. The tape has a leftmost cell, called cell 0, but is potentially infinite to the right. The tape cells all initially hold a special symbol B, the *blank symbol*, with the exception of cells 1 through n for some $n \geq 0$, which hold the input.

The finite control, based on the current symbol being scanned and its own internal state, rewrites the current symbol, moves to either the left or right, and changes its internal state.

There is a special halting state h, and the Turing machine accepts its input if it eventually reaches the halting state. There are two ways a Turing machine can fail to accept its input: it can crash (either by attempting to move left from cell 0, or by entering a state from which no move is defined), or it can enter an infinite loop.

More formally, a Turing machine is a 7-tuple $M = (Q, \Sigma, \Gamma, \delta, q_0, B, h)$. Here Q is a finite set of states, Σ is the finite input alphabet, Γ is the tape alphabet, q_0 is the initial state, B is the special blank symbol, and h is the special halting state. Note that $h \notin Q$ and $B \notin \Sigma$. However, $B \in \Gamma$ and $\Sigma \subseteq \Gamma$. Finally, δ is the transition function, which is a partial function from $Q \times \Gamma$ to $(Q \cup \{h\}) \times \Gamma \times \{L, R\}$. The meaning of a transition such as $\delta(q, a) = (p, b, D)$ is that if the Turing machine is in state q, scanning a tape symbol a, then it rewrites a as b, moves in the direction specified by D, and changes its state to state p.

A *configuration* of a Turing machine is a triple of the form $(p, \alpha, b\beta)$ and represents the current state of the Turing machine. Here p is the state, α represents the contents of the tape to the left of the tape head, and $b\beta$ consists of the symbol currently being scanned, followed by the tape contents to the right of the symbol being scanned. Since the tape is infinite to the right, any suffix of β that consists of infinitely many B's is truncated by convention.

We write $(p, \alpha, b\beta) \vdash (p', \alpha', b'\beta')$ if the second configuration is reached by one move of the Turing machine starting in the first configuration, and $\overset{*}{\vdash}$ is the reflexive, transitive closure of $\vdash$. Then $L(M)$, the language accepted by the Turing machine M, is defined to be

$$\{x \in \Sigma^* : (q_0, \epsilon, Bx) \overset{*}{\vdash} (h, \alpha, b\beta) \text{ for some } \alpha, b, \beta\}.$$

A language L is said to be *recursively enumerable* if $L = L(M)$ for some Turing machine M. A language L is said to be *recursive* if $L = L(M)$ for some Turing machine M that, all inputs, either halts or crashes (never enters an infinite loop).

We now turn to the solvability of decision problems. A *decision problem* is a problem with at least one parameter that takes infinitely many values, and for which the answer is always "yes" or "no".

We can associate a language with a decision problem as follows: we take the set of all encodings of instances of the decision problem for which the answer is "yes". Of course, this raises the question of what encoding to use, but often there is a "natural" encoding that suggests itself.

Example 4.6.1 Consider the decision problem: given an integer n, to decide whether or not n is a prime number. The input n can take infinitely many values (all $n \geq 2$, for example), and the answer is "yes" (the number is prime) or "no" (the number is not).

A natural encoding of an integer n is representation in base 2. The language associated with the previous decision problem is therefore

$$\texttt{PRIMES2} = \{10, 11, 101, 111, \dots\}.$$

We say a decision problem is *solvable* if its associated language is recursive. Note that a solvable decision problem corresponds to what we ordinarily think of as solvable by mechanical means: there exists a finite deterministic procedure that, given an instance of the problem, will halt in a finite amount of time and answer either "yes" or "no".

Turing's fundamental paper of 1936 proved that there exist unsolvable decision problems:

Theorem 4.6.2 *The decision problem "Given a Turing machine T and an input w, does T halt on w?" is unsolvable.*

The problem in the previous theorem is known as the *halting problem*.

Other decision problems P can be proved unsolvable by showing that the halting problem reduces to them, that is, if you could solve problem P, then you could solve the halting problem.

A very famous problem, known as Hilbert's tenth problem, is the following: given a multivariate polynomial equation with integer coefficients, does there exist an integer solution?

Theorem 4.6.3 *Hilbert's tenth problem is unsolvable.*

4.7 Exercises

1. Use the pumping lemma to prove that the following languages are not regular.
   (a) $\{a^{2^n} : n \geq 1\}$.
   (b) $\{a^{n^2} : n \geq 1\}$.
   (c) $\{a^p : p \text{ is prime}\}$.
   (d) $\{a^i b^j : \gcd(i, j) = 1\}$.
   (e) $\{x \in \{a, b, c\}^* : x \text{ is squarefree}\}$.
2. Prove the pumping lemma for context-free languages: if L is a CFL, then there exists a constant n such that for all $z \in L$ with $|z| \geq n$, there exists a way of writing $z = uvwxy$ with $|vwx| \leq n$ and $|vx| \geq 1$ such that $uv^i wx^i y \in L$ for all $i \geq 0$. Apply it to prove the following languages not CFLs:
   (a) $\{a^i b^i c^i : i \geq 1\}$.
   (b) $\{xx : x \in \{a, b\}^*\}$.
   (c) $\{a^i b^j a^i b^j : i, j \geq 1\}$.
3. Complete the proof of Theorem 4.3.8.
4. Prove Corollary 4.1.9.
5. Let $M = (Q, \Sigma, \delta, q_0, F)$ be a DFA. We say two states $p, q \in Q$ are *distinguishable* if there exists $x \in \Sigma^*$ such that exactly one of $\delta(p, x)$, $\delta(q, x)$ is in F; otherwise they are *indistinguishable*. We say p, q are k-*distinguishable* if there exists such an x with $|x| \leq k$; otherwise they are k-indistinguishable.

(a) If p, q are states that are k-indistinguishable, then p, q are $(k + 1)$-indistinguishable if and only if $\delta(p, a)$, $\delta(q, a)$ are k-indistinguishable for all $a \in \Sigma$.

(b) The states p, q are indistinguishable if and only if they are k-indistinguishable, where $k = (\text{Card } Q) - 2$.

(c) If $M_1 = (Q_2, \Sigma, \delta_1, q_1, F_1)$ and $M_2 = (Q_2, \Sigma, \delta_2, q_2, F_2)$ are two DFAs such that $L(M_1) \neq L(M_2)$, then there exists a string $x \in \Sigma^*$ accepted by exactly one of M_1, M_2 such that $|x| \leq (\text{Card } Q_1) + (\text{Card } Q_2) - 2$.

6. Show that if $L_k = \{0, 1\}^* 1 \{0, 1\}^k$, then no DFA with fewer than 2^{k+1} states can accept L_k.

7. Show that the bound of 2^n in Theorem 4.1.3 is best possible, in the sense that for all n there exists an NFA with n states such that the smallest equivalent DFA has 2^n states.

8. Let $\mathbf{w} = a_1 a_2 a_3 \cdots$ be an infinite word. Show that the language L of prefixes of $\mathbf{w}$,

$$L = \{\epsilon, a_1, a_1 a_2, a_1 a_2 a_3, \dots\}$$

is regular if and only if $a_1 a_2 a_3 \cdots$ is ultimately periodic.

9. Let Σ be an alphabet with at least 3 symbols. Show that the set of strings containing a square, i.e., $\{xyyz : x, z \in \Sigma^* \text{ and } y \in \Sigma^+\}$, is not contextfree.

10. Give an example of a context-free language $L \subseteq \{0, 1, 2\}^*$ such that $\overline{L}$, the complement of L, is infinite and contains only squarefree strings.

11. Give an example of a language L that is *factorial* (i.e., each subword of a word in L is also contained in L) and contains no infinite regular subset.

12. Let $L \subseteq \Sigma^*$ be a regular language, and suppose there exists a set of pairs of words $P = \{(x_i, w_i) : 1 \leq i \leq n\}$ such that
(a) $x_i w_i \in L$ for $1 \leq i \leq n$;
(b) $x_j w_i \notin L$ for $1 \leq i, j \leq n$ and $i \neq j$.
Show that any NFA accepting L has at least n states.

13. Use the state elimination technique of Theorem 4.1.5 to find regular expressions for the set of strings over $\{0, 1\}$ having an even (resp., odd) number of occurrences of the subword 11. (As in Section 3.3, these occurrences may be overlapping.)

14. Given a DFA M, describe an efficient procedure for determining $t_n := |\Sigma^n \cap L(M)|$. Also describe an efficient procedure for computing the ith string of length n in $L(M)$, in lexicographic order.

15. Give a context-sensitive grammar, using as few productions as possible, for the language $\{0^{F_n} : n \geq 1\}$, where F_n is the nth Fibonacci number.

16. Give a context-sensitive grammar generating the prime numbers.

17. Prove the following generalization of the Myhill–Nerode theorem. Let f be a finite-state function, and define the equivalence relation R_f as follows: $x R_f y$ if and only if for all z we have $f(xz) = f(yz)$. Show that the minimum number of states in any DFAO computing f is equal to the number of equivalence classes of R_f.

4.8 Open Problems

1. Suppose you are given two distinct words u, v with $|u|$, $|v| \leq n$. What is the size of the smallest DFA that accepts u but rejects v, or vice versa? If u and v are of different lengths, then a simple argument gives a $O(\log n)$ bound. How about if u and v are of the same length? (Remark: It is known that a machine of size $O(n^{2/5}(\log n)^{3/5})$ exists. See Robson [1989].)

2. Determine if the following language is context-free: the set of words over a 4-letter alphabet that contain an abelian square. (Remark: See Main, Bucher, and Haussler [1987].)

3. Is the set of primitive words over $\{0, 1\}$ context-free? (Remarks: This is a long-standing difficult problem in the theory of context-free languages. Some partial results are known; see, for example, M. Ito and Katsura [1991]; Petersen [1994, 1996]; Kászonyi and Katsura [1997]; Dömösi, Hauschildt, Horváth, and Kudlek [1999].)

4.9 Notes on Chapter 4

For a good introduction to finite automata and other models of computation, see Hopcroft and Ullman [1979].

4.1 McCulloch and Pitts [1943] studied finite-state systems that today would be called finite automata. Huffman [1954], Mealy [1955], and E. Moore [1956] discussed simple computational models equivalent to finite automata. Kleene [1956] introduced regular expressions and proved the languages they represented were the same as languages accepted by finite automata. Rabin and Scott [1959] introduced nondeterministic finite automata.

For an historical survey of automata theory, see Perrin [1995a].

Theorem 4.1.6 (closure of regular languages under quotient) is due to Ginsburg and Spanier [1963].

Theorem 4.1.8 is due to Myhill [1957] and Nerode [1958], independently.

Our proof of Lemma 4.1.10 is due to A. Lubiw.

4.2 Lemma 4.2.1, the "pumping lemma", is due to Bar-Hillel, Perles, and Shamir [1961]. In French it is known as the "théorème de l'étoile".

4.3 Mealy [1955] introduced automata with outputs on the transitions, a specific case of our more general concept of finite-state transducer. E. Moore [1956] introduced automata with outputs on the states, similar to our model of DFAO, except that our DFAO model only outputs one symbol per string, based on the last state reached. What we call a finite-state transducer is called a "finitary sequential transformation" by Cobham [1972].

4.4 Context-free grammars were introduced by Chomsky [1956, 1959]. Theorem 4.4.2 was stated by Chomsky and Schützenberger [1963]. The only known complete proof of this theorem, including all details, seems to be in Kuich and Salomaa [1986, Chapter 16]. Also see Kuich [1997, §10].

Oettinger [1961] introduced the pushdown automaton. Theorem 4.4.4 is due to Chomsky [1962] and Evey [1963], independently.

4.5 Context-sensitive languages were introduced by Chomsky [1956, 1959].

Our example, the context-sensitive grammar for $L = \{a^n b^n c^n : n \geq 1\}$, is from Salomaa [1973, p. 11], where it is attributed to M. Soittola.

4.6 The Turing-machine model was introduced by Turing [1936]. For Hilbert's tenth problem, see, for example, Matiyasevich [1993].

5

Automatic Sequences

In this chapter, we introduce the fundamental concept of this book: the automatic sequence.

5.1 Automatic Sequences

In Section 4.3, we introduced the concept of finite-state function. In this book, we will be particularly interested in the case where the input represents a number in base k, i.e., when the input alphabet $\Sigma = \Sigma_k := \{0, 1, \ldots, k - 1\}$ for an integer $k \geq 2$. We call a DFAO a k-DFAO if this is the case.

We are now ready to define the fundamental notion of this book: the concept of k-automatic sequence. Informally, a sequence $(a_n)_{n \geq 0}$ is k-automatic if a_n is a finite-state function of the base-k digits of n. More precisely, we compute a_n by feeding a finite automaton with the base-k representation of n, *starting with the most significant digit*, and then applying an output mapping τ to the last state reached.

Definition 5.1.1 We say the sequence $(a_n)_{n \geq 0}$ over a finite alphabet Δ is k-*automatic* if there exists a k-DFAO $M = (Q, \Sigma_k, \delta, q_0, \Delta, \tau)$ such that $a_n = \tau(\delta(q_0, w))$ for all $n \geq 0$ and all w with $[w]_k = n$.

If M is as above, we say M *generates* the sequence $(a_n)_{n \geq 0}$. Note that our definition requires that the automaton return the correct answer *even if the input possesses leading zeros*; we will relax this requirement in Section 5.2.

Dozens of sequences of mathematical interest turn out to be k-automatic for some integer $k \geq 2$, including some that we already have seen in this book. We consider some examples.

Example 5.1.2 (The Thue–Morse Sequence) This sequence $\mathbf{t} = (t_n)_{n \geq 0}$, already introduced in Section 1.6, counts the number of 1's (mod 2) in the base-2

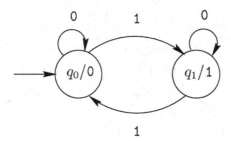

Figure 5.1: Automaton Generating the Thue–Morse Sequence.

representation of n. Here are the first few terms:

$$n = 0\ 1\ 2\ 3\ 4\ 5\ 6\ 7\ 8\ 9\ 10\ 11\ 12\ 13\ 14\ 15\ 16\ 17\ 18\ 19\ 20\ \ldots$$
$$t_n = 0\ 1\ 1\ 0\ 1\ 0\ 0\ 1\ 1\ 0\ 0\ \ 1\ 0\ \ 1\ \ 1\ 0\ 1\ 0\ 0\ 1\ 0\ \ldots$$

The Thue–Morse sequence is 2-automatic, since it can be generated by the DFAO in Figure 5.1.

To see that this DFAO indeed generates **t**, note that being in state q_0 means that the bits of the input seen so far sum to 0 (mod 2), while being in state q_1 means that the bits of the input seen so far sum to 1 (mod 2). This can easily be proved by induction.

The ubiquitous Thue–Morse sequence appears in many different fields of mathematics. Here are just two examples:

Example 5.1.3 (Multigrades) Let A, B be disjoint sets of non-negative integers. A *multigrade* is an identity of the form

$$\sum_{a \in A} a^i = \sum_{b \in B} b^i$$

for $i = 0, 1, 2, \ldots, c$. Suppose we define

$$A_n = \{0 \leq j < 2^{n+1} : t_j = 0\},$$
$$B_n = \{0 \leq j < 2^{n+1} : t_j = 1\}.$$

Then $\sum_{a \in A_n} a^i = \sum_{b \in B_n} b^i$ for $i = 0, 1, 2, \ldots, n$. For example,

$$0^i + 3^i + 5^i + 6^i = 1^i + 2^i + 4^i + 7^i$$

for $i = 0, 1, 2$. See Exercise 1.

Example 5.1.4 (Unusual Infinite Products) We have

$$\prod_{k \geq 0} \left(\frac{2k+1}{2k+2} \right)^{(-1)^{t_k}} = \frac{\sqrt{2}}{2}.$$

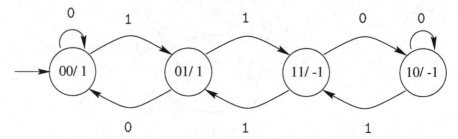

Figure 5.2: DFAO Generating the Rudin–Shapiro Sequence.

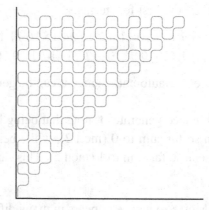

Figure 5.3: Space-filling Curve Derived from the Rudin–Shapiro Sequence.

Now let's look at other examples of automatic sequences.

Example 5.1.5 (The Rudin–Shapiro Sequence) Recall that the *Rudin–Shapiro* sequence $(r_n)_{n\geq0}$, introduced in Example 3.3.1, is defined by the formula $r_n = (-1)^{e_{2;11}(n)}$. In other words, r_n is 1 (or -1) according as the number of (possibly overlapping) occurrences of "11" in the binary expansion of n is even (or odd). Then $(r_n)_{n\geq0}$ is 2-automatic, since it is generated by the DFAO in Figure 5.2.

Here the meaning of a state labeled ab/c is that the running sum of the number of occurrences of "11" so far is congruent to a modulo 2, the last digit input was b, and the output is c. The transitions have been chosen to preserve these meanings.

Here is another interesting feature of the Rudin–Shapiro sequence. Suppose we consider a path visiting lattice points in the plane. We start at the origin and make a first move to $(0, 1)$. At step n, for $n \geq 1$, we decide to turn left (L) or right (R) according to the following rule:

$$d_n = \begin{cases} \text{L} & \text{if } r_n r_{n-1} = (-1)^n, \\ \text{R} & \text{if } r_n r_{n-1} = -(-1)^n. \end{cases} \tag{5.1}$$

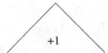

Figure 5.4: Building the Regular Paperfolding Sequence: One Fold.

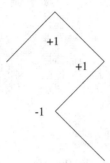

Figure 5.5: Building the Regular Paperfolding Sequence: Two Folds.

The first few terms of this sequence are given below:

$$n = 0\ 1\ 2\ 3\ \ 4\ 5\ \ 6\ \ 7\ 8\ 9\ 10\ 11\ \ 12\ \ 13\ \ 14\ \ 15\ \dots$$
$$r_n = 1\ 1\ 1\ -1\ 1\ 1\ -1\ 1\ 1\ 1\ \ 1\ \ -1\ \ -1\ \ -1\ +1\ -1\ \cdots$$
$$d_{n+1} = \text{R L L}\ \ \text{R}\ \ \text{R R}\ \ \text{L}\ \ \text{L R L L}\ \ \text{L}\ \ \text{R}\ \ \text{R}\ \ \text{L}\ \ \text{R}\ \dots$$

We then get a space-filling curve that visits every lattice point in $\frac{1}{8}$ of the plane; see Figure 5.3.

Example 5.1.6 (The Regular Paperfolding Sequence) First, take a rectangular piece of paper and fold it in half lengthwise, then fold the result in half again, etc., ad infinitum, taking care to make the folds in the same direction each time. Next, unfold the paper. The sequence $(R_i)_{i \geq 1}$ of "hills" $(+1)$ and "valleys" (-1) that results is called the *regular paperfolding sequence*. (For aesthetic reasons, made clear in Section 6.5, we index the sequence starting at 1, not 0.)

For example, after one fold, and unfolding to $90°$, we get the pattern in Figure 5.4. After two folds, and unfolding to $90°$, we get the pattern in Figure 5.5. After twelve folds, we get the interesting "dragon curve" in Figure 5.6 (where corners have been rounded off for clarity).

Here are the first few terms of the limiting sequence **R**:

$$n = 1\ 2\ \ 3\ \ 4\ 5\ \ 6\ \ \ 7\ \ 8\ 9\ 10\ 11\ \ 12\ \ 13\ \ 14\ \ 15\ \ 16\ 17\ 18\ 19\ 20\dots$$
$$R_n = 1\ 1\ -1\ 1\ 1\ -1\ -1\ 1\ 1\ 1\ \ -1\ \ -1\ \ 1\ \ -1\ \ -1\ \ 1\ \ 1\ \ 1\ \ -1\ \ 1\ \dots$$

As we will see in Section 6.5, the regular paperfolding sequence $\mathbf{R} = (R_i)_{i \geq 1}$ is generated by the 2-DFAO in Figure 5.7.

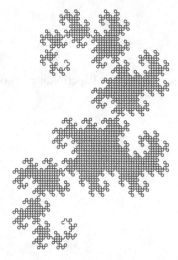

Figure 5.6: Building the Regular Paperfolding Sequence: Twelve Folds.

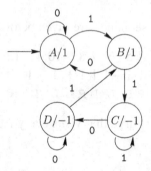

Figure 5.7: 2-DFAO for the Paperfolding Sequence.

There is another interesting method to generate the regular paperfolding sequence, called the *Toeplitz* construction. Start with the following infinite periodic word, of period length 4, over $\{-1, 0, +1\}$,

$$w = +1\ 0\ -1\ 0\ +1\ 0\ -1\ 0\ \cdots.$$

Now replace all the 0's consecutively by the terms of w. We get

$$w' = +1\ +1\ -1\ 0\ +1\ -1\ -1\ 0\ \cdots.$$

Continue iterating this transformation; in the limit we get the paperfolding sequence **R**. For a proof, see Exercise 7.

Example 5.1.7 (The Baum–Sweet Sequence) This sequence $\mathbf{b} = (b_n)_{n\geq 0}$ takes the value 1 if the binary representation of n contains no block of consecutive 0's of

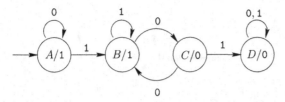

Figure 5.8: 2-DFAO for the Baum–Sweet Sequence.

odd length, and 0 otherwise. Here are the first few terms of this sequence:

$$n = 0\ 1\ 2\ 3\ 4\ 5\ 6\ 7\ 8\ 9\ 10\ 11\ 12\ 13\ 14\ 15\ 16\ 17\ 18\ 19\ 20\ldots$$
$$b_n = 1\ 1\ 0\ 1\ 1\ 0\ 0\ 1\ 0\ 1\ 0\ 0\ 1\ 0\ 0\ 1\ 1\ 0\ 0\ 1\ 0\ldots$$

The 2-DFAO in Figure 5.8 generates this sequence.

Here the meaning of the states is as follows:

A: reading the leading zeros of the input;
B: all blocks of zeros (including current one) are of even length;
C: the last block of zeros seen has odd length so far, but all previous ones have even length;
D: we've seen a block of zeros of odd length.

5.2 Robustness of the Automatic Sequence Concept

In this section we consider some simple variations on the automatic sequence concept, and show that none of these variations change the class of sequences described. In other words, the definition of automatic sequences is robust. Here are some of the variations we consider:

(a) the input number is processed starting with the least significant digit, rather than the most significant digit;
(b) the input number is represented with alternate digit sets; and
(c) the input number is represented in base $-k$.

We start by relaxing a condition in the definition of automatic sequence. Originally, we demanded that our machine M compute a_n correctly no matter which base-k representation of n is input. More precisely, M must give the same answer even if the input has leading zeros. This is a strong requirement, but as the next theorem shows, it is not necessary. In fact, it suffices that the DFAO compute the correct output just for the canonical representations of n in base k (those lacking leading zeros).

Theorem 5.2.1 *The sequence $(a_n)_{n\geq 0}$ is k-automatic if and only if there exists a k-DFAO M such that $a_n = \tau(\delta(q_0, (n)_k))$ for all $n \geq 0$. Moreover, we may choose M such that $\delta(q_0, 0) = q_0$.*

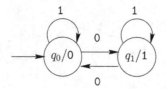

Figure 5.9: Automaton Generating a Variant of the Thue–Morse Sequence.

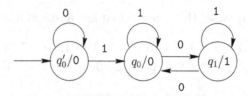

Figure 5.10: Automaton Generating a Variant of the Thue–Morse Sequence.

Proof. $\Longrightarrow$: Trivial.

$\Longleftarrow$: Let $M = (Q, \Sigma_k, \delta, q_0, \Delta, \tau)$. Define $M' = (Q', \Sigma_k, \delta', q_0', \Delta, \tau')$ as follows:

$$Q' = Q \cup \{q_0'\},$$
$$\delta'(q, a) = \delta(q, a) \qquad \text{for all } q \in Q, \ a \in \Sigma_k,$$
$$\delta'(q_0', a) = \begin{cases} \delta(q_0, a) & \text{if } a \neq 0, \\ q_0' & \text{if } a = 0, \end{cases}$$
$$\tau'(q_0') = \tau(q_0),$$
$$\tau'(q) = \tau(q) \qquad \text{for all } q \in Q.$$

Then we claim that

$$\tau'(\delta'(q_0', 0^i(n)_k)) = \tau(\delta(q_0, (n)_k)) \qquad \text{for all } i \geq 0.$$

For we have, for $n \neq 0$, $\delta'(q_0', 0^i(n)_k) = \delta'(q_0', (n)_k) = \delta(q_0, (n)_k)$. ∎

Example 5.2.2 Consider the 2-DFAO in Figure 5.9.

This 2-DFAO computes the sequence $(t_n')_{n \geq 0}$ that counts the number of 0's, modulo 2, in the binary expansion of n:

$$n = 0\ 1\ 2\ 3\ 4\ 5\ 6\ 7\ 8\ 9\ 10\ 11\ 12\ 13\ 14\ 15\ 16\ 17\ 18\ 19\ 20\ \ldots$$
$$t_n' = 0\ 0\ 1\ 0\ 0\ 1\ 1\ 0\ 1\ 0\ 0\ \ \ 0\ \ \ 1\ 0\ \ \ 1\ \ 1\ 0\ 0\ 1\ 1\ 0\ 1\ \ldots$$

Note that the output of the automaton in Figure 5.9 is correct in the case where the input has no leading zeros, but may be incorrect otherwise. To fix this problem, we introduce a new state q_0', as shown in Figure 5.10.

Another variation on the automatic sequence concept is to allow the DFAO to process the base-k representation of the input integer n starting with the *least* significant digit. We have the following theorem.

Theorem 5.2.3 *The following three conditions are equivalent:*

(1) $(a_n)_{n \geq 0}$ *is a k-automatic sequence.*
(2) *There exists a k-DFAO $M = (Q, \Sigma_k, \delta, q_0, \Delta, \tau)$ such that $a_n = \tau(\delta(q_0, w^R))$ for all*
    $n \geq 0$ *and all $w \in \Sigma_k^*$ such that $[w]_k = n$.*
(3) *There exists a k-DFAO $M' = (Q', \Sigma_k, \delta', q_0', \Delta, \tau')$ such that $a_n = \tau(\delta(q_0, (n)_k^R))$ for*
    all $n \geq 0$.

Proof. $(1) \implies (2)$: Let $(a_n)_{n \geq 0}$ be a k-automatic sequence. By Theorem 4.3.3, the function f^R that maps w^R to $a_{[w]_k}$ is a finite-state function, so there exists a DFAO that computes it.

$(2) \implies (3)$: Follows trivially.

$(3) \implies (1)$: Condition (3) implies that there is a finite-state function g that maps $(n)_k^R$ to a_n. Hence by Theorem 4.3.3, the function g^R is a finite-state function, and it maps $(n)_k$ to a_n. Hence there exists a DFAO that computes g^R. By Theorem 5.2.1, this implies that $(a_n)_{n \geq 0}$ is k-automatic. ∎

Yet another variation is to allow DFAOs (processing inputs starting with the least significant digit) that return the correct result when sufficiently many trailing zeros are appended.

Theorem 5.2.4 *Suppose $M = (Q, \Sigma_k, \delta, q_0, \Delta, \tau)$ is a DFAO such that for all w the sequence $(\tau(\delta(q_0, w0^i)))_{i \geq 0}$ is eventually constant. Let $f(w)$ be this constant. Then the sequence $(f((n)_k^R))_{n \geq 0}$ is k-automatic.*

Proof. We may assume without loss of generality that all states of Q are reachable. Since for all w the sequence $(\tau(\delta(q_0, w0^i)))_{i \geq 0}$ is eventually constant, it follows that there exists a map $\tau' : Q \to \Delta$ such that for all $q \in Q$ we have $\tau(\delta(q, 0^i)) = \tau'(q)$ for all sufficiently large i. Now define $M' = (Q, \Sigma_k, \delta, q_0, \Delta, \tau')$. It now follows from Theorem 5.2.3 that $(f((n)_k^R))_{n \geq 0}$ is k-automatic. ∎

Next, we consider cases where the input integer n is expressed in a representation other than ordinary base-k representation. First, we prove two useful lemmas. The first concerns the removal of leading and trailing zeros from strings in a regular language. If

$$w \in \Sigma_k^* = \{0, 1, \ldots, k - 1\}^*,$$

and $w = 0^i a_1 a_2 \cdots a_r$ where $a_1 \neq 0$, then we write $\text{rlz}(w) = a_1 a_2 \cdots a_r$. Similarly, if $w = a_1 a_2 \cdots a_r 0^i$ where $a_r \neq 0$, then we write $\text{rtz}(w) = a_1 a_2 \cdots a_r$.

For a language L we write $\text{rlz}(L) = \{\text{rlz}(w) : w \in L\}$ and $\text{rtz}(L) = \{\text{rtz}(w) : w \in L\}$.

Lemma 5.2.5 *If L is a regular language, then so are $\text{rlz}(L)$ and $\text{rtz}(L)$.*

Proof. We have $\text{rlz}(L) = (L^R/0^*)^R \cap C_k$, where $C_k := \{\epsilon\} \cup (\Sigma_k \setminus \{0\})\Sigma_k^*$, and regular languages are closed under intersection, reversal, and quotient.

Similarly, $\text{rtz}(L) = (L/0^*) \cap (C_k)^R$. ∎

Our second lemma provides a useful alternate characterization of automatic sequences. Let $\mathbf{a} = (a_n)_{n \geq 0}$ be a sequence over Δ, let k be an integer ≥ 2, and let $d \in \Delta$. We define the k-fiber $I_k(\mathbf{a}, d)$ to be the set $\{(n)_k : a_n = d\}$.

Lemma 5.2.6 *Let $\mathbf{a} = (a_n)_{n \geq 0}$. Then $\mathbf{a}$ is k-automatic if and only if each of the fibers $I_k(\mathbf{a}, d)$ is a regular language for all $d \in \Delta$.*

Proof. $\Longrightarrow$: Let $(a_n)_{n \geq 0}$ be generated by the k-DFAO $M = (Q, \Sigma_k, \delta, q_0, \Delta, \tau)$. Then the language

$$J_k(d) = \{w \in \Sigma_k^* : \tau(\delta(q_0, w)) = d\}$$

is evidently regular, as it is the union of regular languages:

$$J_k(d) = \bigcup_{\substack{q \\ \tau(q)=d}} \{w \in \Sigma_k^* : \delta(q_0, w) = q\}.$$

Now $J_k(d) \cap C_k$ is a regular language, by Theorem 4.1.2, and $I_k(\mathbf{a}, d) = J_k(d) \cap C_k$.

$\Longleftarrow$: Suppose that $I_k(\mathbf{a}, d) = \{(n)_k : a_n = d\}$ is regular for all $d \in \Delta = \{1, 2, \ldots, r\}$, (renaming if necessary). Then so is $I'_k(i) := 0^* I_k(\mathbf{a}, i)$, and the $I'_k(i)$ are disjoint and partition Σ_k^*. Using Theorem 4.3.2, we conclude that $(a_n)_{n \geq 0}$ is k-automatic. ∎

We are now ready to prove a theorem on bijective base-k representations.

Theorem 5.2.7 *Let $\langle n \rangle_k$ denote the bijective representation of n in base k using the digits in $\{1, 2, \ldots, k\}$, as discussed in Theorem 3.6.1. Then $\mathbf{a} = (a_n)_{n \geq 0}$ is a k-automatic sequence if and only if there exists a DFAO $M = (Q, \{1, 2, \ldots, k\}, \delta, q_0, \Delta, \tau)$ such that $a_n = \tau(\delta(q_0, \langle n \rangle_k))$ for all $n \geq 0$.*

Proof. Here is an overview of the proof. We use Lemma 5.2.6 to look at the fibers $I_k(\mathbf{a}, d)$ corresponding to the base-k expansions of the integers n for which $a_n = d$.

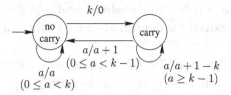

Figure 5.11: Transducer from Bijective to Ordinary Base-k Representation.

We then show that (more or less) the representations using digits $\{1, 2, \ldots, k\}$ can be transduced, using a finite-state transducer, into the ordinary representations using digits $\{0, 1, \ldots, k-1\}$. (Of course this is not exactly true, since the representations can be of different lengths.) We then use Lemma 5.2.6 again to get the desired conclusion.

$\Longrightarrow$: Suppose there exists a DFAO $M = (Q, \{1, 2 \ldots, k\}, \delta, q_0, \Delta, \tau)$ such that $a_n = \tau(\delta(q_0, \langle n \rangle_k))$ for all $n \geq 0$. Then, just as in the proof of Lemma 5.2.6, it follows that each of the languages

$$I'_k(\mathbf{a}, d) := \{\langle n \rangle_k : a_n = d\}$$

is regular.

Now we claim that the transducer T in Figure 5.11 converts from the bijective base-k representation, starting with the *least* significant digit, to the ordinary base-k representation. More precisely, if the input to T is a string of the form $\langle n \rangle_k^R 0$, then the output is a string of the form $(n)_k^R 0^i$ where $i = 0$ or 1. We leave this to the reader to verify.

It now follows that

$$I_k(\mathbf{a}, d) = \{(n)_k : a_n = d\}$$
$$= \mathrm{rlz}(T(I'_k(\mathbf{a}, d)^R\, 0)^R)$$

and hence $I_k(\mathbf{a}, d)$ is regular. By Lemma 5.2.6, $(a_n)_{n \geq 0}$ is a k-automatic sequence.

$\Longleftarrow$: Similarly, if $(a_n)_{n \geq 0}$ is k-automatic, then $I_k(\mathbf{a}, d) = \{(n)_k : a_n = d\}$ is regular for all d. We then have

$$I'_k(\mathbf{a}, d) = \{\langle n \rangle_k : a_n = d\}$$
$$= \mathrm{rlz}(T^{-1}(I_k(\mathbf{a}, d)^R)^R),$$

and it follows from Theorem 4.3.8 that $I'_k(d)$ is regular for all d. In analogy with the proof of Lemma 5.2.6 it follows that there exists a DFAO with $a_n = \tau(\delta(q_0, \langle n \rangle_k))$ for all $n \geq 0$. ∎

5.3 Two-Sided Automatic Sequences

We have defined a sequence $(a_n)_{n \geq 0}$ to be k-automatic if there exists a DFAO such that $a_n = \tau(\delta(q_0, (n)_k))$. As usually defined, automatic sequences are indexed by $\mathbb{N} = \{0, 1, 2, \ldots\}$ and hence are "unidirectional". In this section we examine two different ways to extend the domain of an automatic sequence to $\mathbb{Z}$ and their consequences.

As we have seen in Theorem 3.7.2, if $k \geq 2$ is an integer, then every integer can be uniquely represented (up to leading zeros) in the form $\sum_{0 \leq i \leq r} a_i(-k)^i$, where $a_i \in \{0, 1, \ldots, k - 1\}$. This leads to the following generalization of automatic sequences to the two-sided case:

Definition 5.3.1 We say a two-sided infinite sequence $(a_n)_{n \in \mathbb{Z}}$ is *k-automatic* if there exists a k-DFAO $M = (Q, \Sigma_k, \delta, q_0, \Delta, \tau)$ such that $a_n = \tau(\delta(q_0, (n)_{-k}))$ for all $n \in \mathbb{Z}$.

We then have the following theorem:

Theorem 5.3.2 *A two-sided infinite sequence $(a_n)_{n \in \mathbb{Z}}$ is $(-k)$-automatic if and only if both subsequences $(a_n)_{n \geq 0}$ and $(a_{-n})_{n \geq 0}$ are k-automatic.*

Proof. We first note that an analogue of Lemma 5.2.6 also holds when $k \leq -2$; i.e., a two-sided sequence $(a_n)_{n \in \mathbb{Z}}$ is $(-k)$-automatic ($k \geq 2$) if and only if the each of the fibers

$$I_{-k}(\mathbf{a}, d) = \{(n)_{-k} : a_n = d\}$$

is a regular language for all $d \in \Delta$.

We will now exhibit a finite-state transducer that will convert the representation of a number in base k to its representation in base $-k$. The input will be given starting with the least significant digit. Moreover, since a number's base-$(-k)$ representation can be as much as 2 digits longer than that of its base-k representation, we may have to prefix the base-k representation with 2 leading zeros to obtain the correct result. The output in this case may contain trailing zeros.

More precisely, the input to our transducer will be $(n)_k^R 00$, and the output is in $(n)_{-k}^R \{\epsilon, 0, 00\}$. Figure 5.12 illustrates two transducers in one diagram. The transducer T works for positive inputs, and the transducer T' works for negative ones; they differ only in the specification of the initial states.

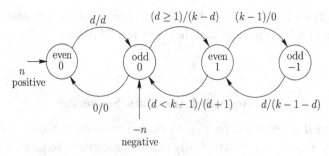

Figure 5.12: Transducers for Converting from Base-k to Base-$(-k)$ Representation.

We leave it to the reader to verify that the transducer T does indeed map $(n)_k^R 00$ to one of $(n)_{-k}^R \{\epsilon, 0, 00\}$, for $n \geq 0$. (It may be helpful to observe that T keeps track of the parity of the position, as well as the "carry" to the next place.) Similarly, by starting at the indicated state, the transducer T' converts $(n)_k^R 00$ to one of $(-n)_{-k}^R \{\epsilon, 0, 00\}$.

Now suppose that $(a_n)_{n \geq 0}$ and $(a_n)_{-n \geq 0}$ are k-automatic. Then the sets

$$I_k(\mathbf{a}, c) = \{(n)_k : a_n = c; n \geq 0\},$$
$$I_k'(\mathbf{a}, c) = \{(n)_k : a_{-n} = c; n \geq 0\}$$

are regular for all $c \in \Delta$. Then so are $I_k(\mathbf{a}, c)^R 00$ and $I_k'(\mathbf{a}, c)^R 00$. Then so are

$$B = T(I_k(\mathbf{a}, c)^R 00)^R \quad \text{and} \quad B' = T'(I_k'(\mathbf{a}, c)^R 00)^R.$$

Now

$$\mathrm{rlz}(B) = \{(n)_{-k} : a_n = c \text{ and } n \geq 0\},$$
$$\mathrm{rlz}(B') = \{(n)_{-k} : a_n = c \text{ and } n \leq 0\}.$$

Thus the sets $\{(n)_{-k} : a_n = c \text{ and } n \geq 0\}$ and $\{(n)_{-k} : a_n = c \text{ and } n \leq 0\}$ are both regular; hence so is their union. It follows that $(a_n)_{n \in \mathbb{Z}}$ is $(-k)$-automatic.

For the converse, assume that

$$K(c) = \{(n)_{-k} : a_n = c; n \in \mathbb{Z}\}$$

is a regular language for all $c \in \Delta$. Then so are

$$K_1(c) = \{(n)_{-k} : a_n = c; n \geq 0\}$$
$$= K(c) \cap ((\Sigma_k \setminus \{0\})(\Sigma_k^2)^* \cup \{\epsilon\})$$

and

$$K_2(c) = \{(n)_{-k} : a_n = c; n \leq 0\}$$
$$= K(c) \cap ((\Sigma_k \setminus \{0\})\Sigma_k(\Sigma_k^2)^* \cup \{\epsilon\}).$$

Now, by an argument similar to the one given above,

$$\mathrm{rlz}((T^{-1}(K_1(c)^R 00))^R) = \{(n)_k : a_n = c; n \geq 0\}$$

is regular, and so is

$$\mathrm{rlz}((T'^{-1}(K_2(c)^R 00))^R) = \{(n)_k : a_{-n} = c; n \geq 0\},$$

which shows that both $(a_n)_{n \geq 0}$ and $(a_{-n})_{n \geq 0}$ are k-automatic. This completes the proof. ∎

Example 5.3.3 We present an analogue to the Thue–Morse sequence in base -2. Define $s_{-2}(n)$ to be the sum of the base-(-2) digits of n. Then $s_{-2}(n) \bmod 2$ is (-2)-automatic, and is generated by the DFAO in Figure 5.13.

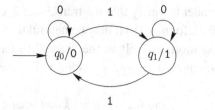

Figure 5.13: Automaton for an Analogue of the Thue–Morse Sequence.

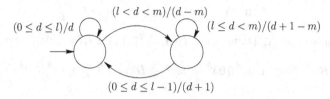

Figure 5.14: Transducer for Positive Numbers.

Here are the first few terms of $s_{-2}(n)$:

n	-9	-8	-7	-6	-5	-4	-3	-2	-1	0	1	2	3	4	5	6	7	8	9	...
$s_{-2}(n)$	3	1	2	3	4	2	3	1	2	0	1	2	3	1	2	3	4	2	3	...

Then the subsequences $(s_{-2}(n) \bmod 2)_{n \geq 0}$ and $(s_{-2}(-n) \bmod 2)_{n \geq 0}$ are both 2-automatic. The construction of the 2-automata is left to the reader in Exercise 3.

Recall the (k, l) numeration systems from Example 3.6.7: these are representations in base $m = k + l + 1$ using the digits in $\Sigma = \{-k, 1 - k, 2 - k, \ldots, -1, 0, 1, 2, \ldots, l - 1, l\}$. We use $(n)_{k,l}$ to denote this type of representation. We say a two-sided infinite sequence $(a_n)_{n \in \mathbb{Z}}$ is (k, l)-automatic if there exists a DFAO $M = (Q, \Sigma, \delta, q_0, \Delta, \tau)$ such that $a_n = \tau(\delta(q_0, (n)_{k,l}))$ for all integers n.

Theorem 5.3.4 *A sequence* $(a_n)_{n \in \mathbb{Z}}$ *is* (k, l)-*automatic if and only if* $(a_n)_{n \geq 0}$ *and* $(a_{-n})_{n \geq 0}$ *are m-automatic.*

Proof. We mimic the proof of Theorem 5.3.2. All the steps should be clear except perhaps the finite-state transducers mapping base-m expansions to (k, l)-expansions. These transducers are given in the accompanying figures. Figure 5.14 illustrates a transducer converting $(n)_m$ to $(n)_{k,l}$. For negative numbers, we simply observe that for $n \geq 0$, $(-n)_{k,l} = -(n)_{l,k}$, where by $-(n)_{k,l}$ we mean negate each term of the (k, l)-expansion. Figure 5.15 gives the transducer converting $(n)_m$ to $(-n)_{k,l}$.

The proof of the result goes through unchanged. (Note that it is easy to deduce whether a (k, l)-expansion denotes a positive or negative number, since this depends only on the sign of the most significant digit.) ∎

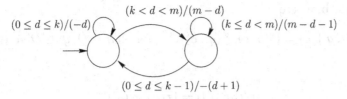

Figure 5.15: Transducer for Negative Numbers.

5.4 Basic Properties of Automatic Sequences

Theorem 5.4.1 *If a sequence $(v_n)_{n \geq 0}$ differs only in finitely many terms from a k-automatic sequence $(a_n)_{n \geq 0}$, then it is k-automatic.*

Proof. This follows easily from Lemma 5.2.6. ∎

Theorem 5.4.2 *If $(a_n)_{n \geq 0}$ is an ultimately periodic sequence, then it is k-automatic for all $k \geq 2$.*

Proof. From Theorem 5.4.1 it suffices to show this is the case where $(a_n)_{n \geq 0}$ is purely periodic of period t, i.e., $a_{tn+i} = a_i$ for $0 \leq i < t$ and $n \geq 0$. Define the k-automaton $M = (Q, \Sigma, \delta, q_0, \Delta, \tau)$, where $\Sigma = \{0, 1, \ldots, k - 1\}$, as follows:

$$Q = \{0, 1, \ldots, t - 1\},$$
$$\delta(q, b) = (kq + b) \bmod t \qquad \text{for all } q \in Q, \ b \in \Sigma,$$
$$\tau(q) = a_q \qquad \text{for } 0 \leq q < t.$$

Then it is easy to see by induction that

$$\delta(q, b_0 b_1 \cdots b_j) = [b_0 b_1 \cdots b_j]_k \bmod t$$

and the result follows. ∎

Theorem 5.4.3 *Let $\mathbf{u} = (u_n)_{n \geq 0}$ be a k-automatic sequence, and let ρ be a coding. Then the sequence $\rho(\mathbf{u})$ is also k-automatic.*

Proof. By the definition of k-automatic, there exists a k-DFAO $M = (Q, \Sigma, \delta, q_0, \Gamma, \tau)$ such that $u_n = \tau(\delta(q_0, (n)_k))$ for all $n \geq 0$. Now consider the k-DFAO $M' = (Q, \Sigma, \delta, q_0, \Gamma, \rho \circ \tau)$. Clearly this DFAO generates $\rho(\mathbf{u})$. ∎

Theorem 5.4.4 *Let $\mathbf{a} = (a_n)_{n \geq 0}$ and $\mathbf{b} = (b_n)_{n \geq 0}$ be two k-automatic sequences with values in Δ and Δ', respectively. Then $\mathbf{a} \times \mathbf{b} = ([a_n, b_n])_{n \geq 0}$ is k-automatic.*

Proof. Let $M = (Q, \Sigma, \delta, q_0, \Delta, \tau)$ generate $(a_n)_{n \geq 0}$ and $M' = (Q', \Sigma, \delta', q'_0, \Delta', \tau')$ generate $(b_n)_{n \geq 0}$. Then $M'' = (Q \times Q', \Sigma, \delta'', [q_0, q'_0], \Delta \times \Delta', \tau'')$

generates $\mathbf{a} \times \mathbf{b}$, where

$$\delta''([q, q'], c) = [\delta(q, c), \delta'(q', c)] \qquad \text{for all } q \in Q, q' \in Q', c \in \Sigma$$

and

$$\tau''([q, q']) = [\tau(q), \tau'(q')]. \qquad \blacksquare$$

Corollary 5.4.5 *Let* $\mathbf{a} = (a_n)_{n \geq 0}$ *and* $\mathbf{b} = (b_n)_{n \geq 0}$ *be two k-automatic sequences with values in finite sets* Δ *and* Δ', *respectively. Let* $f : \Delta \times \Delta' \to \Delta''$ *be any function into a finite set* Δ''. *Then the sequence* $(f(a_n, b_n))_{n \geq 0}$ *is k-automatic.*

Proof. Combine the previous two theorems. $\qquad \blacksquare$

5.5 Nonautomatic Sequences

To prove a sequence is automatic, it suffices to exhibit a DFAO generating it. But how can we prove a sequence nonutomatic? One method is to use the pumping lemma for regular languages in combination with Lemma 5.2.6.

Example 5.5.1 Let us prove that $\mathbf{q} = 1100100001\cdots$, the characteristic sequence of the squares, is not a 2-automatic sequence.

First, we observe that the only solutions in positive integers to the equation

$$y^2 = (2^{2m} - 1)2^{2n+2} + 1$$

are given by $m = n$, $y = 2^{2m+1}$. For if $y^2 - 1 = (2^{2m} - 1)2^{2n+2}$, then $2^{2n+2} \| (y - 1)(y + 1)$. However, if $4 \mid y - 1$, then $2 \| y + 1$. If $4 \mid y + 1$, then $2 \| y - 1$. Hence $2^{2n+1} \| y - 1$ or $2^{2n+1} \| y + 1$. Thus we can write $y = 2^{2n+1}r \pm 1$, where $r \geq 1$ is odd. Now $y^2 - 1 = 2^{4n+2}r^2 \pm r2^{2n+2} = (2^{2m} - 1)2^{2n+2}$. Dividing by 2^{2n+2}, we get $2^{2n}r^2 \pm r = 2^{2m} - 1$. Hence $2^{2m} - 1 \geq 2^{2n}r^2 - r \geq 2^{2n} - 1$, and so $m \geq n$.

Now $m > n$ if and only if $m + n + 2 > 2n + 2$, if and only if $2^{m+n+2} > 2^{2n+2}$, if and only if $2^{2m+2n+2} - 2^{m+n+2} + 1 < 2^{2m+2n+2} - 2^{2n+2} + 1$, if and only if $(2^{m+n+1} - 1)^2 < y^2$, and clearly $y^2 < (2^{m+n+1})^2$. Hence if $m > n$, then y^2 lies strictly between two consecutive squares, a contradiction.

Hence $m = n$, and $y = 2^{2m+1} - 1$.

Now assume that the characteristic sequence of the squares is 2-automatic. It follows from Lemma 5.2.6 that the language

$$\texttt{SQUARES} = \{x \in 1\{0, 1\}^* : [x]_2 \text{ is a square}\}$$

is regular. But by the remarks above,

$$\texttt{SQUARES} \cap (11)^*(00)^*01 = \{1^{2m}0^{2m+1}1 : m \geq 1\},$$

and since regular languages are closed under intersection, $\{1^{2m}0^{2m+1}1 : m \geq 1\}$ must be regular. But an easy argument using the pumping lemma shows this last language is not regular, a contradiction.

The following theorem can often be used to show that some sequences are not automatic.

Theorem 5.5.2 *Let $k \geq 2$ be an integer, and let $\mathbf{a} = (a_n)_{n\geq 0}$ be an automatic sequence generated by the k-DFAO $(Q, \Sigma, \delta, q_0, \Delta, \tau)$. Let $v, w \in \{0, 1, \cdots, k-1\}^*$. Then the sequence $(a_{[vw^i]_k})_{i\geq 0}$ is ultimately periodic.*

Proof. Clearly this is true for $|w| = 0$; hence assume $|w| \geq 1$ and put $w = w_0 w_1 \cdots w_{r-1}$ for some $r \geq 1$. Also put $v = v_0 v_1 \cdots v_{s-1}$ for some $s \geq 0$, and define

$$x_0 x_1 x_2 \cdots := v_0 v_1 v_2 \cdots v_{s-1}(w_0 w_1 \cdots w_{r-1})^{\omega}.$$

Now for $i \geq 1$ define $q_i := \delta(q_0, x_0 x_1 \cdots x_{i-1})$. Since $\mathbf{a}$ is automatic, the set $\{q_i : i \geq 0\}$ is finite. Similarly, the set of pairs $\{(q_i, i \bmod r) : i \geq s\}$ is finite. But for $i \geq s$, the next state q_{i+1} is completely determined by q_i and $i \bmod r$, since

$$\begin{aligned} q_{i+1} &= \delta(q_0, x_0 x_1 \cdots x_i) \\ &= \delta(\delta(q_0, x_0 x_1 x_2 \cdots x_{i-1}), x_i) \\ &= \delta(q_i, x_i) \\ &= \delta(q_i, w_{(i-s)\bmod r}). \end{aligned}$$

There are at most $|Q| \cdot r$ possibilities for $(q_i, i \bmod r)$, so eventually there is a repetition, after which the same sequence repeats (with period at most $|Q| \cdot r$). ∎

Corollary 5.5.3 *If $(a_n)_{n\geq 0}$ is k-automatic, then the subsequences $(a_{k^n})_{n\geq 0}$ and $(a_{k^n-1})_{n\geq 0}$ are ultimately periodic.*

Proof. $(k^n)_k = 1\overbrace{0\cdots 0}^{n}$, which corresponds to $v = 1$, $w = 0$; and $(k^n - 1)_k = \overbrace{1\cdots 1}^{n}$, which corresponds to $v = \epsilon$, $w = 1$. ∎

Example 5.5.4 Consider the sequence $b_n := s_2(\ell_2(n))$, where s_2 is (as usual) the sum-of-digits function and ℓ_2 is the length function introduced in Section 3.2. We show that $(b_n \bmod 2)_{n\geq 0}$ is not 2-automatic. For if it were, by Corollary 5.5.3

we would have $(b_{2^n-1} \bmod 2)_{n \geq 0}$ ultimately periodic. But $b_{2^n-1} = s_2(\ell_2(2^n - 1)) = s_2(n)$, so $b_{2^n-1} \bmod 2 = t_n$, the Thue–Morse sequence. This sequence is not ultimately periodic, since we proved in Theorem 1.6.1 that it is overlap-free.

5.6 k-Automatic Sets

In this section, we define what it means for a set of non-negative integers to be k-automatic, and prove some simple properties of these sets.

Let S be a set of non-negative integers. Then we say that S is k-*automatic* if its characteristic function

$$\chi_S(n) = \begin{cases} 1 & \text{if } n \in S, \\ 0 & \text{otherwise} \end{cases}$$

defines a k-automatic sequence.

Example 5.6.1 The set of powers of 2, $\{1, 2, 4, 8, 16, \dots\}$, forms a 2-automatic set.

Example 5.6.2 By Theorem 5.4.2, any arithmetic progression of integers forms a 2-automatic set.

Theorem 5.6.3 *The class of k-automatic sets is closed under the operations*

(a) intersection;
(b) union;
(c) complement;
(d) set addition, i.e., the operation $R + S = \{r + s : r \in S, s \in S\}$;
(e) non-negative integer multiplication, i.e., the operation $nT = \{nt : t \in T\}$;
(f) the operation defined by $J_{a,b}(S) = \{x : ax + b \in S\}$, where $a \in \mathbb{N}, b \in \mathbb{Z}$.

Proof. Parts (a)–(c) follow immediately from the definition of automatic sequence and Corollary 5.4.5. To see (d), we construct an NFA that, on input $w = (n)_k^R$, accepts w if and only if there exist integers $r \in R$ and $s \in S$ such that $n = r + s$. The idea is to guess the base-k digits of r and s, starting with the least significant digit. We then add these two numbers and compare to the input n, accepting if $r \in R, s \in S$, and $n = r + s$.

Here are more details. Let $M_R = (Q_R, \Sigma_k, \delta_R, q_{R,0}, F_R)$ be a DFA accepting $\{(r)_k^R 0^* : r \in R\}$, and let $M_S = (Q_S, \Sigma_k, \delta_S, q_{S,0}, F_S)$ be a DFA accepting $\{(s)_k^R 0^* : s \in S\}$. We then define an NFA

$$M = (Q_R \times Q_S \times \{0, 1\} \times \{a, r\}, \Sigma_k, \delta, q_0, F)$$

where $q_0 = [q_{R,0}, q_{S,0}, 0, a]$, and

$$\delta([q, q', c, x], v) = \left\{ \left[\delta_R(q, t), \delta_S(q', u), \left\lfloor \frac{t + u + c}{k} \right\rfloor, y \right] : 0 \le t, u < k \right\},$$

where

$$y = \begin{cases} q & \text{if } x = a \text{ and } v = (t + u + c) \bmod k, \\ r & \text{otherwise,} \end{cases}$$

and $F = F_R \times F_S \times \{0\} \times \{a\}$. The meaning of the states of M, represented as 4-tuples, is that the first component records what state we are are currently in for the guessed r, the second component records what state we are currently in for the guesses s, the third component records whether or not there was a carry in adding r to s occurring from the immediately preceding digit, and the fourth component records whether or not the digits of the input seen so far match the digits of the sum $r + s$, where a indicates that it has ("accept") and r indicates it has not ("reject"). This NFA can now be converted, using the subset construction (Theorem 4.1.3), to a DFA that shows that $R + S$ is k-automatic.

For (e), use Lemma 4.3.11 and Lemma 5.2.6. Now (f) is similar to (d); as x is input, it is transduced to $ax + b$, and the result is accepted if it is S. ∎

5.7 1-Automatic Sequences

As we have noted in Chapter 3, it is sometimes convenient to represent integers in "base 1", or unary. The unary representation of n is defined to be the string $1^n = \overbrace{11 \cdots 1}^{n}$. In analogy with our definition for k-automatic sequences ($k \ge 2$), we say a sequence $(a_n)_{n \ge 0}$ is 1-automatic if there exists a DFAO $(Q, \Sigma, \delta, q_0, \Delta, \tau)$, with $\Sigma = \{1\}$, such that $a_n = \tau(\delta(q_0, 1^n))$.

Theorem 5.7.1 *A sequence $(a_n)_{n \ge 0}$ is 1-automatic if and only if it is ultimately periodic.*

Proof. $\Longrightarrow$: We have $a_n = \tau(\delta(q_0, 1^n))$ for some 1-DFAO $M = (Q, \Sigma, \delta, q_0, \Delta, \tau)$. Now there are only finitely many states; hence there exist r, s such that $s > 0$ and $\delta(q_0, 1^r) = \delta(q_0, 1^{r+s})$. Then

$$\delta(q_0, 1^{r+2s}) = \delta(\delta(q_0, 1^{r+s}), 1^s) = \delta(\delta(q_0, 1^r), 1^s) = \delta(q_0, 1^{r+s}),$$

and in a similar fashion it follows that $\delta(q_0, 1^{r+js}) = \delta(q_0, 1^r)$ for all $j \ge 0$. Hence $a_{n+js} = a_n$ for all $n \ge r$ and $j \ge 0$.

$\Longleftarrow$: Suppose $a_{n+js} = a_n$ for all $n \geq r$ and $j \geq 0$. Then we construct a 1-DFAO generating $(a_n)_{n\geq0}$ as follows:

$$Q = \{q_0, q_1, \ldots, q_{r+s-1}\},$$
$$\delta(q_i, 1) = q_{i+1} \quad \text{for } 0 \leq i \leq r+s-2,$$
$$\delta(q_{r+s-1}, 1) = q_r,$$
$$\tau(q_i) = a_i \quad \text{for } 0 \leq i \leq r+s-1.$$

Then it is easy to see that $M = (Q, \Sigma, \delta, q_0, \Delta, \tau)$ generates $(a_n)_{n\geq0}$. ∎

5.8 Exercises

1. Prove the assertion in Example 5.1.3 about multigrades and the Thue–Morse sequence.

2. Let k, m be integers with $k \geq 2$, $m \geq 1$. Show that each of the following sequences defined in Section 3.2 is k-automatic:
   (a) $(s_k(n) \bmod m)_{n\geq0}$;
   (b) $(S_k(n) \bmod m)_{n\geq0}$;
   (c) $(\ell_k(n) \bmod m)_{n\geq0}$;
   (d) $(v_k(n) \bmod m)_{n\geq0}$.

3. Give explicitly the 2-DFAO generating the sequences $(s_{-2}(n) \bmod 2)_{n\geq0}$ and $(s_{-2}(-n) \bmod 2)_{n\geq0}$.

4. Show that the sequence $(d_n)_{n\geq0}$ introduced in Eq. (5.1) is 2-automatic.

5. Show that the following conditions are equivalent:
   (a) The two-sided sequence $(a_n)_{n\in\mathbb{Z}}$ is k-automatic.
   (b) There exists an integer N such that $(a_{n+N})_{n\geq0}$ and $(a_{N-n})_{n\geq0}$ are both k-automatic.
   (c) For all integers N, the sequences $(a_{n+N})_{n\geq0}$ and $(a_{N-n})_{n\geq0}$ are both k-automatic.

6. Give an example of an automatic sequence $(a_n)_{n\geq0}$ such that its summatory function $(\sum_{0\leq i\leq n} a_i)_{n\geq0}$ contains every non-negative integer infinitely often.

7. Prove the Toeplitz characterization of the regular paperfolding sequence given in Example 5.1.6.

8. Let S_n be the nth *Schröder number*, defined to be the number of paths in $\mathbb{Z} \times \mathbb{Z}$ from $(0, 0)$ to (n, n) using the steps $\{(0, 1), (1, 0), (1, 1)\}$, and that contain no points above the line $y = x$.
   (a) Show that $S_{n+1} = 3S_n + \sum_{1\leq k\leq n-1} S_k S_{n-k}$.
   (b) Show that $(S_n \bmod 3)_{n\geq0}$ is a 3-automatic sequence, and characterize this sequence in terms of the base-3 expansion of n.

9. Show that $(s_{k^i}(n) \bmod m)_{n\geq0}$ is a k-automatic sequence for all integers $i, m \geq 1$, where $s_j(n)$ is the sum of the base-j digits of n.

10. Show that the class of k-automatic sets is not closed under set multiplication, i.e., under the operation

$$ST = \{st : s \in S, t \in T\}.$$

11. Let w, x be nonempty strings, and define $w_0 = w$, and $w_{n+1} = w_n x w_n^R$ for $n \geq 0$. Let $z = \lim_{n \to \infty} w_n$ be the infinite string of which $w_0, w_1, \ldots$ are all prefixes.

   (a) Show that the sequence $z = z_0 z_1 z_2 \ldots$ is 2-automatic.

   (b) Show that if x is a palindrome (i.e., $x = x^R$), then z is ultimately periodic.

   (c) Show that if z is ultimately periodic, then x is a palindrome.

12. Show that the characteristic sequence of the prime numbers is not a k-automatic sequence for any $k \geq 2$.

13. Let T be the set of non-negative integers that can be expressed as the sum of three integer squares. Prove that T is a 2-automatic set, and give a 2-DFAO generating the characteristic sequence χ_T.

14. Prove that $(t_{s_2(i)})_{i \geq 0}$ is not a 2-automatic sequence, where $(t_n)_{n \geq 0}$ is the Thue–Morse sequence.

15. Is the sequence $(t'_n)_{n \geq 0}$ of Example 5.2.2 overlap-free?

5.9 Open Problems

1. What is the critical exponent e of the Rudin–Shapiro sequence $\mathbf{r} = (r_n)_{n \geq 0}$? (Remarks: We have $\mathbf{r}[7..10] = 1111$, so $e \geq 4$. On the other hand, Séébold [1986] has shown that $e \leq 5$.)

2. Given a k-DFAO with n states, can one determine in time polynomial in n whether the k-automatic sequence it generates is ultimately periodic? (Remark: For a decision procedure that does not run in polynomial time, see Pansiot [1986]; Honkala [1986]; Harju and Linna [1986].)

3. Explore the theory of *automatic graphs*: graphs on the non-negative integers in which an edge (i, j) exists if and only if the automaton M accepts the input $0^i \# 0^j$, where # is a new symbol (or similar input coded in other ways). What properties (e.g., connectivity) are decidable? (Remark. See, for example, Ly [2000].)

4. Which of the commonly used statistical tests fail to distinguish automatic sequences from truly random sequences? Which succeed?

5.10 Notes on Chapter 5

5.1 The origins of automatic sequences can be found in a paper of Büchi [1960], who attempted to prove (in modern language) that the set of powers of an integer $n \geq 2$ is k-automatic if and only if n and k are multiplicatively dependent. (This is a special case of Cobham's theorem, discussed in Chapter 11.) However, Büchi's proof was flawed; see McNaughton [1963]. Ritchie [1963] proved that the characteristic sequence of the integer squares is not a 2-automatic sequence. Minsky and Papert [1966] studied the characteristic sequence of the primes and showed it was not 2-automatic; also see Allen [1968]; Schützenberger [1968]; Hartmanis and Shank [1968, 1969].

Cobham [1972] was the first to systematically study k-automatic sequences. He called them "uniform tag sequences", in analogy with Post's process of tag; see Post [1965, §3]. Eilenberg [1974] also studied k-automatic sequences, but called them "k-recognizable sequences", terminology that still persists in some circles today.

The first occurrence of the term "automatic sequence" (in French) appears to be in a paper of Deshouillers [1979, pp. 5–21]; also see the *thèse d'état* of Allouche [1983] and Mendès France [1984b]. The review journals *Mathematical Reviews* and *Zentralblatt für Mathematik* have assigned the classification 11B85 to "automata sequences".

For more information on multigrades, see, for example, Lehmer [1947]; Wright [1948]; Roberts [1958]; Wright [1959]; Rokicki and Knuth [1987]; Borwein and Ingalls [1994].

For the unusual infinite products involving $\sqrt{2}$, and some generalizations, see Woods [1978]; Shallit [1985]; Allouche and Cohen [1985]; Allouche, Cohen, Mendès France, and Shallit [1987].

Generalizations of automatic sequences can be found, for example, in Shallit [1988a], Rigo [2000, 2001, 2002], and Lecomte and Rigo [2001].

For the Baum–Sweet sequence, see Baum and Sweet [1976].

For more details about the examples in this section, see, for example, the survey paper of Allouche [1987].

5.2 Theorem 5.2.7 (and much more) is proved in a different way in Allouche, Cateland, Gilbert, Peitgen, Shallit, and Skordev [1997].

5.3 Example 5.3.3 is from Allouche, Cateland, Gilbert, Peitgen, Shallit, and Skordev [1997].

5.4 Theorem 5.4.1 is from Cobham [1972, p. 174]. Theorem 5.4.2 is due to Büchi [1960, p. 88]. Theorem 5.4.4 is from Cobham [1972, p. 177].

5.5 Example 5.5.1, the fact that the squares do not form a 2-automatic set, is due to Ritchie [1963].

5.6 The study of k-automatic sets originated with Büchi [1960], Ritchie [1963], and Minsky and Papert [1966], who called them "recognizable" or "regular". Cobham called such sets "n-regular" [1968b] and "n-recognizable" [1969]. This last term was also used by Eilenberg [1974].

5.7 See, for example, Bruyère, Hansel, Michaux, and Villemaire [1994].

6

Uniform Morphisms and Automatic Sequences

In this chapter we begin our study of the properties of fixed points of morphisms, focusing on the case where the morphism is k-uniform for some integer $k \geq 2$. A fundamental idea is the k-kernel, introduced in Section 6.6. We also study the closure properties of k-automatic sequences.

6.1 Fixed Points of Uniform Morphisms

Recall from Chapter 1 that a morphism is a map φ from Σ^* to Δ^* satisfying $\varphi(xy) = \varphi(x)\varphi(y)$ for all $x, y \in \Sigma^*$. In this chapter, unless otherwise indicated, we suppose that $\Sigma = \Delta$, and that there exists an integer $k \geq 2$ such that $|\varphi(a)| = k$ for all $a \in \Sigma$. In other words, φ is k-uniform.

If there exists a letter a such that $\varphi(a) = ax$ for some $x \in \Sigma^*$ with $|x| = k - 1$, then φ is *prolongable* on a. In this case, the infinite word

$$\mathbf{w} = \varphi^\omega(a) := a\, x\, \varphi(x)\, \varphi^2(x)\, \varphi^3(x) \cdots$$

is the unique infinite fixed point of φ starting with a. As we will see below, the infinite sequences that are images (under a coding) of an infinite fixed point of a k-uniform morphism are precisely the k-automatic sequences.

6.2 The Thue–Morse Infinite Word

As an example, let us recall the Thue–Morse infinite word $\mathbf{t} = t_0 t_1 t_2 \cdots$ introduced in Chapter 1. This word was defined by $t_i = s_2(i) \bmod 2$, where s_2 is the base-2 sum-of-digits function.

However, there are many other ways to define $\mathbf{t}$. For example, the relations $t_0 = 0$, $t_{2n} = t_n$, and $t_{2n+1} = 1 - t_n$, noted in Chapter 1, suffice to define $\mathbf{t}$.

As we saw in Section 5.1, $\mathbf{t}$ is a 2-automatic sequence, and can be generated by the 2-DFAO in Figure 6.1. As we saw in Section 1.7, the Thue–Morse word $\mathbf{t}$ can also be generated as the fixed point of a morphism. Consider the 2-uniform morphism μ defined by $\mu(0) = 01$ and $\mu(1) = 10$. Then μ is prolongable on both

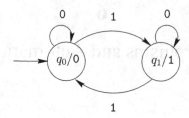

Figure 6.1: Automaton Generating the Thue–Morse Sequence.

0 and 1. We find

$$\mu^{\omega}(0) = \mathbf{t} = 01101001\cdots,$$
$$\mu^{\omega}(1) = \overline{\mathbf{t}} = 10010110\cdots.$$

Yet another definition is as follows. Set $X_0 = 0$, and for $i \geq 0$ define $X_{i+1} = X_i \overline{X_i}$, where $\overline{w}$ is a coding that sends $0 \to 1$ and $1 \to 0$. We find $X_1 = 01$, $X_2 = 0110$, and $\lim_{n \to \infty} X_n = \mathbf{t}$.

Finally, here is a somewhat mysterious way to generate $\mathbf{t}$. As we saw in Section 5.1, we have

$$\prod_{i \geq 0} \left(\frac{2i+1}{2i+2} \right)^{(-1)^{t_i}} = \frac{\sqrt{2}}{2}.$$

Consider trying to determine the t_i iteratively, through the greedy algorithm. More precisely, set $u_0 = 0$, and for $i \geq 0$ set

$$u_{i+1} = \begin{cases} 0 & \text{if } \prod_{0 \leq j \leq i} \left(\frac{2i+1}{2i+2} \right)^{(-1)^{u_i}} > \frac{\sqrt{2}}{2}, \\ 1 & \text{otherwise.} \end{cases}$$

Then $u_i = t_i$ for all $i \geq 0$. We leave the proof to the reader as Exercise 6.

6.3 Cobham's Theorem

In this section, we prove a basic result about fixed points of k-uniform morphisms, known as Cobham's theorem. First, we prove a lemma.

Lemma 6.3.1 *Suppose that* $\mathbf{w} = a_0 a_1 a_2 \cdots$ *is an infinite word with* $\mathbf{w} = \varphi(\mathbf{w})$ *for some k-uniform morphism φ. Then* $\varphi(a_i) = a_{ki} a_{ki+1} a_{ki+2} \cdots a_{ki+k-1}$.

Proof. Since $\mathbf{w} = \varphi(\mathbf{w})$, we have

$$\varphi(a_0 a_1 \cdots a_i) = a_0 a_1 \cdots a_{ki+k-1}.$$

Now

$$\varphi(a_0 a_1 \cdots a_{i-1}) \varphi(a_i) = (a_0 \cdots a_{ki-1})(a_{ki} a_{ki+1} \cdots a_{ki+k-1}),$$

so we get

$$\varphi(a_i) = a_{ki}a_{ki+1}\cdots a_{ki+k-1}.$$ ∎

We are now ready to prove a beautiful description of k-automatic sequences in terms of k-uniform morphisms.

Theorem 6.3.2 (Cobham) *Let $k \geq 2$. Then a sequence $\mathbf{u} = (u_n)_{n\geq 0}$ is k-automatic if and only if it is the image, under a coding, of a fixed point of a k-uniform morphism.*

Proof. $\Longleftarrow$: Suppose $\mathbf{u}$ is the image of a fixed point of a k-uniform morphism. More precisely, suppose $\mathbf{u} = \tau(\mathbf{w})$ for a coding $\tau : \Delta \to \Delta'$, and $\mathbf{w} = \varphi(\mathbf{w})$ for a k-uniform morphism $\varphi : \Delta^* \to \Delta^*$. Write $\mathbf{w} = w_0 w_1 w_2 \cdots$ where each $w_i \in \Delta$, and let $q_0 := w_0$. Now define a k-DFAO $M = (\Delta, \Sigma_k, \delta, q_0, \Delta', \tau)$ where $\delta(q, b) :=$ the bth letter of $\varphi(q)$.

We now claim that $w_n = \delta(q_0, (n)_k)$ for all $n \geq 0$. We prove this by induction on n. Clearly the result is true for $n = 0$. Assume for all $i < n$; we prove it for n. Let $(n)_k = n_1 n_2 \cdots n_t$ where $n = kn' + n_t$, $0 \leq n_t < k$. Then

$$\begin{aligned}
\delta(q_0, (n)_k) &= \delta(q_0, n_1 n_2 \cdots n_t)\\
&= \delta(\delta(q_0, n_1 n_2 \cdots n_{t-1}), n_t)\\
&= \delta(\delta(q_0, (n')_k), n_t)\\
&= \delta(w_{n'}, n_t) \qquad \text{(definition of } \varphi)\\
&= \text{the } n_t\text{'th symbol of } \varphi(w_{n'})\\
&= w_{kn'+n_t} \qquad \text{(by Lemma 6.3.1)}\\
&= w_n.
\end{aligned}$$

It follows that, if $\mathbf{u} = (u_n)_{n\geq 0}$, that

$$u_n = \tau(w_n) = \tau(\delta(q_0, (n)_k)).$$

$\Longrightarrow$: If $(u_n)_{n\geq 0}$ is k-automatic, then it is generated by a k-DFAO $(Q, \Sigma_k, \delta, q_0, \Delta, \tau)$. By Theorem 5.2.1 we may assume without loss of generality that $\delta(q_0, 0) = q_0$. Now define the morphism φ as follows:

$$\varphi(q) := \delta(q, 0)\delta(q, 1)\cdots\delta(q, k-1)$$

for each $q \in Q$.

Let $\mathbf{w} = w_0 w_1 w_2 \cdots$ be the fixed point, starting with q_0, of φ. (It exists because $\delta(q_0, 0) = q_0$.) We now claim that $\delta(q_0, y) = w_{[y]_k}$ for all $y \in \Sigma^*$.

We prove this by induction on $|y|$. When $|y| = 0$ we have $\delta(q_0, \varepsilon) = q_0 = w_0$. Now assume the induction hypothesis is true for all $|y| < i$; we prove it when

$|y| = i$. Write $y = xa$, $a \in \Sigma_k$. Then

$$
\begin{aligned}
\delta(q_0, y) = \delta(q_0, xa) &= \delta(\delta(q_0, x), a) \\
&= \delta(w_{[x]_k}, a) \quad \text{(by induction)} \\
&= \varphi(w_{[x]_k})a \quad \text{(definition of } \varphi) \\
&= w_{k[x]_k + a} \quad \text{(by Lemma 6.3.1)} \\
&= w_{[xa]_k} \\
&= w_{[y]_k}.
\end{aligned}
$$

Hence,

$$
\begin{aligned}
u_n &= \tau(\delta(q_0, (n)_k)) \\
&= \tau(w_n),
\end{aligned}
$$

and so $(u_n)_{n \geq 0}$ is the image (under τ) of the fixed point of φ. ∎

Example 6.3.3 Consider the Baum–Sweet sequence $\mathbf{b} = (b_n)_{n \geq 0}$, which is defined in Example 5.1.7. A DFAO generating this sequence was given in Figure 5.8. From this we deduce that $\mathbf{b} = \tau(h^\omega(\mathrm{A}))$, where

$$
\begin{aligned}
h(\mathrm{A}) &= \mathrm{AB}, & \tau(\mathrm{A}) &= 1, \\
h(\mathrm{B}) &= \mathrm{CB}, & \tau(\mathrm{B}) &= 1, \\
h(\mathrm{C}) &= \mathrm{BD}, & \tau(\mathrm{C}) &= 0, \\
h(\mathrm{D}) &= \mathrm{DD}, & \tau(\mathrm{D}) &= 0.
\end{aligned}
$$

Example 6.3.4 Consider the *period-doubling sequence* $\mathbf{d} = (d_n)_{n \geq 0} = 0100010101000100\cdots$. This sequence is defined by $d_n := \nu_2(n + 1) \bmod 2$, where the function ν_2, introduced in Section 2.1, is the exponent of the highest power of 2 dividing its argument. Alternatively, we have $\mathbf{d} = h^\omega(0)$, where $h(0) = 01$ and $h(1) = 00$. A DFAO generating this sequence is given in Figure 6.2. For more discussion, see Chapter 17.

If $\mathbf{u} = \tau(\varphi^\omega(a))$ for some k-uniform morphism $\varphi : \Delta^* \to \Delta^*$ that is prolongable on a and some coding τ, then $\varphi^\omega(a)$ is sometimes called an *interior sequence* of

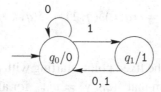

Figure 6.2: DFAO for the Period-Doubling Sequence.

u. If Depth $\varphi = |\Delta|$ is as small as possible, then $\varphi^\omega(a)$ is called a *minimal interior sequence*.

Theorem 6.3.5 *The minimal interior sequence is unique, up to renaming of the symbols.*

Proof. Left to the reader as Exercise 11. ∎

6.4 The Tower of Hanoi and Iterated Morphisms

In this section, we give an detailed example of Cobham's theorem relating automatic sequences to fixed points of morphisms.

The *tower of Hanoi* puzzle consists of three numbered pegs and N disks. Initially, the disks, which have radii $1, 2, \ldots, N$, are all placed on peg 1 in decreasing order of size, so that the smallest disk is on top, and the largest disk is at the bottom. See Figure 6.3.

A *move* of the puzzle consists of taking the top disk off one peg and moving it to another. A move is called *legal* if it does not involve covering a smaller disk with

Figure 6.3: The tower of Hanoi puzzle: reprinted with permission from Ed. Lucas, *Récréations Mathématiques*, Editions Albert Blanchard, Paris.

a larger one. The object of the puzzle is to transfer all of the disks from peg 1 to one of the other two pegs, using only legal moves. We are most interested in the *optimal* solutions to the puzzle; that is, solutions for N disks that use the smallest number of legal moves.

At each step there are at most six possible moves: moving a disk from peg 1 to peg 2, from peg 2 to peg 3, and from peg 3 to peg 1; and their inverses. (Of course, not all of these may be legal at any given time.) We code these moves using the letters a, b, c and $\overline{a}$, $\overline{b}$, $\overline{c}$ respectively.

Thus, for example, we can transfer 3 disks from peg 1 to peg 2 using the following sequence of moves:

$$a\overline{c}bac\overline{b}a.$$

This sequence is optimal – no shorter sequence of legal moves will work. Actually, it is easy to construct an optimal solution for any number of disks; we discuss this further below.

We now describe the classical solution to the puzzle, which has the virtue of being optimal. To move N disks from peg 1 to peg 2, say, first move $N - 1$ disks (recursively) from peg 1 to peg 3, using peg 2 as intermediate storage. Once this is completed, move disk N from peg 1 to peg 2, and then finally move the $N - 1$ disks on peg 3 to peg 2 (recursively), using peg 1 as intermediate storage. Letting T_N be the number of moves in this solution, we see that $T_1 = 1$, and $T_N = 2T_{N-1} + 1$. By induction, one can easily prove that $T_N = 2^N - 1$.

That the classical solution is indeed optimal is not hard to see: to transfer N disks from one peg to another, we must at some point move disk N at least once. In order to move disk N, it must be alone on its peg, and some other peg must be empty; hence the remaining peg must contain the $N - 1$ smaller disks. Hence no algorithm can do better than to (i) move the first $N - 1$ disks to the same peg, using the optimal strategy, (ii) move disk N, and (iii) move the $N - 1$ disks again, using the optimal strategy, and covering disk N. Hence, if $T'(N)$ denotes the total number of moves used by the optimal strategy to transfer N disks, we see that $T'(N) \geq 1 + 2T'(N - 1)$. Since $T'(1) = 1$, we see that $T'(N) \geq 2^N - 1$, as desired.

We now define a coding σ on the alphabet $\{a, b, c, \overline{a}, \overline{b}, \overline{c}\}$ as follows:

$$\sigma(a) = b, \qquad \sigma(\overline{a}) = \overline{b},$$
$$\sigma(b) = c, \qquad \sigma(\overline{b}) = \overline{c},$$
$$\sigma(c) = a, \qquad \sigma(\overline{c}) = \overline{a}.$$

This particular morphism has the following interpretation in terms of the tower of Hanoi puzzle: if a string x has the effect of moving some disks from peg 1 to peg 2, using peg 3 as intermediate storage, then $\sigma(x)$ moves the same configuration of disks from peg 2 to peg 3, using peg 1 as intermediate storage.

We define the coding σ^{-1} as the inverse of σ, namely:

$$\sigma^{-1}(a) = c, \quad \sigma^{-1}(\bar{a}) = \bar{c},$$
$$\sigma^{-1}(b) = a, \quad \sigma^{-1}(\bar{b}) = \bar{a},$$
$$\sigma^{-1}(c) = b, \quad \sigma^{-1}(\bar{c}) = \bar{b}.$$

As we have seen above, for each integer $N \geq 0$, there exists a sequence of $2^N - 1$ moves that constitutes an optimal solution to the tower of Hanoi puzzle with N disks. Actually, there are *two* different solutions: one that results in the disks ending up on peg 2, and another that leaves all the disks on peg 3.

As above, we can code these solutions as strings of symbols, where each symbol represents a move. In what follows, we only consider the solution to the puzzle that takes the disks from peg 1 to peg 2 if N, the number of disks, is odd, and from peg 1 to peg 3 if N is even. This choice might at first seem unnatural, but its advantage is that the sequence of moves for $N + 1$ disks begins with the sequence of moves for N disks. Hence there is an infinite string

$$\mathbf{H} = h_0 h_1 h_2 \cdots = a\bar{c}bac\bar{b}a\bar{c}b\bar{a}cba\bar{c}b\cdots$$

that codes the solution to the puzzle for $N = 1, 2, 3, \ldots$ disks. Another interpretation is that $\mathbf{H}$ solves the puzzle for an *infinite* number of disks.

The infinite string $\mathbf{H}$ can be described as the limit of the sequence of strings $(H_i)_{i \geq 0}$, where each H_i is a string of length $2^i - 1$ that gives the solution to the puzzle for i disks. We can obtain a recursive formula for the H_i using the description of the optimal solution obtained previously:

Proposition 6.4.1 *We have $H_0 = \epsilon$, the empty string, and*

$$H_{2N+1} = H_{2N} \, a \, \sigma^{-1}(H_{2N}) \qquad \text{for } N \geq 0, \qquad (6.1)$$
$$H_{2N} = H_{2N-1} \, \bar{c} \, \sigma(H_{2N-1}) \qquad \text{for } N \geq 1. \qquad (6.2)$$

For example, Eq. (6.2) says in order to solve the puzzle for $2N$ disks, first move $2N - 1$ disks from peg 1 to peg 2. Then, using the move $\bar{c}$, move the $2N$th disk from peg 1 to peg 3. Finally, move $2N - 1$ disks from peg 2 to peg 3; note this is accomplished using the morphism σ.

Using these recursion formulas, we get, for example,

$$H_1 = a,$$
$$H_2 = a\bar{c}b,$$
$$H_3 = a\bar{c}bac\bar{b}a,$$
$$H_4 = a\bar{c}bac\bar{b}a\bar{c}b\bar{a}cba\bar{c}b.$$

We now observe that we can describe the Hanoi sequence **H** as the fixed point of a certain 2-uniform morphism φ, defined below:

$$\varphi(a) = a\overline{c}, \qquad \varphi(\overline{a}) = ac,$$
$$\varphi(b) = c\overline{b}, \qquad \varphi(\overline{b}) = cb,$$
$$\varphi(c) = b\overline{a}, \qquad \varphi(\overline{c}) = ba.$$

First, we prove a lemma:

Lemma 6.4.2 *Let* $\Sigma = \{a, b, c, \overline{a}, \overline{b}, \overline{c}\}$. *Then for* $w \in \Sigma^*$, *we have*

$$\varphi(\sigma(w)) = \sigma^{-1}(\varphi(w)), \tag{6.3}$$
$$\varphi(\sigma^{-1}(w)) = \sigma(\varphi(w)). \tag{6.4}$$

Proof. It suffices to verify the two equations for each letter in Σ. ∎

We now prove the following:

Lemma 6.4.3 *For all* $i \geq 0$, *we have*

$$\varphi(H_{2i})\,a = H_{2i+1},$$
$$\varphi(H_{2i+1})\,b = H_{2i+2}.$$

Proof. By induction on i. The assertions are easily verified for $i = 0$. Then for all $i > 0$,

$$
\begin{aligned}
\varphi(H_{2i})\,a &= \varphi(H_{2i-1}\,\overline{c}\,\sigma(H_{2i-1}))\,a \quad \text{(by (6.2))} \\
&= \varphi(H_{2i-1})\,ba\,\varphi(\sigma(H_{2i-1}))\,a \\
&= \varphi(H_{2i-1})\,b\,a\,\sigma^{-1}(\varphi(H_{2i-1})\,b) \quad \text{(by (6.3))} \\
&= H_{2i}\,a\,\sigma^{-1}(H_{2i}) \quad \text{(by induction)} \\
&= H_{2i+1} \quad \text{(by (6.1))}.
\end{aligned}
$$

Similarly,

$$
\begin{aligned}
\varphi(H_{2i+1})\,b &= \varphi(H_{2i}\,a\,\sigma^{-1}(H_{2i}))\,b \quad \text{(by (6.1))} \\
&= \varphi(H_{2i})\,a\overline{c}\,\varphi(\sigma^{-1}(H_{2i}))\,b \\
&= \varphi(H_{2i})\,a\,\overline{c}\,\sigma(\varphi(H_{2i})\,a) \quad \text{(by (6.4))} \\
&= H_{2i+1}\,\overline{c}\,\sigma(H_{2i+1}) \quad \text{(by above)} \\
&= H_{2i+2} \quad \text{(by (6.2))}.
\end{aligned}
$$

This completes the proof of Lemma 6.4.3. ∎

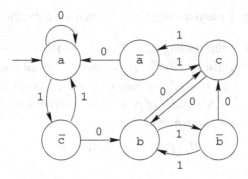

Figure 6.4: The Tower of Hanoi Automaton.

As a corollary, we have

Corollary 6.4.4 *The tower of Hanoi sequence* **H** *is a fixed point of the map* φ*; i.e.,* **H** $= \varphi($**H**$)$.

Proof. It suffices to give an infinite sequence of prefixes of **H** that are mapped by φ into longer prefixes of **H**. But this follows from Lemma 6.4.3, for we see that $\varphi(H_{2i})$ is a prefix of length $2^{2i+1} - 2$ of H_{2i+1}, hence a prefix of **H**. A similar statement holds for $\varphi(H_{2i+1})$. ∎

Of course, by Theorem 6.3.2 it follows that that iterating φ starting with the letter a also produces the infinite sequence **H**. For more details, see Exercise 37. Thus, for example, we have that

$$\varphi(\mathrm{a}) = \mathrm{a}\,\overline{\mathrm{c}} = H_1\,\overline{\mathrm{c}},$$
$$\varphi^2(\mathrm{a}) = \mathrm{a}\,\overline{\mathrm{c}}\,\mathrm{b}\,\mathrm{a} = H_2\,\mathrm{a},$$
$$\varphi^3(\mathrm{a}) = \mathrm{a}\,\overline{\mathrm{c}}\,\mathrm{b}\,\mathrm{a}\,\mathrm{c}\,\overline{\mathrm{b}}\,\mathrm{a}\,\overline{\mathrm{c}} = H_3\,\overline{\mathrm{c}},$$
$$\varphi^4(\mathrm{a}) = \mathrm{a}\,\overline{\mathrm{c}}\,\mathrm{b}\,\mathrm{a}\,\mathrm{c}\,\overline{\mathrm{b}}\,\mathrm{a}\,\overline{\mathrm{c}}\,\overline{\mathrm{b}}\,\mathrm{a}\,\mathrm{c}\,\mathrm{b}\,\mathrm{a}\,\overline{\mathrm{c}}\,\mathrm{b}\,\mathrm{a} = H_4\,\mathrm{a}.$$

Theorem 6.3.2 also tells us that the nth term of the sequence **H** can be computed by the 2-DFAO with six states that is shown in Figure 6.4. The input is $(n)_2$, and the output is the name of the state.

6.5 Paperfolding and Continued Fractions

In this section we discuss paperfolding sequences, and give a remarkable connection with continued fractions.

Recall that in Example 5.1.6 we introduced the notion of iterated paperfolding. In that example, we always folded in such a way to introduce a hill. But in folding n times, we can actually choose our fold to introduce a hill *or* a valley at each step.

If we denote a hill by $+1$ and a valley by -1, then we have the following:

Observation 6.5.1 *Let the folding map* $F_a : \{1, -1\}^* \to \{1, -1\}^*$ *be defined as follows:* $F_a(w) = \mathrm{Concat}(w, a, -w^R)$. *(Note that* F_a *is not a morphism, and that by* $-x$ *for a word* $x \in \{+1, -1\}^*$ *we mean the word obtained by negating each symbol in* x*.) Then if we fold a piece of paper successively with folding instructions* $a_1, a_2, \ldots, a_n$, *we get the pattern of folds*

$$F_{a_1}(F_{a_2}(\cdots(F_{a_n}(\epsilon))\cdots))$$

upon unfolding.

Proof. By induction on n. Clearly the result is true for $n = 1$, since $F_{a_1}(\epsilon) = a_1$. Assume the result is true for $n = m - 1$; we wish to prove it for $n = m$. Fold the paper once with instruction a_1, but pretend it has not been folded. Next fold the folded paper $m - 1$ times with instructions $a_2, a_3, \ldots, a_m$. Finally, unfold it up to, but not including, the first fold. By induction the sequence of folds one sees is

$$w := F_{a_2}(F_{a_3}(\cdots(F_{a_n}(\epsilon))\cdots)).$$

Now unfold the last fold, corresponding to instruction a_1. The new sequence of folds is w, followed by a_1, followed by $-w^R$, which is simply $F_{a_1}(w)$. ∎

We denote $F_{a_1}(F_{a_2}(\cdots(F_{a_n}(\epsilon))\cdots))$ by $\mathrm{Fold}(a_n, a_{n-1}, \ldots, a_1)$, and call the sequence

$$a_n, a_{n-1}, \ldots, a_1$$

the *unfolding instructions*.

Now we would like to fold an *infinite* number of times. In this case, we can specify an infinite sequence $b_0, b_1, b_2, \cdots$ of unfolding instructions. The resulting sequence of folds is then given by

$$\mathrm{Fold}(b_0, b_1, b_2, \ldots).$$

For example, the regular paperfolding sequence is

$$\mathbf{R} := (R_i)_{i \geq 1} = \mathrm{Fold}(1, 1, 1, \ldots) = \mathrm{Concat}(1, 1, -1, 1, 1, -1, -1, \ldots).$$

Our first theorem characterizes the terms of an infinite folding.

Theorem 6.5.2 *Let* $(f_i)_{i \geq 1} = \mathrm{Fold}(b_0, b_1, b_2, \ldots)$. *Let* $n = 2^k(2j + 1)$ *for some integers* $j, k \geq 0$. *Then* $f_n = (-1)^j b_k$.

Proof. By induction on n. Clearly the result is true for $n = 1$, since then $k = j = 0$ and $f_1 = b_0$. Now assume the result is true for all $n < 2^m$; we prove it for $2^m \leq n < 2^{m+1}$. If $n = 2^m$, then from the definition of the folding map it is easy to see that $f_n = b_m$. Otherwise $2^m < n < 2^{m+1}$. In this case, by the folding map it is clear that $f_n = -f_{2^{m+1}-n}$. If $n = 2^k(2j + 1)$, then $2^{m+1} - n = 2^k(2j' + 1)$, where $j' = 2^{m-k} - j - 1$. Now $2^{m+1} - n < n$, so by induction we have $f_{2^{m+1}-n} = (-1)^{j'} b_k$. But $k < m$, so $j' \equiv j + 1 \pmod 2$. Hence $f_n = (-1)^j b_k$, as desired. ∎

This theorem explains the DFAO in Figure 5.7: for the regular paperfolding sequence $(R_i)_{i \geq 1}$, we have $R_i = 1$ if and only if the base-2 representation of i is 10^j or ends in 010^j for some $j \geq 0$.

As a consequence, we can prove

Theorem 6.5.3 *No paperfolding sequence $(f_i)_{i \geq 1}$ is ultimately periodic.*

Proof. Suppose $(f_i)_{i \geq 1}$ is ultimately periodic. Then there exist integers m, N such that $f_n = f_{m+n}$ for all $n \geq N$. Let $m = 2^a(2b + 1)$. Define the sequence $(n_i)_{i \geq 1}$ by $n_i = 2^{a+1}(2i + 1)$. Then $(f_{n_i})_{i \geq 1}$ is purely periodic of period 2 by Theorem 6.5.2. But

$$m + n_i = 2^a(2b + 1) + 2^{a+1}(2i + 1) = 2^a(2(b + 2i + 1) + 1),$$

so $(f_{m+n_i})_{i \geq 1}$ is purely periodic of period 1 by Theorem 6.5.2, a contradiction. ∎

Another consequence is the following:

Theorem 6.5.4 *The paperfolding sequence $(f_i)_{i \geq 1} = \mathrm{Fold}(b_0, b_1, b_2, \dots)$ is 2-automatic if and only if the sequence $(b_i)_{i \geq 0}$ is ultimately periodic.*

Proof. $\Longrightarrow$: By Theorem 6.5.2, we know that $f_{2^i} = b_i$. But, by Corollary 5.5.3, if $(f_i)_{i \geq 1}$ is automatic, then f_{2^i} is ultimately periodic.

$\Longleftarrow$: By Theorem 6.5.2, if $n = 2^j(2k + 1)$, then $f_n = (-1)^k b_j$. Suppose the sequence $(b_j)_{j \geq 0}$ is ultimately periodic. Then it is easy to construct an automaton that on input n determines b_j and the parity of k, and hence f_n. ∎

Now we explain a surprising connection between paperfolding and certain continued fraction expansions of formal power series. We start with the following lemma, known as the *folding lemma*.

Lemma 6.5.5 *Suppose $p_n/q_n = [c_0, c_1, c_2, \dots, c_n]$, and let w represent the word $c_1, c_2, \dots, c_n$. Then*

$$[c_0, w, t, -w^R] = \frac{p_n}{q_n} + \frac{(-1)^n}{t q_n^2}.$$

Proof. By Eq. (2.6) we have

$$\begin{bmatrix} c_0 & 1 \\ 1 & 0 \end{bmatrix} \begin{bmatrix} c_1 & 1 \\ 1 & 0 \end{bmatrix} \cdots \begin{bmatrix} c_n & 1 \\ 1 & 0 \end{bmatrix} = \begin{bmatrix} p_n & p_{n-1} \\ q_n & q_{n-1} \end{bmatrix}.$$

Similarly, it is easily proved by induction that

$$
\begin{bmatrix} -c_0 & 1 \\ 1 & 0 \end{bmatrix} \begin{bmatrix} -c_1 & 1 \\ 1 & 0 \end{bmatrix} \cdots \begin{bmatrix} -c_n & 1 \\ 1 & 0 \end{bmatrix} = \begin{bmatrix} (-1)^{n+1} p_n & (-1)^n p_{n-1} \\ (-1)^n q_n & (-1)^{n-1} q_{n-1} \end{bmatrix}. \quad (6.5)
$$

Now take the transpose of both sides of Eq. (6.5); we get

$$
\begin{bmatrix} -c_n & 1 \\ 1 & 0 \end{bmatrix} \begin{bmatrix} -c_{n-1} & 1 \\ 1 & 0 \end{bmatrix} \cdots \begin{bmatrix} -c_0 & 1 \\ 1 & 0 \end{bmatrix} = \begin{bmatrix} (-1)^{n+1} p_n & (-1)^n q_n \\ (-1)^n p_{n-1} & (-1)^{n-1} q_{n-1} \end{bmatrix}. \quad (6.6)
$$

Now multiply Eq. (6.6) on the right by

$$
\begin{bmatrix} -c_0 & 1 \\ 1 & 0 \end{bmatrix}^{-1} = \begin{bmatrix} 0 & 1 \\ 1 & c_0 \end{bmatrix}
$$

to get

$$
\begin{bmatrix} -c_n & 1 \\ 1 & 0 \end{bmatrix} \begin{bmatrix} -c_{n-1} & 1 \\ 1 & 0 \end{bmatrix} \cdots \begin{bmatrix} -c_1 & 1 \\ 1 & 0 \end{bmatrix} = \begin{bmatrix} (-1)^n q_n & * \\ (-1)^{n-1} q_{n-1} & * \end{bmatrix}, \quad (6.7)
$$

where the asterisks represent entries that do not concern us. Hence we find

$$
\begin{bmatrix} c_0 & 1 \\ 1 & 0 \end{bmatrix} \begin{bmatrix} c_1 & 1 \\ 1 & 0 \end{bmatrix} \cdots \begin{bmatrix} c_n & 1 \\ 1 & 0 \end{bmatrix} \begin{bmatrix} t & 1 \\ 1 & 0 \end{bmatrix} \begin{bmatrix} -c_n & 1 \\ 1 & 0 \end{bmatrix} \begin{bmatrix} -c_{n-1} & 1 \\ 1 & 0 \end{bmatrix} \cdots \begin{bmatrix} -c_1 & 1 \\ 1 & 0 \end{bmatrix}
$$

$$
= \begin{bmatrix} p_n & p_{n-1} \\ q_n & q_{n-1} \end{bmatrix} \begin{bmatrix} t & 1 \\ 1 & 0 \end{bmatrix} \begin{bmatrix} (-1)^n q_n & * \\ (-1)^{n-1} q_{n-1} & * \end{bmatrix}
$$

$$
= \begin{bmatrix} (tp_n + p_{n-1})(-1)^n q_n + (-1)^{n-1} p_n q_{n-1} & * \\ (tq_n + q_{n-1})(-1)^n q_n + (-1)^{n-1} q_n q_{n-1} & * \end{bmatrix}.
$$

It follows that

$$
[c_0, c_1, \ldots, c_n, t, -c_n, -c_{n-1}, \ldots, -c_1] = \frac{(-1)^n (tp_n q_n + p_{n-1} q_n - p_n q_{n-1})}{(-1)^n (tq_n^2 + q_{n-1} q_n - q_n q_{n-1})}
$$

$$
= \frac{tp_n q_n + (-1)^n}{tq_n^2}
$$

$$
= \frac{p_n}{q_n} + \frac{(-1)^n}{tq_n^2},
$$

where we have used Theorem 2.4.2.   ∎

Theorem 6.5.6 *Let $(e_i)_{i \geq 0}$ be an infinite sequence of ± 1's, with $e_0 = 1$. Let $f(X)$ be the formal power series $X \sum_{i \geq 0} e_i X^{-2^i}$. Then the continued fraction expansion of f is given by* $[1, \text{Fold}(e_1 X, -e_2 X, -e_3 X, -e_4 X, \ldots)]$.

Proof. It suffices to prove that

$$
X \sum_{0 \leq i \leq n} e_i X^{-2^i} = [1, \text{Fold}(e_1 X, -e_2 X, -e_3 X, \ldots, -e_n X)].
$$

This is easily established by induction on n, using the folding lemma.   ∎

Example 6.5.7 Take $e_i = 1$ for all $i \geq 0$. Then we have

$$1 + X^{-1} + X^{-3} + X^{-7} + X^{-15} + \cdots$$
$$= [1, \ X, \ -X, \ -X, \ -X, \ X, \ X, \ -X, \ -X, \ X, \ -X, \ -X, \ X,$$
$$X, \ X, \ -X, \ldots]. \tag{6.8}$$

6.6 The k-Kernel

Let $\mathbf{u} = (u_n)_{n \geq 0}$ be an infinite sequence. We define the *k-kernel* of u to be the set of subsequences

$$K_k(\mathbf{u}) = \{(u_{k^i \cdot n + j})_{n \geq 0} : i \geq 0 \text{ and } 0 \leq j < k^i\}.$$

Example 6.6.1 Let $\mathbf{t} = (t_n)_{n \geq 0}$ be the Thue–Morse sequence. The 2-kernel consists of two sequences: $\mathbf{t}$ and $\bar{\mathbf{t}}$, since we have $t_{2n} = t_n$ and $t_{2n+1} = (t_n + 1) \bmod 2$.

Theorem 6.6.2 *Let $k \geq 2$. The sequence $\mathbf{u} = (u_n)_{n \geq 0}$ is k-automatic if and only if $K_k(\mathbf{u})$ is finite.*

Proof. $\Longrightarrow$: By Theorem 5.2.3 it follows that there exists a k-DFAO $(Q, \Sigma_k, \delta, q_0, \Delta, \tau)$ such that

$$u_n = \tau(\delta(q_0, (n)_k^R 0^t)) \qquad \forall t \geq 0.$$

Now let $q = \delta(q_0, w^R)$ where $|w| = i$ and $[w]_k = j$. Since

$$(k^i \cdot n + j)_k = (n)_k w$$

except possibly when $n = 0$, it follows that, for $n > 0$,

$$\delta(q_0, (k^i \cdot n + j)_k^R) = \delta(\delta(q_0, w^R), (n)_k^R)$$
$$= \delta(q, (n)_k^R).$$

In the case when $n = 0$ we have

$$(k^i \cdot n + j)_k = (j)_k,$$

and $w = 0^t (j)_k$ for some $t \geq 0$. Then

$$\delta(q_0, (k^i \cdot n + j)_k^R) = \delta(q_0, (j)_k^R)$$
$$= \delta(q_0, (j)_k^R 0^t)$$
$$= \delta(q_0, w^R)$$
$$= q$$
$$= \delta(q, (0)_k^R).$$

It follows that the subsequence $(u_{k^i \cdot n + j})_{n \geq 0}$ is generated by the k-DFAO $(Q, \Sigma_k, \delta, q, \Delta, \tau)$. Since there are only finitely many choices for q, the finiteness of $K_k(\mathbf{u})$ follows.

$\Longleftarrow$: Suppose that $K_k(\mathbf{u})$ is finite. Then Σ_k^* is partitioned into a finite number of disjoint equivalence classes under the equivalence relation

$$w \equiv x \quad \text{if and only if} \quad u_{k^{|w|} \cdot n + [w]_k} = u_{k^{|x|} \cdot n + [x]_k}$$

for all $n \geq 0$.

Make a k-DFAO as follows:

$$Q = \{[x] : x \in \Sigma_k^*\},$$
$$\delta([x], a) = [ax],$$
$$\tau([w]) = u_{[w]_k},$$
$$q_0 = [\varepsilon],$$

where $[x]$ is the equivalence class containing x. We need to see that this definition is meaningful, i.e., if $[x] = [w]$, then $\delta([x], a) = \delta([w], a)$ and $\tau([x]) = \tau([w])$. For the first, we need to see that $[ax] = [aw]$. Now if $[x] = [w]$ then

$$u_{k^{|w|} \cdot n + [w]_k} = u_{k^{|x|} \cdot n + [x]_k} \qquad \forall n \geq 0.$$

Now upon setting $n = km + a$ it follows that

$$u_{k^{|aw|} \cdot m + [aw]_k} = u_{k^{|ax|} \cdot m + [ax]_k} \qquad \forall m \geq 0.$$

For the second assertion we need to see that if

$$u_{k^{|w|} \cdot n + [w]_k} = u_{k^{|x|} \cdot n + [x]_k}$$

then

$$u_{[w]_k} = u_{[x]_k}.$$

To do this, set $n = 0$. We now claim that $\tau(\delta(q_0, w^R)) = u_{[w]_k}$ for all $w \in \Sigma_k^*$. First we show that $\delta(q_0, w^R) = [w]$. This is an easy induction on $|w|$, and is omitted. By definition of τ the result now follows. $\blacksquare$

Example 6.6.3 Let's compute the 2-kernel for the Rudin–Shapiro sequence $\mathbf{r} = (r(n))_{n \geq 0}$. Since $r(n)$ is $+1$ or -1 according to whether the number of occurrences of 11 in $(n)_2$ is even or odd, we clearly have

$$r(2n) = r(n),$$
$$r(4n + 1) = r(n),$$
$$r(8n + 7) = r(2n + 1),$$
$$r(16n + 3) = r(8n + 3),$$
$$r(16n + 11) = r(4n + 3).$$

It follows that

$$K_2(\mathbf{r}) = \{(r(n))_{n \geq 0}, (r(2n + 1))_{n \geq 0}, (r(4n + 3))_{n \geq 0}, (r(8n + 3))_{n \geq 0}\}.$$

Theorem 6.6.4 *For all $m \geq 1$, a sequence $\mathbf{a} = (a_i)_{i \geq 0}$ is k-automatic if and only if it is k^m-automatic.*

Proof. $\Longrightarrow$: We know from Theorem 6.3.2 that if $\mathbf{a}$ is k-automatic, it can be expressed as the image (under a coding τ) of the fixed point of a k-uniform morphism φ. Thus

$$a_0 a_1 a_2 \cdots = \tau(\varphi^\infty(a))$$

for some letter a. Defining $\gamma = \varphi^m$, we easily see that

$$a_0 a_1 a_2 \cdots = \tau(\gamma^\infty(a)).$$

Hence $(a_i)_{i \geq 0}$ is k^m-automatic.

$\Longleftarrow$: If $\mathbf{a}$ is k^m-automatic, then by Lemma 5.2.6, each of the fibers

$$I_k(\mathbf{a}, d) = \{(n)_{k^m} : a_n = d\}$$

is regular. Now consider the morphism defined by $\beta(j) = w$ for all integers j, $0 \leq j < k^m$, where $|w| = m$ and $[w]_k = j$. Then by Corollary 4.3.7, $\beta(I_d)$ is regular for all d. But

$$\left((\beta(I_d))^R / 0^*\right)^R \cap C_k = \{(n)_k : a_n = d\},$$

and so $(a_i)_{i \geq 0}$ is k-automatic. ∎

6.7 Cobham's Theorem for (k, l)-Numeration Systems

In Section 6.3 we proved Cobham's theorem relating k-automatic sequences and the images of fixed points of k-uniform morphisms. In this section we show how to do the same thing for (k, l)-morphisms, which upon iteration generate two-sided infinite sequences. If $k, l \geq 1$ are integers, then a (k, l)-*morphism* is a uniform morphism h for which there exists a letter a such that $h(a) = wax$, for some strings w, x such that $|w| = k$ and $|x| = l$. In this case, anticipating a notion discussed more fully in Section 7.4, we define

$$\overleftrightarrow{h^\omega}(a) := \cdots h^2(w)\, h(w)\, w.a\, x\, h(x)\, h^2(x) \cdots,$$

a two-sided infinite word.

Example 6.7.1 Consider the map defined by

$$h(0) = 201,$$
$$h(1) = 012,$$
$$h(2) = 120.$$

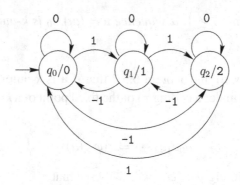

Figure 6.5: A (1, 1)-Automaton for $\overleftrightarrow{h^\omega}(0)$.

Then

$$\overleftrightarrow{h^\omega}(0) = \cdots a_{-3}a_{-2}a_{-1}.a_0a_1a_2a_3 \cdots$$
$$= \cdots 0121202011202.01012201012120, \cdots.$$

Furthermore, this sequence is $(1, 1)$-automatic, since it is generated by the $(1, 1)$-automaton in Figure 6.5.

From this it is easy to see that the fixed point

$$\cdots a_{-3}a_{-2}a_{-1}.a_0a_1a_2a_3 \cdots = \cdots 202.0101 \cdots$$

counts, modulo 3, the sum of digits in the balanced ternary representation of n.

More generally we have

Theorem 6.7.2 *Let* $\cdots a_{-3}a_{-2}a_{-1}.a_0a_1a_2 \cdots$ *be a two-sided infinite sequence with values in a finite set* Δ'. *Then* $\cdots a_{-3}a_{-2}a_{-1}.a_0a_1a_2 \cdots$ *is* (k, l)-*automatic if and only if it can be expressed as the image, under a coding* $\tau : \Delta \to \Delta'$, *of*

$$\overleftrightarrow{h^\omega}(a)$$

where $h : \Delta^* \to \Delta^*$ *is a* (k, l)-*morphism.*

Proof. Suppose that $\cdots a_{-3}a_{-2}a_{-1}.a_0a_1a_2 \cdots$ is (k, l)-automatic, generated by the DFAO $M = (Q, \Sigma, \delta, q_0, \Delta', \tau)$, where $\Sigma = \{-k, 1-k, \ldots, -1, 0, 1, \ldots, l\}$. Furthermore assume that M reads its input starting with the most significant digit, and that it tolerates leading zeros. (By that we mean $\delta(q_0, 0) = q_0$.) Then define

$$h(q) = \delta(q, -k)\delta(q, -(k-1)) \cdots \delta(q, 0) \cdots \delta(q, l)$$

for all $q \in Q$. It is now easy to see that $\cdots a_{-3}a_{-2}a_{-1}.a_0a_1a_2 \cdots$ is the image (under the coding τ) of $\overleftrightarrow{h^\omega}(q_0.)$

For the converse, assume that $\cdots a_{-3}a_{-2}a_{-1}.a_0a_1a_2 \cdots$ is the image, under the coding $\tau : \Delta \to \Delta'$, of $\overleftrightarrow{h^{\omega}}(a)$, where $h : \Delta^* \to \Delta^*$ such that $h(a) = wax$, $|w| = k$, $|x| = l$. Then define $M = (\Delta, \Sigma, \delta, q_0, \Delta', \tau)$ by setting $\Sigma = \{-k, \ldots, l\}$, $q_0 = a$, and letting $\delta(q, i) =$ the $(i + k)$th symbol of $h(q)$. The result now follows. ∎

6.8 Basic Closure Properties

In this section we explore some of the basic kinds of modifications to automatic sequences that result in automatic sequences, in other words, closure properties.

Our first theorem shows that certain subsequences of k-automatic sequences are still k-automatic.

Theorem 6.8.1 *Let* $\mathbf{u} = (u(n))_{n \geq 0}$ *be a* k-*automatic sequence. Then for all integers* $a, b \geq 0$, *the subsequence* $(u(an + b))_{n \geq 0}$ *is also* k-*automatic.*

Proof. If $a = 0$ then $u(an + b) = u(b)$, so the sequence $(u(an + b))_{n \geq 0}$ is constant and hence trivially k-automatic. Now assume $a \geq 1$. We use the characterization in terms of the k-kernel. Since $\mathbf{u} = (u(n))_{n \geq 0}$ is k-automatic, it has a finite k-kernel, say

$$K_k(\mathbf{u}) = \{(u_1(n))_{n \geq 0}, (u_2(n))_{n \geq 0}, \ldots, (u_r(n))_{n \geq 0}\}.$$

Now consider the set of $r(a + b)$ sequences

$$S = \{(u_i(an + c))_{n \geq 0} : 1 \leq i \leq r, \ 0 \leq c < a + b\}.$$

We claim that each element of the k-kernel of $\mathbf{v} = (v(n))_{n \geq 0}$ is in S, where $v(n) := u(an + b)$. To see this, consider $(v(k^e \cdot n + j))_{n \geq 0}$ where $0 \leq j < k^e$ and $e \geq 0$. Using the division algorithm, write

$$ja + b = d \cdot k^e + f, \qquad 0 \leq f < k^e, \quad 0 \leq d < a + b.$$

Then, for all $n \geq 0$, we have

$$\begin{aligned}
v(k^e \cdot n + j) &= u(a(k^e \cdot n + j) + b) \\
&= u(k^e(an + d) + f).
\end{aligned}$$

Now

$$(u(k^e \cdot m + f))_{m \geq 0} = (u_i(m))_{m \geq 0}$$

for some i, $1 \leq i \leq r$, so (putting $m = an + d$) we get

$$(u(k^e(an + d) + f))_{n \geq 0} = (u_i(an + d))_{n \geq 0}$$

and hence

$$(v(k^e \cdot n + j))_{n \geq 0} = (u_i(an + d))_{n \geq 0}.$$

Thus

$$(v(k^e \cdot n + j))_{n \geq 0} \in S$$

and so $K_k(\mathbf{v}) \subseteq S$. Thus $K_k(\mathbf{v})$ is finite, and the result follows.   ∎

The next theorem is a kind of converse to Theorem 6.8.1. Roughly speaking, it says that the a-way merge, for an integer $a \geq 1$, of k-automatic sequences is still k-automatic.

Theorem 6.8.2 *Let $a > 0$ be an integer, and let $(u(n))_{n \geq 0}$ be a sequence taking values in Δ such that $(u(an + i))_{n \geq 0}$ is k-automatic for $0 \leq i < a$. Then $(u(n))_{n \geq 0}$ is itself k-automatic.*

Proof. Define $t_i(n) = u(an + i)$ for $0 \leq i < a$, $n \geq 0$. It then follows from Lemma 5.2.6 that each of the fibers

$$\{(n)_k : t_i(n) = d\}$$

is regular for all $d \in \Delta$, $0 \leq i < a$. Then using Lemmas 4.3.9 and 4.3.11 with Corollary 4.3.10 it follows that

$$\begin{aligned} X_{i,d} &= \{(an + i)_k : t_i(n) = d\} \\ &= \{(an + i)_k : u(an + i) = d\} \end{aligned}$$

is a regular language for all $d \in \Delta$, $0 \leq i < a$. Now define, for each $d \in \Delta$,

$$y_d = \bigcup_{0 \leq i < a} X_{i,d}.$$

Since y_d is a union of regular languages, it is itself a regular language, and $y_d = \{(n)_k : u(n) = d\}$. Thus by Lemma 5.2.6, $(u(n))_{n \geq 0}$ is a k-automatic sequence.   ∎

As a consequence we get the following corollary:

Corollary 6.8.3 *Let $\mathbf{u} = (u(n))_{n \geq 0}$ be a k-automatic sequence, and let h be an a-uniform morphism for some $a \geq 1$. Then $h(\mathbf{u})$ is also k-automatic.*

Proof. It is easy to see that $h(\mathbf{u})$ is the a-way merge of sequences obtained by applying a coding to $\mathbf{u}$. Each of these sequences is automatic, by Theorem 5.4.3. The result now follows by Theorem 6.8.2.   ∎

The next theorem shows that the class of k-automatic sequences is closed under right shifts.

Theorem 6.8.4 *Let* $\mathbf{u} = (u(n))_{n \geq 0}$ *be a k-automatic sequence. Then so is* $\mathbf{v} = (v(n))_{n \geq 0}$, *where*

$$v(n) = \begin{cases} u(n-1) & \text{if } n \geq 1, \\ a & \text{if } n = 0. \end{cases}$$

Proof. Let $K_k(\mathbf{u}) = \{(u_1(n))_{n \geq 0}, (u_2(n))_{n \geq 0}, \dots, (u_r(n))_{n \geq 0}\}$ be the k-kernel. Then for $1 \leq i \leq r$ define

$$v_i(n) = \begin{cases} u_i(n-1) & \text{if } n \geq 1, \\ a & \text{if } n = 0. \end{cases}$$

Then we claim that

$$K_k(\mathbf{v}) \subseteq \{(u_1(n))_{n \geq 0}, \dots (u_r(n))_{n \geq 0}, (v_1(n))_{n \geq 0}, (v_2(n))_{n \geq 0}, \dots, (v_r(n))_{n \geq 0}\}.$$

To see this, consider the subsequences

$$(v(k^e \cdot n + j))_{n \geq 0}, \qquad k \geq 0, \quad 0 \leq j < k^e.$$

If $j \geq 1$, then

$$v(k^e \cdot n + j) = u(k^e \cdot n + j - 1) \qquad \text{for all } n \geq 0.$$

But $u(k^e \cdot n + j - 1) = u_i(n)$ for some i, $1 \leq i \leq r$. Hence $v(k^e \cdot n + j) = u_i(n)$. If $j = 0$, then we observe that, for some i, $1 \leq i \leq r$ we have

$$u(k^e \cdot n + k^e - 1) = u_i(n)$$

for all $n \geq 0$ and hence

$$u(k^e \cdot n - 1) = u_i(n-1)$$

for all $n \geq 1$. Thus

$$v(k^e \cdot n) = \begin{cases} u(k^e \cdot n - 1) & \text{if } n \geq 1, \\ a & \text{if } n = 0 \end{cases}$$

$$= \begin{cases} u_i(n-1) & \text{if } n \geq 1, \\ a & \text{if } n = 0 \end{cases}$$

$$= v_i(n). \qquad \blacksquare$$

Corollary 6.8.5 *Let* $\mathbf{u}$ *be a k-automatic sequence. Then so is the shifted sequence* $S^i(\mathbf{u})$, *for any integer* $i \in \mathbb{Z}$.

Proof. For $i \geq 0$, we use Theorem 6.8.1. For $i < 0$, we apply Theorem 6.8.4 i times. $\blacksquare$

Finally, we show that if $(a_n)_{n \geq 0}$ is k-automatic and generated by a k-DFAO M, then so is the sequence obtained by applying a transducer to the base-k digits of n before processing by M.

Theorem 6.8.6 *Let* $\Sigma_k = \{0, 1, \ldots, k-1\}$, *and let* $T = (Q', \Sigma_k, \delta', q_0', \Sigma_k, \tau')$ *be a finite-state transducer. Let* $M = (Q, \Sigma_k, \delta, q_0, \Delta, \tau)$ *be a* k-*DFAO such that* $a_n = \tau(\delta(q_0, (n)_k))$ *for all* $n \geq 0$. *Then the sequence* $(b_n)_{n \geq 0}$ *defined by* $b_n = a_{[T((n)_k)]_k}$ *is also* k-*automatic.*

Proof. We construct a new k-DFAO $M' = (Q \times Q', \Sigma_k, \delta'', [q_0, q_0'], \Delta, \tau'')$ as follows: we let $\delta''([p, q], a) = [\delta(p, \tau'(a)), \delta'(q, a)]$ for all $p \in Q$, $q \in Q'$, and $a \in \Sigma_k$, and we let $\tau''([p, q]) = \tau(p)$ for all $p \in Q$, $q \in Q'$.

An easy induction now shows that on input $c_1 c_2 \cdots c_i$, the machine M' computes the output

$$\tau(\delta(q_0, T(c_1 c_2 \cdots c_i))),$$

as desired.　　　　　　　　　　　　　　　　　　　　　　　　■

6.9 Uniform Transduction of Automatic Sequences

We have seen that some automatic sequences, such as the Thue–Morse sequence, can be represented as fixed points of uniform morphisms. For others, however, such as the Rudin–Shapiro sequence, we applied a coding to a fixed point to obtain the sequence. This naturally raises the question of whether the coding can be dispensed with.

In general, the answer is no. It can be shown, for example, that the Rudin–Shapiro sequence is not the pure fixed point of any uniform morphism; see Exercise 24.

However, suppose we group the terms of the Rudin–Shapiro sequence into blocks of two consecutive symbols, treating each block as a single symbol. Then it is not hard to see (see Exercise 25) that the resulting sequence *is* a fixed point of a morphism, namely:

$$[1, \ 1] \to [1, \ 1] \, [1, \ -1],$$
$$[1, \ -1] \to [1, \ 1] \, [-1, \ 1],$$
$$[-1, \ 1] \to [-1, \ -1] \, [1, \ -1],$$
$$[-1, \ -1] \to [-1, \ -1] \, [-1, \ 1].$$

One might conjecture that every k-automatic sequence can be so represented, that is, that such a sequence can be represented as a fixed point of a map that sends blocks of size k^a into blocks of size k^b, for some $b > a \geq 0$. The following lemma shows this is true.

Lemma 6.9.1 *Let Δ' be a finite set of symbols and let $\mathbf{x} = (x_i)_{i \geq 0}$ be a sequence taking values in Δ'. Then $\mathbf{x}$ is k-automatic if and only if there exist integers q, r with $0 \leq q < r$ such that for all non-negative integers i, j we have*

$$\mathbf{x}[ik^q..(i+1)k^q - 1] = \mathbf{x}[jk^q..(j+1)k^q - 1]$$

implies

$$\mathbf{x}[ik^r..(i+1)k^r - 1] = \mathbf{x}[jk^r..(j+1)k^r - 1].$$

Proof. $\Rightarrow$: Let $\mathbf{x} = (x_i)_{i \geq 0}$ be a k-automatic sequence. Then by Theorem 6.3.2 we know that $(x_i)_{i \geq 0}$ is the image, under a coding $\tau : \Delta \to \Delta'$, of the fixed point of a k-uniform morphism $\varphi : \Delta \to \Delta^k$; in other words, $\mathbf{x} = \tau(\varphi^\infty(b))$ for some symbol $b \in \Delta$. Now each k^s-aligned subword of $\mathbf{x}$ has the form

$$\tau(\varphi^s(c)) \qquad \text{for some} \quad c \in \Delta.$$

(Recall the definition of k-aligned subword from Section 1.1.)

Let s be fixed and for $a, a' \in \Delta$ define $a \sim_s a'$ if and only if $\tau(\varphi^s(a)) = \tau(\varphi^s(a'))$. Clearly $\sim_s$ is an equivalence relation on Δ, and since there are only a finite number of such relations (see Exercise 26) we know there exist integers q, r with $q < r$ and $\sim_q$ coincides with $\sim_r$. Now if

$$\mathbf{x}[ik^q..(i+1)k^q - 1] = \mathbf{x}[jk^q..(j+1)k^q - 1],$$

then, letting $y = \varphi^\infty(b)$ (so that $\mathbf{x} = \tau(y)$), we have

$$\tau(\varphi^q(y_i)) = \tau(\varphi^q(y_j)),$$

so $y_i \sim_q y_j$. But then $y_i \sim_r y_j$, so

$$\tau(\varphi^r(y_i)) = \tau(\varphi^r(y_j))$$

and hence

$$\mathbf{x}[ik^r..(i+1)k^r - 1] = \mathbf{x}[jk^r..(j+1)k^r - 1].$$

$\Longleftarrow$: Now suppose that $\mathbf{x} = (x_i)_{i \geq 0}$ is a sequence taking values in Δ', such that there exist integers q, r with the stated properties. For non-negative integers i, j we say $i \sim j$ if

$$\mathbf{x}[ik^s..(i+1)k^s - 1] = \mathbf{x}[jk^s..(j+1)k^s - 1] \tag{6.9}$$

for $s = 0, 1, 2, \ldots, r - 1$. Clearly $\sim$ is an equivalence relation on $\mathbb{N}$, and since Δ' is finite, the number of distinct equivalence classes given by $\sim$ is also finite. (Each class is determined by its values at $1 + k + \cdots + k^{r-1} = \frac{k^r - 1}{k - 1}$ different positions, giving a bound of $|\Delta'|^{(k^r - 1)/(k-1)}$ on the number of equivalence classes.)

Let C be this set of equivalence classes, and for $i \in \mathbb{N}$ let $\bar{i}$ denote the equivalence class to which i belongs. If $i \sim j$, then (6.9) holds and so it holds for $s = q$. Then by hypothesis, Eq. (6.9) also holds for $s = r$. Hence

$$\mathbf{x}[ik^{s+1}..(i+1)k^{s+1} - 1] = \mathbf{x}[jk^{s+1}..(j+1)k^{s+1} - 1]$$

for $s = 0, 1, 2, \ldots, r - 1$. Now for any l with $0 \leq l < k$,

$$\mathbf{x}[(ik+l)k^s..(ik+l+1)k^s - 1]$$

and

$$\mathbf{x}[(jk+l)k^s..(jk+l+1)k^s - 1]$$

are subwords in corresponding positions of

$$\mathbf{x}[ik^{s+1}..(i+1)k^{s+1} - 1]$$

and

$$\mathbf{x}[jk^{s+1}..(j+1)k^{s+1} - 1],$$

respectively, and hence equal for $0 \leq s < r$. Therefore $ik + l \sim jk + l$. In other words, we have shown that if $\bar{i} = \bar{j}$ then $\overline{ik+l} = \overline{jk+l}$ for $0 \leq l < k$. Now if we set

$$\varphi(\bar{i}) = \overline{ik}\; \overline{ik+1} \cdots \overline{(i+1)k - 1},$$

then we have a well-defined map φ from C into C^k, and $\varphi(\bar{0})$ begins with $\bar{0}$. Similarly, if we define $\tau(\bar{i}) = \mathbf{x}[i]$, then if $\bar{i} = \bar{j}$ we have (setting $s = 0$ in Eq. (6.9)) $\mathbf{x}[i] = \mathbf{x}[j]$, and so τ is also well defined.

We now claim that $\mathbf{x} = \tau(\varphi^\infty(\bar{0}))$. We prove this by induction. Our induction hypothesis is: if $\mathbf{y} = \varphi^\infty(\bar{0})$ then $\mathbf{y}[i] = \bar{i}$ for all $i < k^s$. Clearly this is true for $s = 0$. Now assume it is true for s; we prove it for $s + 1$. Then we have

$$\mathbf{y}[ik \cdots (i+1)k - 1] = \varphi(\mathbf{y}[i])$$
$$= \varphi(\bar{i})$$
$$= (\overline{ik}) \cdots \overline{(i+1)k - 1}$$

for all $i < k^s$; hence $\mathbf{y}[i] = \bar{i}$ for all $i < k^{s+1}$. By induction, $\mathbf{y}[i] = \bar{i}$ for all $i \geq 0$, and $\mathbf{x}[i] = \tau(y_i) = \tau(\bar{i})$. This completes the proof. ∎

We are now ready to prove the main theorem of this section: that the class of automatic sequences is closed under uniform transductions, that is, under the transducers where every transition outputs a string of the same length. This result generalizes the theorems of Section 6.8. Without loss of generality we may assume that the transducer T is 1-uniform; if not, let us say it is a-uniform. Then, using the result proved below, the outputs of the a different transducers T_i on input x (an automatic sequence) that mimic T except only for the ith symbol ($0 \leq i < a$) are all k-automatic, and we can put them together to get an automatic, sequence using Theorem 6.8.2.

Now let us prove:

Theorem 6.9.2 *Automatic sequences are closed under 1-uniform transducers.*

Proof. Let $(x_i)_{i \geq 0}$ be a k-automatic sequence, given by $\mathbf{x} = \tau(\varphi^\infty(a))$ for $\varphi : \Delta \to \Delta^k$, $a \in \Delta$, and $\tau : \Delta \to \Delta'$. Let $T = (Q, \Delta', \delta, q_0, \Delta'', \lambda)$ be a transducer as in Chapter 4. We must show that

$$\mathbf{x}' := T(\mathbf{x}) = \lambda(q_0, x_0)\lambda(\delta(q_0, x_0), x_1) \ldots$$

is a k-automatic sequence.

Without loss of generality, by "absorbing" the coding τ into T, we may assume that τ is actually the identity map. More formally, set

$$\delta'(q, a) = \delta(q, \tau(a)),$$
$$\lambda'(q, a) = \lambda(q, \tau(a)),$$

and then replace δ, λ in the transducer T by δ', λ'. As a result, we can assume $\mathbf{x} = \varphi^\infty(a)$.

We may also assume without loss of generality that $\Delta'' = Q \times \Delta$ and that λ is the identity map. (For if not, we set $\Delta'' = Q \times \Delta$ and let λ be the identity map, and then apply a coding to the result.) Hence we may assume that

$$T = (Q, \Delta, \delta, q_0, Q \times \Delta, \text{identity}).$$

Let $\mathbf{x}' = T(\mathbf{x})$. Now define $\delta_k(q, a) = \delta(q, \varphi^k(a))$ for $k \in \mathbb{N}$, $q \in Q$, and $a \in \Delta$. Since there are only finitely many functions from $Q \times \Sigma$ into Q, there must be integers $r, s \geq 0$ with $0 \leq r < s$ such that $\delta_r = \delta_s$, i.e., $\delta(q, \varphi^r(a)) = \delta(q, \varphi^s(a))$.

Hence, in particular,

$$\delta(q_0, \varphi^r(\mathbf{x}[0..i-1])) = \delta(q_0, \varphi^s(\mathbf{x}[0..i-1]))$$

for all $i \geq 0$, and so

$$\delta(q_0, \mathbf{x}[0..ik^r - 1]) = \delta(q_0, \mathbf{x}[0..ik^s - 1]) \tag{6.10}$$

for all $i \geq 0$. The idea is to use Lemma 6.9.1. Suppose that (i, j) is a pair such that

$$\mathbf{x}'[ik^r..(i+1)k^r - 1] = \mathbf{x}'[jk^r..(j+1)k^r - 1].$$

Then, since

$$\mathbf{x}[k'] = \langle \delta(q_0, \mathbf{x}[0..k-1]), \mathbf{x}[k] \rangle,$$

we have, by projection on the first component, that

$$\delta(q_0, \mathbf{x}[0..ik^r - 1]) = \delta(q_0, \mathbf{x}[0..jk^r - 1]). \tag{6.11}$$

Furthermore, by projection on the second component, we see that

$$\mathbf{x}'[ik^r..(i+1)k^r - 1] = \mathbf{x}[jk^r..(j+1)k^r - 1]. \tag{6.12}$$

Now

$$
\begin{aligned}
\delta(q_0, \mathbf{x}[0..ik^s - 1]) &= \delta(q_0, \mathbf{x}[0..ik^r - 1]) &\text{(by (6.10))}\\
&= \delta(q_0, \mathbf{x}[0..jk^r - 1]) &\text{(by (6.11))}\\
&= \delta(q_0, \mathbf{x}[0..jk^s - 1]) &\text{(by (6.10)).} \tag{6.13}
\end{aligned}
$$

Applying φ^{s-r} to both sides of (6.12), we get

$$\mathbf{x}[ik^s..(i+1)k^s - 1] = \mathbf{x}[jk^s..(j+1)k^s - 1]. \tag{6.14}$$

Combining (6.13) and (6.14), we get

$$\delta(q_0, \mathbf{x}[0..ik^s + t - 1]) = \delta(q_0, \mathbf{x}[0..jk^s + t - 1]) \tag{6.15}$$

for $0 \le t \le k^s$. Now, combining (6.14) and (6.15), we get

$$\mathbf{x}'[ik^s + t] = \mathbf{x}'[jk^s + t]$$

for $0 \le t < k^s$; hence

$$\mathbf{x}'[ik^s..(i+1)k^s - 1] = \mathbf{x}'[jk^s..(j+1)k^s - 1].$$

Thus, by Lemma 6.9.1, $\mathbf{x}'$ is k-automatic.   ∎

Corollary 6.9.3 *Let* $\mathbf{a} = (a_i)_{i\ge0}$ *be a k-automatic sequence taking values in* $\Sigma_r = \{0, 1, \ldots, r - 1\}$ *for some integer* $r \ge 2$. *Then the following sequences are also k-automatic:*

(a) the running sum sequence $((\sum_{0\le j\le i} a_j) \bmod r)_{i\ge0}$;
(b) the running product sequence $((\prod_{0\le j\le i} a_j) \bmod r)_{i\ge0}$.

Proof. It is easy to construct 1-uniform transducers that on input $\mathbf{a}$ will compute the desired sequences.   ∎

Here is another application of Theorem 6.9.2. Given a k-automatic set S, of positive integers and 0, we may enumerate its members as $S = \{s_1, s_2, \ldots\}$ where $s_1 < s_2 < \cdots$. The following theorem says that "periodic selection" still results in a k-automatic set:

Theorem 6.9.4 *The set* $\{s_{aj+b} : j \ge 0\}$ *is also k-automatic if* $S = \{s_1, s_2, \ldots\}$ *is k-automatic.*

Proof. Use Theorem 6.9.2 on the underlying (automatic) characteristic sequence. It is easy to construct a transducer that changes the 1's to 0's (except the 1's whose occurrence number is congruent to b modulo a) in this sequence.   ∎

6.10 Sums of Digits, Polynomials, and Automatic Sequences

In Section 5.5, we explored some techniques for proving sequences not automatic. In this section, we discuss a new technique, which is exemplified in the following theorem.

Theorem 6.10.1 *Let k, m be integers with $k \geq 2$, $m \geq 1$, and let $p(X)$ be a polynomial in X with non-negative integer coefficients. Then the sequence*

$$(s_k(p(n)) \bmod m)_{n \geq 0}$$

is k-automatic if and only if either $\deg p \leq 1$ *or* $m \mid k - 1$.

Proof. Suppose $\deg p \leq 1$. Then $p(X) = aX + b$ for integers $a, b \geq 0$. The sequence $(s_k(n) \bmod m)_{n \geq 0}$ is k-automatic by Exercise 5.2(a). Hence, by Theorem 6.8.1, the subsequence $(s_k(an + b) \bmod m)_{n \geq 0}$ is k-automatic.

Now suppose $m \mid k - 1$. By Exercise 3.6 we know $s_k(n) \equiv n \pmod{k - 1}$. It follows that $s_k(p(n)) \equiv p(n) \pmod{k - 1}$. Let $r \geq 1$ be an integer. By the binomial theorem $p(n + cr) \equiv p(n) \pmod{r}$ for all $c \geq 0$, so setting $r = k - 1$ and reducing modulo m, we see that $(s_k(p(n)) \bmod m)_{n \geq 0}$ is periodic with period length dividing $k - 1$. By Theorem 5.4.2 we know $(s_k(p(n)) \bmod m)_{n \geq 0}$ is k-automatic.

The converse is harder. If $\deg p \leq 1$ we are done, so assume $\deg p \geq 2$. Here is a rough outline of the proof: we show in Lemma 6.10.3 that if $\mathbf{s} := (s_k(p(n)) \bmod m)_{n \geq 0}$ is k-automatic, then the coefficients of $p(X)$ must satisfy a certain relation. However, any affine subsequence $(s_k(p(un + v)) \bmod m)_{n \geq 0}$ of $\mathbf{s}$ would also be k-automatic, and hence the coefficients of $p(uX + v)$ must satisfy the same relation. By combining enough of these relations we show $m \mid k - 1$. We begin by stating a technical lemma.

Lemma 6.10.2 *Suppose $(a(n))_{n \geq 0}$ is a k-automatic sequence. Then there exist integers $n_0 \geq 0$, $T \geq 1$ such that $a(k^n j + 1) = a(k^{n+Tc} j + 1)$ for all integers $n \geq n_0$, $j \geq 0$, $c \geq 0$.*

Proof. Clearly the result is true for $j = 0$. Otherwise assume $j \geq 1$. Assume $M = (Q, \Sigma_k, \delta, q_0, \Delta, \tau)$ is a k-DFAO that, on input the reversed base-k expansion of n, produces a_n. (Such a machine exists by Theorem 5.2.3.)

Now for $j \geq 1$ we have

$$(k^n j + 1)_k = [j]_k \, 0^{n-1} \, 1,$$
$$(k^{n+Tc} j + 1)_k = [j]_k \, 0^{n+Tc-1} \, 1.$$

Take the machine M, and consider what happens when we feed it with 1 followed by a string of 0's: eventually some state is repeated. In other words, defining $q_1 := \delta(q_0, 1)$, there exist integers $r \geq 0$, $T \geq 1$ such that $\delta(q_1, 0^r) = \delta(q_1, 0^{r+T})$.

Hence

$$\delta(q_0, 1\,0^{n-1}\,[j]_k) = \delta(q_0, 1\,0^{n+Tc-1}\,[j]_k)$$

for $n \geq r + 1$, $c \geq 0$. Since the automaton M reaches the same state on these two inputs, it produces the same output, and the lemma follows.   ∎

The next lemma is crucial for our argument.

Lemma 6.10.3 *Suppose* $p(X) = \sum_{0 \leq i \leq d} a_i X^i$ *for some non-negative integers* a_i *with* $a_d \neq 0$, *and* $d \geq 2$. *Let* k, m *be integers with* $k \geq 2$, $m \geq 1$. *If* $(s_k(p(n)) \bmod m)_{n \geq 0}$ *is* k-*automatic, then* $s_k(a_{d-1} + (d+1)a_d) \equiv s_k(a_d) + s_k(a_{d-1} + da_d) \pmod{m}$.

Proof. Define $j = k^{(d-1)n} + k^{(d-2)n}$. Since $d \geq 2$, the number j is an integer. Using the binomial theorem twice, we have

$$s_k(p(k^n j + 1)) = s_k\left(\sum_{0 \leq i \leq d} a_i(k^{dn} + k^{(d-1)n} + 1)^i\right)$$

$$= s_k\left(\sum_{0 \leq i \leq d} a_i \sum_{0 \leq l \leq i} \binom{i}{l} k^{(d-1)nl}(k^n + 1)^l\right)$$

$$= s_k\left(\sum_{0 \leq i \leq d} \sum_{0 \leq l \leq i} \sum_{0 \leq r \leq l} a_i \binom{i}{l}\binom{l}{r} k^{(d-1)nl+nr}\right)$$

$$= s_k\left(\sum_{0 \leq l \leq d} \sum_{0 \leq r \leq l} b_{l,r} k^{(d-1)nl+nr}\right), \tag{6.16}$$

where $b_{l,r} := \sum_{l \leq i \leq d} a_i \binom{i}{l}\binom{l}{r}$.

Now the sum (6.16) is over various powers of k that could coincide. Let us determine when this happens. Suppose

$$(d-1)nl + nr = (d-1)nl' + nr'$$

for $(l, r) \neq (l', r')$, $l' \leq l$, $0 \leq r \leq l \leq d$, and $0 \leq r' \leq l' \leq d$. This holds if and only if $(d-1)(l-l') = (r'-r)$, which holds if and only if $r = 0$, $l = d$, $r' = d - 1$, $l' = d - 1$. It follows that

$$\sum_{0 \leq l \leq d} \sum_{0 \leq r \leq l} b_{l,r} k^{(d-1)nl+nr} = (b_{d,0} + b_{d-1,d-1})k^{(d-1)dn}$$

$$+ \sum_{\substack{0 \leq l \leq d \\ 0 \leq r \leq l \\ (l,r) \notin \{(d,0),(d-1,d-1)\}}} b_{l,r} k^{(d-1)nl+nr}. \tag{6.17}$$

Note that all the exponents in the right-hand side are distinct, and since they are multiples of n, they differ by at least n. It follows that

$$
s_k\left((b_{d,0} + b_{d-1,d-1})k^{(d-1)dn} + \sum_{\substack{0 \le l \le d \\ 0 \le r \le l \\ (l,r) \notin \{(d,0),(d-1,d-1)\}}} b_{l,r} k^{(d-1)nl+nr}\right)
$$

$$
= \left(\sum_{\substack{0 \le l \le d \\ 0 \le r \le l}} s_k(b_{l,r})\right) - s_k(b_{d,0}) - s_k(b_{d-1,d-1}) + s_k(b_{d,0} + b_{d-1,d-1}) \quad (6.18)
$$

for all n sufficiently large so that $k^n > 2\max_{l,r} b_{l,r}$. Hence, combining Eqs. (6.16)–(6.18), we find

$$
s_k(p(k^n j + 1)) = \left(\sum_{\substack{0 \le l \le d \\ 0 \le r \le l}} s_k(b_{l,r})\right) - s_k(b_{d,0}) - s_k(b_{d-1,d-1}) + s_k(b_{d,0} + b_{d-1,d-1})
$$

$$
(6.19)
$$

for all n sufficiently large.

In a similar fashion we can expand $s_k(p(k^{n+Tc} j + 1))$ to find

$$
s_k(p(k^{n+Tc} j + 1)) = s_k\left(\sum_{0 \le l \le d} \sum_{0 \le r \le l} b_{l,r} k^{((d-1)n+Tc)l+nr}\right). \quad (6.20)
$$

Once again we try to determine which exponents of k coincide. Suppose $(l, r) \neq (l', r')$ with $0 \le r \le l \le d, 0 \le r' \le l' \le d$. Now

$$
((d-1)n + Tc)l + nr = ((d-1)n + Tc)l' + nr'
$$

if and only if

$$
Tc(l - l') = n((r' - r) + (d-1)(l' - l)).
$$

If $(r' - r) + (d-1)(l' - l) = 0$, then $(l, r) = (l', r')$, a contradiction. Hence $(r' - r) + (d-1)(l' - l) \neq 0$. It follows that for all n sufficiently large we have

$$
|Tc(l - l') - n((r' - r) + (d-1)(l' - l))| \ge n - |Tc(l - l')|
$$
$$
\ge n - Tcd
$$
$$
\ge Tc.
$$

Now choose a c sufficiently large so that $k^{Tc} > \max_{l,r} b_{l,r}$. We find

$$
s_k\left(\sum_{0 \le l \le d} \sum_{0 \le r \le l} b_{l,r} k^{((d-1)n+Tc)l+nr}\right) = \sum_{\substack{0 \le l \le d \\ 0 \le r \le l}} s_k(b_{l,r}) \quad (6.21)
$$

for all n sufficiently large. Hence

$$s_k(p(k^{n+Tc}j + 1)) = \sum_{\substack{0 \leq l \leq d \\ 0 \leq r \leq l}} s_k(b_{l,r}) \qquad (6.22)$$

for all n sufficiently large.

Now by Lemma 6.10.2 we know $s_k(k^n j + 1) \equiv s_k(k^{n+Tc}j + 1)$ (mod m) for all sufficiently large n. By Eqs. (6.19) and (6.22) we obtain $s_k(b_{d,0} + b_{d-1,d-1}) \equiv s_k(b_{d,0}) + s_k(b_{d-1,d-1})$ (mod m). Now a simple computation shows that $b_{d,0} = a_d$ and $b_{d-1,d-1} = a_{d-1} + da_d$, and the result follows. ∎

Lemma 6.10.4 *Let a, b, b' be non-negative integers with $a \geq 1$ and $b - b' \geq a$. Then the sequence*

$$((s_k(an + b) - s_k(an + b')) \bmod m)_{n \geq 0}$$

is eventually constant if and only if $m \mid k - 1$.

Proof. Suppose $m \mid k - 1$. From Exercise 3.6 we have $s_k(n) \equiv n$ (mod m). Hence

$$s_k(an + b) - s_k(an + b') \equiv b - b' \pmod{m}.$$

For the converse, suppose $((s_k(an + b) - s_k(an + b')) \bmod m)_{n \geq 0}$ is eventually constant. Since $b - b' \geq a$, there exists an integer $r \geq 1$ such that $b' < ar \leq b$. Let $n = k^t - r$. For all t large enough we have

$$s_k(an + b) - s_k(an + b') = s_k(ak^t + b - ar) - s_k(ak^t + b' - ar)$$
$$= s_k(a) + s_k(b - ar) - s_k(ak^t - (ar - b')).$$

Now fix an integer $s < t$ such that $k^s > ar - b'$. We obtain

$$s_k(a) + s_k(b - ar) - s_k(ak^t - (ar - b'))$$
$$= s_k(a) + s_k(b - ar) - s_k((a - 1)k^t + k^t - k^s + k^s - (ar - b'))$$
$$= s_k(a) + s_k(b - ar) - s_k(a - 1) - s_k(k^t - k^s) - s_k(k^s - (ar - b')). \quad (6.23)$$

Note that all the quantities in this last expression are constants, with the possible exception of $s_k(k^t - k^s) = (k - 1)(t - s)$. But $((k - 1)(t - s) \bmod m)_{t \geq 0}$ is eventually constant if and only if $k - 1 \equiv 0$ (mod m). ∎

We can now complete the proof of Theorem 6.10.1. Let u, v be positive integers with $u > d$. If $(s_k(p(n)) \bmod m)_{n \geq 0}$ is k-automatic, then by Theorem 6.8.1 so is the affine subsequence $(s_k(p(un + v)) \bmod m)_{n \geq 0}$. Then Lemma 6.10.3 holds for the polynomial $p(uX + v)$. But $p(uX + v) = \sum_{0 \le i \le d} a_i (uX + v)^i$, and by the binomial theorem the coefficient of X^d in $p(uX + v)$ is $a_d u^d$, while the coefficient of X^{d-1} is $a_d u^{d-1} v + a_{d-1} u^{d-1}$. By Lemma 6.10.3 we have $s_k(av + b) - s_k(av + b') = e$, where

$$a = a_d u^{d-1},$$
$$b = a_{d-1} u^{d-1} + (d + 1) a_d u^d,$$
$$b' = a_{d-1} u^{d-1} + d a_d u^d,$$
$$e = s_k(a_d u^d).$$

Since $u > d$, we have $b - b' \ge a$, and so by Lemma 6.10.4 we have $m \mid k - 1$, as desired. ∎

Exercise 32 provides a small generalization of Theorem 6.10.1.

6.11 Exercises

1. Consider the first differences of the Thue–Morse sequence, taken modulo 2. More precisely, define $d_n = (t_{n+1} - t_n) \bmod 2$ for $n \ge 0$. Show that $\mathbf{d} = d_0 d_1 d_2 \cdots$ is the fixed point of the morphism that sends $1 \to 10$, $0 \to 11$.

2. Show that the sequence $\mathbf{s} = (s_n)_{n \ge 0}$ consisting of the first digit in the base-k representation of n is k-automatic, for all integers $k \ge 2$, and explicitly write down a k-uniform morphism h such that $\mathbf{s} = h^\omega(0)$.

3. Show that $\mathbf{H}$, the tower of Hanoi sequence on six symbols, is squarefree.

4. Find a 2-uniform morphism φ and coding τ such that $\tau(\varphi^\omega(0)) = (S_2(n) \bmod 2)_{n \ge 1}$, where $S_2(n)$ is the function defined in Section 3.2.

5. (Niederreiter and Vielhaber) Given a sequence $(a_n)_{n \ge 1}$, define $B(h)$ for $h \ge 0$ to be the set of all finite subsequences of the form

$$(a_k, a_{2k}, a_{2k+1}, a_{4k}, a_{4k+1}, a_{4k+2}, a_{4k+3}, \ldots, a_{2^h k}, \ldots, a_{2^h k + 2^h - 1})$$

where k ranges over all integers ≥ 1. Also define $K(h) = |B(h)|$. Show that $(a_n)_{n \ge 1}$ is 2-automatic if and only if $K(h) = O(1)$.

6. Define $u_0 = 0$, and for $i \ge 0$ set

$$u_{i+1} = \begin{cases} 0 & \text{if } \prod_{0 \le j \le i} \left(\frac{2i+1}{2i+2} \right)^{(-1)^{u_i}} > \frac{\sqrt{2}}{2}, \\ 1 & \text{otherwise.} \end{cases}$$

Show that $u_i = t_i$ for all $i \ge 0$, where $\mathbf{t} = t_0 t_1 t_2 \cdots$ is the Thue–Morse sequence.

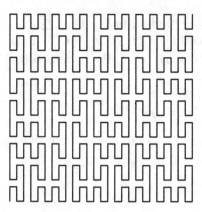

Figure 6.6: Third Iteration in the Construction of Peano's Space-Filling Curve.

7. Consider the morphism f defined as follows:

$$S_1 \rightarrow S_1 R_1 R_2,$$
$$S_2 \rightarrow S_2 L_1 L_2,$$
$$R_1 \rightarrow S_2 L_1 L_2,$$
$$R_2 \rightarrow S_1 S_2 S_1,$$
$$L_1 \rightarrow S_1 R_1 R_2,$$
$$L_2 \rightarrow S_2 S_1 S_2,$$

and let τ be a coding that removes the subscripts. Let $w_i = \tau(f^{2i}(S_1))$ with the last two characters removed, and interpret the letters in w_i as drawing instructions, where S means move straight one unit in the direction you are already traveling, R means turn right and move one unit, and L means turn left and move one unit. Show that under this interpretation, the words w_i code the ith iteration in the construction of Peano's famous space-filling curve. For example, Figure 6.6 shows the graphic interpretation of w_3.

8. Consider the morphism g defined as follows:

$$R \rightarrow RLLSRRLR,$$
$$L \rightarrow RLLSRRLL,$$
$$S \rightarrow RLLSRRLS.$$

Define $x_i = g^i(RRRR)$, and consider the figures specified by x_i, where the interpretations are as in the previous exercise. Show that x_i codes the ith iteration in the construction of von Koch's famous snowflake curve. The first four iterations are shown in Figure 6.7.

9. Consider the sequence $(a_n)_{n \geq 0} = 1264224288 \cdots$ that gives the least significant nonzero digit in the base-10 expansion of $n!$.

   (a) Show that $(a_n)_{n \geq 0}$ is a 5-automatic sequence.
   (b) Give a 5-uniform morphism h and a coding τ such that $(a_n)_{n \geq 0} = \tau(h^\omega(0))$.
   (c) Show that $(a_n)_{n \geq 0}$ is not ultimately periodic.

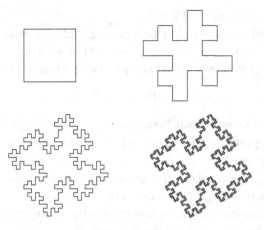

Figure 6.7: Four Iterations in the Construction of the von Koch Curve.

10. Let $\mathbf{u} = (u_n)_{n \geq 0}$ be an infinite sequence over a finite alphabet. Suppose $\mathbf{u}$ is k-automatic and k'-automatic, for integers $k, k' \geq 2$. Show that $\mathbf{u}$ is kk'-automatic. (Note: A much stronger result is proved in Chapter 11.)

11. Prove Theorem 6.3.5. Hint: Use a suitable generalization of Corollary 4.1.9.

12. Let $\mathbf{t} = (t_n)_{n \geq 0}$ be the Thue–Morse sequence. We have seen in Example 6.6.1 that its 2-kernel consists of two sequences, $\mathbf{t}$ and $\bar{\mathbf{t}}$. Show that as $k \to \infty$, the size of the 2-kernel of the shifted sequence $\mathcal{S}^k(\mathbf{t})$ goes to ∞.

13. Let $\mathbf{R} = (R_n)_{n \geq 1}$ be the regular paperfolding sequence.
    (a) Show that $\sum_{1 \leq k \leq n} R_k$ is always positive.
    (b) Show the converse: if a paperfolding sequence $(v_k)_{k \geq 1}$ has the property that $\sum_{1 \leq k \leq n} v_k > 0$ for all $n \geq 1$, then $(v_k)_{k \geq 1}$ is the regular paperfolding sequence.

14. Consider the paperfolding curves formed by iterated folding, followed by unfolding to 90°. Show that the resulting curves are always self-avoiding; that is, they may meet at a point, but the curve never crosses itself.

15. Let $\mathbf{x} = \mathrm{Fold}(1, 2, 3, 4, \ldots) = (12\bar{1}31\bar{2}1\bar{4} \ldots)$, where by $\bar{x}$ we mean $-x$. Show that this sequence is the fixed point of the map

$$
\begin{aligned}
1 &\to 1\,2, & \bar{1} &\to 1\,\bar{2}, \\
2 &\to \bar{1}\,3, & \bar{2} &\to \bar{1}\,\bar{3}, \\
3 &\to \bar{1}\,4, & \bar{3} &\to \bar{1}\,\bar{4}, \\
4 &\to \bar{1}\,5, & \bar{4} &\to \bar{1}\,\bar{5}, \\
&\;\vdots & &\;\vdots
\end{aligned}
$$

16. Let $(f_i)_{i \geq 1}$ be any paperfolding sequence with $f_1 = 1$. Show that f is multiplicative; that is, $f_{mn} = f_m f_n$ for all $m, n \geq 1$ with $\gcd(m, n) = 1$.

17. (a) Suppose we are looking for a partition of the set

$$
\{0, 1, \ldots, 2^n - 1\}
$$

into two disjoint sets S and T such that $\sum_{s \in S} s^i = \sum_{t \in T} t^i$ for $0 \le i < n$. Explain how this problem is related to the following: find a univariate polynomial $p_n(X)$ of degree $2^n - 1$ and coefficients in $\{-1, +1\}$ such that $(X - 1)^n$ divides $p_n(X)$.

(b) Prove or disprove: there exists exactly one way to partition the set

$$\{0, 1, \ldots, 2^n - 1\}$$

into two disjoint sets, S and T, such that $\sum_{s \in S} s^i = \sum_{t \in T} t^i$ for $0 \le i < n$.

18. Show that a paperfolding sequence is k-automatic for some $k \ge 2$ if and only if it is 2-automatic.

19. (a) Let $I_n = \int_0^{\pi/2} (\sin x)^n \, dx$. Using integration by parts, prove that $I_{n+2} = \frac{n+1}{n+2} I_n$. Deduce that $(n + 2)I_{n+2}I_{n+1} = (n + 1)I_{n+1}I_n = \cdots = \frac{\pi}{2}$.

(b) Noting that the sequence $(I_n)_{n \ge 0}$ is nonincreasing, prove that $I_{n+2}/I_n \le I_{n+1}/I_n \le 1$. Deduce that $I_{n+1} \sim I_n$, and that $I_n \sim \sqrt{\pi/2n}$.

(c) Using the recurrence above, prove that $I_{2n} = (\pi/2)1 \cdot 3 \cdots (2n - 1)/2 \cdot 4 \cdots 2n$. Deduce that

$$\prod_{n \ge 1} \frac{(2n - 1)(2n + 1)}{(2n)^2} = \frac{4}{\pi}.$$

(d) Let $\mathbf{t} = (t_n)_{n \ge 0} = 01101001 \cdots$ be the Thue–Morse sequence. Prove that

$$\prod_{n \ge 0} \left(\frac{2n + 1}{2n + 2} \right)^{2t_n} \left(\frac{2n + 3}{2n + 2} \right) = \frac{4\sqrt{2}}{\pi},$$

$$\prod_{n \ge 0} \left(\frac{2n + 1}{2n + 2} \right)^{2(1 - t_n)} \left(\frac{2n + 3}{2n + 2} \right) = \frac{2\sqrt{2}}{\pi}.$$

20. With any paperfolding sequence $(v_k)_{k \ge 1}$ we can associate the real number $\sum_{k \ge 1} v_k 2^{-k}$. Show that the set of real numbers specified by all paperfolding sequences is of Lebesgue measure 0.

21. Let $\mathbf{w} = a_0 a_1 a_2 \cdots$ be an infinite word. Define its prefix language Pref($\mathbf{w}$) to be the set $\{a_0 a_1 \cdots a_i : i \ge 0\}$.

(a) Let h be a morphism that is prolongable on a, and let $\mathbf{w} = h^\omega(a)$. Show that the prefix language Pref($\mathbf{w}$) is a co-CFL, that is, $\overline{\text{Pref}(\mathbf{w})}$ is a CFL.

(b) Give an explicit example of an infinite word $\mathbf{w}$ such that Pref($\mathbf{w}$) is not a co-CFL.

22. Define

$$g(n) = \begin{cases} g(3n + 1) + 1 & \text{if } n > 1 \text{ is odd}, \\ g(n/2) & \text{if } n \text{ is even}, \\ 0 & \text{if } n = 1. \end{cases}$$

The *Collatz conjecture* states that g is well defined for each $n \ge 1$. For $i \ge 0$ define $S_i = \{k \ge 1 : g(k) = i\}$. Show that for each $i \ge 0$, S_i is a 2-automatic set.

23. Show that the class of k-automatic sequences is not closed under arbitrary finite-state transductions.

24. Show that the Rudin–Shapiro sequence is not the fixed point of any nontrivial morphism.

25. Let $(r_n)_{n\geq 0}$ be the Rudin–Shapiro sequence. Show that the sequence $(v_n)_{n\geq 0}$ defined by $v_i = [r_{2i}, r_{2i+1}]$ is the fixed point of a 2-uniform morphism.

26. The proof of Lemma 6.9.1 used the fact that the number B_n of distinct equivalence relations on a finite set is finite. In this exercise we explore the properties of this sequence.

    (a) Show B_n is the total number of ways to distribute n labeled balls onto labeled plates, such that each plate contains at least one ball. The numbers B_k are sometimes called *Bell numbers*.
    (b) Show that $B_1 = 1$, $B_2 = 2$, $B_3 = 5$, $B_4 = 15$, and $B_5 = 52$.
    (c) Show that if

    $$e^{e^X - 1} = \sum_{n \geq 0} \frac{a_n X^n}{n!}$$

    then $a_n = B_n$.
    (d) Define $B_0 = 1$. Show that

    $$B_{n+1} = \sum_{0 \leq i \leq n} \binom{n}{i} B_i$$

    for all $n \geq 0$.
    (e) Show that

    $$\sum_{n \geq 0} \frac{n^k}{n!} = B_k \cdot e.$$

    (f) Show that B_k can be efficiently computed using the *Bell triangle*:

$$
\begin{array}{ccccccc}
1 & & & & & \\
1 & 2 & & & & \\
2 & 3 & 5 & & & \\
5 & 7 & 10 & 15 & & \\
15 & 20 & 27 & 37 & 52 & \\
52 & 67 & 87 & 114 & 151 & 203
\end{array}
$$

    Determine the rule behind this triangle, and prove that it works.
    (g) Show that the sequence $(v_2(B_k))_{k \geq 0}$ is periodic with period 12.

27. (von Haeseler, Allouche) Let $(R_n)_{n \geq 1}$ be the regular paperfolding sequence. Let s be a complex number such that $\Re(s) > 0$. Prove that

$$\sum_{n \geq 1} \frac{R_n}{n^s} = \frac{2^s}{2^s - 1} \sum_{n \geq 0} \frac{(-1)^n}{(2n + 1)^s}.$$

Deduce as a special case that

$$\sum_{n \geq 1} \frac{R_n}{n} = \frac{\pi}{2}.$$

28. Let $(R_n)_{n \geq 1}$ denote the regular paperfolding sequence. Define $g_n = \sum_{1 \leq i \leq n} R_i$. Show that $g_n = e_{2;10}(n) + e_{2;01}(n)$. Conclude that $g_n = O(\log n)$.

29. Let $(R_n)_{n \geq 1}$ denote the regular paperfolding sequence. Find the expansion of R_n in terms of pattern functions (see Section 3.3).

30. The generalized Arshon word of order n is an infinite word over the n-letter alphabet $\Sigma_n = \{0, 1, \ldots, n - 1\}$ that is generated as follows: for a single letter $i \in \Sigma_n$ define

$$\alpha_0(i) = (i, i + 1, i + 2, \ldots, n - 2, n - 1, 0, 1, \ldots, i - 1)$$

and

$$\alpha_1(i) = \alpha_0(i)^R,$$

where the entries are taken mod n. Given a string $w = c_0 c_1 c_2 \cdots c_{r-1}$ we define

$$\alpha(w) = \prod_{0 \leq i < r} \alpha_{i \bmod 2}(c_i).$$

(Note that α is not a morphism, in general.) Finally, define $\mathbf{a}_n = \alpha^\omega(0)$.

(a) Show that $\mathbf{a}_n$ is n-automatic.

(b) Show that $\mathbf{a}_n$ is pure morphic (i.e., the fixed point of a morphism without coding) if n is even, but not pure morphic if n is odd.

(c) Show that $\mathbf{a}_3$ avoids $\frac{7}{4}$ powers.

31. (Gunturk) Define the running sum map $RS(x_1, x_2, \ldots, x_r)$ to be $(x_1, x_1 + x_2, \ldots, x_1 + x_2 + \cdots + x_r)$, and let $RS^t(x)$ be the map iterated t times. Prove or disprove: the first 2^n symbols of the Thue–Morse sequence on ± 1 achieve the minimum value of $|RS^n(x)|_\infty$, where $|x|_\infty$ is the L^∞ norm of the vector x, defined by $\max_{1 \leq i \leq r} |x_i|$.

32. Prove that Theorem 6.10.1 holds if we allow the coefficients of $p(X)$ to be rational numbers, provided $p(\mathbb{N}) \subseteq \mathbb{N}$.

33. Let $x \geq 1$ be a real number, and for $n \geq 1$ define $x_n = \lfloor x^n \rfloor - x \lfloor x^{n-1} \rfloor$. Show that $(x_n)_{n \geq 1}$ is ultimately periodic if and only if x is an integer.

34. Give a 4-uniform morphism h and coding τ such that $\tau(h^\omega(0))$ generates the characteristic sequence for those numbers that are the sums of three squares.

35. Show that the following number-theoretic functions are not k-automatic for any $k \geq 2$:

(a) $(\sigma(n) \bmod v)_{n \geq 0}$ for integers $v \geq 3$, where $\sigma(n)$ denotes the sum of the divisors of n;

(b) $(\varphi(n) \bmod v)_{n \geq 0}$ for integers $v \geq 3$, where $\varphi(n)$ denotes the Euler φ-function;

(c) $(\mu(n) \bmod v)_{n \geq 0}$ for integers $v \geq 3$, where $\mu(n)$ denotes the Möbius function.

36. The classical *van der Waerden theorem* asserts that for every mapping $\alpha : \mathbb{N} \to \Sigma$, where Σ is a finite alphabet, and every finite $S \subseteq \mathbb{N}$, there exist integers $a \geq 1$, $b \geq 0$ and a letter $t \in \Sigma$ such that $\alpha(as + b) = t$ for all $s \in S$. Consider the following variant on this: let S be a finite subset of $\mathbb{N}$, and let $k \geq 1$. We say the property $VW(S, k)$ holds if for every mapping $\alpha : \mathbb{N} \to \Sigma$ with $|\Sigma| = k$ there exist integers $a \geq 1$, $b \geq 0$ and a letter $t \in \Sigma$ such that $\alpha(a(s + b)) = t$ for all $s \in S$. Show that $VW(\{0, 1, 2\}, 2)$ does not hold. Hint: Consider the mapping $\alpha : i \to \mathbf{w}[i]$, where $\mathbf{w} = 001001101 \cdots$ is a fixed point of the morphism $h(0) = 001$ and $h(1) = 101$.

37. Prove that for all $i \geq 0$ we have

$$\varphi^{2i}(\mathbf{a}) = H_{2i}\, \mathbf{a},$$
$$\varphi^{2i+1}(\mathbf{a}) = H_{2i+1}\, \overline{\mathbf{c}},$$

where φ and H_j are defined as in Section 6.4.

38. Let $(a(n))_{n \geq 1}$ be a sequence over a finite alphabet. Let k be an integer ≥ 2. Then $(a(v_k(n)))_{n \geq 1}$ is k-automatic if and only if $(a(n))_{n \geq 1}$ is ultimately periodic.

39. Let $(a(n))_{n \geq 0}$ be a sequence over a finite alphabet. Let k be an integer ≥ 2. For $n \geq 0$ define $b(k^n) = a(n)$, and $b(i) = 0$ for i not a power of k. Show that $(a(n))_{n \geq 0}$ is ultimately periodic if and only if $(b(n))_{n \geq 0}$ is k-automatic.

40. Let $\mathbf{t} = (t_n)_{n \geq 0}$ be the Thue–Morse sequence. Define $Z(n)$ to be the number of real zeros of the polynomial $\sum_{0 \leq i < n} t_i X^i$. Show that $\lim_{n \to \infty} \frac{1}{n} \sum_{0 \leq t < n} Z(t) = \frac{11}{4}$.

41. Let $(d_i)_{i \geq 0}$ be the period-doubling sequence introduced in Example 6.3.4. Let $(a_2(n))_{n \geq 0}$ be the alternating sum of the binary digits, as discussed in Exercise 3.11. Show that $\sum_{0 \leq i < n} d_i = (n - a_2(n))/3$.

42. Define $g(n) = \sum_{0 \leq i < n} \binom{n}{i+1}\binom{2i}{i}/(i + 1)$. Show that $g(n) \equiv t_n \pmod{2}$, where $(t_n)_{n \geq 0}$ is the Thue–Morse sequence.

6.12 Open Problems

1. Decide whether or not the number

$$\prod_{n \geq 1} \left(\frac{2n}{2n + 1}\right)^{(-1)^{t_n}} \doteq 1.6281601297189$$

is algebraic.

2. Discuss the speed of convergence of

$$R(m) := \prod_{0 \leq n < m} \left(\frac{2n + 1}{2n + 2}\right)^{(-1)^{t_n}}$$

to $\sqrt{2}/2$. In particular, give a sharp estimate on $\sqrt{2}/2 - R(2^m)$ as a function of m. (Remark: From Allouche and Cohen [1985] a weak estimate can be derived.)

3. As we have seen in Lemma 5.2.6, automatic sequences may be defined in terms of k-fibers. This suggests a generalization to what we might call k-context-free sequences. We say a sequence $\mathbf{a} = (a_n)_{n\geq 0}$ taking values in Δ is k-context-free if for all $a \in \Delta$ the k-fiber $I_k(\mathbf{a}, a)$ is a context-free language. Explore the properties of these sequences, particularly with regard to subword complexity.

4. Find a simple closed form for the determinants of the Hankel matrices given by $M_n = (m_{i,j})_{0\leq i,j<n}$, where $m_{i,j} = t_{i+j}$ and $t_0 t_1 t_2 \cdots$ is the Thue–Morse sequence, and $M'_n = (m'_{i,j})_{0\leq i,j<n}$, where $m_{i,j} = s_2(i + j)$, the sum-of-bits function.

5. Prove or disprove: The Thue–Morse partition of Example 5.1.3 is the unique partition of $\{0, 1, \ldots, 2^n - 1\}$ for which $\sum_{s\in S} s^i = \sum_{t\in T} t^i$ for $0 \leq i < n$ and for which $\sum_{s\in S} s^n - \sum_{t\in T} t^n$ is minimized.

6. Show that the limit of the sequence specified by $x_1 = 12$, $x_{n+1} = x_n x_n x_n^R x_n^R 2$ is not 2-automatic. Is it k-automatic for any k? (Remark: The sequence is a fixed point of the map $1 \to 12122121, 2 \to 2$.)

7. Let $w \in \{0, 1\}^* \setminus 0^*$. Consider the morphism defined by $f(0) = 0w$, $f(1) = 1$. Give simple necessary and sufficient conditions for the fixed point of f starting with 0 to be a 2-automatic sequence.

8. Does the result of Exercise 21(a) above still hold if we define $\mathbf{w} = \tau(h^\omega(a))$ for some coding τ?

9. The *Liouville function* $\lambda(n)$ is defined as follows for integers $n \geq 1$: if the prime factorization of n is $p_1^{e_1} \cdots p_t^{e_t}$, then $\lambda(n) = (-1)^{e_1+\cdots+e_t}$. For which $k \geq 2$ is $(\lambda(n))_{n\geq 1}$ a k-automatic sequence? (Remark: See Yazdani [2001].)

6.13 Notes on Chapter 6

6.1 Morphisms are sometimes called "substitutions" in the literature, especially the ergodic-theory literature. We have avoided this terminology in this book because "substitution" has a different meaning in the computer science literature; namely, a substitution is a language-valued homomorphism. In the physics literature morphisms are often called "inflation rules".

6.2 The Thue–Morse sequence was introduced by Thue [1912] and Morse [1921] independently; it appears implicitly in an earlier paper of Prouhet [1851]. For a translation of Thue's work into English, see Berstel [1995].

   The Thue–Morse sequence appears in many areas of mathematics, such as construction of magic squares (Adler and Li [1977]); bifurcation theory (Parry [1981]); partially ordered sets (Trotter and Winkler [1987]); hook numbers (Grassl and Mullhaupt [1989]); the maximum of the Knopp function (Dubuc and Elqortobi [1990]); and group theory (Boffa and Point [1991, 1992]; Wantiez [1994, 1995]).

   For generalizations of the Thue–Morse sequence, see Keane [1968]; J. Martin [1976, 1977]; Nürnberg [1983]; Černý [1984]; Séébold [1986, 2000]; and Yao [1997b].

Allouche, Peyrière, Wen, and Wen [1998] studied Hankel determinants formed from the terms of the Thue–Morse sequence. (For other work on Hankel determinants, see Tamura [1999]; Kamae, Tamura, and Wen [1999].)

Allouche and Shallit [1999] and Mauduit [2001] are surveys about the Thue–Morse sequence.

For other papers on the Thue–Morse sequence, see Pirillo [1991]; Dekking [1992]; Ferenczi [1992]; Fredricksen [1992].

6.3 For Cobham's theorem, see Cobham [1972].

For the Baum–Sweet sequence, see Baum and Sweet [1976].

Properties of the period-doubling sequence were studied by Damanik [2000]. The name "period-doubling sequence" is derived from the relationship of this sequence to the "period-doubling scenario" (also called the Feigenbaum cascade). When iterating the function $f_\lambda : [0, 1] \to \mathbb{R}$, where $(f_\lambda)_{\lambda \in \Lambda}$ is a family of continuous unimodal (i.e., increasing then decreasing) functions, the following universal scenario occurs for many families of such functions – including the "typical" family $f_\lambda(x) := \lambda x(1 - x)$: there exist successive intervals for the parameter λ such that f_λ has first an attractive fixed point (i.e., a point p_λ such that $f(p_\lambda) = p_\lambda$, and $\lim_{n \to \infty} f_\lambda^{(n)}(x) = p_\lambda$ for all $x \in (0, 1)$), then an attractive 2-cycle (i.e., two points p_λ and q_λ such that $f_\lambda(p_\lambda) = q_\lambda$ and $f_\lambda^{(2)}(p_\lambda) = p_\lambda$, and the limit set of the sequence $f_\lambda^{(n)}(x)$ is the set $\{p_\lambda, q_\lambda\}$), then an attractive 4-cycle, then an attractive 8-cycle, and so on. Now a binary coding of the orbit $f_\lambda^{(n)}(x_\lambda)$ (where x_λ is the point where f_λ takes its maximum) gives a periodic sequence that tends to the period-doubling sequence as λ increases. This phenomenon was discovered by Feigenbaum [1978]. Also see Collet and Eckmann [1980]; Peitgen, Jürgens, and Saupe [1992, Chap. 11].

6.4 This section is taken largely verbatim from the article of Allouche, Astoorian, Randall, and Shallit [1994], with permission.

The tower of Hanoi puzzle was invented by Lucas [1884]. The results in this section first appeared in Allouche, Bétréma, and Shallit [1989] and Allouche and Dress [1990].

For other papers on the tower of Hanoi, see de Parville [1884]; Scorer, Grundy, and Smith [1944]; Crowe [1956]; Hayes [1977]; Buneman and Levy [1980]; Wood [1981, 1983]; Walsh [1982, 1983b, 1998]; Er [1982, 1983a, 1983b, 1984a, 1985b, 1985c, 1986b, 1987b, 1987c, 1989]; Leiss [1983a, 1983b, 1984]; Lavallée [1985]; Pettorossi [1985]; Cull and Ecklund [1985]; Cull and Gerety [1985]; Gault and Clint [1987]; Chan [1989]; Hinz [1989a, 1992a, 1992b, 1999]; Fournier [1990]; Noland [1990]; Minsker [1991]; J. Wu and Chen [1992]; Poole [1992, 1994]; C. Klein and Minsker [1993]; Lu and Dillon [1994]; Ray and Majumdar [1995a].

For generalizations to $n > 3$ pegs, see Dudeney [1908]; B. Stewart [1939]; T. Roth [1974]; Brousseau [1976]; Cull and Ecklund [1982]; Bendisch [1985]; Boardman, Garrett, and Robson [1986]; Rohl and Gedeon [1986]; Lunnon

[1986]; Er [1987a]; Lu [1988, 1989]; Hinz [1989b]; I. Chu and Johnsonbaugh [1991]; Gedeon [1992]; van de Liefvoort [1992]; Stockmeyer [1994]; Majumdar [1994a, 1994b, 1995, 1996]; Lu and Dillon [1995, 1996]; Mallol, López, and Serrato [1996]; Klavžar and Milutinović [1997]; Sarkar [2000].

In 1975, Knuth (unpublished) suggested a variant of the tower of Hanoi problem in which disks are permitted to move only in one cyclic direction; see Lunnon [1986]. It was first published by Atkinson [1981]. For more on this variant, see Walsh [1983a]; Er [1984b, 1984c, 1984d, 1985a, 1986a, 1986c]; J. Wu and Chen [1993]; Allouche [1994b]; Ray and Majumdar [1995b].

6.5　The paperfolding sequence was apparently first discovered by J. E. Heighway around 1966; see Gardner [1967b]. Davis and Knuth [1970] gave a detailed study of these sequences. The three-part survey of Dekking, Mendès France, and van der Poorten [1982] provides a delightful introduction to the topic.

For more on paperfolding, see Mendès France [1981, 1984c, 1989, 1990a]; Mendès France and Tenenbaum [1981]; Mendès France and van der Poorten [1981]; Blanchard and Mendès France [1982]; Dekking, Mendès France, and van der Poorten [1982]; and Bercoff [1995, 1997].

The folding lemma is from van der Poorten and Shallit [1992]. Earlier versions include Mendès France [1973a]; Shallit [1979, 1982a, 1982b]; Kmošek [1979]; and Dekking, Mendès France, and van der Poorten [1982]. The papers of Leighton and Scott [1939] and Scott and Wall [1940] contain similar results, but in a disguised form. The connection between paperfolding and continued fractions was noticed first by Mendès France [1981]. The convergents of folded continued fractions were studied by Allouche, Lubiw, Mendès France, van der Poorten, and Shallit [1996] and Mendès France, van der Poorten, and Shallit [1999].

For other interesting papers on continued fractions of formal power series, see Baum and Sweet [1976]; Mills and Robbins [1986]; Buck and Robbins [1995]; Lasjaunias [1997, 1999, 2000a, 2000b]; Thakur [1999]; W. Schmidt [2000]. Mkaouar [1995] and Yao [1997a] showed that the partial quotients of the continued fraction for the Baum–Sweet power series do not form an automatic sequence.

Mendès France and Shallit [1989] considered the analogue of paperfolding sequences in three dimensions.

Davis and Knuth [1970] discussed paperfolding curves resulting from folding into three pieces or more at each step, instead of folding in half. Other papers along these lines include Razafy Andriamampianina [1989, 1992, 1996]; Bercoff [1996]; Koskas [1996].

Prodinger and Urbanek [1979] showed that the regular paperfolding sequence contains no squares w^2 with $|w| > 5$; hence it is of bounded repetition. Allouche [1984] observed that in fact all paperfolding sequences have this property; for a detailed proof see Allouche and Bousquet-Mélou [1994b].

Allouche [1992] discussed the subword complexity of paperfolding sequences. For other aspects of this topic, see Allouche and Bousquet-Mélou [1994a].

6.6 Theorem 6.6.2 is due to Eilenberg [1974, Prop. V.3.3] and Christol [1979].

Theorem 6.6.4 is due to Eilenberg [1974, Prop. V.3.5].

The term "k-kernel" is due to O. Salon.

6.7 The results in this section can be found, in a much more general setting, in Allouche, Cateland, Gilbert, Peitgen, Shallit, and Skordev [1997].

6.8 Theorem 6.8.1 is due to Cobham [1972, p. 174]. Our proof is based on that in Allouche and Shallit [1992]; also see Allouche [1982].

6.9 Theorem 6.9.2 is due to Cobham [1972].

6.10 Theorem 6.10.1 is essentially due to Allouche [1982], who proved it for the case $k = p$, a prime number. For a generalization, see Allouche and Salon [1993].

7

Morphic Sequences

In Chapter 6 we began our study of sequences that are fixed points of morphisms. There we concentrated on uniform morphisms, where each letter is sent to a word of the same length. In this chapter, we study the more general case where the morphism need not be uniform.

Recall from Section 1.4 that if there exists an integer $j \geq 1$ such that $h^j(a) = \epsilon$, then the letter a is said to be *mortal*, and the set of mortal letters associated with a morphism h is denoted by M_h. Recall that the mortality exponent of a morphism h is defined to be the least integer $t \geq 0$ such that $h^t(a) = \epsilon$ for all $a \in M_h$; we write the mortality exponent as $\exp(h) = t$. Also recall the notion of morphic sequence from Section 1.4: let $h : \Sigma^* \to \Sigma^*$ be a morphism, and suppose there exists a letter $a \in \Sigma$ with $h(a) = ax$, $x \notin M_h^*$. Then h is *prolongable* on a, and the sequence of words $a, h(a), h^2(a) \ldots$ converges, in the limit, to the infinite word

$$h^\omega(a) := a \, x \, h(x) \, h^2(x) \cdots,$$

which is a fixed point of h, that is, $h(h^\omega(a)) = h^\omega(a)$. If $\mathbf{w} = h^\omega(a)$, then $\mathbf{w}$ is a *pure morphic sequence*. If there is a coding $\tau : \Sigma \to \Delta$ and $\mathbf{w} = \tau(h^\omega(a))$, then $\mathbf{w}$ is a *morphic sequence*.

In order to build intuition, the next section explores some of the basic properties of one of the most famous morphic sequences, the infinite Fibonacci word.

7.1 The Infinite Fibonacci Word

Let φ be the morphism that maps $0 \to 01$, $1 \to 0$; we call φ the *Fibonacci morphism*. One of the simplest examples of a pure morphic sequence is the infinite Fibonacci word

$$\mathbf{f} := \varphi^\omega(0) = 01001010010010100101 0 \cdots,$$

which is the infinite fixed point of φ. In this section, we explore some of the basic properties of this word.

First, we observe that **f** can be computed as the limit of a certain sequence of words.

Theorem 7.1.1 *Define* $\Phi_1 = 1$, $\Phi_2 = 0$, *and* $\Phi_n = \Phi_{n-1}\Phi_{n-2}$ *for* $n \geq 3$. *Then* $\varphi^n(1) = \Phi_{n+1}$ *and* $\varphi^n(0) = \Phi_{n+2}$ *for* $n \geq 0$. *Hence* **f** $= \lim_{n \to \infty} \Phi_n$. *Furthermore, for* $n \geq 1$ *we have* $|\Phi_n| = F_n$, *the* n *th Fibonacci number.*

The words Φ_n are sometimes called the (finite) Fibonacci words.

Proof. We prove the two assertions about φ^n by induction on n. They are clearly true for $n = 0, 1$. Now assume they are true for all $m < n$; we prove them for $m = n$:

$$\varphi^n(1) = \varphi^{n-1}(0) = \Phi_{n+1},$$
$$\varphi^n(0) = \varphi^{n-1}(01) = \varphi^{n-1}(0)\varphi^{n-1}(1)$$
$$= \Phi_{n+1}\Phi_n = \Phi_{n+2}.$$

It now follows that $\mathbf{f} = \lim_{n \to \infty} \Phi_n$.

The assertion about $|\Phi_n|$ also follows easily by induction. ∎

One of the most useful features of the finite Fibonacci words is the following "almost commutative" property:

Theorem 7.1.2 *For any string* $w \in \{0, 1\}^*$ *with* $|w| \geq 2$, *define the map* $c(w)$ *that interchanges the last two characters of* w. *Then* $\Phi_n\Phi_{n+1} = c(\Phi_{n+1}\Phi_n)$ *for all* $n \geq 1$.

Proof. By induction on n. The base cases $n = 1, 2$ are left to the reader. Assume the result is true for all $m \leq n$; we prove it for $m = n + 1$.

We have

$$\Phi_{n+1}\Phi_{n+2} = \Phi_{n+1}(\Phi_{n+1}\Phi_n) = \Phi_{n+1}c(\Phi_n\Phi_{n+1})$$
$$= (\Phi_{n+1}\Phi_n)c(\Phi_{n+1}) = \Phi_{n+2}c(\Phi_{n+1}) = c(\Phi_{n+2}\Phi_{n+1}),$$

where we have used the fact that $c(xy) = xc(y)$ for $|y| \geq 2$. ∎

7.2 Finite Fixed Points

As we have seen above, a morphic sequence is the image, under a coding, of a word $h^\omega(a)$ that is the fixed point of a morphism. This raises the natural question, are there fixed points of morphisms *not* of the form $h^\omega(a)$? In this section we characterize the finite fixed points of a morphism.

Lemma 7.2.1 *Let $h : \Sigma^* \to \Sigma^*$ be a morphism. Let $w \in \Sigma^+$ be a finite nonempty word such that w is a subword of $h(w)$. Then there exists a letter $a \in \Sigma$ occurring in w such that a occurs in $h(a)$.*

Proof. Let $w = c_1 c_2 \cdots c_n$, where $c_i \in \Sigma$ for $1 \leq i \leq n$. For $0 \leq i \leq n$ define $s_w(i) = |h(c_1 c_2 \cdots c_i)|$. (If the context is clear, we omit the subscript on s.) In particular, $s(0) = 0$.

Let $h(w) = d_1 d_2 \cdots d_{s(n)}$, where $d_i \in \Sigma$ for $1 \leq i \leq s(n)$. Hence

$$h(c_i) = d_{s(i-1)+1} \cdots d_{s(i)}$$

for $1 \leq i \leq n$. Since w is a subword of $h(w)$, we know there must exist an integer $t, 0 \leq t \leq s(n) - n$, such that $w = d_{t+1} \cdots d_{t+n}$. Hence $c_i = d_{t+i}$ for $1 \leq i \leq n$.

Consider the least index $j \geq 1$ for which $s(j) \geq t + j$. Such an index must exist, since the inequality holds for $j = n$. There are now two cases to consider.

Case 1: $j = 1$: Then $h(c_1) = d_1 d_2 \cdots d_{s(1)}$ and $s(1) \geq t + 1$. Hence $h(c_1)$ contains $d_{t+1} = c_1$. Let $a = c_1$.

Case 2: $j > 1$: Then by the definition of j we must have $s(j-1) < t + j - 1$. Hence $s(j-1) + 1 < t + j \leq s(j)$, and since $h(c_j) = d_{s(j-1)+1} \cdots d_{s(j)}$, we know $h(c_j)$ contains $d_{t+j-1} d_{t+j} = c_{j-1} c_j$ as a subword. Let $a = c_j$. ∎

Corollary 7.2.2 *If $w \in \Sigma^+$ is a nonempty finite word with $h(w) = w$, then there exist words $w_1, w_2, w_3, w_4 \in \Sigma^*$ and a letter $a \in \Sigma$ such that $w = w_1 w_2 a w_3 w_4$, $h(w_1 w_2) = w_1$, $h(a) = w_2 a w_3$, and $h(w_3 w_4) = w_4$.*

Proof. If $h(w) = w$, then, using the notation in the proof of Lemma 7.2.1, we have $t = 0$ and $s(n) = n$. Define

$$w_1 := d_1 \cdots d_{s(j-1)},$$
$$w_2 := d_{s(j-1)+1} \cdots d_{j-1},$$
$$a := d_j,$$
$$w_3 := d_{j+1} \cdots d_{s(j)},$$
$$w_4 := d_{s(j)+1} \cdots d_n.$$

The reader can now easily verify that $h(w_1 w_2) = w_1$, $h(a) = w_2 a w_3$, and $h(w_3 w_4) = w_4$. ∎

Now define

$$A_h = \{a \in \Sigma : \exists\, x, y \in M_h^* \text{ such that } h(a) = xay\}$$

and

$$F_h = \{h^t(a) : a \in A_h \text{ and } t = \exp(h)\}.$$

Note that there is at most one way to write $h(a)$ in the form xay with $x, y \in M_h^*$.

Theorem 7.2.3 *Let $h : \Sigma^* \to \Sigma^*$ be a morphism. Then a finite word $w \in \Sigma^*$ has the property that $w = h(w)$ if and only if $w \in F_h^*$.*

Proof. $\Longleftarrow$: Suppose $w \in F_h^*$. Then we can write $w = w_1 w_2 \cdots w_r$, where each $w_i \in \Sigma^*$, and there exist letters $a_1, a_2, \ldots, a_r \in A_h$ such that $w_i = h^t(a_i)$, with $t = \exp(h)$.

Since $a_i \in A_h$, we know that there exist x_i, y_i with $x_i, y_i \in M_h^*$ such that $h(a_i) = x_i a_i y_i$. Since $t = \exp(h)$, we have $h^t(x_i) = h^t(y_i) = \epsilon$. Hence

$$h^{t+1}(a_i) = h^t(x_i)\, h^t(a_i)\, h^t(y_i) = h^t(a_i).$$

Thus $h(w_i) = w_i$ for $1 \le i \le r$, and so $h(w) = w$.

$\Longrightarrow$: We prove the result by contradiction. Suppose $h(w) = w$, and assume w is the shortest such word with $w \notin F_h^*$. Clearly $w \neq \epsilon$.

By Corollary 7.2.2 there exist w_1, w_2, w_3, w_4, a such that $w = w_1 w_2 a w_3 w_4$, $h(w_1 w_2) = w_1$, $h(a) = w_2 a w_3$, and $h(w_3 w_4) = w_4$.

Now a is a subword of w, so $h(a)$ is a subword of $h(w) = w$, and hence by an easy induction, it follows that

$$h^i(a) \text{ is a subword of } w \text{ for all } i \ge 0. \tag{7.1}$$

Then we must have $w_2 w_3 \in M_h^*$, since otherwise the length of

$$h^i(a) = h^{i-1}(w_2) \cdots h(w_2)\, w_2\, a\, w_3\, h(w_3) \cdots h^{i-1}(w_3)$$

would grow without bound as $i \to \infty$, contradicting (7.1). It follows that $h^t(w_2 w_3) = \epsilon$, where $t = \exp(h)$.

Now we have $w_1 = h(w_1 w_2)$, so by applying h^t to both sides, we see

$$h^t(w_1) = h^{t+1}(w_1 w_2) = h^{t+1}(w_1)\, h^{t+1}(w_2) = h^{t+1}(w_1).$$

Hence, defining $y_1 = h^t(w_1)$, we have $h(y_1) = y_1$. In a similar fashion, if we set $y_2 = h^t(w_4)$, then $h(y_2) = y_2$. Since $|y_1|, |y_2| < |w|$, it follows by the minimality of w that $y_1, y_2 \in F_h^*$. Now

$$w = h^t(w) = h^t(w_1)\, h^t(w_2)\, h^t(a)\, h^t(w_3)\, h^t(w_4) = y_1\, h^t(a)\, y_2,$$

and hence $w \in F_h^*$, a contradiction. ∎

7.3 Morphic Sequences and Infinite Fixed Points

Let $\mathbf{w} = c_1 c_2 c_3 \cdots$ be a one-sided right-infinite word over Σ, and let h be a morphism. In this section we characterize those $\mathbf{w}$ for which $h(\mathbf{w}) = \mathbf{w}$.

Recall that Σ^ω denotes the set of all right-infinite words over the alphabet Σ. If $\mathbf{w} = a_1 a_2 a_3 \cdots$, then $h(\mathbf{w}) = h(a_1) h(a_2) h(a_3) \cdots$. If $L \subseteq \Sigma^*$ is a language, then we define

$$L^\omega = \{w_1 w_2 w_3 \cdots : w_i \in L \setminus \{\epsilon\} \text{ for all } i \geq 1\}.$$

Perhaps slightly less obviously, we can also define the word $h^\omega(a)$ for a letter a, provided $h(a) = xay$ and $x \in M_h^*$. In this case, there exists $t \geq 0$ such that $h^t(x) = \epsilon$. Then we define

$$h^\omega(a) = h^{t-1}(x) \cdots h(x) \, x \, a \, y \, h(y) \, h^2(y) \cdots,$$

which is infinite if and only if $y \notin M_h^*$.

Theorem 7.3.1 *The right-infinite word* **w** *is a fixed point of h if and only if at least one of the following two conditions holds:*

(a) **w** $\in F_h^\omega$; *or*
(b) **w** $\in F_h^* h^\omega(a)$ *for some $a \in \Sigma$, and there exist $x \in M_h^*$ and $y \notin M_h^*$ such that $h(a) = xay$.*

Note that there is at most one way to write $h(a) = xay$ with $x \in M_h^*$ and $y \notin M_h^*$.

Proof. $\Longleftarrow$: First, suppose condition (a) holds. Then we can write **w** $= w_1 w_2 w_3 \cdots$, where each $w_i \in F_h$. Then by Theorem 7.2.3 we have $h(w_i) = w_i$. It follows that $h(\mathbf{w}) = \mathbf{w}$.

Second, suppose condition (b) holds. Then we can write **w** $= v\,\mathbf{z}$, where $v \in F_h^*$ and $\mathbf{z} = h^\omega(a)$, where $h(a) = xay$ for some $x \in M_h^*$, $y \notin M_h^*$. Then from Theorem 7.2.3, we have $h(v) = v$.

Since $x \in M_h^*$, we have $h^t(x) = \epsilon$ where $t = \exp(h)$, and hence

$$\mathbf{z} = h^\omega(a) = h^{t-1}(x) \cdots h(x) \, x \, a \, y \, h(y) \, h^2(y) \, h^3(y) \cdots.$$

Since $y \notin M_h^*$, it follows that $|h^i(y)| \geq 1$ for all $i \geq 0$, and hence **z** is indeed an infinite word. We then have

$$h(\mathbf{z}) = h^t(x) \cdots h(x) \, x \, a \, y \, h(y) \, h^2(y) \, h^3(y) \cdots = \mathbf{z}$$

and so $h(\mathbf{w}) = h(v\mathbf{z}) = v\mathbf{z} = \mathbf{w}$.

$\Longrightarrow$: Now suppose **w** $= c_1 c_2 c_3 \cdots$ is an infinite word, with $c_i \in \Sigma$ for $i \geq 1$, and $h(\mathbf{w}) = \mathbf{w}$. As before, we define $s_\mathbf{w}(i) = |h(c_1 c_2 \cdots c_i)|$ for $i \geq 0$, and if the context is clear, we omit the subscript on s. There are several cases to consider.

Case 1: $s_\mathbf{w}(i) = i$ for infinitely many integers $i \geq 1$. Suppose $s(i) = i$ for $i = i_0, i_1, i_2, \ldots$. Clearly we may take $i_0 = 0$. Then we can write

$$\mathbf{w} = y_1 y_2 y_3 \cdots,$$

where $y_j = c_{i_{j-1}+1} \cdots c_{i_j}$ and $h(y_j) = y_j$ for $j \geq 1$. It follows that **w** $\in F_h^\omega$.

Case 2: $s_{\mathbf{w}}(i) = i$ for finitely many $i \geq 1$, and at least one such i. Let $i_0 = 0, i_1, \ldots, i_r, r \geq 1$, be all the indices i for which $s(i) = i$. Then we can write

$$\mathbf{w} = y_1 y_2 y_3 \cdots y_r \, \mathbf{x},$$

where $y_j = c_{i_{j-1}+1} \cdots c_{i_j}$, and $h(y_j) = y_j$ for $1 \leq j \leq r$, and $h(\mathbf{x}) = \mathbf{x}$. Furthermore, if we write $\mathbf{x} = d_1 d_2 d_3 \cdots$ for $d_i \in \Sigma, i \geq 1$, then

$$s(i) \neq i \qquad \text{for all } i \geq 1. \tag{7.2}$$

If we can show that (7.2) implies that $\mathbf{x} = h^{\omega}(a)$, where $h(a) = xay$ for some $x \in M_h^*, y \notin M_h^*$, we will be done. This leads to case 3.

Case 3: $s_{\mathbf{w}}(i) \neq i$ for all $i \geq 1$. Suppose there exist i, j with $1 \leq i < j$ and

$$s(i) > i \quad \text{but} \quad s(j) < j. \tag{7.3}$$

Among all pairs (i, j) with $1 \leq i < j$ satisfying (7.3), choose one with $j - i$ minimal. Suppose there exists an integer k with $i < k < j$. If $s(k) < k$, then (i, k) has a smaller difference, while if $s(k) > k$, then (k, j) has a smaller difference. It follows that $j = i + 1$. Then $s(i) > i$, but $s(i + 1) < i + 1$, a contradiction, since $s(i) \leq s(i + 1)$. It follows that either (a) $s(i) < i$ for all $i \geq 1$, or (b) there exists an integer $r \geq 1$ such that $s(i) < i$ for $1 \leq i < r$ and $s(i) > i$ for all $i \geq r$.

Case 3a: $s_{\mathbf{w}}(i) < i$ for all $i \geq 1$. Since this is true for $i = j_0 := 1$, in particular we see that $h(c_1) = \epsilon$. Now let j_1 be the least index such that

$$h(c_{j_1}) \text{ contains } c_1; \tag{7.4}$$

such an index must exist, since $h(\mathbf{w}) = \mathbf{w}$. We then have $h(c_2) = h(c_3) = \cdots = h(c_{j_1-1}) = \epsilon$, so the first occurrence of c_{j_1} in $\mathbf{w}$ is at position j_1.

Now inductively assume that we have constructed a strictly increasing sequence $j_0 < j_1 < \cdots < j_t$ such that the first occurrence of c_{j_i} in $\mathbf{w}$ is at position j_i, for $1 \leq i \leq t$. Let j_{t+1} be the least index such that $h(c_{j_{t+1}})$ contains c_{j_t}. Assume $j_t \geq j_{t+1}$. Since $s(i) < i$ for all i, we have $h(c_{j_{t+1}}) = c_k \cdots c_l$ with $l < j_{t+1} \leq j_t$. Since $h(c_{j_{t+1}})$ contains c_{j_t}, this implies that c_{j_t} occurs to the left of position j_t, a contradiction. Hence $j_t < j_{t+1}$.

Thus we can construct an infinite strictly increasing sequence $j_0 < j_1 < \cdots$ such that the first occurrence of c_{j_i} in $\mathbf{w}$ is at position j_i. It follows that the letters $c_{j_0}, c_{j_1}, \ldots$ in Σ are all distinct. But Σ is finite, a contradiction. Hence this case cannot occur.

Case 3b: There exists an integer $r \geq 1$ such that

$$s_{\mathbf{w}}(i) < i \quad \text{for } 1 \leq i < r \quad \text{and} \quad s_{\mathbf{w}}(i) > i \quad \text{for all } i \geq r. \tag{7.5}$$

Put $a = c_r$. Then $h(a) = xay$ for some $x, y \in \Sigma^*$, and $|y| \geq 1$. Furthermore, the conditions (7.5) on s imply that we can write $\mathbf{w} = u \, a \, \mathbf{v}$ and $h(\mathbf{w}) = h(u) \, x \, a \, y \, h(\mathbf{v})$

such that $u = h(u)x$. An easy induction now gives

$$h^i(\mathbf{w}) = h^i(u)\,h^{i-1}(x)\cdots h(x)\,x\,a\,y\,h(y)\cdots h^{i-1}(y)\,h^i(\mathbf{v}) \qquad (7.6)$$

and

$$u = h^i(u)\,h^{i-1}(x)\cdots h(x)\,x \qquad (7.7)$$

for all $i \geq 0$. Since $|u| < \infty$, it follows from letting $i \to \infty$ in Eq. (7.7) that there exists an integer $j \geq 0$ such that $h^j(x) = \epsilon$. Hence $x \in M_h^*$, and so $h^t(x) = \epsilon$, where $t = \exp(h)$.

Now $u = h(u)x$, so $h^t(u) = h^{t+1}(u)h^t(x) = h^{t+1}(u)$. Define $u' = h^t(u)$; then $h(u') = u'$. Hence, putting $j = |u'|$, it follows that $s(j) = j$. Hence $j = 0$ and $u' = \epsilon$.

Now, to get a contradiction, suppose that $y \in M_h^*$. Then $h^t(y) = \epsilon$. Define $z = h^t(a)$. Then

$$h(z) = h^{t+1}(a) = h^t(h(a)) = h^t(xay) = h^t(x)\,h^t(a)\,h^t(y) = h^t(a) = z.$$

Hence, putting $j = |z|$, we see that $s(j) = j$, a contradiction since $|z| \geq 1$. Hence $y \notin M_h^*$.

Now, letting $i \to \infty$ in (7.6), we see that $\mathbf{w} = h^\omega(a)$.   ∎

We can also characterize the left-infinite words that are fixed points of morphisms. Suppose $h(a) = xay$ with $x \notin M_h^*$ and $y \in M_h^*$. Then there exists $t \geq 0$ such that $h^t(x) = \epsilon$. We define

$$\overset{\leftarrow}{h^\omega}(a) = h^2(x)\,h(x)\,x\,a\,y\,h(y)\cdots h^{t-1}(y).$$

Theorem 7.3.2 *The left-infinite word $\mathbf{w}$ is a fixed point of h if and only if at least one of the following two conditions holds:*

(a) $\mathbf{w} \in {}^\omega F_h$; *or*

(b) $\mathbf{w} \in \overset{\leftarrow}{h^\omega}(a)F_h^*$ *for some $a \in \Sigma$, and there exist $x \notin M_h^*$ and $y \in M_h^*$ such that $h(a) = xay$.*

Proof. Exactly like the proof of Theorem 7.3.1.   ∎

7.4 Two-Sided Infinite Fixed Points

Recall $\Sigma^{\mathbb{Z}}$ denotes the set of all two-sided infinite words over the alphabet Σ, which are of the form $\cdots c_{-2}c_{-1}c_0.c_1c_2\cdots$. (We use a decimal point to the left of the character c_1, to indicate how the word is indexed.) Recall that S is the shift function, where

$$\mathcal{S}^k(\cdots c_{-2}c_{-1}c_0.c_1c_2c_3\cdots) = \cdots c_{k-1}c_k.c_{k+1}c_{k+2}\cdots$$

for all $k \in \mathbb{Z}$. If $\mathbf{w}, \mathbf{x}$ are 2 two-sided infinite words, and there exists an integer k such that $\mathbf{x} = \mathcal{S}^k(\mathbf{w})$, then we call $\mathbf{w}$ and $\mathbf{x}$ *conjugates*, and we write $\mathbf{w} \sim \mathbf{x}$. It is easy to see that $\sim$ is an equivalence relation. We extend this notation to sets of infinite words, as follows: if L is a set of two-sided infinite words, then by $\mathbf{w} \sim L$ we mean there exists $\mathbf{x} \in L$ such that $\mathbf{w} \sim \mathbf{x}$.

If $i = |wa|$, $h(a) = wax$, and $w, x \notin M_h^*$, then we define

$$\overleftrightarrow{h^{\omega;i}}(a) := \cdots h^2(w)\,h(w)\,w.a\,x\,h(x)\,h^2(x)\cdots,$$

a two-sided infinite word. Note that in this case the factorization of $h(a)$ as wax is *not* necessarily unique, and we use the superscript i to indicate which a is being chosen.

We assume $h : \Sigma^* \to \Sigma^*$ is a morphism that is extended to the domain $\Sigma^{\mathbb{Z}}$ as follows:

$$h(\cdots c_{-2}c_{-1}c_0.c_1c_2\cdots) = \cdots h(c_{-2})h(c_{-1})h(c_0).h(c_1)h(c_2)\cdots.$$

We first consider the equation $h(\mathbf{w}) = \mathbf{w}$ for two-sided infinite words.

Proposition 7.4.1 *The equation $h(\mathbf{w}) = \mathbf{w}$ has a solution if and only if at least one of the following conditions holds:*

(a) $\mathbf{w} \in F_h^{\mathbb{Z}}$; *or*

(b) $\mathbf{w} \in \overleftarrow{h^{\omega}}(a)F_h^*.F_h^{\omega}$ *for some $a \in \Sigma$, and there exist $x \notin M_h^*$, $y \in M_h^*$ such that $h(a) = xay$; or*

(c) $\mathbf{w} \in {}^{\omega}F_h.F_h^*h^{\omega}(a)$ *for some $a \in \Sigma$, and there exist $x \in M_h^*$, $y \notin M_h^*$ such that $h(a) = xay$; or*

(d) $\mathbf{w} \in \overleftarrow{h^{\omega}}(a)\,F_h^*.F_h^*h^{\omega}(b)$ *for some $a, b \in \Sigma$, and there exist $x, z \notin M_h^*$, $y, w \in M_h^*$, such that $h(a) = xay$ and $h(b) = wbz$.*

Proof. Let $\mathbf{w} = \cdots c_{-2}c_{-1}c_0.c_1c_2c_3\cdots$. By definition, we have

$$h(\mathbf{w}) = \cdots h(c_{-2})h(c_{-1})h(c_0).h(c_1)h(c_2)h(c_3)\cdots,$$

so if $h(\mathbf{w}) = \mathbf{w}$, then we have $h(c_1c_2c_3\cdots) = c_1c_2c_3\cdots$ and $h(\cdots c_{-2}c_{-1}c_0) = \cdots c_{-2}c_{-1}c_0$.

We may now apply Theorem 7.3.1 (Theorem 7.3.2) to $R(\mathbf{w}) = c_1c_2c_3\cdots$ (to $L(\mathbf{w}) = \cdots c_{-2}c_{-1}c_0$). There are 2 cases to consider for each side, giving $2 \times 2 = 4$ total cases. ∎

Example 7.4.2 Let μ be the Thue–Morse morphism, which maps $0 \to 01$ and $1 \to 10$. Define $g = \mu^2$. Then $g(0) = 0110$, $g(1) = 1001$. Let $\mathbf{t} =$

01101001 $\cdots$, the one-sided Thue–Morse infinite word. Then there are exactly 4 two-sided infinite fixed points of g, as follows:

$$\mathbf{t}^R.\mathbf{t} = \cdots 10010110.01101001 \cdots,$$
$$\bar{\mathbf{t}}^R.\mathbf{t} = \cdots 01101001.01101001 \cdots,$$
$$\bar{\mathbf{t}}^R.\bar{\mathbf{t}} = \cdots 01101001.10010110 \cdots,$$
$$\mathbf{t}^R.\bar{\mathbf{t}} = \cdots 10010110.10010110 \cdots.$$

All of these fall under case (d) of Proposition 7.4.1. Incidentally, all four of these words are overlap-free.

Now we characterize the two-sided infinite fixed points of a morphism in the "unpointed" case. That is, our goal is to characterize the solutions to $h(\mathbf{w}) \sim \mathbf{w}$. The following theorem is the first of our two main results.

Theorem 7.4.3 *Let h be a morphism. Then the two-sided infinite word $\mathbf{w}$ satisfies the relation $h(\mathbf{w}) \sim \mathbf{w}$ if and only if at least one of the following conditions holds:*

(a) $\mathbf{w} \sim F_h^{\mathbb{Z}}$; or

(b) $\mathbf{w} \sim \overset{\leftarrow}{h^{\omega}}(a) . F_h^{\omega}$ for some $a \in \Sigma$, and there exist $x \notin M_h^*$ and $y \in M_h^*$ such that $h(a) = xay$; or

(c) $\mathbf{w} \sim {}^{\omega}F_h . h^{\omega}(a)$ for some $a \in \Sigma$, and there exist $x \in M_h^*$ and $y \notin M_h^*$ such that $h(a) = xay$; or

(d) $\mathbf{w} \sim \overset{\leftarrow}{h^{\omega}}(a) . F_h^* h^{\omega}(b)$ for some $a, b \in \Sigma$ and there exist $x, z \notin M_h^*$, $y, w \in M_h^*$, such that $h(a) = xay$ and $h(b) = wbz$; or

(e) $\mathbf{w} \sim \overset{\longleftrightarrow}{h^{\omega;i}}(a)$ for some $a \in \Sigma$, and there exist $x, y \notin M_h^*$ such that $h(a) = xay$ with $|xa| = i$; or

(f) $\mathbf{w} = (xy)^{\mathbb{Z}}$ for some $x, y \in \Sigma^+$ such that $h(xy) = yx$.

Before we begin the proof of Theorem 7.4.3, we state three useful lemmas.

Lemma 7.4.4 *Suppose $\mathbf{w}, \mathbf{x}$ are 2 two-sided infinite words with $\mathbf{w} \sim \mathbf{x}$, and suppose h is a morphism. Then $h(\mathbf{w}) \sim h(\mathbf{x})$.*

Proof. Left to the reader as Exercise 2.   ■

Our second lemma concerns periodicity of infinite words. We say a two-sided infinite word

$$\mathbf{w} = \cdots c_{-2}c_{-1}c_0.c_1c_2 \cdots$$

is *periodic* if there exists a nonempty word x such that $\mathbf{w} = x^{\mathbb{Z}}$, i.e., if there exists an integer $p \geq 1$ such that $\mathbf{w} = S^p(\mathbf{w})$. The integer p is called a *period* of $\mathbf{w}$.

Lemma 7.4.5 *Suppose* $\mathbf{w} = \cdots c_{-2}c_{-1}c_0.c_1c_2\cdots$ *is a two-sided infinite word such that there exists a one-sided right-infinite word* $\mathbf{x}$ *and infinitely many negative indices* $0 > i_1 > i_2 > \cdots$ *such that*

$$\mathbf{x} = c_{i_j}c_{i_j+1}c_{i_j+2}\cdots$$

for $j \geq 1$. *Then* $\mathbf{w}$ *is periodic.*

Proof. Left to the reader as Exercise 1.5. ∎

Our third lemma concerns the growth functions of iterated morphisms.

Lemma 7.4.6 *Let* $h : \Sigma^* \to \Sigma^*$ *be a morphism. Then*

(a) *there exist integers* i, j *with* $0 \leq i < j$ *and* $|h^i(w)| \leq |h^j(w)|$ *for all* $w \in \Sigma^*$; *and*

(b) *there exists an integer* M, *depending only on* $k = \operatorname{Card} \Sigma$, *such that for all* $h : \Sigma^* \to \Sigma^*$, *there exists* i, j *with* $0 \leq i < j < M$ *such that* $|h^i(w)| \leq |h^j(w)|$ *for all* $w \in \Sigma^*$.

Proof. (a): Suppose $\Sigma = \{a_1, a_2, \ldots, a_k\}$. First, choose $i_{1,1}$ to be the least index such that $|h^{i_{1,1}}(a_1)| = \min_{i \geq 0}|h^i(a_1)|$. Next, successively choose $i_{1,2}, i_{1,3}, i_{1,4}, \ldots$ such that $|h^{i_{1,n+1}}(a_1)| = \min_{i > i_{1,n}}|h^i(a_1)|$ for $n \geq 1$. Clearly $|h^{i_{1,n}}(a_1)| \leq |h^{i_{1,n+1}}(a_1)|$ for all $n \geq 1$. Let $S_1 = \{i_{1,1}, i_{1,2}, i_{1,3}, \ldots\}$.

Now, choose $i_{2,1}$ to be the least index $i \in S_1$ such $|h^{i_{2,1}}(a_2)| = \min_{i \in S_1}|h^i(a_2)|$. Next, successively choose $i_{2,2}, i_{2,3}, i_{2,4}, \ldots \in S_1$ such that $|h^{i_{2,n+1}}(a_2)| = \min_{i \in S_1; i > i_{2,n}}|h^i(a_2)|$ for $n \geq 1$. Clearly $|h^{i_{2,n}}(a_j)| \leq |h^{i_{2,n+1}}(a_j)|$ for $j = 1, 2$ and all $n \geq 1$. Let $S_2 = \{i_{2,1}, i_{2,2}, i_{2,3}, \ldots\}$. Note that $S_2 \subseteq S_1$.

Continuing in this fashion, we produce an infinite sequence of indices $i_{k,1}, i_{k,2}, i_{k,3}, \ldots$ such that $|h^{i_{k,n}}(a_j)| \leq |h^{i_{k,n+1}}(a_j)|$ for $j = 1, 2, \ldots, k$ and all $n \geq 1$. We can then choose $i = i_{k,1}$ and $j = i_{k,2}$.

(b): In fact, we can take $M = 2^k$; see Exercise 8.9. ∎

Now we can prove Theorem 7.4.3.

Proof. $\Longleftarrow$: Suppose case (a) holds, and $\mathbf{w} \sim F_h^{\mathbb{Z}}$. Then there exists $\mathbf{x} \in F_h^{\mathbb{Z}}$ with $\mathbf{w} \sim \mathbf{x}$. Since $\mathbf{x} \in F_h^{\mathbb{Z}}$, we can write $\mathbf{x} = \cdots x_{-2}x_{-1}x_0.x_1x_2\cdots$, where $x_i \in F_h$ for all $i \in \mathbb{Z}$. Since $x_i \in F_h$, we have $h(x_i) = x_i$ for all $i \in \mathbb{Z}$. It follows that $h(\mathbf{x}) = \mathbf{x}$. Now, applying Lemma 7.4.4, we conclude that $h(\mathbf{w}) \sim h(\mathbf{x}) = \mathbf{x} \sim \mathbf{w}$.

Next, suppose case (b) holds, and $\mathbf{w} \sim \overset{\leftarrow}{h^\omega}(a).F_h^\omega$. Then $\mathbf{w} \sim \mathbf{x}$ for some $\mathbf{x}$ of the form

$$\mathbf{x} = \overset{\leftarrow}{h^\omega}(a).x_1x_2x_3\cdots,$$

where $x_i \in F_h$ for all $i \geq 1$, and $h(a) = xay$ with $x \notin M_h^*$ and $y \in M_h^*$. Then we have $h(\mathbf{x}) = \mathbf{x}$, and by Lemma 7.4.4, we conclude that $h(\mathbf{w}) \sim h(\mathbf{x}) = \mathbf{x} \sim \mathbf{w}$.

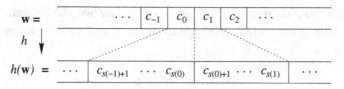

Figure 7.1: Interpretation of the Function s.

Cases (c) and (d) are similar to case (b).

For case (e), if $\mathbf{w} \sim \overleftrightarrow{h^{\omega;i}}(a)$, then by Lemma 7.4.4 we have

$$
\begin{aligned}
h(\mathbf{w}) &\sim h(\overleftrightarrow{h^{\omega;i}}(a)) \\
&= \cdots h^3(w)\,h^2(w)\,h(w)\,.\,h(a)h(x)h^2(x)h^3(x)\cdots \\
&= \cdots h^3(w)\,h^2(w)\,h(w)\,.\,xayh(x)h^2(x)h^3(x)\cdots \\
&\sim \cdots h^3(w)\,h^2(w)\,h(w)\,x\,.\,ayh(x)h^2(x)h^3(x)\cdots \\
&= \overleftrightarrow{h^{\omega;i}}(a) \\
&\sim \mathbf{w}.
\end{aligned}
$$

Finally, if case (f) holds, then

$$
h(\mathbf{w}) = h(\cdots xyxy.xyxy\cdots) = \cdots yxyx.yxyx\cdots,
$$

and so $h(\mathbf{w}) = \mathcal{S}^k(\mathbf{w})$ for $k = |x|$.

$\Longrightarrow$: Suppose $\mathbf{w} = \cdots c_{-2}c_{-1}c_0.c_1c_2\cdots$, and there exists k such that $h(\mathbf{w}) = \mathcal{S}^k(\mathbf{w})$. Define

$$
s(i) := \begin{cases} |h(c_1c_2\cdots c_i)| + k & \text{if } i \geq 0, \\ k - |h(c_{i+1}c_{i+2}\cdots c_0)| & \text{if } i < 0. \end{cases} \tag{7.8}
$$

Then it is not hard to see that

$$
h(c_i) = c_{s(i-1)+1}\cdots c_{s(i)} \tag{7.9}
$$

for $i \in \mathbb{Z}$; see Figure 7.1. Note that $s(0) = k$.

We define the set C as follows: $C = \{i \in \mathbb{Z} : s(i) = i\}$. Our argument is divided into two major cases, depending on whether or not C is empty.

Case 1: $C \neq \emptyset$. In this case, there exists j such that $s(j) = j$. Now consider the pointed word $\mathbf{x} = \cdots c_{j-2}c_{j-1}c_j.c_{j+1}c_{j+2}\cdots$. We have $\mathbf{x} \sim \mathbf{w}$, and by Eq. (7.9) we have $h(\mathbf{x}) = \mathbf{x}$. Then, by Proposition 7.4.1, one of cases (a)–(d) must hold.

Case 2: $C = \emptyset$. There are several subcases to consider.

Case 2a: There exist integers i, j with $i < j$ such that

$$
s(i) > i \quad \text{but} \quad s(j) < j. \tag{7.10}
$$

Among all pairs (i, j) satisfying (7.10), choose one with $j - i$ minimal. Suppose there exists an integer k with $i < k < j$. If $s(k) < k$, then (i, k) is a pair satisfying (7.10) with smaller difference, while if $s(k) > k$, then (k, j) is a pair satisfying (7.10) with smaller difference. Hence $s(k) = k$. But this is impossible by our assumption. It follows that $j = i + 1$. Then $s(i) > i$, but $s(i + 1) < i + 1$, a contradiction, since $s(i) \leq s(i + 1)$. Hence this case cannot occur.

Case 2b: There exists an integer r such that $s(i) < i$ for all $i < r$, and $s(i) > i$ for all $i \geq r$. Then $h(c_r) = c_{s(r-1)+1} \cdots c_{s(r)}$, which by the inequalities contains $c_{r-1}c_r c_{r+1}$ as a subword. Therefore, letting $a = c_r$, it follows that

$$\mathbf{w} \sim \mathbf{u} \, x \, . \, a \, y \, \mathbf{v},$$

where $\mathbf{u} = \cdots c_{s(r-1)-1} c_{s(r-1)}$ is a left-infinite word, $x = c_{s(r-1)+1} \cdots c_{r-1}$ and $y = c_{r+1} \cdots c_{s(r)}$ are finite words, and $\mathbf{v} = c_{s(r)+1} c_{s(r)+2} \cdots$ is a right-infinite word. Furthermore, we have $h(\mathbf{u}x) = \mathbf{u}$, $h(a) = xay$, and $h(y\mathbf{v}) = \mathbf{v}$.

Now the equation $h(y\mathbf{v}) = \mathbf{v}$ implies that $h(y)$ is a prefix of $\mathbf{v}$, and by an easy induction we have $h(y)h^2(y)h^3(y) \cdots$ is a prefix of $\mathbf{v}$. Suppose this prefix is finite. Then $y \in M_h^*$, and so $h(y)h^2(y)h^3(y) \cdots = h(y)h^2(y) \cdots h^t(y)$, where $t = \exp(h)$. Define $z = h(y)h^2(y) \cdots h^t(y)$; then $h(yz) = z$. Now $\mathbf{w}[1..r + |y| + |z|] = \mathbf{w}[1..s(r-1)]xayz$, and $h(\mathbf{w}[1..s(r-1)]xayz) = \mathbf{w}[k+1..s(r-1)]xayz$. It follows that $s(r + |y| + |z|) = r + |y| + |z|$, a contradiction, since we have assumed $C = \emptyset$. It follows that $\mathbf{z} := h(y)h^2(y)h^3(y) \cdots$ is right-infinite and hence $y \notin M_h^*$.

By exactly the same reasoning, we find that $\cdots h^3(x) \, h^2(x) \, h(x)$ is a left-infinite suffix of $\mathbf{u}$. We conclude that $\mathbf{w} \sim \overleftrightarrow{h^{\omega;i}} (a)$, and hence case (e) holds.

Case 2c: $s(i) > i$ for all $i \in \mathbb{Z}$. Let $\mathbf{w} = \cdots c_{-2} c_{-1} c_0 . c_1 c_2 \cdots$.

Now consider the following factorization of certain conjugates of $\mathbf{w}$, as follows: for $i \leq 0$, we have $\mathbf{w} \sim \mathbf{x}_i \, y_i \, . \, \mathbf{z}_i$, where $\mathbf{x}_i = \cdots c_{i-2} c_{i-1}$ (a left-infinite word), $y_i = c_i \cdots c_{s(i-1)}$ (a finite word), and $\mathbf{z}_i = c_{s(i-1)+1} c_{s(i-1)+2} \cdots$ (a right-infinite word). Note that $i - 1 < s(i-1)$ by assumption, so $i \leq s(i-1)$; hence y_i is nonempty. Evidently we have

$$h(\mathbf{x}_i) = \mathbf{x}_i \, y_i,$$
$$h(y_i \mathbf{z}_i) = \mathbf{z}_i. \tag{7.11}$$

Now the equation $h(y_i \mathbf{z}_i) = \mathbf{z}_i$ implies that $h(y_i)$ is a prefix of $\mathbf{z}_i$. Now an easy induction, as in case 2b, shows that $v := h(y_i)h^2(y_i)h^3(y_i) \cdots$ is a prefix of $\mathbf{z}_i$. If v were finite, then we would have $y_i \in M_h^*$, and so $h(\mathbf{w}[i..j]) = h(y_i v) = v$. Thus $s(j) = j$ for $j = s(i-1) + |v|$, a contradiction, since $C = \emptyset$. Hence v is right-infinite, and so $y_i \notin M_h^*$. There are now two further subcases to consider: (i) $\sup_{i \leq 0}(s(i) - i) < +\infty$, and (ii) $\sup_{i \leq 0}(s(i) - i) = +\infty$.

Case 2c(i): Suppose $\sup_{i \leq 0}(s(i) - i) = d < +\infty$. It then follows that $|y_i| \leq d$. Hence there is a finite word u such that $y_i = u$ for infinitely many indices $i \leq 0$. From the above argument we see that the right-infinite word $h(u)h^2(u)h^3(u) \cdots$ is a suffix of $\mathbf{w}$, beginning at position $s(i-1) + 1$, for infinitely many indices $i \leq 0$. We now use Lemma 7.4.5 to conclude that $\mathbf{w}$ is periodic.

Thus we can write $\mathbf{w} = \cdots c_{-2}c_{-1}c_0.c_1c_2\cdots$, and $\mathbf{w} = \cdots vvv.vvv\cdots$, where $v = c_1c_2\cdots c_p$ for some integer $p \geq 1$. Without loss of generality, we may assume p is minimal.

We claim $|h(v)| = p$. For if not we must have $|h(v)| = q$, for $q \neq p$, and then, since $h(\mathbf{w}) \sim \mathbf{w}$, we would have that $\mathbf{w}$ is periodic with periods p and q, hence periodic of period $\gcd(p, q)$. But since p was minimal, we must have $p \mid q$. Hence $q \geq 2p$. Now let $s(p) = l$; since $s(i) > i$ for all i, we must have $l > 0$. Then

$$h(c_1c_2\cdots c_p) = c_{s(0)+1}\cdots c_{s(p)} = c_{l-q+1}\cdots c_l.$$

It now follows that

$$s(ip) = l - q + iq \tag{7.12}$$

for all integers i. Now $p < q$, so $p \leq q - 1$, and hence $p < q - 1 + q/l$. Hence, multiplying by $-l$, we get $-lp > l - ql - q$. Now take $i = -l$ in Eq. (7.12); we obtain

$$s(-lp) = l - q - lq < -lp,$$

a contradiction, since $s(i) > i$ for all i. It follows that $|h(v)| = p$.

Recall that $h(c_1c_2\cdots c_p) = c_{k+1}c_{k+2}\cdots c_{k+p}$. Using the division theorem, write $k = jp + r$, where $0 \leq r < p$. Define

$$y = c_{k+1}\cdots c_{(j+1)p} = c_{r+1}\cdots c_p,$$
$$x = c_{(j+1)p+1}\cdots c_{k+p} = c_1\cdots c_r.$$

We have $h(xy) = yx$ and $v = xy$. Then $\mathbf{w} = v^{\mathbb{Z}} = (xy)^{\mathbb{Z}}$.

By the above we know $|v| \geq 1$, so $xy \neq \epsilon$. Suppose $y = \epsilon$. Then $h(x) = x$, and so $x \in F_h^*$. It follows that $\mathbf{w} \in F_h^{\mathbb{Z}}$. A similar argument applies if $x = \epsilon$. However, if $\mathbf{w} \in F_h^{\mathbb{Z}}$, then $C \neq \emptyset$, a contradiction. Thus $x, y \neq \epsilon$, and case (f) holds.

Case 2c(ii): $\sup_{i \leq 0}(s(i) - i) = +\infty$. Recall that $s(i) > i$ for all $i \in \mathbb{Z}$ and $\mathbf{w} = \cdots c_{-2}c_{-1}c_0.c_1c_2\cdots$. Define

$$\mathbf{x} := \cdots c_{-2}c_{-1}c_0,$$
$$y := c_1c_2\cdots c_{s(0)},$$
$$\mathbf{z} := c_{s(0)+1}c_{s(0)+2}\cdots.$$

Then $\mathbf{w} = \mathbf{x}.y\mathbf{z}$ and $h(\mathbf{x}) = xy$, $h(y\mathbf{z}) = \mathbf{z}$.

For all integers n define $B_j(n) := s^j(n) - s^{j-1}(n)$, where s^j denotes the j-fold composition of the function s with itself. First we prove the following technical lemma.

Lemma 7.4.7 *For all integers $r, t \geq 1$ there exists an integer $n \leq 0$ such that $B_j(n) > r$ for $1 \leq j \leq t$.*

Proof. By induction on t. For $t = 1$ the result follows, since

$$\sup_{i \leq 0} B_1(i) = \sup_{i \leq 0} (s(i) - i) = +\infty.$$

Now assume the result is true for t; we prove it for $t + 1$. Define $m := \max_{a \in \Sigma} |h(a)|$. By induction there exists an integer n_1 such that $B_j(n_1) > mr + m^{t+1}$ for $1 \leq j \leq t$. Then, by the definition of m there exist an integer $n_2 \leq n_1$ with $n_1 - n_2 < m$, and an integer n_3 such that $s(n_3) = n_2$.

Now $h(c_{n_3+1} \cdots c_{n_2}) = c_{s(n_3)+1} \cdots c_{s(n_2)}$, so $s(n_2) - s(n_3) \leq m(n_2 - n_3)$. Similarly, we have

$$s^j(n_2) - s^j(n_3) \leq m^j(n_2 - n_3) \tag{7.13}$$

for all $j \geq 0$. By the same reasoning, we have

$$s^j(n_1) - s^j(n_2) \leq m^j(n_1 - n_2) \leq m^j(m - 1) \tag{7.14}$$

for all $j \geq 0$. Thus we find

$$
\begin{aligned}
B_1(n_3) &= s(n_3) - n_3 \\
&= n_2 - n_3 \\
&\geq \frac{s(n_2) - s(n_3)}{m} \qquad \text{(by Eq. (7.13))} \\
&= \frac{s(n_2) - n_2}{m} \\
&= \frac{(s(n_1) - n_1) - ((s(n_1) - s(n_2)) - (n_1 - n_2))}{m} \\
&= \frac{B_1(n_1) - ((s(n_1) - s(n_2)) - (n_1 - n_2))}{m} \\
&> \frac{mr + m^{t+1} - m(m - 1)}{m} \qquad \text{(by induction and Eq. (7.14))} \\
&> r.
\end{aligned}
$$

Similarly, for $2 \leq j \leq t + 1$, we have

$$
\begin{aligned}
B_j(n_3) &= s^j(n_3) - s^{j-1}(n_3) \\
&= s^{j-1}(n_2) - s^{j-2}(n_2) \\
&= (s^{j-1}(n_1) - s^{j-2}(n_1)) \\
&\quad - ((s^{j-1}(n_1) - s^{j-1}(n_2)) - (s^{j-2}(n_1) - s^{j-2}(n_2))) \\
&= B_{j-1}(n_1) - ((s^{j-1}(n_1) - s^{j-1}(n_2)) - (s^{j-2}(n_1) - s^{j-2}(n_2))) \\
&> mr + m^{t+1} - m^{j-1}(m - 1) \qquad \text{(by Eq. (7.14))} \\
&\geq r.
\end{aligned}
$$

It thus follows that we can take $n = n_3$. This completes the proof of Lemma 7.4.7. ∎

Now let M be the integer specified in Lemma 7.4.6, and define $r :=$ $\sup_{1 \le i \le M} B_i(0)$. By Lemma 7.4.7 there exists an integer $n \le 0$ such that $B_j(n) > r$ for $1 \le j \le M$. Define $w := c_{n+1} \cdots c_0$. We have

$$|h^j(w)| = s^j(0) - s^j(n),$$
$$|h^{j-1}(w)| = s^{j-1}(0) - s^{j-1}(n).$$

It follows that

$$
\begin{aligned}
|h^j(w)| &= (s^j(0) - s^{j-1}(0)) - (s^j(n) - s^{j-1}(n)) + |h^{j-1}(w)| \\
&= B_j(0) - B_j(n) + |h^{j-1}(w)| \\
&< B_j(0) - r + |h^{j-1}(w)| \\
&\le |h^{j-1}(w)|
\end{aligned}
$$

for $1 \le j \le M$. But this contradicts Lemma 7.4.6. This contradiction shows that case 2c(ii) cannot occur.

Case 2d: $s(i) < i$ for all $i \in \mathbb{Z}$. This case is the mirror image of case 2c,[1] and the proof is identical. The proof of Theorem 7.4.3 is complete. ∎

Example 7.4.8 Consider the morphism f defined by a $\to$ bb, b $\to \epsilon$, c $\to$ aad, d $\to$ c. Let

$$\mathbf{w} = \cdots \text{aadbbbbcaadbbbbc.aadbbbbcaadbbbbc} \cdots.$$

Then

$$f(\mathbf{w}) = \cdots \text{bbbbcaadbbbbcaad.bbbbcaadbbbbcaad} \cdots.$$

This falls under case (f) of Theorem 7.4.3, with $x = $ bbbbc and $y = $ aad.

Example 7.4.9 Consider the morphism φ defined by $0 \to 201$, $1 \to 012$, and $2 \to 120$. Then if

$$\mathbf{w} = \overleftrightarrow{\varphi^{\omega;2}}(0) = \cdots c_{-2}c_{-1}.c_0 c_1 c_2 \cdots = \cdots 1202.01012 \cdots,$$

we have $\varphi(\mathbf{w}) \sim \mathbf{w}$. This falls under case (e) of Theorem 7.4.3. Incidentally, c_i equals the sum of the digits, modulo 3, in the balanced ternary representation of i; see Example 6.7.1.

7.5 More on Infinite Fixed Points

We start with the following theorem, which shows that every infinite fixed point of the form $g^\omega(a)$ is the image (under a coding) of a nonerasing morphism.

---

[1] Note that $s(i) > i$ for all i implies that $s(i - 1) > i - 1$. Therefore $s(i - 1) + 1 > i$, and hence case 2d really *is* the mirror image of case 2c.

Theorem 7.5.1 *Suppose* $g : \Sigma^* \to \Sigma^*$ *is a morphism prolongable on the letter* $a \in \Sigma$. *Then there exists an alphabet* Δ, *a letter* $b \in \Delta$, *a nonerasing morphism* $f : \Delta^* \to \Delta^*$, *and a coding* $h : \Delta \to \Sigma$ *such that* $g^\omega(a) = h(f^\omega(b))$.

Proof. Let $\Sigma = \{a_1, a_2, \ldots, a_r\}$ and let $a = a_1$.

By Lemma 7.4.6 there exist non-negative integers j, k with $j < k$ such that $|g^j(a_i)| \leq |g^k(a_i)|$ for $1 \leq i \leq r$. Now for $1 \leq i \leq r$ define $t_i = |g^j(a_i)|$. Let $\Delta = \{b_{i,m} : 1 \leq i \leq r, \ 1 \leq m \leq t_i\}$, where the $b_{i,m}$ are new symbols.

Let $\varphi : \Sigma^* \to \Delta^*$ be the map sending a_i to $b_{i,1}b_{i,2} \cdots b_{i,t_i}$. Then $|\varphi(w)| = |g^j(w)|$ for all $w \in \Sigma^*$. Now $g^k(a_i) = g^j(g^{k-j}(a_i))$, so

$$|\varphi(g^{k-j}(a_i))| = |g^j(g^{k-j}(a_i))| = |g^k(a_i)| \geq |g^j(a_i)| = t_i.$$

Hence $\varphi(g^{k-j}(a_i))$ can be arbitrarily decomposed as the product of t_i nonempty words, say

$$\varphi(g^{k-j}(a_i)) = w_{i,1}w_{i,2} \cdots w_{i,t_i}.$$

Since $|g^j(a_1)| < |g^k(a_1)|$, we may choose $w_{1,1}$ such that $|w_{1,1}| \geq 2$. Now let $f : \Delta^* \to \Delta^*$ be the map that sends $b_{i,m}$ to $w_{i,m}$. Note that $b_{1,1}$ is a prefix of $w_{1,1}$, and so f is prolongable on $b_{1,1}$. By our construction, f is nonerasing, and we see that $f \circ \varphi = \varphi \circ g^{k-j}$. Hence, applying f to both sides, we see that

$$\begin{aligned} f^2 \circ \varphi &= f \circ \varphi \circ g^{k-j} \\ &= \varphi \circ g^{k-j} \circ g^{k-j} \\ &= \varphi \circ g^{2(k-j)}. \end{aligned}$$

Similarly, we get

$$f^n \circ \varphi = \varphi \circ g^{n(k-j)} \tag{7.15}$$

for all $n \geq 0$.

Now define the coding $h : \Delta \to \Sigma$ to send $b_{i,m}$ to the mth symbol of $g^j(a_i)$. We see that

$$h \circ \varphi = g^j. \tag{7.16}$$

Hence we have

$$\begin{aligned} h \circ f^n \circ \varphi &= h \circ \varphi \circ g^{n(k-j)} \quad \text{(by Eq. (7.15))} \\ &= g^j \circ g^{n(k-j)} \quad \text{(by Eq. (7.16))} \\ &= g^{j+n(k-j)} \end{aligned}$$

for all $n \geq 0$. In particular, then,

$$h(f^n(\varphi(a_1))) = g^{j+n(k-j)}(a_1). \tag{7.17}$$

for all $n \geq 0$. But $\varphi(a_1) = b_{1,1} \cdots b_{1,t_1}$. Hence, letting $n \to \infty$ in (7.17), and recalling that f is nonerasing and $|f(b_{1,1})| \geq 2$, we see

$$h(f^{\omega}(b_{1,1})) = g^{\omega}(a_1).$$

Take $b = b_{1,1}$. This completes the proof.   ∎

Example 7.5.2 Consider the morphism g sending $0 \to 01, 1 \to 202$, and $2 \to \epsilon$. Hence

$$g^{\omega}(0) = 012020101202 \cdots.$$

We find

i	$g^0(i)$	$g^1(i)$	$g^2(i)$	$g^3(i)$
0	0	01	01202	0120201
1	1	202	01	01202
2	2	ϵ	ϵ	ϵ

Thus we may take $j = 1, k = 3$. We then have $\varphi(0) = b_{01}b_{02}$, $\varphi(1) = b_{11}b_{12}b_{13}$, and $\varphi(2) = \epsilon$. Then $\varphi(g^2(0)) = b_{01}b_{02}b_{11}b_{12}b_{13}b_{01}b_{02}$, so we can (arbitrarily) choose $w_{01} = b_{01}b_{02}b_{11}$ and $w_{02} = b_{12}b_{13}b_{01}b_{02}$. Similarly, $\varphi(g^2(1)) = b_{01}b_{02}b_{11}b_{12}b_{13}$, so we can choose $w_{11} = b_{01}b_{02}$, $w_{12} = b_{11}$, and $w_{13} = b_{12}b_{13}$. Now define $f(b_{01}) = w_{01}$, $f(b_{02}) = w_{02}$, $f(b_{11}) = w_{11}$, $f(b_{12}) = w_{12}$, and $f(b_{13}) = w_{13}$. Also set $h(b_{01}) = 0$, $h(b_{02}) = 1$, $h(b_{11}) = 2$, $h(b_{12}) = 0$, $h(b_{13}) = 2$. We have

$$f^{\omega}(b_{01}) = b_{01}b_{02}b_{11}b_{12}b_{13}b_{01}b_{02}b_{01}b_{02} \cdots$$

and hence

$$h(f^{\omega}(b_{01})) = 012020101 \cdots.$$

7.6 Closure Properties

We next turn to the closure properties of morphic sequences. In particular, this class of words is closed under shifts. We first consider the case of a left shift.

Theorem 7.6.1 *Let $g : \Sigma^* \to \Sigma^*$ be a morphism prolongable on a letter $a \in \Sigma$. Let $\tau : \Sigma \to \Delta$ be a coding, and write $\tau(g^{\omega}(a)) = a_0a_1a_2 \cdots$. Then there exists an alphabet Γ, a morphism $h : \Gamma^* \to \Gamma^*$ prolongable on a letter $c \in \Gamma$, and a coding $\rho : \Gamma \to \Delta$ such that $\rho(h^{\omega}(c)) = a_1a_2a_3 \cdots$.*

Proof. We have $g^{\omega}(a) = a \, x \, g(x) \, g^2(x) \, g^3(x) \cdots$ for some $x \notin M_g^*$. Let $x = by$ with $b \in \Sigma$ and $y \in \Sigma^*$.

Let $c \notin \Sigma$ be a new symbol, and define $\Gamma = \Sigma \cup \{c\}$. Let $h : \Gamma^* \to \Gamma^*$ be defined by

$$h(d) = \begin{cases} g(d) & \text{if } d \in \Sigma, \\ c \, y \, g(b) & \text{if } d = c. \end{cases}$$

Then we have

$$h^\omega(c) = c \, y \, g(b) \, g(y) \, g^2(b) \, g^2(y) \, g^3(b) \cdots$$
$$= c \, y \, g(x) \, g^2(x) \, g^3(x) \cdots.$$

Note that $x \notin M_g^*$, so either $b \notin M_g$ or $y \notin M_g^*$. Hence $yg(b) \notin M_g^*$, so h is prolongable on c.

Now define $\rho : \Gamma \to \Delta$ as follows:

$$\rho(d) = \begin{cases} \tau(d) & \text{if } d \in \Sigma, \\ \tau(b) & \text{if } d = c. \end{cases}$$

It follows that

$$\rho(h^\omega(c)) = \mathcal{S}(\tau(g^\omega(a))),$$

where $\mathcal{S}$ is the shift function. ∎

Example 7.6.2 Consider the Thue–Morse word **t**, generated by the map $0 \to 01$, $1 \to 10$. Then $\mathcal{S}(\mathbf{t})$ can be generated as the image under ρ of the fixed point of $2 \to 210, 0 \to 01, 1 \to 10$, where ρ maps $2 \to 1, 1 \to 1$, and $0 \to 0$.

Next, we consider the case of a right shift.

Theorem 7.6.3 *Let* $g : \Sigma^* \to \Sigma^*$ *be a morphism prolongable on a letter* $a \in \Sigma$. *Let* $\tau : \Sigma \to \Delta$ *be a coding, and write* $\tau(g^\omega(a)) = \mathbf{x}$. *Let* $w \in \Sigma^*$. *Then there exists an alphabet* Γ, *a morphism* $h : \Gamma^* \to \Gamma^*$ *prolongable on a letter* $c \in \Gamma$, *and a coding* $\rho : \Gamma \to \Delta$ *such that* $\rho(h^\omega(c)) = w \, \mathbf{x}$.

Proof. Let $w = b_1 b_2 \cdots b_r$. Introduce $r + 1$ new symbols $c_1, c_2, \ldots, c_r, c \notin \Sigma$, and set $\Gamma = \Sigma \cup \{c_1, c_2, \ldots, c_r, c\}$. Define

$$h(d) = \begin{cases} c_1 c_2 \cdots c_r \, c & \text{if } d = c_1, \\ \epsilon & \text{if } d = c_i \text{ with } 2 \le i \le r, \\ x & \text{if } d = c, \\ g(d) & \text{if } d \in \Sigma. \end{cases}$$

Since g is prolongable on a, we have

$$g^\omega(a) = a \, x \, g(x) \, g^2(x) \cdots$$

for some $x \notin M_g^*$, and by construction

$$h^\omega(c_1) = c_1 c_2 \cdots c_r \, c \, x \, g(x) \, g^2(x) \cdots.$$

Finally, define $\rho : \Gamma \to \Delta$ as follows:

$$\rho(d) = \begin{cases} b_i & \text{if } d = c_i, \\ \tau(a) & \text{if } d = c, \\ \tau(d) & \text{if } d \in \Sigma. \end{cases}$$

It follows that

$$\rho(h^\omega(c_1)) = w_1 w_2 \cdots w_r \tau(g^\omega(a)) = w \, \mathbf{x}. \qquad \blacksquare$$

Finally, we show that the class of morphic words is closed under direct product with a shift of the original word.

Theorem 7.6.4 *Let $g : \Sigma^* \to \Sigma^*$ be a morphism prolongable on the letter $a \in \Sigma$. Let $\tau : \Sigma \to \Delta$ be a coding, and write $\tau(g^\omega(a)) = a_0 a_1 a_2 \cdots$. Then there exists an alphabet Γ, a morphism $h : \Gamma^* \to \Gamma^*$ prolongable on a letter $c \in \Gamma$, and a coding $\rho : \Gamma \to \Delta \times \Delta$ such that $\rho(h^\omega(c)) = [a_0, a_1][a_1, a_2][a_2, a_3] \cdots$.*

Proof. By Theorem 7.5.1 we may, without loss of generality, assume that g is non-erasing. Suppose $g^\omega(a) = b_0 b_1 b_2 \cdots$.

Let $\Gamma = \Sigma \times \Sigma$. Intuitively speaking, the meaning of the ordered pair $[c, d]$ is "the symbol c is followed by d". If $g(c) = c_1 c_2 \cdots c_r$ and $g(d) = d_1 d_2 \cdots d_s$, then we define $h([c, d]) = [c_1, c_2][c_2, c_3] \cdots [c_r, d_1]$. Let $\rho([c, d]) = [\tau(c), \tau(d)]$, and define $a' = [b_0, b_1]$.

Now it is easy to see that if $g^n(a) = b_0 b_1 b_2 \cdots b_t$ for some $n \geq 1$, then

$$h^n(a') = [b_0, b_1][b_1, b_2][b_2, b_3] \cdots [b_t, b_{t+1}].$$

From this the result easily follows. $\blacksquare$

Example 7.6.5 Consider the Fibonacci word on $\{a, b\}$. It is generated by $a \to ab$ and $b \to a$. Writing the pairs $[c, d]$ in the proof of Theorem 7.6.4 as c_d, we get the following morphism h:

$$\begin{aligned} a_b &\to a_b b_a, \\ b_a &\to a_a, \\ a_a &\to a_b b_a. \end{aligned}$$

If we now define $\tau(c_d) = d$ for $c, d \in \{a, b\}$, then $\tau(h^\omega(a_b))$ generates the shifted Fibonacci word.

Theorem 7.6.6 *The class of words that are images of fixed points of morphisms is not closed under direct product.*

Proof. (Sketch.) Let α, β, γ be any three linearly independent quadratic irrationals with purely periodic continued fraction expansion, and consider the associated Sturmian words (see Chapter 9). These are generated as the fixed points of morphisms. However, the direct product of the three words has subword complexity $(n + 1)^3$, which means it cannot be generated as the image of a fixed point of a morphism; see Corollary 10.4.9. ■

7.7 Morphic Images of Morphic Words

In this section we prove a result that generalizes Theorem 7.5.1: namely, that if f and g are arbitrary morphisms with $f(g^{\omega}(a))$ an infinite word, then there exists a nonerasing morphism h and a coding τ such that $f(g^{\omega}(a)) = \tau(h^{\omega}(a))$. This is Corollary 7.7.5.

Theorem 7.7.1 *Let* $\mathbf{u} = (u_n)_{n \geq 0}$ *be a sequence defined on a finite alphabet* Σ. *Let* $e = \mathrm{Card}\ \Sigma$. *Then the following conditions are equivalent:*

(a) the sequence $\mathbf{u}$ *is morphic;*

(b) the sequence $\mathbf{u}$ *is morphic, and the morphism that generates it can be taken to be nonerasing;*

(c) there exist an integer $d > 1$ *and* d-*automatic sets* $S_1, \ldots, S_e$ *such that if* $T = \bigcup_{j=1}^{e} S_j = \{t_1, t_2, \ldots\}$ *and* $t_1 < t_2 < \cdots$, *then* T *is infinite, and for all* $a \in \Sigma$, $u_n = a$ *if and only if* $t_n \in S_a$;

(d) there exist a finite alphabet $\Sigma' = \Sigma \cup \{\ell\}$ *with* $\ell \notin \Sigma$, *an integer* $d > 1$, *and a* d-*automatic sequence* $\mathbf{z}$ *(defined over* Σ'*) that is not ultimately equal to* ℓ, *and such that the sequence* $\mathbf{u}$ *equals the sequence obtained from* $\mathbf{z}$ *by erasing all occurrences of the letter* ℓ.

Proof. (a) $\Leftrightarrow$ (b): proved in Theorem 7.5.1.

(c) $\Rightarrow$ (d): We note that the set $\mathbb{N} \setminus T$ is d-automatic (see Theorem 5.6.3). Next, we introduce a new letter ℓ that does not belong to Σ, and we define the sequence $\mathbf{z} = (z_n)_{n \geq 0}$ on the alphabet $\Sigma' = \Sigma \cup \{\ell\}$ as follows:

$$z_n = \begin{cases} u_n & \text{for all } n \in T, \\ \ell & \text{for all } n \in \mathbb{N} \setminus T. \end{cases}$$

Since the sets S_i and $\mathbb{N} \setminus T$ are d-automatic, it follows that the sequence $\mathbf{z}$ is d-automatic.

(d) $\Rightarrow$ (c): It suffices to define the sets S_i as the sets where the sequence $\mathbf{z}$ is constant and not equal to the letter ℓ.

(a) $\Rightarrow$ (d): It suffices to prove that every pure morphic sequence $\mathbf{v}$ can be obtained by erasing all occurrences of a given letter in some automatic sequence. Let Θ be a finite alphabet, μ be a morphism prolongable on a letter $a \in \Theta$, and let $\mathbf{v} = \mu^\omega(a)$. Let $h = \max\{|\mu(b)| : b \in \Theta\}$. Take a new letter ℓ not in Θ, and define the morphism φ on $\Theta' = \Theta \cup \{\ell\}$ as follows:

$$\varphi(b) = \begin{cases} \mu(b)\ell^{h-|\mu(b)|} & \text{if } b \in \Theta, \\ b^h & \text{if } b \notin \Theta. \end{cases}$$

Then φ is a morphism of length h that is prolongable on a. Furthermore, the sequence obtained from $\varphi^\omega(a)$ by erasing all occurrences of the letter ℓ is equal to the sequence $\mathbf{v}$.

(d) $\Rightarrow$ (b): Proved below in the more general Theorem 7.7.4. ∎

Before stating the next theorem, we need some definitions and a lemma.

Definition 7.7.2 Let μ be a morphism on the finite alphabet Σ, and let $\Gamma \subseteq \Sigma$. We classify the letters in Σ into three types, as follows. We say that a letter $x \in \Sigma$ is

(μ, Γ)-*dead* if for all $k \geq 0$ the word $\mu^k(x) \in \Gamma^*$. (Of course, any dead letter belongs to Γ, and for any dead letter x and for any $k \geq 0$, all letters of $\mu^k(x)$ are also (μ, Γ)-dead.)

(μ, Γ)-*moribund* if there exists $k \geq 0$ such that the word $\mu^k(x)$ contains at least one letter in $\Sigma \setminus \Gamma$, and for all $j > k$, the word $\mu^j(x) \in \Gamma^*$.

(μ, Γ)-*robust* if there exist infinitely many $k \geq 0$ such that the word $\mu^k(x)$ contains at least one letter in $\Sigma \setminus \Gamma$.

Remark. If $\Gamma = \emptyset$, the definition above gives:

there is no $(\mu, \emptyset)$-dead letter;
a letter is $(\mu, \emptyset)$-moribund if and only if it is mortal (see Section 1.4);
a letter is $(\mu, \emptyset)$-robust if and ony if it is immortal (see Section 1.4).

Lemma 7.7.3 *Let μ be a morphism defined on the alphabet Σ. Let $\Gamma \subseteq \Sigma$. Then there exists an integer $T \geq 1$ such that the morphism $\varphi = \mu^T$ satisfies the following properties:*

(a) *If x is (φ, Γ)-moribund, then for all $j > 0$ the word $\varphi^j(x)$ contains only (φ, Γ)-dead letters. Furthermore $x \in \Sigma \setminus \Gamma$.*

(b) *If x is (φ, Γ)-robust, then for all $j > 0$ the word $\varphi^j(x)$ contains at least one letter in $\Sigma \setminus \Gamma$.*

Proof. For all $j \geq 0$ let $U(j) = \{x \in \Sigma : \mu^j(x) \text{ contains at least one letter in } \Sigma \setminus \Gamma\}$. We note that $x \in U(j + 1)$ if and only if $\mu(x)$ contains at least one letter that belongs to $U(j)$. Hence, if $U(j) = U(k)$, then $U(j + 1) = U(k + 1)$. Since all the sets $U(j)$ are subsets of Σ, the sequence of sets $(U(j))_{j \geq 0}$ is ultimately periodic.

Then there exists some multiple T of the period such that $U(T) = U(2T) = \cdots = U(kT) = \cdots$. Define $\varphi = \mu^T$.

(a): If x is (φ, Γ)-moribund, then there exists k such that $\varphi^k(x)$ contains at least one letter in $\Sigma \setminus \Gamma$, and for all $j > k$ the word $\varphi^j(x)$ contains only letters in Γ. Since $\varphi^k = \mu^{kT}$, we see that $x \in U(kT)$. If $k \neq 0$, this implies $x \in U(T) = U(2T) = \cdots$. Hence for all $j > k$, the word $\varphi^j(x) = \mu^{jT}(x)$ contains at least one letter in $\Sigma \setminus \Gamma$, which is not the case. Hence $k = 0$, and thus $x \in \Sigma \setminus \Gamma$.

(b): If x is (φ, Γ)-robust, then there exist infinitely many k's such that $\varphi^k(x)$ contains at least one letter that belongs to $\Sigma \setminus \Gamma$. Choose such a $k > 0$. Then $\mu^{kT}(x)$ contains at least one letter belonging to $\Sigma \setminus \Gamma$. This means $x \in U(kT)$. Hence, as above, $x \in U(jT)$ for all $j > 0$, which is exactly our claim. ∎

We are now ready for our next theorem.

Theorem 7.7.4 *Let* **u** *be a morphic sequence defined on the finite alphabet* Σ. *Let* $\Gamma \subseteq \Sigma$. *Let* **v** *be the sequence obtained from* **u** *by erasing all occurrences of the letters belonging to* Γ. *Then the sequence* **v** *is either finite or morphic.*

Proof. First we give two reductions of the problem, and next we give the proof of the restricted case.

First reduction. We can suppose that the sequence **u** is itself a pure morphic sequence, and the first letter of **u** does not belong to Γ. For if all letters of the sequence **u** are in Γ, the sequence obtained by erasing them is empty, hence finite, and we are done. Otherwise the sequence begins with, say, k letters that are in Γ, followed by a letter not in Γ. The sequence obtained by shifting the sequence **u** k times is still morphic (by Theorem 7.6.1). Thus it is the image under a coding, say τ, of some pure morphic sequence, say **v**. It now suffices to prove that the sequence obtained from **v** by erasing the letters in the set $\tau^{-1}(\Gamma)$ is morphic.

Second reduction. Up to replacing the morphism by a power (that is still prolongable on the same letter, and generates the same fixed point), we can suppose that the morphism μ that generates **u** has the properties listed for φ in Lemma 7.7.3 above.

Proof of the restricted case. We now suppose we have a morphism μ on the finite set Σ, and it is prolongable on $a_0 \in \Sigma$. We have a subset $\Gamma \subseteq \Sigma$ that does not contain the letter a_0. The morphism μ and the set Γ have the properties listed in Lemma 7.7.3 above. We define the morphisms λ_Δ, λ_M, and λ_Γ, by

> if b is (μ, Γ)-dead, then $\lambda_\Delta(b) = \epsilon$, otherwise $\lambda_\Delta(b) = b$;
> if b is (μ, Γ)-moribund, then $\lambda_M(b) = \epsilon$, otherwise $\lambda_M(b) = b$;
> if $b \in \Gamma$, then $\lambda_\Gamma(b) = \epsilon$, otherwise $\lambda_\Gamma(b) = b$.

We know from the hypotheses on μ that, for any b that is (μ, Γ)-robust, the word $\mu(b)$ contains at least one letter in $\Sigma \setminus \Gamma$. Since any (μ, Γ)-dead letter clearly belongs to Γ, we thus have that $\lambda_\Delta(\mu(b))$ contains at least one letter in $\Sigma \setminus \Gamma$. Define, for each b that is (μ, Γ)-robust,

$$\lambda_\Delta(\mu(b)) := w(b, 1)w(b, 2) \cdots w(b, \ell_b),$$

where each of the words $w(b, i)$ contains exactly one letter in $\Sigma \setminus \Gamma$. Of course $\ell_b \geq 1$. Now any of the words $\lambda_\Gamma(w(b, i))$ consists of a single letter (and this letter belongs to $\Sigma \setminus \Gamma$). Let $a(b, i) := \lambda_\Gamma(w(b, i))$ be this letter.

For each (μ, Γ)-robust letter b, and for each i with $1 \leq i \leq \ell_b$, write

$$\lambda_M(w(b, i)) = c(b, i, 1)c(b, i, 2) \cdots c(b, i, n_{b,i}),$$

where the $c(b, i, j)$'s are letters in Σ. Since a moribund letter is in $\Sigma \setminus \Gamma$, we know that $\lambda_M(w(b, i))$ either is equal to $w(b, i)$, or is obtained from $w(b, i)$ by erasing exactly one moribund letter. Note that $\lambda_M(w(b, i))$ may be empty; this is the case if and only if $w(b, i)$ contains exactly one letter that is moribund.

We note the easy commutation relations

$$\lambda_\Gamma \circ \lambda_\Delta = \lambda_\Gamma,$$
$$\lambda_\Delta \circ \mu^n \circ \lambda_M = \lambda_\Delta \circ \mu^n \qquad \text{for all } n > 0,$$
$$\lambda_\Gamma \circ \mu^n \circ \lambda_M = \lambda_\Gamma \circ \mu^n \qquad \text{for all } n > 0,$$
$$\lambda_\Gamma \circ \mu^k \circ \lambda_\Delta = \lambda_\Gamma \circ \mu^k \qquad \text{for all } k \geq 0.$$

The first one is a consequence of the fact that any dead letter belongs to Γ. The second one is clear for any non-moribund letter, whereas both sides give ϵ for a moribund letter (remember that μ satisfies the hypotheses of the second reduction step). The third relation is an easy consequence of the first two. The last relation is clear for any non-dead letter, whereas both sides give ϵ for a dead letter.

We now define an alphabet $\mathfrak{A}$ of new symbols associated with the words $w(b, i)$, by

$$\mathfrak{A} = \{\alpha(b, i) : b \text{ is } (\mu, \Gamma)\text{-robust}, 1 \leq i \leq \ell_b\}.$$

Note that Card $\mathfrak{A} = \prod_b \ell_b$, where b runs through the (μ, Γ)-robust letters.

We define, for each (μ, Γ)-robust letter b, for each i and j with $1 \leq i \leq \ell_b$ and $1 \leq j \leq n_{b,i}$, the words $z(b, i, j)$ on the alphabet $\mathfrak{A}$ by

$$z(b, i, j) = \alpha(c(b, i, j), 1)\alpha(c(b, i, j), 2) \cdots \alpha(c(b, i, j), \ell_{c(b,i,j)}).$$

We finally define the morphism θ on $\mathfrak{A}$, and the map p from $\mathfrak{A}$ to Σ as follows: for all b that is (μ, Γ)-robust, for all i with $1 \leq i \leq \ell_b$,

$$\theta(\alpha(b, i)) = z(b, i, 1)z(b, i, 2) \cdots z(b, i, n_{b,i}),$$
$$p(\alpha(b, i)) = a(b, i).$$

We prove by induction on n that

$$(p \circ \theta^n)(\alpha(b, 1)\alpha(b, 2) \cdots \alpha(b, \ell_b)) = (\lambda_\Gamma \circ \mu^{n+1})(b) \qquad (7.18)$$

for all (μ, Γ)-robust letters b. We have $p(\alpha(b, i)) = a(b, i) = \lambda_\Gamma(w(b, i))$. Hence

$$p(\alpha(b, 1) \cdots \alpha(b, \ell_b)) = a(b, 1) \cdots a(b, \ell_b) = \lambda_\Gamma(w(b, 1)) \cdots \lambda_\Gamma(w(b, \ell_b))$$
$$= \lambda_\Gamma(w(b, 1) \cdots w(b, \ell_b)) = (\lambda_\Gamma \circ \lambda_\Delta \circ \mu)(b)$$
$$= (\lambda_\Gamma \circ \mu)(b),$$

which gives the case $n = 0$ of the induction. Suppose now that, for each (μ, Γ)-robust letter b, we have

$$(p \circ \theta^n)(\alpha(b, 1)\alpha(b, 2) \cdots \alpha(b, \ell_b)) = (\lambda_\Gamma \circ \mu^{n+1})(b).$$

Then, for all b, i, j for which $z(b, i, j)$ is defined, we have

$$(p \circ \theta^n)(z(b, i, j)) = (p \circ \theta^n)(\alpha(c(b, i, j), 1) \cdots \alpha(c(b, i, j), \ell_{c(b,i,j)}))$$
$$= (\lambda_\Gamma \circ \mu^{n+1})(c(b, i, j)).$$

Hence

$$(p \circ \theta^n)(z(b, i, 1) \cdots z(b, i, n_{b,i})) = (\lambda_\Gamma \circ \mu^{n+1})(c(b, i, 1) \cdots c(b, i, n_{b,i}))$$
$$= (\lambda_\Gamma \circ \mu^{n+1} \circ \lambda_M)(w(b, i))$$
$$= (\lambda_\Gamma \circ \mu^{n+1})(w(b, i)).$$

This in turn gives

$$(p \circ \theta^{n+1})(\alpha(b, 1)\alpha(b, 2) \cdots \alpha(b, \ell_b)) = (p \circ \theta^n)(\theta(\alpha(b, 1)) \cdots \theta(\alpha(b, \ell_b)))$$
$$= (p \circ \theta^n)(z(b, 1, 1) \cdots z(b, 1, n_{b,1})z(b, 2, 1) \cdots z(b, 2, n_{b,2}) \cdots$$
$$z(b, \ell_b, 1) \cdots z(b, \ell_b, n_{b,\ell_b}))$$
$$= [(\lambda_\Gamma \circ \mu^{n+1})(w(b, 1))][(\lambda_\Gamma \circ \mu^{n+1})(w(b, 2))] \cdots [(\lambda_\Gamma \circ \mu^{n+1})(w(b, \ell_b))]$$
$$= (\lambda_\Gamma \circ \mu^{n+1} \circ \lambda_\Delta)(\mu(b))$$
$$= (\lambda_\Gamma \circ \mu^{n+1})(\mu(b)) = (\lambda_\Gamma \circ \mu^{n+2})(b).$$

Now, taking $b = a_0$ in (7.18), we obtain

$$(p \circ \theta^n)(\alpha(a_0, 1)\alpha(a_0, 2) \cdots \alpha(a_0, \ell_{a_0})) = (\lambda_\Gamma \circ \mu^{n+1})(a_0).$$

We know that $\mu(a_0)$ begins with a_0, and we clearly have that a_0 is (μ, Γ)-robust. Hence $w(a_0, 1)$ begins with a_0. This gives $a(a_0, 1) = a_0$ and $c(a_0, 1, 1) = a_0$. Hence

$$z(a_0, 1, 1) = \alpha(c(a_0, 1, 1), 1)\alpha(c(a_0, 1, 1), 2) \cdots \alpha(c(a_0, 1, 1), \ell_{c(a_0,1,1)})$$
$$= \alpha(a_0, 1)\alpha(a_0, 2) \cdots \alpha(a_0, \ell_{a_0}).$$

Let

$$\theta(\alpha(a_0, 1)) = z(a_0, 1, 1)z(a_0, 1, 2) \cdots z(a_0, 1, n_{a_0,1})$$
$$:= z(a_0, 1, 1)\tilde{z}.$$

We can write

$$
\begin{aligned}
p(\theta^{n+1}(\alpha(a_0, 1))) \\
&= [p(\theta^n(z(a_0, 1, 1)))][p(\theta^n(\tilde{z}))] \\
&= [p(\theta^n(\alpha(c(a_0, 1, 1), 1) \cdots \alpha(c(a_0, 1, 1), \ell_{c(a_0,1,1)})))][p(\theta^n(\tilde{z})))] \\
&= [\lambda_\Gamma(\mu^{n+1}(a_0))][p(\theta^n(\tilde{z}))].
\end{aligned}
$$

Since we know that $\lambda_\Gamma(\mu^{n+1}(a_0))$ tends to a limit that is exactly the (infinite) sequence obtained from $\mathbf{u}$ by erasing the letters that belong to Γ, we deduce that $p(\theta^{n+1}(\alpha(a_0, 1)))$ has a limit, and that

$$
p(\theta^\omega(\alpha(a_0, 1))) = \lambda_\Gamma(\mu^\omega(a_0)).
$$ ∎

Corollary 7.7.5 *If a sequence is morphic, then the image of this sequence by any morphism is either finite or morphic.*

Proof. Let $\mathbf{u}$ be a morphic sequence defined on the finite alphabet Σ. Let Λ be a finite alphabet, and let φ be a morphism $\varphi : \Sigma^* \to \Lambda^*$. We want to prove that the sequence $\varphi(\mathbf{u})$ is finite or morphic.

Using Theorem 7.7.1, we know that there exist a finite alphabet $\Sigma' = \Sigma \cup \{\ell\}$, with $\ell \notin \Sigma$, an integer $d > 1$, and a d-automatic sequence $\mathbf{z}$ on Σ' that is not ultimately equal to ℓ and is such that $\mathbf{u} = \lambda_\ell(\mathbf{z})$, where λ_ℓ is the morphism defined on Σ' by

$$
\begin{aligned}
\lambda_\ell(a) &= a \quad \text{for all } a \in \Sigma, \\
\lambda_\ell(\ell) &= \epsilon.
\end{aligned}
$$

Let $e = \max\{|(\varphi \circ \lambda_\ell)(b)| : b \in \Sigma'\}$. Let $\ell' \notin \Lambda$ and let $\Lambda' := \Lambda \cup \{\ell'\}$. Define the morphism $\psi : \Sigma'^* \to \Lambda'^*$ by, for all $a \in \Sigma'$,

$$
\psi(a) = [(\varphi \circ \lambda_\ell)(a)][(\ell')^{e-|(\varphi \circ \lambda_\ell)(a)|}].
$$

Note that ψ is a morphism of constant length e. Denoting by $\lambda_{\ell'}$ the morphism defined by

$$
\begin{aligned}
\lambda_{\ell'}(a) &= a \quad \text{for all } a \in \Lambda, \\
\lambda_{\ell'}(\ell') &= \epsilon,
\end{aligned}
$$

we clearly have $\lambda_{\ell'} \circ \psi = \varphi \circ \lambda_\ell$.

Now, $\varphi(\mathbf{u}) = (\varphi \circ \lambda_\ell)(\mathbf{z}) = (\lambda_{\ell'} \circ \psi)(\mathbf{z})$. The sequence $\psi(\mathbf{z})$ is the image of a d-automatic sequence by a morphism of constant length; hence it is d-automatic (see Corollary 6.8.3). Hence the sequence $(\lambda_{\ell'} \circ \psi)(\mathbf{z})$ is finite or morphic from Theorem 7.7.4.   ∎

7.8 Locally Catenative Sequences

In this section we prove a theorem that shows that a large class of sequences generated by recursion is morphic.

Definition 7.8.1 We say a sequence of words $(X_i)_{i \geq 0}$ over Σ is *locally catenative* if there exist an integer $N \geq 0$, $N + 1$ nonempty words $X_0, X_1, \ldots, X_N$, an integer $t \geq 1$, t integers $1 \leq c_1, \ldots, c_t \leq N + 1$, and t codings $\mu_1, \mu_2, \ldots, \mu_t$ such that

$$X_n = X_{n-1} \mu_1(X_{n-c_1}) \cdots \mu_t(X_{n-c_t})$$

for all $n > N$.

Thus a sequence is locally catenative if it satisfies the analogue of a linear recurrence over the rationals, where the codings μ_s play the role of the constant coefficients.

Example 7.8.2 Let the Fibonacci representation of n be $(n)_F = e_{t-1} \cdots e_0$. We define $g_n = (\sum_{0 \leq i < t} e_i) \bmod 2$, i.e., the parity of the sum of digits of n expressed in the Fibonacci representation. Then $\mathbf{g} = (g_n)_{n \geq 0} = 01110100100011 \cdots$ is an analogue of the Thue–Morse sequence for Fibonacci representation.

Define $X_0 = 0$, $X_1 = 01$, and $X_n = X_{n-1} \overline{X_{n-2}}$ for $n \geq 2$, where the overbar is a morphism sending $0 \to 1$ and $1 \to 0$. Then it is easy to see that $\lim_{n \to \infty} X_n = \mathbf{g}$. The sequence of words $(X_i)_{i \geq 0}$ is locally catenative, with $N = 1$, $t = 1$, $c_1 = 2$, and μ_1 defined by the map $0 \to 1, 1 \to 0$.

Theorem 7.8.3 *Suppose $(X_i)_{i \geq 0}$ is a locally catenative sequence of words. Then $\lim_{n \to \infty} X_n$ exists and is a morphic infinite word.*

Proof. Clearly $\mathbf{X} = \lim_{n \to \infty} X_n$ exists and is infinite. Define $u_i = |X_i|$ for $i \geq 0$. To see that $\mathbf{X}$ is morphic, we construct a morphism φ, a coding τ, and a letter a such that $\mathbf{X} = \tau(\varphi^\omega(a))$.

We can write

$$X_n = \prod_{0 \leq s \leq t} \mu_s(X_{n-c_s}) \tag{7.19}$$

where $c_0 = 1$ and μ_0 is defined to be the identity map.

First, consider the monoid M generated by all the codings μ_s under composition. Clearly M is finite, since the set of all functions from Σ to Σ is of finite cardinality. Then we can write $M = \{\lambda_0, \lambda_1, \ldots, \lambda_r\}$, where λ_0 is the identity map. We also adopt the following notation for composition of the codings: We write $\mu_{h_0 h_1 \cdots h_q}$ in place of $\mu_{h_0} \circ \mu_{h_1} \circ \cdots \circ \mu_{h_q}$, and we define μ_ϵ to be the identity map.

We now introduce some new symbols not in Σ, namely those in the alphabet

$$\Gamma = \{a_{i,j,k} : 0 \le i \le N, \; 0 \le j < u_i, \; 0 \le k \le r\}.$$

We also introduce "aliases" for the $a_{i,j,k}$ as follows: we let $a_{i,j,k} = a_{i,j}^{(h)}$ for $h \in T^*$ if $\lambda_k = \mu_h$, where $T = \{0, 1, \dots, t\}^*$. Note that each symbol has infinitely many aliases.

We next extend the domain of the codings μ_s to $\Sigma \cup \Gamma$ as follows: we define

$$\mu_l(a_{i,j}^{(h)}) = a_{i,j}^{(lh)}$$

for $0 \le i \le N, 0 \le j < u_i$, and $0 \le l \le t$.

Now define

$$Y_i^{(h)} = a_{i,0}^{(h)} \cdots a_{i,u_i-1}^{(h)}$$

for $0 \le i \le N$ and $h \in T^*$. It follows that $\mu_l(Y_i^{(h)}) = Y_i^{(lh)}$ for all $l, h \in T^*, 0 \le i \le N$.

We define φ on Γ implicitly by requiring that

$$\varphi(Y_i^{(h)}) = Y_{i+1}^{(h)}, \qquad 0 \le i < N, \quad h \in T^*, \tag{7.20}$$

$$\varphi(Y_N^{(h)}) = \prod_{0 \le s \le t} Y_{N-c_s+1}^{(hs)}, \qquad h \in T^*. \tag{7.21}$$

Of course, since we are using the aliased representation, we must check that these definitions are consistent. This is implied by the associativity of composition. The definitions are meaningful because each $Y_i^{(h)}$ is nonempty, so we are free to choose any one of a number of possible definitions of φ on members of Γ such that (7.20) and (7.21) hold; the particular choice of how the $Y_i^{(h)}$ are partitioned is irrelevant. Furthermore, we can also require that $|\varphi(a_{N,0}^{(h)})| \ge 2$ for all $h \in T^*$.

Define $\tau(Y_i^{(h)}) = \mu_h(X_i)$ for $h \in T^*, 0 \le i \le N$. We claim τ commutes with μ_l for $l \in T^*$. For we have $\tau(\mu_l(Y_i^{(h)})) = \tau(Y_i^{(lh)}) = \mu_{lh}(X_i)$, while $\mu_l(\tau(Y_i^{(h)})) = \mu_l(\mu_h(X_i)) = \mu_{lh}(X_i)$.

We now claim that $\tau(\varphi^n(Y_0^{(h)})) = \mu_h(X_n)$ for all $n \ge 0, h \in T^*$. We prove this by induction on n. For $0 \le n \le N$, we have $\tau(\varphi^n(Y_0^{(h)})) = \tau(Y_n^{(h)}) = \mu_h(X_n)$. Now assume $n \ge N$. Then

$$\varphi^{N+1}(Y_0^{(h)}) = \varphi(\varphi^N(Y_0^{(h)}))$$
$$= \varphi(Y_N^{(h)}) \quad \text{(by (7.20))}$$
$$= \prod_{0 \le s \le t} Y_{N-c_s+1}^{(hs)} \quad \text{(by (7.21))}$$
$$= \prod_{0 \le s \le t} \varphi^{N-c_s+1}(Y_0^{(hs)}) \quad \text{(by (7.20))},$$

so, applying φ^{n-N} to both sides, we get

$$\varphi^{n+1}(Y_0^{(h)}) = \prod_{0 \leq s \leq t} \varphi^{n-c_s+1}(Y_0^{(hs)}) \tag{7.22}$$

for all $n \geq N$. Thus

$$\tau(\varphi^{n+1}(Y_0^{(h)})) = \tau\left(\prod_{0 \leq s \leq t} \varphi^{n-c_s+1}(Y_0^{(hs)})\right) \quad \text{(by (7.22))}$$

$$= \prod_{0 \leq s \leq t} \tau(\varphi^{n-c_s+1}(Y_0^{(hs)}))$$

$$= \prod_{0 \leq s \leq t} \mu_{hs}(X_{n-c_s+1}) \quad \text{(by induction)}$$

$$= \mu_h\left(\prod_{0 \leq s \leq t} \mu_s(X_{n-c_s+1})\right)$$

$$= \mu_h(X_{n+1}) \quad \text{(by (7.19))}.$$

Now set $h = \epsilon$, and write $Y_N^{(\epsilon)} = a_{N,0}^{(\epsilon)} \cdots a_{N,u_N-1}^{(\epsilon)}$. We get

$$\lim_{n \to \infty} \tau(\varphi^n(Y_0^{(\epsilon)})) = \lim_{n \to \infty} \tau(\varphi^{n-N}(Y_N^{(\epsilon)}))$$

$$= \lim_{n \to \infty} \tau(\varphi^{n-N}(a_{N,0}^{(\epsilon)}))\tau(\varphi^{n-N}(a_{N,1}^{(\epsilon)} \cdots a_{N,u_N-1}^{(\epsilon)}))$$

$$= \lim_{n \to \infty} \tau(\varphi^n(a_{N,0}^{(\epsilon)})),$$

so, taking $a = a_{N,0}^{(\epsilon)}$, we get the desired result. ∎

Example 7.8.4 Let us continue Example 7.8.2. We find $M = \{\lambda_0, \mu_1\}$. We define

$$\Gamma = \{a_{0,0}, a_{1,0}, a_{1,1}, a_{0,0}', a_{1,0}', a_{1,1}'\},$$

and φ as follows:

$$a_{0,0} \to a_{1,0}a_{1,1},$$
$$a_{1,0} \to a_{1,0}a_{1,1},$$
$$a_{1,1} \to a_{0,0}',$$
$$a_{0,0}' \to a_{1,0}'a_{1,1}',$$
$$a_{1,0}' \to a_{1,0}'a_{1,1}',$$
$$a_{1,1}' \to a_{0,0}.$$

Define $\tau(a_{0,0}) = \tau(a_{1,0}) = \tau(a_{1,1}') = 0$ and $\tau(a_{1,1}) = \tau(a_{1,0}') = \tau(a_{0,0}') = 1$. Then we have $\mathbf{g} = \tau(\varphi^\omega(a_{1,0}))$.

7.9 Transductions of Morphic Sequences

In this section we prove the analogue of Theorem 6.9.2 for morphic sequences: namely, that the finite-state transduction of a morphic sequence is still morphic, even if the transducer is nonuniform. More precisely, we prove the following:

Theorem 7.9.1 *Let $M = (Q, \Sigma, \delta, q_0, \Delta, \lambda)$ be a finite-state transducer. Let $\mathbf{x}$ be a morphic sequence. Then $M(\mathbf{x})$ is morphic or finite.*

Proof. Without loss of generality, we may assume that $\mathbf{x}$ is pure morphic (by absorbing the coding into M, if necessary). Let $h : \Sigma^* \to \Sigma^*$ be a morphism, prolongable on $a \in \Sigma$, such that $\mathbf{x} = h^\omega(a)$. Let $\mathbf{x} = a_1 a_2 a_3 \cdots$. For $w \in \Sigma^*$, define $\tau_w(q) = \delta(q, w)$. We thus have

$$\tau_{wx}(q) = \delta(q, wx) = \delta(\delta(q, w), x) = \tau_x(\tau_w(q)) = (\tau_x \circ \tau_w)(q),$$

so $\tau_{wx} = \tau_x \circ \tau_w$ for all $w, x \in \Sigma^*$.

Define $\theta_h(w) = (\tau_w, \tau_{h(w)}, \tau_{h^2(w)}, \ldots)$. Let

$$\Sigma' = \{(c, \theta_h(w)) : c \in \Sigma, \ w \in \Sigma^*\}.$$

The alphabet Σ' is not finite, but we will see below that Σ' can be replaced by a finite alphabet.

Now define the morphism $g : \Sigma'^* \to \Sigma'^*$ as follows:

$$g((c, \theta_h(w))) = d_1 d_2 \cdots d_r, \qquad (7.23)$$

where $h(c) = c_1 c_2 \cdots c_r$, and $d_i = (c_i, \theta_h(h(w)c_1 \cdots c_{i-1}))$ for $1 \le i \le r$. (Note that we could have $\theta_h(w) = \theta_h(w')$ for $w \ne w'$, but this does not affect the definition of the d_i.) Furthermore, set $\mathbf{y} = g^\omega((a, \theta_h(\epsilon)))$ and assume $\mathbf{y} = b_1 b_2 b_3 \cdots$. We claim

$$b_j = (a_j, \theta_h(a_1 a_2 \cdots a_{j-1})) \qquad (7.24)$$

for $j \ge 1$.

We first prove that Eq. (7.24) holds for $1 \le j \le |h^n(a)|$ by induction on n. For $n = 1$ this follows from the definition (7.23) and the fact that $a = a_1$. Now suppose $n \ge 1$. Since $h^{n+1}(a) = h(h^n(a))$, it follows that for each j with $1 \le j \le |h^{n+1}(a)|$ there exist integers i, k, such that $1 \le i \le |h^n(a)|$, $1 \le k \le |h(a_i)|$ such that

$$a_1 a_2 \cdots a_j = h(a_1 a_2 \cdots a_{i-1})(h(a_i))[1..k]. \qquad (7.25)$$

Hence, by the definition of g, the fact that $\mathbf{y} = g(\mathbf{y})$, and the fact that $|g((c, \theta_h(w)))| = |h(c)|$ for $c \in \Sigma$, $w \in \Sigma^*$, we get

$$b_1 b_2 \cdots b_j = g(b_1 b_2 \cdots b_{i-1})(g(b_i))[1..k]. \qquad (7.26)$$

Then we have

$$b_j = g(b_i)[k]$$
$$= (g((a_i, \theta_h(a_1 a_2 \cdots a_{i-1}))))[k] \quad \text{(by (7.24) and induction)}$$
$$= (h(a_i)[k], \theta_h(h(a_1 a_2 \cdots a_{i-1})(h(a_i))[1..k-1])) \quad \text{(by the definition of } g\text{)}$$
$$= (a_j, \theta_h(a_1 a_2 \cdots a_{j-1})) \quad \text{(by (7.25))},$$

which completes the proof.

Recall that $\mathbf{y}$ is pure morphic (modulo the inconvenience that Σ' is not finite). Now apply the coding ρ defined by $\rho((c, \theta_h(w))) = (c, \tau_w(q_0))$. Then

$$\rho(b_j) = \rho((a_j, \theta_h(a_1 a_1 \cdots a_{j-1})))$$
$$= (a_j, \tau_{a_1 a_2 \cdots a_{j-1}}(q_0))$$
$$= (a_j, \delta(q_0, a_1 a_2 \cdots a_{j-1})).$$

It follows that the sequence

$$((a_j, \delta(q_0, a_1 a_2 \cdots a_{j-1})))_{j \geq 1}$$

is morphic. Now apply the morphism λ', where $\lambda'((c, b)) = \lambda(b, c)$ for $b, c \in Q$. Then

$$\lambda'((a_j, \delta(q_0, a_1 a_2 \cdots a_{j-1}))) = \lambda(\delta(q_0, a_1 a_2 \cdots a_{j-1}), a_j).$$

By Corollary 7.7.5, the resulting sequence is morphic or finite.

It now remains to see that Σ' may be replaced with a finite alphabet.

Lemma 7.9.2 *The sequence $\theta_h(w) = (\tau_w, \tau_{h(w)}, \tau_{h^2(w)}, \dots)$ is uniformly ultimately periodic, i.e., there exist integers $p \geq 1$, $N \geq 0$ such that for all $w \in \Sigma^*$ we have $\tau_{h^i(w)} = \tau_{h^{i+p}(w)}$ for all $i \geq N$.*

Proof. Without loss of generality we may assume $\Sigma = \{0, 1, \dots, s-1\}$. Define a map $H : (Q^Q)^s \to (Q^Q)^s$ as follows:

$$H((f_0, f_1, \dots, f_{s-1})) = (f_{h(0)}, \dots, f_{h(s-1)}).$$

Then $H^n(\tau_0, \dots, \tau_{s-1}) = (\tau_{h^n(0)}, \dots, \tau_{h^n(s-1)})$. Since H is a map on a finite set, H^n is ultimately periodic with period length p and preperiod length N. Hence each of the sequences $(\tau_{h^n(j)})_{n \geq 0}$ with $0 \leq j < s$ is ultimately periodic with the same period length p and preperiod length N. Now if $w = c_1 c_2 \cdots c_t$, then

$$\tau_{h^i(c_1 \cdots c_t)} = \tau_{h^i(c_t)} \circ \cdots \circ \tau_{h^i(c_1)}$$
$$= \tau_{h^{i+p}(c_t)} \circ \cdots \circ \tau_{h^{i+p}(c_1)}$$
$$= \tau_{h^{i+p}(c_1 c_2 \cdots c_t)},$$

so the result follows. It follows that we may replace the infinite alphabet Σ' by the finite alphabet

$$\hat{\Sigma} = \{(c, \tau_w, \tau_{h(w)}, \ldots, \tau_{h^{N+p-1}(w)}) : c \in \Sigma,\ w \in \Sigma^*\}. \qquad \blacksquare$$

The proof of Theorem 7.9.1 is now complete. $\blacksquare$

7.10 Exercises

1. Let $h : \Sigma^* \to \Sigma^*$ be a morphism. Consider the following algorithm for computing M_h. Define $S_1 = \{a \in \Sigma : h(a) = \epsilon\}$. Compute S_i for $i \geq 2$ inductively as follows: $S_{i+1} = \{a \in \Sigma : h(a) \in S_i^*\}$. Eventually we must have $S_n = S_{n+1}$. Show that $M_h = S_n$, $\exp(h) = n$, and $a \in S_i$ if and only if $h^i(a) = \epsilon$.

2. Prove Lemma 7.4.4: if $\mathbf{w}, \mathbf{x}$ are 2 two-sided infinite words with $\mathbf{w} \sim \mathbf{x}$, then $h(\mathbf{w}) \sim h(\mathbf{x})$.

3. Define $v_0 = \text{ab}$ and $v_{i+1} = \text{a}v_0 v_1 \cdots v_i \text{b}$ for $i \geq 0$.
   (a) Show that $h^i(v_0) = v_i$, where h is the morphism defined by $\text{a} \to \text{aab}, \text{b} \to \text{b}$.
   (b) Show that
   $$h^i(a) = \text{ab}^{v_2(1)} \text{ab}^{v_2(2)} \text{ab}^{v_2(3)} \cdots \text{ab}^{v_2(2^i)}$$
   for $i \geq 0$, where v_2 is defined as in Section 2.1.
   (c) Let $\mathbf{w} = w_0 w_1 w_2 \cdots$ be the infinite fixed point of h starting with the symbol a, so $\mathbf{w} = \text{aabaabb} \cdots$. Show that the following are equivalent:
   (1) $w_n = \text{a}$;
   (2) there exists $e_j \in \{0, 1\}$ such that $n = \sum_{j \geq 1} e_j (2^j - 1)$;
   (3) there exists an integer $m \geq 0$ such that $n = m - s_2(m)$;
   (4) there exists an integer $r \geq 0$ such that $n = r + v_2(r!)$.
   (d) Prove that $\mathbf{w}$ is not 2-automatic.

4. Consider the mapping $\varphi : \text{c} \to \text{cab}, \text{a} \to \text{aa}, \text{b} \to \text{b}$. Show that $|\varphi^n(\text{c})| = 2^n + n$.

5. Give an explicit example of a sequence (s_n) for which all the sequences in the 2-kernel are distinct.

6. Recall that in the proof of Theorem 7.5.1 we showed there exist j, k with $j < k$ such that $|g^j(a_i)| \leq |g^k(a_i)|$ for $1 \leq i \leq r$.
   (a) Show how to construct g such that $k = e^{\Omega(\sqrt{r \log r})}$.
   (b) Show that $k \leq 2^r$.

7. Show the following: given a morphism $h : \Sigma^* \to \Sigma^*$ that is prolongable on a letter $a \in \Sigma$, and a subset $\Delta \subseteq \Sigma$, it is decidable if there exists an $e \geq 1$ such that $h^e(a) \in \Delta^*$.

8. We say that a word $w \in \Sigma^*$ is expressible as a *straight-line program* or *word chain* if there exist words $X_1, X_2, \ldots, X_r = w$ such that either $X_i = b$ for some $b \in \Sigma$ or $X_i = X_j X_k$ with $1 \leq j, k < i$. The *length* of the program is defined to be r.
   (a) Let φ be a morphism prolongable on the letter a, and let τ be a coding. Show that, for any $n \geq 0$, the word $\tau(\varphi^n(a))$ is expressible as a straight-line program of length $O(n)$.

(b) Show that the shortest straight-line program computing $aba^2ba^3b \cdots a^nb$ has length $\Theta(n)$.

9. Show that if $h : \{0, 1\}^* \to \Delta^*$ is any morphism, then $h(\mathbf{t})$ is 2-automatic, where $\mathbf{t} = 0110 \cdots$ is the Thue–Morse sequence. Show, more generally, that the same result is true if $\mathbf{t}$ is replaced by any fixed point $\mathbf{p}$ of a morphism $f : \Sigma^* \to \Sigma^*$ in which $|f(a)|_b$ is a constant for all letters $a, b \in \Sigma^*$. Is the converse true?

10. Give an example of a morphism (other than the identity) with an uncountable number of infinite fixed points.

11. (Hamm) Show that it is possible for the number of subwords of length r in the fixed point of a morphism of depth m and width n to be larger than any polynomial in r, m, n.

12. (de Luca) Show that, for $n \geq 5$, each finite Fibonacci word Φ_n can be expressed as the product of two nonempty palindromes, and this expression is unique. The lengths of these palindromes are $F_{n-1} - 2$ and $F_{n-2} + 2$.

13. (Cummings, Moore, and Karhumäki) Recall that a *border* of a word w is a proper subword that is both a prefix and suffix of w. Show that the longest border of Φ_n is Φ_{n-2}, where Φ_n is the nth Fibonacci word.

14. (Allouche) Consider the tower of Hanoi problem as discussed in Section 6.4, with the following variation, known as the *cyclic tower of Hanoi*: we demand that only the moves a, b, and c be used. Show that a solution of minimal length is given by the prefixes of $\tau(f^\omega(a_1))$, where f is the nonuniform morphism that sends

$$a_1 \to a_1 b_2 a_1,$$
$$b_1 \to b_1 c_2 b_1,$$
$$c_1 \to c_1 a_2 c_1,$$
$$a_2 \to b_1 a_1,$$
$$b_2 \to c_1 b_1,$$
$$c_2 \to a_1 c_1,$$

and τ is a coding that removes the subscripts.

15. Suppose $h : \Sigma^* \to \Sigma^*$ is a morphism, and suppose there exist a word $w \in \Sigma^*$ and a constant c such that $c = |w| = |h(w)| = \cdots = |h^n(w)|$, where $n = |\Sigma|$.
    (a) Show that $c = |h^i(w)|$ for all $i \geq 0$.
    (b) Show that the result does not necessarily hold if n is replaced by $n - 1$.

16. Let φ be a morphism, and let a be a growing letter (i.e., $|\varphi^i(a)|$ is unbounded as $i \to \infty$). Show that the characteristic sequence of the sequence $(|\varphi^n(a)|)_{n \geq 0}$ is morphic.

17. In the proof of Theorem 7.8.3, show that φ commutes with the μ_l on the symbols of Γ.

18. Show that every purely periodic sequence can be written as $h^\omega(a)$ for some primitive morphism h.

19. Suppose h possesses a nonempty finite fixed point w. How long can the shortest w be, as a function of Width h and Depth h?

20. Let $h : \Sigma^* \to \Sigma^*$ be a morphism such that there exists a letter a with $h(a) = xayaz$, with $x, z \notin M_h^*$. In this case there are at least two distinct two-sided infinite words of the form $h^{\omega;i}(a)$, one corresponding to $i = |xa|$ and the other to $i = |xaya|$. Give an example where these two words are conjugate, and give an example where they are not.

21. Characterize all solutions to the equation $h(\mathbf{y}) = x\mathbf{y}$, where h is a morphism, $\mathbf{y}$ is a one-sided (right-)infinite word, and x is a finite word.

22. Suppose h is a k-uniform morphism satisfying $h(\mathbf{y}) = x\mathbf{y}$, where $\mathbf{y}$ is an infinite word and x is a finite word. Show that $\mathbf{y}$ is k-automatic.

23. Give an example of an infinite word $\mathbf{x}$ that is a fixed point of an erasing morphism, but not the fixed point of any nonerasing morphism.

24. Give an algorithm for the following problem: given two morphisms g and h, decide if there exist integers $i, j \geq 1$ such that $g^i = h^j$.

25. Let k be an integer ≥ 2. Let $a_0, a_1, \cdots, a_{k-1}$ be distinct symbols and let $\Sigma = \{a_0, a_1, \cdots, a_{k-1}\}$. Consider the morphism $\varphi : \Sigma^* \to \Sigma^*$ defined below, which generalizes the Fibonacci morphism:

$$\varphi(a_0) = a_0 a_1,$$
$$\varphi(a_1) = a_0 a_2,$$
$$\vdots$$
$$\varphi(a_{k-2}) = a_0 a_{k-1},$$
$$\varphi(a_{k-1}) = a_0.$$

(a) Define $A_j := |\varphi^j(a_0)|$. Show that $A_0 = 1, A_1 = 2, A_2 = 4, \ldots, A_{k-1} = 2^{k-1}$, and $A_n = A_{n-1} + A_{n-2} + \cdots + A_{n-k}$ for $n \geq k$.

(b) Define $X_j = \varphi^j(a_0)$ for $j = 0, 1, 2, \ldots, k-1$. Thus, for example, $X_0 = a_0$ and $X_1 = a_0 a_1$. Define $X_n = X_{n-1} X_{n-2} \cdots X_{n-k}$ for $n \geq k$. Prove that $X_n = \varphi^n(a_0)$ for all $n \geq 0$.

(c) Show that every integer $n \geq 0$ can be written uniquely in the form $n = \sum_{i \geq 0} \epsilon_i(n) A_i$ where $\epsilon_i(n) \in \{0, 1\}$, subject to the condition $\epsilon_i(n)\epsilon_{i+1}(n) \cdots \epsilon_{i+k-1}(n) = 0$.

(d) Define $X_m = b_0 b_1 b_2 \cdots b_{A_m - 1}$. for $m \geq 0$. Prove that, for all $n \geq 0$, $b_n = a_j$ if and only if $\epsilon_0(n) = 1$, $\epsilon_1(n) = 1, \ldots, \epsilon_{j-1}(n) = 1$, and $\epsilon_j(n) = 0$.

7.11 Open Problems

1. Given two morphisms f and g, prolongable on a and b, respectively, and two codings τ, ρ, is it decidable whether $\tau(f^\omega(a)) = \rho(g^\omega(b))$? (Remarks: For the same problem without codings, see Culik and Harju [1984].)

2. Is the following problem decidable: given a morphism φ prolongable on a, is $\varphi^\omega(a)$ k-automatic for some k? How about if k is given as part of the input?

3. Is the following problem decidable: given a morphism h, and a letter a, do there exist integers $i_1, i_2, \ldots, i_t$ and an integer n such that $h^n(a) = h^{n-i_1}(a) \cdots h^{n-i_t}(a)$? (Remarks: For partial results see Choffrut [1992].)

Figure 7.2: Melodic transcription of the Thue–Morse sequence, © 1996 Tom Johnson. Used by permission.

7.12 Notes on Chapter 7

Morphic sequences are sometimes called *substitutive* in the literature.

Iterated morphisms and their properties are studied in the literature under several different names, including D0L (and CD0L) sequences and tag systems.

A *D0L sequence* is specified a triple (Σ, h, w) where Σ is a finite alphabet, $h : \Sigma^* \to \Sigma^*$ is a morphism, and w is a finite word. Given such a triple, the corresponding D0L sequence is $w, h(w), h^2(w), \ldots$. A *CD0L sequence* is specified by a 4-tuple (Σ, h, w, τ), where τ is a coding, and the corresponding sequence is $\tau(w), \tau(h(w)), \tau(h^2(w)), \ldots$. If $\lim_{n\to\infty} h^n(w)$ exists, it is sometimes called the ω-*word* of the D0L system. Many of the properties of morphisms we discuss in this book appear, in suitably translated form, in papers devoted to D0L systems. Incidentally, the term "D0L system" is an abbreviation for "deterministic Lindenmayer system with 0 symbols of context", named after the mathematical biologist A. Lindenmayer.

For tag systems, see Cobham [1968b, 1972].

Culik and Karhumäki [1994b] explored alternative methods for generating infinite words. Also see Lepistö [1996].

For the use of iterated morphisms in musical composition, see Allouche and Johnson [1995, 1996]; Laakso [1996]. In Figure 7.2 we display some music written by the Paris-based composer Tom Johnson, inspired by the Thue–Morse sequence. Here 1 is coded as a descending semitone, and 0 as a rising semitone.

7.1 It is hard to pin down precisely who first introduced the Fibonacci words. The infinite Fibonacci word **f** is a simple example of a Sturmian word, discussed further in Chapter 9. Sturmian words were studied prior to 1900 by Bernoulli, Christoffel, Smith, and Markoff.

The earliest reference we know where Fibonacci words appear explicitly is Recht and Rosenman [1947]. Also see Pennington [1956], Knuth [1968, Exercise 1.2.8-36], Cohn [1974], and Stolarsky [1976]. In the paper of Knuth, Morris, and Pratt [1977], Fibonacci strings appear as the worst case of a pattern-matching algorithm.

Berstel [1980b, 1986b] was the first to systematically study Fibonacci words. Crochemore [1981] used Fibonacci words to show the optimality of an algorithm for computing square subwords.

Karhumäki [1983] proved that **f** contains cubes, but no fourth powers. This result was strengthened by Mignosi and Pirillo [1992], who showed that **f** has a critical exponent of $(5 + \sqrt{5})/2 \doteq 3.61803$. Pirillo [1993] studied maximal powers in Fibonacci strings. Iliopoulos, Moore, and Smyth [1996, 1997] characterized the squares appearing in Fibonacci strings.

Factorizations of Fibonacci words were studied by Wen and Wen [1994a], de Luca [1995], and Pirillo [1999].

Other papers on Fibonacci words include Higgins [1987]; Turner [1988]; Séébold [1991]; Chuan [1992, 1993a, 1993b, 1995a, 1995b]; Hendel and Monteferrante [1994]; Droubay [1995]; Grytczuk [1996a].

For generalizations of the Fibonacci word, see Pansiot [1983b] and Exercise 25. Chekhova, Hubert, and Messaoudi [2001] studied the properties of the so-called "Tribonacci" morphism, a generalization of the Fibonacci morphism.

For generalizations to two-dimensional arrays, see Apostolico and Brimkov [2000].

7.2 Theorem 7.2.3 is due to Head [1981].

7.3 Theorem 7.3.1 is due to Head and Lando [1986]. Our presentation is from Hamm and Shallit [1999].

Shallit and Swart [1999] showed how, given a prolongable morphism h, to compute the ith letter of $h^n(a)$ in time polynomial in $\log i$, $\log n$, and the description size of h. A similar problem for fixed points of finite-state transducers, however, is EXPTIME-hard.

7.4 Theorem 7.4.3 is due to Shallit and Wang [2002]; our presentation is taken more or less verbatim from that paper, with permission from Elsevier Science.

7.5 Theorem 7.5.1 was proved by Cobham [1968b].

7.6 The construction in Theorem 7.6.4 is apparently due to B. Klein [1972, p. 115]. A similar construction was given by Peyrière [1978].

7.7 Cobham [1968b] was apparently the first to observe that the morphic image of a morphic word is still morphic, although his proof is rather sketchy. Another proof was given by Pansiot [1983a]. Our presentation is loosely based on Cobham's ideas.

7.8 Locally catenative sequences were introduced by Rozenberg and Lindenmayer [1973]. Their definition was less general than ours and did not include codings. Our presentation is based on Shallit [1988a].

Other papers on locally catenative formulas include Herman, Lindenmayer, and Rozenberg [1975]; Kobuchi [1977]; Ehrenfeucht and Rozenberg [1978]; Kobuchi and Wood [1981]; Seki and Kobuchi [1991]; Choffrut [1992].

7.9 Theorem 7.9.1 is from Dekking [1994]. For an analogous result on primitive morphic words, see Holton and Zamboni [2000].

8

Frequency of Letters

Our goal in this chapter is to prove theorems about the frequency of letters in automatic and morphic sequences. These theorems can be used, for example, to show that certain sequences are not k-automatic for any k.

8.1 Some Examples

Recall that the frequency of a letter a in an infinite word $\mathbf{t}$ was defined in Section 1.1 as follows:

$$\text{Freq}_\mathbf{t}(a) = \lim_{n \to \infty} \frac{|\mathbf{t}[0..n-1]|_a}{n},$$

if this limit exists.

We start with some examples:

Example 8.1.1 Let $\mathbf{t} = t_0 t_1 t_2 \cdots$ be the Thue–Morse sequence – the fixed point, starting with 0, of the morphism μ which maps $0 \to 01, 1 \to 10$. Then it is easy to see that, for all $k \geq 0$, the subword $\mathbf{t}[2k..2k+1]$ has one 0 and one 1. Hence $\text{Freq}_\mathbf{t}(0) = \text{Freq}_\mathbf{t}(1) = \frac{1}{2}$.

Example 8.1.2 The frequency of a letter in a k-automatic sequence may not always exist. Consider the following morphism:

$$h : \mathbf{a} \to \mathbf{ab}$$
$$\mathbf{b} \to \mathbf{cc}$$
$$\mathbf{c} \to \mathbf{bb}.$$

It is easy to see that

$$|h^n(\mathbf{a})|_\mathbf{a} = 1,$$

$$|h^n(\mathbf{a})|_\mathbf{b} = \begin{cases} \frac{2^n - 1}{3} & \text{if } n \text{ is even,} \\ \frac{2^{n+1} - 1}{3} & \text{if } n \text{ is odd,} \end{cases}$$

$$|h^n(\mathbf{a})|_\mathbf{c} = \begin{cases} \frac{2^{n+1} - 2}{3} & \text{if } n \text{ is even,} \\ \frac{2^n - 2}{3} & \text{if } n \text{ is odd.} \end{cases}$$

247

Hence

$$\lim_{n\to\infty} \frac{|h^n(\mathrm{a})|_{\mathrm{b}}}{2^n} = \begin{cases} \frac{1}{3} & \text{if } n \text{ ranges over even integers only,} \\ \frac{2}{3} & \text{if } n \text{ ranges over odd integers only,} \end{cases}$$

and so the limit does not exist.

Example 8.1.3 The frequency of a letter in a morphic sequence may be irrational. For example, consider the fixed point $\mathbf{f}$ of

$$\varphi : 1 \to 10$$
$$0 \to 1.$$

i.e., the infinite Fibonacci word of Section 7.1. Then

$$\varphi^\omega(1) = 10110101\cdots = f_1 f_2 f_3 \cdots;$$

and, as we will see in Section 9.1, we have

$$f_i = \lfloor (i+1)\alpha \rfloor - \lfloor i\alpha \rfloor,$$

where $\alpha = \frac{1}{2}(\sqrt{5}-1) \doteq 0.61803$. Now it is easy to see that $|\mathbf{f}[1..n]|_1 = \sum_{1\le i\le n} f_i = \lfloor (n+1)\alpha \rfloor$. Hence $\mathrm{Freq}_{\mathbf{f}}(1) = \frac{\sqrt{5}-1}{2}$.

8.2 The Incidence Matrix Associated with a Morphism

Given a morphism $\varphi : \Sigma^* \to \Sigma^*$ for some finite set $\Sigma = \{a_1, a_2, \ldots, a_d\}$, we define the *incidence matrix* $M = M(\varphi)$ as follows:

$$M = (m_{i,j})_{1\le i,j\le d},$$

where $m_{i,j}$ is the number of occurrences of a_i in $\varphi(a_j)$, i.e., $m_{i,j} = |\varphi(a_j)|_{a_i}$.

Example 8.2.1 Consider the morphism φ defined by

$$\varphi : a \to ab$$
$$b \to cc$$
$$c \to bb.$$

Then

$$M(\varphi) = \begin{array}{c} \\ a \\ b \\ c \end{array} \begin{array}{ccc} a & b & c \\ \left[\begin{array}{ccc} 1 & 0 & 0 \\ 1 & 0 & 2 \\ 0 & 2 & 0 \end{array}\right] \end{array}$$

The matrix $M(\varphi)$ is useful because of the following proposition.

Proposition 8.2.2

$$
\begin{bmatrix}
|\varphi(w)|_{a_1} \\
|\varphi(w)|_{a_2} \\
\vdots \\
|\varphi(w)|_{a_d}
\end{bmatrix}
= M(\varphi)
\begin{bmatrix}
|w|_{a_1} \\
|w|_{a_2} \\
\vdots \\
|w|_{a_d}
\end{bmatrix}.
$$

Proof. Clearly we have

$$
|\varphi(w)|_{a_i} = \sum_{1 \le j \le d} |\varphi(a_j)|_{a_i} |w|_{a_j}.
$$

From this, the desired equation easily follows. ∎

Now an easy induction gives $M(\varphi)^n = M(\varphi^n)$, and hence

Corollary 8.2.3

$$
\begin{bmatrix}
|\varphi^n(w)|_{a_1} \\
|\varphi^n(w)|_{a_2} \\
\vdots \\
|\varphi^n(w)|_{a_d}
\end{bmatrix}
= (M(\varphi))^n
\begin{bmatrix}
|w|_{a_1} \\
|w|_{a_2} \\
\vdots \\
|w|_{a_d}
\end{bmatrix}.
$$

Hence we find

Corollary 8.2.4

$$
|\varphi^n(w)| = \begin{bmatrix} 1 & 1 & 1 & \cdots & 1 \end{bmatrix} M(\varphi)^n
\begin{bmatrix}
|w|_{a_1} \\
|w|_{a_2} \\
\vdots \\
|w|_{a_d}
\end{bmatrix}.
$$

8.3 Some Results on Non-negative Matrices

In this section we recall some theorems concerning the eigenvalues of non-negative matrices. We also address the asymptotic behavior of the sequence $(M^n)_{n \ge 0}$, where M is a non-negative matrix. We begin with some definitions. In what follows we let I denote the identity matrix whose dimension is clear from the context.

Definition 8.3.1 Let $M = (m_{i,j})_{1 \le i, j \le d}$ be a $d \times d$ matrix over a field. We define the *adjoint* matrix adj(M) to be the transpose of the $d \times d$ matrix whose (i, j)th term is equal to $(-1)^{i+j}$ times the determinant of the matrix obtained from M by erasing the ith row and jth column. In particular $M \cdot \mathrm{adj}(M) = (\det M)I$.

If $MV = \lambda V$ for a vector $V \neq 0$ and a number λ, we call V an *eigenvector* of M and λ an *eigenvalue* of M.

The *characteristic polynomial* of the matrix M is defined by $Q(\lambda) := \det(\lambda I - M)$. The *minimal polynomial* of M is defined as the monic polynomial $q(\lambda)$ of smallest degree such that $q(M) = 0$.

Let $g(\lambda)$ be the monic greatest common divisor of all the (polynomial) coefficients of the matrix $\operatorname{adj}(\lambda I - M)$. The *reduced adjoint* of $\lambda I - M$ is defined by $C(\lambda) := \operatorname{adj}(\lambda I - M)/g(\lambda)$. Note that $C(\lambda)$ is also a matrix with coefficients that are polynomials in λ.

If M is a $d \times d$ matrix, and $P(\lambda)$ is a polynomial: $P(\lambda) = \alpha_0 + \alpha_1 \lambda + \alpha_2 \lambda^2 + \cdots + \alpha_e \lambda^e$, we denote by $P(M)$ the matrix obtained by replacing λ^j by M^j for $j = 0, 1, 2, \ldots, e$, i.e., $P(M) := \alpha_0 I + \alpha_1 M + \alpha_2 M^2 + \cdots + \alpha_e M^e$.

A *permutation matrix* is a matrix of 0's and 1's such that there is exactly one 1 in each row, and exactly one 1 in each column.

Let us now consider matrices and vectors with rational (or real) coefficients. A matrix M is said to be *non-negative* (*positive*) if all its entries are non-negative (positive); we write $M \geq 0 \, (M > 0)$. Similarly, a vector v is said to be *non-negative* (*positive*) if all its entries are non-negative (positive); we write $v \geq 0 \, (v > 0)$. A non-negative matrix M is said to be *reducible* if there exists a permutation matrix S such that the matrix $S^{-1} M S$ can be written in the form

$$S^{-1}MS = \begin{bmatrix} B & C \\ 0 & D \end{bmatrix}$$

where B and D are square matrices, and C and 0 are rectangular matrices such that all dimensions fit. A non-negative matrix M is said to be *irreducible* if it is not reducible. A non-negative matrix M is said to be *primitive* if there exists an integer $k \geq 1$ such that all the entries of M^k are positive. Finally, a matrix M is said to be (row-)*stochastic* if it is non-negative and if the sum of the entries of each row of M is equal to 1.

Lemma 8.3.2 *Let M be a $d \times d$ matrix with entries in a (commutative) field K. Let $P(\lambda) = \alpha_0 + \alpha_1 \lambda + \alpha_2 \lambda^2 + \cdots + \alpha_e \lambda^e$ be a polynomial with coefficients in K. Then the following conditions are equivalent:*

(i) $P(M) = 0$.

(ii) *There exists a $d \times d$ matrix $T(\lambda)$ whose entries are polynomials in λ such that the relation $(\lambda I - M)T(\lambda) = P(\lambda)I$ holds.*

Proof. First note that a $d \times d$ matrix $T(\lambda)$ whose entries are polynomials in λ can be written $T(\lambda) = T_0 + \lambda T_1 + \lambda^2 T_2 + \cdots + \lambda^{e-1} T_{e-1}$ where the T_j's are $d \times d$ matrices with constant coefficients. The existence of a $d \times d$ matrix $T(\lambda)$ such that $(\lambda I - M)T(\lambda) = P(\lambda)I$ is equivalent to the existence of $d \times d$ matrices

$T_0, T_1, \ldots, T_{e-1}$ such that

$$(\lambda I - M)(T_0 + \lambda T_1 + \lambda^2 T_2 + \cdots + \lambda^{e-1} T_{e-1})$$
$$= (\alpha_0 + \alpha_1 \lambda + \alpha_2 \lambda^2 + \cdots + \alpha_e \lambda^e) I,$$

i.e.,

$$
\begin{aligned}
-MT_0 &= \alpha_0 I, \\
T_0 - MT_1 &= \alpha_1 I, \\
T_1 - MT_2 &= \alpha_2 I, \\
&\vdots \\
T_{e-2} - MT_{e-1} &= \alpha_{e-1} I, \\
T_{e-1} &= \alpha_e I.
\end{aligned}
$$

Starting from the last equation, we find successively $T_{e-1} = \alpha_e I$, $T_{e-2} = \alpha_e M + \alpha_{e-1} I, \ldots$, up to the second equation, which gives $T_0 = \alpha_e M^{e-1} + \alpha_{e-1} M^{e-2} + \cdots + \alpha_1 I$. Hence the first equation is satisfied if and only if $\alpha_0 I = -MT_0 = \alpha_e M^e + \alpha_{e-1} M^{e-1} + \cdots + \alpha_1 M$, which is equivalent to $P(M) = 0$. ∎

For our next proposition, we need a definition. If $\| \ \|$ is a norm on the vector space of $n \times n$ complex matrices that satisfies $\|MN\| \le \|M\| \, \|N\|$ for all matrices M, N, then we say that $\| \ \|$ is *submultiplicative*. Operator norms, which are norms on the vector space of complex $n \times n$ matrices given by $\|M\| = \sup_{\|X\|=1} \|MX\|$, where $\|X\|$ is a norm on $\mathbb{C}^n$, are submultiplicative.

Proposition 8.3.3 *Let M be a matrix with complex coefficients. Let $Q(\lambda)$ be its characteristic polynomial, and let $q(\lambda)$ be its minimal polynomial. Let $g(\lambda)$ be the greatest common divisor of the (polynomial) coefficients of the matrix $\mathrm{adj}(\lambda I - M)$, and let $C(\lambda) = \mathrm{adj}(\lambda I - M)/g(\lambda)$. Then:*

(a) *The complex number λ is an eigenvalue of M if and only if $Q(\lambda) = 0$.*

(b) *$Q(M) = 0$ (the Cayley–Hamilton theorem).*

(c) *The minimal polynomial $q(\lambda)$ divides the polynomial $Q(\lambda)$. Furthermore, any root of $Q(\lambda)$ in $\mathbb{C}$, i.e., any eigenvalue of M, is a root of $q(\lambda)$.*

(d) *$q(\lambda) = Q(\lambda)/g(\lambda)$, and hence $C(\lambda)/q(\lambda) = \mathrm{adj}(\lambda I - M)/Q(\lambda)$. Furthermore, the relation $(\lambda I - M)C(\lambda) = C(\lambda)(\lambda I - M) = q(\lambda) I$ holds.*

(e) *$Q'(\lambda) = \mathrm{tr}(\mathrm{adj}(\lambda I - M))$, where Q' is the derivative of Q.*

(f) *Let $\| \ \|$ be a submultiplicative norm on the vector space of $n \times n$ complex matrices. Let M be a complex matrix such that $\|M\| \le 1$. If the minimal polynomial of M has a root of modulus 1, this root is simple.*

Proof. (a): The complex number λ is an eigenvalue of the matrix M if and only if there exists a nonzero vector V such that $(M - \lambda I)V = 0$. This happens if and only if the matrix $M - \lambda I$ is not invertible, which is equivalent to having $\det(M - \lambda I) = 0$.

(b): $(\lambda I - M) \operatorname{adj}(\lambda I - M) = \det(\lambda I - M) I = Q(\lambda)I$. Since $\operatorname{adj}(\lambda I - M)$ is clearly a matrix whose entries are polynomials in λ of degree at most $d - 1$, there exist $d \times d$ matrices $B_0, B_1, \ldots, B_{d-1}$ such that $\operatorname{adj}(\lambda I - M) = B_0 + \lambda B_1 + \lambda^2 B_2 + \cdots + \lambda^{d-1} B_{d-1}$. Hence

$$(\lambda I - M)(B_0 + \lambda B_1 + \lambda^2 B_2 + \cdots + \lambda^{d-1} B_{d-1}) = Q(\lambda)I.$$

It now suffices to apply Lemma 8.3.2.

(c): Dividing the characteristic polynomial Q by the minimal polynomial q, we have $Q(\lambda) = S(\lambda)q(\lambda) + T(\lambda)$, where either $\deg T < \deg q$ or $T = 0$. Replacing λ by the matrix M gives $0 = 0 + T(M)$. If T were different from 0, we would have $\deg T < \deg q$ and $T(M) = 0$, contradicting the minimality of q. Hence $T = 0$, and q divides Q. Note that the same proof shows that *any* polynomial q_1 such that $q_1(M) = 0$ is divisible by the minimal polynomial q.

Let λ_0 be an eigenvalue, i.e., a root of $Q(\lambda) = 0$. Dividing q by the polynomial $\lambda - \lambda_0$, we have $q(\lambda) = R(\lambda)(\lambda - \lambda_0) + q(\lambda_0)$. Replacing λ by M gives

$$0 = q(M) = R(M)(M - \lambda_0 I) + q(\lambda_0)I.$$

Now, let $V \neq 0$ be an eigenvector of the matrix M for the eigenvalue λ_0. Applying the matrix equality above to the vector V we obtain

$$0 = R(M)(MV - \lambda_0 V) + q(\lambda_0)V = 0 + q(\lambda_0)V.$$

Hence $q(\lambda_0) = 0$.

(d): Expanding the determinant $\det(\lambda I - M) = Q(\lambda)$ with respect to the first row, we see easily that the polynomial g divides Q. Let q_1 be the polynomial defined by $q_1(\lambda) := Q(\lambda)/g(\lambda)$. Dividing the equality

$$(\lambda I - M) \operatorname{adj}(\lambda I - M) = \det(\lambda I - M) I = Q(\lambda)I$$

by $g(\lambda)$, we get

$$(\lambda I - M)C(\lambda) = q_1(\lambda)I.$$

Using Lemma 8.3.2 we have $q_1(M) = 0$. Hence, as in the proof of the first assertion of (c), the polynomial q_1 is divisible by q, say $q_1 = q q_2$. Hence

$$(\lambda I - M)C(\lambda) = q(\lambda)q_2(\lambda)I.$$

On the other hand, since $q(M) = 0$, we have from Lemma 8.3.2 that there exists a matrix with polynomial entries $\widetilde{C}(\lambda)$ such that $(\lambda I - M)\widetilde{C}(\lambda) = q(\lambda)I$. Hence for all sufficiently large real numbers λ we have

$$\frac{\widetilde{C}(\lambda)}{q(\lambda)} = (\lambda I - M)^{-1} = \frac{C(\lambda)}{q(\lambda)q_2(\lambda)}.$$

Hence, for all sufficiently large real numbers λ,

$$q_2(\lambda)\widetilde{C}(\lambda) = C(\lambda).$$

This equality is thus true when we consider both sides as matrices with polynomial entries. In particular, $q_2(\lambda)$ must be a constant polynomial, since the gcd of the (polynomial) entries of $C(\lambda)$ is 1. Looking at the leading coefficients shows that $q_2 = 1$. Hence $q_1 = q$, which gives $(\lambda I - M)C(\lambda) = C(\lambda)(\lambda I - M) = q(\lambda)I$, $q(\lambda) = Q(\lambda)/g(\lambda)$, and $C(\lambda)/q(\lambda) = \mathrm{adj}(\lambda I - M)/Q(\lambda)$.

(e): The proof is straightforward when M is a diagonal matrix; see Exercise 5. We deduce from this the result when M is diagonalizable: suppose $M = PDP^{-1}$ where D is a diagonal matrix. Take λ to be a large real number. Then

$$(\lambda I - M)^{-1} = (P(\lambda I - D)P^{-1})^{-1} = P(\lambda I - D)^{-1}P^{-1}.$$

Hence by multiplying by $Q(\lambda)$, which is also equal to the characteristic polynomial of $\lambda I - D$, we get

$$P\,\mathrm{adj}(\lambda I - M)\,P^{-1} = P\,\mathrm{adj}(\lambda I - D)\,P^{-1}.$$

Hence

$$Q'(\lambda) = \mathrm{tr}(\mathrm{adj}(\lambda I - D)) = \mathrm{tr}(\mathrm{adj}(\lambda I - M)).$$

To end the proof it suffices to remember that the set of diagonalizable matrices is dense (see Exercise 6) in the set of all matrices, and that the trace, the adjoint, and the determinant are all continuous functions of the entries of a matrix.

(f): Let λ be a complex number of modulus 1. To show that λ cannot be a multiple root of the minimal polynomial of the matrix M, it suffices to show that for any matrix B we have

$$(M - \lambda I)^2 B = 0 \quad \text{implies that} \quad (M - \lambda I)B = 0,$$

and to apply this statement when B is a polynomial in M. Note that this implication can also be written

$$M(M - \lambda I)B = \lambda(M - \lambda I)B \quad \text{implies that} \quad (M - \lambda I)B = 0.$$

Suppose that $M(M - \lambda I)B = \lambda(M - \lambda I)B$ holds. Then

$$
\begin{aligned}
MB &= (M - \lambda I)B + \lambda B, \\
M^2 B &= M(M - \lambda I)B + \lambda MB = \lambda(M - \lambda I)B + \lambda((M - \lambda I)B + \lambda B) \\
&= 2\lambda(M - \lambda I)B + \lambda^2 B, \\
&\ \ \vdots \\
M^k B &= k\lambda^{k-1}(M - \lambda I)B + \lambda^k B \qquad \text{for any integer } k.
\end{aligned}
$$

Hence

$$k\|(M - \lambda I)B\| = \|k\lambda^{k-1}(M - \lambda I)B\|$$
$$= \|M^k B - \lambda^k B\| \leq \|M\|^k \|B\| + \|B\| \leq 2\|B\|.$$

Since the right-hand term is bounded, the left-hand term is bounded; hence we have $\|(M - \lambda I)B\| = 0$, i.e., $(M - \lambda I)B = 0$. ∎

Proposition 8.3.4

(a) *If M is an irreducible non-negative $d \times d$ matrix ($d \geq 2$), and if y is a non-negative vector of dimension d with exactly k positive entries ($1 \leq k \leq d - 1$), then $(I + M)y$ has strictly more than k positive entries.*

(b) *If M is an irreducible non-negative $d \times d$ matrix, and if $y \neq 0$ is a non-negative vector of dimension d, then all the entries of $(I + M)^{d-1}y$ are positive.*

(c) *A non-negative $d \times d$ matrix is irreducible if and only if all the entries of the matrix $(I + M)^{d-1}$ are positive.*

(d) *A non-negative eigenvector of a non-negative irreducible matrix must have positive entries.*

(e) *Let $M = (m_{i,j})_{i,j}$ be a non-negative $d \times d$ matrix, and let $M^k = (m_{i,j}^{(k)})_{i,j}$ be its kth power. The matrix M is irreducible if and only if for each i, j there exists $k = k(i, j)$ such that $m_{i,j}^{(k)} > 0$ (and there is such a k with $k \leq d$). In particular, any primitive matrix is irreducible.*

Proof. (a): Let P be a permutation matrix such that the first k entries of Py are positive and the others are zero. We have $(I + M)y = y + My$, and $M \geq 0$; hence $(I + M)y$ cannot have strictly more than $d - k$ zero entries. If $(I + M)y$ has exactly $d - k$ zero entries, then $y_i = 0$ implies that $(My)_i = 0$. Hence $(Py)_i = 0$ implies that $(PMy)_i = 0$, and so we have $(PMy)_i = 0$ for $i = k + 1, k + 2, \ldots, d$. Let N be the non-negative matrix defined by $N := (n_{i,j})_{i,j} := PMP^{-1}$, and $z := Py$. We thus have $(Nz)_i = (PMy)_i = 0$ for $i = k + 1, k + 2, \ldots, d$, i.e., $\sum_{1 \leq j \leq d} n_{i,j}z_j = 0$ for $i = k + 1, k + 2, \ldots, d$. Since $z_1, z_2, \ldots, z_k > 0$, we have $n_{i,j} = 0$ for $i = k + 1, k + 2, \ldots, d$ and $j = 1, 2, \ldots, k$, and hence M is reducible.

(b): It suffices to iterate property (a) above.

(c): If M is irreducible, then for each vector e_j of the canonical basis, we have $(I + M)^{d-1}e_j > 0$. Hence all columns of $(I + M)^{d-1}$ are positive. The converse is straightforward: the reader can easily prove that, if a non-negative matrix N is such that N^q is irreducible for some integer $q \geq 1$, then N itself is irreducible, and that a non-negative matrix N is irreducible if and only if the matrix $I + N$ is irreducible.

(d): Let $Mx = \lambda x$ where M is non-negative and irreducible, $x \geq 0$ and $x \neq 0$. Since $(I + M)x = (1 + \lambda)x$, if x had $k \geq 1$ zero entries, then $(1 + \lambda)x$ would also have k zero entries, although $(I + M)x$ would have strictly less than k zero entries by property (a) above.

(e): Suppose that M is irreducible. We know by property (c) above that all the entries of the matrix $(I + M)^{d-1}$ are positive. Hence all the entries of the matrix $M(I + M)^{d-1}$ are positive (if an entry were zero, that would imply that a row of the matrix M is zero and M would be reducible). It now suffices to look at the entry (i, j) of the matrix $M(I + M)^{d-1}$ to conclude that there exists a k (and this k is at most equal to d) such that $m_{i,j}^{(k)} > 0$.

The converse is easy: if the matrix M is not irreducible, then there exists a permutation matrix S such that

$$S^{-1}MS = \begin{bmatrix} B & C \\ 0 & D \end{bmatrix}.$$

Hence, for all $k \geq 0$,

$$S^{-1}M^k S = \begin{bmatrix} B^k & C_k \\ 0 & D^k \end{bmatrix}$$

where C_k is some rectangular matrix. In particular the entry in the lower left corner of the matrix $S^{-1}M^k S$ is equal to 0 for any $k \geq 0$. But this entry is equal to $m_{i,j}^{(k)}$ for some index i, j independent of k. ∎

Definition 8.3.5 Let M be an irreducible $d \times d$ non-negative matrix. The *Collatz–Wielandt function* associated with M is the function f_M defined on the vectors $x \neq 0$ of dimension d with non-negative entries by

$$f_M(x) := \min_{x_i \neq 0} \frac{(Mx)_i}{x_i}.$$

Proposition 8.3.6 *Let M be an irreducible $d \times d$ non-negative matrix. Let f_M be its Collatz–Wielandt function.*

(a) *The function f_M satisfies $f_M(tx) = f_M(x)$ for all $t > 0$.*

(b) *Let $x \neq 0$ be a vector of dimension d with non-negative entries. Then $f_M(x)$ is the largest real number ρ for which the vector $(M - \rho I)x$ has non-negative entries.*

(c) *Let $x \neq 0$ be a vector of dimension d with non-negative entries. Let $y = (I + M)^{d-1}x$; then $f_M(y) \geq f_M(x)$.*

(d) *The function f_M attains its maximum on the set $\mathcal{E}$ of vectors of dimension d with non-negative entries and such that the sum of their entries is equal to 1.*

Proof. (a) and (b): The proofs are straightforward and left to the reader.

(c): All the entries of $Mx - f_M(x)x$ are non-negative. Multiplying on the left by the matrix $(I + M)^{d-1}$, we see that all the entries of $(I + M)^{d-1}Mx - f_M(x)(I + M)^{d-1}x$ are non-negative. But M and $(I + M)^{d-1}$ commute; hence all entries of $M(I + M)^{d-1}x - f_M(x)(I + M)^{d-1}x$ are non-negative, i.e., all the entries of $My - f_M(x)y$ are non-negative. Using property (b), this implies that $f_M(y) \geq f_M(x)$.

(d): Let $\mathcal{F} = \{(I + M)^{d-1}x : x \in \mathcal{E}\}$, where $\mathcal{E}$ is defined in the statement of property (d). The set $\mathcal{F}$ is compact, and by Proposition 8.3.4(b), all the entries of any vector in $\mathcal{F}$ are positive. This implies that f_M is continuous on $\mathcal{F}$. Since $\mathcal{F}$ is compact, the function f_M attains its maximum on $\mathcal{F}$ at some point z. Using property (a), we may suppose that the sum of the entries of z is equal to 1, i.e., that z belongs to $\mathcal{E}$. Now if x is any vector in $\mathcal{E}$ and if $y = (I + M)^{d-1}x$, then:

$f_M(x) \leq f_M(y)$ by property (c),
$f_M(y) \leq f_M(z)$ by the maximality of $f_M(z)$ on $\mathcal{F}$.

This implies that f_M attains its maximum on $\mathcal{E}$ at z.   ∎

Theorem 8.3.7 (Perron–Frobenius) *Let $M = (m_{i,j})_{i,j}$ be a $d \times d$ irreducible non-negative matrix. Let $r = \max\{f_M(x),\ x \in \mathcal{E}\}$, where f_M is defined in Definition 8.3.5, and $\mathcal{E}$ is defined in Proposition 8.3.6(d) above. Then the number r, called the Perron–Frobenius eigenvalue of M, has the following properties.*

(a) *r is a positive eigenvalue of the matrix M. Furthermore, any (complex) eigenvalue of M satisfies $|\lambda| \leq r$. If M has exactly h eigenvalues of modulus r, the number h is called the index of imprimitivity of the matrix M.*

(b) *There exists an eigenvector corresponding to the eigenvalue r which has positive entries. More precisely, any vector $\xi \in \mathcal{E}$ for which $f_M(\xi) = r$ is an eigenvector for the eigenvalue r, and it has positive entries. Furthermore, any eigenvector corresponding to the eigenvalue r is a multiple of this vector.*

(c) *r is a simple root of the characteristic polynomial of M.*

Proof. (a): We know by Proposition 8.3.6(d) that there exists a vector $z \in \mathcal{E}$ such that, for each vector $x \in \mathcal{E}$, we have $f_M(z) \geq f_M(x)$. Let $r := f_M(z)$ ($= \max\{f_M(x),\ x \in \mathcal{E}\}$). Let u be the d-dimensional vector having all its entries equal to $1/d$. Now

$$r \geq f_M(u) = \min_i \frac{(Mu)_i}{u_i} = \min_i \sum_{0 \leq j \leq d} m_{i,j} > 0,$$

since the matrix M cannot have a row of zeros.

Now, for z as above we have that $Mz - rz = Mz - f_M(z)z$ has all its entries non-negative. If $Mz - rz \neq 0$, then, from Proposition 8.3.4(b), the vector $(I + M)^{d-1}(Mz - rz)$ has all its entries positive. Let $y := (I + M)^{d-1}z$. We thus have that $My - ry$ has all its entries positive. Hence there exists $\varepsilon > 0$ such that the vector $My - (r + \varepsilon)y$ has all its entries non-negative. This implies that $f_M(y) \geq r + \varepsilon$ by Proposition 8.3.6(b). Hence $r < f_M(y)$, which contradicts the maximality of r. Hence $Mz - rz = 0$. Thus r is an eigenvalue of M, and z is a non-negative eigenvector of M corresponding to the eigenvalue r. We know by Proposition 8.3.4(d) that z must have positive entries.

Suppose now λ is an eigenvalue of the matrix M. Then there exists a vector $x \neq 0$ such that $Mx = \lambda x$. This implies, for all $i = 1, 2, \ldots, d$, that $\lambda x_i = \sum_{1 \leq j \leq d} m_{i,j} x_j$. Hence $|\lambda||x_i| \leq \sum_{1 \leq j \leq d} m_{i,j} |x_j|$. In other words, if we denote by x' the vector having the $|x_i|$'s as entries, the vector $Mx' - |\lambda| x'$ has non-negative entries. Hence

$$|\lambda| \leq f_M(x') \leq r.$$

(b): We have just seen that there exists an eigenvector z corresponding to the eigenvalue r that has positive entries. To prove that the eigenspace of the matrix M corresponding to the eigenvalue r has dimension 1, we take the vector z as in (a): this is an eigenvector of the matrix M corresponding to the eigenvalue r, and z has positive entries. Now let y be an eigenvector of M for the eigenvalue r. Let y' be the vector whose entries are the moduli of the entries of y. Note that, by applying the inequalities at the end of the proof of (a) above, we have $f_M(y') = r$. Since $My = ry$ and $y \neq 0$, we have that $My' - ry'$ is non-negative and $y' \neq 0$. Then, as in the proof of (a), the vector $My' - ry'$ equals 0, and y' must have positive entries. Hence all entries of y are nonzero. Hence *all entries of any eigenvector of M for the eigenvalue r are nonzero.* Now take x and y two eigenvectors of M for the eigenvalue r. Let x_i and y_i be their ith entries. The vector $y_1 x - x_1 y$ either is equal to 0 or is an eigenvector of M for the eigenvalue r. But its first entry is 0: it thus cannot be an eigenvector for r. Hence $y_1 x - x_1 y = 0$, i.e., the vectors x and y are linearly dependent.

(c): Let $Q(\lambda) = \det(\lambda I - M)$ be the characteristic polynomial of M. We want to show that $Q'(r) \neq 0$. We have by Proposition 8.3.3(e) the relation $Q'(r) = \mathrm{tr}(\mathrm{adj}(rI - M))$. Let $B(\lambda) := \mathrm{adj}(\lambda I - M)$. We have

$$(\lambda I - M) \, \mathrm{adj}(\lambda I - M) = \det(\lambda I - M) \, I = Q(\lambda) I;$$

hence $(rI - M)B(r) = Q(r)I = 0$. Hence each column of $B(r)$ is either 0 or an eigenvector of M for the eigenvalue r, and thus a real multiple of a vector having all its entries positive. Hence each column of $B(r)$ either is 0 or has all its entries positive or all its entries negative. Since the transpose of M is irreducible and has r as its Perron–Frobenius eigenvalue, each row of $B(r)$ either is 0 or has all its entries positive or all its entries negative. Furthermore at least one of the columns and one of the rows of $B(r)$ is nonzero: we have seen in (b) that the eigenspace of M corresponding to the eigenvalue r has dimension 1, hence the matrix $rI - M$ has rank $d - 1$, and its adjoint $B(r)$ cannot be zero. Finally, this implies that all entries of $B(r)$ are positive or all entries of $B(r)$ are negative. Hence, using Proposition 8.3.3(e), we have $Q'(r) = \mathrm{tr}(B(r)) \neq 0$. (Note that looking at $B(\lambda)$ for $\lambda \to \infty$ shows that actually all entries of $B(r)$ are positive.) ∎

Theorem 8.3.8 *Let $M = (m_{i,j})_{i,j}$ be a $d \times d$ irreducible non-negative matrix. Let $B = (b_{i,j})_{i,j}$ be a $d \times d$ complex matrix, such that, for all i, j, we have $|b_{i,j}| \leq m_{i,j}$*

(M is said to dominate B). Let r be the Perron–Frobenius eigenvalue of M. Then, for every eigenvalue s of B, we have $|s| \leq r$.

Furthermore, there exists an eigenvalue s of B such that $|s| = r$ if and only if $B = e^{i\varphi} D^{-1} M D$, where $s = re^{i\varphi}$ and D is a diagonal matrix with all its diagonal entries of modulus 1.

Proof. Let $y \neq 0$ such that $By = sy$. Hence for all i we have $\sum_{1 \leq j \leq d} b_{i,j} y_j = sy_i$, which implies

$$|s||y_i| \leq \sum_{1 \leq j \leq d} |b_{i,j}||y_j| \leq \sum_{1 \leq j \leq d} m_{i,j}|y_j|. \tag{8.1}$$

Hence, denoting by y' the vector with entries $|y_i|$, we have $My' - |s|y' \geq 0$. Using Proposition 8.3.6(b), Theorem 8.3.7, and Proposition 8.3.6(a), this implies the inequalities

$$|s| \leq f_M(y') \leq r. \tag{8.2}$$

Suppose that $B = e^{i\varphi} D^{-1} M D$. Then B and $e^{i\varphi} M$ are similar. Since r is an eigenvalue of M, then $re^{i\varphi}$ is an eigenvalue of B of modulus r.

Suppose now that $s = re^{i\varphi}$ is an eigenvalue of the matrix B. We deduce from Eq. (8.2) that

$$|s| = f_M(y') = r. \tag{8.3}$$

Hence y' is an eigenvector of M for the Perron–Frobenius eigenvalue r (use Theorem 8.3.7(b) and Proposition 8.3.6(a)), i.e., $My' = ry'$. Hence, using Eq. (8.1), we have for every i

$$r|y_i| = |s||y_i| \leq \sum_{1 \leq j \leq d} |b_{i,j}||y_j| \leq \sum_{1 \leq j \leq d} m_{i,j}|y_j| = r|y_i|,$$

which implies

$$\sum_{1 \leq j \leq d} (m_{i,j} - |b_{i,j}|)|y_j| = 0.$$

Since $m_{i,j} \geq |b_{i,j}|$ by hypothesis, and since y' is a positive vector, we then have $m_{i,j} = |b_{i,j}|$ for all i, j. Let D be the diagonal matrix whose diagonal entries are

$$\frac{y_1}{|y_1|}, \frac{y_2}{|y_2|}, \ldots, \frac{y_d}{|y_d|},$$

and let $G = (g_{i,j})_{i,j} := e^{-i\varphi} D^{-1} B D$. Since $By = sy$ and $Dy' = y$, we have

$$Gy' = e^{-i\varphi} D^{-1} B D y' = e^{-i\varphi} D^{-1} By = s e^{-i\varphi} D^{-1} y = ry' = My'.$$

Hence, for all i,

$$\sum_{1 \leq j \leq d} (g_{i,j} - m_{i,j})|y_j| = 0. \tag{8.4}$$

Since

$$g_{i,j} = e^{-i\varphi} \frac{|y_i|}{y_i} b_{i,j} \frac{y_j}{|y_j|} \qquad \text{for all } i, j,$$

we have $|g_{i,j}| = |b_{i,j}|$. But we have seen that $m_{i,j} = |b_{i,j}|$, hence $m_{i,j} = |g_{i,j}|$. Using Eq. (8.4), we thus have for all i

$$\sum_{1 \leq j \leq d} (|g_{i,j}| - g_{i,j})|y_j| = 0.$$

Taking the real part and the imaginary part, and using that $|y_j| > 0$ for all j, we deduce that $|g_{i,j}| = g_{i,j}$ for all i, j. Hence $g_{i,j} = m_{i,j}$ for all i, j, i.e., $B = e^{i\varphi} DMD^{-1}$. ∎

Theorem 8.3.9 *Let M be a $d \times d$ irreducible non-negative matrix, with index of imprimitivity h. Let $\lambda^d + a_1 \lambda^{d_1} + a_2 \lambda^{d_2} + \cdots + a_k \lambda^{d_k}$, with $d > d_1 > d_2 > \cdots > d_k$ and $a_j \neq 0$ for all j, be its characteristic polynomial.*

(a) The relation

$$\begin{aligned} h &= \gcd(d - d_1, d - d_2, d - d_3, \ldots, d - d_k) \\ &= \gcd(d - d_1, d_1 - d_2, d_2 - d_3, \ldots, d_{k-1} - d_k) \end{aligned}$$

holds.
(b) If the trace of the matrix M is positive, then its index of imprimitivity h is equal to 1.

Proof. (a): Let θ be a complex number. The matrices M and θM have the same spectrum if and only if their characteristic polynomials are equal, i.e.,

$$\begin{aligned} \lambda^d &+ a_1 \lambda^{d_1} + a_2 \lambda^{d_2} + \cdots + a_k \lambda^{d_k} \\ &= \lambda^d + a_1 \theta^{d-d_1} \lambda^{d_1} + a_2 \theta^{d-d_2} \lambda^{d_2} + \cdots + a_k \theta^{d-d_k} \lambda^{d_k}. \end{aligned}$$

Hence $a_j = a_j \theta^{d-d_j}$ for all j. Now taking $\theta = e^{2i\pi/m}$ with $m \geq 1$ integer, the matrices M and $e^{2i\pi/m} M$ have the same spectrum for $m = h$ and not for $m > h$ (Theorem 8.3.10(a)). Since $a_j = a_j e^{2i\pi(d-d_j)/m}$ for all j, we see that

$$h = \gcd(d - d_1, d - d_2, d - d_3, \ldots, d - d_k).$$

Of course this gives immediately

$$h = \gcd(d - d_1, d_1 - d_2, d_2 - d_3, \ldots, d_{k-1} - d_k).$$

(b): If the trace of M is > 0, then $d_1 = d - 1$. Hence, from (a), $h = \gcd(d - (d - 1), \ldots) = 1$. ∎

Theorem 8.3.10 (Perron–Frobenius) *Let M be a d × d non-negative matrix.*

(a) If M is irreducible, and if its Perron–Frobenius eigenvalue is equal to r and its index of imprimitivity is equal to h, then the h eigenvalues of M of modulus r are the numbers $e^{2i\ell\pi/h}r$, where $\ell = 0, 1, 2, \ldots, h - 1$. These eigenvalues are simple. Furthermore the whole set of eigenvalues of M is closed under multiplication by $e^{2i\pi/h}$, and not closed under multiplication by $e^{i\alpha}$, where α is any positive number smaller than $2\pi/h$.

(b) The matrix M is primitive if and only if it is irreducible and its index of imprimitivity h is equal to 1.

Proof. (a): Let $\lambda_t = re^{i\varphi_t}$, with $t = 1, 2, \ldots, h$, be the eigenvalues of M of modulus r. Using Theorem 8.3.8 with $B = M$, there exists for each $t = 1, 2, \ldots, h$ a diagonal matrix D_t (with diagonal entries of modulus 1) such that

$$M = e^{i\varphi_t} D_t M D_t^{-1}.$$

Hence M and $e^{i\varphi_t} M$ are similar. Since r is a simple eigenvalue of M, the complex number $\lambda_t = re^{i\varphi_t}$ is a simple eigenvalue of $e^{i\varphi_t} M$, hence of M. Now for each t and s we have

$$M = e^{i\varphi_t} D_t M D_t^{-1} = e^{i\varphi_t} D_t(e^{i\varphi_s} D_s M D_s^{-1}) D_t^{-1} = e^{i(\varphi_t + \varphi_s)}(D_t D_s) M (D_t D_s)^{-1}.$$

This shows that M and $e^{i(\varphi_t + \varphi_s)} M$ are similar. Hence $re^{i(\varphi_t + \varphi_s)}$ is an eigenvalue of M (of modulus r), which implies that $e^{i(\varphi_t + \varphi_s)} \in S := \{e^{i\varphi_1}, e^{i\varphi_2}, \ldots, e^{i\varphi_h}\}$. Hence the set S is closed under multiplication. This implies that this set is exactly the set of hth roots of 1.

Since we have just seen that M and $e^{2i\pi/h} M$ are similar, the spectrum of M is closed under multiplication by $e^{2i\pi/h}$. It cannot be closed under multiplication by $e^{i\alpha}$ where α is any positive number smaller than $2\pi/h$, since what precedes shows that the set of eigenvalues of M of modulus r itself is not closed under multiplication by $e^{i\alpha}$.

(b): Suppose that M^m has positive entries for some integer m. Then M is irreducible (if it were reducible, M^m would also be reducible and would not have positive entries). Since M is irreducible, it cannot have a row of zeros; hence the matrix $M^{m+1} = M M^m$ also has positive entries. Let r be the Perron–Frobenius eigenvalue of M, and let h be its index of imprimitivity. We know by Theorem 8.3.10(a) that the eigenvalues of M of modulus r are $re^{2i\pi t/h}$ for $t = 1, 2, \ldots, h - 1$. Hence the numbers $r^m e^{2im\pi t/h}$ are eigenvalues of M^m of modulus r^m. Since M^m is irreducible and has positive entries, its trace is positive; hence its index of imprimitivity is equal to 1 (Theorem 8.3.9(b)). The eigenvalues $r^m e^{2im\pi t/h}$ must then all be equal to r^m. Hence h divides m. But we have seen that M^{m+1} also has positive entries, and so $h \mid m + 1$. Hence $h = 1$.

Suppose now that the matrix M is irreducible, and that its index of imprimitivity is equal to 1. Up to dividing M by its Perron–Frobenius value r, we may suppose that $r = 1$. Then, from properties (c) in Theorem 8.3.7 and (a) in Theorem 8.3.10, we know that this eigenvalue is simple and that no other eigenvalue of M has modulus 1. Hence M admits a Jordan form of the type $M = S(J_1 + B)S^{-1}$ where J_1 is the matrix defined by $(J_1)_{1,1} = 1$ and $(J_1)_{i,j} = 0$ for $(i, j) \neq (1, 1)$, and B is a matrix such that, in particular, $(J_1 + B)^k = J_1^k + B^k = J_1 + B^k$ and $\lim_{k \to \infty} B^k = 0$. Furthermore the first column of S is an eigenvector of M corresponding to the Perron–Frobenius eigenvalue 1, and the first row of S^{-1} is an eigenvector of the transpose of M for the Perron–Frobenius eigenvalue 1. Hence this column and this row have no zero entry. Then $\lim_{k \to \infty} M^k = S J_1 S^{-1}$. For all i, j the entry of index (i, j) of $S J_1 S^{-1}$ is $(S^{-1})_{i,1} S_{1,j}$ and hence is not zero. Since $\lim_{k \to \infty} M^k$ is non-negative, we see that all the entries of this limit are actually positive. Hence M^k has positive entries for large k. ∎

Theorem 8.3.11 *Let M be a non-negative $d \times d$ matrix. Then there exists a real number $r \geq 0$ (called the Perron–Frobenius eigenvalue of M) such that:*

(a) r is an eigenvalue of M. Furthermore, any (complex) eigenvalue of M satisfies $|\lambda| \leq r$.
(b) There exists a non-negative eigenvector corresponding to the eigenvalue r.
(c) There exists a positive integer h such that any eigenvalue of M with $|\lambda| = r$ satisfies $\lambda^h = r^h$.

Proof. Take any matrix N with positive entries (for example the matrix whose all entries are equal to 1), and consider the matrix $M + \varepsilon N$, where ε is a positive real number. The matrix $M + \varepsilon N$ has clearly positive entries. We apply Theorem 8.3.7 to $M + \varepsilon N$. Hence $M + \varepsilon N$ has a Perron–Frobenius eigenvalue r_ε, all its other eigenvalues λ_ε satisfy $|\lambda_\varepsilon| \leq r_\varepsilon$, and there exists a positive vector $x_\varepsilon \in \mathcal{E}$ such that $(M + \varepsilon N)x_\varepsilon = r_\varepsilon x_\varepsilon$. Now, letting ε tend to zero, and using the fact that eigenvalues and eigenvectors are continuous functions of the entries of a matrix, we see that the eigenvalues λ_ε tend to the eigenvalues of M (and all eigenvalues of M are limits of eigenvalues of M_ε), that r_ε tends to a real non-negative eigenvalue r of M, and that all the moduli of all eigenvalues of M are at most equal to r. Furthermore x_ε tends to a vector $x \in \mathcal{E}$: hence $x \neq 0$, $Mx = rx$, and hence x is a non-negative vector of the matrix M for the eigenvalue r. This gives assertions (a) and (b). For assertion (c), see Exercise 7. ∎

Theorem 8.3.12 *Let M be a primitive $d \times d$ matrix, and r be its Perron–Frobenius eigenvalue. Let $C(\lambda)$ be the reduced adjoint matrix of $\lambda I - M$, and $q(\lambda)$ be the minimal polynomial of M. Then*

(a) The columns of $C(r)$ are all proportional. More precisely, let x be an eigenvector of M for the eigenvalue r, and similarly let y be a (row) left eigenvector for the matrix M and the eigenvalue r. Suppose that x and y are normalized by $yx = 1$. Then we have

$$C(r) = q'(r)xy.$$

In particular, all the entries of the matrix $C(r)/q'(r)$ are positive.

(b) The relation

$$\lim_{n \to \infty} \frac{M^n}{r^n} = \frac{C(r)}{q'(r)}$$

holds. In particular, if M is a matrix with rational coefficients, then $\lim_{n \to \infty} M^n / r^n$ *also has rational coefficients.*

Proof. (a): First we write, using Proposition 8.3.3(d),

$$(\lambda I - M)C(\lambda) = C(\lambda)(\lambda I - M) = q(\lambda)I.$$

Hence $(rI - M)C(r) = 0$. In other words, $MC(r) = rC(r)$. This means that each column of the matrix $C(r)$ is an eigenvector of M associated with the Perron–Frobenius eigenvalue r. Hence all columns of $C(r)$ are proportional to some positive eigenvector x for the eigenvalue r (Theorem 8.3.7(b)).

Similarly each row of the matrix $C(r)$ is a left eigenvector for the matrix M and the (Perron–Frobenius) eigenvalue r. Hence all the rows of $C(r)$ are proportional to some positive (row) eigenvector y for the eigenvalue r.

We suppose that x and y are normalized so that $yx = 1$. The properties of the columns and rows of $C(r)$ show that $C(r) = cxy$, where c is some positive constant. Now from Proposition 8.3.3(d), we have $(\lambda I - M)C(\lambda) = q(\lambda)I$. Hence, by differentiating with respect to λ,

$$C(\lambda) + (\lambda I - M)C'(\lambda) = q'(\lambda)I.$$

Multiplying by y on the left and putting $\lambda = r$ gives

$$yC(r) = yC(r) + y(rI - M)C'(r) = q'(r)y.$$

Now replacing $C(r)$ by cxy, we obtain

$$cy(xy) = q'(r)y.$$

But $yx = 1$ and $y \neq 0$, and hence $q'(r) = c$.

(b): We have seen in the proof of Theorem 8.3.10(b) that the limit $\lim_{n \to \infty} M^k / r^k := L$ exists. This implies that, for each real number $\lambda > 1$, the matrix $(M/r\lambda)^k$ converges (elementwise) to 0 at least geometrically. Hence the series $\sum_{k \geq 0} M^k / r^k \lambda^k$ converges. We claim that the following limit exists and equals L:

$$\lim_{\lambda \to 1_+} \frac{\lambda - 1}{\lambda} \sum_{k \geq 0} \frac{M^k}{r^k \lambda^k}.$$

Namely, given $\varepsilon > 0$, take k_0 such that all entries of $M^k / r^k - L$ are smaller than $\varepsilon/2$ for $k \geq k_0$. Then all entries of

$$\frac{\lambda - 1}{\lambda} \sum_{k \geq k_0} \frac{\frac{M^k}{r^k} - L}{\lambda^k}$$

are bounded by

$$\frac{\varepsilon}{2}\left(\frac{\lambda-1}{\lambda}\sum_{k\geq k_0}\frac{1}{\lambda^k}\right)\leq\frac{\varepsilon}{2}\left(\frac{\lambda-1}{\lambda}\sum_{k\geq 0}\frac{1}{\lambda^k}\right)=\frac{\varepsilon}{2}.$$

Now note that M^k/r^k-L has all its entries uniformly bounded by some constant c_1 independent of k, since all its entries converge to 0. Then, k_0 being fixed, take $\alpha>0$ such that when $|\lambda-1|\leq\alpha$, then

$$\left|\frac{\lambda-1}{\lambda}\right|\leq\frac{\varepsilon}{2c_1k_0},$$

so that for such λ's we have that all entries of

$$\frac{\lambda-1}{\lambda}\sum_{0\leq k<k_0}\frac{\frac{M^k}{r^k}-L}{\lambda^k}$$

are bounded by

$$\frac{\varepsilon}{2c_1k_0}c_1\sum_{0\leq k<k_0}\frac{1}{\lambda^k}\leq\frac{\varepsilon}{2}\qquad\text{(remember that }\lambda>1\text{)}.$$

Hence

$$\left(\frac{\lambda-1}{\lambda}\sum_{k\geq 0}\frac{M^k}{r^k\lambda^k}\right)-L=\frac{\lambda-1}{\lambda}\sum_{k\geq 0}\frac{\frac{M^k}{r^k}-L}{\lambda^k}$$

$$=\frac{\lambda-1}{\lambda}\sum_{0\leq k<k_0}\frac{\frac{M^k}{r^k}-L}{\lambda^k}+\frac{\lambda-1}{\lambda}\sum_{k\geq k_0}\frac{\frac{M^k}{r^k}-L}{\lambda^k}$$

has all its entries bounded by ε.

Now, using again Proposition 8.3.3(d), $(\lambda I-M)C(\lambda)=q(\lambda)I$, hence, for all $\lambda>1$, we have $(r\lambda I-M)C(r\lambda)=q(r\lambda)I$. Hence

$$\frac{\lambda-1}{\lambda}\sum_{k\geq 0}\frac{M^k}{r^k\lambda^k}=\frac{\lambda-1}{\lambda}\left(I-\frac{M}{r\lambda}\right)^{-1}$$

$$=r(\lambda-1)(r\lambda I-M)^{-1}=\frac{r(\lambda-1)}{q(r\lambda)}C(r\lambda).$$

Letting λ tend to 1, we deduce

$$\lim_{k\to\infty}\frac{M^k}{r^k}=L=\lim_{\lambda\to 1_+}\frac{\lambda-1}{\lambda}\sum_{k\geq 0}\frac{M^k}{r^k\lambda^k}=\frac{1}{q'(r)}C(r).\qquad\blacksquare$$

Theorem 8.3.13 *Let M be a stochastic $d\times d$ matrix. Then its Perron–Frobenius eigenvalue is equal to 1. This eigenvalue, and all other eigenvalues of modulus 1 if any, are simple roots of the minimal polynomial of M. If furthermore M has no* *eigenvalue $\lambda \neq 1$ such that $|\lambda| = 1$, then*

$$\lim_{n \to \infty} M^n = \frac{C(1)}{q'(1)},$$

where $C(\lambda)$ is the reduced adjoint matrix of $\lambda I - M$, and $q(\lambda)$ the minimal polynomial of M. In particular, if M is a matrix with rational coefficients, then $\lim_{n \to \infty} M^n$ also has rational coefficients.

Proof. Let $M = (m_{i,j})_{i,j}$ be a stochastic matrix. Then 1 is an eigenvalue of M, since the vector V whose entries are all equal to 1 clearly satisfies $MV = V$. Now, if λ is an eigenvalue associated with the vector W whose entries are $w_1, w_2, \ldots, w_d$, then the relation $MW = \lambda W$ gives

$$|\lambda||w_i| = |\lambda w_i| = \left| \sum_{1 \leq j \leq d} m_{i,j} w_j \right|$$

$$\leq \sum_{1 \leq j \leq d} |m_{i,j}||w_j| = \sum_{1 \leq j \leq d} m_{i,j}|w_j|$$

$$\leq \max_j |w_j| \sum_{1 \leq j \leq d} m_{i,j} = \max_j |w_j|.$$

Taking $i = i_0$ such that $|w_{i_0}| = \max_j |w_j|$, and noting that $|w_{i_0}| \neq 0$ (since the vector W is not zero), we have

$$|\lambda| \leq 1.$$

Hence 1 is the Perron–Frobenius value of the matrix M.

We now prove that 1, as well as all eigenvalues of M of modulus 1 (if any), are simple roots of the minimal polynomial of M. Define $\|B\|$ for a $d \times d$ matrix B by

$$\|B\| := \max_{1 \leq i \leq d} \sum_{1 \leq j \leq d} |b_{i,j}|.$$

It is straightforward to prove that $\|\ \|$ is a submultiplicative norm. Furthermore $\|M\| = 1$ since M is stochastic. Since 1 is an eigenvalue of M, it is a root of the minimal polynomial (see Proposition 8.3.3(c)). By Proposition 8.3.3(f) it is a simple root. The same result shows that any eigenvalue of M of modulus 1 (if any) is a root and hence a simple root of the minimal polynomial.

The assertion about M^n, with the extra hypothesis that M has no eigenvalue $\lambda \neq 1$ of modulus 1, is proved exactly as in Theorem 8.3.12 for the existence of the limit and the value $C(1)/q'(1)$. ∎

Theorem 8.3.14 *Let $M = (m_{i,j})_{i,j}$ be a non-negative $d \times d$ matrix, and let $M^n = (m_{i,j}^{(n)})_{i,j}$ be its n th power. Denote by r the Perron–Frobenius eigenvalue of M.*

(a) *If M has no eigenvalue* λ *such that* $|\lambda| = r$ *and* $\lambda \neq r$, *then there exist an integer* $p > 0$ *and constants* $e_{i,j} \geq 0$, *such that, as* $n \to \infty$,

$$m_{i,j}^{(n)} = e_{i,j} n^{p-1} r^n + o(n^{p-1} r^n).$$

Furthermore there exist i, j *such that* $e_{i,j} > 0$.

(b) *If the matrix M has its entries in* $\mathbb{Q}$, *then the quantities* $e_{i,j}$ *and* r *are algebraic numbers.*

Proof. We already know by Theorem 8.3.11 that M admits a Perron–Frobenius eigenvalue r, and (with the extra hypothesis that M has no eigenvalue λ such that $|\lambda| = r$ and $\lambda \neq r$) that any other eigenvalue λ of M satisfies $|\lambda| < r$. We now reduce M to Jordan normal form. There exists an invertible matrix T such that the matrix $J = T^{-1}MT$ is a Jordan matrix, i.e., J has square blocks $M_1, \ldots, M_t$ on its diagonal, and 0's elsewhere, M_k is a $d_k \times d_k$ matrix

$$M_k = \begin{bmatrix} \lambda_k & 1 & 0 & \cdots & 0 \\ 0 & \lambda_k & 1 & \cdots & 0 \\ \vdots & & \ddots & \ddots & \vdots \\ 0 & 0 & \cdots & \lambda_k & 1 \\ 0 & 0 & 0 & \cdots & \lambda_k \end{bmatrix},$$

and the set of λ_k's with multiplicities d_k is exactly the set of eigenvalues of M with their multiplicities.

Since $M^n = T J^n T^{-1}$, the entries of M^n are linear combinations with coefficients independent of n of the entries of J^n. Since J^n has the blocks $M_1^n, \ldots, M_t^n$ on its diagonal and 0's elsewhere, and since

$$M_k^n = \begin{bmatrix} \lambda_k^n & n\lambda_k^{n-1} & \binom{n}{2}\lambda_k^{n-2} & \cdots & \binom{n}{d_k-2}\lambda_k^{n-d_k+2} & \binom{n}{d_k-1}\lambda_k^{n-d_k+1} \\ 0 & \lambda_k^n & n\lambda_k^{n-1} & \cdots & \binom{n}{d_k-3}\lambda_k^{n-d_k+2} & \binom{n}{d_k-2}\lambda_k^{n-d_k+1} \\ \vdots & \vdots & \vdots & & \vdots & \vdots \\ 0 & 0 & 0 & \cdots & \lambda_k^n & n\lambda_k^{n-1} \\ 0 & 0 & 0 & \cdots & 0 & \lambda_k^n \end{bmatrix},$$

we see that there exist constants $c_{i,j}^{k,\ell}$ independent of n, such that, for all $n \geq 1$,

$$m_{i,j}^{(n)} = \sum_{1 \leq k \leq t} \sum_{0 \leq \ell < d_k} c_{i,j}^{k,\ell} \binom{n}{\ell} \lambda_k^{n-\ell}.$$

Let $p = \max\{d_k : \lambda_k = r\}$. Then there exist constants $e_{i,j}$, independent of n, such that

$$m_{i,j}^{(n)} = e_{i,j} n^{p-1} r^n + o(n^{p-1} r^n).$$

Since $m_{i,j}^{(n)} \geq 0$, we have $e_{i,j} \geq 0$.

To prove that there exist i, j such that $e_{i,j} > 0$, let us suppose that $e_{i,j} = 0$ for all i, j. Then all entries of the matrix M^n are $o(n^{p-1}r^n)$ as $n \to \infty$. This implies that all entries of the matrix J^n are also $o(n^{p-1}r^n)$ as $n \to \infty$. But this is not possible, since one of the entries of J^n is equal to $\binom{n}{p-1}r^{n-p+1}$, which is not $o(n^{p-1}r^n)$ as $n \to \infty$.

The assertion that r and the $e_{i,j}$ are algebraic is a straightforward consequence of the fact that all computations in the proof take place in the field generated by $\mathbb{Q}$ and the eigenvalues of the matrix M. ∎

8.4 Frequencies of Letters and Words in a Morphic Sequence

Before studying frequencies of letters in a morphic sequence, we begin with a proposition about the growth of the length of the nth iterate of a primitive morphism. The proof is a simple example of the methods used later in this section.

Proposition 8.4.1 *Let $\varphi : \Sigma^* \to \Sigma^*$ be a primitive morphism on the finite set $\Sigma = \{a_1, a_2, \ldots, a_d\}$. Let $M(\varphi)$ be the incidence matrix of φ, and let r be the Perron–Frobenius eigenvalue of $M(\varphi)$. Then*

(a) the incidence matrix $M(\varphi)$ is primitive;
(b) for each nonempty word w on Σ, there exists a positive constant $c(w)$ such that

$$\lim_{n \to \infty} \frac{|\varphi^n(w)|}{r^n} = c(w).$$

Proof. The first assertion is clear. In order to prove the second assertion, we have, using Corollary 8.2.4 and Theorem 8.3.12(b), that for each nonempty word $w \in \Sigma^*$ the limit $\lim_{n \to \infty} |\varphi^n(w)|/r^n$ exists and

$$c(w) := \lim_{n \to \infty} \frac{|\varphi^n(w)|}{r^n} = \begin{bmatrix} 1 & 1 & 1 & \cdots & 1 \end{bmatrix} \frac{C(r)}{q'(r)} \begin{bmatrix} |w|_{a_1} \\ |w|_{a_2} \\ \vdots \\ |w|_{a_d} \end{bmatrix}.$$

Since the morphism φ is primitive, there exists an integer $k > 0$ such that, for each $i = 1, 2, \ldots, d$, we have $|\varphi^k(w)|_{a_i} > 0$. Hence

$$c(w) = \lim_{n \to \infty} \frac{|\varphi^{n+k}(w)|}{r^{n+k}} = \lim_{n \to \infty} \frac{|\varphi^n(\varphi^k(w))|}{r^{n+k}}$$

$$= \begin{bmatrix} 1 & 1 & 1 & \cdots & 1 \end{bmatrix} \frac{C(r)}{r^k q'(r)} \begin{bmatrix} |\varphi^k(w)|_{a_1} \\ |\varphi^k(w)|_{a_2} \\ \vdots \\ |\varphi^k(w)|_{a_d} \end{bmatrix}.$$

Since we know by Theorem 8.3.12(a) that $C(r)/q'(r)$ is positive, we thus have $c(w) > 0$. ∎

The ordinary frequency of letters was defined in Section 1.1. A related concept, the *logarithmic frequency*, often proves useful.

Definition 8.4.2 We say that a letter a occurs in the sequence $\mathbf{x} = x_1x_2x_3\cdots$ with *logarithmic frequency* $\text{LogFreq}_{\mathbf{x}}(a)$ if the sequence

$$\frac{1}{\log N} \sum_{\substack{1 \le n \le N \\ x_n = a}} \frac{1}{n} \to \text{LogFreq}_{\mathbf{x}}(a) \quad \text{as} \quad N \to \infty.$$

Example 8.4.3 Consider the word $\mathbf{x} = x_1x_2x_3\cdots$ defined by taking x_i to be the most significant digit in the base-3 expansion of i. Thus $\mathbf{x} = 12111222111111111\cdots$. We prove that the logarithmic frequency of the letter 1 in $\mathbf{x}$ exists and is equal to $\log 2 / \log 3$. Let us define $E = \{i \ge 1 : x_i = 1\}$. Let

$$B_\ell = \sum_{3^\ell \le j < 3^{\ell+1}, \ j \in E} \frac{1}{j}.$$

Clearly

$$B_\ell = \sum_{3^\ell \le j < 2 \cdot 3^\ell} \frac{1}{j}.$$

When n goes to infinity, we have

$$\sum_{1 \le j \le n} \frac{1}{j} = \log n + \gamma + o(1),$$

where γ is the Euler constant. An easy consequence is that, when ℓ goes to infinity, B_ℓ goes to $\log 2$.

Let $n \ge 1$ be an integer, and let k be the integer such that $3^k \le n < 3^{k+1}$. Then

$$\frac{1}{\log n} \sum_{1 \le j \le n, \ j \in E} \frac{1}{j} = \frac{1}{\log n} \sum_{1 \le j < 3^k, \ j \in E} \frac{1}{j} + \frac{1}{\log n} \sum_{3^k \le j \le n, \ j \in E} \frac{1}{j}$$

$$= \frac{1}{\log n} \sum_{1 \le \ell < k} B_\ell + \frac{1}{\log n} \sum_{3^k \le j \le n, \ j \in E} \frac{1}{j}.$$

The last sum is bounded by $\frac{1}{\log n} B_k$, and hence tends to 0 when n tends to ∞. The first sum is equivalent to

$$\frac{1}{\log 3^k} \sum_{1 \le \ell < k} B_\ell = \frac{1}{\log 3} \left(\frac{1}{k} \sum_{1 \le \ell < k} B_\ell \right).$$

This sum tends to $\frac{\log 2}{\log 3}$. More precisely, B_ℓ tends to $\log 2$ when ℓ tends to infinity; hence its Cesáro mean (see Exercise 2.32) also converges to $\log 2$.

We now state a proposition, without proof, about the existence of the logarithmic frequency.

Proposition 8.4.4

(a) *If the frequency of a in the sequence $\mathbf{x}$ exists, then the logarithmic frequency of a in $\mathbf{x}$ also exists, and these two frequencies are equal.*

(b) *The letter a occurs in the sequence $\mathbf{x}$ with logarithmic frequency if and only if*

$$\lim_{s \to 1^+} \sum_{n \geq 1;\ x_n = a} \frac{1}{n^s}.$$

exists. Furthermore the logarithmic frequency and this limit are then equal.

We are now ready to give results on frequencies of letters in a morphic sequence.

Theorem 8.4.5

(a) *Let $\mathbf{x} = (x_n)_{n \geq 0}$ be a morphic sequence. If the frequency of a letter exists, then it is an algebraic number.*

(b) *Let $\mathbf{x} = (x_n)_{n \geq 0}$ be an automatic sequence. If the frequency of a letter exists, then it is a rational number.*

Proof. Let $\mathbf{x} = (x_n)_{n \geq 0}$ be a morphic sequence on the alphabet Σ. From Theorem 7.5.1 we can suppose that there exists an alphabet Γ, a nonerasing morphism μ on Γ, prolongable on $a_1 \in \Gamma$, and a coding $\tau : \Gamma \to \Sigma$, such that

$$\mathbf{x} = \tau(\mu^\omega(a_1)).$$

Let $\Gamma = \{a_1, \ldots, a_d\}$. We suppose that all letters of Γ occur in $\mu^\omega(a_1)$. Let M be the incidence matrix of the morphism μ. Up to replacing M by a power of M, we can suppose, using Theorem 8.3.11, that every eigenvalue λ of M (if any) that is different from the Perron–Frobenius eigenvalue r satisfies $|\lambda| < r$.

Suppose that the frequency of some letter $a \in \Sigma$ exists. Let $\Gamma = \Gamma' \cup \Gamma''$, where

$$\Gamma'' = \{b \in \Gamma : \text{for all } n \geq 0,\ \tau(\mu^n(b)) \in (\Sigma \setminus \{a\})^*\}$$

and $\Gamma' = \Gamma \setminus \Gamma''$.

Up to replacing M by $T^{-1}MT$, where T is a permutation matrix, we can suppose that the letters in Γ are ordered in such a way that the first d' letters are precisely the letters in Γ' and the remaining letters are the letters in Γ''. Hence the matrix M

can be written

$$M = \begin{bmatrix} M' & 0 \\ N & M'' \end{bmatrix}$$

where M' is a $d' \times d'$ matrix, M'' is a $d'' \times d''$ matrix, $d' + d'' = d$, and N and 0 are rectangular matrices of appropriate size. We also have

$$M^n = \begin{bmatrix} (M')^n & 0 \\ N_n & (M'')^n \end{bmatrix}$$

where N_n is some rectangular matrix.

We apply Theorem 8.3.14 to the matrix M. Let $M = (m_{i,j})_{i,j}$, and let $M^n = (m_{i,j}^{(n)})_{i,j}$ be its nth power. There exist an integer $p > 0$ and constants $e_{i,j} \geq 0$, such that, as $n \to \infty$,

$$m_{i,j}^{(n)} = e_{i,j} n^{p-1} r^n + o(n^{p-1} r^n).$$

We can also apply Theorem 8.3.14 to the matrix M'. Since $(m'_{i,j})^{(n)} = m_{i,j}^{(n)}$ for all i, j for which $m'_{i,j}$ is defined, we obtain exactly the above asymptotic behavior. But, since we are now dealing *with the matrix M'*, we can say that there exist $i_0, j_0 \in \Gamma'$ such that $e_{i_0, j_0} > 0$. Since a_1 belongs to Γ', and since a_{j_0} occurs in $\mu^\omega(a_1)$, there exists $v \geq 1$ such that $m_{j_0,1}^{(v)} \geq 1$. Now, since $i_0 \in \Gamma'$, there exist $\ell \in \Gamma'$ and $u \geq 1$ such that ℓ occurs in $\mu^u(a_{i_0})$. Hence $m_{\ell,i_0}^{(u)} \geq 1$. For all $n \geq 0$ we have

$$m_{\ell,1}^{(u+n+v)} \geq m_{\ell,i_0}^{(u)} m_{i_0,j_0}^{(n)} m_{j_0,1}^{(v)} \geq m_{i_0,j_0}^{(n)}.$$

Now, replacing the m's by their asymptotic behaviors given above, we see that $e_{\ell,1} > 0$. The number of a occurring in $\tau(\mu^n(a_1))$ is exactly $\sum_{1 \leq i \leq d',\ \tau(a_i)=a} m_{i,1}^{(n)}$ and the total number of letters in $\tau(\mu^n(a_1))$ is equal to $\sum_{1 \leq i \leq d} m_{i,1}^{(n)}$. The first sum satisfies

$$\sum_{1 \leq i \leq d',\ \tau(a_i)=a} m_{i,1}^{(n)} = e n^{p-1} r^n + o(n^{p-1} r^n),$$

where $e = \sum_{1 \leq i \leq d'} e_{i,1}$. The second sum satisfies

$$\sum_{1 \leq i \leq d} m_{i,1}^{(n)} = \tilde{e} n^{p-1} r^n + o(n^{p-1} r^n),$$

where $\tilde{e} = \sum_{1 \leq i \leq d} e_{i,1}$, which is bounded from below by $e_{\ell,1} > 0$. Hence $\tilde{e} > 0$. Finally we see that the frequency of a in $\mathbf{x}$ equals

$$\lim_{n \to \infty} \frac{|\tau(\mu^n(a_1))|_a}{|\tau(\mu^n(a_1))|} = \frac{e}{\tilde{e}}.$$

This frequency is an algebraic number by Theorem 8.3.14.

Now let $\mathbf{x}$ be an automatic sequence on the alphabet Σ. Then there exist an alphabet Γ, a morphism of constant length μ on Γ, a letter $b \in \Gamma$, and a coding

$\tau : \Gamma \to \Sigma$, such that $\mathbf{x} = \tau(\mu^\omega(b))$. Let a be a letter in Σ. We suppose that the frequency of a in $\mathbf{x}$ exists and is equal to α. This implies that

$$\lim_{n\to\infty} \frac{|\tau(\mu^n(b))|_a}{|\tau(\mu^n(b))|} = \alpha.$$

Let $h \geq 1$ be an integer to be fixed later on. We clearly have

$$\lim_{n\to\infty} \frac{|\tau(\mu^{hn}(b))|_a}{|\tau(\mu^{hn}(b))|} = \alpha.$$

Let $b_{i_1}, \ldots, b_{i_t}$ be all the letters of Γ such that $\tau(b_{i_j}) = a$. It suffices to prove that

$$\lim_{n\to\infty} \frac{|\mu^{hn}(b)|_{b_{i_j}}}{|\mu^{hn}(b)|}$$

exists and is rational for each i_j to conclude that α, which is the sum of all these limits, is rational. Let r be the length of the morphism μ (i.e., the maximum over $|\mu(c)|$ for $c \in \Gamma$). Let M be the incidence matrix of μ. The matrix M^T/r is a stochastic matrix. Hence its Perron–Frobenius eigenvalue is 1 by Theorem 8.3.13. From Theorem 8.3.11 there exists an integer $h \geq 1$ such that $(M^T/r)^h$ has no eigenvalue λ such that $|\lambda| = 1$ and $\lambda \neq 1$. Using Theorem 8.3.13 once more, we have

$$\lim_{n\to\infty} \frac{((M^T)^h)^n}{r^{hn}} = \frac{\tilde{C}(1)}{\tilde{q}'(1)},$$

where $\tilde{C}(\lambda)$ is the reduced adjoint of $\lambda I - (M^T/r)$ and $\tilde{q}$ the minimal polynomial of M^T/r. Hence

$$\lim_{n\to\infty} \frac{(M^h)^n}{r^{hn}} = \frac{\tilde{C}(1)^T}{\tilde{q}'(1)}.$$

Furthermore this matrix has rational coefficients by Theorem 8.3.13. Finally, if $\Gamma = \{b_1, b_2, \ldots, b_d\}$, we have

$$\frac{1}{r^{hn}} M^{hn} \begin{bmatrix} 1 \\ 0 \\ \vdots \\ 0 \end{bmatrix} = \frac{1}{|\mu^{hn}(b)|} \begin{bmatrix} |\mu^{hn}(b)|_{b_1} \\ |\mu^{hn}(b)|_{b_2} \\ \vdots \\ |\mu^{hn}(b)|_{b_d} \end{bmatrix}.$$

Hence all the limits $\lim\limits_{n\to\infty} \dfrac{|\mu^{hn}(b)|_{b_i}}{|\mu^{hn}(b)|}$ exist, and they are rational, since

$$
\begin{bmatrix}
\lim\limits_{n\to\infty} \dfrac{|\mu^{hn}(b)|_{b_1}}{|\mu^{hn}(b)|} \\
\lim\limits_{n\to\infty} \dfrac{|\mu^{hn}(b)|_{b_2}}{|\mu^{hn}(b)|} \\
\vdots \\
\lim\limits_{n\to\infty} \dfrac{|\mu^{hn}(b)|_{b_d}}{|\mu^{hn}(b)|}
\end{bmatrix}
= \lim\limits_{n\to\infty} \dfrac{1}{r^{hn}} M^{hn}
\begin{bmatrix} 1 \\ 0 \\ \vdots \\ 0 \end{bmatrix}
= \dfrac{\tilde{C}(1)^T}{\tilde{q}'(1)}
\begin{bmatrix} 1 \\ 0 \\ \vdots \\ 0 \end{bmatrix}.
$$

∎

Theorem 8.4.6 *Let* $\mathbf{x} = (x_n)_{n\geq 0}$ *be a pure morphic sequence, generated by a morphism* μ. *If the frequencies of all letters exist, then the vector of frequencies is a non-negative normalized eigenvector of the incidence matrix of* μ, *associated with the Perron–Frobenius eigenvalue of this matrix.*

Proof. Let $\mu : \Gamma^* \to \Gamma^*$ be a morphism prolongable on $a_1 \in \Gamma$, such that $\mathbf{x} = \mu^\omega(a_1)$. Write $\mathbf{x} = (x_n)_{n\geq 0}$. Let $M = (m_{i,j})_{i,j}$ be the incidence matrix of μ. Since all frequencies of letters occurring in $\mathbf{x}$ exist, we have in particular

$$
\lim\limits_{n\to\infty} \dfrac{1}{|\mu^n(a_1)|} M^n
\begin{bmatrix} 1 \\ 0 \\ \vdots \\ 0 \end{bmatrix}
=
\begin{bmatrix}
\mathrm{Freq}_{\mathbf{x}}(a_1) \\
\mathrm{Freq}_{\mathbf{x}}(a_2) \\
\vdots \\
\mathrm{Freq}_{\mathbf{x}}(a_d)
\end{bmatrix}.
$$

Hence

$$
M
\begin{bmatrix}
\mathrm{Freq}_{\mathbf{x}}(a_1) \\
\mathrm{Freq}_{\mathbf{x}}(a_2) \\
\vdots \\
\mathrm{Freq}_{\mathbf{x}}(a_d)
\end{bmatrix}
= M \lim\limits_{n\to\infty} \dfrac{1}{|\mu^n(a_1)|} M^n
\begin{bmatrix} 1 \\ 0 \\ \vdots \\ 0 \end{bmatrix}
$$

$$
= \lim\limits_{n\to\infty} \dfrac{|\mu^{n+1}(a_1)|}{|\mu^n(a_1)|} \dfrac{1}{|\mu^{n+1}(a_1)|} M^{n+1}
\begin{bmatrix} 1 \\ 0 \\ \vdots \\ 0 \end{bmatrix}.
$$

Since we have

$$
\lim\limits_{n\to\infty} \dfrac{1}{|\mu^{n+1}(a_1)|} M^{n+1}
\begin{bmatrix} 1 \\ 0 \\ \vdots \\ 0 \end{bmatrix}
=
\begin{bmatrix}
\mathrm{Freq}_{\mathbf{x}}(a_1) \\
\mathrm{Freq}_{\mathbf{x}}(a_2) \\
\vdots \\
\mathrm{Freq}_{\mathbf{x}}(a_d)
\end{bmatrix}
$$

and since at least one of the $\text{Freq}_x(a_i)$'s is not zero (since the $\text{Freq}_x(a_i)$'s sum to 1), we see that $|\mu^{n+1}(a_1)|/|\mu^n(a_1)|$ tends to a limit, say λ, for $n \to \infty$, and that

$$
M \begin{bmatrix} \text{Freq}_x(a_1) \\ \text{Freq}_x(a_2) \\ \vdots \\ \text{Freq}_x(a_d) \end{bmatrix} = \lambda \begin{bmatrix} \text{Freq}_x(a_1) \\ \text{Freq}_x(a_2) \\ \vdots \\ \text{Freq}_x(a_d) \end{bmatrix}.
$$

Let r be the Perron–Frobenius eigenvalue of the matrix M. Using Theorem 8.3.11, there exists an integer $h \geq 1$ such that M^h (whose Perron–Frobenius eigenvalue is r^h) has no eigenvalue ξ satisfying $|\xi| = r^h$ and $\xi \neq r^h$. We then easily obtain from the proof of the first part of Theorem 8.4.5 that there exist an integer $p \geq 1$ and a number $\tilde{e} > 0$ such that $|\mu^{hn}(a_1)| = \tilde{e}n^{p-1}r^{hn} + o(n^{p-1}r^{hn})$. Hence

$$
\lambda = \lim_{n \to \infty} \frac{|\mu^{n+1}(a_1)|}{|\mu^n(a_1)|} = \lim_{n \to \infty} \left(\frac{|\mu^{h(n+1)}(a_1)|}{|\mu^{hn}(a_1)|} \right)^{1/h} = r. \qquad \blacksquare
$$

Theorem 8.4.7 *Let* $\mathbf{x} = (x_n)_{n \geq 0}$ *be a primitive morphic sequence, that is the image by a coding of a fixed point of a primitive morphism. Then the frequencies of all letters exist, and are nonzero. Furthermore, the vector of frequencies of the letters occurring in the fixed point itself is the positive normalized vector associated with the Perron–Frobenius eigenvalue of the incidence matrix and is equal to*

$$
\frac{C(r) \begin{bmatrix} 1 \\ 0 \\ \vdots \\ 0 \end{bmatrix}}{\begin{bmatrix} 1 & 1 & \cdots & 1 \end{bmatrix} C(r) \begin{bmatrix} 1 \\ 0 \\ \vdots \\ 0 \end{bmatrix}},
$$

where r *is the Perron–Frobenius eigenvalue of the incidence matrix of the morphism, and* $C(\lambda)$ *the reduced adjoint matrix.*

Proof. We can first clearly restrict ourselves to the case where the sequence is a fixed point of a prolongable primitive morphism. Using Theorem 8.3.12 we see that, if the frequencies exist, then they are given by the formula above, and hence are positive. Let us now prove that the frequencies exist for the fixed point of a prolongable primitive morphism.

Let $\mathbf{x} = (x_n)_{n \geq 0}$ be a sequence over the alphabet Γ. We suppose that $\mathbf{x}$ is the fixed point of a primitive morphism μ prolongable on x_0. Let $\Gamma = \{a_0, a_1, \ldots, a_d\}$ with $a_0 = x_0$. For any letter $a_j \in \Gamma$ and for any $n \geq 0$, we have, defining E_j to be the vector whose all entries are 0 except the jth, which is equal to 1,

$$\frac{1}{|\mu^n(a_j)|} \begin{bmatrix} |\mu^n(a_j)|_{a_1} \\ |\mu^n(a_j)|_{a_2} \\ \vdots \\ |\mu^n(a_j)|_{a_d} \end{bmatrix} = \frac{M^n E_j}{\begin{bmatrix} 1 & 1 & \cdots & 1 \end{bmatrix} M^n E_j} = \frac{\frac{M^n}{r^n} E_j}{\begin{bmatrix} 1 & 1 & \cdots & 1 \end{bmatrix} \frac{M^n}{r^n} E_j}.$$

Hence the limit for $n \to \infty$ of the left-hand side exists and

$$\lim_{n\to\infty} \frac{1}{|\mu^n(a_j)|} \begin{bmatrix} |\mu^n(a_j)|_{a_1} \\ |\mu^n(a_j)|_{a_2} \\ \vdots \\ |\mu^n(a_j)|_{a_d} \end{bmatrix} = \frac{C(r)E_j}{\begin{bmatrix} 1 & 1 & \cdots & 1 \end{bmatrix} C(r)E_j} := D_j.$$

Furthermore, this limit does not depend on j: by Theorem 8.3.12 the columns of $C(r)$, i.e., the vectors $C(r)E_j$, are all proportional. Hence the normalized vectors D_j are all equal to

$$D_1 := \begin{bmatrix} \alpha_1 \\ \alpha_2 \\ \vdots \\ \alpha_d \end{bmatrix},$$

and we have, for all i, j, $\lim_{n\to\infty} |\mu^n(a_j)|_{a_i}/|\mu^n(a_j)| = \alpha_i$.

Define, for $n \geq 0$, $\ell_n := \max_{b\in\Gamma} |\mu^n(b)|$. If $N \geq \ell_n$, then there exist $j(n, N) \geq 0$ and a word $B_{n,N}$ such that $|B_{n,N}| < \ell_n$ and

$$x_0 \ldots x_{N-1} = \mu^n(x_0)\mu^n(x_1)\ldots \mu^n(x_{j(n,N)})B_{n,N}.$$

This implies

$$N = |x_0 \ldots x_{N-1}| = \sum_{0\leq i \leq j(n,N)} |\mu^n(x_i)| + |B_{n,N}|$$

and for any letter $a_i \in \Gamma$

$$|x_0 \ldots x_{N-1}|_{a_i} = \sum_{0\leq i \leq j(n,N)} |\mu^n(x_i)|_{a_i} + |B_{n,N}|_{a_i}.$$

We know from what precedes that for all $\varepsilon > 0$ and for all $b \in \Gamma$, there exists $n_0(b, \varepsilon)$ such that for all $n \geq n_0(b, \varepsilon)$

$$\left| |\mu^n(b)|_{a_i} - \alpha_i |\mu^n(b)| \right| \leq \varepsilon |\mu^n(b)|.$$

Let us fix $\varepsilon > 0$. Since Γ is finite, there exists $n_1 = n_1(\varepsilon)$ such that for all $b, a_t \in \Gamma$ we have

$$\left| |\mu^{n_1}(b)|_{a_t} - \alpha_t |\mu^{n_1}(b)| \right| \leq \varepsilon |\mu^{n_1}(b)|.$$

Hence

$$\left| |x_0 \ldots x_{N-1}|_{a_t} - \alpha_t |x_0 \ldots x_{N-1}| \right|$$

$$= \left| \sum_{0 \leq i \leq j(n_1, N)} (|\mu^{n_1}(x_i)|_{a_t} - \alpha_t |\mu^{n_1}(x_i)|) + |B_{n_1, N}|_{a_t} - \alpha_t |B_{n_1, N}| \right|,$$

which gives

$$\left| |x_0 \ldots x_{N-1}|_{a_t} - \alpha_t N \right| \leq \sum_{0 \leq i \leq j(n_1, N)} \left| |\mu^{n_1}(x_i)|_{a_t} - \alpha_t |\mu^{n_1}(x_i)| \right| + 2|B_{n_1, N}|$$

$$\leq \varepsilon \left(\sum_{0 \leq i \leq j(n_1, N)} |\mu^{n_1}(x_i)| \right) + 2|B_{n_1, N}| \leq \varepsilon N + 2\ell_{n_1}.$$

Thus

$$\left| \frac{1}{N} |x_0 \ldots x_{N-1}|_{a_t} - \alpha_t \right| \leq \varepsilon + 2\frac{\ell_{n_1}}{N}.$$

It then suffices to choose $N > \lfloor 2\ell_{n_1}/\varepsilon \rfloor$ to have $\left| \frac{1}{N} |x_0 \ldots x_{N-1}|_{a_t} - \alpha_t \right| \leq 2\varepsilon$. Hence we have that $\frac{1}{N} |x_0 \ldots x_{N-1}|_{a_t}$ tends to α_t for $N \to \infty$. ∎

We now prove that the logarithmic frequencies always exist for an automatic sequence. Actually, we prove a more general theorem.

Theorem 8.4.8 *Let* $\mathbf{x} = (x_n)_{n \geq 0}$ *be an automatic sequence with values in* $\mathbb{C}$. *Then the limit*

$$\lim_{s \to 1+} (s - 1) \sum_{n \geq 1} \frac{x_n}{n^s}$$

exists. Furthermore this limit is explicitly given by the following converging series: let $\mathbf{x}$ *be* k-*automatic, and let* $\mathbf{x}^{(1)} = \mathbf{x} = (x_n^{(1)})_{n \geq 0}, \mathbf{x}^{(2)} = (x_n^{(2)})_{n \geq 0}, \ldots, \mathbf{x}^{(d)} = (x_n^{(d)})_{n \geq 0}$ *be the (finitely many) sequences of the* k-*kernel of* $\mathbf{x}$ *(see Theorem 6.6.2). Defining* $V(n)^T := (x_n^{(1)}, x_n^{(2)}, \ldots, x_n^{(d)})$, *there exist* k *matrices of size* $d \times d$, *say* $M_0, \ldots, M_{k-1}$ *such that, for any* $j = 0, 1, \ldots, k - 1$ *we have* $V(kn + j) = M_j V(n)$. *Let* $M = (\sum_{0 \leq j \leq k} M_j)/k$, *let* $C(\lambda)$ *be the reduced adjoint matrix of* $\lambda I - M$, *and let* $q(\lambda)$ *be the minimal polynomial of* M. *Then*

$$\lim_{s \to 1_+} (s - 1) \sum_{n \geq 1} \frac{x_n}{n^s}$$

$$= \frac{C(1)}{q'(1) \log k} \left(\sum_{1 \leq j < k} \frac{V(j)}{j} - \sum_{1 \leq j < k} j M_j \sum_{n \geq 1} V(n) \frac{1}{kn(kn + j)} \right).$$

Proof. Since the sequence $\mathbf{x}$ is automatic, say k-automatic, its k-kernel is finite by Theorem 6.6.2. In particular, there exists a finite set of sequences $\{\mathbf{x}^{(1)} = (x_n^{(1)})_{n\geq 0}, \ldots, \mathbf{x}^{(d)} = (x_n^{(d)})_{n\geq 0}\}$ such that $\mathbf{x}^{(1)} = \mathbf{x}$, and for any sequence $\mathbf{y} = (y_n)_{n\geq 0}$ belonging to this set and for any $j = 0, 1, \ldots, k-1$, the sequence $(y_{kn+j})_{n\geq 0}$ also belongs to this set. In other words, there exist k matrices of size $d \times d$, say $M_0, \ldots, M_{k-1}$, such that for any $j = 0, 1, \ldots, k-1$ we have

$$V(kn + j) = M_j V(n), \qquad \text{where } V(n) := \begin{bmatrix} x_n^{(1)} \\ x_n^{(2)} \\ \vdots \\ x_n^{(d)} \end{bmatrix}.$$

Let $G(s) := \sum_{n\geq 1} V(n)/n^s$. The Dirichlet vector $G(s)$ converges for $\Re(s) > 1$, since $V(n)$ takes only finitely many values. For such an s we can write

$$G(s) = \sum_{n\geq 1} \frac{V(n)}{n^s} = \sum_{n\geq 1} \frac{V(kn)}{k^s n^s} + \sum_{1\leq j<k} \sum_{n\geq 0} \frac{V(kn+j)}{(kn+j)^s}$$

$$= k^{-s} M_0 \sum_{n\geq 1} \frac{V(n)}{n^s} + \sum_{1\leq j<k} M_j \sum_{n\geq 0} \frac{V(n)}{(kn+j)^s}$$

$$= k^{-s} M_0 \sum_{n\geq 1} \frac{V(n)}{n^s} + \sum_{1\leq j<k} \frac{V(j)}{j^s} + \sum_{1\leq j<k} M_j \sum_{n\geq 1} \frac{V(n)}{(kn+j)^s}.$$

Hence

$$\left(I - k^{-s} \sum_{0\leq j<k} M_j\right) G(s)$$

$$= \sum_{1\leq j<k} \frac{V(j)}{j^s} + \sum_{1\leq j<k} M_j \sum_{n\geq 1} V(n) \left(\frac{1}{(kn+j)^s} - \frac{1}{k^s n^s}\right).$$

Let $W(s)$ be the right-hand member of this equality, and let $M := \frac{1}{k} \sum_{0\leq j<k} M_j$. Since for each $j = 0, 1, \ldots, k-1$ the matrix M_j has only 0's and 1's and exactly one 1 in each row, the matrix M is stochastic. Hence its Perron–Frobenius eigenvalue is 1, and 1 is a simple root of the minimal polynomial (see Theorem 8.3.13 above). Denoting by $C(\lambda)$ the reduced adjoint matrix of $\lambda I - M$, and by $Q(\lambda)$ and $q(\lambda)$ the characteristic and the minimal polynomial of M, we have for real $s > 1$, using Proposition 8.3.3,

$$G(s) = k^{s-1}(k^{s-1}I - M)^{-1} W(s) = k^{s-1} \frac{\text{adj}(k^{s-1}I - M)}{Q(k^{s-1})} W(s)$$

$$= k^{s-1} \frac{C(k^{s-1})}{q(k^{s-1})} W(s).$$

But as $s \to 1+$, the vector $W(s)$ tends to

$$W(1) = \sum_{1 \le j < k} \frac{V(j)}{j} - \sum_{1 \le j < k} jM_j \sum_{n \ge 1} V(n)\frac{1}{kn(kn+j)}.$$

Furthermore

$$\frac{s-1}{q(k^{s-1})} = \frac{s-1}{k^{s-1}-1} \frac{k^{s-1}-1}{q(k^{s-1})-q(1)};$$

hence $\lim\limits_{s \to 1+} \dfrac{s-1}{q(k^{s-1})}$ exists and

$$\lim_{s \to 1+} \frac{s-1}{q(k^{s-1})} = \frac{1}{q'(1)\log k}.$$

Hence the limit of $(s-1)G(s)$ for $s \to 1+$ exists, and

$$\lim_{s \to 1+} (s-1)G(s) = \frac{C(1)}{q'(1)\log k} W(1). \qquad \blacksquare$$

Corollary 8.4.9 *Let* $\mathbf{x}$ *be an automatic sequence. Then all letters occurring in* $\mathbf{x}$ *have a logarithmic frequency. Furthermore, the logarithmic frequencies are explicitly given by a converging series analogous to the one above.*

Proof. Let $\mathbf{x}$ be an automatic sequence on the alphabet Σ. Let $a \in \Sigma$. Define $\tau : \Sigma \to \{0, 1\}$ by $\tau(b) = 0$ if $b \in \Sigma \setminus \{a\}$, and $\tau(a) = 1$. Then the sequence $\mathbf{y} := \tau(\mathbf{x})$ is also automatic. It then suffices to apply Theorem 8.4.8 above to $\mathbf{y}$. $\qquad \blacksquare$

8.5 An Application

In Exercise 7.14 we considered a variation on the tower of Hanoi problem called the cyclic tower of Hanoi. This gives rise to an infinite sequence $\mathbf{c} = \tau(f^{\omega}(\mathsf{a}_1))$, where f is the nonuniform morphism defined by

$$\begin{aligned}
f(\mathsf{a}_1) &= \mathsf{a}_1\mathsf{b}_2\mathsf{a}_1, \\
f(\mathsf{b}_1) &= \mathsf{b}_1\mathsf{c}_2\mathsf{b}_1, \\
f(\mathsf{c}_1) &= \mathsf{c}_1\mathsf{a}_2\mathsf{c}_1, \\
f(\mathsf{a}_2) &= \mathsf{b}_1\mathsf{a}_1, \\
f(\mathsf{b}_2) &= \mathsf{c}_1\mathsf{b}_1, \\
f(\mathsf{c}_2) &= \mathsf{a}_1\mathsf{c}_1,
\end{aligned}$$

and τ is the coding that removes the subscripts. We now prove that $\mathbf{c}$ is not an automatic sequence.

Theorem 8.5.1 *The sequence* $\tau(f^{\omega}(\mathsf{a}_1))$ *is not d-automatic for any $d \ge 1$.*

Proof. First, we note that the word aba in the sequence $\tau(f^\omega(a_1))$ can only occur as $\tau(a_1 b_2 a_1)$. For aba is necessarily of the form $\tau(a_i b_j a_k)$, where $i, j, k \in \{1, 2\}$. Looking at the images of each letter under f^2, we see that the only subword of the form $a_i b_j a_k$ that occurs in $\tau(f^\omega(a_1))$ is $a_1 b_2 a_1$. This implies that the number of times the subword aba $= \tau(a_1 b_2 a_1)$ occurs in $\tau(f^\omega(a_1))[0..N-1]$ is equal to the number of times the subword $a_1 b_2 a_1$ occurs in $f^\omega(a_1)[0..N-1]$.

Since the letter b_2 must be preceded and followed by a_1, we also see that the number of times the subword $a_1 b_2 a_1$ occurs in $f^\omega(a_1)[0..N-1]$ is equal either to the number of times the letter b_2 occurs in $f^\omega(a_1)[0..N-1]$ or to this number minus 1.

Suppose now that the sequence $\mathbf{c} = c_0 c_1 c_2 \cdots$ is automatic. Then, by combining the results of Theorems 5.4.4 and 6.8.1, we see that the sequence

$$\mathbf{c}' := [c_0, c_1, c_2][c_1, c_2, c_3][c_2, c_3, c_4]\cdots$$

of its 3-letter subwords is also automatic. Hence, if the frequency of a "letter" of this new sequence $\mathbf{c}'$ exists, it is rational by Theorem 8.4.5 above. In particular, if the frequency of the "letter" aba exists, it must be rational. But from the argument above, the frequency of this "letter" exists if and only if the frequency of the letter b_2 exists in the sequence $f^\omega(a_1)$, and the two frequencies are equal.

To conclude, we note that the incidence matrix of the morphism f is

$$M := \begin{bmatrix} 2 & 0 & 0 & 1 & 0 & 1 \\ 0 & 2 & 0 & 1 & 1 & 0 \\ 0 & 0 & 2 & 0 & 1 & 1 \\ 0 & 0 & 1 & 0 & 0 & 0 \\ 1 & 0 & 0 & 0 & 0 & 0 \\ 0 & 1 & 0 & 0 & 0 & 0 \end{bmatrix}.$$

It is easy to compute the characteristic polynomial of this matrix: $Q(\lambda) = (\lambda - 1)^4(\lambda^2 - 2\lambda - 2)$. Hence the Perron–Frobenius eigenvalue of this matrix is equal to $1 + \sqrt{3}$. Furthermore, the matrix M is primitive, since $M^4 > 0$. Therefore, we know that all frequencies exist, and the vector of frequencies is the normalized non-negative vector associated with the Perron–Frobenius value of the matrix. It is easy to find this vector:

$$V = \begin{bmatrix} \frac{\sqrt{3}-1}{3} \\ \frac{\sqrt{3}-1}{3} \\ \frac{\sqrt{3}-1}{3} \\ \frac{2-\sqrt{3}}{3} \\ \frac{2-\sqrt{3}}{3} \\ \frac{2-\sqrt{3}}{3} \end{bmatrix}.$$

Hence the frequency of b_2 equals $\frac{2-\sqrt{3}}{3}$, which is an irrational number. This yields the desired contradiction. ∎

8.6 Gaps

In this section we discuss the gaps between consecutive occurrences of a letter in an automatic sequence. If $\mathbf{x} = (x_n)_{n \geq 0}$ is an infinite sequence over the alphabet Δ, we define the *occurrence function*

$$\operatorname{occ}_{\mathbf{x}}(a, i) = (\sup\{n \geq -1 : |\mathbf{x}[0..n]|_a < i\}) + 1,$$

where $a \in \Delta$ and $i \geq 1$. Thus $\operatorname{occ}_{\mathbf{x}}(a, i)$ gives the position of the ith occurrence of the letter a in $\mathbf{x}$ (or $+\infty$ if a does not occur i times).

Our first theorem says that if $\mathbf{x}$ is a k-automatic sequence and a is a letter that occurs sufficiently often, then the smallest gaps between successive occurrences of a cannot be too large.

Theorem 8.6.1 *Let $\mathbf{x}$ be a k-automatic sequence taking values in Δ, and let $a \in \Delta$. If*

$$\limsup_{n \to \infty} \frac{|\mathbf{x}[0..n-1]|_a}{\log n} = \infty,$$

then

$$\liminf_{j \to \infty} (\operatorname{occ}_{\mathbf{x}}(a, j+1) - \operatorname{occ}_{\mathbf{x}}(a, j)) < \infty.$$

Proof. Since $\mathbf{x}$ is k-automatic, we can write $\mathbf{x} = \tau(\mathbf{y})$, where $\mathbf{y} = \varphi^{\omega}(b)$, where τ is a coding, b is a single symbol, and φ is a k-uniform morphism prolongable on b.

Let u be the smallest integer such that every letter that appears in $\mathbf{y}' := \mathbf{y}[k^u..\infty]$ occurs infinitely often in $\mathbf{y}'$. Now suppose for all integers i, r with $r \geq 0$ and $k^u \leq i < k^{u+1}$ we have $|\tau(\varphi^r(\mathbf{y}[i]))|_a \leq 1$. Then for all $r \geq 0$ we have

$$|\tau(\varphi^r(\mathbf{y}[k^u..k^{u+1}-1]))|_a \leq k^{u+1} - k^u$$

and

$$\mathbf{x}[0..k^{u+r+1} - 1] = \mathbf{x}[0..k^u - 1] \prod_{0 \leq t \leq r} \tau(\varphi^t(\mathbf{y}[k^u..k^{u+1} - 1])).$$

Combining these, we have

$$|\mathbf{x}[0..k^{u+r+1} - 1]|_a \leq k^u + (r + 1)(k^{u+1} - k^u)$$

for all $r \geq 0$. Now take n such that $k^{u+r} \leq n < k^{u+r+1}$. Then

$$\frac{|\mathbf{x}[0..n-1]|_a}{\log n} \leq \frac{k^u + (r + 1)(k^{u+1} - k^u)}{(u + r) \log k}.$$

Now, letting $r \to \infty$, we find

$$\limsup_{n \to \infty} \frac{|\mathbf{x}[0..n-1]|_a}{\log n} \leq \frac{k^{u+1} - k^u}{\log k} < \infty.$$

This contradicts the hypothesis.

So it must be that in fact there exist integers i, r with $|\tau(\varphi^r(\mathbf{y}[i]))|_a \geq 2$ with $k^u \leq i < k^{u+1}$ and $r \geq 0$. But $\mathbf{y}[i]$ occurs infinitely often in $\mathbf{y}$, so $\tau(\varphi^r(\mathbf{y}[i]))$ occurs infinitely often in $\mathbf{x} = \tau(\varphi^r(\mathbf{y}))$. Thus there are infinitely many consecutive occurrences of a in $\mathbf{x}$ that are separated by at most $k^r < \infty$. This is the desired result. ∎

Example 8.6.2 In Example 5.5.1 we proved that $\mathbf{q} = 1100100001\cdots$, the characteristic sequence of the squares, was not 2-automatic. Using Theorem 8.6.1, however, it easily follows that this sequence is not k-automatic for any $k \geq 2$. For we have

$$|\mathbf{q}[0..n-1]|_1 = 1 + \lfloor \sqrt{n-1} \rfloor,$$

but

$$\mathrm{occ}_{\mathbf{q}}(a, j+1) - \mathrm{occ}_{\mathbf{q}}(a, j) = 2j - 1.$$

Our next theorem discusses when the frequency of a symbol in an automatic sequence can be 0. Recall the definition of $e_{k;P}$ from Section 3.3: $e_{k;P}(n)$ counts the number of occurrences of the word P in the word $0^{|P|-1}(n)_k$.

Theorem 8.6.3 *Let* $\mathbf{x}$ *be a k-automatic sequence over an alphabet Δ, and let $a \in \Delta$. Then* $\mathrm{Freq}_{\mathbf{x}}(a) = 0$ *if and only if there exists a word $P \in \Sigma_k^+$ such that $\mathbf{x}[n] \neq a$ for all n with $e_{k;P}(n) > 0$.*

Proof. Suppose there exists such a word P. Then by Exercise 2, the symbol a is of frequency 0.

Otherwise there is no such word P. Suppose $\mathbf{x}$ is generated by the k-DFAO $M = (Q, \Sigma_k, \delta, q_0, \Delta, \tau)$. By Theorem 5.2.1 we may assume $\delta(q_0, 0) = q_0$. Then for all $P \in \Sigma_k^+$ there exist words $y = y(P)$, $z = z(P)$ such that $\tau(\delta(q_0, yPz)) = a$. Now we claim that in fact we can choose y, z such that $|y|, |z| < |Q|$. For consider the shortest prefix y' of y such that $\delta(q_0, y') = \delta(q_0, y)$. By the pigeonhole principle, we have $|y'| < |Q|$. Similarly, by considering the shortest prefix z' of z such that $\delta(q_0, yPz') = \delta(q_0, ypz)$ we get $|z'| < |Q|$. Hence $\delta(q_0, yPz) = \delta(q_0, y'Pz') = a$, as desired.

Now let us estimate $|\mathbf{x}[0..n-1]|_a$ for large n. Write $k^{u+2|Q|-2} \leq n < k^{u+2|Q|-1}$, and for each integer m with $k^{u-1} \leq m < k^u$ consider the word $P = (m)_k$ of length u. From the above, there exist words $y = y(P)$, $z = z(P)$ such that $\tau(\delta(q_0, yPz)) = a$ and $|y|, |z| < |Q|$. Without loss of generality we may assume y has no leading zeros. Consider the map sending P to $P' := yPz$; then $|P'| \leq u + 2|Q| - 2$ and P' is the base-k representation of an integer $< n$. This map could be many–one, but since P is a length-u subword of P', and there are at most $2|Q| - 1$ such subwords, at most $2|Q| - 1$ different words P can get mapped to the same P'. It follows that

$|\mathbf{x}[0..n-1]|_a \geq (k^u - k^{u-1})/(2|Q| - 1)$. Hence

$$\frac{|\mathbf{x}[0..n-1]|_a}{n} \geq \frac{k^u - k^{u-1}}{k^{u+2|Q|-1}(2|Q|-1)}$$

$$= \frac{k-1}{k^{2|Q|}(2|Q|-1)},$$

and so $\mathrm{Freq}_{\mathbf{x}}(a) > 0$.   ∎

Example 8.6.4 Consider $\mathbf{p} = (p_n)_{n \geq 1} = 01101010001\cdots$, the characteristic sequence of the prime numbers introduced in Example 1.1.3. From the prime number theorem we know the number of primes $\leq n$ is asymptotically $n/\log n$, and hence $\mathrm{Freq}_{\mathbf{p}}(1) = 0$. On the other hand, for any $P \in \Sigma_k^+$, there exists a prime $q \equiv k[P]_k + 1 \pmod{k^{|P|+1}}$ by Dirichlet's theorem. For this q we have that $\mathbf{p}[q] = 1$ and P is a subword of $(q)_k$. Hence by Theorem 8.6.3, $\mathbf{p}$ is not k-automatic for any $k \geq 2$.

Corollary 8.6.5 *Let $\mathbf{x}$ be a k-automatic sequence over the alphabet Δ, and let $a \in \Delta$. If $\mathrm{Freq}_{\mathbf{x}}(a) = 0$ and a occurs infinitely often in $\mathbf{x}$, then*

$$\limsup_{n \to \infty} \frac{\mathrm{occ}_{\mathbf{x}}(a, n+1)}{\mathrm{occ}_{\mathbf{x}}(a, n)} > 1.$$

Proof. By Theorem 8.6.3, if $\mathrm{Freq}_{\mathbf{x}}(a) = 0$, then there exists a word $P \in \Sigma_k^+$ such that $\mathbf{x}[n] \neq a$ if P is a subword of $(n)_k$. By replacing P with $1P$ if necessary, we may assume P has no leading zeros. Let $y = [P]_k$. Then $\mathbf{x}[n] \neq a$ for $yk^r \leq n < (y+1)k^r$. Setting $j = |\mathbf{x}[0..yk^r]|_a$, we have

$$\frac{\mathrm{occ}_{\mathbf{x}}(a, j+1)}{\mathrm{occ}_{\mathbf{x}}(a, j)} \geq \frac{y+1}{y}.$$

Hence

$$\limsup_{n \to \infty} \frac{\mathrm{occ}_{\mathbf{x}}(a, n+1)}{\mathrm{occ}_{\mathbf{x}}(a, n)} \geq \frac{y+1}{y} > 1.$$ ∎

8.7 Exercises

1. Prove that the characteristic sequence of the squarefree numbers is not morphic.
2. Let $P \in \Sigma_k^+$, and let $\mathbf{a}_P = (a_n)_{n \geq 0}$ be a sequence over Σ_2 defined as follows: $a_n = 1$ if $e_{k;P}(n) > 0$, and 0 otherwise. (See Section 3.3 for the definition of $e_{k;P}$.) For example, if $k = 2$ and $P = 11$, then $\mathbf{a}_P = 0001001100011111\cdots$.

(a) Show that $\mathbf{a}_P$ is k-automatic for all P.

(b) Show that $\mathrm{Freq}_{\mathbf{a}_P}(0) = 0$.

3. Suppose a sequence is the image of a fixed point of a (not necessarily uniform) morphism, and all its subwords for which a frequency exists occur with rational frequency. Is it true that the sequence is k-automatic for some k?

4. Let M be a $d \times d$ matrix with non-negative entries. Suppose that M is irreducible. Let r be the Perron–Frobenius eigenvalue of M. Supose that x is an eigenvector of the matrix M for an eigenvalue λ. Prove that if the entries of x are positive, then $\lambda = r$.

5. Let M be a diagonal complex $d \times d$ matrix. Let $m_1, \ldots, m_d$ be its diagonal terms. Prove that the adjoint of $\lambda I - M$ is the $d \times d$ diagonal matrix whose diagonal entries are $\frac{Q(\lambda)}{\lambda - m_1}, \ldots, \frac{Q(\lambda)}{\lambda - m_d}$, where $Q(\lambda) = \prod_{j=1}^{d}(\lambda - m_j)$ is the characteristic polynomial of the matrix D. Deduce that $Q'(\lambda) = \mathrm{tr}(\mathrm{adj}(\lambda I - M))$.

6. Prove that the set of diagonalizable complex matrices is dense in the set of all complex matrices.

7. Let M be a non-negative $d \times d$ complex matrix. Let r be its Perron–Frobenius eigenvalue. Prove that there exists a positive integer h such that any eigenvalue λ of M such that $|\lambda| = r$ satisfies $\lambda^h = r^h$.

8. Let $M = (m_{i,j})_{1 \le i,j \le d}$ be a $d \times d$ matrix with non-negative entries.

(a) Define $G(M)$ to be the directed graph with vertices $\{1, 2, \ldots, d\}$ such that there is a directed edge from i to j if and only if $m_{i,j} > 0$. Show that M is primitive if and only if there exists an integer e such that for all vertices i, j there exists a walk of length exactly e from i to j.

(b) Suppose M is primitive and there exists i such that $m_{i,i} > 0$. Show that all the entries of M^e are positive for $e = 2(d - 1)$, and this bound is best possible. Conclude that if h is a primitive morphism prolongable on a, and Depth $h = d$, then $h^{2(d-1)}$ $(b) \in \Sigma^* c \Sigma^*$ for all $b, c \in \Sigma$.

(c) Let $\gamma(M)$ be the least integer e such that M^e has all positive entries. Prove that, for all primitive matrices M, we have $\gamma(M) \le (d - 1)^2 + 1$.

(d) Prove that the bound $\gamma(M) \le (d - 1)^2 + 1$ is best possible.

9. In this exercise we outline a proof of the result mentioned in the proof of Lemma 7.4.6: if A is an $n \times n$ matrix with non-negative integer entries then there exist integers r, s with $0 \le r < s \le 2^n$ such that $A^r \le A^s$. (By $M \le M'$ we mean that each entry of M is $\le$ the corresponding entry of M'.)

(a) Let A be an irreducible $n \times n$ matrix. Define $\beta(A)$ to be the least integer $e \ge 1$ such that the entries of the diagonal of A^e are all positive. Define $\beta(n) = \sup \beta(A)$ where the supremum is taken over all irreducible $n \times n$ matrices A. Show that $\beta(n) \le n(n - 1)$ for $n \ge 2$.

(b) Suppose A is an $n \times n$ non-negative integer matrix of the form

$$\begin{bmatrix} B & C \\ 0 & D \end{bmatrix}$$

with B, D square matrices and D has nonzero entries on its diagonal. For $l \ge 0$

define C_l by

$$A^l = \begin{bmatrix} B^l & C_l \\ 0 & D^l \end{bmatrix}.$$

Show that $C_l \leq C_{l+1}$ and $D^l \leq D^{l+1}$.

(c) Now prove the result by induction on n.

8.8 Open Problems

1. For a morphic sequence **w**, must the logarithmic frequencies of each letter exist?
2. Must the logarithmic frequency of a letter occurring in an automatic sequence be of the form $\log \alpha / \log \beta$, where α and β are rational numbers?

8.9 Notes

8.2 What we call "incidence matrix" is sometimes called "transition matrix" or "substitution matrix" in the literature.

8.3 For the original papers on the Perron–Frobenius theorem, see Perron [1907a, 1907b] and Frobenius [1908, 1912]. Our presentation is based on the books of Gantmacher [1960] and Minc [1988]. For applications of the Perron–Frobenius theorem, see MacCluer [2000].

   The proof of the assertion in Proposition 8.3.3(f) is a simplification by Allouche of a proof due to B. Randé (personal communication).

8.4 For an interesting discussion of logarithmic frequencies and the first-digit problem, see Raimi [1976].

   For the proof of Proposition 8.4.4, see, for example, Tenenbaum [1995, Theorem 2 (p. 272) and Theorem 3 (p. 274)].

   Theorem 8.4.5(b) is due to Cobham [1972]. In the particular case that the automatic sequence is uniformly recurrent, this result is also due to B. Klein [1972].

   Theorem 8.4.7 is due to Michel [1975, 1976a].

   For Theorem 8.4.8 see Allouche, Mendès France, and Peyrière [2000]. For Corollary 8.4.9, see Cobham [1972] and Allouche, Mendès France, and Peyrière [2000].

8.5 Theorem 8.5.1 is due to Allouche [1994b].

8.6 Theorems 8.6.1 and 8.6.3 is due to Cobham [1972]. Corollary 8.6.5 is due to Minsky and Papert [1966].

9

Characteristic Words

In this chapter we introduce a special class of infinite words over $\{0, 1\}$ with many remarkable properties, the so-called *characteristic words*. These infinite words are of great number-theoretic interest, and can be viewed as generalizations of the infinite Fibonacci word introduced in Section 7.1.

Additional properties of characteristic words (and their generalizations) are discussed in Sections 10.5 and 10.6.

9.1 Definitions and Basic Properties

Let θ be a real number with $0 < \theta < 1$. For $n \geq 1$, define

$$f_\theta(n) := \lfloor (n+1)\theta \rfloor - \lfloor n\theta \rfloor \tag{9.1}$$

and

$$\mathbf{f}_\theta = f_\theta(1) f_\theta(2) f_\theta(3) \cdots .$$

Then $\mathbf{f}_\theta$ is called a *characteristic word* with slope θ. Note that $\mathbf{f}_\theta$ is an infinite word over the alphabet $\{0, 1\}$.

Alternatively, we could define

$$f_\theta(n) := \begin{cases} 1 & \text{if } \{n\theta\} \in [1 - \theta, 1), \\ 0 & \text{otherwise.} \end{cases}$$

It is easy to see that the two definitions are identical.

By telescoping cancellation applied to Eq. (9.1), we have

$$\sum_{1 \leq i \leq n} f_\theta(i) = \lfloor (n+1)\theta \rfloor. \tag{9.2}$$

Example 9.1.1 Consider the case where $\theta = (\sqrt{5} - 1)/2 \doteq .61803$. Then we have

$$\mathbf{f}_\theta = 1011010110 \cdots ,$$

which, as we will see below, is (up to a coding) the infinite Fibonacci word of Section 7.1.

There is a natural generalization of characteristic words called *Sturmian words*. If θ, ρ are real numbers with $0 < \theta < 1$, we define, for $n \geq 1$,

$$s_n := \lfloor (n+1)\theta + \rho \rfloor - \lfloor n\theta + \rho \rfloor,$$
$$s_n' := \lceil (n+1)\theta + \rho \rceil - \lceil n\theta + \rho \rceil.$$

Then

$$\mathbf{s}_{\theta,\rho} := s_1 s_2 s_3 \cdots,$$
$$\mathbf{s}_{\theta,\rho}' := s_1' s_2' s_3' \cdots$$

are said to be *Sturmian words* with *slope* θ. Again, both $\mathbf{s}_{\theta,\rho}$ and $\mathbf{s}_{\theta,\rho}'$ are words over the alphabet $\{0, 1\}$.

Define the coding r as follows: $r(0) = 1$ and $r(1) = 0$. Then we have

Lemma 9.1.2 *If $0 < \theta < 1$ is an irrational real number, then*

$$\mathbf{f}_{1-\theta} = r(\mathbf{f}_\theta).$$

Proof. We know that

$$f_\theta(n) = \lfloor (n+1)\theta \rfloor - \lfloor n\theta \rfloor$$

and

$$f_{1-\theta}(n) = \lfloor (n+1)(1 - \theta) \rfloor - \lfloor n(1 - \theta) \rfloor$$
$$= \lfloor -(n+1)\theta \rfloor - \lfloor -n\theta \rfloor + 1,$$

and so

$$f_\theta(n) + f_{1-\theta}(n) = \lfloor x \rfloor + \lfloor -x \rfloor - \lfloor y \rfloor - \lfloor -y \rfloor + 1,$$

where $x = (n+1)\theta$, $y = n\theta$. Now, using Exercise 9.1, we find $f_\theta(n) + f_{1-\theta}(n) = 1$. ∎

We now define a related infinite word. Let α be a real number, and define

$$g_\alpha(n) := \begin{cases} 1 & \text{if } n = \lfloor k\alpha \rfloor \text{ for some integer } k, \\ 0 & \text{otherwise.} \end{cases}$$

Also define

$$\mathbf{g}_\alpha := g_\alpha(1) g_\alpha(2) g_\alpha(3) \cdots.$$

Lemma 9.1.3 *Let $\alpha > 1$ be an irrational real number. Then $\mathbf{g}_\alpha = \mathbf{f}_{1/\alpha}$.*

Proof.

$$g_\alpha(n) = 1 \iff \exists k \text{ such that } n = \lfloor k\alpha \rfloor$$
$$\iff \exists k \text{ such that } n \leq k\alpha < n + 1$$
$$\iff \exists k \text{ such that } \frac{n}{\alpha} \leq k < \frac{n+1}{\alpha}$$
$$\iff \exists k \text{ such that } \left\lfloor \frac{n}{\alpha} \right\rfloor = k - 1 \text{ and } \left\lfloor \frac{n+1}{\alpha} \right\rfloor = k$$
$$\iff \left\lfloor \frac{n+1}{\alpha} \right\rfloor - \left\lfloor \frac{n}{\alpha} \right\rfloor = 1$$
$$\iff f_{1/\alpha}(n) = 1.$$

∎

We now define the *characteristic morphisms* $h_n : \{0, 1\}^* \to \{0, 1\}^*$, as follows:

$$h_n(0) = 0^{n-1}1; \quad h_n(1) = 0^{n-1}10$$

for $n \geq 1$.

Lemma 9.1.4 *Let α be an irrational real number with $0 < \alpha < 1$, and let k be an integer ≥ 1. Then $h_k(\mathbf{f}_\alpha) = \mathbf{f}_{1/(k+\alpha)}$.*

Proof. Define the words $d_i = h_k(f_\alpha(i))$ for $i \geq 1$. Then clearly

$$h_k(\mathbf{f}_\alpha) = d_1 d_2 d_3 \cdots .$$

Let m be arbitrary, and let n be the position of the mth 1 in $h_k(\mathbf{f}_\alpha)$. Since each d_i contains exactly one 1, this means that we are interested in the 1 in d_m. Then

$$|d_1 d_2 \cdots d_{m-1}| = (m - 1)k + f_\alpha(1) + f_\alpha(2) + \cdots + f_\alpha(m - 1)$$
$$= (m - 1)k + \lfloor m\alpha \rfloor,$$

where we have used Eq. (9.2). It follows that

$$n = |d_1 d_2 \cdots d_{m-1}| + k = mk + \lfloor m\alpha \rfloor = \lfloor m(k + \alpha) \rfloor.$$

Hence

$$(h_k(\mathbf{f}_\alpha))(n) = 1 \iff \exists m \text{ such that } n = \lfloor m(k + \alpha) \rfloor$$
$$\iff g_{k+\alpha}(n) = 1,$$

and the result follows by Lemma 9.1.3. ∎

Theorem 9.1.5 *Let α be an irrational real number, $0 < \alpha < 1$, with continued fraction expansion $\alpha = [0, a_1, a_2, \ldots]$. For $n \geq 1$ define $\beta_n = [0, a_n, a_{n+1}, a_{n+2}, \ldots]$. Then for all $n \geq 0$ we have*

$$\mathbf{f}_\alpha = (h_{a_1} \circ h_{a_2} \circ \cdots \circ h_{a_n})(\mathbf{f}_{\beta_{n+1}}).$$

Proof. By induction on n. Since $\alpha = \beta_1$, the theorem is clearly true for $n = 0$.

Now assume the theorem is true for all $m < n$; we prove it for $m = n$. From Lemma 9.1.4, we have

$$h_{a_n}(\mathbf{f}_{\beta_{n+1}}) = \mathbf{f}_{1/(a_n+\beta_{n+1})} = \mathbf{f}_{\beta_n}.$$

It follows that

$$\begin{aligned}
\mathbf{f}_\alpha &= (h_{a_1} \circ h_{a_2} \circ \cdots \circ h_{a_{n-1}})(\mathbf{f}_{\beta_n}) \\
&= (h_{a_1} \circ h_{a_2} \circ \cdots \circ h_{a_{n-1}})(h_{a_n}(\mathbf{f}_{\beta_{n+1}})) \\
&= (h_{a_1} \circ h_{a_2} \circ \cdots \circ h_{a_n})(\mathbf{f}_{\beta_{n+1}}).
\end{aligned}$$
∎

Recall from Section 2.4 that $[a_0, a_1, \ldots, a_n, \overline{a_{n+1}, \ldots, a_{n+t}}]$ denotes a continued fraction with an ultimately periodic sequence of partial quotients.

Corollary 9.1.6 *Let* $0 < \alpha < 1$ *be an irrational real number with purely periodic continued fraction expansion, i.e.,*

$$\alpha = [0, \overline{a_1, a_2, \ldots, a_n}].$$

Then the characteristic word $\mathbf{f}_\alpha$ *is a fixed point of the morphism*

$$h_{a_1} \circ h_{a_2} \cdots \circ h_{a_n}.$$

As an example, consider the case where $\alpha = [0, \overline{1}] = [0, 1, 1, 1, \ldots] = (\sqrt{5} - 1)/2$. In this case

$$\mathbf{f}_\alpha = 10110 \cdots$$

and $\mathbf{f}_\alpha$ is a fixed point of the map $1 \to 10$, $0 \to 1$. Hence $\mathbf{f}_\alpha$ is, up to a coding reversing 0 and 1, the infinite Fibonacci word introduced in Section 7.1.

We now turn to examining the prefixes of f_α. For $n \geq 0$ define

$$X_n = (h_{a_1} \circ h_{a_2} \circ \cdots \circ h_{a_n})(0),$$
$$Y_n = (h_{a_1} \circ h_{a_2} \circ \cdots \circ h_{a_n})(1).$$

The string X_n is sometimes called the nth *characteristic block*.

Lemma 9.1.7 *For* $n \geq 1$ *we have* $Y_n = X_n X_{n-1}$.

Proof.

$$\begin{aligned}
Y_n &= (h_{a_1} \circ h_{a_2} \circ \cdots \circ h_{a_n})(1) \\
&= (h_{a_1} \circ h_{a_2} \circ \cdots \circ h_{a_{n-1}})(h_{a_n}(1)) \\
&= (h_{a_1} \circ h_{a_2} \circ \cdots \circ h_{a_{n-1}})(h_{a_n}(0)0) \\
&= (h_{a_1} \circ h_{a_2} \circ \cdots \circ h_{a_n})(0)(h_{a_1} \circ h_{a_2} \circ \cdots \circ h_{a_{n-1}})(0) \\
&= X_n X_{n-1}.
\end{aligned}$$
∎

Theorem 9.1.8 *We have $X_0 = 0$, $X_1 = 0^{a_1-1}1$, and $X_n = X_{n-1}^{a_n} X_{n-2}$ for $n \geq 2$.*

Proof. We prove the result by induction on n. Clearly it is correct for $n = 0, 1$. Now assume the result is true for all $m < n$; we prove it for n:

$$
\begin{aligned}
X_n &= (h_{a_1} \circ h_{a_2} \circ \cdots \circ h_{a_n})(0) \\
&= (h_{a_1} \circ h_{a_2} \circ \cdots \circ h_{a_{n-1}})(h_{a_n}(0)) \\
&= (h_{a_1} \circ h_{a_2} \circ \cdots \circ h_{a_{n-1}})(0^{a_n-1}1) \\
&= X_{n-1}^{a_n-1} Y_{n-1} \\
&= X_{n-1}^{a_n-1} X_{n-1} X_{n-2} \qquad \text{(by Lemma 9.1.7)} \\
&= X_{n-1}^{a_n} X_{n-2}.
\end{aligned}
$$
∎

Lemma 9.1.9 *For $n \geq 0$ we have*

(1) $|X_n|_0 = q_n - p_n$, $\quad |X_n|_1 = p_n$,
(2) $|Y_n|_0 = q_n - p_n + q_{n-1} - p_{n-1}$, $\quad |Y_n|_1 = p_n + p_{n-1}$,

where $p_n/q_n = [0, a_1, a_2, \ldots, a_n]$.

Proof. The result is clearly true for $n = 0$, since $X_0 = 0$ and $Y_0 = 1$, while $p_0 = q_{-1} = 0$ and $p_{-1} = q_0 = 1$.

Similarly, the result is true for $n = 1$, since $X_1 = 0^{a_1-1}1$ and $Y_1 = 0^{a_1-1}10$, while $p_1 = 1, q_1 = a_1$.

Now assume the result is true for all $k < n$, where $n \geq 2$. We prove it for n. By Theorem 9.1.8 we have $X_n = X_{n-1}^{a_n} X_{n-2}$. It follows that $|X_n|_0 = a_n|X_{n-1}|_0 + |X_{n-2}|_0$. Hence, by induction, $|X_n|_0 = a_n(q_{n-1} - p_{n-1}) + (q_{n-2} - p_{n-2}) = q_n - p_n$, as desired. Similarly, $|X_n|_1 = a_n|X_{n-1}|_1 + |X_{n-2}|_1$, so by induction $|X_n|_1 = a_n p_{n-1} + p_{n-2} = p_n$.

Finally, using Lemma 9.1.7, we have $|Y_n|_0 = |X_n X_{n-1}|_0 = q_n - p_n + q_{n-1} - p_{n-1}$ and $|Y_n|_1 = |X_n X_{n-1}|_1 = p_n + p_{n-1}$.
∎

The next theorem explains the importance of the characteristic blocks.

Theorem 9.1.10 *For $n \geq 1$, X_n is exactly the prefix of $\mathbf{f}_\alpha$ of length q_n.*

Proof. By Theorem 9.1.5, either

$$X_n = (h_{a_1} \circ h_{a_2} \circ \cdots \circ h_{a_n})(0)$$

or

$$Y_n = (h_{a_1} \circ h_{a_2} \circ \cdots \circ h_{a_n})(1)$$

is a prefix of $\mathbf{f}_\alpha$. By Lemma 9.1.7, X_n is a prefix of both X_n and Y_n. Hence X_n is a prefix of $\mathbf{f}_\alpha$, and by Lemma 9.1.9, $|X_n| = q_n$.
∎

Note that Theorem 9.1.10 is not necessarily true for $n = 0$. In fact, it is easy to see that X_0 is a prefix of $\mathbf{f}_\alpha$ if and only if $\alpha < \frac{1}{2}$.

In Section 7.1, we saw that the consecutive Fibonacci words satisfied an "almost commutative" property. The same is true for the words X_n. Recall that, if $|w| \geq 2$, $c(w)$ is the map that interchanges the last two symbols of w and leaves the other symbols unchanged.

Theorem 9.1.11 *For $n \geq 1$ we have $X_n X_{n-1} = c(X_{n-1} X_n)$.*

Proof. We prove the result by induction on n. The result is clearly true for $n = 1$, since then

$$X_1 X_0 = 0^{a_1-1} 1\, 0 = c(0^{a_1-1} 01) = c(0\, 0^{a_1-1} 1) = c(X_0 X_1).$$

Now assume the result is true for all k with $1 \leq k < n$, and $n \geq 2$. We have

$$X_n X_{n-1} = (X_{n-1}^{a_n} X_{n-2}) X_{n-1} = X_{n-1}^{a_n} c(X_{n-1} X_{n-2})$$
$$= c(X_{n-1}^{a_n} X_{n-1} X_{n-2}) = c(X_{n-1} X_{n-1}^{a_n} X_{n-2}) = c(X_{n-1} X_n). \qquad \blacksquare$$

An interesting corollary of the previous result is that X_n starts with a large power.

Corollary 9.1.12 *For $n \geq 5$, X_{n-2}^e is a prefix of X_n, where $e = a_{n-1} + 1 + (q_{n-3} - 2)/q_{n-2}$.*

Proof. We have $X_n = X_{n-1}^{a_n} X_{n-2} = (X_{n-2}^{a_{n-1}} X_{n-3})^{a_n} X_{n-2}$. Since $a_n \geq 1$, it follows that $X_{n-2}^{a_{n-1}} X_{n-3} X_{n-2}$ is a prefix of X_n. Since $n \geq 5$, we have $|X_{n-3}| = q_{n-3} \geq 2$, and so

$$X_{n-2}^{a_{n-1}} X_{n-3} X_{n-2} = X_{n-2}^{a_{n-1}} c(X_{n-2} X_{n-3})$$
$$= X_{n-2}^{a_{n-1}} X_{n-2} c(X_{n-3})$$
$$= X_{n-2}^{a_{n-1}+1} c(X_{n-3}).$$

Now X_{n-3} is a prefix of X_{n-2}, so the result follows. $\blacksquare$

Finally, we observe an interesting connection between the characteristic words $\mathbf{f}_\alpha$ and the Ostrowski α-representation introduced in Section 3.9. The next theorem gives a factorization of every prefix of $\mathbf{f}_\alpha$ in terms of characteristic blocks.

Theorem 9.1.13 *Let $0 < \alpha < 1$ be an irrational real number with continued fraction expansion $\alpha = [0, a_1, a_2, \dots]$. Let $m \geq 0$ be an integer, and let $b_s b_{s-1} \cdots b_0$ be the Ostrowski α-representation of m. Then*

$$f_\alpha(1) f_\alpha(2) \cdots f_\alpha(m) = X_s^{b_s} X_{s-1}^{b_{s-1}} \cdots X_0^{b_0}.$$

Proof. By induction on m. The result is clearly true for $m = 0$, since then both sides are ϵ. It is also true if $0 < m < q_1 = a_1$, since then $b_0 = m$, $X_0 = 0$, and $f_\alpha(1) \cdots f_\alpha(q_1) = 0^{a_1 - 1} 1$, so $f_\alpha(1) \cdots f_\alpha(m) = 0^m = X_0^{b_0}$.

Now fix $s \geq 1$, and suppose the result is true for all $m < q_s$. We prove it for all $m < q_{s+1}$. Suppose $q_s \leq m < q_{s+1}$. Write $m = b_s q_s + r$, where $1 \leq b_s \leq a_{s+1}$, and $0 \leq r < q_s$.

By induction we have $r = \sum_{0 \leq i \leq s-1} b_i q_i$, and $f_\alpha(1) \cdots f_\alpha(r) = X_{s-1}^{b_{s-1}} \cdots X_0^{b_0}$. There are two cases to consider: $b_s < a_{s+1}$ and $b_s = a_{s+1}$.

Case 1: $b_s < a_{s+1}$. Then $f_\alpha(1) \cdots f_\alpha(m)$ is a prefix of $X_{s+1} = X_s^{a_{s+1}} X_{s-1}$. But $m = b_s q_s + r$, so $m < q_{s+1}$. Hence $f_\alpha(1) \cdots f_\alpha(m)$ is a prefix of $X_s^{b_s + 1} = X_s^{b_s} X_s$. Then

$$
\begin{aligned}
f_\alpha(1) \cdots f_\alpha(m) &= X_s^{b_s} f_\alpha(1) \cdots f_\alpha(r) \\
&= X_s^{b_s} X_{s-1}^{b_{s-1}} \cdots X_0^{b_0}.
\end{aligned}
$$

Case 2: $b_s = a_{s+1}$. Now $m < q_{s+1}$ by hypothesis, and so $m = b_s q_s + r = a_{s+1} q_s + r < q_{s+1}$. Thus $r < q_{s-1}$, and hence $f_\alpha(1) \cdots f_\alpha(r)$ is a prefix of X_{s-1}. But $f_\alpha(1) \cdots f_\alpha(m)$ is a prefix of $X_{s+1} = X_s^{a_{s+1}} X_{s-1}$. Therefore

$$
\begin{aligned}
f_\alpha(1) \cdots f_\alpha(m) &= X_s^{a_{s+1}} f_\alpha(1) \cdots f_\alpha(r) \\
&= X_s^{b_s} X_{s-1}^{b_{s-1}} \cdots X_0^{b_0}.
\end{aligned}
$$
∎

Corollary 9.1.14 *Let $0 < \alpha < 1$ be an irrational real number with continued fraction expansion $\alpha = [0, a_1, a_2, \ldots]$. Let $m \geq 1$ be an integer, and let $b_s b_{s-1} \cdots b_0$ be the Ostrowski α-representation of m. Then*

$$
\lfloor (m+1)\alpha \rfloor = \sum_{0 \leq i \leq s} p_i b_i,
$$

where p_i is the ith numerator in the continued fraction expansion for α.

Proof. Combine Theorem 9.1.13 with Lemma 9.1.9. ∎

Theorem 9.1.15 *For $n \geq 1$, we have that $f_\alpha(n) = 1$ if and only if $b_s b_{s-1} \cdots b_0$, the Ostrowski α-representation for n, ends in an odd number of zeros.*

Proof. It is clear, from the definition of h_k and Theorem 9.1.10, that for $n \geq 1$ we have

$$
f_\alpha(q_n) = \begin{cases} 1 & \text{if } n \text{ is odd}, \\ 0 & \text{if } n \text{ is even}. \end{cases}
$$

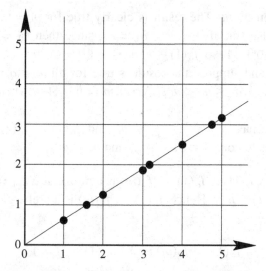

Figure 9.1: A Cutting Sequence for $\theta = (\sqrt{5} - 1)/2$.

Let i be the least index for which $b_i > 0$. Then by Theorem 9.1.13, $f_\alpha(n)$ is the last symbol of X_i. But the last symbol of X_i is $f_\alpha(q_i)$. Hence $f_\alpha(n) = 1$ if and only if i is odd if and only if $b_{i-1} \cdots b_0 = 0^i$ is an odd number of zeros. $\blacksquare$

9.2 Geometric Interpretation of Characteristic Words

In this section we examine a useful geometric characterization of characteristic words.

Let $\theta > 0$ be an irrational real number, and consider the line $L_\theta : y = \theta x$ through the origin with slope θ, as pictured in Figure 9.1 for $\theta = (\sqrt{5} - 1)/2$.

Now consider this line for $x > 0$, and as it travels to the right, write

$$c_i = \begin{cases} 0 & \text{if } L \text{ intersects a vertical line,} \\ 1 & \text{if } L \text{ intersects a horizontal line.} \end{cases}$$

Call the resulting infinite word

$$\mathbf{c}_\theta = c_1 c_2 c_3 \cdots.$$

The infinite word $\mathbf{c}_\theta$ is sometimes called a *cutting sequence*.

For example, for $\theta = (\sqrt{5} - 1)/2$ we get

$$\mathbf{c}_\theta = 01001010 \cdots,$$

which can be factored as

$$0 \cdot 10 \cdot 0 \cdot 10 \cdot 10 \cdots.$$

Now assume $0 < \theta < 1$. Note that, as x increases from n to $n + 1$, we get

an extra 0 if $\lfloor (n+1)\theta \rfloor = \lfloor n\theta \rfloor$;
an extra 10 if $\lfloor (n+1)\theta \rfloor > \lfloor n\theta \rfloor$.

It follows that, for $0 < \theta < 1$, we have

$$\mathbf{c}_\theta = h(0\mathbf{f}_\theta)$$

where $h(0) = 0$, $h(1) = 10$. Hence, by Exercise 9.2, we have

$$\mathbf{c}_\theta = h'(\mathbf{f}_\theta),$$

where $h'(0) = 0$, $h'(1) = 01$.

Now notice that $h' = r \circ h_1$, where r is the map $r(0) = 1, r(1) = 0$ introduced in Lemma 9.1.2, and h_1 is the characteristic morphism introduced after Lemma 9.1.3. Hence we have $\mathbf{c}_\theta = r(h_1(\mathbf{f}_\theta))$. Now use Lemma 9.1.4, and we have proved

Theorem 9.2.1 *Let* $0 < \theta < 1$ *be an irrational real number. Then*

$$\mathbf{c}_\theta = \mathbf{f}_{\theta/(\theta+1)}.$$

9.3 Application: Unusual Continued Fractions

In this section, we apply our theory of characteristic words to prove a theorem on continued fractions. The simplest example of our theorem is the following identity:

$$\sum_{n \geq 1} 2^{-\lfloor n/\alpha \rfloor} = [0, \, 2^0, \, 2^1, \, 2^1, \, 2^2, \, 2^3, \, 2^5, \, 2^8, \, 2^{13}, \, 2^{21}, \, 2^{34}, \ldots].$$

Here $\alpha = (\sqrt{5} - 1)/2$, and the exponents in the right-hand side are the Fibonacci numbers.

The following definitions will hold throughout this section. Let α be an irrational real number with $0 < \alpha < 1$. Let the continued fraction expansion of α be $\alpha = [0, a_1, a_2, \ldots]$, and let $p_n/q_n = [0, a_1, a_2, \ldots, a_n]$ be the nth convergent. Let $\mathbf{f}_\alpha = f_\alpha(1)f_\alpha(2)f_\alpha(3) \cdots$ be the characteristic word with slope α.

We now define x_n as the integer whose base-b representation is given by the string X_n of the previous section, where b is an integer ≥ 2; that is, $x_n = [X_n]_b$. In other words, for $n \geq 1$ define

$$x_n = f_\alpha(1)b^{q_n-1} + f_\alpha(2)b^{q_n-2} + \cdots + f_\alpha(q_n)b^0$$

$$= b^{q_n} \sum_{1 \leq k \leq q_n} f_\alpha(k)b^{-k}.$$

Also define $y_n = (b^{q_n} - 1)/(b - 1)$.

Lemma 9.3.1 *Define* $t_0 = 0$ *and*

$$t_n = \frac{b^{q_n} - b^{q_n-2}}{b^{q_n-1} - 1}$$

for $n \geq 1$. *Then*

$$[0, t_1, t_2, \ldots, t_n] = \frac{x_n}{y_n}.$$

Proof. The result is easily verified for $n = 0, 1$. For $n \geq 2$, the result will follow if we can show that $x_n = t_n x_{n-1} + x_{n-2}$ and $y_n = t_n y_{n-1} + y_{n-2}$ for $n \geq 2$. We have

$$
\begin{aligned}
x_n &= [X_n]_b \\
&= [X_{n-1}^{a_n} X_{n-2}]_b \\
&= b^{q_{n-2}} x_{n-1}(1 + b^{q_{n-1}} + \cdots + b^{q_{n-1}(a_n-1)}) + x_{n-2} \\
&= b^{q_{n-2}} \cdot \frac{b^{q_{n-1} a_n} - 1}{b^{q_{n-1}} - 1} \cdot x_{n-1} + x_{n-2} \\
&= \frac{b^{q_n} - b^{q_{n-2}}}{b^{q_{n-1}} - 1} \cdot x_{n-1} + x_{n-2} \\
&= t_n x_{n-1} + x_{n-2},
\end{aligned}
$$

where we have used Theorem 9.1.8.

Similarly,

$$
\begin{aligned}
t_n y_{n-1} + y_{n-2} &= \frac{b^{q_n} - b^{q_{n-2}}}{b^{q_{n-1}} - 1} \cdot y_{n-1} + y_{n-2} \\
&= \frac{b^{q_n} - b^{q_{n-2}}}{b^{q_{n-1}} - 1} \cdot \frac{b^{q_{n-1}} - 1}{b - 1} + \frac{b^{q_{n-2}} - 1}{b - 1} \\
&= \frac{(b^{q_n} - b^{q_{n-2}})(b^{q_{n-1}} - 1) + (b^{q_{n-2}} - 1)(b^{q_{n-1}} - 1)}{(b^{q_{n-1}} - 1)(b - 1)} \\
&= \frac{b^{q_n + q_{n-1}} - b^{q_n} - b^{q_{n-1}} + 1}{(b^{q_{n-1}} - 1)(b - 1)} \\
&= \frac{b^{q_n} - 1}{b - 1} \\
&= y_n.
\end{aligned}
$$

■

Theorem 9.3.2 *We have*

$$
(b - 1) \sum_{k \geq 1} f_\alpha(k) b^{-k} = [0, t_1, t_2, t_3, \ldots].
$$

Proof. From the previous lemma, we have

$$
\begin{aligned}
[0, t_1, t_2, \ldots, t_n] &= \frac{x_n}{y_n} \\
&= \frac{b^{q_n} \sum_{1 \leq k \leq q_n} f_\alpha(k) b^{-k}}{\frac{b^{q_n} - 1}{b - 1}} \\
&= \frac{b^{q_n}}{b^{q_n} - 1} \cdot (b - 1) \cdot \sum_{1 \leq k \leq q_n} f_\alpha(k) b^{-k}.
\end{aligned}
$$

To obtain the result, let $n \to \infty$.

■

We will show in Section 13.6 that the numbers $(b-1)\sum_{k\geq 1} f_\alpha(k)b^{-k}$ are all transcendental.

Corollary 9.3.3 *Let $b \geq 2$ be an integer, and let F_i denote the ith Fibonacci number. Let $\mathbf{f} = f_1 f_2 \cdots$ be the Fibonacci word. Then $[0, b^{F_0}, b^{F_1}, b^{F_2}, \ldots] = (b-1)\sum_{k\geq 1} f_k b^{-k}$.*

Proof. Choose $\alpha = [0, 1, 1, 1, \ldots] = (\sqrt{5}-1)/2$. Then in Theorem 9.3.2 we have $q_n = F_{n+1}$, and an easy calculation gives $t_n = b^{F_{n-1}}$. ∎

9.4 Exercises

1. Show that
$$\lfloor x \rfloor + \lfloor -x \rfloor = \begin{cases} 0 & \text{if } x \in \mathbb{Z}, \\ -1 & \text{if } x \notin \mathbb{Z}. \end{cases}$$

2. Let $h(0) = 0, h(1) = 10$, and $h'(0) = 0, h'(1) = 01$. Prove that $h(0w) = h'(w0)$ for all $w \in \{0, 1\}^*$. Conclude that $h(0w) = h'(w)$ for all $w \in \{0, 1\}^\omega$.

3. Let $0 < \theta, \rho < 1$ be real numbers. Show that $\mathbf{s}_{\theta,\rho}$ is ultimately periodic if and only if θ is a rational number. Conclude that there exists k such that $\mathbf{s}_{\theta,\rho}$ is k-automatic if and only if θ is rational.

4. Let $\alpha > 1$ be an irrational number, let $0 < \beta < 1$ be a real number, and let n be an integer. Show that there exists an integer k such that $n = \lfloor k\alpha + \beta \rfloor$ if and only if $\{(\beta - n)/\alpha\} < 1/\alpha$.

5. Let $0 < \alpha, \beta, \gamma, \rho < 1$ be real numbers with α, β irrational. Show that $\mathbf{s}_{\alpha,\gamma} = \mathbf{s}_{\beta,\rho}$ if and only if $(\alpha, \gamma) = (\beta, \rho)$.

6. Can the nth term of the characteristic sequence of α be computed in time polynomial in $\log n$, where α is an algebraic number?

7. Suppose you are given an oracle that returns the jth character in the continued fraction for α with cost proportional to j. Show how to determine the nth term of the characteristic sequence for α in time polynomial in $\log n$.

8. Let α and β be positive irrational numbers satisfying the equation
$$\frac{1}{\alpha} + \frac{1}{\beta} = 1.$$

Define $A = \{\lfloor \alpha n \rfloor : n \geq 1\}$ and $B = \{\lfloor \beta n \rfloor : n \geq 1\}$. Show that $A \cap B = \emptyset$ and $A \cup B = \{1, 2, 3, \ldots\}$.

9. Suppose $\alpha_1, \alpha_2, \ldots, \alpha_r$ are r distinct positive real numbers such that every positive integer occurs in exactly one of the sequences $(\lfloor n\alpha_i \rfloor)_{n\geq 1}$. Show that $r < 3$.

10. *Wythoff's game* is a game for two players who take turns. Two piles of counters are placed on a table, each with an arbitrary number of counters. At each turn,

a player can remove any number of counters from a single pile, or the same number of counters from both piles. A player wins when, after completing a turn, there are no counters left. Prove that if $\tau = (1 + \sqrt{5})/2$, then the second player has a forced win if and only if the numbers of counters in the piles are initially $\lfloor n\tau \rfloor$ and $\lfloor n\tau^2 \rfloor$.

11. Define $g(0) = 0$, and $g(n) = n - g(g(n-1))$ for $n \geq 1$. Prove that $g(n) = \lfloor (n+1)\alpha \rfloor$ for $\alpha = (\sqrt{5} - 1)/2$.

12. Let $0 < \alpha < 1$ be an irrational real number, and let X be an indeterminate. Prove the following power series identities:

$$\sum_{k \geq 1} f_\alpha(k) X^{-k} = \sum_{n \geq 1} X^{-\lfloor n/\alpha \rfloor} = (X - 1) \sum_{j \geq 1} \lfloor j\alpha \rfloor X^{-j}.$$

13. (Komatsu) Let a be a positive integer, and let $0 < \alpha < 1$ be an irrational real number. Determine the continued fraction for $\sum_{n \geq 1} X^{-\lceil n/\alpha \rceil}$. Hint: Exercise 2.33 may be useful.

14. Define a sequence of linear polynomials as follows: $g_1(x) = x/2$, $g_2(x) = (x+1)/2$, and $g_n(x) = g_{n-1}(g_{n-2}(x))$ for $n \geq 3$. Find an expression for $g_n(x)$, and compute $\lim_{n \to \infty} g_n(0)$.

15. Show that if $0 < \alpha < 1$ is an irrational real number, then the characteristic sequence $\mathbf{f}_\alpha$ has arbitrarily large prefixes that are palindromes.

16. Recall that $\ell_k(n)$ denotes the number of digits in the base-k representation of n. Is it true that

$$\sum_{n \geq 1} \frac{\ell_{10}(2^n)}{2^n} = \frac{1169}{1023}?$$

17. Develop an efficient algorithm that, given a finite sequence of integers $(a_i)_{1 \leq i \leq n}$, will determine whether there exist real numbers α and β such that $a_i = \lfloor i\alpha + \beta \rfloor$ for $1 \leq i \leq n$.

18. Define a function $\sigma : \mathbb{N} \times \mathbb{N} \to \{0, 1\}^*$ as follows: for $0 \leq h \leq k$ we have

$$\sigma(h, k) := \begin{cases} 0^k & \text{if } h = 0, \\ 0^{k-1}1 & \text{if } h \mid k, \\ \sigma(r, h)^q \sigma(h, r) & \text{if } k = qh + r, 0 \leq r < h. \end{cases}$$

For $h > k$ we define $\sigma(h, k) := \sigma(h \bmod k, k)$.

(a) Show that $\sigma(h, k)$ is a string of length k.

(b) For $h, k \geq 1$, show that $\sigma(h, k)^\omega$ and $\sigma(k, h)^\omega$ agree on the first $h + k - \gcd(h, k) - 1$ symbols, but differ at the $h + k - \gcd(h, k)$th symbol.

(c) In parts (c)–(f) assume $\gcd(h, k) = 1$. Let $h/k = [0, a_1, a_2, \ldots, a_t]$. Show that

$$\sigma(h, k) = (h_{a_t} \circ h_{a_{t-1}} \circ \cdots \circ h_{a_1})(0),$$

where the h_n are the characteristic morphisms introduced in Section 9.1.

(d) Compute the number of 1's in $\sigma(h, k)$. Hint: The number depends on $h^{-1} \bmod k$ and the length of the continued fraction for h/k.

(e) (Jenkinson and Zamboni) Consider the $k \times k$ array A formed by taking the conjugates of $\sigma(h, k)$ in increasing lexicographic order. Show that successive rows of A can be obtained by cyclically shifting by the same amount. Prove the same thing for the columns.

(f) (Gambaudo, Lanford, and Tresser) Show that the lexicographically least conjugate of $\sigma(h, k)$ is lexicographically $\geq$ the lexicographically least conjugate of any other word over $\{0, 1\}$ of the same length with the same number of 1's.

19. Let α be irrational with continued fraction expansion $\alpha = [0, a_1, a_2, \dots]$, and let p_n/q_n be the nth convergent. Show that $\sum_{1 \leq i \leq q_n} \lfloor i\alpha \rfloor = \frac{1}{2}(p_n q_n - q_n + p_n + (-1)^n)$.

9.5 Open Problems

1. Let $0 < \alpha < 1$ be an irrational real number, and let $n \geq 1$ be an integer. Consider the permutation that orders 0, $\{\alpha\}$, $\{2\alpha\}$, $\dots$, $\{n\alpha\}$, 1 into ascending order. What can you say about the cycle structure, the number of cycles, or the number of inversions in this permutation? (Remark. Ellis and Steele [1981] and Devroye and Goudjil [1998] have studied certain properties of this permutation.)

9.6 Notes on Chapter 9

9.1 Johann (Jean) Bernoulli III (1744–1807) was apparently the first to investigate what are now generally called Sturmian words. In our notation, he studied $\mathbf{s}_{\alpha, 1/2}$ and found a connection with continued fractions; see Bernoulli [1772]. However, he did not provide any proofs.

Christoffel [1875, 1888] and Smith [1876] independently found similar results for $\mathbf{s}_{\alpha, 0}$.

Markoff [1882] proved Bernoulli's assertions about $\mathbf{s}_{\alpha, 1/2}$. An exposition of Markoff's work in English can be found in Venkov [1970, pp. 65–68].

The term "Sturmian" was introduced by Morse and Hedlund [1940] in their work on symbolic dynamics. (The term is rather unfortunate in that Sturm apparently never worked on these sequences.) Also see Hedlund [1944].

Uspensky published a four-part paper on Bernoulli's problem [1946a, 1946b]. This paper is a detailed study of the sequences $\lfloor n\alpha \rfloor$ for both rational and irrational α.

Stolarsky [1976] wrote an influential paper that gave several different descriptions of characteristic words. This paper also has a large bibliography. For related papers, see Fraenkel, Mushkin, and Tassa [1978]; Porta and Stolarsky [1990]; and Lunnon and Pleasants [1992].

Sturmian words appear in other areas of mathematics. For an application to linear filters, see Kieffer [1988]. For an application to routing in networks, see Altman, Gaujal, and Hordijk [2000].

Lemma 9.1.3, the fact that $\mathbf{g}_\alpha = \mathbf{f}_{1/\alpha}$, appears in Fraenkel, Mushkin, and Tassa [1978].

The "characteristic morphisms" h_k were introduced by Stolarsky [1976]. Similar maps had been previously studied by Markoff [1882]; also see Venkov [1970, pp. 65–68]. Cohn [1974] studied similar maps. Lemma 9.1.4 is due to T. Brown [1991].

Corollary 9.1.6, the fact that reduced quadratic irrationals have characteristic words that are fixed points of morphisms, is due independently to S. Ito and Yasutomi [1990], S. Ito [1991], T. Brown [1991], and Shallit [1991b]. Corollary 9.1.6 does not completely characterize the quadratic irrationals whose characteristic words are fixed points of morphisms; the complete characterization has been given by Crisp, Moran, Pollington, and Shiue [1993]. Other proofs were given by Berstel and Séébold [1993b] and Komatsu and van der Poorten [1996]. Also see Allauzen [1998].

Theorem 9.1.8, the recurrence relation for prefixes of $\mathbf{f}_\alpha$, is due to Smith [1876]. Also see Shallit [1991b]; Nishioka, Shiokawa, and Tamura [1992, Thm. 2].

Borel and Laubie [1993] linked morphisms on characteristic words (which they call infinite Christoffel words) with the action of homographies on the projective line; also see Borel and Laubie [1991] and Laubie [1991]. Laubie and Laurier [1995] and Borel [1997, 2001] studied homographic transformations and sums of the slopes of characteristic words.

Theorem 9.1.13, giving a factorization of prefixes of the characteristic word $\mathbf{f}_\alpha$ in terms of the Ostrowski α-representation, is due to T. Brown [1993]. Corollary 9.1.14 can be found in Fraenkel, Levitt, and Shimshoni [1972]. Our proof follows T. Brown [1993]. Theorem 9.1.15 is due to Fraenkel, Levitt, and Shimshoni [1972].

Berstel and Pocchiola [1996] showed how to generate a "random" Sturmian word.

A *Sturmian morphism* essentially maps Sturmian words to Sturmian words. For more on this concept, see Berstel and Séébold [1993b, 1994a, 1994b]; Mignosi and Séébold [1993b]; de Luca [1996, 1997]; Séébold [1998]; Chuan [1999]; Richomme [1999].

9.2 For cutting sequences, see Series [1985] and Lunnon and Pleasants [1992].

Because of the geometric interpretation, Sturmian words have received some attention in the computer graphics and image processing literature. Four problems have received particular attention: (1) efficient methods for drawing an approximation to a line on a raster display, (2) efficient specifications for digitized straight lines, (3) characterization and recognition of digitized straight lines, and (4) counting the number of essentially different digitized straight lines. For surveys, see Bruckstein [1990, 1991].

For the first problem, Bresenham [1965] gave an algorithm for digital line drawing. For variations on Bresenham's method, see Bresenham [1982,

1985]; Pitteway and Green [1982]; Sproull [1982]; Pitteway [1985]; Castle and Pitteway [1985]; Angel and Morrison [1991]. Berstel [1990] explained Bresenham's algorithm and its variations in terms of formal language theory.

Earlier, Freeman [1961a, 1961b] had described a method for encoding a figure made up of straight lines that he called a "chain code". Also see Freeman [1970]; Bongiovanni, Luccio, and Zorat [1975].

For the second problem (efficient specification of digitized straight lines), see Brons [1974, 1985] and Rothstein and Weiman [1976].

For the third problem (characterization and recognition of digitized straight lines), see Rosenfeld [1974]; Gaafar [1977]; Arcelli and Massarotti [1978]; L. Wu [1980, 1982]; Rosenfeld and Kim [1982]; Dorst and Smeulders [1984]; McIlroy [1984]; Ronse [1985]; Voss [1991]; Lindenbaum and Bruckstein [1993].

The fourth problem (counting the number of essentially different digitized straight lines) corresponds to counting the number of subwords of length n in all Sturmian words. Dulucq and Gouyou-Beauchamps [1990] studied this problem and conjectured a formula that was later proven by Mignosi [1991]. For other proofs, see Berstel and Pocchiola [1993] and de Luca and Mignosi [1994].

In the graphics literature, less complete analysis was done by Sundar Raj and Koplowitz [1986]; Berenstein, Kanal, Lavine, and Olson [1987]; Berenstein and Lavine [1988]; Koplowitz, Lindenbaum, and Bruckstein [1990]; Lindenbaum and Koplowitz [1991].

9.3 Theorem 9.3.2 is originally due to Böhmer [1926]. It was independently rediscovered by Danilov [1972], Davison [1977], Adams and Davison [1977], Bullett and Sentenac [1994], and Shiu [1999]. Our presentation is based on the paper of Anderson, Brown, and Shiue [1995]. Also see Gould, Kim, and Hoggatt [1977]; Bundschuh [1980]; Adams [1985]; Bowman [1988]; Graham, Knuth, and Patashnik [1989, pp. 293–295]. Tamura [1992, 1995] generalized Theorem 9.3.2 to multidimensional continued fractions.

10

Subwords

10.1 Introduction

An infinite word **u** may be partially understood by studying its finite subwords. Among the types of natural questions that arise are:

How many distinct subwords of **u** of length n are there, and what is the growth rate of this quantity as n tends to infinity?
Does every subword of **u** occur infinitely often in **u**, and if so, how big are the gaps between successive occurrences?

We start with the first question, which refers to a measure of complexity for infinite words, called *subword complexity*. This measure is of particular interest because automatic sequences and, more generally, morphic sequences have relatively low subword complexity, while the typical "random" sequence has high subword complexity.

Definition 10.1.1 Let $\mathbf{w} = a_0 a_1 a_2 \cdots$ be an infinite word over a finite alphabet Σ. We define $\text{Sub}_{\mathbf{w}}(n)$ to be the set of all subwords of length n of **w**. We define $\text{Sub}(\mathbf{w})$ to be the set of all finite subwords of **w**. Finally, we define $p_{\mathbf{w}}(n)$, the *subword complexity* function of **w**, to be the function counting the number of distinct length-n subwords of **w**.

Example 10.1.2 If **w** is an ultimately periodic word, then it is easy to see (Theorem 10.2.6 below) that $p_{\mathbf{w}}(n) = O(1)$.

Example 10.1.3 Consider $\mathbf{f} = 0100101001001 \cdots$, the Fibonacci word introduced in Section 7.1. Then a short computation shows

$$
\begin{aligned}
\text{Sub}_{\mathbf{f}}(0) &= \{\epsilon\}, & p_{\mathbf{f}}(0) &= 1, \\
\text{Sub}_{\mathbf{f}}(1) &= \{0, 1\}, & p_{\mathbf{f}}(1) &= 2, \\
\text{Sub}_{\mathbf{f}}(2) &= \{00, 01, 10\}, & p_{\mathbf{f}}(2) &= 3, \\
\text{Sub}_{\mathbf{f}}(3) &= \{001, 010, 100, 101\}, & p_{\mathbf{f}}(3) &= 4,
\end{aligned}
$$

298

and this suggests the conjecture $p_{\mathbf{f}}(n) = n + 1$. We prove this conjecture – and much more – in Section 10.5 below.

Example 10.1.4 The Barbier infinite word $\mathbf{B} =$
$123456789101112131415161718192021\cdots$, introduced in Exercise 3.26, consists of the concatenation of the decimal expansions of the positive integers, written in increasing order. For this word we have $p_{\mathbf{B}}(n) = 10^n$ for all $n \geq 0$.

Example 10.1.5 The infinite word $\mathbf{p} =$
$235711131719232931374143475359 6167\cdots$ consists of the concatenation of the decimal expansions of the prime numbers, written in increasing order. In Exercise 2 you are asked to prove that $p_{\mathbf{p}}(n) = 10^n$ for all $n \geq 0$.

The subword complexity in Examples 10.1.4 and 10.1.5 is actually typical of a "random" infinite word, as we now prove.

Theorem 10.1.6 *Almost all sequences* $\mathbf{w}$ *over a finite alphabet* Σ *satisfy* $p_{\mathbf{w}}(n) = |\Sigma|^n$ *for all* $n \geq 0$.

Proof. Let $k = |\Sigma|$. Without loss of generality, assume $\Sigma = \{0, 1, \ldots, k - 1\}$. Let

$$Y = \{\mathbf{w} \in \Sigma^\omega : \exists i \geq 1 \text{ such that } p_{\mathbf{w}}(i) < k^i\}.$$

We need to show that $m(Y) = 0$, where m is the measure introduced in Section 1.2.
Let $d \in \Sigma$, and define

$$T_d = \{\mathbf{w} \in \Sigma^\omega : \mathbf{w}[i] \neq d \ \forall i \geq 0\}.$$

Hence

$$\overline{T_d} = \Sigma^\omega \setminus T_d = \{\mathbf{w} \in \Sigma^\omega : \exists i \text{ such that } \mathbf{w}[i] = d\}.$$

Also define

$$U_d(i) = \{\mathbf{w} \in \Sigma^\omega : \min\{j : \mathbf{w}[j] = d\} = i\}.$$

Then $\overline{T_d} = \bigcup_{i \geq 0} U_d(i)$, and this union is disjoint.
Now

$$m(U_d(i)) = \left(\frac{k-1}{k}\right)^i \cdot \frac{1}{k},$$

since the first factor gives the measure of those sequences $\mathbf{w}$ for which the symbols $\mathbf{w}[0..i - 1]$ all differ from d, while the second factor gives the measure of those sequences for which $\mathbf{w}[i] = d$. Hence

$$m(\overline{T_d}) = \sum_{i \geq 0} m(U_d(i)) = \sum_{i \geq 0} \left(\frac{k-1}{k}\right)^i \cdot \frac{1}{k} = \left(\frac{1}{1 - \frac{k-1}{k}}\right) \cdot \frac{1}{k} = 1.$$

Thus we have $m(T_d) = 0$. That is, almost all sequences contain an occurrence of d somewhere.

Now let $x \in \Sigma^+$, and, generalizing the definition of T_d, let

$$T_x = \{\mathbf{w} \in \Sigma^\omega : \mathbf{w}[i..i + |x| - 1] \neq x \; \forall i \geq 0\}.$$

Then we can compute $m(T_x)$ by grouping the terms of $\mathbf{w}$ into blocks of size $|x|$.

If $\mathbf{w} = a_0 a_1 a_2 \cdots$, then let $B(\mathbf{w}, r) = b_0 b_1 b_2 \cdots$ where $b_i = [a_{ir} a_{ir+1} \cdots a_{ir+r-1}]_k$. In other words, $B(\mathbf{w}, r)$ is the infinite word over $\{0, 1, \ldots, k^r - 1\}$ resulting from grouping the terms of $\mathbf{w}$ into blocks of size r, and treating each block as a base-k number. Let $|x| = r$, and define $d = [x]_k$, so $0 \leq d < k^r$. For $0 \leq j < r$ define

$$V_{r,j} = \{\mathbf{w} \in \Sigma^\omega : B(\mathcal{S}^j(\mathbf{w}), r)[i] \neq d \; \forall i \geq 0\},$$

where $\mathcal{S}$ denotes the shift map. Thus $V_{r,j}$ is the set of those $\mathbf{w}$ not containing any occurrences of the word x that occur beginning at a position congruent to $j \pmod r$. Hence

$$T_x = \bigcup_{0 \leq j < r} V_{r,j};$$

this union is not disjoint. But, by analogy with T_d, we have $m(V_{r,j}) = 0$ for $0 \leq j < r$. Hence $m(T_x) = 0$.

Finally, we have

$$Y = \bigcup_{x \in \Sigma^+} T_x.$$

Then Y is the countable union of sets of measure 0, and hence $m(Y) = 0$. ∎

10.2 Basic Properties of Subword Complexity

Let $\mathbf{w}$ be an infinite word over a finite alphabet Σ. In this section we study the basic properties of $p_{\mathbf{w}}(n)$.

Theorem 10.2.1 *For all $n \geq 0$ we have $p_{\mathbf{w}}(n) \geq 1$ and $p_{\mathbf{w}}(n) \leq |\Sigma|^n$.*

Proof. Clear. ∎

Theorem 10.2.2 *Let $\mathbf{w}$ be an infinite word over an alphabet Σ of cardinality k. Then $p_{\mathbf{w}}(n) \leq p_{\mathbf{w}}(n + 1) \leq k p_{\mathbf{w}}(n)$ for all $n \geq 0$.*

Proof. Let $X = \{x_1, x_2, \ldots, x_j\}$ be the set of all subwords of length n in **w**. For $1 \le i \le j$, let a_i be any letter that follows some occurrence of x_i in **w**. Then the words $x_i a_i$ for $1 \le i \le j$ are all distinct, and each is of length $n + 1$. Hence $p_{\mathbf{w}}(n) \le p_{\mathbf{w}}(n + 1)$.

On the other hand, any subword of length $n + 1$ must be of the form $x_i a$ for some $x_i \in X$ and $a \in \Sigma$. Hence $p_{\mathbf{w}}(n + 1) \le k p_{\mathbf{w}}(n)$. ∎

Theorem 10.2.3 *Let* **w** *be an infinite word over an alphabet* Σ *of cardinality* k. *Then* $p_{\mathbf{w}}(n + 1) - p_{\mathbf{w}}(n) \le k(p_{\mathbf{w}}(n) - p_{\mathbf{w}}(n - 1))$ *for all* $n \ge 1$.

Proof. For $1 \le i \le k$ and $n \ge 0$, define $T_i(n)$ to be the set of all subwords x of length n of **w** for which there exist at least i distinct symbols $a \in \Sigma$ such that xa is a subword of **w**. Also define $t_i(n) = \operatorname{Card} T_i(n)$.

Each $x \in T_i(n)$ can be written as $x = bw$, where $|b| = 1$ and $|w| = n - 1$. Furthermore, $w \in T_i(n - 1)$. Since there are k choices for b, it follows that $t_i(n) \le k t_i(n - 1)$.

Now $p_{\mathbf{w}}(n) = t_1(n)$ and $p_{\mathbf{w}}(n + 1) = t_1(n) + t_2(n) + \cdots + t_k(n)$. Hence

$$p_{\mathbf{w}}(n + 1) - p_{\mathbf{w}}(n) = t_2(n) + \cdots + t_k(n)$$

and, substituting $n - 1$ for n, we get

$$p_{\mathbf{w}}(n) - p_{\mathbf{w}}(n - 1) = t_2(n - 1) + \cdots + t_k(n - 1).$$

Since $t_i(n) \le k t_i(n - 1)$, the desired result follows. ∎

Our next theorem describes how subword complexity is affected by nonerasing morphisms. The case of erasing morphisms is discussed in Exercise 32.

Theorem 10.2.4 *Let* **u** *be an infinite word over* Σ. *Let* $h : \Sigma^* \to \Delta^*$ *be a nonerasing morphism, and define* $\mathbf{v} = h(\mathbf{u})$. *Then* $p_{\mathbf{v}}(n) \le W p_{\mathbf{u}}(n)$, *where* $W = \operatorname{Width} h$.

Proof. Let $x \in \Delta^*$ be a subword of length n of $\mathbf{v} = b_0 b_1 b_2 \cdots$, say $x = b_i b_{i+1} \cdots b_{i+n-1}$. Let us now define $w = a_j a_{j+1} \cdots a_{j+n-1}$, where j is chosen to be the largest integer such that $|h(a_0 a_1 \cdots a_{j-1})| \le i$. Since h is nonerasing, it follows that x is a subword of $h(w)$. Further, the string x is completely determined by the pair (w, k), where $k := i - |h(a_0 a_1 \cdots a_{j-1})|$. Now $0 \le k < |h(a_j)| \le \operatorname{Width} h$, so it follows that $p_{\mathbf{v}}(n) \le W p_{\mathbf{u}}(n)$, as desired. ∎

In particular, this shows that if **v** is the image of **u** under a coding, then $p_{\mathbf{v}}(n) \le p_{\mathbf{u}}(n)$.

Lemma 10.2.5 *Let* **w** *be an infinite word. Then for integers* $m, n \geq 0$ *we have* $p_{\mathbf{w}}(m + n) \leq p_{\mathbf{w}}(m) p_{\mathbf{w}}(n)$.

Proof. Every subword of length $m + n$ can be expressed as a subword of length m followed by a subword of length n. There are at most $p_{\mathbf{w}}(m) p_{\mathbf{w}}(n)$ possibilities. ∎

Theorem 10.2.6 *Let* $\mathbf{w} = b_0 b_1 b_2 \cdots$ *be an infinite word over a finite alphabet* Σ. *Then the following are equivalent:*

(a) *There exists a non-negative integer N such that for all integers $n \geq 0$ we have $p_{\mathbf{w}}(n) \leq N$.*

(b) *There exists a non-negative integer n_0 such that for all integers $n \geq n_0$ we have $p_{\mathbf{w}}(n) = p_{\mathbf{w}}(n_0)$.*

(c) *There exists a non-negative integer k such that $p_{\mathbf{w}}(k) \leq k$.*

(d) *There exists a non-negative integer m such that $p_{\mathbf{w}}(m) = p_{\mathbf{w}}(m + 1)$.*

(e) **w** *is ultimately periodic.*

Proof. (a) $\implies$ (b): Choose N as small as possible; then there exists n_0 such that $p_{\mathbf{w}}(n_0) = N$. By Theorem 10.2.2, $p_{\mathbf{w}}(n) \geq N$ for all $n \geq n_0$. But $p_{\mathbf{w}}(n) \leq N$ for $n \geq n_0$ by hypothesis. Hence $p_{\mathbf{w}}(n) = N = p_{\mathbf{w}}(n_0)$ for all $n \geq n_0$.

(b) $\implies$ (c): Let $k = p_{\mathbf{w}}(n_0)$. If $k \leq n_0$, then by Theorem 10.2.2, $p_{\mathbf{w}}(k) \leq p_{\mathbf{w}}(n_0) = k$. If $k > n_0$, then by hypothesis $p_{\mathbf{w}}(k) = p_{\mathbf{w}}(n_0) = k$.

(c) $\implies$ (d): If $p_{\mathbf{w}}(k) \leq k$, then since $p_{\mathbf{w}}(0) = 1$, by Theorem 10.2.2 and the pigeonhole principle there must exist an integer $m \leq k$ for which $p_{\mathbf{w}}(m) = p_{\mathbf{w}}(m + 1)$.

(d) $\implies$ (e): Suppose $p_{\mathbf{w}}(m) = p_{\mathbf{w}}(m + 1)$, and let $p_{\mathbf{w}}(m) = r$. Consider a fixed subword of length m in **w**, say $x = a_1 a_2 \cdots a_m$. Then

$$\text{every occurrence of } x \text{ in } \mathbf{w} \text{ must be followed by the same letter;} \quad (10.1)$$

for if not, we would have $p_{\mathbf{w}}(m + 1) > p_{\mathbf{w}}(m)$. Now consider the $r + 1$ strings

$$b_0 b_1 \cdots b_{m-1},$$
$$b_1 b_2 \cdots b_m,$$
$$b_2 b_3 \cdots b_{m+1},$$
$$\vdots$$
$$b_r b_{r+1} \cdots b_{r+m-1}.$$

Since $p_{\mathbf{w}}(m) = r$, these $r + 1$ subwords cannot all be distinct; thus there must exist indices i, j such that $0 \leq i < j \leq r$ and

$$b_i \cdots b_{i+m-1} = b_j \cdots b_{j+m-1}.$$

By our earlier observation (10.1), we must have $b_{i+m} = b_{j+m}$. Hence

$$b_{i+1} \cdots b_{i+m} = b_{j+1} \cdots b_{j+m}.$$

Again by (10.1), we have $b_{i+m+1} = b_{j+m+1}$, and continuing in this fashion, we get that $b_{i+l} = b_{j+l}$ for all $l \geq 0$. It follows that $\mathbf{w}$ is ultimately periodic, and its period length is $j - i \leq k$.

(e) $\implies$ (a): If $\mathbf{w}$ is an ultimately periodic word, say $\mathbf{w} = xy^{\omega}$ with $x \in \Sigma^*$, $y \in \Sigma^+$, then $p_{\mathbf{w}}(n) \leq |x| + |y|$ for all $n \geq 0$. This follows because any subword begins at a position inside x or a position inside a repetition of y, and there are at most $|x| + |y|$ such distinct positions. $\blacksquare$

Subword complexity is related to the concept of (topological) entropy. Let $\mathbf{w}$ be an infinite word over the finite alphabet Δ. The *entropy* $h(\mathbf{w})$ of $\mathbf{w}$ is defined as follows:

$$h(\mathbf{w}) := \lim_{n \to \infty} \frac{\log_{|\Delta|} p_{\mathbf{w}}(n)}{n}. \tag{10.2}$$

To see that this definition makes sense, we first need to see that this limit exists. First, a definition: we say a sequence $(A(n))_{n \geq 0}$ is *subadditive* if for all $m, n \geq 0$ we have $A(m + n) \leq A(m) + A(n)$.

Lemma 10.2.7 *Let* $(A(n))_{n \geq 0}$ *be a sequence such that* $A(n) \geq 0$ *for all* $n \geq 0$, *and* A *is subadditive. Then* $\lim_{n \to \infty} \frac{1}{n} A(n)$ *exists and equals* $\inf_{n \geq 1} \frac{1}{n} A(n)$.

Proof. Define $t = \inf_{n \geq 1} \frac{A(n)}{n}$. Note that t exists and is non-negative. Fix $b \geq 1$; then for all N there exist n, r such that $N = bn + r$ with $0 \leq r < b$. Now

$$A(N) = A(bn + r) \leq A(bn) + A(r) \leq nA(b) + A(r).$$

Hence

$$\frac{A(N)}{N} \leq \frac{nA(b)}{bn + r} + \frac{A(r)}{bn + r}.$$

Then

$$\limsup_{N \to \infty} \frac{A(N)}{N} \leq \frac{A(b)}{b}.$$

But this is true for every b, so

$$\limsup_{N \to \infty} \frac{A(N)}{N} \leq t.$$

But clearly

$$t \leq \liminf_{N \to \infty} \frac{A(N)}{N},$$

so

$$\limsup_{N \to \infty} \frac{A(N)}{N} = \liminf_{N \to \infty} \frac{A(N)}{N},$$

and the limit exists. ∎

Now to see that the limit (10.2) exists, put $A(n) = \log_{|\Delta|} p_u(n)$. Then $A(n)$ satisfies the hypothesis of Lemma 10.2.7 by Lemma 10.2.5.

10.3 Results for Automatic Sequences

In this section we give a linear upper bound on the subword complexity of an automatic sequence.

Theorem 10.3.1 *Let* $h : \Sigma^* \to \Sigma^*$ *be a* t-*uniform morphism prolongable on a* $\in$ Σ, *and let* $\tau : \Sigma \to \Delta$ *be a coding. Let* $\mathbf{u} = \tau(h^\omega(a))$. *Then* $p_{\mathbf{u}}(n) \leq tk^2 n$ *for all* $n \geq 1$, *where* $k = \mathrm{Card}\,\Sigma$.

Proof. By Theorem 10.2.4, it suffices to prove the upper bound for $\mathbf{v} = h^\omega(a)$. Let $\mathbf{v} = v_0 v_1 v_2 \cdots$. Let $n \geq 1$, let r be such that $t^{r-1} \leq n < t^r$, and let $v_i v_{i+1} \cdots v_{i+n-1}$ be any subword of $\mathbf{v}$ of length n. Let $j = \lfloor i/t^r \rfloor$; then $v_i v_{i+1} \cdots v_{i+n-1}$ is a subword of $v_{jt^r} \cdots v_{(j+1)t^r} \cdots v_{(j+2)t^r-1}$. But $v_{jt^r} \cdots v_{(j+2)t^r-1} = h^r(v_j v_{j+1})$, so $v_i \cdots v_{i+n-1}$ is completely determined by $i \pmod{t^r}$, v_j and v_{j+1}. There are t^r possibilities for $i \pmod{t^r}$, and k^2 possibilities for v_j and v_{j+1}. Hence $p_{\mathbf{v}}(n) \leq t^r k^2 \leq tk^2 n$. ∎

Now, combining this result with Theorem 10.2.6, we get

Corollary 10.3.2 *If* $\mathbf{u}$ *is a* k-*automatic sequence that is not ultimately periodic, then* $p_{\mathbf{u}}(n) = \Theta(n)$.

Example 10.3.3 Let us now compute the subword complexity for a particular automatic sequence. Let $\mathbf{a} = (a_n)_{n \geq 0}$ be the sequence, over the alphabet $\{A, B, C, D\}$, that is the fixed point of the map defined by

$$\varphi(A) = AB,$$
$$\varphi(B) = CB,$$
$$\varphi(C) = AD,$$
$$\varphi(D) = CD.$$

(This is the interior sequence for the paperfolding sequence.) We have

$$n = 0\ 1\ 2\ 3\ 4\ 5\ 6\ 7\ 8\ 9\ 10\ 11\ 12\ 13\ 14\ 15\ 16\ 17\ 18\ 19\ 20 \ldots$$
$$a_n = A\ B\ C\ B\ A\ D\ C\ B\ A\ B\ C\ D\ A\ D\ C\ B\ A\ B\ C\ B\ A \ldots$$

We find

$$\text{Sub}_\mathbf{a}(1) = \{\text{A, B, C, D}\}, \qquad\qquad p_\mathbf{a}(1) = 4,$$
$$\text{Sub}_\mathbf{a}(2) = \{\text{AB, AD, BA, BC, CB, CD, DA, DC}\}, \qquad p_\mathbf{a}(2) = 8,$$
$$\text{Sub}_\mathbf{a}(3) = \{\text{ABC, ADC, BAB, BAD, BCB, BCD, CBA, CDA,}$$
$$\text{DAB, DAD, DCB, DCD}\}, \qquad\qquad p_\mathbf{a}(3) = 12.$$

We might guess that $p_\mathbf{a}(n) = 4n$ for all $n \geq 1$. This is in fact correct.

Theorem 10.3.4 *Let* $\mathbf{a} = (a_n)_{n \geq 0}$ *be the fixed point of* φ, *where* $\varphi(\text{A}) = \text{AB}$, $\varphi(\text{B}) = \text{CB}$, $\varphi(\text{C}) = \text{AD}$, $\varphi(\text{D}) = \text{CD}$. *Then* $p_\mathbf{a}(n) = 4n$ *for* $n \geq 1$.

Proof. We split the set of subwords into two groups: the *odd* and *even* subwords. An even subword $f = a_r a_{r+1} \cdots a_{r+k}$ corresponds to an even r, and an odd subword is defined analogously. Note that since the letters in $\mathbf{a}$ alternate, that is, since $\mathbf{a} \in ((\text{A} + \text{C})(\text{B} + \text{D}))^\omega$, the sets of odd and even subwords are disjoint. Let o_n denote the number of odd subwords of length n, and e_n the number of even subwords of length n. We claim that

$$\text{(i)} \qquad o_n = e_{n+1} \qquad (n \geq 3);$$
$$\text{(ii)} \qquad e_{2n} = e_n + o_n \qquad (n \geq 1);$$
$$\text{(iii)} \qquad e_{2n} = e_{2n+1} \qquad (n \geq 1).$$

(i): Let f be an odd subword,

$$f = a_{2r-1} a_{2r} a_{2r+1} \cdots a_{2r+n-2} \qquad (r \geq 1),$$

of length $n \geq 3$. We claim that a_{2r-2} is uniquely determined by (a_{2r}, a_{2r+1}), according to the following table:

$a_{2r} a_{2r+1}$	a_{2r-2}
AB	C
CB	A
AD	C
CD	A

For example, suppose that $a_{2r} a_{2r+1} = \text{AB}$; then $a_r = \text{A}$. Hence $a_{r-1} \in \{\text{B, D}\}$. Then $\varphi(a_{r-1}) = a_{2r-2} a_{2r-1} \in \{\text{CB, CD}\}$, and in any event $a_{2r-2} = \text{C}$. The other cases are similar. It follows that the odd subword f, of length n, can be uniquely extended (on the left) to an even subword of length $n + 1$. Similarly, by deleting the first symbol, an even subword g, of length $n + 1$, corresponds to a unique odd subword of length n. Hence $o_n = e_{n+1}$.

(ii): An even subword $f = a_{2r} a_{2r+1} \cdots a_{2r+2n-1}$ of length $2n$ is the image of a subword

$$a_r a_{r+1} \cdots a_{r+n-1}$$

of length n in exactly one way.

(iii): We claim that an even subword $f = a_{2r}a_{2r+1}\cdots a_{2r+2n-1}$ of length $2n$ can be extended uniquely on the right to an even subword of length $2n + 1$, for $n \geq 1$. For we have

$$\varphi(a_r a_{r+1} \cdots a_{r+n-1}) = a_{2r}a_{2r+1}\cdots a_{2r+2n-1},$$

and $a_r a_{r+1} \cdots a_{r+n-1}$ is unique. We claim that a_{2r+2n} is uniquely determined by $a_{2r+2n-2}a_{2r+2n-1}$, as given in the following table:

$a_{2r+2n-2}a_{2r+2n-1}$	a_{2r+2n}
AB	C
CB	A
AD	C
CD	A

For example, suppose that $a_{2r+2n-2}a_{2r+2n-1} = \mathrm{AB}$; then $a_{r+n-1} = \mathrm{A}$. Hence $a_{r+n} \in \{\mathrm{B}, \mathrm{D}\}$. Then $\varphi(a_{r+n}) = a_{2r+2n}a_{2r+2n+1} \in \{\mathrm{CB}, \mathrm{CD}\}$, and in any event $a_{2r+2n} = \mathrm{C}$. Hence, as before, $e_{2n} = e_{2n+1}$. Using (i),(ii),(iii), and the fact that $e_1 = o_1 = 2$ and $e_2 = o_2 = 4$, we can easily prove by induction that

$$o_{2n} = 4n,$$
$$o_{2n+1} = 4n + 4,$$
$$e_{2n} = 4n,$$
$$e_{2n+1} = 4n,$$
$$o_n + o_{n+1} = 4n + 4,$$
$$e_n + o_n = 4n,$$

for $n \geq 1$. ∎

10.4 Subword Complexity for Morphic Words

Our goal in this section is to prove a bound on the subword complexity of morphic words. Our arguments are based on a certain infinite tree associated with the iteration of a morphism.

Let $h : \Sigma^* \to \Sigma^*$ be a nonerasing morphism, and let $w \in \Sigma^+$. For $i \geq 0$, define $h^i(w) = a_{i,1} \cdots a_{i,l_i}$, where $l_i := |h^i(w)|$. We think of the words $h^i(w)$ arranged in an infinite tree $T(h, w)$ (or more precisely, a forest) rooted at the letters of w. If $v' := |h(a_{i,1} \cdots a_{i,v-1})|$ and $v'' := |h(a_{i,v})|$, then the children of the node $a_{i,v}$ are $a_{i+1,v'+1}, \ldots, a_{i+1,v'+v''}$.

Example 10.4.1 Consider the morphism h defined by

$$a \to abd,$$
$$b \to bb,$$
$$c \to c,$$
$$d \to dc.$$

Then the first few levels of $T(h, \mathrm{a})$ are shown in Figure 10.1.

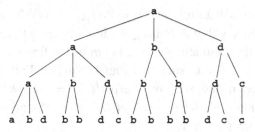

Figure 10.1: The Tree $T(h, \text{a})$.

Definition 10.4.2 Let $h : \Sigma^* \to \Sigma^*$ be a morphism. We say a letter $c \in \Sigma$ is *growing* if for each integer $n \geq 0$ there exists an integer i such that $|h^i(c)| \geq n$.

Definition 10.4.3 Let $G(n)$ denote Landau's function, the maximum order of a permutation on n elements.

Since every permutation may be written as the product of disjoint cycles, we have

$$G(n) = \max_{\substack{c_1, c_2, \dots, c_i \geq 1 \\ c_1 + c_2 + \cdots + c_i = n}} \text{lcm}(c_1, c_2, \dots, c_i).$$

Then we have the following theorem, due to Landau, which we state without proof:

Theorem 10.4.4 *We have* $\log G(n) \sim \sqrt{n \log n}$.

First, we prove a useful result on the prefixes and suffixes of words arising by iterating a morphism.

Definition 10.4.5 If $w = a_1 a_2 \cdots a_s$, where each $a_i \in \Sigma$, then for $t \geq 0$ we define $\text{Pref}_t(w) = a_1 a_2 \cdots a_{\min(s,t)}$ and $\text{Suff}_t(w) = a_{\max(1,s-t+1)} \cdots a_s$.

Lemma 10.4.6 *Let* $h : \Sigma^* \to \Sigma^*$ *be a nonerasing morphism, and let* $k = \text{Card } \Sigma$. *Let* $w \in \Sigma^+$. *Then there exist integers* g, g' *with* $1 \leq g, g' \leq G(k)$ *such that for all integers* $t \geq 1$, $j \geq tk$, *and* $n \geq 0$ *we have*

$$\text{Pref}_t(h^j(w)) = \text{Pref}_t(h^{j+ng}(w)), \tag{10.3}$$

$$\text{Suff}_t(h^j(w)) = \text{Suff}_t(h^{j+ng'}(w)). \tag{10.4}$$

Proof. We prove the result (10.3) for prefixes. The result (10.4) for suffixes follows by a similar argument.

In fact, we prove a slightly stronger statement by induction on t. For each letter $c \in \Sigma$, consider the sequence $(\text{Pref}_1(h^j(c)))_{j \geq 0}$. Clearly this sequence is ultimately periodic; let its period length be p_c. We have $1 \leq p_c \leq k$. Further, there exists an integer $0 \leq i_c < k$ such that $\mathrm{Pref}_1(h^j(c)) = \mathrm{Pref}_1(h^{j+p_c}(c))$ for all $j \geq i_c$. Then we prove that (10.3) holds with $g = \mathrm{lcm}_{c \in \Sigma} \, p_c$. It is now easy to see that $g \leq G(k)$.

The base case of the argument is $t = 1$. Let a_1 be the first letter of w. Then there exists an integer i, $0 \leq i < k$, such that $\mathrm{Pref}_1(h^j(a_1)) = \mathrm{Pref}_1(h^{j+p_{a_1}}(a_1))$ for all $j \geq i$. Since h is nonerasing, we have $\mathrm{Pref}_1(h^j(w)) = \mathrm{Pref}_1(h^j(a_1))$ for all $j \geq 0$. Moreover $p_{a_1} \mid g$. Thus $\mathrm{Pref}_1(h^j(w)) = \mathrm{Pref}_1(h^{j+ng}(w))$ for all $j \geq k$ and all $n \geq 0$. Hence the result holds for $t = 1$.

Now assume the result is true for $t = 1, 2, \ldots, r$. We prove it for $t = r + 1$.

By induction we have $\mathrm{Pref}_r(h^j(w)) = \mathrm{Pref}_r(h^{j+ng}(w))$ for $j \geq rk, n \geq 0$. There are three cases to consider. Recall that $l_i := |h^i(w)|$.

Case 1: $l_{(r+1)k} < r + 1$. Since h is nonerasing, this means that $a_{u,r+1}$ does not exist for $u \leq (r + 1)k$. Then by Exercise 33 we have $l_j < r + 1$ for all $j \geq 0$, and so, trivially, we have $\mathrm{Pref}_{r+1}(h^j(w)) = \mathrm{Pref}_{r+1}(h^{j+ng}(w))$.

Case 2: There exists u with $rk < u \leq (r + 1)k$ such that $a_{u,r+1}$ exists, and the parent of $a_{u,r+1}$ is one of $a_{u-1,1}, \ldots, a_{u-1,r}$. Then by induction

$$a_{u-1,1} \cdots a_{u-1,r} = a_{u-1+ng,1} \cdots a_{u-1+ng,r}$$

for $n \geq 0$, so applying h^e to both sides, and recalling that h is nonerasing, we get

$$\mathrm{Pref}_{r+1}(h^j(w)) = \mathrm{Pref}_{r+1}(h^{j+ng}(w))$$

for $j \geq u$.

Case 3: For all u with $rk < u \leq (r + 1)k$ the parent of $a_{u,r+1}$ exists and is equal to $a_{u-1,r+1}$. Then, by the pigeonhole principle, among $a_{rk,r+1}, \ldots, a_{(r+1)k,r+1}$ there must be a repeated letter. Choose i', m' minimal with $0 \leq i' < m' \leq k$ such that $a_{rk+i',r+1} = a_{rk+m',r+1}$. Put $s' = m' - i'$. Then by an argument similar to that in case 2, we get $\mathrm{Pref}_{r+1}(h^j(w)) = \mathrm{Pref}_{r+1}(h^{j+n\hat{g}}(w))$ where $\hat{g} = \mathrm{lcm}(g, s')$. But $\mathrm{lcm}(g, s') = g$. This completes the proof. ∎

We are now ready to prove the main result of this section:

Theorem 10.4.7 *Let $h : \Sigma^* \to \Sigma^*$ be a nonerasing morphism, prolongable on the letter a. Let $\mathbf{u} = h^\omega(a)$. Let $k = \mathrm{Card}\, \Sigma$, $W = \max_{c \in \Sigma} |h(c)|$. Then for all $n > W$, we have*

$$p_{\mathbf{u}}(n) < k \frac{W(W - 1)}{2}(kn^2 + 2nG(k)^2 + 2kn(n - 1) + W^k(G(k) + k)).$$

Proof. Let $n > W$ and let u be a subword of length n of $h^\omega(a)$. Then u is a subword of $h^r(a)$ for some $r \geq 1$.

Once more, as in the proof of Lemma 10.4.6, we think of the words $a = h^0(a), h^1(a), \ldots, h^r(a)$ arranged in a tree $T(h, a)$ with root a.

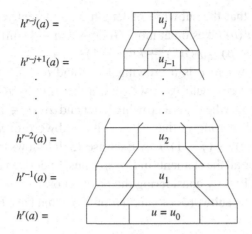

$h^{r-j}(a) =$

$h^{r-j+1}(a) =$

$h^{r-2}(a) =$

$h^{r-1}(a) =$

$h^r(a) =$

Figure 10.2: Tree of Substrings, T.

We now define a finite sequence of subwords of $h^i(a)$, $i \leq r$, as follows: let $u_0 = u$, and for $i \leq r$, let u_i be the subword of $h^{r-i}(a)$ corresponding to the parents of u_{i-1} in $T(h, a)$. Thus $h(u_i)$ covers u_{i-1}, and in fact u_i is the minimal subword of $h^{r-i}(a)$ that does so. We continue this procedure, ascending the tree $T(h, a)$, until we discover an index j such that $|u_j| = 1$. We then get a picture as in Figure 10.2. Since we chose $|u| > W$, we must have $u_1 \geq 2$ and so $j \geq 2$.

Now let

$$B = \{\beta \in \Sigma^* : |\beta| > 1 \text{ and } \beta \text{ is a subword of } h(c) \text{ for some } c \in \Sigma\}.$$

For each $\beta \in B$, define $L(n, \beta)$ as follows:

$$L(n, \beta) = \{u \in \Sigma^* : u \text{ is a subword of length } n \text{ of } h^\omega(a) \text{ and } u_{j-1} = \beta\}.$$

Here j is the index defined above.

Now we estimate Card $L(n, \beta)$. There are four cases to consider:

Case 1: β contains no growing letter. Then $S = \{h^i(\beta) : i \geq 0\}$ must be finite. By Exercise 35(b) we have Card $S \leq G(k) + k$, and by Exercise 35(a) we know that each element of S is bounded in length by $|\beta|W^{k-1}$. It follows that Card $L(n, \beta) \leq W^k(G(k) + k)$.

Case 2: $\beta = c\alpha$ where c is a growing letter and $\alpha \in \Sigma^+$. Write $\beta = cb\alpha'$, where $\alpha' \in \Sigma^*$. By construction each $u \in L(n, \beta)$ contains a descendant of c and a descendant of b; if not, β would not be minimal. Hence any subword of length n of $h^i(\beta)$ in $L(n, \beta)$, for $i \geq 0$, must straddle the boundary between $h^i(c)$ and $h^i(b)$. It follows that $h^i(\beta)$ contains at most n subwords of length n in $L(n, \beta)$. Now by Lemma 10.4.6 we know that the sequence $(\text{Suff}_{n-1}(h^i(c)))_{i \geq 0}$ is ultimately periodic with preperiod length bounded by $k(n - 1)$ and period length bounded by $G(k)$. Similarly, $(\text{Pref}_{n-1}(h^i(b\alpha')))_{n \geq 0}$ is ultimately periodic with preperiod length bounded by $k(n - 1)$ and period length bounded by $G(k)$. Putting these results together, it follows that the subword of length $2n - 2$ centered at the boundary between $h^i(c)$ and $h^i(b)$ takes on at most $G(k)^2 + k(n - 1)$ different values. Thus in this case Card $L(n, \beta) \leq nG(k)^2 + kn(n - 1)$.

Case 3: $\beta = \alpha c$ where c is a growing letter and $\alpha \in \Sigma^+$. This case can be handled similarly to case 2, and again we get Card $L(n, \beta) \leq nG(k)^2 + kn(n - 1)$.

Case 4: $\beta = \alpha_1 c \alpha_2$, where c is a growing letter and $\alpha_1, \alpha_2 \in \Sigma^+$. Since $\alpha_1, \alpha_2 \neq \epsilon$, and β is minimal, it follows that $h^{j-1}(c)$ is a subword of u. Thus $n = |u| > |h^{j-1}(c)|$. But $|h^{j-1}(c)| > (j - 1)/k$ by Exercise 33. It follows that $n > (j - 1)/k$, and so $j \leq kn$. Hence all the elements in $L(n, \beta)$ must be derived from β by applying h at most kn times. But, by considering the position of $h^i(c)$ within $h^i(\beta)$, we see that there are at most n subwords of length n of $h^i(\beta)$ that have $h^i(c)$ as a subword. Hence Card $L(n, \beta) \leq kn^2$.

It follows that Card $L(n, \beta) \leq kn^2 + 2nG(k)^2 + 2kn(n - 1) + W^k(G(k) + k)$. Now Card $B \leq k(1 + 2 + \cdots + W - 1) = kW(W - 1)/2$. It follows that for $n > W$ we get $p_u(n) < k\frac{W(W-1)}{2}(kn^2 + 2nG(k)^2 + 2kn(n - 1) + W^k(G(k) + k))$. ∎

An estimate on $p_u(n)$ not involving G can be obtained using the following result of Massias, which we state without proof:

Theorem 10.4.8 *For all integers $n \geq 1$ we have $G(n) \leq 2.87^{\sqrt{n \log n}}$.*

As a corollary to Theorem 10.4.7, we get

Corollary 10.4.9 *If $\mathbf{u}$ is a morphic word, then $p_{\mathbf{u}}(n) = O(n^2)$.*

Proof. If $\mathbf{u}$ is morphic, then by Theorem 7.5.1 there exist a letter b, a nonerasing morphism g prolongable on b, and a coding λ such that $\mathbf{u} = \lambda(g^\omega(b))$. By Theorem 10.4.7 the subword complexity of $g^\omega(b)$ is $O(n^2)$, and hence the subword complexity of $\mathbf{u} = \lambda(g^\omega(b))$ is also $O(n^2)$. ∎

Example 10.4.10 Let us continue with Example 10.4.1. Let

$$\mathbf{u} = h^\omega(a) = a \prod_{i \geq 0} b^{2^i} d c^i.$$

Then it is easy to see that $\mathbf{u}$ contains a subword of the form $b^i d c^j b \{b, c, d\}^{n-(i+j+2)}$ for $0 \leq i \leq j \leq (n - 2)/2$, and all of these subwords are distinct. It follows that $p_{\mathbf{u}}(n) = \Omega(n^2)$.

The form of the upper bound in Theorem 10.4.7 suggests that it should be possible to construct morphic infinite words that have extremely high subword complexity for short subwords. (Of course, as $n \to \infty$ this behavior cannot persist, since the subword complexity is $O(n^2)$.) This is true, as the following example shows.

Example 10.4.11 Let $p_1 = 2$, p_2, p_3, ... be the sequence of prime numbers. Let $k \geq 1$ be a fixed integer, and define a morphism h as follows: $c \to c\, c_{1,0}\, c_{2,0} \cdots c_{k,0}$, and $c_{i,j} \to c_{i,j+1 \bmod p_i}$ for $1 \leq i \leq k$ and $0 \leq j < p_i$. Then h is defined over an alphabet with $A_k := 1 + \sum_{1 \leq i \leq k} p_i$ symbols. By the prime number theorem, it follows that $A_k \sim (k^2 \log k)/2$. On the other hand, if $\mathbf{u} = h^\omega(a)$, then it is easy to see that $p_{\mathbf{u}}(k) \geq \prod_{1 \leq i \leq k} p_i$, which by the prime number theorem is $e^{k(1+o(1))}$.

While general morphic words have subword complexity $O(n^2)$, for certain types of morphic words this bound can be improved.

Theorem 10.4.12 *Let $h : \Sigma^* \to \Sigma^*$ be a primitive morphism, prolongable on the letter $a \in \Sigma$. Let $\mathbf{u} = h^\omega(a)$. Then $p_{\mathbf{u}}(n) = O(n)$.*

Proof. Let $k = \operatorname{Card} \Sigma$. The result is clearly true if $k = 1$, so we assume $k \geq 2$.

Since h is a primitive morphism, there exists an integer $e \geq 1$ such that for all $c, d \in \Sigma$ we have $h^e(c) \in \Sigma^* d \Sigma^*$. (In fact, by Exercise 8.8(b), we may take $e = 2(k-1)$.) Let $g = h^e$, and define $\hat{W} := \operatorname{Width} g$ and $W := \operatorname{Width} h$. Then $\hat{W} \leq W^{2(k-1)}$.

Since $g(c)$ contains at least one occurrence of each symbol in Σ, we have $|g(c)| \geq k$ for all $c \in \Sigma$. By Theorem 1.4.3 we have $|g^n(c)| \leq \hat{W} |g^n(d)|$ for all $c, d \in \Sigma$ and $n \geq 0$.

Let u be a subword of length ≥ 4 of $\mathbf{u}$. Then u is a subword of $g^r(a)$ for some integer $r \geq 1$. As in the proof of Theorem 10.4.7, consider forming the sequence of subwords $u = u_0, u_1, \ldots$, where u_{i+1} is the shortest word such that $h(u_{i+1})$ covers u_i. Let j be the largest index such that $|u_j| \geq 4$, and write $u_j = \alpha v \beta$ for letters $\alpha, \beta \in \Sigma$ and $v \in \Sigma^*$. Since $|u_j| \geq 4$, we have $|v| \geq 2$.

Now $|u_{j+1}| \leq 3$, and u is a subword of $g^{j+1}(u_{j+1})$, so $|u| \leq |g^{j+1}(u_{j+1})| \leq 3|g^{j+1}(c)|$ for some $c \in \Sigma$. On the other hand, $g^j(v)$ is a subword of u, so $|u| \geq 2|g^j(d)|$ for some $d \in \Sigma$. Thus u is completely specified by u_{j+1} (for which there are at most k^3 distinct choices) and the position of u within $g^{j+1}(u_{j+1})$. But we have $|g^{j+1}(u_{j+1})| \leq 3|g^{j+1}(c)| \leq 3\hat{W}|g^j(c)| \leq 3\hat{W}^2|g^j(d)| \leq \frac{3}{2}\hat{W}^2|u|$. It follows that $p_{\mathbf{u}}(n) \leq \frac{3}{2}W^{4k-2}k^3 n$ for $n \geq 4$. ∎

Example 10.4.13 We continue with Examples 10.4.1 and 10.4.10. Let

$$\mathbf{u} = h^\omega(\mathrm{a}) = \mathrm{a} \prod_{i \geq 0} \mathrm{b}^{2^i} \mathrm{d}\, \mathrm{c}^i.$$

Then from Example 10.4.10, we know that $p_{\mathbf{u}}(n) = \Omega(n^2)$. By Theorem 10.4.12, $\mathbf{u}$

cannot be primitive morphic (i.e., the image, under a coding, of a word of the form $g^\omega(a)$ where g is a primitive morphism prolongable on a), since then its subword complexity would be $O(n)$.

10.5 Sturmian Words

In this section we continue our study of Sturmian words, begun in Chapter 9, and we prove a basic result about the subword complexity of these words.

Up to now infinite words were indexed by the integers $\{0, 1, 2, \ldots\}$. However, Sturmian characteristic words are traditionally indexed by the positive integers $\{1, 2, \ldots\}$. It will be understood here that two infinite words are equal if and only if their letters are respectively equal. More precisely, if $\mathbf{w} = a_0 a_1 a_2 \cdots$, and $\mathbf{s} = s_1 s_2 s_3 \cdots$, then $\mathbf{w} = \mathbf{s}$ means that $a_i = s_{i+1}$ for all $i \geq 0$.

Let α be an irrational real number. For integers $i \geq 0$ and real numbers α and x, define $v_i(x) = \lfloor (i+1)\alpha + x \rfloor - \lfloor i\alpha + x \rfloor$. Then, as a function of x, the function $v_i(x)$ is periodic of period 1. Note that if

$$\mathbf{s}_{\alpha,\theta} = s_1 s_2 s_3 \cdots,$$

then

$$v_i(j\alpha + \theta) = s_{i+j}.$$

Now consider the $n + 2$ numbers

$$0, \{-\alpha\}, \{-2\alpha\}, \ldots, \{-n\alpha\}, 1,$$

where (as in Chapter 2) $\{x\}$ denotes the fractional part of the number x and α is irrational. Arrange these numbers in increasing order:

$$0 = c_0(n) < c_1(n) < \cdots < c_n(n) < c_{n+1}(n) = 1.$$

For $0 \leq j \leq n$, define the half-open interval $L_j(n) = [c_j(n), c_{j+1}(n))$.

Lemma 10.5.1 *The word-valued function of x*

$$V_n(x) = v_0(x)v_1(x) \cdots v_{n-1}(x)$$

is constant on $L_j(n)$.

Proof. It suffices to show that for each integer i with $0 \leq i \leq n$, the quantity $i\alpha + x$ is never an integer for any x in the range $c_j(n) < x < c_{j+1}(n)$. If this is the case, then for $0 \leq i \leq n$, $\lfloor i\alpha + x \rfloor$ is constant on $L_j(n)$. Hence for $0 \leq i \leq n - 1$, the word-valued function $v_i(x)$ is constant on $L_j(n)$.

So assume, contrary to what we want to prove, that there exists x_0, $0 < x_0 < 1$, such that $i\alpha + x_0 = r$ for some integer r, with $c_j(n) < x_0 < c_{j+1}(n)$. Then $-i\alpha = x_0 - r$, so $\{-i\alpha\} = \{x_0\}$. But this equality contradicts our definition of $c_j(n)$ and

$c_{j+1}(n)$. It follows that if $x \in L_j(n)$, then $V_n(x) = v_0(x)v_1(x) \cdots v_{n-1}(x)$ has a fixed value $B_j(n)$ depending only on j and n, not on x. ∎

We now apply our result to prove

Theorem 10.5.2 *Let α, θ be real numbers with α irrational and $0 < \alpha < 1$. Then every Sturmian word $\mathbf{s}_{\alpha,\theta} = s_1 s_2 s_3 \cdots$ has $n + 1$ distinct subwords of length n; that is, $p_{\mathbf{s}_{\alpha,\theta}}(n) = n + 1$.*

Proof. Let $w = \mathbf{s}_{\alpha,\theta}[m..m + n - 1]$ be a subword of length n of $\mathbf{s}_{\alpha,\theta}$. Then we have $w = v_0(x)v_1(x) \cdots v_{n-1}(x)$ for $x = \{m\alpha + \theta\}$. By Lemma 10.5.1, we have $w = B_j(n)$ for some j, $0 \leq j \leq n$, where $x \in L_j(n)$. Hence $p_{\mathbf{s}_{\alpha,\theta}}(n) \leq n + 1$ for all $n \geq 0$.

To see that $p_{\mathbf{s}_{\alpha,\theta}}(n) \geq n + 1$ for all $n \geq 0$, assume the contrary. Then there exists an n such that $p_{\mathbf{s}_{\alpha,\theta}}(n) < n + 1$, i.e., $p_{\mathbf{s}_{\alpha,\theta}}(n) \leq n$. But then, by Theorem 10.2.6, the sequence $\mathbf{s}_{\alpha,\theta}$ is ultimately periodic. But, by Exercise 1.6, every letter in an ultimately periodic sequence has rational density. However, the density of 1's in $\mathbf{s}_{\alpha,\theta}$ is easily seen to be α, which by assumption is irrational. This is a contradiction. It follows that $p_{\mathbf{s}_{\alpha,\theta}}(n) = n + 1$ for all $n \geq 0$, and a similar argument proves the same result for $p_{\mathbf{s}'_{\alpha,\theta}}(n)$. ∎

Finally, we observe that the set of subwords of $\mathbf{s}_{\alpha,\theta}$ is independent of θ.

Theorem 10.5.3 *Let $0 < \alpha < 1$ be irrational, and θ, ρ be any real numbers. The finite subwords of $\mathbf{s}_{\alpha,\theta}$ coincide with the finite subwords of $\mathbf{s}_{\alpha,\rho}$.*

Proof. By the remarks above, the length-n subwords of $\mathbf{s}_{\alpha,\theta}$ are given by $\{B_j(n) : 0 \leq j \leq n\}$. But this depends only on n and not on θ. ∎

We now introduce the notion of binary balanced infinite words. These words have the property that the numbers of occurrences of a given letter in any two subwords of the same length differ at most by 1. We first establish some properties of these words. The section ends with Theorem 10.5.8 showing that the set of Sturmian words, the set of non-ultimately-periodic balanced words, and the set of words of subword complexity $p(n) = n + 1$ for all $n \geq 0$ are identical.

Definition 10.5.4 An infinite word $\mathbf{u} = (u(n))_{n \geq 0}$ on $\Sigma_2 = \{0, 1\}$ is called *balanced* if and only if for any two subwords x and y of $\mathbf{u}$, we have

$$|x| = |y| \implies \left| |x|_1 - |y|_1 \right| \leq 1.$$

Remark. The condition above is clearly equivalent to the condition

$$|x| = |y| \implies \big||x|_0 - |y|_0\big| \leq 1.$$

We first give an equivalent definition of binary balanced infinite words.

Proposition 10.5.5 *An infinite word* $\mathbf{u} = (u(n))_{n\geq 0}$ *on* $\Sigma_2 = \{0, 1\}$ *is balanced if and only if for any two nonempty subwords x and y of $\mathbf{u}$, we have*

$$\big||x|_1|y| - |y|_1|x|\big| \leq |x| + |y| - \gcd(|x|, |y|). \tag{10.5}$$

Proof. First suppose the inequality (10.5) holds. Then if x and y are subwords of the word $\mathbf{u}$, and $|x| = |y|$, either x and y are empty and $\big||x|_1 - |y|_1\big| = 0$, or they are nonempty and we have, letting $\ell = |x| = |y|$,

$$\big||x|_1\ell - |y|_1\ell\big| \leq \ell + \ell - \ell = \ell, \qquad \text{i.e.,} \qquad \big||x|_1 - |y|_1\big| \leq 1.$$

Suppose now that the word $\mathbf{u}$ is balanced. We prove by induction on $|x| + |y|$ that the inequality (10.5) holds for any nonempty words x and y that are subwords of $\mathbf{u}$. It clearly holds if $|x| + |y| = 2$. Suppose it holds for any nonempty subwords x and y such that $|x| + |y| \leq k$, and let x and y be two subwords of $\mathbf{u}$ such that $|x| + |y| = k + 1$. If $|x| = |y|$, the inequality (10.5) is an immediate consequence of the definition of balanced words. Otherwise, let us suppose that $|x| > |y|$. Hence there exist two nonempty words z and t (that are subwords of $\mathbf{u}$) such that $x = zt$ and $|z| = |y|$. We then have

$$\begin{aligned}
&\big||x|_1|y| - |y|_1|x|\big| \\
&= \big|(|z|_1 + |t|_1)|y| - |y|_1(|z| + |t|)\big| \\
&= \big|(|z|_1|y| - |y|_1|z|) + (|t|_1|y| - |y|_1|t|)\big| \\
&= \big|(|z|_1 - |y|_1)|z| + (|t|_1|y| - |y|_1|t|)\big| \qquad \text{(since $|y| = |z|$)} \\
&\leq \big||z|_1 - |y|_1\big||z| + \big||t|_1|y| - |y|_1|t|\big| \qquad \text{(by the triangle inequality)} \\
&\leq |z| + \big||t|_1|y| - |y|_1|t|\big| \qquad \text{(since $|z| = |y|$ and $\mathbf{u}$ is balanced)} \\
&\leq |z| + (|y| + |t| - \gcd(|y|, |t|)) \qquad \text{(from the induction hypothesis)} \\
&= |z| + |t| + |y| - \gcd(|y|, |y| + |t|) \\
&= |x| + |y| - \gcd(|y|, |z| + |t|) \\
&= |x| + |y| - \gcd(|y|, |x|).
\end{aligned}$$

∎

Corollary 10.5.6 *Let* **u** *be an infinite balanced word on the alphabet* $\{0, 1\}$. *Then the following properties hold:*

(a) *The frequency of* 1 *and the frequency of* 0 *in the word* **u** *exist.*

(b) *Let* α *be the frequency of* 1 *in* **u**. *Then either for each subword* z *of the word* **u** *we have* $-1 < |z|_1 - \alpha|z| \leq 1$, *or for each subword* z *of the word* **u** *we have* $-1 \leq |z|_1 - \alpha|z| < 1$.

(c) *The word* **u** *is ultimately periodic if and only if* α *is rational.*

(d) *For all real* β *we have either* $|w_n|_1 \leq \lfloor \alpha n + \beta \rfloor$ *for all* $n \geq 0$ *or* $|w_n|_1 \geq \lfloor \alpha n + \beta \rfloor$ *for all* $n \geq 0$, *where* $w_n := \mathbf{u}[0..n - 1]$.

Proof. We first prove (a). For each $n \geq 0$, let $w_n := \mathbf{u}[0..n - 1]$ be the prefix of length n of the word **u**. From Proposition 10.5.5 we have, for any $m, n \geq 1$,

$$\big||w_n|_1|w_m| - |w_m|_1|w_n|\big| \leq |w_m| + |w_n| - \gcd(|w_m|, |w_n|) < |w_m| + |w_n|.$$

Hence

$$\left|\frac{|w_n|_1}{|w_n|} - \frac{|w_m|_1}{|w_m|}\right| \leq \frac{1}{|w_n|} + \frac{1}{|w_m|} = \frac{1}{n} + \frac{1}{m}.$$

This implies that the sequence $|w_n|_1/n$ is a Cauchy sequence, and hence it converges. Denote its limit by α. Hence the frequency of 1 in **u** exists and is equal to α. Clearly the frequency of 0 in **u** also exists, and it is equal to $1 - \alpha$.

To prove (b), we first apply Proposition 10.5.5 to the words z and w_n, where z is any subword of **u**, and w_n is defined as above. This yields the inequality

$$\big||z|_1|w_n| - |w_n|_1|z|\big| \leq |z| + |w_n| - \gcd(|z|, |w_n|).$$

Hence, dividing by $|w_n| = n$,

$$\left||z|_1 - \frac{|w_n|_1}{n}|z|\right| \leq \frac{|z|}{n} + 1 - \frac{\gcd(|z|, n)}{n}.$$

Letting n tend to ∞ gives

$$\big||z|_1 - \alpha|z|\big| \leq 1,$$

i.e.,

$$-1 \leq |z|_1 - \alpha|z| \leq 1.$$

If (b) does not hold, this implies that there exist two subwords z and w of **u** such that

$$|z|_1 - \alpha|z| = -1 \quad \text{and} \quad |w|_1 - \alpha|w| = 1.$$

Hence

$$\big||w|_1|z| - |z|_1|w|\big| = \big|(\alpha|w|+1)|z| - (\alpha|z|-1)|w|\big| = |z| + |w|,$$

which contradicts Proposition 10.5.5.

We prove (c): first note that if the word **u** is ultimately periodic, then trivially the frequency of 1 in **u** is rational. Let us now suppose that the frequency of 1 in **u** is rational, say equal to p/q. We suppose, from (b), that for each subword z of the word **u**, we have

$$-1 < |z|_1 - \frac{p}{q}|z| \le 1,$$

(the other case is analogous). Hence, for any subword z of **u** of length q, we have $-1 < |z|_1 - p \le 1$. Hence $|z|_1 \in \{p, p+1\}$. We claim that there exist a finite word w and an infinite word **v** such that $\mathbf{u} = w\mathbf{v}$ and all subwords z of **v** of length q satisfy $|z|_1 = p$. For if not, there would exist infinitely many subwords z of **u** such that $|z| = q$ and $|z|_1 = p+1$. In particular there would exist two non-overlapping such words, i.e., there would exist a subword zxz' of **u** such that $|z| = |z'| = q$ and $|z|_1 = |z'|_1 = p+1$. Hence

$$2(p+1) + |x|_1 = |zxz'|_1 \le \frac{p}{q}|zxz'| + 1 = 2p + \frac{p}{q}|x| + 1.$$

This would imply

$$|x|_1 \le \frac{p}{q}|x| - 1,$$

which contradicts

$$-1 < |x|_1 - \frac{p}{q}|x| \le 1.$$

Now, since all subwords z of length q of **v** satisfy $|z|_1 = p$, we have that the first and the last symbols of each subword of length $q+1$ of **v** must be equal. Namely, let ayb be any subword of **v** such that a and b are letters and y is a word of length $q-1$. Since $|ay| = |yb| = q$, we have $|ay|_1 = |yb|_1 = p$; hence $a = b$. It clearly follows that **v** is periodic and q is a period of **v**. Hence **u** is ultimately periodic (see also Exercise 52).

To prove (d) we suppose to the contrary that there exist two integers $m > n$ such that

$$\text{either} \quad \begin{cases} |w_m|_1 < \lfloor \alpha m + \beta \rfloor, \text{ i.e., } |w_m|_1 \le \lfloor \alpha m + \beta \rfloor - 1 \text{ and} \\ |w_n|_1 > \lfloor \alpha n + \beta \rfloor, \text{ i.e., } |w_n|_1 \ge \lfloor \alpha n + \beta \rfloor + 1, \end{cases}$$

$$\text{or} \quad \begin{cases} |w_m|_1 > \lfloor \alpha m + \beta \rfloor, \text{ i.e., } |w_m|_1 \ge \lfloor \alpha m + \beta \rfloor + 1 \text{ and} \\ |w_n|_1 < \lfloor \alpha n + \beta \rfloor, \text{ i.e., } |w_n|_1 \le \lfloor \alpha n + \beta \rfloor - 1. \end{cases}$$

Defining the word z by $w_m = w_n z$, this implies that either

$$|z|_1 = |w_m|_1 - |w_n|_1 \leq \lfloor \alpha m + \beta \rfloor - \lfloor \alpha n + \beta \rfloor - 2$$
$$< (\alpha m + \beta) - (\alpha n + \beta - 1) - 2$$

and hence

$$|z|_1 < \alpha(m - n) - 1 = \alpha|z| - 1,$$

or

$$|z|_1 = |w_m|_1 - |w_n|_1 \geq \lfloor \alpha m + \beta \rfloor - \lfloor \alpha n + \beta \rfloor + 2$$
$$> (\alpha m + \beta - 1) - (\alpha n + \beta) + 2$$

and hence

$$|z|_1 > \alpha(m - n) + 1 = \alpha|z| + 1.$$

In either case this contradicts assertion (b). ∎

The following proposition will prove useful. It relates balanced binary infinite words to palindromes.

Proposition 10.5.7 *Let* **u** *be an infinite word with values in* $\Sigma_2 = \{0, 1\}$. *Then the following properties are equivalent:*

(i) The word **u** *is balanced.*
(ii) For each subword w of **u**, *either* $0w0$ *or* $1w1$ *is not a subword of* **u**.
(iii) For each palindromic subword w of **u**, *either* $0w0$ *or* $1w1$ *is not a subword of* **u**.

Proof. The proof of (i) $\Longrightarrow$ (ii) follows from the definition of balanced words. The implication (ii) $\Longrightarrow$ (iii) is trivial. Let us prove that (iii) $\Longrightarrow$ (i) by proving that, if a word **u** is not balanced, then there exists a palindromic subword of w such that both $0w0$ and $1w1$ are subwords of **u**. Since **u** is not balanced there exist two subwords y and z of **u**, such that $\big||y|_1 - |z|_1\big| \geq 2$ and such that $|y| = |z| = \ell$. We can suppose that these subwords are chosen so that ℓ is minimal. Since ℓ is minimal, the words y and z cannot begin with the same letter, nor can they end with the same letter. Suppose y begins with 0 and z begins with 1. Let w be the longest word (possibly empty) such that $0w$ is a prefix of y and $1w$ a prefix of z. Since the last letters of y and z are different, we see that $y \neq 0w$, $z \neq 1w$. Since w is the longest word such that $0w$ is a prefix of y and $1w$ a prefix of z, the letters following w in y and z must be different. Then, there exist two words (possibly empty) y' and z', and two distinct letters a, b in Σ_2, such that

$$y = 0way' \quad \text{and} \quad z = 1wbz'.$$

Of course $|y'| = |z'|$. If $a = 1$ and $b = 0$, then $\big||y'|_1 - |z'|_1\big| = \big||y|_1 - |z|_1\big| \geq 2$, which would contradict the minimality of ℓ. Hence $a = 0$ and $b = 1$. Hence

$$y = 0w0y' \quad \text{and} \quad z = 1w1z'.$$

Using the minimality of ℓ again we see that y' and z' must be empty, i.e.,

$$y = 0w0 \quad \text{and} \quad z = 1w1.$$

It remains to prove that w is a palindrome. Let x be the (possibly empty) longest prefix of w such that x^R is a suffix of w. If w is not a palindrome, then $|x| \neq |w|$, and there exists a letter $a \in \Sigma_2$ such that xa is a prefix of w and $\bar{a}x^R$ is a suffix of w, where, as usual, $\bar{0} = 1$ and $\bar{1} = 0$. Hence $0xa$ is a prefix of y, and $\bar{a}x^R 1$ is a suffix of z.

If $a = 0$, then $0x0$ and $1x^R 1$ are two subwords of $\mathbf{u}$ of length $< \ell$ such that $\big||0x0|_1 - |1x^R 1|_1\big| = 2$, which contradicts the minimality of ℓ. Hence $a = 1$, so $0x1$ is a prefix of y and $0x^R 1$ is a suffix of z. Their common length is $< \ell$; hence there exist two nonempty words y'' and z'' such that

$$y = 0x1y'' \quad \text{and} \quad z = z''0x^R 1.$$

We clearly have $\big||y''|_1 - |z''|_1\big| = \big||y|_1 - |z|_1\big| \geq 2$, contradicting once more the minimality of ℓ. Hence w is a palindrome, and we are done. ∎

We now give the main result of this section.

Theorem 10.5.8 *Let* $\mathbf{u} = (u_n)_{n \geq 0}$ *with values in* $\Sigma_2 = \{0, 1\}$. *Then the following conditions are equivalent:*

(i) the subword complexity of the word $\mathbf{u}$ *satisfies* $p_{\mathbf{u}}(n) = n + 1$ *for all* $n \geq 1$;
(ii) the word $\mathbf{u}$ *is balanced and non-ultimately-periodic;*
(iii) the word $\mathbf{u}$ *is Sturmian, i.e.,* $\mathbf{u}$ *is equal either to* $\mathbf{s}_{\alpha, \theta}$ *or to* $\mathbf{s}'_{\alpha, \theta}$ *for some irrational* α *in* $(0, 1)$ *and* θ *in* $[0, 1)$.

Furthermore, the frequency of 1 *in the word* $\mathbf{u}$ *exists, and it is equal to the number* α *that occurs in (iii).*

Proof. We first prove that (i) $\implies$ (ii). Since the complexity of $\mathbf{u}$ satisfies $p_{\mathbf{u}}(n) = n + 1$ for all n, the word $\mathbf{u}$ cannot be ultimately periodic from Theorem 10.2.6. We suppose that $\mathbf{u}$ is not balanced, and get a contradiction. If $\mathbf{u}$ is not balanced, we have, using Proposition 10.5.7, that there exists a palindromic subword w of $\mathbf{u}$ such that both $0w0$ and $1w1$ are subwords of $\mathbf{u}$. We see in particular that both words $w0$ and $w1$ are subwords of $\mathbf{u}$. Since $p_{\mathbf{u}}(n) = n + 1$ for all n, then for each word $z \neq w$ such that $|z| = |w|$, each occurrence of z in $\mathbf{u}$ must be always followed by the same letter, say $a_z \in \Sigma_2$. For the same reason there exists exactly one subword of $\mathbf{u}$ of length $|w| + 1$ whose occurrences in $\mathbf{u}$ can be followed either by 0 or by 1. Furthermore w must be a suffix of this word. Without loss of generality, suppose this word is $0w$. Hence $0w0$, $0w1$, and $1w1$ are subwords of **u**, and $1w0$ is not. In other words, each occurrence of $1w$ in **u** must be followed by 1.

Let $1w1x$ be a subword of **u** such that $|x| = |w|$. We claim that $0w$ is not a subword of $1w1x$. For if it is, let $1w1x = z'0wz''$. The word z' must begin with 1, say $z' = 1y$. This gives $1w1x = 1y0wz''$, and hence $w1x = y0wz''$. Note that $|y| + |z''| = |x| = |w|$.

If z'' is empty, then $w1x = y0w$ and $|y| = |w|$. This is impossible, because $1 \neq 0$.

If z'' is not empty, then $|y| = |w| - |z''| \leq |w| - 1$, and $w1x = y0wz''$ shows that there exists a word t (prefix of w) such that $w = y0t$. Hence $y0t1x = y0wz''$, i.e., $t1x = wz''$. Since $|t| < |w|$, we see that the $(|t|+1)$th letter of w is 1. But $w = w^R = (y0t)^R = t^R 0 y^R$. Hence the $(|t|+1)$th letter of w is 0. This gives the desired contradiction, and the word $1w1x$ does not contain the subword $0w$. Hence each subword of $1w1x$ of length $|w| + 1$ is a subword of **u** whose every occurrence in **u** is always followed by the same letter. This implies that **u** is ultimately periodic (see Exercise 52), which is not the case.

We now prove that (ii) $\implies$ (iii). For each $n \geq 0$ let w_n be the prefix of length n of the word **u**. Using Corollary 10.5.6, we know that the frequency of 1 in the word **u** exists and is irrational. Let α be this frequency. Let θ be defined by

$$\theta := \inf\{\beta : |w_n|_1 \leq \lfloor \alpha n + \beta \rfloor \text{ for all } n \geq 0\}.$$

We have that θ is finite: more precisely, θ belongs to $[0, 1]$ from its definition (take $n = 0$) and from assertion (b) in Corollary 10.5.6. Also note that, by taking in the set defining θ a sequence $(\beta_j)_j$ that converges to θ, we have $|w_n|_1 \leq \lfloor \alpha n + \theta \rfloor \leq \alpha n + \theta$ for all $n \geq 0$. On the other hand, we have $|w_n|_1 \geq \alpha n + \theta - 1$ for all $n \geq 0$. For if not, there would be an n_0 such that $|w_{n_0}|_1 < \alpha n_0 + (\theta - 1)$. Let $\theta' := |w_{n_0}|_1 + 1 - \alpha n_0$. Hence $\theta' < \theta$ and $|w_{n_0}|_1 < \alpha n_0 + \theta'$. From assertion (d) in Corollary 10.5.6 this would imply that the inequality $|w_n|_1 \leq \alpha n + \theta'$ holds for all $n \geq 0$, which would contradict the definition of θ. Hence we have

$$|w_n|_1 \leq \alpha n + \theta \leq |w_n|_1 + 1 \qquad \text{for all } n \geq 0. \tag{10.6}$$

Since α is irrational, the quantity $\alpha n + \theta$ can be an integer for at most one value of n. Hence either

$(\alpha n + \theta)$ is different from $|w_n|_1 + 1$ for all $n \geq 0$ and Eq. (10.6) implies

$$\forall n \geq 0, \qquad |w_n|_1 \leq \alpha n + \theta < |w_n|_1 + 1,$$

i.e., $\forall n \geq 0$, $|w_n|_1 = \lfloor \alpha n + \theta \rfloor$, which clearly implies that $\mathbf{u} = \mathbf{s}_{\rho,\theta}$; or there exists a unique integer n_0 such that $\alpha n_0 + \theta = |w_{n_0}| + 1$, and then, for any other integer n, the quantity $\alpha n + \theta$ cannot be an integer and Eq. (10.6) implies

$$\forall n \geq 0, \ n \neq n_0, \qquad |w_n|_1 < \alpha n + \theta < |w_n|_1 + 1,$$

i.e., finally, $|w_n|_1 = \lceil \alpha n + \theta \rceil - 1$ for all $n \geq 0$, which clearly implies $\mathbf{u} = \mathbf{s}'_{\rho,\theta}$.

Finally, (iii) $\implies$ (i) was proved in Theorem 10.5.2 above. ∎

10.6 Sturmian Words and *k*th-Power-Freeness

In this section, we apply the results of Section 10.5 to prove the following beautiful theorem due to F. Mignosi about repetitions in Sturmian words.

Theorem 10.6.1 *Let $0 < \alpha < 1$ be irrational, and let θ be a real number. Then there exists an integer $k \geq 2$ such that the Sturmian word $\mathbf{s}_{\alpha,\theta}$ is kth-power-free if and only if α has bounded partial quotients.*

Proof. The proof consists of several steps. First, by the result of Theorem 10.5.3, the set of subwords of $\mathbf{s}_{\alpha,\theta}$ coincides with the set of finite subwords of $\mathbf{s}_{\alpha,0}$. Thus we may restrict our attention to characteristic words. Next, it is an easy consequence of the results in Section 9.1 that if α has unbounded partial quotients, then $f_\alpha = \mathbf{s}_{\alpha,0}$ contains arbitrarily large powers. Finally, in the most difficult part of the proof, we show that if α has bounded partial quotients, then there exists an integer k such that f_α is kth-power-free.

Lemma 10.6.2 *If the continued fraction for α has unbounded partial quotients, then the characteristic word $\mathbf{f}_\alpha = f(1)f(2)f(3)\cdots$ contains arbitrarily large powers.*

Proof. Let k be a given integer. Since $\alpha = [0, a_1, a_2, \ldots]$ has unbounded partial quotients, there exists an index $i \geq 2$ such that $a_i \geq k$. Then by Theorem 9.1.8, we have $X_i = X_{i-1}^{a_i} X_{i-2}$, where $X_i = f(1)f(2)\cdots f(q_i)$ is a prefix of $\mathbf{f}_\alpha$. It follows that $\mathbf{f}_\alpha$ begins with a kth power. ∎

Lemma 10.6.3 *Let x, y be real numbers such that $\{x\}$, $\{x + y\}$, and $\{x + 2y\}$ are all contained in some interval I, where $|I| < \frac{1}{2}$. Then either*

$$\{x\} \leq \{x + y\} \leq \{x + 2y\}$$

or

$$\{x\} \geq \{x + y\} \geq \{x + 2y\}.$$

Proof. Case (i): $\{x\} \leq \{x + y\}$. Then

$$\{y\} = \{x + y\} - \{x\} < \tfrac{1}{2}$$

by Exercise 2.3. Suppose, contrary to what we want to prove, that $\{x + 2y\} <$

$\{x + y\}$. Then

$$\{-y\} = \{x + y\} - \{x + 2y\} < \tfrac{1}{2},$$

again by Exercise 2.3. Hence $\{y\} < \tfrac{1}{2}$ and $\{-y\} < \tfrac{1}{2}$. It follows that $\{y\} = 0$, which contradicts our assumption that $\{x + 2y\} < \{x + y\}$. Hence $\{x + 2y\} \geq \{x + y\}$.

Case (ii), where $\{x\} > \{x + y\}$, is handled similarly. ∎

Lemma 10.6.4 *Let k be a positive integer, and let m, n be positive integers. Define $x_i = \{(n + im)\alpha\}$. Let h be an integer with $0 \leq h \leq m$. Suppose $x_i \in L_h(m)$ for $0 \leq i < k$, and suppose $|L_h(m)| < \tfrac{1}{2}$. Then either*

$$x_0 \leq x_1 \leq \cdots \leq x_{k-1}$$

or

$$x_0 \geq x_1 \geq \cdots \geq x_{k-1}.$$

Proof. By induction on k. For $k = 1, 2$ the result is vacuously true. For $k = 3$ it follows from Lemma 10.6.3. Now assume true for $k \geq 3$; we prove it for $k + 1$. By induction we have either

$$x_0 \leq x_1 \leq \cdots \leq x_{k-1} \tag{10.7}$$

or

$$x_0 \geq x_1 \geq \cdots \geq x_{k-1}. \tag{10.8}$$

Similarly, by applying the induction hypothesis to the range $1 \leq i \leq k$, we have either

$$x_1 \leq x_2 \leq \cdots \leq x_k \tag{10.9}$$

or

$$x_1 \geq x_2 \geq \cdots \geq x_k. \tag{10.10}$$

Since $k \geq 3$, Eqs. (10.7) and (10.10) cannot hold simultaneously. Similarly, Eqs. (10.8) and (10.9) cannot hold simultaneously. The desired conclusion now follows. ∎

Lemma 10.6.5 *Let $w = B_h(m)$, the length-m subword of $\mathbf{f}_\alpha$ associated with the interval $L_h(m)$, $0 \leq h \leq m$. Suppose $|L_h(m)| < \tfrac{1}{2}$. If w^k is a subword of $\mathbf{f}_\alpha$, then*

$$k - 1 \leq \frac{|L_h(m)|}{\min(\{-m\alpha\}, \{m\alpha\})}.$$

Proof. Suppose w^k is a subword of $\mathbf{f}_\alpha = f(1)f(2)f(3) \cdots$, and $|w| = m$. Then there exists an integer $n \geq 1$ such that $w^k = f(n)f(n + 1) \cdots f(n + km - 1)$.

Since the words $B_h(m)$ $(0 \le h \le m)$ are distinct, it follows by Lemma 10.5.1 that $\{(n + im)\alpha\} \in L_h(m)$ for $0 \le i < k$. Then, by Lemma 10.6.4, either

$$\{n\alpha\} < \{(n + m)\alpha\} < \cdots < \{(n + (k - 1)m)\alpha\}$$

or

$$\{n\alpha\} > \{(n + m)\alpha\} > \cdots > \{(n + (k - 1)m)\alpha\}.$$

Without loss of generality, assume the former holds. Since all these points are in $L_h(m)$, it follows that at least one of the implied intervals, say $I = [\{(n + im)\alpha\}, \{(n + (i + 1)m)\alpha\})$, is of length $\le |L_h(m)|/(k - 1)$. The result now follows.   ∎

We define the *index* of a finite subword w of an infinite word $\mathbf{x}$ to be k, where w^k is the highest power of w that occurs as a subword of $\mathbf{x}$. If arbitrarily high powers of w occur in $\mathbf{x}$, then we say w is of infinite index.

Corollary 10.6.6 *Every finite subword of* $\mathbf{s}_{\alpha,0}$ *is of finite index.*

Proof. Suppose, contrary to what we want to prove, that w is a subword of $\mathbf{s}_{\alpha,0}$ of infinite index. Choose an integer $j \ge 1$ sufficiently large so that $|L_h(m)| < \frac{1}{2}$ for all h with $0 \le h \le m$ for $m := j|w|$.

Then, by assumption, w^{jk} is a subword of $\mathbf{s}_{\alpha,0}$ for all $k \ge 1$. It then follows from Lemma 10.6.5 that

$$jk - 1 \le \frac{\max_{0 \le h \le m} |L_h(m)|}{\min(\{-m\alpha\}, \{m\alpha\})}$$

for all $k \ge 1$. But this is clearly absurd, since the quantity on the right is finite.   ∎

We are now ready to complete the proof of Mignosi's theorem.

Suppose the partial quotients in the continued fraction for α are bounded; say $a_i \le K$ for all $i \ge 1$. Let w be a subword of f_α. By Corollary 10.6.6, w has finite index, so let k be the largest integer such that w^k is a subword of f_α. Let $m := |w|$; as in the proof of Theorem 10.5.2, we know that $w = B_h(m)$ for some h with $0 \le h \le m$.

Define

$$C(\alpha) = \begin{cases} a_1/2 & \text{if } a_1 > 1, \\ (a_2 + 1)/2 & \text{if } a_1 = 1. \end{cases} \tag{10.11}$$

Then by Exercise 28 we know that $\max_{0 \le h \le m} |L_h(m)| < \frac{1}{2}$ if $m > C(\alpha)$. Hence Lemma 10.6.5 applies, and we have

$$k - 1 \le \frac{|L_h(m)|}{\min(\{-m\alpha\}, \{m\alpha\})}. \tag{10.12}$$

But clearly

$$\min(\{-m\alpha\}, \{m\alpha\}) \geq \min_{0 \leq h \leq m} |L_h(m)|,$$

so we may use Corollary 2.6.4 to conclude that

$$\frac{|L_h(m)|}{\min(\{-m\alpha\}, \{m\alpha\})} \leq a_{k+1} + 2 \leq K + 2$$

if $q_k \leq m < q_{k+1}$. Combining this inequality with (10.12), we have $k - 1 \leq K + 2$, and so $k \leq K + 3$.

On the other hand, if $m \leq C(\alpha)$, then there are only a finite number of different subwords of length m contained in $\mathbf{s}_{\alpha,0}$, and by Corollary 10.6.6, each of these subwords has finite index. The proof of Mignosi's theorem is now complete. ∎

10.7 Subword Complexity of Finite Words

Up to now we have been largely concerned with the subword complexity of infinite words. However, it is also of interest to study the subword complexity of finite words. Here, the main questions are the following (let $\mathbf{u}$ be an infinite word):

What is the length of the shortest finite word $w = w(n)$ for which $p_w(i) = p_\mathbf{u}(i)$ for $i = 0, 1, 2, \ldots, n$?
How many such words are there?
What is an efficient method for generating such words?

In this section we examine the first question in one simple case: where the alphabet is $\{0, 1\}$ and the infinite word $\mathbf{u}$ has maximum complexity.

Theorem 10.7.1 *For all $n \geq 0$ there exists a word $w = w(n)$ over $\{0, 1\}$, of length $2^n + n - 1$, such that $p_w(i) = 2^i$ for $i = 0, 1, 2, \ldots, n$.*

Such a word is often called a *de Bruijn sequence* in the literature.

Let $\mathbf{u}$ be an infinite word over the alphabet Σ. In order to prove Theorem 10.7.1, we introduce a useful family of directed graphs $(G_\mathbf{u}(n))_{n \geq 1}$, called *word graphs*. The vertices of $G_\mathbf{u}(n)$ are the subwords of $\mathbf{u}$ of length n. The edges of $G_\mathbf{u}(n)$ correspond to the subwords of $\mathbf{u}$ of length $n + 1$: if $a, b \in \Sigma$, $z \in \Sigma^{n-1}$, and azb is a subword of length $n + 1$, then we create a directed edge (az, zb).

In the particular case of sequences of maximum complexity over $\{0, 1\}$, we easily see that $B(n) := G_\mathbf{u}(n)$ is a directed graph with 2^n vertices and 2^{n+1} edges. Furthermore, the indegree (number of edges entering a vertex) and outdegree (number of edges leaving a vertex) of each vertex is 2. For example, Figure 10.3 illustrates $B(3)$.

We can now prove Theorem 10.7.1.

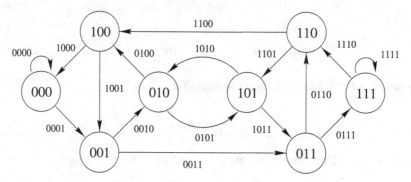

Figure 10.3: The Word Graph $B(3)$.

Proof. It suffices to show that $B(n)$ possesses an *Eulerian cycle*: a path that traverses every edge exactly once and begins and ends at the same vertex. For, given such a cycle, we can form a word $a_1 a_2 \cdots a_m$ of length $m = 2^{n+1} + n$ where $a_i a_{i+1} \cdots a_{i+n-1}$ is the label of the ith vertex visited. We use the following lemma:

Lemma 10.7.2 *A directed graph G possesses an Eulerian cycle if and only if it is strongly connected (i.e., there is a directed path connecting any two vertices) and the indegree of every vertex equals its outdegree.*

Proof. Suppose G has an Eulerian cycle. Then it is clearly strongly connected. Also, if the indegree of a vertex v differed from its outdegree, then our Eulerian cycle would be forced to repeat an edge.

Now suppose G is strongly connected and satisfies the condition on the degrees of the vertices. Let P be a longest closed path with the property that every edge it traverses, it traverses at most once. If there is an edge of G not in P, then by strong connectivity P must omit some edge incident to a vertex of P. Consider such an edge leaving a vertex x on P and entering a vertex v; by the condition on degrees there must be an edge not in P leaving v. Follow this edge. Continuing in this fashion, we encounter other edges not in P, and again, by the degree condition, we must return eventually to x. We have now found a new path P', and we can add P' to P to obtain a longer closed path, a contradiction. ∎

Example 10.7.3 Let us use the graph $B(3)$ in Figure 10.3 to create a word w of length 19 such that $p_w(4) = 16$. We find `0000111101001011000`.

10.8 Recurrence

We now begin our study of the second of the two questions mentioned in the introduction: namely, whether all of the finite subwords of a given infinite word appear infinitely often. A subword that occurs infinitely often is said to be *recurrent*.

Example 10.8.1 Consider the characteristic sequence **c** of the powers of 2, i.e.,

$$\mathbf{c} = 1101000100000001 \cdots.$$

Then it is clear that any subword of **c** that contains at least two 1's appears only once in **c**. Thus, **c** possesses infinitely many subwords that are not recurrent. On the other hand, the subword 0 is clearly recurrent.

Example 10.8.2 Every subword of the Thue–Morse sequence **t** is recurrent. For let w be any subword of **t**. Then w is a subword of some prefix of **t** of the form $\mu^n(0)$, where μ maps $0 \to 01$ and $1 \to 10$. But $\mathbf{t} = \mu^n(\mathbf{t})$, and 0 clearly occurs infinitely often in **t**. Therefore $\mu^n(0)$ occurs infinitely often, and hence so does w.

We say an infinite word **x** is *recurrent* if every finite subword of **x** is recurrent.

Example 10.8.3 Every purely periodic infinite word is recurrent. On the other hand, if a recurrent infinite word is ultimately periodic, then it is purely periodic; see Exercise 37.

Proposition 10.8.4 *An infinite word* **x** *is recurrent if and only if every finite subword of* **x** *occurs at least twice.*

Proof. One direction is clear. For the other direction, assume every finite subword of **x** occurs at least twice. Let w be such a subword. Assume, contrary to what we want to prove, that w occurs only finitely many times in **x**. Let w' be the shortest prefix of **x** containing all occurrences of w in **x**. Then w' is itself a subword of **x**, so it must occur twice in **x**. Let w'' be the shortest prefix of **x** containing two occurrences of w'. Then w'' contains at least one occurrence of w not contained in w', a contradiction. ■

Let $r_\mathbf{w}(n)$ denote the number of distinct recurrent subwords of **w** of length n. Note that every recurrent subword x of length m is a prefix of some recurrent subword of length $m + 1$; for if not, x would not be recurrent, because the total number of occurrences of xa, for a letter a, would be finite. Keeping this observation in mind, it is not hard to prove the following result.

Theorem 10.8.5 *Theorems 10.2.1–10.2.6 hold for $r_\mathbf{w}(n)$ in place of $p_\mathbf{w}(n)$.*

Proof. Left to the reader. (See Exercise 8.) ■

We now examine some transformations which, when applied to recurrent words, produce recurrent results.

Theorem 10.8.6 *Let $\mathbf{x} \in \Sigma^\omega$ be a recurrent infinite word, and let $h : \Sigma^* \to \Sigma^*$ be a morphism that is nontrivial on the alphabet of $\mathbf{x}$, i.e., there exists at least one letter a occurring in $\mathbf{x}$ for which $h(a) \neq \epsilon$. Then $h(\mathbf{x})$ is a recurrent infinite word.*

Proof. Since $\mathbf{x}$ is recurrent, every letter occurring in $\mathbf{x}$ occurs infinitely often. Since h is nontrivial on the alphabet of $\mathbf{x}$, $h(\mathbf{x})$ is infinite. Now let w be a finite subword of $h(\mathbf{x})$. Then there exists a finite subword $y = \mathbf{x}[m..n]$ that covers w, i.e., w is a subword of $h(y)$. Since $\mathbf{x}$ is recurrent, y occurs infinitely often in $\mathbf{x}$, and hence w occurs infinitely often in $h(\mathbf{x})$. ∎

We now consider the case of k-block compression. Recall that the k-block compression of a sequence is formed by grouping the terms into blocks of size k, and recoding this new sequence with a symbol for each distinct block.

Theorem 10.8.7 *Let $\mathbf{w}$ be an infinite recurrent sequence, and let $k \geq 2$ be an integer. Then the k-block compression of $\mathbf{w}$ is recurrent.*

Proof. First we need the following lemma:

Lemma 10.8.8 *Let a, k be integers with $k \geq 2$ and $0 \leq a < k$. Suppose $\mathbf{w} = c_0 c_1 c_2 \cdots$ is a recurrent infinite word, and suppose u is a finite subword of $\mathbf{w}$ that occurs at least once in $\mathbf{w}$ at some position $i \equiv a \pmod{k}$. Then u occurs infinitely often in $\mathbf{w}$ at positions congruent to $a \pmod{k}$.*

Proof. Suppose, contrary to what we want to prove, that u occurs only finitely many times in $\mathbf{w}$ at positions congruent to $a \pmod{k}$. Suppose the last such occurrence of u in $\mathbf{w}$ is $c_j c_{j+1} \cdots c_{j+r-1}$. Define

$$S = \{i : 0 \leq i < k \text{ and } u \text{ occurs infinitely often}$$
$$\text{in } \mathbf{w} \text{ in positions congruent to } i \pmod{k}\}.$$

Since $\mathbf{w}$ is recurrent, we know that S is nonempty, and by above $a \notin S$. Let v be a subword of $\mathbf{w}$ beginning at position j and extending to the right sufficiently far to include at least one occurrence of u starting at positions congruent to each of the elements of S, modulo k.

Since $\mathbf{w}$ is recurrent, v itself must occur infinitely often in $\mathbf{w}$, and hence must occur infinitely often starting at a position congruent to some fixed $b \pmod{k}$. It follows that $b \in S$, and further that S is stable under the operation of adding $b - a$ to each element of S. Hence $b + (k-1)(b-a) \in S$. But $b + (k-1)(b-a) \equiv a \pmod{k}$, so $a \in S$, a contradiction. ∎

Now we can prove Theorem 10.8.7. Let $\mathbf{x}$ be the k-block compression of $\mathbf{w}$, and let t be a finite subword of $\mathbf{x}$. Then t corresponds to a subword of $\mathbf{w}$ that occurs starting at a position congruent to 0 (mod k). By Lemma 10.8.8 this subword must occur infinitely often at positions congruent to 0 (mod k), and hence t must occur infinitely often in **x**. ∎

We now turn to a topological interpretation of recurrence. Recall that the orbit of a one-sided infinite word $\mathbf{x} = a_0 a_1 a_2 \cdots$ is the set of all shifted words $\{S^k(\mathbf{x}) : k \geq 0\}$, i.e., the set of all suffixes of **x**. We denote this set by Orb(**x**). The *orbit closure*, Cl(Orb(**x**)), is the closure of Orb(**x**) under the usual topology. The following simple proposition gives two alternative characterizations of Cl(Orb(**x**)):

Proposition 10.8.9 *Let* **x**, **y** *be infinite words over a finite alphabet* Σ. *Then the following properties are equivalent:*

(a) $\mathbf{y} \in$ Cl(Orb(**x**));
(b) *every finite prefix of* **y** *is a subword of* **x**;
(c) Sub(**y**) $\subseteq$ Sub(**x**).

Proof. (a) $\Longleftrightarrow$ (b) follows easily from the definition of the topology on infinite words.
   (b) $\Longrightarrow$ (c): Every subword of **y** is contained in some prefix of **y**.
   (c) $\Longrightarrow$ (b): Every prefix of **y** is a subword of **y**. ∎

Example 10.8.10 Continuing Example 10.8.1, let $\mathbf{c} = 011010001 \cdots$ be the characteristic sequence of the powers of 2. As we have seen, **c** is not recurrent. The orbit closure of **c** consists of all suffixes of **c**, together with the set $\{0^i 10^\omega : i \geq 0\}$.

The orbit closure of an infinite word may at first sight be a somewhat mysterious object. The following interpretation may make it somewhat easier to understand. Suppose we are given an infinite word **x**. We now construct an infinite tree $T = T(\mathbf{x})$ as follows: the root of T is labeled ϵ. If a node of T is labeled w, then it has children labeled wa for each subword wa occurring in **x**. Furthermore, we arrange the children from left to right so that if the children of the node labeled w are labeled $wa_1, wa_2, \ldots, wa_k$, then the first occurrence of wa_1 in **x** precedes the first occurrence of wa_2, which precedes the first occurrence of wa_3, etc. We label the edge from w to wa_i with the symbol a_i. We say an infinite word **y** belongs to T if there is an infinite path, descending from the root, with edge labels given by the symbols of **y**. Then Cl(Orb(**x**)), the orbit closure of **x**, is the set of all infinite words belonging to $T(\mathbf{x})$.

Example 10.8.11 Consider the Thue–Morse sequence **t**. As we have seen above in Example 10.8.2, **t** is recurrent. Figure 10.4 gives the first few levels in the tree $T(\mathbf{t})$.
   In particular, $\bar{\mathbf{t}}$ is a sequence that is in the orbit closure of **t**, but not in the orbit of **t**.

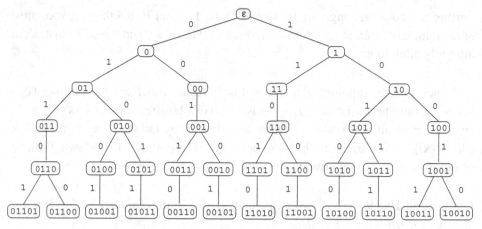

Figure 10.4: First Few Levels of the Tree $T(\mathbf{t})$.

One consequence of this interpretation is the following:

Theorem 10.8.12 *Suppose* $\mathbf{x}$ *is a recurrent infinite word. Then* $\mathrm{Cl}(\mathrm{Orb}(\mathbf{x}))$ *is uncountable if and only if* $\mathbf{x}$ *is not purely periodic.*

Proof. One direction is clear. Now suppose $\mathbf{x}$ is not purely periodic. Since $\mathbf{x}$ is recurrent, by Exercise 15 every prefix p of $\mathbf{x}$ can be extended to a (possibly longer) prefix p' such that $p'a$ and $p'b$ are both subwords of $\mathbf{x}$, for some $a \neq b$. This means every path in $T(\mathbf{x})$ encounters infinitely many levels where at least two distinct choices can be made, each of which results in a distinct member of the orbit closure. It follows that $\mathrm{Cl}(\mathrm{Orb}(\mathbf{x}))$ is uncountable. ∎

10.9 Uniform Recurrence

As we have seen in the previous section, recurrent words have several pleasant properties that make them worthy of study. In this section, we focus on infinite words which, in addition to being recurrent, have the property that successive occurrences of words are not widely separated.

We say an infinite word $\mathbf{x}$ is *uniformly recurrent* if, corresponding to every finite subword w of $\mathbf{x}$, there exists an integer k such that every subword of length k of $\mathbf{x}$ contains w. Clearly a uniformly recurrent word is recurrent. We define $R_{\mathbf{x}}(w)$, the recurrence function, to be the smallest such k, or ∞, if no such k exists. If $\mathbf{x}$ is clear from the context, we may omit the subscript on R. We also define $R_{\mathbf{x}}(n) = \max_{w \in \mathrm{Sub}_{\mathbf{x}}(n)} R_{\mathbf{x}}(w)$. Thus $R_{\mathbf{x}}(n)$ is the least integer m such that every block of m consecutive symbols in $\mathbf{x}$ contains at least one occurrence of every length-n subword that appears anywhere in $\mathbf{x}$.

Example 10.9.1 We saw in Example 10.8.3 that every purely periodic infinite word is recurrent. Suppose $\mathbf{x}$ is purely periodic of period length k. Let w be a subword of

x of length n. By the periodicity of **x**, for any $a \geq 0$ there exists an occurrence of w beginning at one of the positions $a, a + 1, \ldots a + k - 1$ of **x**. Thus w is a subword of $\mathbf{x}[a..a + n + k - 2]$, and so

$$R_{\mathbf{x}}(n) \leq n + k - 1. \tag{10.13}$$

Hence purely periodic words are uniformly recurrent, and recurrence can be viewed as a generalization of periodicity.

Notice that the inequality (10.13) cannot, in general, be replaced by an equality. For example, if $\mathbf{x} = (\text{abba})^{\omega}$, then $k = 4$, but $R_{\mathbf{x}}(1) = 3$. However, see Exercise 53.

The following theorem gives some of the basic properties of the recurrence function.

Theorem 10.9.2 *Let* **x** *be a uniformly recurrent infinite word. Then we have*

(a) $R_{\mathbf{x}}(n + 1) > R_{\mathbf{x}}(n)$ *for all* $n \geq 0$;
(b) $R_{\mathbf{x}}(n) \geq p_{\mathbf{x}}(n) + n - 1$ *for all* $n \geq 0$;
(c) $R_{\mathbf{x}}(n) \geq 2n$ *for all* n *if and only if* **x** *is not purely periodic.*

Proof. (a): Let $t = R_{\mathbf{x}}(n)$. Then from the definition of $R_{\mathbf{x}}(n)$ there exists an integer $j \geq 0$ such that $\mathbf{x}[j..j + t - 1]$ contains all of the length-n subwords of **x**, but $\mathbf{x}[j..j + t - 2]$ fails to contain at least one length-n subword. In fact, it fails to contain $w := \mathbf{x}[j + t - n..j + t - 1]$. Let $a = \mathbf{x}[j + t]$. Then wa does not appear in $\mathbf{x}[j..j + t - 1]$, for if it did, w would be a subword of $\mathbf{x}[j..j + t - 2]$, a contradiction. It follows that $R_{\mathbf{x}}(n) \geq t + 1$.

(b): A subword of any length $\ell \geq n$ contains at most $\ell - n + 1$ distinct subwords of length n. But **x** contains $p_{\mathbf{x}}(n)$ subwords of length n. Therefore $p_{\mathbf{x}}(n) \leq R_{\mathbf{x}}(n) - n + 1$, which gives the desired inequality.

(c): Suppose **x** is not periodic. Since it is recurrent, it cannot be ultimately periodic, by Exercise 37. Thus by Theorem 10.2.6 we have $p_{\mathbf{x}}(n) \geq n + 1$ for all n. Hence by part (b) we have $R_{\mathbf{x}}(n) \geq 2n$ for all n. On the other hand, if **x** is periodic with period length p, then as we remarked in Example 10.9.1, $R_{\mathbf{x}}(n) \leq n + p - 1$. Hence for $n = p$ we have $R_{\mathbf{x}}(n) \leq 2n - 1 < 2n$. ∎

Example 10.9.3 As usual, let $\mathbf{t} = t_0 t_1 t_2 \cdots$ denote the Thue–Morse sequence. We show

$$R_{\mathbf{t}}(n) = \begin{cases} 3 & \text{if } n = 1, \\ 9 & \text{if } n = 2, \\ 9 \cdot 2^k + n - 1 & \text{if } k \geq 0 \text{ and } 2^k + 2 \leq n \leq 2^{k+1} + 1. \end{cases}$$

First, we observe that the subwords of **t** of length ≥ 4 can be classified into two disjoint sets: those that occur only at an even-indexed position in **t**, and those that occur only at an odd-indexed position in **t**. (We call the former "even" and the latter "odd". To see this, note that the even subwords of length 4 are $\{0110, 1010,$ $1001, 0101\}$ and the odd subwords of length 4 are $\{1101, 0100, 0011,$ $1100, 0010, 1011\}$. Then no even subword can be odd, or vice versa, since if one were, we would either have 11 or 00 as the image of a letter under μ (the Thue–Morse morphism), or have 000 or 111 as a subword of $\mathbf{t}$, contradicting Theorem 1.6.1. For words of length > 4, simply consider a prefix of length 4.

We start by showing that $R(2n) \leq 2R(n+1) - 1$ for $n \geq 2$. It suffices to show that every subword x of length $2n$ of $\mathbf{t}$ appears in every subword B of $\mathbf{t}$ of length $2R(n+1) - 1$. There are two cases to consider: x is even and x is odd.

Case 1: If x is even, then we can write $x = \mu(y)$, where y is a subword of $\mathbf{t}$ of length n. Then $|B| = 2R(n+1) - 1 \geq 2R(n) + 1$ by Theorem 10.9.2(a). Thus B contains an even subword z of length $2R(n)$. Then we can write $z = \mu(w)$, where w is a subword of $\mathbf{t}$ of length $R(n)$. By the definition of R, w must contain every subword of $\mathbf{t}$ of length n, and so it contains y. Then $x = \mu(y)$ is a subword of $z = \mu(w)$, and hence a subword of B.

Case 2: If x is odd, then there exist $a, b \in \{0, 1\}$ and a subword y of $\mathbf{t}$ of length $n + 1$ such that $axb = \mu(y)$.

If B itself is odd, then let c be a symbol that immediately precedes B in $\mathbf{t}$. Then cB is an even subword of $\mathbf{t}$ of length $2R(n+1)$. It follows that we can write $cB = \mu(w)$ where w is a subword of $\mathbf{t}$ of length $R(n+1)$. By the definition of R, w must contain every subword of $\mathbf{t}$ of length $n + 1$, and so it contains y. Then $cB = \mu(w)$ contains $axb = \mu(y)$, and since B and x are both odd, we have that B contains x.

If B itself is even, then we let d be a symbol that immediately follows B in $\mathbf{t}$, and repeat the previous argument with Bd in place of cB.

We have now shown that every subword of length $2n$ of $\mathbf{t}$ appears in every subword of length $2R(n+1) - 1$ of $\mathbf{t}$, and so the inequality $R(2n) \leq 2R(n+1) - 1$ follows.

Next, we show $R(2n+1) \leq 2R(n+1)$ for $n \geq 2$. As before, it suffices to show that every subword x of length $2n+1$ of $\mathbf{t}$ appears in every subword B of $\mathbf{t}$ of length $2R(n+1)$. There are two cases to consider:

Case 1: B is even. If x is odd, then $x = \overline{c}\mu(y)$ for some subword cy of $\mathbf{t}$ of length $n + 1$. Now $|B| = 2R(n+1)$, so $B = \mu(w)$ for a subword w of $\mathbf{t}$ of length $R(n+1)$. By the definition of R, w must contain every subword of $\mathbf{t}$ of length $n + 1$, and so it contains cy. Thus $B = \mu(w)$ contains $\mu(cy) = c\overline{c}\mu(y) = cx$.

If x is even, then we write $x = \mu(y)c$ and repeat the previous argument.

Case 2: B is odd. If x is odd, then $x = \overline{c}\mu(y)$ for some subword cy of $\mathbf{t}$ of length $n + 1$. As before $B = \overline{a}\mu(w)b$ for some subword awb of $\mathbf{t}$ of length $R(n+1) + 1$. Then aw contains every subword of $\mathbf{t}$ of length $n + 1$; hence it contains cy. Then $\mu(cy)$ is a subword of $\mu(aw)$; since both are even, it follows that $x = \overline{c}\mu(y)$ is a subword of $\overline{a}\mu(w)$, hence a subword of B.

If x is even, then we write $x = \mu(y)c$ and repeat the previous argument.

We have now shown that every subword of length $2n + 1$ of **t** appears in every subword of length $2R(n + 1)$ of **t**, and so the inequality $R(2n + 1) \leq 2R(n + 1)$ follows.

Next, we argue that $R(2n) \geq 2R(n + 1) - 1$ for $n \geq 2$. Call a subword u of **t** an *n-minimax block* if u contains all subwords of length n of **t**, but no proper subword of u does. It follows that $|u| = R(n)$. Now let g be an $(n + 1)$-minimax block of length $R(n + 1)$, and write $g = g'axb$, where $|x| = n - 1$ and $a, b \in \{0, 1\}$. Then axb is not a subword of $g'ax$, for otherwise g would not be minimax. We claim the subword $\mu(g'ax)$ omits the subword $v := \overline{a}\mu(x)b$. Now v is clearly odd, and since $|v| \geq 4$, it cannot also be even. If v appears in $\mu(g'ax)$, then since $\mu(g'ax)$ is even, it must be that $\mu(g'ax)$ contains $av\overline{b} = a\overline{a}\mu(x)b\overline{b} = \mu(axb)$. Hence $g'ax$ contains axb, a contradiction. It follows that $R(2n) > 2(R(n + 1) - 1)$, and so $R(2n) \geq 2R(n + 1) - 1$. Now $R(2n + 1) > R(2n)$ by Theorem 10.9.2(a), and so $R(2n + 1) \geq R(2n) + 1 \geq 2R(n + 1)$.

We have now shown

$$R(2n) = 2R(n + 1) - 1,$$
$$R(2n + 1) = 2R(n + 1)$$

for $n \geq 2$. In view of the equalities $R(1) = 3$, $R(2) = 9$, $R(3) = 11$, the proofs of which are left to the reader, the desired result now follows by an easy induction.

Uniformly recurrent infinite words have many interesting properties, and can be defined in several equivalent ways. Suppose T is a set of infinite words. We say T is *closed under the shift* if, for all $\mathbf{x} \in T$ and all $k \geq 0$, we have $\mathcal{S}^k(\mathbf{x}) \in T$.

An infinite word $\mathbf{x}$ is said to be *minimal* if the only closed subsets of $\mathrm{Cl}(\mathrm{Orb}(\mathbf{x}))$ that are closed under the shift are $\emptyset$ and $\mathrm{Cl}(\mathrm{Orb}(\mathbf{x}))$ itself. We show in a moment that an infinite word is minimal if and only if it is uniformly recurrent.

Suppose $\mathbf{x} = a_0a_1a_2\cdots$ is recurrent, and w is an arbitrary subword of $\mathbf{x}$ beginning at positions $i_1 < i_2 < \cdots$. We define $\gamma_{\mathbf{x}}(w) = \sup_{n \geq 0}(i_{n+1} - i_n)$, where $i_0 := -1$. Thus, $\gamma_{\mathbf{x}}(w)$ measures the maximum distance between successive occurrences of w in $\mathbf{x}$. We define $\gamma_{\mathbf{x}}(n) = \max_{w \in \mathrm{Sub}_{\mathbf{x}}(n)} \gamma_{\mathbf{x}}(w)$. If $\gamma_{\mathbf{x}}(n) < \infty$ for all n, then we say $\mathbf{x}$ has the *bounded gap property*.

We now show that a sequence is uniformly recurrent if and only if it is minimal, if and only if it has the bounded gap property.

Theorem 10.9.4 *Let* $\mathbf{x} = a_0a_1a_2\cdots$ *be an infinite sequence over a finite alphabet. Then the following properties are equivalent:*

(a) *The sequence* $\mathbf{x}$ *is minimal.*
(b) *For every sequence* $\mathbf{y}$, *if* $\mathbf{y} \in \mathrm{Cl}(\mathrm{Orb}(\mathbf{x}))$, *then* $\mathbf{x} \in \mathrm{Cl}(\mathrm{Orb}(\mathbf{y}))$.

(c) *For every sequence* **y**, *if* **y** *has arbitrarily long prefixes in* Sub(**x**), *then* **x** *has arbitrarily long prefixes in* Sub(**y**).

(d) *For every sequence* **y**, *if* Sub(**y**) $\subseteq$ Sub(**x**), *then* Sub(**x**) $\subseteq$ Sub(**y**).

(e) *For every sequence* **y**, *if* Sub(**y**) $\subseteq$ Sub(**x**), *then* Sub(**x**) $=$ Sub(**y**).

(f) *For every sequence* **y**, *we have* **y** $\in$ Cl(Orb(**x**)) $\Leftrightarrow$ Sub(**y**) $=$ Sub(**x**).

(g) *The sequence* **x** *has the bounded gap property.*

(h) *The sequence* **x** *is uniformly recurrent.*

Proof. To prove that (a) $\Leftrightarrow$ (b), we note that the sequence **x** is minimal if and only if for any sequence **y** $\in$ Cl(Orb(**x**)) we have Cl(Orb(**y**)) $=$ Cl(Orb(**x**)). This is equivalent to saying **x** $\in$ Cl(Orb(**y**)), since Cl(Orb(**y**)) is closed, and closed under the shift.

The equivalences between (b), (c), and (d) are easy consequences of Proposition 10.8.9. The equivalence between (d) and (e) is trivial.

To prove that (e) $\Leftrightarrow$ (f), we first note that (e) can be rephrased under the equivalent form: Sub(**y**) $\subseteq$ Sub(**x**) if and only if Sub(**y**) $=$ Sub(**x**). It then suffices to apply Proposition 10.8.9.

We now prove (g) $\Leftrightarrow$ (h). In fact, we prove that $\gamma_{\mathbf{x}}(n) = R_{\mathbf{x}}(n) - n + 1$ for all $n \geq 0$.

Suppose **x** has the bounded gap property, and let $g = \gamma_{\mathbf{x}}(n)$. Then $g < \infty$. Now consider any subword y of **x** of length $g + n - 1$, say $y = \mathbf{x}[i..i + g + n - 2]$. Let w be a subword of length n of **x**. We claim w occurs in y. For if not, there can be no occurrence of w in **x** beginning at any position $p \in \{i, i + 1, i + 2, \ldots, i + g - 1\}$. It follows that $\gamma_{\mathbf{x}}(n) \geq g + 1$, a contradiction. Thus $R_{\mathbf{x}}(n) \leq g + n - 1$.

Now suppose **x** is uniformly recurrent, and let $r = R_{\mathbf{x}}(n)$. Then $r < \infty$. Let w be a subword of length n of **x**, and let $\mathbf{x}[i..i + n - 1]$, $\mathbf{x}[i'..i' + n - 1]$ be any two consecutive occurrences of w with $i < i'$; we permit $i = -1$, in which case $\mathbf{x}[i'..i' + n - 1]$ is the first occurrence of w in **x**. Then $\mathbf{x}[i + 1..i' + n - 2]$ contains no occurrences of w, so it follows that $R_{\mathbf{x}}(n) \geq i' - i + n - 1$. Since $R_{\mathbf{x}}(n) < \infty$, it follows that $i' - i < \infty$; since i, i' were the starting positions of arbitrary consecutive occurrences of w, we have $\gamma_{\mathbf{x}}(n) \leq R_{\mathbf{x}}(n) - n + 1$.

To prove that (h) implies (d), let **y** be a sequence such that Sub(**y**) $\subseteq$ Sub(**x**), and let $x \in$ Sub(**x**). Take any subword y of **y**, such that $|y| = R_{\mathbf{x}}(|x|)$. Since Sub(**y**) $\subseteq$ Sub(**x**), the word y is also a subword of **x**. Since it has length $R_{\mathbf{x}}(|x|)$, it contains all subwords of **x** of length $|x|$, in particular x itself. Hence x is a subword of y, and so x is a subword of **y**.

It remains to prove, for example, that (d) implies (g). We prove that if the sequence **x** does not have the bounded gap property, then it does not satisfy condition (d). If there is a subword x of **x** that occurs only a finite number of times in **x**, then there exists a sequence **y** obtained by applying a power of the shift to **x**, and that does not contain the word x. Hence condition (d) is not satisfied. Suppose now that a word x occurs infinitely often in **x**, but with unbounded gaps. Then there exists an increasing sequence $(n_j)_j$ such that, denoting the shift by S, any sequence $S^{n_j}(\mathbf{x})$

begins with a word $w_j x$, where w_j does not contain x as a subword, and the lengths $|w_j|$ are not bounded. Up to replacing the sequence $(n_j)_j$ by a subsequence, we can suppose that the sequence of lengths $(|w_j|)_j$ tends to infinity. Now the sequence $(S^{n_j}(\mathbf{x}))_j$ admits an accumulation point, whose subwords are all in $\mathrm{Sub}(\mathbf{x})$, but that misses the word x, contradicting (d). ∎

Theorem 10.9.5 *Let* $h : \Sigma^* \to \Sigma^*$ *be a primitive morphism, prolongable on* a. *Then* $h^\omega(a)$ *is uniformly recurrent.*

Proof. Let $k = \mathrm{Card}\, \Sigma$. Since h is primitive, there exists an integer e such for all $c, d \in \Sigma$, we have $h^e(c) \in \Sigma^* d \Sigma^*$. (In fact, by Exercise 8.8(b), we may take $e = 2(k-1)$.)

Let $\mathbf{u} = h^\omega(a)$, and let w be a subword of $\mathbf{u}$. Then w is a subword of $h^n(a)$ for some $n \geq 1$. Now $\mathbf{u}$ is a fixed point of h, so if $\mathbf{u} = a_0 a_1 a_2 \cdots$ with each $a_i \in \Sigma$, then we also have $\mathbf{u} = h^{n+e}(a_0) h^{n+e}(a_1) h^{n+e}(a_2) \cdots$. However, $h^{n+e}(a_i) = h^n(h^e(a_i)) = h^n(x_i a y_i)$ for all $i \geq 0$. It follows that every subword of length $2W^{n+e}$ of $\mathbf{u}$ contains an occurrence of $h^n(a)$, and hence an occurrence of w. ∎

10.10 Appearance

There are two other related functions of interest dealing with infinite words. The first is called *appearance*, and denotes the starting position of the last subword of length n to appear for the first time in $\mathbf{w} = a_0 a_1 a_2 \cdots$. More precisely, we define $\alpha_{\mathbf{w}}(n)$ to be the least index i such that every length-n subword of $\mathbf{w}$ appears in the prefix $a_0 a_1 \cdots a_i \cdots a_{i+n-1}$.

The second function is similar to recurrence, except instead of handling all blocks, we consider only prefixes. If $\mathbf{x}$ is an infinite word, and w is a subword of $\mathbf{x}$, we define $R'_{\mathbf{x}}(w)$ to be the smallest integer k such that w is a subword of $\mathbf{x}[0..k-1]$. Notice that R' is defined on all infinite words $\mathbf{x}$, not just uniformly recurrent ones. Further, we define

$$R'_{\mathbf{x}}(n) = \max_{w \in \mathrm{Sub}_{\mathbf{x}}(n)} R'_{\mathbf{x}}(w).$$

Hence $R'_{\mathbf{x}}(n)$ is the length of the shortest prefix of $\mathbf{x}$ that contains at least one occurrence of each length-n subword.

Note that R' and α are related by $R'_{\mathbf{x}}(n) = \alpha_{\mathbf{x}}(n) + n$.

Example 10.10.1 Let $\mathbf{x}$ be an ultimately periodic word, with preperiod of length k and period of length p. Then $\alpha_{\mathbf{x}}(n) = k + p - 1$ and $R'_{\mathbf{x}}(n) = k + p + n - 1$.

Theorem 10.10.2 *Let* $\mathbf{x}$ *be an infinite word. Then*

(a) $R'_{\mathbf{x}}(n+1) > R'_{\mathbf{x}}(n)$ *for all* $n \geq 0$;
(b) $R'_{\mathbf{x}}(n) \geq p_{\mathbf{x}}(n) + n - 1$ *for all* $n \geq 0$;
(c) $R'_{\mathbf{x}}(n) \geq 2n$ *for all* n *if and only if* $\mathbf{x}$ *is not ultimately periodic.*

Proof. (a): Let $t = R'_{\mathbf{x}}(n)$. Then $\mathbf{x}[0..t-1]$ contains all subwords appearing in $\mathbf{x}$ of length n. Furthermore, if we let $w = \mathbf{x}[t-n..t-1]$, then the first occurrence of w in $\mathbf{x}$ begins at position $t-n$ of $\mathbf{x}$; if not, we would have $R'_{\mathbf{x}}(n) < t$. Hence, if we let $a = \mathbf{x}[t]$, then wa is a word of length $n+1$ that occurs for the first time at position $t-n$ of $\mathbf{x}$, and so $R'_{\mathbf{x}}(n+1) > t$.

(b): If there are $p_{\mathbf{x}}(n)$ distinct subwords of length n in $\mathbf{x}$, then in the "worst case" they could appear consecutively (overlapping) in $\mathbf{x}$, in which case we would need $p_{\mathbf{x}} + n - 1$ positions to cover them all.

(c): Suppose $\mathbf{x}$ is not ultimately periodic. Then by Theorem 10.2.6 we have $p_{\mathbf{x}}(n) \geq n+1$ for all n. Hence by part (b) we have $R'_{\mathbf{x}} \geq n+1+n-1 = 2n$. On the other hand, if $\mathbf{x}$ is ultimately periodic, then by Example 10.10.1 we have $R'_{\mathbf{x}}(n) = C + n - 1$ for some integer constant C, so taking $n = C$ we get $R'_{\mathbf{x}}(n) = 2n - 1 < 2n$. ∎

Example 10.10.3 We compute α and R' for $\mathbf{t} = t_0 t_1 t_2 \cdots$, the Thue–Morse sequence.

We use the terminology of "even" and "odd" subwords introduced in Example 10.9.3.

Now consider a subword of length $2n \geq 4$ of $\mathbf{t}$. If it is even, it must be the image under μ of a unique subword of length n of $\mathbf{t}$. If it is odd, it must be formed by dropping the first and last characters of the image under μ of a unique subword of length $n+1$. It now easily follows that $\alpha_{\mathbf{t}}(2n) = 2\alpha_{\mathbf{t}}(n+1) + 1$ for $n \geq 2$. A similar argument shows $\alpha_{\mathbf{t}}(2n+1) = 2\alpha_{\mathbf{t}}(n+1) + 1$ for ≥ 2.

Now by induction we get, for $n \geq 3$, that $\alpha_{\mathbf{t}}(n) = 3 \cdot 2^{k+1} - 1$ if $2^k + 2 \leq n \leq 2^{k+1} + 1$. Thus, for $n \geq 3$, we have $R'_{\mathbf{t}}(n) = 3 \cdot 2^{k+1} + n - 1$ if $2^k + 2 \leq n \leq 2^{k+1} + 1$.

Note that for $k \geq 1$, the above result implies that $R'_{\mathbf{t}}(2^k + 1) = 2^{k+2}$. Furthermore by Exercise 10 we know that $p_{\mathbf{t}}(2^k + 1) = 3 \cdot 2^k$. It follows that the first 2^{k+2} symbols of $\mathbf{t}$ contain all the $3 \cdot 2^k$ subwords of length $2^k + 1$, and each subword appears exactly once. For example, the first 16 symbols of $\mathbf{t}$ are `0110100110010110`, and the 12 length-5 subwords of $\mathbf{t}$ are `01101`, `11010`, `10100`, `01001`, `10011`, `00110`, `01100`, `11001`, `10010`, `00101`, `01011`, and `10110`.

10.11 Exercises

1. What is the subword complexity of the word `1101000100000001`$\cdots$, the characteristic sequence of the powers of 2?

2. Show that $\mathbf{p} = $ `235711131719232931`$\cdots$, the concatenation of the decimal expansions of the prime numbers, has subword complexity $p_{\mathbf{p}}(n) = 10^n$. Hint: Use Theorem 2.13.3.

3. In Theorem 10.3.1, we proved that if $\mathbf{u} = h^\omega(a)$ for some t-uniform morphism h of depth k, then $p_{\mathbf{u}}(n) \leq tk^2 n$ for all $n \geq 1$. Is this bound attainable?

4. Suppose that $|\Sigma| \geq 2$. Give an explicit example of an infinite word $\mathbf{w} \in \Sigma^\omega$ such that $p_\mathbf{w}(n) = |\Sigma|^n$ for all $n \geq 0$.

5. Suppose that $w_n := \lfloor (n+1)\alpha + \beta \rfloor - \lfloor n\alpha + \beta \rfloor$. Show that $w_n = 0 \Leftrightarrow \{n\alpha + \beta\} \in [0, 1 - \alpha)$.

6. Show how, for all real numbers $0 \leq \alpha \leq 1$, to construct a sequence over $\{0, 1\}$ with entropy α.

7. Show how to construct an infinite word $\mathbf{w}$ over $\{0, 1\}$ with complexity $p_\mathbf{w}(n) = F_{n+2}$, where F_n is the nth Fibonacci number.

8. Prove Theorem 10.8.5.

9. A three-dimensional analogue to Sturmian words is the following. Let $\alpha_1, \alpha_2, \alpha_3$ be totally irrational (rationally independent). Consider the sequence on three letters $\{0, 1, 2\}$ defined by the path of the ray with parameterization $x = \alpha_1 t$, $y = \alpha_2 t, z = \alpha_3 t$ for $t \geq 0$ through the integer lattice in $\mathbb{R}^3$. Show that these words have subword complexity $n^2 + n + 1$.

10. Let $\mathbf{t}$ be the Thue–Morse infinite word. Show that for $n \geq 2$

$$p_\mathbf{t}(n+1) = \begin{cases} 4n - 2^a & \text{if } n = 2^a + b, \text{ where } a \geq 1, 0 \leq b < 2^{a-1}, \\ 4n - 2^a - 2b & \text{if } n = 2^a + 2^{a-1} - b, \text{ where } a \geq 1, 0 \leq b < 2^{a-1}. \end{cases}$$

11. Show that if $\mathbf{t}$ is the Thue–Morse sequence, then $(p_\mathbf{t}(n) - p_\mathbf{t}(n-1))_{n \geq 1}$ is a 2-automatic sequence, and give a DFAO computing it.

12. Let $k \geq 2$ be an integer. The generalized Thue–Morse sequence $\mathbf{t}_k = t_k(0)t_k(1)\cdots$ is defined as follows: for $n \geq 0$, set $t_k(n) = s_2(n) \bmod k$. Show that the generalized Thue–Morse word has subword complexity

$$p_{\mathbf{t}_k}(n+1) = \begin{cases} k & \text{if } n = 0, \\ k^2 & \text{if } n = 1, \\ k(kn - 2^{a-1}) & \text{if } n = 2^a + b, \text{ where } a \geq 1, 0 \leq b < 2^{a-1}, \\ k(kn - 2^{a-1} - b) & \text{if } n = 2^a + 2^{a-1} + b, \text{ where } a \geq 1, 0 \leq b < 2^{a-1}. \end{cases}$$

13. Show that there is no infinite binary word w for which $p_\mathbf{w}(0) = 1$, $p_\mathbf{w}(1) = 2$, $p_\mathbf{w}(2) = 3$, $p_\mathbf{w}(3) = 5$, and $p_\mathbf{w}(4) = 9$. Conclude that the conditions in Theorems 10.2.2 and 10.2.3 are necessary, but not sufficient, for a function to be the subword complexity function of an infinite word.

14. Suppose a language L is infinite and factorial, i.e., every subword of a word in L is also in L. Show there is a two-sided infinite word whose subwords are all contained in L.

15. Suppose $\mathbf{x}$ is a recurrent infinite word such that if $\mathbf{x}[0..n-1] = \mathbf{x}[k..k + n-1]$, then $\mathbf{x}[n] = \mathbf{x}[k+n]$. Show that $\mathbf{x}$ is periodic.

16. Let p_k/q_k be the convergents of the golden ratio $\frac{1}{2}(\sqrt{5}-1)$, and let **w** be the infinite word defined by

$$\mathbf{w} = \prod_{k=1}^{\infty} 0^{p_k} 1^{q_k}.$$

Show that $p_\mathbf{w}(n) = O(n)$, and determine when $2 \mid p_\mathbf{w}(n)$.

17. Exercise 16 may be generalized as follows. Let $0 < \alpha < 1$ be irrational. Let p_n/q_n be the partial quotients of α. Let

$$\mathbf{w}_\alpha = \prod_{k=1}^{\infty} 0^{p_k} 1^{q_k}$$

Show that $p_{\mathbf{w}_\alpha}(n) = O(n)$.

18. Consider the generalized Fibonacci word **w** generated by the morphism φ defined by $0 \to 0^m 1^n$ and $1 \to 0$, where m and n are fixed positive integers. Let $x \in \Sigma^*$ (here $\Sigma = \{0, 1\}$), and let $P(x)$ denote the frequency of occurrence of the word x in the word generated by φ. In other words,

$$P(x) = \lim_{n \to \infty} \frac{\text{\# of times } x \text{ occurs in the first } n \text{ letters of } \mathbf{w}}{n - |x| + 1}.$$

Define $P(a \mid y)$ (where $y \in \Sigma^*, a \in \Sigma$) to be the frequency of the subword y of **w** occurring somewhere in **w** immediately followed by the letter a. Determine for **w**:

   (a) $P(0), P(1), P(00), P(11), P(01), P(10)$;
   (b) $P(1 \mid 1), P(1 \mid 0), P(0 \mid 1), P(0 \mid 0)$;
   (c) $P(1 \mid 11), P(0 \mid 11)$.

19. Do the previous exercise, but for the Thue–Morse sequence **t**.

20. Consider the Barbier infinite word

$$\mathbf{B} = 1234567891011121314151617181920212 2 \cdots$$

defined in Exercise 3.26.

   (a) Show **B** is not morphic.
   (b) Show that the orbit closure of **B** includes every sequence over Σ_{10}.

21. Let $A = \{1, 2\}$, and let h_1 and h_2 be two morphisms on A^* defined by

$$h_1 : 1 \to 1$$
$$2 \to 11,$$

$$h_2 : 1 \to 2$$
$$2 \to 22.$$

Define the mapping h from A^* into itself as follows: For $w = a_1 a_2 \cdots a_n$, where

$a_i \in A$,

$$h(w) = h_{i_1}(a_1)h_{i_2}(a_2) \cdots h_{i_n}(a_n),$$

where $i_k \equiv k \pmod 2$. Now starting at the word $w = 2$ and iterating the mapping h, we obtain the sequence

$$w, h(w), h^2(w), \ldots.$$

Now the word $h^2(w) = 12$ is a proper prefix of the word $h^3(w)$, and therefore

$$\mathbf{Kol} = \lim_{n \to \infty} h^n(w)$$

exists; it is referred to as the *Kolakoski word*.

(a) Show that in **Kol** all square subwords have lengths in $\{2, 4, 6, 18, 54\}$, and each of these possibilities can occur.

(b) Show that the Kolakoski word is cubefree.

(c) Show that the Kolakoski word is not pure morphic.

(d) Does there exist a morphism φ such that $\mathbf{Kol} = \varphi(\mathbf{t})$, where $\mathbf{t}$ is the Thue–Morse sequence?

22. Prove Furstenberg's theorem: Let $\mathbf{t}$ be an infinite word over a finite alphabet Σ. Then there exists a uniformly recurrent infinite word $\mathbf{r}$ such that $\mathrm{Sub}(\mathbf{r}) \subseteq \mathrm{Sub}(\mathbf{t})$.

23. Suppose $\mathbf{x}$ is a uniformly recurrent infinite word such that there is a constant C such that successive occurrences of *any* subword are separated by distance $\leq C$. Show that $\mathbf{x}$ is purely periodic.

24. Give an example of a uniformly recurrent word that is not k-automatic for any $k \geq 2$.

25. Let $0 < \theta < 1$ be an irrational real number. Show that the characteristic word $\mathbf{f}_\theta$ is uniformly recurrent.

26. Let $\mathbf{u}$ be a binary sequence, and suppose there exists k such that $p_\mathbf{u}(k) < 2^k$. Show that there exists a real number $\alpha < 2$ such that $p_\mathbf{u}(n) = O(\alpha^n)$.

27. Show that for any uniformly recurrent word $\mathbf{x}$ over an alphabet of size $k \geq 2$, there exists a constant c such that $p_\mathbf{x}(n) < k^n$ for all $n \geq c$.

28. Recall the definition of C from Eq. (10.11). Define

$$C(\alpha) = \begin{cases} a_1/2 & \text{if } a_1 > 1, \\ (a_2 + 1)/2 & \text{if } a_1 = 1. \end{cases}$$

Show that $\max_{0 \leq h \leq m} |L_h(m)| < \frac{1}{2}$ if $m > C(\alpha)$.

29. For a word $w \in \{0, 1\}^*$, define $T_0(w) = w\,w$ and $T_1(w) = w\,\overline{w}$, where $\overline{w}$ is the coding that sends $0 \to 1$, $1 \to 0$. Extend the domain of T as follows: $T_\epsilon(w) = w$, and if $x = ya$, with $a \in \{0, 1\}$, define $T_x(w) = T_a(T_y(w))$. For an

infinite word $\mathbf{z}$, define $T_{\mathbf{z}}(w) = \lim T_x(w)$, where x runs through the prefixes of $\mathbf{z}$.

(a) Show that $|T_x(w)| = |w| \cdot 2^{|x|}$.

(b) Show that $T_{\mathbf{z}}(w)$ is ultimately periodic if and only if there exists a finite word y with $\mathbf{z} = y\, 0^\omega$.

(c) Let $x = a_{r-1} \cdots a_1 a_0$ for some integer $r \geq 0$, and let $T_x(0) = b_0 b_1 \cdots b_{2^r-1}$. Let $0 \leq n < 2^r$, and let the base-2 representation of n be $c_{r-1} \cdots c_1 c_0$, where c_0 is the least significant digit. Then $b_n \equiv \sum_{0 \leq i < r} a_i c_i \pmod 2$.

(d) The lexicographically greatest word of the form $T_{\mathbf{x}}(0)$ is the Thue–Morse word $\mathbf{t}$, corresponding to the case $\mathbf{x} = 111 \cdots$. The lexicographically least word of the form $T_{\mathbf{x}}(1)$ is $\bar{\mathbf{t}}$, corresponding to the case $\mathbf{x} = 111 \cdots$.

(e) For all words w and infinite words $\mathbf{x}$, the infinite word $T_{\mathbf{x}}(w)$ is recurrent.

(f) $T_{\mathbf{x}}(w)$ is uniformly recurrent if and only if the distance between consecutive occurrences of the symbol 1 in $\mathbf{x}$ is bounded.

30. Show that the orbit closure of any morphic word has measure 0.

31. (Hamm) Show that it is possible for the first occurrence of a particular subword of a pure morphic word to be exponentially far out (in terms of the description size of the morphism).

32. Give an example of an infinite word $\mathbf{w}$ over $\{0, 1, 2\}$ with subword complexity $O(n)$, such that if the 2's are deleted from $\mathbf{w}$, the resulting word has subword complexity 2^n.

33. Suppose $h : \Sigma^* \to \Sigma^*$ is a morphism and a is a growing letter. Suppose $k = $ Card Σ. Show that $h^{rk}(a)$ contains at least $r + 1$ immortal letters for all integers $r \geq 0$.

34. Suppose $h : \Sigma^* \to \Sigma^*$ is a morphism and $a \in \Sigma$ is a letter such that the set $D_h = \{h^i(a) : i \geq 0\}$ is finite.

(a) Show that if $k = $ Card Σ and $W = \max_{c \in \Sigma} |h(c)|$, then $|h^i(a)| \leq W^{k-1}$ for all $i \geq 0$. Also show this bound is best possible.

(b) Show that Card $D_h \leq G(k) + k$, where G is Landau's function. Also show this bound is essentially optimal.

35. Give an example of a morphic word $\mathbf{u}$ with subword complexity $O(n)$ and a morphism g such that $g(\mathbf{u})$ has subword complexity $\Omega(n^2)$.

36. Give an example of a nonprimitive morphism h that is prolongable on a letter a, yet such that $h^\omega(a)$ is uniformly recurrent.

37. Suppose $\mathbf{x}$ is a recurrent infinite word that is ultimately periodic. Show that $\mathbf{x}$ is purely periodic.

38. Suppose $\mathbf{x}$ is a recurrent (respectively, uniformly recurrent) infinite word. Let h be a morphism. Show that $h(\mathbf{x})$ is finite or recurrent (respectively, finite or uniformly recurrent). Is the converse true?

39. Suppose we consider choosing the terms of an infinite sequence $\mathbf{x}$ over $\{0, 1\}$ uniformly at random. What is the expected value of $R'_{\mathbf{x}}(n)$?

40. Show that for any sequence of positive integers $a_1, a_2, a_3, \ldots$ there exists an infinite word $\mathbf{u}$ over the alphabet $\{0, 1, 2\}$ and an erasing morphism g such that $R'_{\mathbf{u}}(n) = O(n)$, but $R'_{g(\mathbf{u})}(n) \geq a_n$ for all $n \geq 1$.

41. Let θ be an irrational real number, and define a sequence $\mathbf{x} = a_0 a_1 a_2 \cdots$ as follows:

$$a_n = \begin{cases} 0 & \text{if } \{n\theta\} < \frac{1}{2}, \\ 1 & \text{if } \{n\theta\} > \frac{1}{2}. \end{cases}$$

Here $\{\theta\}$ means, as usual, the fractional part of x. Show that $p_{\mathbf{x}}(n) = 2n$ for all $n \geq 1$.

42. Consider the following generalization of the Toeplitz construction mentioned in Example 5.1.6. Let Σ be a finite alphabet, and let $\diamond$ be a new symbol not in Σ. For infinite words $\mathbf{u} \in (\Sigma \cup \{\diamond\})^\omega$ and $x \in \Sigma(\Sigma \cup \{\diamond\})^*$, define $F_x(\mathbf{u})$ to be the word that results from replacing the occurrences of $\diamond$ in $\mathbf{u}$ successively with the elements of x^ω. Define $T_0(x) = \diamond^\omega$, and for $i \geq 1$ define $T_{i+1}(x) = F_x(T_i(x))$. Clearly $T(x) = \lim_{i \to \infty} T_i(x)$ exists. It is called the *Toeplitz word* generated by x. Show that Toeplitz words are uniformly recurrent.

43. Let $\mathbf{t}$ be the Thue–Morse sequence and μ the Thue–Morse morphism. Let $k \geq 1$ be an integer, and let $a = k \bmod 2$. Show that the subword w of length $2^k + 2$ that maximizes $R'_{\mathbf{t}}(w)$ is $a \mu^k(0)\mathbf{1}$.

44. Let F_i denote the ith Fibonacci number. Show that the shortest finite word $w = w(n)$ for which $p_w(i) = F_{i+2}$ for $0 \leq i \leq n$ is of length $F_{n+2} + F_n$, and give an efficient algorithm to construct such words.

45. Let $B(n)$ be the de Bruijn graph on 2^n vertices.
    (a) Show that for each i, $2^n \leq i \leq 2^{n+1}$, there exists a closed walk of length i in $B(n)$, with no repeated edges and visiting every vertex at least once.
    (b) Find a tight upper bound on

$$\max_{|w|=n} \sum_{0 \leq i \leq n} p_w(i),$$

the largest total number of subwords of a binary string of length n.

46. For each of the following sequences, determine if they are recurrent, uniformly recurrent, or neither. Find an expression for $R'(n)$ and, if the sequence is uniformly recurrent, an expression for $R(n)$:
    (a) the Rudin–Shapiro sequence;
    (b) the regular paperfolding sequence;
    (c) the infinite Fibonacci word;
    (d) the Barbier infinite word.

47. Prove that there exists a constant C such that every paperfolding sequence $\mathbf{f}$ (see Section 6.5) is uniformly recurrent with $R_{\mathbf{f}}(n) \leq Cn$.

48. Let $\mathbf{w}$ be a uniformly recurrent infinite word. Show that if Sub($\mathbf{w}$) is a context-free language, then $\mathbf{w}$ is periodic.

49. Let $\mathbf{p} = 0110101000101 \cdots$ be the characteristic sequence of the prime numbers. Estimate $p_{\mathbf{p}}(n)$ and show that $\mathbf{p}$ has entropy 0.

50. We say an infinite word $\mathbf{x}$ is *mirror-invariant* if whenever w is a subword of $\mathbf{x}$, then so is w^R.

(a) Show that if an infinite word is mirror-invariant, then it is recurrent.

(b) Show that the converse is not true.

51. The *arithmetic complexity* $a_{\mathbf{w}}(n)$ of an infinite word $\mathbf{w} = a_0 a_1 a_2 \cdots$ is the cardinality of the set $\{a_r a_{r+k} a_{r+2k} \cdots a_{r+(n-1)k} : r \geq 0, \ k \geq 1\}$. In other words, the function $a_{\mathbf{w}}(n)$ counts the number of distinct length-n words obtained by choosing from $\mathbf{w}$ the symbols at positions given by an arithmetic progression.

   (a) Show that if $\mathbf{t} = 0110 \cdots$ is the infinite Thue–Morse sequence, then $a_{\mathbf{t}}(n) = 2^n$ for $n \geq 0$.

   (b) Show that if $\mathbf{f}$ is the paperfolding sequence, then $a_{\mathbf{f}}(n) = 8n + 4$ for $n \geq 14$.

52. Let $\mathbf{u}$ be a word having exactly m subwords of length n. Suppose each of these m subwords has the property that any of its occurrences in $\mathbf{u}$ is always followed by the same letter. Further suppose that these m words are subwords of a subword x of $\mathbf{u}$ of length $m + n$. Prove that $\mathbf{u}$ is ultimately periodic.

53. Suppose $\mathbf{x}$ is purely periodic of period k. Show that $R_{\mathbf{x}}(n) = n + k - 1$ for $n \geq k$.

10.12 Open Problems

1. Determine the subword complexity of **Kol**, the Kolakoski word.
2. Is it true that **Kol** contains exactly two palindromic subwords of each length?
3. Is there a k-context-free sequence with bounded gaps that is not k-automatic? (For the definition of k-context-free, see Open Problem 6.3.)
4. Improve the upper bound in Theorem 10.4.7.
5. Give necessary and sufficient conditions for a morphism h, prolongable on a, to be such that $h^\omega(a)$ is uniformly recurrent.
6. Compute good upper and lower bounds for the subword complexity of the characteristic word of the squarefree numbers $111011100110 \cdots$.
7. Consider the infinite word $\mathbf{u} = (t_{n^2})_{n \geq 0}$, where $\mathbf{t} = t_0 t_1 t_2 \cdots$ is the Thue–Morse sequence. Is it true that $p_{\mathbf{u}}(n) = 2^n$?

10.13 Notes on Chapter 10

10.1 In this chapter we have focused on the subword complexity of infinite words. However, it is also possible to study the subword complexity of *languages*; Rozenberg [1981] is a brief survey of this area. For other papers in this vein, see de Luca and Varricchio [1989b] and Baron and Urbanek [1989].

   Subword complexity for infinite words over a finite alphabet was apparently first studied by Morse and Hedlund [1938, p. 825 et seq.].

   Although subword complexity has been actively studied, the terminology and notation have not been standardized. For example, Morse and Hedlund [1938] used the terms "symbolic trajectory" and "ray" for what we call an infinite word, and "permutation number" for what we call subword complexity.

Many papers in the literature refer to subword complexity simply as "complexity".

Theorem 10.1.6 tells only a small part of the story. For other results on the distribution of subwords of a "random" infinite word, see, for example, Bender and Kochman [1993].

For generalizations of subword complexity, see Avgustinovich, Fon-Der-Flaass, and Frid [2001]; Kamae [2001]; Kamae and Zamboni [2001].

10.2 Theorem 10.2.4 essentially appears, in a more general form, in Ehrenfeucht and Rozenberg [1982a].

For Theorem 10.2.6, see Morse and Hedlund [1938], Bush [1955], and Coven and Hedlund [1973]. For a generalization to subword complexity of languages, see Ehrenfeucht and Rozenberg [1982a].

After the pioneering work of Morse and Hedlund, the subject of subword complexity lay dormant until the independent work of Cobham [1972] (who proved Theorem 10.3.1), Coven and Hedlund [1973], and Ehrenfeucht and Rozenberg [1973].

Allouche [1994a] and Ferenczi and Kása [1999] are surveys on subword complexity, the former in French. Berthé [2000b] surveys sequences of low complexity. For a survey explaining connections to dynamical systems, see Ferenczi [1999].

Tijdeman [1999] proved an interesting general lower bound on the complexity of infinite words.

For other papers on subword complexity, see Tapsoba [1987, 1994]; Bleuzen-Guernalec and Blanc [1989]; Mouline [1990]; Allouche and Shallit [1993]; Ferenczi [1995, 1996]; Arnoux and Mauduit [1996]; Mossé [1996a]; Yasutomi [1996]; Cassaigne [1996, 1997a]; Frid [1997a, 1998]; Frid and Avgustinovich [1999]; Fraenkel, Seeman, and Simpson [2001]; Aberkane [2001]; Heinis [2002].

Theorem 10.2.3 appears to be new. For the second difference of the subword complexity function, see Cassaigne [1997a].

10.3 The exact subword complexity of the Thue–Morse sequence **t** was computed independently by Brlek [1989], de Luca and Varricchio [1989a], and Avgustinovich [1994]. For a generalization, see Tromp and Shallit [1995] and Exercise 12. Another generalization was given by Frid [1997b]. The "bispecial" subwords of **t** (those subwords x such that $0x$, $1x$, $x0$, and $x1$ are also all subwords of **t**) were discussed by de Luca and Mione [1994].

Theorem 10.3.1 is due to Cobham [1972]. Bleuzen-Guernalec [1985] generalized this result to the case of fixed points of uniform transducers.

10.4 The function $G(n)$ was first studied by Landau [1903], who proved Theorem 10.4.4. Nicolas [1969a] gave an algorithm for computing $G(n)$. Nathanson [1972] gave an elementary proof that $G(n)$ is not bounded by any power of n.

Theorem 10.4.8, the estimate on $G(n)$, is due to Massias [1984].

Miller [1987] gives a nice survey of work on $G(n)$.

Corollary 10.4.9 is due to Ehrenfeucht, Lee, and Rozenberg [1975]. Actually, they proved a more general result on the subword complexity of a D0L language. The proof can also be found in Rozenberg and Salomaa [1980, Thm. IV.4.4, p. 210]. The explicit version with constants (Theorem 10.4.7) is new.

Ehrenfeucht and Rozenberg [1981a, 1981c] proved that if a morphic word **x** is squarefree, then $p_\mathbf{x}(n) = O(n \log n)$. Sajo [1984] discussed the subword complexity of context-free languages.

Cassaigne and Karhumäki [1995a, 1995b], and independently Koskas [1998], discussed the subword complexity of Toeplitz words. Toeplitz sequences were introduced by Jacobs and Keane [1969] and were based on a construction of Toeplitz [1928].

Other papers on subword complexity of D0L and related languages include Ehrenfeucht, Lee, and Rozenberg [1975, 1976]; Ehrenfeucht and Rozenberg [1981b, 1981c, 1983a, 1983b, 1983c].

Theorem 10.4.12 (subword complexity for primitive morphic words) can be essentially found in Michel [1975, 1976a]. Our proof is based on the paper of Pansiot [1984a]. Also see Queffélec [1987b, Prop. V.19, pp. 105–106].

Primitive morphic words were also studied by Queffélec [1987b]; Ferenczi, Mauduit, and Nogueira [1996]; Holton and Zamboni [1999].

Pansiot [1984a, 1984b] characterized the subword complexity functions for morphic words. Also see Pansiot [1985].

10.5 The subword complexity of Sturmian words was first considered by Morse and Hedlund [1940], and later, independently, by Bloom [1971]. Coven and Hedlund [1973] and Coven [1975] proved many combinatorial properties. Also see Stolarsky [1976], Fraenkel, Mushkin, and Tassa [1978], and Porta and Stolarsky [1990].

Our treatment of balanced sequences, in particular the equivalence given in Theorem 10.5.8, was inspired by Morse and Hedlund [1940], Coven and Hedlund [1973], Parvaix [1998], and Berstel and Séébold [2002]. This last paper is an excellent survey of Sturmian words.

For generalizations of the concept of balance, see, for example, Hubert [2000]; Berthé and Tijdeman [2002]; and Fagnot and Vuillon [2002].

Arnoux and Rauzy [1991] constructed infinite words with subword complexity $2n + 1$. For more on the Arnoux–Rauzy sequences, see Risley and Zamboni [2000]; Cassaigne, Ferenczi, and Zamboni [2000]; Wozny and Zamboni [2001]; Mignosi and Zamboni [2002]. Rote [1994] constructed infinite words with subword complexity $2n$. Both constructions have interesting similarities to Sturmian words. Arnoux, Mauduit, Shiokawa, and Tamura [1994a, 1994b] constructed sequences with subword complexity $n^2 + n + 1$; these have an interesting geometric interpretation in terms of billiards in the cube. They conjectured a formula for billiards in higher dimensions, which was later proved by Baryshnikov [1995]. For other papers on billiards and subword complexity, see Hubert [1995a, 1995b]. For two-dimensional generalization of Sturmian words, see Vuillon [1998]; Berthé and Vuillon [2001].

As we have seen, Sturmian words can be generated through rotations on the unit circle. For more on rotations, see Keane [1970]; Alessandri [1993]; Blanchard and Kurka [1998]; Didier [1998a, 1998b]; Arnoux, Ferenczi, and Hubert [1999]; Chekhova [2000]; Berthé and Vuillon [2000b]. Alternatively, Sturmian words can be generated through interval exchange transformations. Interval exchange was introduced by Katok and Stepin [1967] and later studied by Keane [1975], Rauzy [1979], and Veech [1984a, 1984b, 1984c]. For discussions of interval exchange and its relation to symbolic sequences, see Arnoux and Rauzy [1991]; Santini-Bouchard [1997]; Didier [1997]; Lopez and Narbel [2001]; Ferenczi, Holton, and Zamboni [2001].

Instead of complexity $n + 1$, one can study infinite words of complexity $n + c$ for a fixed constant c. It is known that $p_{\mathbf{u}}(n) = n + c$ for all $n \geq n_0$ if and only if $\mathbf{u} = wh(\mathbf{v})$, where w is a finite word, $\mathbf{v}$ is a Sturmian sequence on the alphabet $\{0, 1\}$, and $h : \{0, 1\}^* \to \Sigma^*$ is a morphism with $h(01) \neq h(10)$. This result is due to Paul [1975] in the case of uniformly recurrent words, and Coven [1975] for the general case. Also see Cassaigne [1998] and Didier [1999].

10.6 Theorem 10.6.1 is due to Mignosi [1989]. Another proof was given by Berstel [1999]. For more information about repetitions in Sturmian sequences, see Franěk, Karaman, and Smyth [2000]; Vandeth [2000]; Justin and Pirillo [2001]; Damanik and Lenz [2002].

10.7 The existence of de Bruijn sequences was apparently first proved by Flye Sainte-Marie [1894], and later rediscovered by M. Martin [1934], de Bruijn [1946], and Good [1946]. The technical report of de Bruijn [1975] gives an interesting history of this problem.

There is a vast literature on de Bruijn sequences; see, for example, Fredricksen [1982].

The graphs we have called word graphs are also called "factor graphs" or "Rauzy graphs" in the literature, the latter usage arising from Rauzy [1983]. For more on word graphs, see, for example, Frid [1999a].

10.8 The important paper of Morse and Hedlund [1938] introduced many concepts dealing with recurrence.

For an advanced treatment of recurrence in number theory, see Furstenberg [1981].

Theorem 10.9.5, the uniform recurrence of primitive morphic words, is due to Gottschalk [1963].

Rauzy [1983] is an influential survey on subword complexity and recurrence.

10.9   Theorem 10.9.2 is due to Morse and Hedlund [1938, §7].

Theorem 10.9.4 is due to Morse and Hedlund [1938, Thm. 7.2]. Note: they used the term "recurrent" for what we call uniformly recurrent. Other terms for uniformly recurrent in the literature include "almost periodic", "minimal", and (infrequently) "repetitive" and "primitive".

For more on uniform recurrence in number theory, see, for example, Furstenberg [1981].

For the recurrence function of Sturmian sequences, see Cassaigne [1999a].

10.10 For more on the appearance function, see Section 15.3 and Allouche and Bousquet-Mélou [1995]. Frid [1999b] computed the appearance function of a large class of infinite words.

11

Cobham's Theorem

As we have seen (Theorem 5.4.2), every ultimately periodic sequence is k-automatic for all integers $k \geq 2$. In this chapter we prove a beautiful and deep theorem due to Cobham, which states that if a sequence $\mathbf{s} = (s(n))_{n \geq 0}$ is both k-automatic and l-automatic and k and l are multiplicatively independent, then $\mathbf{s}$ is ultimately periodic. (Recall that Theorem 2.5.7 discusses when two integers are multiplicatively independent.)

11.1 Syndetic and Right Dense Sets

In this section, we prove some useful preliminary results.

We say that a set $X \subseteq \Sigma^*$ is *right dense* if for any word $u \in \Sigma^*$ there exists a $v \in \Sigma^*$ such that $uv \in X$ (that is, any word appears as a prefix of some word in X).

Lemma 11.1.1 *Let $k, l \geq 2$ be multiplicatively independent integers, and let X be an infinite k-automatic set of integers. Then $0^*(X)_l = 0^* \{(n)_l : n \in X\}$ is right dense.*

Proof. Since X is infinite and k-automatic, by the pumping lemma there exist strings t, u, v with u nonempty such that $tu^*v \subseteq (X)_k$. Let $x \in \{0, 1, \ldots, l-1\}^*$. Our goal is to construct y such that $xy \in 0^*(X)_l$. Then define

$$g = |u|,$$
$$h = |v|,$$
$$a = k^g,$$
$$b = l,$$
$$d = k^h \left([t]_k + \frac{[u]_k}{k^g - 1} \right),$$
$$k_1 = \frac{[x]_l + \frac{1}{4}}{d},$$
$$k_2 = \frac{[x]_l + \frac{1}{2}}{d}.$$

345

By Lemma 2.5.9 there exist arbitrarily large integers $p, q > 0$ such that $k_1 < a^p/b^q < k_2$. Hence there exist arbitrarily large integers $p, q > 0$ with

$$[x]_l + \frac{1}{4} < \frac{k^{gp+h}}{l^q}\left([t]_k + \frac{[u]_k}{k^g - 1}\right) < [x]_l + \frac{1}{2}. \tag{11.1}$$

But for q sufficiently large we have

$$-\frac{1}{4} < \frac{[v]_k - \frac{k^h[u]_k}{k^g-1}}{l^q} < \frac{1}{2}. \tag{11.2}$$

Adding Eqs. (11.1) and (11.2) we get, after multiplying by l^q,

$$[x]_l l^q < k^{gp+h}[t]_k + [u]_k \cdot \frac{k^{gp} - 1}{k^g - 1} \cdot k^h + [v]_k < ([x]_l + 1)\,l^q,$$

or, in other words,

$$[x]_l l^q < [tu^p v]_k < ([x]_l + 1)l^q.$$

Thus there exists an integer j, $0 < j < l^q$ such that

$$[x]_l l^q + j = [tu^p v]_k.$$

Hence there exists a word y such that $[xy]_l = [tu^p v]_k \in X$, and hence $xy \in 0^*(X)_l$. ∎

We now discuss another property of sets of integers. A set $X \subseteq \mathbb{N}$ is called *d-syndetic* if $X \cap [n, n + d) \neq \emptyset$ for all sufficiently large integers $n \geq 0$. If there exists a d such that X is d-syndetic, then X is said to be *syndetic*.

Lemma 11.1.2 *If X is a k-automatic set and $0^*(X)_k$ is right dense, then X is syndetic.*

Proof. Let X be a k-automatic set such that $0^*(X)_k$ is right dense. Then, by definition of what it means to be right dense, for all integers $n \geq 0$ there exist integers p and t, $0 \leq t < k^p$, such that $nk^p + t \in X$. Since X is k-automatic, we can choose p so it is bounded above by the number s of states in any DFA accepting $0^*(X)_k$. Thus X is $2k^s$-syndetic. ∎

Corollary 11.1.3 *Let $k, l \geq 2$ be multiplicatively independent integers. If an infinite set of integers is both k- and l-automatic, then it is syndetic.*

Proof. Combine Lemma 11.1.1 and Lemma 11.1.2. ∎

Now we prove a sort of converse to Lemma 11.1.2.

Lemma 11.1.4 *If X is a syndetic set, then $0^*(X)_k$ is right dense.*

Proof. Assume that X is d-syndetic. Then it is k^p-syndetic for any p with $k^p \geq d$. Choose an integer q large enough such that for all $n \geq 0$ the interval $[nk^q, (n+1)k^q)$ contains an element of X. Hence there exists a t such that $nk^q + t \in X$, and so $0^*(X)_k$ is right dense. ∎

Next we prove a technical result on syndetic sets.

Lemma 11.1.5 *Let X be a d-syndetic set of integers. Then for all positive integers K, L, h and each real $a > 0$ such that $K < L < K + a$, there exists an $x \in X$ and an integer y such that $yL \leq xK + h \leq yL + ad$.*

Proof. Let r be the smallest integer such that $rK + h < rL$. Then for all integers $i \geq 1$, we have $(r - i)L \leq (r - i)K + h$ by the minimality of r, and also

$$(r - i)K + h = rK + h - iK < rL - iK < rL - iL + ia = (r - i)L + ia.$$

Thus, for $1 \leq i \leq d$, we have

$$(r - i)L \leq (r - i)K + h < (r - i)L + da. \tag{11.3}$$

Let j be an integer such that $jL + r - d \geq x$ for some $x \in X$, and $jK + r - d \geq 0$. Add jKL to Eq. (11.3) to obtain

$$(jK + r - i)L \leq (jL + r - i)K + h < (jK + r - i)L + da$$

for all i, $1 \leq i \leq d$. Since X is d-syndetic, we may choose j sufficiently large that there is an $x \in X$ of the form $x = jL + r - i$ for some i with $1 \leq i \leq d$, and the proof is complete. ∎

11.2 Proof of Cobham's Theorem

We are now ready to prove the theorem

Theorem 11.2.1 *Suppose $k, l \geq 2$ are multiplicatively independent integers. If a set of integers X is both k- and l-automatic, then its characteristic word is ultimately periodic.*

Proof. If X is finite, the symbols of its characteristic word are all 0 from some point on.

Hence assume X is infinite and both k- and l-automatic. Then by Theorem 5.6.3(f) we know that the set

$$E_{tj} = \{y \in \mathbb{Z} : yk^j + t \in X\}$$

is l-automatic for all $t, j \geq 0$. Now, choosing $u \in \{0, 1, \ldots, k - 1\}^*$ such that $|u| = j$ and $[u]_k = t$, we have

$$(E_{tj})_k = (X)_k/u$$

so $[(X)_k/u]_k$ is l-automatic.

Now define ρ_k to be the equivalence relation on $\mathbb{N}$ given by the minimal automaton accepting $(X)_k$. In other words, $x \sim y \pmod{\rho_k}$ if and only if for all $w \in \Sigma^*$ we have $(x)_k w \in (X)_k \Leftrightarrow (y)_k w \in (X)_k$. It follows from Lemma 4.1.10 that the equivalence classes of ρ_k can be written as a Boolean combination of the sets $(X)_k/u$; hence each equivalence class E of ρ_k is l-automatic. Now X is the union of some of the equivalence classes of ρ_k.

Now we can construct a single automaton with input alphabet $\Sigma_l = \{0, 1, \ldots, l - 1\}$ that sends two inputs to different states if they are the base-l representations of elements of different equivalence classes of ρ_k. This automaton has a natural equivalence relation θ' on Σ_l^* associated with it:

$$w \sim x \pmod{\theta'} \quad \text{if and only if} \quad \delta(q_0, w) = \delta(q_0, x),$$

and θ' is right-invariant. Hence there is an equivalence relation θ that is a *refinement* of ρ_k and which is l-*stable*, i.e., which satisfies

$$x \sim y \pmod{\theta} \quad \Rightarrow \quad xl^j + t \sim yl^j + t \pmod{\theta}$$

for $j \geq 0, 0 \leq y < l^j$. Let c denote the number of equivalence classes of θ. Let $\mathbf{v}$ be the infinite word

$$\mathbf{v} = v_0 v_1 v_2 \cdots,$$

where v_n is the equivalence class of $n \pmod{\rho_k}$. Let $\mathbf{u}$ be the infinite word

$$\mathbf{u} = u_0 u_1 u_2 \cdots,$$

where u_n is the equivalence class of $n \pmod{\theta}$.

To show that the characteristic word of X is ultimately periodic, it suffices to show that $\mathbf{v}$ is ultimately periodic, because X is the union of some of the equivalence classes of ρ_k. We will prove this by showing that there exists an integer m such that the number of recurrent subwords of length m of $\mathbf{v}$ is bounded by m. The result will then follow from Theorem 10.8.5.

Now for all recurrent subwords $w = w_1 w_2$ of length 2 of $\mathbf{v}$, the set of indices n such that $v_n = w_1$ and $v_{n+1} = w_2$ is both k- and l-automatic, and therefore syndetic. Hence there exists an integer $d > 0$ such that any recurrent subword of length 2 of $\mathbf{v}$ has a second occurrence at distance at most d. Now choose a real number ϵ such that $0 < \epsilon < 1$ and $c\epsilon/(1 - \epsilon) < \frac{1}{2}$. By Lemma 2.5.9 we can find integers $p, q \geq 0$ such that

$$1 < \frac{l^q}{k^p} < 1 + \frac{\epsilon}{d}.$$

Let $K = k^p$, $L = l^q$, and $m = \lfloor K(1 - \epsilon) \rfloor$.

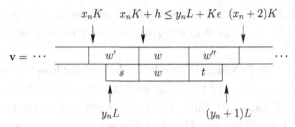

Figure 11.1: Illustration of the Lemma.

We now claim that for any recurrent subword w (of length m) of $\mathbf{v}$, there exists an integer y such that

$$v_{yL}v_{yL+1} \cdots v_{(y+1)L-1} = \mathbf{v}[yL..(y+1)L-1] = swt$$

and $|s| \le \epsilon K$.

To see this, observe that w is a recurrent subword of length at most K, and so appears infinitely often at the same position in subwords of the form $\mathbf{v}[xK..(x+2)K-1]$, $x \ge 0$. Since ρ_k is k-stable, $\mathbf{v}[xK..(x+2)K-1]$ is completely determined by $\mathbf{v}[x..x+1]$. By definition of d, every recurrent subword of length 2 of $\mathbf{v}$ has a second occurrence at distance at most d, so there exists a strictly increasing sequence of integers $(x_n)_{n \ge 1}$ such that $x_{n+1} - x_n \le d$ and

$$\mathbf{v}[x_n K..(x_n+2)K-1] = w'ww''.$$

Let $h = |w'|$. Now apply Lemma 11.1.5 with $a = K\epsilon/d$ and $X = \{x_1, x_2, \dots\}$. The lemma says there exist x_n, y_n such that $y_n L \le x_n K + h \le y_n L + K\epsilon$. Thus we have a situation as in Figure 11.1. (Note that $x_n K + h \le y_n L + K\epsilon$, so

$$x_n K + h + K(1 - \epsilon) \le y_n L + K < (y_n + 1)L,$$
$$x_n K + h + m < (y_n + 1)L,$$

so the picture is accurate.) Hence there exists a y_n such that $\mathbf{v}[y_n L..(y_n + 1)L - 1] = swt$ with $|s| \le da = K\epsilon$.

This proves the claim. Now let us count the number of subwords of $\mathbf{v}$ of the form $\mathbf{v}[yL..(y+1)L-1]$. We claim this number is bounded by the number of subwords of $\mathbf{u}$ of the form $\mathbf{u}[yL..(y+1)L-1]$. But this number is bounded by c, since this subword is completely determined by the equivalence class of y, since θ is an l-stable refinement of ρ_k. Thus the number of recurrent subwords of length m of $\mathbf{v}$ is at most equal to

$$(K\epsilon)c \le \frac{1}{2}K(1 - \epsilon) \le \frac{1}{2}(m + 1) \le m,$$

which by Theorem 10.8.5 shows $\mathbf{v}$ is ultimately periodic. ∎

As a corollary we get

Theorem 11.2.2 *Let $k, l \geq 2$ be multiplicatively independent integers, and suppose the sequence* $\mathbf{s} = (s_n)_{n \geq 0}$ *is both k- and l-automatic. Then* $\mathbf{u}$ *is ultimately periodic.*

Proof. Suppose $\mathbf{s}$ takes its values in a finite set Δ. Without loss of generality we may assume $\Delta = \{0, 1, \ldots, e - 1\}$ for some integer $e \geq 1$. For each $a \in \Delta$, consider the set $S_a = \{n \in \mathbb{Z} : s_n = a\}$. Since $\mathbf{s}$ is k- and l-automatic, each set S_a is both k- and l-automatic. From Theorem 11.2.1, it follows that the characteristic word $\mathbf{w}_a$ of S_a is ultimately periodic. From this, it follows that $\mathbf{s} = \sum_{a \in \Delta} a \mathbf{w}_a$ is ultimately periodic (see Exercise 1.8). ∎

11.3 Exercises

1. Give a purely "algebraic" proof that a formal power series with coefficients taking only finitely many distinct integer values is either rational or transcendental over $\mathbb{Q}(X)$.

2. Let j, k be integers ≥ 2. Prove that the characteristic sequence of the powers of k is j-automatic if and only if j and k are multiplicatively dependent.

11.4 Notes on Chapter 11

11.1 The term "syndetic" comes from Gottschalk and Hedlund [1955].

11.2 Cobham's original paper, in which Theorem 11.2.1 is proved, is Cobham [1969]. The proof is quite difficult. Earlier, Büchi [1960] found connections with mathematical logic, but McNaughton [1963] pointed out several errors in his work.

Our proof is based on one by Hansel; see Hansel [1982] and particularly Perrin [1990, pp. 39–43].

There are some very interesting connections with logic. For these, see Semenov [1977]; Hodgson [1983]; Michaux and Point [1986]; Villemaire [1992a, 1992b, 1992c]; Michaux and Villemaire [1993, 1996a, 1996b]; Bès [1997]; Maes [1999]. The survey by Bruyère, Hansel, Michaux, and Villemaire [1994] is particularly useful.

Randé [1993] extended Cobham's theorem to the case of certain Mahler equations. Fabre [1994] found an extension to certain morphic sequences. His results were improved by Durand [1998a]. For other generalizations, see Point and Bruyère [1997]; Fagnot [1997]; Hansel [1998]; Durand [1998c].

12

Formal Power Series

Recall the definitions of formal power series and formal Laurent series from Section 2.10: The ring $K[[X]]$ of formal power series with coefficients in a field K is defined by

$$K[[X]] := \left\{ \sum_{n \geq 0} a_n X^n : a_n \in K \right\},$$

where

$$\left(\sum_{n \geq 0} a_n X^n \right) + \left(\sum_{n \geq 0} b_n X^n \right) := \sum_{n \geq 0} (a_n + b_n) X^n,$$

$$\left(\sum_{n \geq 0} a_n X^n \right) \times \left(\sum_{n \geq 0} b_n X^n \right) := \sum_{n \geq 0} \left(\sum_{i+j=n} a_i b_j \right) X^n.$$

The ring $K[[X]]$ is a subring of the field $K((X))$ of formal Laurent series defined by

$$K((X)) = \left\{ \sum_{n \geq -n_0} a_n X^n : n_0 \in \mathbb{Z}, \, a_n \in K \right\},$$

where the addition and the multiplication are defined in a similar way.

Since this field contains the field of rational functions $K(X)$, we can define algebraicity over $K(X)$. We recall that a formal Laurent series $F(X) = \sum_{n \geq -n_0} a_n X^n$ is said to be *algebraic over the field $K(X)$*, or just *algebraic*, if there exist an integer $d \geq 1$ and polynomials $A_0(X), A_1(X), \ldots, A_d(X)$, with coefficients in K and not all zero, such that

$$A_0 + A_1 F + A_2 F^2 + \cdots + A_d F^d = 0.$$

This chapter is devoted to studying algebraicity and transcendence of formal power series when the ground field K is finite, and their relations with finite

351

automata. The basic result is a theorem of Christol (Theorem 12.2.5), which gives an equivalence between an algebraic property of a formal power series on a finite field and a combinatorial property of the sequence of its coefficients. We begin with some examples.

12.1 Examples

Example 12.1.1 Let f be the formal power series on $GF(2)$ defined by

$$f(X) = X + X^2 + X^4 + \cdots = \sum_{i \geq 0} X^{2^i}.$$

This series is algebraic over $GF(2)(X)$, since

$$f(X^2) = f(X) - X,$$

which implies, over $GF(2)$, that

$$f(X)^2 + f(X) + X = 0.$$

Example 12.1.2 Let $T(X) = \sum_{n \geq 0} t_n X^n$ where $(t_n)_{n \geq 0}$ is the Thue–Morse sequence. Then

$$T(X) = X + X^2 + X^4 + X^7 + X^8 + X^{11} + \cdots.$$

Now

$$T(X) = \sum_{n \geq 0} t_n X^n = \sum_{n \geq 0} t_{2n} X^{2n} + \sum_{n \geq 0} t_{2n+1} X^{2n+1}$$

$$= \sum_{n \geq 0} t_n X^{2n} + X \sum_{n \geq 0} (t_n + 1) X^{2n}$$

$$= T(X^2) + X T(X^2) + X \frac{1}{1 - X^2}.$$

Hence we have, over $GF(2)$,

$$(1 + X)^3 T(X)^2 + (1 + X)^2 T(X) + X = 0.$$

Thus $T(X)$ is algebraic over $GF(2)(X)$.

Example 12.1.3 Generalizing the previous example, let p be a prime number, let $t_p(n) = s_p(n) \bmod p$, and let $T_p(X) = \sum_{n \geq 0} t_p(n) X^n$.

Then we have

$$T_p(X) = \sum_{0 \le a < p} \sum_{n \ge 0} t_p(pn + a) X^{pn+a} = \sum_{0 \le a < p} \sum_{n \ge 0} (t_p(n) + a) X^{pn+a}$$

$$= \left(\sum_{0 \le a < p} X^a T_p(X)^p \right) + \left(\sum_{0 \le a < p} a X^a \left(\sum_{n \ge 0} X^n \right)^p \right)$$

$$= T_p(X)^p \left(\sum_{0 \le a < p} X^a \right) + \frac{1}{(1-X)^p} \left(\sum_{0 \le a < p} a X^a \right)$$

$$= T_p(X)^p \frac{1 - X^p}{1 - X} + \frac{X}{(1-X)^2}.$$

It follows that

$$(1 - X)^{p+1} T_p(X)^p - (1 - X)^2 T_p(X) + X = 0,$$

which proves that T_p is algebraic over $GF(p)(X)$. Exercise 35 asks you to prove that $(1 - X)^{p+1} Y^p - (1 - X)^2 Y + X$ is irreducible.

Example 12.1.4 Let $(s_n)_{n \ge 0}$ denote the Rudin–Shapiro sequence on $\{0, 1\}$, i.e., $s_n = e_{2;11}(n) \bmod 2$. Let $S(X) = \sum_{n \ge 0} s_n X^n$ and $T(X) = \sum_{n \ge 0} s_{2n+1} X^n$. Splitting these sums into even and odd indices and recalling that the equalities are taken modulo 2, we have

$$S(X) = \sum_{n \ge 0} s_n X^{2n} + X \sum_{n \ge 0} s_{2n+1} X^{2n} = S(X)^2 + X T(X)^2, \tag{12.1}$$

$$T(X) = \sum_{n \ge 0} s_{4n+1} X^{2n} + \sum_{n \ge 0} s_{4n+3} X^{2n+1} = \sum_{n \ge 0} s_n X^{2n} + X \sum_{n \ge 0} (s_{2n+1} + 1) X^{2n}$$

$$= S(X)^2 + X T(X)^2 + \frac{X}{(1+X)^2}.$$

Adding together the two relations for S and T, we get $S(X) + T(X) = X/(1 + X)^2$. Squaring, we get $S(X)^2 + T(X)^2 = X^2/(1 + X)^4$. Substituting back in (12.1), we get $S(X) = S(X)^2 + X(S(X)^2 + X^2/(1 + X)^4)$. Therefore $(1 + X)^5 S(X)^2 + (1 + X)^4 S(X) + X^3 = 0$.

Example 12.1.5 Here is an example over $GF(3)$. Let $(c_i)_{i \ge 0}$ be the Cantor sequence, i.e., $c_i = 1$ if the base-3 expansion of i contains only 0's and 2's, and 0 otherwise. Define the Cantor formal power series $C(X) = \sum_{i \ge 0} c_i X^i$. Then we have

$$C(X) = \sum_{i \geq 0} c_i X^{3i} + \sum_{i \geq 0} c_i X^{3i+2} = C(X^3) + X^2 C(X^3),$$

and so we get $(1 + X^2)C(X)^3 - C(X) = 0$. Since $C(X) \neq 0$, this gives $(1 + X^2)$ $C(X)^2 - 1 = 0$, and so $C(X) = \sqrt{1/(1 + X^2)}$.

These examples are not isolated and a general result explains them, as shown in the next section.

12.2 Christol's Theorem

Before stating the main theorem of this chapter, we need to define operators on the formal power series and prove two lemmas.

Definition 12.2.1 We define, for $0 \leq r < q$, a linear transformation

$$\Lambda_r : GF(q)[[X]] \rightarrow GF(q)[[X]]$$

as follows:

$$\Lambda_r \left(\sum_{i \geq 0} a_i X^i \right) = \sum_{i \geq 0} a_{qi+r} X^i.$$

Lemma 12.2.2 *We have the following properties*

(a) Let A be a formal power series in $GF(q)[[X]]$. Then

$$A(X) = \sum_{i \geq 0} a_i X^i = \sum_{0 \leq r < q} X^r \Lambda_r(A(X))^q. \tag{12.2}$$

(b) Let G and H be two formal power series in $GF(q)[[X]]$. Then

$$\Lambda_r(G^q H) = G \Lambda_r(H). \tag{12.3}$$

Proof. (a): We have

$$A(X) = \sum_{i \geq 0} a_i X^i = \sum_{0 \leq r < q} \sum_{i \geq 0} a_{qi+r} X^{qi+r} = \sum_{0 \leq r < q} X^r \sum_{i \geq 0} a_{qi+r} X^{iq}.$$

Hence

$$A(X) = \sum_{0 \leq r < q} X^r \left(\sum_{i \geq 0} a_{qi+r} X^i \right)^q = \sum_{0 \leq r < q} X^r \Lambda_r(A(X))^q.$$

(b): We have

$$\Lambda_r(G^q H) = \Lambda_r \left(\left(\sum_{k \geq 0} g_k X^k \right)^q \left(\sum_{j \geq 0} h_j X^j \right) \right)$$

$$= \Lambda_r \left(\left(\sum_{k \geq 0} g_k X^{qk} \right) \left(\sum_{j \geq 0} h_j X^j \right) \right).$$

Hence

$$\Lambda_r(G^q H) = \Lambda_r \left(\sum_{i \geq 0} X^i \sum_{\substack{k,j \geq 0 \\ qk+j=i}} g_k h_j \right) = \sum_{i \geq 0} X^i \left(\sum_{\substack{k,j \geq 0 \\ qk+j=qi+r}} g_k h_j \right)$$

$$= \sum_{i \geq 0} X^i \left(\sum_{0 \leq k \leq i} g_k h_{q(i-k)+r} \right) = \sum_{k \geq 0} g_k X^k \left(\sum_{i \geq k} h_{q(i-k)+r} X^{i-k} \right)$$

$$= \sum_{k \geq 0} g_k X^k \left(\sum_{i \geq 0} h_{qi+r} X^i \right) = \left(\sum_{k \geq 0} g_k X^k \right) \left(\sum_{i \geq 0} h_{qi+r} X^i \right)$$

$$= G \cdot \Lambda_r(H). \qquad \blacksquare$$

Lemma 12.2.3 *Let $A(X) = \sum_{i \geq 0} a_i X^i$ be a formal power series with coefficients in $GF(q)$, with $q = p^n$. Then A is algebraic over $GF(q)(X)$ if and only if there exist polynomials $B_1(X), \ldots, B_t(X)$, not all equal to zero, such that*

$$B_0 A + B_1 A^q + B_2 A^{q^2} + \cdots + B_t A^{q^t} = 0.$$

Furthermore we can suppose that $B_0 \neq 0$.

Proof. The first part of this lemma is known as Ore's theorem. If A is algebraic, the series $A, A^q, A^{q^2}, \ldots$, cannot be all linearly independent. Hence there exists a nontrivial linear relation

$$B_0 A + B_1 A^q + B_2 A^{q^2} + \cdots + B_t A^{q^t} = 0.$$

On the other hand, if such a nontrivial relation holds, the series A is clearly algebraic.

Now let us prove that we can find such a relation with $B_0 \neq 0$. Assume that

$$B_0 A + B_1 A^q + B_2 A^{q^2} + \cdots + B_t A^{q^t} = 0$$

with t minimal, and let j be the smallest non-negative integer such that $B_j(X) \neq 0$. We show that $j = 0$. Since

$$B_j = \sum_{0 \leq r < q} X^r (\Lambda_r(B_j))^q$$

by (12.2), it follows that there exists an r with $\Lambda_r(B_j) \neq 0$. Now, since $\sum_{j \leq i \leq t} B_i A(X)^{q^i} = 0$, we have

$$\sum_{j \leq i \leq t} \Lambda_r(B_i A^{q^i}) = 0,$$

and, using (12.3), we see that, if $j \neq 0$,

$$\sum_{j \leq i \leq t} \Lambda_r(B_i) A^{q^{i-1}} = 0,$$

which gives a new relation with the coefficient of $A^{q^{j-1}} \neq 0$, a contradiction. Hence $j = 0$. We thus have the relation

$$\sum_{0 \leq i \leq t} B_i A^{q^i} = 0,$$

with $B_0 \neq 0$.   ■

Lemma 12.2.4 *Suppose* $\mathbf{a} = (a_i)_{i \geq 0}$ *is a sequence over* $GF(q)$. *Then* $\mathbf{a}$ *is* q-*automatic if and only if there exists a finite collection of formal power series* $\mathcal{F}$ *such that (a)* $A \in \mathcal{F}$, *where* $A(X) := \sum_{i \geq 0} a_i X^i$; *and (b) for all* $g \in \mathcal{F}$, $0 \leq r < q$, *we have* $\Lambda_r(g) \in \mathcal{F}$.

Proof. Let $K_q(\mathbf{a}) = \{\mathbf{a}^{(1)}, \mathbf{a}^{(2)}, \ldots, \mathbf{a}^{(r)}\}$ be the q-kernel of the sequence $\mathbf{a}$, with $\mathbf{a} = \mathbf{a}^{(1)}$. Write $\mathbf{a}^{(i)} = (a_n^{(i)})_{n \geq 0}$.

$\Longrightarrow$: It is easy to see that we may take

$$\mathcal{F} = \left\{ \sum_{n \geq 0} a_n^{(i)} X^n : 1 \leq i \leq r \right\}.$$

$\Longleftarrow$: It is easy to see that $\sum_{n \geq 0} a_n^{(i)} X^n \in \mathcal{F}$. It follows that $|K_q(\mathbf{a})| \leq |\mathcal{F}|$, so the q-kernel is finite.   ■

We are now ready for one of the most important theorems of this book.

Theorem 12.2.5 (Christol) *Let* Δ *be a (nonempty) finite set, and let* $\mathbf{a} = (a_i)_{i \geq 0}$ *be a sequence over* Δ. *Let* p *be a prime number. Then* $\mathbf{a}$ *is* p-*automatic if and only if there exists an integer* $n \geq 1$ *and an injective map* $\beta : \Delta \to GF(p^n)$ *such that the formal power series* $\sum_{i \geq 0} \beta(a_i) X^i$ *is algebraic over* $GF(p^n)(X)$.

Proof. $\Longrightarrow$: Choose an integer n sufficiently large that $|\Delta| \leq p^n$, and an injection $\beta : \Delta \to GF(p^n)$. We may therefore assume, without loss of generality, that $\Delta \subseteq GF(p^n)$. Let us then show that $\sum_{i \geq 0} a_i X^i$ is algebraic over $GF(p^n)(X)$.

Since $(a_i)_{i \geq 0}$ is p-automatic, we know from Theorem 6.6.4 that it is q-automatic, where $q = p^n$. By Theorem 6.6.2 we know that the q-kernel $K_q(\mathbf{a})$ is finite, say

$K_q(\mathbf{a}) = \{\mathbf{a}^{(1)}, \mathbf{a}^{(2)}, \ldots, \mathbf{a}^{(d)}\}$, with $\mathbf{a}^{(1)} = \mathbf{a}$. Write $\mathbf{a}^{(i)} = (a_n^{(i)})_{n \geq 0}$. Define

$$A_j(X) = \sum_{n \geq 0} a_n^{(j)} X^n \quad \text{for } 1 \leq j \leq d.$$

Then, for $1 \leq j \leq d$,

$$A_j(X) = \sum_{0 \leq r \leq q-1} \sum_{m \geq 0} a_{qm+r}^{(j)} X^{qm+r}$$

$$= \sum_{0 \leq r \leq q-1} X^r \sum_{m \geq 0} a_{qm+r}^{(j)} X^{qm}.$$

But the sequence $(a_{qm+r}^{(j)})_{m \geq 0}$ is one of the $a^{(i)}$'s, which shows that $A_j(X)$ is a $GF(q)[X]$-linear combination of the power series $A_i(X^q)$. In other words $A_j(X)$ belongs to the $GF(q)(X)$-vector space generated by the series $A_i(X^q)$:

$$\forall j \in [1, d], \qquad A_j(X) \in \langle A_1(X^q), A_2(X^q), \ldots, A_d(X^q) \rangle.$$

But this implies that

$$\forall j \in [1, d], \qquad A_j(X^q) \in \langle A_1(X^{q^2}), A_2(X^{q^2}), \ldots, A_d(X^{q^2}) \rangle,$$

and also, by transitivity, that

$$\forall j \in [1, d], \qquad A_j(X) \in \langle A_1(X^{q^2}), A_2(X^{q^2}), \ldots, A_d(X^{q^2}) \rangle.$$

Hence

$$\forall j \in [1, d], \qquad A_j(X), A_j(X^q) \in \langle A_1(X^{q^2}), A_2(X^{q^2}), \ldots, A_d(X^{q^2}) \rangle.$$

This implies that

$$\forall j \in [1, d], \qquad A_j(X^q), A_j(X^{q^2}) \in \langle A_1(X^{q^3}), A_2(X^{q^3}), \ldots, A_d(X^{q^3}) \rangle.$$

Hence

$$\forall j \in [1, d], \qquad A_j(X), A_j(X^q), A_j(X^{q^2}) \in \langle A_1(X^{q^3}), A_2(X^{q^3}), \ldots, A_d(X^{q^3}) \rangle.$$

Iterating the reasoning, we finally have

$$\forall j \in [1, d], \ \forall k \in [0, d], \qquad A_j(X^{q^k}) \in \langle A_1(X^{q^{d+1}}), A_2(X^{q^{d+1}}), \ldots, A_d(X^{q^{d+1}}) \rangle.$$

But the dimension of $\langle A_1(X^{q^{d+1}}), A_2(X^{q^{d+1}}), \ldots, A_d(X^{q^{d+1}}) \rangle$ as a vector space over $GF(q)(X)$ is at most d, the number of generators, so for any $j \in [1, d]$, the formal power series

$$A_j(X), A_j(X^q), \ldots, A_j(X^{q^d})$$

are linearly related over $GF(p^n)(X)$. In particular for $j = 1$, this gives that $A_1(X) = \sum_{i \geq 0} a_i^{(1)} X^i$ is algebraic over $GF(p^n)(X)$.

$\Longleftarrow$: Suppose that $A(X) = \sum_{i \geq 0} a_i X^i$ is algebraic over $GF(q)(X)$. Then there exist polynomials $B_0(X), \ldots, B_t(X)$ such that

$$\sum_{0 \leq i \leq t} B_i(X)A(X)^{q^i} = 0,$$

and $B_0 \neq 0$ (Lemma 12.2.3). Put $G = A(X)/B_0(X)$; then

$$G = \sum_{1 \leq i \leq t} C_i G^{q^i}, \qquad \text{where } C_i = -B_i B_0^{q^i - 2}.$$

Now let

$$N = \max(\deg B_0, \max_i \deg C_i),$$

and let $\mathcal{H}$ be defined as follows:

$$\mathcal{H} = \left\{ H \in GF(q)[[X]] : H = \sum_{0 \leq i \leq t} D_i G^{q^i} \text{ with} \right.$$

$$\left. D_i \in GF(q)[X] \text{ and } \deg D_i \leq N \right\}.$$

Now $\mathcal{H}$ is a finite set and $A = B_0 G$ belongs to $\mathcal{H}$. We now show that $\mathcal{H}$ is mapped into itself by Λ_r. Let $H \in \mathcal{H}$. Then

$$\Lambda_r(H) = \Lambda_r \left(D_0 G + \sum_{1 \leq i \leq t} D_i G^{q^i} \right) = \Lambda_r \left(\sum_{1 \leq i \leq t} (D_0 C_i + D_i) G^{q^i} \right)$$

$$= \sum_{1 \leq i \leq t} \Lambda_r(D_0 C_i + D_i) G^{q^{i-1}}.$$

Since $\deg D_0, \deg D_i, \deg C_i \leq N$, it follows that $\deg(D_0 C_i + D_i) \leq 2N$, and so

$$\deg(\Lambda_r(D_0 C_i + D_i)) \leq \frac{2N}{q} \leq N.$$

By Lemma 12.2.4, the sequence $(a_i)_{i \geq 0}$ is q-automatic. ∎

Here is an application of Christol's theorem. Suppose $F = \sum_{i \geq 0} f_i X^i$ and $G = \sum_{i \geq 0} g_i X^i$ are two formal power series in $K[[X]]$. The *Hadamard product* is defined to be $F \odot G = \sum_{i \geq 0} f_i g_i X^i$.

Theorem 12.2.6 *If F, G are algebraic over $GF(q)(X)$, then so is $F \odot G$.*

Proof. Suppose F and G are algebraic over $GF(q)(X)$. Then, by Christol's theorem, the sequences $(f_i)_{i \geq 0}$ and $(g_i)_{i \geq 0}$ are q-automatic. Then, by Corollary 5.4.5, the sequence $(f_i g_i)_{i \geq 0}$ is q-automatic. Hence, by Christol's theorem, $F \odot G$ is algebraic over $GF(q)(X)$. ∎

Notice that Theorem 12.2.6 is not true, in general, for characteristic 0. For example,

$$F = \sum_{n \geq 0} \binom{2n}{n} X^n = (1 - 4X)^{-1/2}$$

is algebraic over any field K, but it can be shown that $F \odot F$ is transcendental over $\mathbb{Q}(X)$.

12.3 First Application to Transcendence Results

Let $(a_n)_{n \geq 0}$ be a sequence with values in a finite set $\mathcal{A}$, let $GF(q_1)$ and $GF(q_2)$ be two finite fields, and let φ_1 (respectively, φ_2), be an injective map from A to $GF(q_1)$ (respectively, $GF(q_2)$). Now suppose that the formal power series $\sum \varphi_1(a_n)X^n$ is algebraic over the field $GF(q_1)(X)$. Under what conditions can the series $\sum \varphi_2(a_n)X^n$ also be algebraic, i.e., algebraic over the field $GF(q_2)(X)$?

To make the question more concrete, let $(t_n)_{n \geq 0}$ be the Thue–Morse sequence. The formal power series $\sum_{n \geq 0} t_n X^n$ can be considered both as a series with coefficients in $GF(2)$ and as a series with coefficients in $GF(3)$. Can these two series both be algebraic? Here, by algebraic we mean, of course, over $GF(2)(X)$ for the first one, and over $GF(3)(X)$ for the second one.

The answer is no. Using Christol's theorem (Theorem 12.2.5) above, we can reformulate the theorems of Cobham previously given, i.e., Theorems 11.2.2 and 6.6.4, and state the following.

Theorem 12.3.1 *Let $(a_n)_{n \geq 0}$ be a sequence with values in a finite set $\mathcal{A}$, let $GF(q_1)$ and $GF(q_2)$ be two finite fields, and let φ_1 (respectively, φ_2) be an injective map from A to $GF(q_1)$ (respectively, $GF(q_2)$).*

(a) *If q_1 and q_2 are multiplicatively dependent, then the series $\sum \varphi_1(a_n)X^n$ is algebraic over the field $GF(q_1)(X)$ if and only if the series $\sum \varphi_2(a_n)X^n$ is algebraic over the field $GF(q_2)(X)$.*

(b) *If q_1 and q_2 are multiplicatively independent, and the series $\sum \varphi_k(a_n)X^n$ is algebraic over the field $GF(q_k)(X)$ for $k = 1, 2$, then both series are rational, i.e., the sequence $(a_n)_{n \geq 0}$ is ultimately periodic.*

12.4 Formal Laurent Power Series and Carlitz Functions

In this section we consider expressions of the form

$$\sum_{n \geq -t} a_n \left(\frac{1}{X}\right)^n,$$

where a_n belongs to a finite field $GF(q)$ of characteristic $p > 0$. These are formal Laurent series in the variable X^{-1}. Changing the variable X into X^{-1} in Christol's theorem (Theorem 12.2.5), we easily see that the series $\sum_{n \geq -t} a_n \left(\frac{1}{X}\right)^n$ is algebraic over $GF(q)(X)$ if and only if the sequence $(a_{n-t})_{n \geq 0}$ is p-automatic.

Carlitz introduced functions in positive characteristic that resemble the usual exponential function, the logarithm function, and the Riemann zeta function. We describe the Carlitz zeta function. Recall the definition of the Riemann zeta function for $\Re(s) > 1$:

$$\zeta(s) = \sum_{n \geq 1} \frac{1}{n^s}.$$

It is known that if s is even, then $\zeta(s) = \pi^s \cdot r$ where r is a rational number. For example, $\zeta(2) = \pi^2/6$ and $\zeta(4) = \pi^4/90$. No such expression is known for odd s; however, Apéry showed that $\zeta(3)$ is irrational.

Consider the following analogue in formal power series. Define for $n \geq 1$

$$\zeta_q(n) = \sum_{\substack{P \text{ monic} \\ P \in GF(q)[X]}} \frac{1}{P^n}.$$

Thus, for example,

$$\zeta_2(1) = \frac{1}{1} + \frac{1}{X} + \frac{1}{X+1} + \frac{1}{X^2} + \frac{1}{X^2+1} + \frac{1}{X^2+X} + \frac{1}{X^2+X+1} + \cdots$$
$$= 1 + X^{-2} + X^{-3} + X^{-4} + X^{-5} + X^{-9} + X^{-10} + \cdots$$
$$\in GF(2)[[X^{-1}]].$$

This function ζ_q, now called the Carlitz zeta function, has many properties similar to those of the Riemann zeta function. For example, it admits the following Euler product:

$$\zeta_q(n) = \prod_{\substack{P \text{ monic and irreducible} \\ P \in GF(q)[X]}} \frac{1}{1 - \frac{1}{P^n}}.$$

Carlitz also showed that if $q - 1 \mid n$, then $\zeta_q(n) = \pi_q^n \cdot r$ where r is a rational function and

$$\pi_q := \prod_{k \geq 1} \left(1 - \frac{X^{q^k} - X}{X^{q^{k+1}} - X}\right).$$

Note, by the way, that $q - 1 = |GF(q)^*|$, just as in the real case we have $2 = |\mathbb{Z}^*|$. A proof that π_q is transcendental was given by Wade. We now present a proof of Wade's result based on Christol's theorem (Theorem 12.2.5).

Theorem 12.4.1 *Let* $q = p^k$, p *prime*, $k \geq 1$. *Then* π_q *is transcendental over* $GF(q)(X)$.

Proof. By Exercise 17 we have

$$\frac{\pi_q'}{\pi_q} = \sum_{k \geq 1} \left(\frac{1}{1 - \frac{X^{q^k} - X}{X^{q^{k+1}} - X}} \right) \left(\frac{(X^{q^{k+1}} - X) - (X^{q^k} - X)}{(X^{q^{k+1}} - X)^2} \right) = \sum_{k \geq 1} \frac{1}{X^{q^{k+1}} - X}$$

$$(12.4)$$

$$= \left(\sum_{k \geq 1} \frac{1}{X^{q^k} - X} \right) - \frac{1}{X^q - X}. \qquad (12.5)$$

Now suppose that π_q is algebraic over $GF(q)(X)$. Then, by Exercise 18, so is π_q'. Hence, by Theorem 2.9.1, so is π_q'/π_q. But then, by (12.5) so is

$$\mathcal{B} := \sum_{k \geq 1} \frac{1}{X^{q^k} - X} = \sum_{k \geq 1} \frac{1}{[k]},$$

the so-called *bracket series* introduced by Wade, who defined $[k] := X^{q^k} - X$.

Thus it suffices to show that $\mathcal{B}$ is transcendental. We have

$$\mathcal{B} = \sum_{k \geq 1} \frac{1}{X^{q^k} - X} = \sum_{k \geq 1} \frac{1}{X^{q^k} \left(1 - \left(\frac{1}{X} \right)^{q^k - 1} \right)} = \sum_{k \geq 1} \frac{1}{X^{q^k}} \sum_{n \geq 0} \left(\frac{1}{X} \right)^{n(q^k - 1)}.$$

Hence

$$\mathcal{B} = \frac{1}{X} \sum_{k \geq 1} \frac{1}{X^{q^k - 1}} \sum_{n \geq 0} \left(\frac{1}{X} \right)^{n(q^k - 1)} = \frac{1}{X} \sum_{\substack{k \geq 1 \\ n \geq 0}} \left(\frac{1}{X} \right)^{(n+1)(q^k - 1)}$$

$$= \frac{1}{X} \sum_{\substack{k \geq 1 \\ n \geq 1}} \left(\frac{1}{X} \right)^{n(q^k - 1)} = \frac{1}{X} \sum_{m \geq 1} \left(\frac{1}{X} \right)^m \sum_{\substack{k,n \geq 1 \\ n(q^k - 1) = m}} 1$$

$$= \frac{1}{X} \sum_{m \geq 1} \left(\frac{1}{X} \right)^m \sum_{\substack{k \geq 1 \\ q^k - 1 \mid m}} 1 = \frac{1}{X} \sum_{m \geq 1} \left(\frac{1}{X} \right)^m c(m),$$

where

$$c(m) := \sum_{\substack{k \geq 1 \\ q^k - 1 \mid m}} 1.$$

Now, by virtue of Christol's theorem (Theorem 12.2.5), in order to show that $\sum_{k \geq 1} 1/(X^{q^k} - X)$ is transcendental over $GF(q)(X)$, it suffices to prove that $(c(m))_{m \geq 1}$ is not q-automatic.

If the sequence $(c(m))_{m \geq 1}$ were q-automatic, then by Corollary 5.5.3, the subsequence $(c(q^n - 1))_{n \geq 0}$ would be ultimately periodic. But

$$c(q^n - 1) = \sum_{\substack{k \geq 1 \\ q^k - 1 \mid q^n - 1}} 1 = \sum_{\substack{k \geq 1 \\ k \mid n}} 1 = d(n)$$

by Exercise 2.22, where $d(n)$ is the number of positive integral divisors of n.

Now $q = p^j$ for some $j \geq 1$, where p is a prime, so it suffices to show that $(d(n) \bmod p)_{n \geq 1}$ is not ultimately periodic. If it were, then there would exist integers $t \geq 1$, $n_0 \geq 0$ such that for all $n \geq n_0$ and $i \geq 1$ we have

$$d(n + it) \equiv d(n) \pmod{p}.$$

Choose $i = ni'$. Then

$$d(n(1 + i't)) \equiv d(n) \pmod{p}.$$

for all $i' \geq 1$. Now by Dirichlet's theorem we can find $i' \geq 1$ such that $p' = 1 + i't$ is a prime. Choose $n = p'$; we find

$$d(p'^2) \equiv d(p') \pmod{p}.$$

and hence $3 \equiv 2 \pmod{p}$. This contradiction completes the proof. ∎

12.5 Transcendence of Values of the Carlitz–Goss Gamma Function

In this section, we prove the transcendence of certain values of the Carlitz–Goss gamma function, which is a p-adic interpolation of the ordinary gamma function.

Let $q = p^k$, with p prime and k an integer ≥ 1. For $j \geq 1$ define

$$D_j = (X^{q^j} - X)(X^{q^j} - X^q) \cdots (X^{q^j} - X^{q^{j-1}});$$

note that $\deg D_j = jq^j$.

Let n be a p-adic integer, i.e., a formal expression of the form

$$n = \sum_{j \geq 0} a_j p^j \tag{12.6}$$

with $0 \leq a_j < p$. Note that n is a natural number if and only if there exists an integer $R \geq 0$ such that $a_j = 0$ for all $j \geq R$.

We can group the expansion (12.6) into k bits at a time:

$$n = \sum_{j \geq 0} n_j q^j,$$

where $0 \leq n_j < q$. Then we make the following definition:

$$n! = \prod_{j \geq 1} \left(\frac{D_j}{X^{\deg D_j}} \right)^{n_j} \in GF(q)[[X^{-1}]].$$

As an example, consider the case $p = q = 2$, and set $n = -1$, i.e., $n_j = 1$ for all $j \geq 0$. Then over $GF(2)$ we find

$$
(-1)! = \frac{X^2 - X}{X^2} \cdot \frac{(X^4 - X^2)(X^4 - X)}{X^8} \cdot \frac{(X^8 - X^4)(X^8 - X^2)(X^8 - X)}{X^{24}} \cdots
$$

$$
= (1 + X^{-1})(1 + X^{-2} + X^{-3} + X^{-5})
$$

$$
\times (1 + X^{-4} + X^{-6} + X^{-7} + X^{-10} + X^{-11} + \cdots) \cdots
$$

$$
= 1 + X^{-1} + X^{-2} + X^{-6} + X^{-9} + X^{-12} + X^{-14} + X^{-15} + \cdots.
$$

We now state the theorem of Mendès France and Yao:

Theorem 12.5.1 *Let n be a p-adic integer. Then $n!$ is transcendental over $GF(q)(X)$ if and only if n is not a natural number.*

We need two technical lemmas:

Lemma 12.5.2 *Let a, b, c, q be integers with $a, b, c \geq 1$ and $q \geq 2$. Then $q^c - 1$ $\mid q^a(q^b - 2) + 1$ if and only if $c \mid \gcd(a, b)$.*

Proof. $\Longleftarrow$: Suppose $c \mid \gcd(a, b)$. Then $q^c \equiv 1 \pmod{q^c - 1}$, and since a/c is an integer, we can raise both sides to the a/c power to get $q^a \equiv 1 \pmod{q^c - 1}$. Similarly, $q^b \equiv 1 \pmod{q^c - 1}$. Hence

$$
q^a(q^b - 2) + 1 \equiv (1)(1 - 2) + 1 \equiv 0 \pmod{q^c - 1}.
$$

$\Longrightarrow$: Suppose $q^c - 1 \mid q^a(q^b - 2) + 1$. Define $c_1 = (a + b) \bmod c$ and $a_1 = a \bmod c$, so that $0 \leq a_1, c_1 < c$. Then there exist natural numbers s, t such that $a + b = sc + c_1$ and $a = tc + a_1$. It follows that

$$
q^{c_1} - 2q^{a_1} + 1 = (q^a(q^b - 2) + 1) - q^{c_1}(q^{sc} - 1) + 2q^{a_1}(q^{tc} - 1).
$$

Since $q^c - 1$ divides the right-hand side of this equation, it follows that $q^c - 1$ $\mid q^{c_1} - 2q^{a_1} + 1$.

Now if $c_1 > a_1$, we have

$$
0 \leq q^{c_1} - 2q^{a_1} + 1 \leq q^{c_1} - 2 + 1 < q^{c-1} - 1,
$$

while if $a_1 \geq c_1$ we have

$$
0 \leq -(q^{c_1} - 2q^{a_1} + 1) \leq 2(q^{a_1} - 1) < 2(q^{c-1} - 1).
$$

Hence $|q^{c_1} - 2q^{a_1} + 1| < q^c - 1$. But $q^c - 1 \mid q^{c_1} - 2q^{a_1} + 1$; hence $q^{c_1} - 2q^{a_1} + 1 = 0$. Considering both sides modulo q, we see that $a_1 = c_1 = 0$. It follows that $c \mid a$ and $c \mid b$. $\blacksquare$

Lemma 12.5.3 *Let $q = p^k$, and let $n \in \mathbb{Z}_p \setminus \mathbb{N}$ be a p-adic integer with q-adic expansion $\sum_{l \geq 0} n_l q^l$. For $1 \leq j \leq k$ define*

$$E_j(n) = \{l \geq 0 : n_l \not\equiv 0 \pmod{p^j}\}.$$

Then there exists $m \in \mathbb{Z}_p \setminus \mathbb{N}$ such that $E_1(m)$ is infinite, and $n!$ is transcendental over $GF(q)(X)$ if and only if $m!$ is transcendental over $GF(q)(X)$.

Proof. Let $n \in \mathbb{Z}_p \setminus \mathbb{N}$. Let h be the least integer j, with $1 \leq j < k$, such that $E_j(n)$ is infinite. Note that h exists because by hypothesis $n \notin \mathbb{N}$ implies that $E_k(n)$ is infinite.

If $h = 1$, we may take $m = n$.

Otherwise $h \geq 2$. Then $E_{h-1}(n)$ is finite. Let $L \geq 0$ be the greatest element of $E_{h-1}(n)$. Hence for all $l > L$, the quantity $m_l = n_l / p^{h-1}$ is an integer. Now define

$$m = \sum_{L < i < \infty} m_l q^l.$$

Then we have

$$n! = \prod_{1 \leq i \leq L} \left(\frac{D_i}{X^{\deg D_i}} \right)^{n_i} \prod_{L < i < \infty} \left(\frac{D_i}{X^{\deg D_i}} \right)^{n_i} = \left(\prod_{1 \leq i \leq L} \left(\frac{D_i}{X^{\deg D_i}} \right)^{n_i} \right) (m!)^{p^{h-1}}.$$

Also, it is clear that $E_1(m)$ is infinite. ∎

Now we are ready to prove Theorem 12.5.1.

Proof. By Lemma 12.5.3 we may suppose that $E_1(n)$ is infinite. Now for all integers $j \geq 1$ we have

$$\frac{D_j'}{D_j} = \sum_{0 \leq i < j} \frac{(X^{q^j} - X^{q^i})'}{X^{q^j} - X^{q^i}} = \frac{-1}{X^{q^j} - X}.$$

From this we deduce

$$\frac{(n!)'}{n!} = -\sum_{j \geq 1} \frac{n_j}{X^{q^j} - X} = -\sum_{j \geq 1} \frac{n_j}{X^{q^j}} \left(1 - \left(\frac{1}{X} \right)^{q^j - 1} \right)^{-1}$$

$$= -\sum_{j \geq 1} \frac{n_j}{X^{q^j}} \left(\sum_{l \geq 0} \left(\frac{1}{X} \right)^{l(q^j - 1)} \right) = -\sum_{j \geq 1} \sum_{l \geq 0} \frac{n_j}{X^{q^j}} \left(\frac{1}{X} \right)^{l(q^j - 1)}.$$

Hence we find

$$-X \frac{(n!)'}{n!} = \sum_{j, l \geq 1} n_j \left(\frac{1}{X} \right)^{l(q^j - 1)} = \sum_{m \geq 1} c(m) X^{-m},$$

where

$$c(m) = \left(\sum_{\substack{j,l \geq 1 \\ l(q^j-1)=m}} n_j \right) \bmod p = \left(\sum_{\substack{j \geq 1 \\ q^j-1 \mid m}} n_j \right) \bmod p.$$

Suppose $n!$ is algebraic. Then, by Exercise 18, so is $(n!)'$. Then, by Theorem 2.9.1, so is $-X\frac{(n!)'}{n!}$. Hence, if we can show $-X\frac{(n!)'}{n!}$ is transcendental, it follows that $n!$ is too. We do this by showing that $K_q(c)$, the q-kernel of the sequence c, is infinite.

For $m \geq 0$, define

$$d_l(m) = c(q^l m + 1).$$

Now choose any two distinct elements $a, b \in E_1(n)$. Without loss of generality we may suppose $a > b$. We show that the sequences d_a and d_b are different. Define

$$E = \{l \geq 0 : n_l \not\equiv 0 \,(\bmod\, p),\ l \mid a,\ l \nmid b\}.$$

Note $a \in E$, so $E \neq \emptyset$. Let $h = \min_{e \in E} e$. Suppose there exists t, $1 \leq t < h$, with $t \mid a$ and $t \nmid b$. Then from the definition of E, we must have $n_t \equiv 0 \,(\bmod\, p)$. It follows that, using Lemma 12.5.2,

$$d_a(q^h - 2) - d_b(q^h - 2) \equiv \sum_{\substack{l \geq 1 \\ q^l-1 \mid q^a(q^h-2)+1}} n_l - \sum_{\substack{l \geq 1 \\ q^l-1 \mid q^b(q^h-2)+1}} n_l$$

$$\equiv \sum_{\substack{l \geq 1 \\ l \mid (a,h)}} n_l - \sum_{\substack{l \geq 1 \\ l \mid (b,h)}} n_l,$$

where the congruences are mod p. Hence

$$d_a(q^h - 2) - d_b(q^h - 2) \equiv \sum_{\substack{l \mid \gcd(a,h) \\ l \nmid \gcd(b,h)}} n_l - \sum_{\substack{l \nmid \gcd(a,h) \\ l \mid \gcd(b,h)}} n_l \equiv \sum_{\substack{l \mid h \\ l \nmid b}} n_l \equiv n_h \not\equiv 0 \,(\bmod\, p).$$

Thus the subsequences d_a and d_b differ at $q^h - 2$, and hence are different. But a, b were arbitrary elements of $E_1(n)$, which we supposed to be infinite. Hence the q-kernel of the sequence c is infinite, and thus c is not q-automatic. By Theorem 12.2.5, it follows that $n!$ is transcendental. ∎

12.6 Application to Transcendence Proofs over $\mathbb{Q}(X)$

Christol's theorem may also be fruitfully applied to give easy proofs of transcendence of formal power series over $\mathbb{Q}(X)$. The basic principle is the following:

Theorem 12.6.1 *Let* $F(X) = \sum_{i \geq 0} f_i X^i \in \mathbb{Z}[[X]]$ *be a formal power series with integer coefficients, and let* p *be a prime number. Let* $F_p(X) = \sum_{i \geq 0}(f_i \bmod p)X^i \in GF(p)[[X]]$ *be the reduction of* F *modulo* p. *If* F *is algebraic over* $\mathbb{Q}(X)$, *then* F_p *is algebraic over* $GF(p)(X)$.

Proof. Suppose F is algebraic over $\mathbb{Q}(X)$. Then there exists an integer $d \geq 1$ and $d + 1$ polynomials $B_0, B_1, \ldots, B_d \in \mathbb{Q}[X]$ such that

$$B_0 + B_1 F + \cdots + B_d F^d = 0. \tag{12.7}$$

By clearing denominators, we may in fact assume that $B_i \in \mathbb{Z}[X]$ for $0 \leq i \leq d$. We can also assume that the greatest common divisor e of the coefficients of all the B_i is 1; for if not, we could divide (12.7) by e. Now consider (12.7) modulo p. Since not all the coefficients of the polynomials are divisible by p, the reduction of (12.7) modulo p gives a nontrivial relation for F_p (of possibly lower degree). It then follows that F_p is algebraic over $GF(p)(X)$. ∎

We now apply Theorem 12.6.1 to some series of interest. Define

$$\theta_3(X) = \sum_{-\infty < n < \infty} X^{n^2},$$

the classical theta series.

Theorem 12.6.2 $\theta_3(X)$ *is transcendental over* $\mathbb{Q}(X)$.

Proof. Suppose $\theta_3(X)$ is algebraic over $\mathbb{Q}(X)$. Then since $\theta_3(X) = 1 + 2\sum_{n \geq 1} X^{n^2}$, the series $g(X) = \sum_{n \geq 1} X^{n^2}$ would be algebraic over $\mathbb{Q}(X)$. By Theorem 12.6.1, g would be algebraic over $GF(2)(X)$. But by Theorem 12.2.5, this means that the characteristic sequence of the squares is 2-automatic, which we have seen in Example 5.5.1 is not the case. ∎

For our second application, we consider the following problem. Recall from Section 1.5 that a nonempty word w is said to be primitive if it cannot be written in the form $w = x^k$ where $k \geq 2$ is an integer. Let P_k be the set of all primitive words over $\{0, 1, \ldots, k - 1\}$. For example,

$$P_2 = \{0, 1, 01, 10, 001, 010, 011, 100, 101, 110, \ldots\}.$$

It is currently a major open problem in formal languages to show that P_k is not context-free for any $k \geq 2$; see Open Problem 4.3. Although this problem is unsolved, it is possible to prove the following weaker result:

Theorem 12.6.3 *For $k \geq 2$, if P_k is context-free, then it is inherently ambiguous.*

Proof. Let $\psi_k(n)$ be the number of primitive words of length n over an alphabet of size k. From Theorem 1.5.5 we have

$$\psi_k(n) = \sum_{d \mid n} \mu(d) k^{n/d}.$$

Suppose P_k is context-free and possesses an unambiguous grammar. Then by the theorem of Chomsky-Schützenberger (Theorem 4.4.2), we know that

$$R(X) := \sum_{n \geq 1} \psi_k(n) X^n$$

is algebraic over $\mathbb{Q}(X)$. Then

$$\widetilde{R}(X) := \sum_{n \geq 1} \frac{\psi_k(n)}{k} X^n$$

is also algebraic over $\mathbb{Q}(X)$. Note that $\psi_k(n)/k$ is an integer for any $n \geq 1$.

Let p be a prime dividing k. Then by Theorem 12.6.1, we know that

$$\widetilde{R}_p(X) = \sum_{n \geq 1} \left(\frac{\psi_k(n)}{k} \bmod p \right) X^n$$

is algebraic over $GF(p)(X)$. But

$$\frac{\psi_k(n)}{k} = \sum_{d \mid n} \mu(d) k^{n/d-1} = \mu(n) + \sum_{\substack{d \mid n \\ d \neq n}} \mu(d) k^{n/d-1} \equiv \mu(n) \pmod{p}.$$

It follows that

$$\widetilde{R}_p(X) = \sum_{n \geq 1} \mu(n) X^n$$

and so the sequence $(\mu(n) \bmod p)_{n \geq 0}$ must be p-automatic. But then $(\mu(n)^2 \bmod p)_{n \geq 0}$ would be p-automatic. However, $\mu(n)^2 \equiv 1 \pmod{p}$ if and only if n is squarefree. By a classical theorem (see Exercise 2.21), the density of the squarefree numbers exists and is equal to $6/\pi^2$, an irrational number. But by Theorem 8.4.5, the frequency of a symbol in automatic sequences (if it exists) must be rational, a contradiction. ∎

12.7 Furstenberg's Theorem

We now show that algebraic series can be easily obtained from rational power series in higher dimension. Let us begin with an example.

Example 12.7.1 Consider the rational function in two variables in $GF(2)(X)$ defined by

$$R(X, Y) = \frac{Y}{1 + Y(1 + XY) + \frac{X}{(1+XY)^2}}.$$

Since the relation $X/(1 + XY)^2 = X \sum_{i \geq 0} (XY)^{2i}$ holds over $GF(2)(X)$, we have

$$R(X, Y) = \frac{Y}{1 + Y(1 + XY) + X \sum_{i \geq 0}(XY)^{2i}}$$

$$= \sum_{j \geq 0} Y \left(Y(1 + XY) + X \sum_{i \geq 0}(XY)^{2i} \right)^j$$

$$= \sum_{j,k \geq 0} \binom{j}{k} Y^{k+1}(1 + XY)^k X^{j-k} \left(\sum_{i \geq 0}(XY)^{2i} \right)^{j-k}$$

$$:= \sum_{m,n \geq 0} a_{m,n} X^m Y^n.$$

Now let us compute the diagonal of R, i.e., the formal power series $(\mathcal{D}R)(X) = \sum_{n \geq 0} a_{n,n} X^n$. In the expression

$$\sum_{j,k \geq 0} \binom{j}{k} Y^{k+1}(1 + XY)^k X^{j-k} \left(\sum_{i \geq 0}(XY)^{2i} \right)^{j-k}$$

the only indices j, k that can give equal exponents for X and Y should satisfy $k + 1 = j - k$. Hence j is odd, and $j = 2k + 1$. Then

$$(\mathcal{D}R)(X) = \sum_{k \geq 0} \binom{2k + 1}{k} X^{k+1}(1 + X)^k \left(\sum_{i \geq 0} X^{2i} \right)^{k+1}$$

$$= \sum_{k \geq 0} \binom{2k + 1}{k} X^{k+1}(1 + X)^{-k-2}.$$

Now, splitting into odd and even indices, we obtain

$$(\mathcal{D}R)(X) = \sum_{k \geq 0} \binom{4k + 1}{2k} X^{2k+1}(1 + X)^{-2k-2}$$

$$+ \sum_{k \geq 0} \binom{4k + 3}{2k + 1} X^{2k+2}(1 + X)^{-2k-3}.$$

Using Exercise 12, we have

$$\forall k \geq 0, \qquad \binom{4k + 3}{2k + 1} \equiv \binom{2k + 1}{k} \pmod 2$$

and

$$\forall k \geq 1, \qquad \binom{4k + 1}{2k} \equiv \binom{2k}{k} \equiv 0 \pmod 2.$$

Hence

$$(\mathcal{D}R)(X) = X(1+X)^{-2} + \sum_{k \geq 0} \binom{2k+1}{k} X^{2k+2}(1+X)^{-2k-3}$$
$$= X(1+X)^{-2} + (1+X)(\mathcal{D}R)^2(X).$$

Hence $(\mathcal{D}R)(X)$ satisfies the same quadratic equation as the Thue–Morse series $T(X)$ in Example 12.1.2. Since both series have 0 as constant term, they coincide. In other words, we have obtained the Thue–Morse series as the diagonal of a rational function in two variables.

Definition 12.7.2 Let

$$F(X, Y) = \sum_{\substack{m \geq m_0 \\ n \geq n_0}} a_{m,n} X^m Y^n,$$

where $m_0, n_0 \in \mathbb{Z}$, be a formal Laurent series in the two variables X and Y. The diagonal $\mathcal{D}F$ is the formal Laurent series in one variable defined by

$$\mathcal{D}F(X) = \mathcal{D}\left(\sum_{m \geq m_0, \, n \geq n_0} a_{m,n} X^m Y^n\right) := \sum_{k \geq \max\{m_0, n_0\}} a_{k,k} X^k.$$

Theorem 12.7.3 (Furstenberg) *A formal Laurent series over a finite field is algebraic if and only if it is the diagonal of a rational formal Laurent series in two variables over that field.*

Proof. The first part of the proof consists of showing that the diagonal of a double rational formal power series on a finite field is algebraic. This proof is postponed until Chapter 14, where we will prove that, more generally, the diagonal of a double algebraic formal power series in two variables is algebraic (Theorem 14.4.2).

Let us give the second part of the proof. Let p be a prime and $n \geq 1$ be an integer, and set $q = p^n$. Let $F(X) = \sum_i a_i X^i$ be a formal Laurent series with coefficients in $GF(q)$ that is algebraic over $GF(q)(X)$. We can suppose that F is a formal power series, i.e., the exponents are all non-negative, up to subtracting a rational function from F. Then, from Lemma 12.2.3, we can find polynomials $B_j(X)$, not all equal to zero, such that

$$B_0 F + B_1 F^q + \cdots + B_t F^{q^t} = 0,$$

with $B_0 \neq 0$. This relation shows that there exist relations of the form

$$C_0 H + C_1 H^q + \cdots + C_t H^{q^t} + E(X) = 0, \tag{12.8}$$

with E a polynomial and $F(X) = R(X) + X^h H(X)$, where $h \geq 0$, R is a polynomial, H is a formal power series, and each of the C_i are polynomials. We claim that we can find such a relation (12.8) with $C_0(0) \neq 0$, and $H(0) = 0$. (The notation $H(0)$ stands for the constant term of H.) Suppose that we have such a relation with r the largest integer for which X^r divides C_0 and X^{r+1} does not divide C_0, and r minimal among all such relations. Then let $H(X) = \sum_{j \geq 0} \lambda_j X^j = \lambda_0 + XK(X)$. We have

$$C_0 X K + C_1 X^q K^q + \cdots + C_t X^{q^t} K^{q^t} + \widetilde{E} = 0, \qquad (12.9)$$

with $\widetilde{E} = E + \lambda_0 C_0 + \lambda_0^q C_1 + \cdots + \lambda_0^{q^t} C_t$ and $F(X) = R(X) + \lambda_0 X^h + X^{h+1} K$. Now, defining $s = \min(r + 1, q)$, we see that X^s divides all terms other than $\widetilde{E}$ of the equality (12.9); hence it also divides $\widetilde{E}$. If $r \neq 0$, then $0 \leq r - s + 1 < r$. Dividing (12.9) by X^s we obtain an equality of type (12.8) such that X^{r-s+1} divides the coefficient analogous to C_0, and X^{r-s+2} does not divide this coefficient, a contradiction. The claim follows if $H(0) = 0$. If $H(0) \neq 0$, putting $\widetilde{H} = H - H(0)$, we obtain a relation of type (12.8), with H replaced by $\widetilde{H}$, E replaced by $E + H(0)C_0 + \cdots + C_t H(0)^{q^t}$ and the C_i's (in particular C_0) unchanged. Furthermore, $F = (R + H(0)X^h) + X^h \widetilde{H}$, and we are done.

Now, define

$$P(X, Y) = C_0(X)Y + C_1(X)Y^q + \cdots + C_t(X)Y^{q^t} + E(X), \qquad (12.10)$$

so that $P(X, H(X)) = 0$. Noticing that $P(X, Y)$ can be considered as a polynomial in $GF(q)((X))[Y]$, and using factorization of polynomials over the field $GF(q)((X))$, we can write

$$P(X, Y) = (Y - H(X))Q(X, Y), \qquad (12.11)$$

where $Q(X, Y)$ is a polynomial in Y with coefficients in $GF(q)((X))$. Now notice from Eq. (12.10) that $\frac{\partial P}{\partial Y}(0, 0) = C_0(0) \neq 0$. But from Eq. (12.11) we have $\frac{\partial P}{\partial Y}(0, 0) = Q(0, 0) - H(0)\frac{\partial Q}{\partial Y}(0, 0) = Q(0, 0)$. Hence $Q(0, 0) \neq 0$. Now, taking the partial logarithmic derivative with respect to Y of Eq. (12.11), we get

$$\frac{1}{P}\frac{\partial P}{\partial Y}(X, Y) = \frac{1}{Y - H(X)} + \frac{1}{Q}\frac{\partial Q}{\partial Y}(X, Y).$$

Now, since $Q(0, 0) \neq 0$, the term $\frac{1}{Q}\frac{\partial Q}{\partial Y}(X, Y)$ can be expanded as a formal Laurent series in two variables. Multiplying by Y^2, replacing X by XY, and taking diagonals, we have

$$\mathcal{D}\left(\frac{Y^2}{P(XY, Y)}\frac{\partial P}{\partial Y}(XY, Y)\right) = \mathcal{D}\left(\frac{Y^2}{Y - H(XY)}\right)$$
$$+ \mathcal{D}\left(\frac{Y^2}{Q(XY, Y)}\frac{\partial Q}{\partial Y}(XY, Y)\right).$$

The term $\mathcal{D}\left(\frac{Y^2}{Q(XY,Y)}\frac{\partial Q}{\partial Y}(XY, Y)\right)$ is the diagonal of a series in XY and Y multiplied by Y^2 and so is zero. On the other hand

$$\mathcal{D}\left(\frac{Y^2}{Y - H(XY)}\right) = \mathcal{D}\left(Y\left(1 - Y^{-1}H(XY)\right)^{-1}\right)$$

$$= \mathcal{D}\left(\sum_{i \geq 0} Y^{1-i} H(XY)^i\right) = H(X),$$

and

$$\frac{Y^2}{P(XY, Y)} \frac{\partial P}{\partial Y}(XY, Y)$$

is a rational function, which concludes the proof. ∎

12.8 Exercises

1. What is the power series solution to the functional equation $f(X)f(X^2) = 1/(1 - X)$?

2. Consider the sequence $(a_n)_{n \geq 0}$ that is the fixed point of the morphism $0 \to 01$, $1 \to 00$. Define $A(X) = \sum_{n \geq 0} a_n X^n$. Find an equation over $GF(2)$ that the power series $A(X)$ satisfies.

3. Consider the sequence $\mathbf{a}_{11} = (a_n)_{n \geq 0}$ introduced in Exercise 8.2. Find the equation over $GF(2)$ satisfied by $\sum_{n \geq 0} a_n X^n$.

4. Let the Taylor series expansion of

$$\frac{\sqrt{\frac{5x+1}{x+1}} - 1}{2(x + 1)}$$

at $x = 0$ be $\sum_{n \geq 0} a_n x^n$. Show that $a_n \equiv t_n \pmod 2$, where $(t_n)_{n \geq 0}$ is the Thue–Morse sequence.

5. Show that if R is a commutative ring with unit element, then $R[[X]]$, the ring of formal power series over R, is a commutative ring with unit element. (The only nontrivial part is the associativity of multiplication.)

6. Give an example of two rings, R and S with $R \subseteq S$ such that the cardinality of the algebraic elements of S over R is uncountable.

7. Suppose $f(X)$ and $g(X)$ are two algebraic formal power series in $K[[X]]$, and suppose $g(0) = 0$. Show that $f(g(X))$ is algebraic.

8. Let α be a real number, and let g be a polynomial of degree ≥ 1. Then $\sum_{n \geq 0} g(\lfloor \alpha n \rfloor)X^n$ is a rational function if and only if α is rational.

9. Observe that $\prod_{n \geq 0}(1 - X^{2^n}) = \sum_{k \geq 0}(-1)^{s_2(k)} X^k$. Let $f(X) = \sum_{k \geq 0}(-1)^{s_2(k)} X^k$. Show that

$$\frac{X}{1 - X} - \frac{Xf'(X)}{f(X)} = 2\sum_{k \geq 1} 2^{v_2(k)} X^k.$$

10. Show that $\prod_{k \geq 0}(1 + X^{2^k})^{-(k+1)} = \sum_{i \geq 0}(-1)^{s_2(i)} X^i$. Conclude that

$$\int_0^1 \frac{dX}{\prod_{k \geq 0}(1 + X^{2^k})^{k+1}} = \sum_{n \geq 1} \frac{(-1)^{s_2(n-1)}}{n}.$$

11. Prove Lucas's formula: if $m = pa + i$ and $n = pb + j$, where p is a prime number, and $0 \leq i, j \leq p - 1$, then

$$\binom{m}{n} \equiv \binom{a}{b}\binom{i}{j} \pmod{p}.$$

Deduce that, if the base-p expansions of m and n are $m = \sum m_i p^i$ and $n = \sum n_i p^i$ with the m_i's and the n_i's in $[0, p - 1]$ and ultimately equal to zero, then

$$\binom{m}{n} \equiv \prod_i \binom{m_i}{n_i} \pmod{p}.$$

12. Prove that $\binom{2k}{k} \equiv 0 \pmod{2}$ for every $k \geq 1$. Exercise 2.28 may prove useful.

13. (Allouche and Laubie) Let $(a_n)_{n \geq 0}$ be a sequence with values in $GF(q)$. Prove that the formal power series $\sum_{n \geq 0} a_n X^n$ is a rational function (i.e., belongs to $GF(q)(X)$) if and only if the formal power series $\sum_{n \geq 0} a_n X^{q^n}$ is algebraic over $GF(q)[X]$.

14. (C. Moore) Consider the generating function of Pascal's triangle, taken modulo 2, with variables X and Y indicating the position of each term:

$$f(X, Y) = 1$$
$$+ Y + XY$$
$$+ Y^2 + X^2 Y^2$$
$$+ Y^3 + XY^3 + X^2 Y^3 + X^3 Y^3$$
$$+ Y^4 + X^4 Y^4 + \cdots$$

Show that $f(X, Y)$ is algebraic over $GF(2)[X, Y]$, but not algebraic over $\mathbb{Z}[X, Y]$.

15. Prove Euler's formula relating π^{2n}, $\zeta(2n)$, and the Bernoulli numbers. That is, show that if n is a positive number, then

$$2(2n)! \zeta(2n) = (-1)^{n+1}(2\pi)^{2n} B_{2n},$$

where B_t is the tth Bernoulli number.

16. Prove that $e^X = \sum_{k \geq 0} X^k / k!$ is transcendental over $\mathbb{Q}[X]$.

17. Let $f(X) = \prod_{k\geq 1} f_k(X)$. Compute $f'(X)/f(X)$, the formal logarithmic derivative.

18. Show that if $f(X) \in GF(p^n)[[X]]$ is a formal power series that is algebraic over $GF(p^n)(X)$, then so is $f'(X)$, its formal derivative. Give a proof that works for the finite field $GF(p^n)$, and give another proof that works when $GF(p^n)$ is replaced by any commutative field K.

19. (Thakur) Let p be a prime number, and define $f = \sum_{n\geq 0} X^{p^n}$. Show that f is the diagonal of the rational expression

$$\frac{X}{1 - (X^{p-1} + Y)}.$$

20. Show that the power series $f(z) = \sum_{n\geq 0} z^{2^n}$ has the unit circle as a natural boundary. Show that f has precisely two real zeros, and give good approximations to them. Find some complex zeros of f. Does f have zeros in every neighborhood of the unit circle?

21. Give a counterexample to the converse of Theorem 12.6.1. More precisely, give an example of a formal power series in $\mathbb{Q}[[X]]$ transcendental over $\mathbb{Q}[X]$, such that for all primes p, its reduction modulo p is algebraic over $GF(p)[X]$.

22. For each integer $n \geq 0$ give an example of a formal power series $\sum_{i\geq 0} a_i X^i$ such that the coefficients of $(\sum_{i\geq 0} a_i X^i)^k$ are bounded above by a constant for $0 \leq k < n$, but unbounded for $k = n$.

23. As in Section 12.6, let

$$\theta_3(X) = \sum_{-\infty < n < \infty} X^{n^2}.$$

Show that

$$\theta_3(x)^2 = 1 + 4\sum_{n\geq 0}(-1)^n \frac{X^{2n+1}}{1 - X^{2n+1}}.$$

24. Prove that the Hadamard product of two algebraic series over $GF(p)$ need not be of bounded degree. More precisely, prove that for all integers $a \geq 1$, there exists a prime p and a formal power series modulo p that is the Hadamard product of two quadratic series and whose degree is $> a$.

25. Define an analogue of the Baum–Sweet sequence as follows: $a_n = 1$ if there is a nonempty block of 0's of even length in $(n)_2$, and 0 otherwise. Thus, for example, $a_{12} = 1$ and $a_{49} = 0$. It is easy to see that $(a_n)_{n\geq 0}$ is a 2-automatic sequence. Define $A(X) = \sum_{n\geq 0} a_n X^n$, so

$$A(X) = X^4 + X^9 + X^{12} + \cdots.$$

Find an algebraic equation satisfied by A over $GF(2)[X]$.

26. Show that for all $n \geq 1$ there exist 2-automatic sequences $\mathbf{u} = (u_i)_{i \geq 0}$ such that Card $K_2(\mathbf{u}) \geq n$, while the degree of the algebraic equation of which $\sum_{i \geq 0} u_i X^i$ is a root is 2.

27. (Bowman) Let $t(X) = \sum_{n \geq 0} (-1)^{s_2(n)} X^n$ be the Thue–Morse power series on $\{-1, +1\}$. Show that $t(X)$ has continued fraction expansion

$$\frac{1}{1} + \frac{X}{1} - \frac{X + X^2}{1 + X + X^2} - \frac{1 + X^2}{2 + X^2} - \frac{X^2 + X^4}{1 + X^2 + X^4} - \frac{1 + X^4}{2 + X^4} - \frac{1 + X^4}{2 + X^4} -$$

$$\frac{X^4 + X^8}{1 + X^4 + X^8} - \frac{1 + X^8}{2 + X^8} - \frac{1 + X^8}{2 + X^8} - \frac{1 + X^8}{2 + X^8} - \cdots$$

Hint: Apply Exercise 2.13.

28. Let $T(X) = \sum_{n \geq 0} t_n X^{-n}$ be the Thue–Morse power series over $GF(2)[[1/X]]$.
    (a) Show that the continued fraction expansion of T is

$$T(X) = [0, \ X + 1, \ \overline{X, \ X, \ X^3 + X, \ X}]$$

    where the vinculum denotes the periodic portion.
    (b) Similarly, let $U(X) := \sum_{n \geq 0} (1 + t_n) X^{-n}$. Show that

$$U(X) = [1, \ X^3 + X + 1, \ \overline{X, \ X, \ X, \ X^3 + X}].$$

29. Let $(s_n)_{n \geq 0}$ denote the Rudin–Shapiro sequence on $\{0, 1\}$, i.e., $s_n = e_{2;11}(n)$ mod 2. Let $S(X) = \sum_{n \geq 0} s_n X^{-n}$ be the Rudin–Shapiro power series over $GF(2)[[1/X]]$. Compute the continued fraction expansion of $S(X)$.

30. Let $C(X) = c_i X^{-i}$ be the Cantor formal power series in $GF(3)[[1/X]]$, where $c_i = 1$ if the base-3 expansion of i contains only 0's and 2's, and $c_i = 0$ otherwise. Compute the continued fraction expansion of $C(X)$.

31. Let p be a prime. In this exercise we give an algebraic construction of finite words of length $p^r + r - 1$ over the alphabet $GF(p)$ that contain every subword of length r. Let f be an irreducible polynomial of degree r over $GF(p)$.
    (a) Show that f divides $X^{p^r} - X$.
    (b) Prove that for every prime p and $r \geq 1$ there exists a *primitive polynomial* g, i.e., an irreducible polynomial for which a root of g generates $GF(p^r)$.
    (c) Suppose g is primitive, of degree r. Show that the coefficients of the formal power series for $1/g$ are purely periodic with period $p^r - 1$, and contain every subword of length r except 0^r.
    (d) Show how to modify the word consisting of the first $p^r + r - 2$ coefficients of the formal power series for $1/g$ to get the desired word.

32. Let $a_0 = 0$, $a_1 = 1$, and for $n \geq 0$ define $a_{4n} = a_n$, $a_{2n+1} = a_{n+1}$, and $a_{4n+2} = 0$.

(a) Show that $(a_i)_{i \geq 0}$ is 2-automatic and is the image under $n \to n \pmod 2$ of the fixed point of φ, where φ is defined as follows:

$$0 \to 01,$$
$$1 \to 20,$$
$$2 \to 34,$$
$$3 \to 24,$$
$$4 \to 44.$$

(b) Show that $\sum_{i \leq 2^n} a_i = F_{n+1}$, the $(n+1)$st Fibonacci number.

(c) Let $f(X) = \sum_{i \geq 0} a_i X^{-i} \in GF(2)[[X^{-1}]]$. Show that $f^3 + Xf + 1 = 0$.

(d) Prove that the continued fraction expansion of f is $[0, X, X^2, X^4, X^8, \ldots]$.

(e) Show that if $g^{2^n+1} + Xg + 1 = 0$, then the continued fraction expansion of g is $[0, X, X^{2^n}, X^{2^{2n}}, X^{2^{3n}}, \ldots]$.

33. Use the folding lemma (Lemma 6.5.5) to describe the continued fraction expansion for the formal power series $\sum_{i \geq 0} X^{-a_i}$ over $GF(2)[[X^{-1}]]$, provided that $a_{i+1} \geq 2a_i$. In particular, deduce that

$$[0, \ X, \ X^2, \ X, \ X^3, \ X, \ X^2, \ X, \ X^4, \ldots, \ X^{1+\nu_2(i)}, \ldots] = \sum_{n \geq 2} X^{n+1-2^n}$$

and

$$[0, \ X, \ X^2, \ X, \ X^4, \ X, \ X^2, \ X, \ X^8, \ldots, \ X^{2^{\nu_2(i)}}, \ldots] = \sum_{n \geq 1} X^{-(n+1)2^n}.$$

34. Let $C(n) = \binom{2n}{n}/(n+1)$ be the nth Catalan number. Let p be a prime ≥ 2. Is the sequence $(C(n) \bmod p)_{n \geq 0}$ p-automatic?

35. Show that $(1-X)^{p+1}Y^p - (1-X)^2 Y + X$ is irreducible for all primes p.

36. Suppose $\sum_{n \geq 0} f_n X^n \in GF(q)[[X]]$ is algebraic over $GF(q)(X)$. Show that for all $a \geq 1$, $b \geq 0$ the formal power series $\sum_{n \geq 0} f_{an+b} X^n$ is algebraic over $GF(q)(X)$.

12.9 Open Problems

1. Prove or disprove: the continued fraction $[0, \ X, \ X^2, \ X^3, \ X^4, \ldots]$ is transcendental over $GF(q)(X)$ for all prime powers q.

2. Prove or disprove: the continued fraction $[0, \ X, \ X^3, \ X^7, \ X^{15}, \ X^{31}, \ X^{63}, \ldots]$ is transcendental over $GF(2)(X)$.

3. (Thakur) Let p be a prime number, and consider

$$F(X) = \exp(X + X^p/p + X^{p^2}/p^2 + X^{p^3}/p^3 + \cdots)$$
$$= \prod_{p \nmid n} (1 - X^n)^{-\mu(n)/n},$$

where μ is the Möbius function. It can be shown that, writing $F(X) = \sum a_k X^k$, no a_k has a denominator congruent to zero modulo p. Hence it makes sense to consider $F(X)$ reduced modulo p. Is it algebraic over $GF(p)(X)$?

12.10 Notes on Chapter 12

12.1 In the literature, the series $\sum_{n \geq 0} X^{2^n}$ in Example 12.1.1 is sometimes called the Fredholm series, but Fredholm apparently never studied it.

Many other examples of algebraic formal power series can be found in the paper of Christol, Kamae, Mendès France and Rauzy [1980]. Also see Allouche [1987].

12.2 The operators introduced in Definition 12.2.1 are called *Cartier operations* or *Cartier operators*; see Christol [1970, 1972].

Theorem 12.2.5 was first given by Christol [1979] for sequences taking only the values 0 and 1, then completed by Christol, Kamae, Mendès France, and Rauzy [1980].

Algebraic formal power series whose coefficients are p-adic integers and their diagonals were studied by Christol [1986] and by Denef and Lipshitz [1987].

Christol's theorem (Theorem 12.2.5) in the case of a finite field was generalized to the multidimensional case by Salon [1987, 1989b]. The generalization to an infinite field of positive characteristic for formal power series with a finite number of variables was done independently by Sharif and Woodcock [1988] and Harase [1988]. Both generalizations are discussed in Chapter 14.

The Hadamard product was introduced by Hadamard [1899]. Theorem 12.2.6 was proved by Furstenberg [1967]. Fliess [1969, 1974a] noted that Furstenberg's proof works for perfect fields of positive characteristic, i.e., fields of characteristic p that contain pth roots of all of their elements.

Hadamard products and diagonals of power series are related; see the notes on Section 12.7 below, in particular those concerning papers of Deligne [1984], Denef and Lipshitz [1987], Harase [1988], and Sharif and Woodcock [1988].

Harase [1989] found bounds for the size of minimal polynomial of the Hadamard product of two algebraic functions in positive characteristic.

Borel showed that the Hadamard product of two rational functions in one variable over a field of characteristic 0 is again a rational function; see Jungen [1931, Thm. 7]. (This result is actually true over any field.) Woodcock and Sharif [1990] considered the Hadamard product of two rational functions in several variables.

For surveys of these results and others, see Allouche [1989] and Sharif [1993].

Beals and Thakur [1998] generalized automatic formal power series in finite characteristic to the case where the set of base-q expansions of the coefficients lies in some language classes other than the class of regular languages.

12.3 It is not known whether the real-number analogue of Theorem 12.3.1 holds. See Open Problem 13.11 for more information.

12.4 Allouche [1982] used Christol's theorem to prove the transcendence of the formal power series $\sum_{n\geq 0} s_p(P(n))X^n$, where $s_p(n)$ is the value modulo p of the sum of digits of n in base p and P is any polynomial of degree ≥ 2 such that $P(\mathbb{N}) \subseteq \mathbb{N}$; see Section 6.10.

   Allouche [1990] used Christol's theorem to give a new proof of the transcendence of the formal power series π_q (first proved by Wade [1941]). Other results for the Carlitz functions are described below. We mention a result due to Becker [1993]: let $g, h \geq 2$ be natural integers, and let $(a_n(g, h))_{n\geq 1}$ be the sequence obtained by concatenating the base-h expansions of the numbers $1, g, g^2, g^3, \ldots$ in order. Then the formal power series $\sum_{n\geq 1} a_n(g, h)X^n$ is transcendental over $GF(p)(X)$ for $p > h$. This result is a natural analogue of a result of Mahler [1937a, 1937b, 1976b] in characteristic zero.

12.5 Berthé [1992, 1993] proved (via automata) that $\zeta(n)/\pi_q^n$ is transcendental for $1 \leq n \leq q - 2$; the result was previously proved by J. Yu [1991] for every n such that $(q - 1){\not|}n$, but Berthé's proof is elementary. Recher [1992] obtained transcendence results, via automata, for periods of generalized Carlitz exponentials, i.e., of generalizations of π_q. Berthé [1994], still using finite automata, also proved transcendence results for the Carlitz logarithm. Berthé [1995] also gave results on linear expressions in $\zeta(n)/\pi_q^n$ for $1 \leq n \leq q - 2$.

   To prove the transcendence of values of Carlitz functions four methods were used. The first goes back to Wade [1941, 1943, 1944, 1946] and mimics one of the methods for proving transcendence of real numbers over $\mathbb{Q}$. Also see Spencer [1952] and Damamme and Hellegouarch [1988, 1991]. A second method uses Drinfeld modules and can be compared to studying periods of elliptic functions. For a survey of these topics, see J. Yu [1992]. A third method uses Diophantine approximation; see, for example, the paper of Chérif and de Mathan [1993]. The last method, explained in this chapter, uses automata theory and Christol's theorem. These methods are competing, with the most powerful seeming to be the Drinfeld module method.

   The following results, however, are not known to be obtainable using the Drinfeld module approach: Allouche [1996] proved, via automata, the transcendence of the values of the Carlitz–Goss gamma function for all p-adic rational arguments that are not natural numbers; also see Thakur [1996c]. Mendès France and Yao [1997] extended the result to *all* the values of the Carlitz–Goss gamma function at p-adic arguments that are not natural numbers. Wen and Yao [2002] proved that a value of the T-adic Carlitz–Goss gamma function is transcendental over $GF(q)(T)$ if and only if the q-adic coefficients of the argument are not ultimately constant.

   Recent work of Koskas shows that the Diophantine approximation method and the automaton method can be unified; in particular a paper of Fresnel, Koskas, and de Mathan [2000] gives a quantitative version of Christol's theorem. Finally, we cite a proof via automata that the period of the Tate elliptic curve is transcendental, due to Thakur [1996b]. Also see Thakur [1998]; Allouche and Thakur [1999].

12.6 Theorem 12.6.3 was first proved by Petersen [1994, 1996]. Our proof is due to Allouche [1997a].

For the use of the Chomsky–Schützenberger theorem to prove that certain context-free languages are inherently ambiguous, see Flajolet [1985, 1987]. For related results see Allouche [1999].

Cucker and Gabarró [1989] proved that if $\sum_{n\geq0} f(n)X^n$ is algebraic over $\mathbb{Q}(X)$, then f is primitive recursive.

For other transcendence results obtained by reducing formal power series modulo primes see Woodcock and Sharif [1989] and Allouche, Gouyou-Beauchamps, and Skordev [1998].

12.7 Furstenberg [1967] was basically interested in the algebraicity of the Hadamard product of two formal power series, which is defined by (see the definition preceding Theorem 12.2.6)

$$\left(\sum_{n\geq0} a_n X^n\right) \odot \left(\sum_{n\geq0} b_n X^n\right) = \sum_{n\geq0} a_n b_n X^n.$$

Of course, this question is related to diagonals of formal power series. We have the following relations:

$$\mathcal{D}\left(\left(\sum_{n\geq0} a_n X^n\right)\left(\sum_{m\geq0} b_m Y^m\right)\right) = \sum_{n\geq0} a_n b_n X^n$$

$$= \left(\sum_{n\geq0} a_n X^n\right) \odot \left(\sum_{n\geq0} b_n X^n\right)$$

and

$$\left(\frac{1}{1-XY}\right) \odot \left(\left(\sum_{n\geq0} a_n X^n\right)\left(\sum_{m\geq0} b_m Y^m\right)\right) = \sum_{n\geq0} a_n b_n (XY)^n.$$

Deligne [1984] proved that any diagonal of an algebraic Laurent series in several variables is itself algebraic provided the ground field has positive characteristic. Christol [1986] studied algebraic series and diagonals of rational series. Denef and Lipshitz [1987] generalized Furstenberg's theorem. In the same paper they gave an elementary proof of Deligne's result, and they proved the corresponding result in the p-adic case. Harase [1988] and Sharif and Woodcock [1988] also gave an elementary proof of Deligne's result.

Other papers dealing with diagonals of formal power series include Hautus and Klarner [1971], Gessel [1981], and Fagnot [1996].

13

Automatic Real Numbers

In his famous 1936 paper, Alan Turing discussed the computability of real numbers. Roughly speaking, a real number α is computable if there is a Turing machine that, on input i, will compute a rational approximation to α good to within 2^{-i}.

In this chapter, we discuss an analogous concept: the *automatic real numbers*. More precisely, we consider associating a single real number x with a DFAO as follows: on input n represented in base k, the automaton outputs the nth digit of the base-b expansion of the fractional part of x. We then investigate the properties of such numbers. It turns out that the set of such numbers forms a vector space over the rational numbers. It is also conjectured, but not yet proved, that no irrational algebraic real number can be an automatic real.

13.1 Basic Properties of the Automatic Reals

In this section we prove some basic properties of the automatic real numbers. In particular, we show the automatic reals form a vector space over $\mathbb{Q}$.

Let k, b be integers ≥ 2. Let r be a real number, and suppose

$$r = a_0 + \sum_{i \geq 1} a_i b^{-i}$$

for $a_i \in \mathbb{Z}$ for $i \geq 0$ and $0 \leq a_i < b$ for $i \geq 1$. We say r is (k, b)-*automatic* if the sequence of digits $(a_i)_{i \geq 0}$ is a k-automatic sequence. We let $L(k, b)$ denote the set of all (k, b)-automatic reals.

Our first result shows that $\mathbb{Q} \subseteq L(k, b)$ for all $k, b \geq 2$.

Theorem 13.1.1 *If r is rational, then $r \in L(k, b)$ for all $k, b \geq 2$.*

Proof. Let r be a rational number. Then by Theorem 3.4.2 the expansion of r in base b is ultimately periodic. Hence by Theorem 5.4.2, the sequence $(a_i)_{i \geq 0}$ is k-automatic. ∎

The following theorem is a sort of converse to the preceding one.

Theorem 13.1.2 *Let $j, k \geq 2$ be multiplicatively independent integers. Then $L(j, b) \cap L(k, b) = \mathbb{Q}$.*

Proof. For one direction, we use Theorem 13.1.1. The other direction follows immediately from Cobham's theorem, Theorem 11.2.2. ■

Theorem 13.1.3 *Let k, b be integers ≥ 2. If $x \in L(k, b)$, then $-x \in L(k, b)$.*

Proof. The result is clearly true for $x \in \mathbb{Z}$. For $x \notin \mathbb{Z}$, write $x = a_0 + \sum_{i \geq 1} a_i b^{-i}$ with $0 \leq a_i < b$ for $i \geq 1$. Then by hypothesis $(a_i)_{i \geq 0}$ is a k-automatic sequence. Now consider the coding $h : \{0, 1, 2, \ldots, b - 1\} \to \{0, 1, 2, \ldots, b - 1\}$ defined as follows: $h(i) = b - 1 - i$. Let $c_i = h(a_{i+1})$ for $i \geq 0$, and define $y = \sum_{i \geq 0} c_i b^{-(i+1)}$. A simple calculation now gives $y = a_0 + 1 - x$.

By Theorem 6.8.1, the shifted sequence $(a_{i+1})_{i \geq 0}$ is k-automatic. By Theorem 5.4.3, $(c_i)_{i \geq 0}$ is k-automatic. Now define

$$d_i = \begin{cases} -(a_0 + 1) & \text{if } i = 0, \\ c_{i-1} & \text{if } i \geq 1. \end{cases}$$

Then $(d_i)_{i \geq 0}$ is k-automatic by Theorem 6.8.4, and $\sum_{i \geq 0} d_i b^{-i} = -x$. ■

We now prove a useful normalization lemma.

Lemma 13.1.4 *Let C be a positive integer, and let $(a_i)_{i \geq 0}$ be a k-automatic sequence of integers with $0 \leq a_i \leq C$ for all $i \geq 1$. Then $y := \sum_{i \geq 0} a_i b^{-i}$ is a (k, b)-automatic real number.*

Proof. Without loss of generality we may assume $a_0 = 0$.

The result is trivial if $C < b$, for then the digits in the base-b expansion of y are precisely a_i. The only difficulty occurs when the carries are taken into account, since carries may come from arbitrarily far to the right.

The idea of the proof is as follows: first, in a bounded number of steps, we rewrite $y = \sum_{i \geq 1} a_i' b^{-i}$ in such a way that $0 \leq a_i' \leq b$. Next, we show how to perform the potential carries resulting from the digits equal to b.

For the first step, define $g_i = a_i \bmod b$ and $h_i = \lfloor a_{i+1}/b \rfloor$ for $i \geq 0$. Then clearly $y = \sum_{i \geq 0} a_i' b^{-i}$, where $a_i' = g_i + h_i$. Now $(g_i)_{i \geq 0}$ is easily seen to be k-automatic, and the fact that $(h_i)_{i \geq 0}$ is k-automatic follows from Theorems 5.4.3 and 6.8.1. Hence $(a_i')_{i \geq 0}$ is k-automatic.

Now if $a_i \leq C$ for all $i \geq 1$, then $a_i' \leq b - 1 + \lfloor C/b \rfloor$ for all $i \geq 1$. If $C \geq b + 2$, then $C \geq b^2/(b - 1)$. Hence $b^2 \leq bC - C$, and so, adding $-b + C$ to both sides, we get $b^2 - b + C \leq bC - b$. Now, dividing by b, we get $a_i' \leq b - 1 + C/b \leq$

$C - 1$. On the other hand, if $C = b + 1$, then $a_i' \leq b$. Finally, if $C \leq b$, then $a_i' \leq b$. Hence applying this transformation repeatedly, we eventually reach a k-automatic sequence, say $(e_i)_{i \geq 0}$, whose terms are all $\leq b$, and $y = \sum_{i \geq 0} e_i b^{-i}$.

The second step of the construction involves determining the carry bits that arise from the terms of e_i that equal b. Define the *carry sequence* $(c_i)_{i \geq 0}$ as follows:

$$c_i = \begin{cases} 1 & \text{if there exists } j > i \text{ with } e_t = b - 1 \text{ for } i < t < j, \text{ and } e_j = b, \\ 0 & \text{otherwise.} \end{cases}$$

Then it is easy to see that if $f_i = ((e_i + c_i) \bmod b)_{i \geq 0}$, then $y = \sum_{i \geq 0} f_i b^{-i}$, and $0 \leq f_i < b$. Thus it suffices to create a DFAO M that generates $(c_i)_{i \geq 0}$.

Our construction of $M = M_5$ goes in several stages. Let $M_0 = (Q, \Sigma, \delta, q_0, \Delta, \tau)$ be a DFAO generating $(e_i)_{i \geq 0}$; here $\Sigma = \{0, 1, \ldots, k - 1\}$. First, we create a nondeterministic finite automaton (NFA) $M_1 = (Q', \Sigma \times \Sigma, \delta', q_0', F)$ that, roughly speaking, has two non-negative integer inputs, i and j, and accepts if there exists n, $i < n < j$, such that $e_n \neq b - 1$. The inputs i and j are, of course, provided in base k, starting with the most significant digit, with the shorter padded by 0's in the front if necessary to make the lengths of the expansions the same. The NFA M_1 functions by nondeterministically guessing the base-k digits of n, and maintaining the relationship of the current guessed n with i and j.

The states of M_1 are triples of the form $[q, u, v]$, where $q \in Q$, and $u, v \in \{<, =\}$. The meaning of the state $[q, u, v]$ is that the guessed expansion of n seen so far would take us to state q in M_0, and furthermore the relationship of n to the currently seen inputs i and j is given by $i \, u \, n \, v \, j$ (e.g., $i < n = j$). The start state of M_1 is $q_0' = [q_0, =, =]$. The transition function δ' is given by

$$\delta'([q, u, v], [c, d])$$
$$= \begin{cases} [\delta(q, c), =, =] & \text{if } (u, v) = (=, =) \text{ and } c = d, \\ [\delta(q, c), =, <] \cup [\delta(q, d), <, =] & \text{if } (u, v) = (=, =) \text{ and } c < d, \\ \quad \cup \bigcup_{c < z < d} [\delta(q, z), <, <] & \\ [\delta(q, d), <, =] \cup \bigcup_{0 \leq z < d} [\delta(q, z), <, <] & \text{if } (u, v) = (<, =), \\ [\delta(q, c), =, <] \cup \bigcup_{c < z < k} [\delta(q, z), <, <] & \text{if } (u, v) = (=, <), \\ \bigcup_{0 \leq z < k} [\delta(q, z), <, <] & \text{if } (u, v) = (<, <). \end{cases}$$

Here c should be thought of as the next base-k digit of i; d, as the next digit of j; and z, as the "guessed" next digit of n. The set of final states is given by

$$F = \{[q, <, <] : \tau(q) \neq b - 1\}.$$

We leave it to the reader to verify that M_1 really behaves as we have claimed.

Now, using the standard construction, we convert M_1 to a deterministic finite automaton (DFA) M_2 accepting the same set. Then, by interchanging accepting and nonaccepting states of M_2, we get a DFA M_3 that accepts the base-k representations of pairs (i, j) such that for all n, with $i < n < j$, we have $e_n = b - 1$.

Next, we create a new NFA M_4 that, on input i, "guesses" the base-k digits of j and simulates M_3 on input (i, j). Our NFA M_4 also simulates M_0 on input j, and accepts if and only if M_3 accepts (i, j) and M_0 outputs b on input j. Now M_4 can be easily converted to a DFAO M_5 that (essentially) generates the carry sequence $(c_i)_{i \geq 0}$. We say "essentially" because the base-k representation of j may have substantially more digits than that of i; hence only those base-k representations of i that have sufficiently many leading zeros will result in the correct output. However, this problem may be easily dealt with; see Exercise 22. This completes the proof of the lemma.   ∎

We can now prove that $L(k, b)$ is closed under addition.

Theorem 13.1.5 (Lehr) *If $r, s \in L(k, b)$, then so is $r + s$.*

Proof. We can add the base-b expansions of r and s digit by digit, using Theorem 5.4.4 and the function $f(a, c) = a + c$. This gives us an "unnormalized" base-b expansion $\sum_{n \geq 0} u_n b^{-n}$ with $u_i \in \{0, 1, \ldots, 2b - 2\}$ for $i \geq 1$. The result then follows from Lemma 13.1.4.   ∎

Theorem 13.1.6 *Let $x \in L(k, b)$, i.e., the set of numbers whose base-b expansions are k-automatic. If c is a nonzero integer, then $x/c \in L(k, b)$.*

Proof. We construct a 1-uniform transducer that transforms the sequence x_1, $x_2, \ldots, x_i, \ldots$ into the sequence $y_1, y_2, \ldots, y_j, \ldots$, where $x = .x_1 x_2 \cdots$, $y = x/c = y_1 y_2 \cdots$ in base b. It then follows from Theorem 6.9.2 that $y = x/c$ is in $L(k, b)$.

Define the transducer $T = (Q, \Sigma, \delta, q_0, \Delta, p)$ by $Q = \{0, 1, 2, \ldots, c - 1\}$, $\Sigma = \{0, 1, 2, \ldots, b - 1\}$, $\delta(d, a) = (bd + a) \bmod c$ for $d \in Q, a \in \Sigma$, $q_0 = 0$, $\Delta = \{0, 1, \ldots, b - 1\}$, $p(d, a) = \left\lfloor \frac{bd + a}{c} \right\rfloor$. Then this transducer essentially divides its input by c using the ordinary pencil-and-paper method of long division.   ∎

Corollary 13.1.7 *The set $L(k, b)$ forms a vector space over the rational numbers.*

13.2 Non-closure Properties of $L(k, b)$

It is natural to wonder, after seeing the results of the previous section, whether $L(k, b)$ forms a multiplicative group. It does not, as the following result shows.

Theorem 13.2.1 *$L(k, b)$ is not closed under multiplication.*

Proof. Let $k \geq 2$ be an integer, and define

$$f(X) = \sum_{r \geq 0} X^{k^r} = X + X^k + X^{k^2} + \cdots$$

and

$$g(X) = \sum_{m \geq 1, \, n \geq 0} X^{(k^m - 1)k^n}.$$

Let $b \geq 2$ be an integer, and define $y = f(1/b)$, $z = g(1/b)$. Then $y \in L(k, b)$, since the base-b representation of y has 1's in those positions whose base-k representation is given by the regular language 10^*. Similarly, $z \in L(k, b)$, since the base-b representation of z has 1's in those positions whose base-k representation is given by the regular language $(k - 1)^+ 0^*$. We will show that $yz \notin L(k, b)$.

First, note that

$$f(X)g(X) = \sum_{\substack{m \geq 1, \, n \geq 0, \, r \geq 0}} X^{k^r} \cdot X^{(k^m - 1)k^n} = \sum_{\substack{m \geq 1, \, n \geq 0, \, r \geq 0}} X^{k^r + (k^m - 1)k^n}$$

$$= \sum_{\substack{r < n \\ m \geq 1, \, n \geq 0, \, r \geq 0}} X^{k^r + (k^m - 1)k^n} + \sum_{\substack{r = n \\ m \geq 1, \, n \geq 0, \, r \geq 0}} X^{k^r + (k^m - 1)k^n}$$

$$+ \sum_{\substack{r > n \\ m \geq 1, \, n \geq 0, \, r \geq 0}} X^{k^r + (k^m - 1)k^n}$$

$$= S(X) + T(X) + U(X).$$

Second, note that

$$S(X) = \sum_{\substack{r < n \\ m \geq 1, \, n \geq 0, \, r \geq 0}} X^{k^r + (k^m - 1)k^n} = \sum_{\substack{r < n \\ m \geq 1, \, n \geq 0, \, r \geq 0}} X^{k^r (1 + k^{n-r}(k^m - 1))}$$

$$= \sum_{\substack{m \geq 1, \, p \geq 1, \, r \geq 0}} X^{k^r (1 + k^p (k^m - 1))} = \sum_{\substack{m \geq 1, \, p \geq 1, \, r \geq 0}} X^{k^r (k^{p+m} - k^p + 1)}.$$

Now $(k^r(k^{p+m} - k^p + 1))_k = (k - 1)^m \, 0^{p-1} \, 1 \, 0^r$, so it follows that $S(X) = \sum_{i \geq 0} s_i X^i$, where

$$s_i = \begin{cases} 1 & \text{if } (i)_k \in (k - 1)^+ \, 0^* \, 1 \, 0^*, \\ 0 & \text{otherwise.} \end{cases}$$

Hence $(s_i)_{i \geq 0}$ is a k-automatic sequence, and therefore $S(1/b) \in L(k, b)$.

Third, note that

$$U(X) = \sum_{\substack{r > n \\ m \geq 1, \, n \geq 0, \, r \geq 0}} X^{k^r + (k^m - 1)k^n} = \sum_{\substack{r > n \\ m \geq 1, \, n \geq 0, \, r \geq 0}} X^{k^n (k^{r-n} + k^m - 1)}$$

$$= \sum_{\substack{m \geq 1, \, n \geq 0, \, q \geq 1}} X^{k^n (k^q + k^m - 1)}.$$

Now

$$(k^n(k^q + k^m - 1))_k = \begin{cases} 1 \ 0^{q-m} \ (k-1)^m \ 0^n & \text{if } m < q, \\ 1 \ (k-1)^m \ 0^n & \text{if } m = q, \\ 1 \ 0^{m-q} \ (k-1)^q \ 0^n & \text{if } m > q. \end{cases}$$

Hence $U(x) = \sum_{i \geq 0} u_i X^i$, where

$$u_i = \begin{cases} 2 & \text{if } (i)_k \in 1 \ 0^+ \ (k-1)^+ \ 0^*, \\ 1 & \text{if } (i)_k \in 1 \ (k-1)^+ \ 0^*, \\ 0 & \text{otherwise.} \end{cases}$$

It follows that $(u_i)_{i \geq 0}$ is a k-automatic sequence, and by the normalization lemma we have $U(1/b) \in L(k, b)$.

Finally, note that

$$T(X) = \sum_{\substack{r=n \\ m \geq 1, \ n \geq 0, \ r \geq 0}} X^{k^r + (k^m - 1)k^n}$$

$$= \sum_{r \geq 0, \ m \geq 1} X^{k^{m+r}} = \sum_{n \geq 1} n X^{k^n}.$$

Now consider the base-b expansion of $T(1/b)$, say $(T(1/b))_b = 0.c_0 c_1 c_2 \cdots$. Evidently the base-$b$ digits immediately to the left of position k^n are just $(n)_b$. It follows that every element of $\{0, 1, \ldots, b-1\}^*$ eventually appears as a subword of the infinite sequence

$$\mathbf{c} = c_0 c_1 c_2 \cdots.$$

Recall that the subword complexity $p_\mathbf{d}(n)$ of an infinite sequence $\mathbf{d} = (d_i)_{i \geq 0}$ is the number of distinct subwords of length n that appear in $\mathbf{d}$. Then we have shown that $p_\mathbf{c}(n) = b^n$. But, by Theorem 10.3.1, if $\mathbf{c}$ were k-automatic, we would have $p_\mathbf{c}(n) = O(n)$. This gives a contradiction, and so $T(1/b) \notin L(k, b)$.

It follows that $yz \notin L(k, b)$, since $yz = S(1/b) + T(1/b) + U(1/b)$. ∎

Corollary 13.2.2 *The set $L(k, b)$ is not closed under the map $x \to x^2$.*

Proof. Suppose it were. Then, since for all $y, z \in L(k, b)$ we have

$$yz = \frac{1}{4}((y + z)^2 - (y - z)^2),$$

it would follow that $L(k, b)$ is closed under multiplication, a contradiction. ∎

Actually, using the same method as used to prove Theorem 13.2.1, we can prove the following:

Theorem 13.2.3 *We have $g(1/b)^2 \notin L(k, b)$, where $g(X) = \sum_{m \geq 1, \ n \geq 0} X^{(k^m - 1)k^n}$.*

Proof. Left to the reader as Exercise 13. ∎

Furthermore, the set $L(k, b)$ is not closed under reciprocal.

Corollary 13.2.4 *The set $L(k, b)$ is not closed under the map $x \to 1/x$.*

Proof. Suppose it were. Then, since

$$y^2 = y + \frac{1}{\frac{1}{y-1} - \frac{1}{y}}$$

for all $y \in L(k, b) \setminus \{0, 1\}$, we would have that $L(k, b)$ is closed under squaring, a contradiction. ∎

13.3 Transcendence: An Ad Hoc Approach

We now turn to questions about transcendence of real numbers. We begin gently, using an ad hoc method to prove a simple transcendence result about the automatic real number $F = \sum_{n \geq 0} B^{-2^n}$. More general techniques are given in the next section.

Theorem 13.3.1 *The real number $F = \sum_{n \geq 0} B^{-2^n}$ is transcendental for all integers $B \geq 2$.*

Proof. Assume F is algebraic and satisfies the polynomial equation

$$c_e F^e + \cdots + c_1 F + c_0 = 0 \tag{13.1}$$

with $c_i \in \mathbb{Z}$ for $0 \leq i \leq e$ and $c_e > 0$. Let $H = \max_{0 \leq i \leq e} |c_i|$.

Now rewrite (13.1) as follows:

$$c_e F^e + \cdots = b_s F^s + \cdots, \tag{13.2}$$

where the coefficients on both sides are non-negative and $0 \leq s < e$.

Now define $f(X) = \sum_{n \geq 0} X^{2^n}$. For $r, k \geq 0$ let $a(r, k)$ denote the coefficient of X^r in $f(X)^k$. Note that $a(r, k)$ is the number of ways that r can be written as a sum of k powers of 2, where different orderings are counted as distinct.

Lemma 13.3.2 *Let e, m be fixed integers, and let k be an integer with $1 \leq k \leq e$. Define $N = (2^e - 1) \cdot 2^m$. Then for $N - (2^{m-1} - 1) \leq r \leq N + 2^m - 1$ we have*

$$a(r, k) = \begin{cases} e! & \text{if } r = N \text{ and } k = e, \\ 0 & \text{otherwise.} \end{cases} \tag{13.3}$$

Proof. We have $(N)_2 = 1^e\, 0^m$. Then for $N - (2^{m-1} - 1) \le r < N$ we have

$$(r)_2 = 1^{e-1}\, 01\, x,$$

where the string x contains at least one 1.

For $N < r \le N + (2^m - 1)$ we have

$$(r)_2 = 1^e\, x'$$

where the string x' also has at least one 1. Hence for all $r \ne N$ in the specified range, r has at least $e + 1$ 1's in its binary expansion, and hence $a(r, k) = 0$.

If, on the other hand, $r = N$, then $a(r, k) = 0$ for $1 \le k < e$. If $k = e$, then $a(r, k) = e!$, since then N can be written as the sum of e distinct powers of 2, and all $e!$ permutations of these will work.   ∎

Now consider Eq. (13.2) as a number in base B, with both sides thought of for the moment without carries. The left-hand side will have, in digit positions specified by the interval $I := [N - (2^{m-1} - 1), N + (2^m - 1)]$, all zeros, except at position N. The right-hand side will have all zeros in these positions.

It now remains to consider the effect of the carries.

Lemma 13.3.3 *For integers $k, r \ge 1$ we have*

$$a(r, k) \le (1 + \log_2 r)^k.$$

Proof. We can use powers from 2^0 up to $2^{\lfloor \log_2 r \rfloor}$ in the summands to represent r, which gives $1 + \lfloor \log_2 r \rfloor$ different choices; each choice can be used at most k times.   ∎

We now show that, for m sufficiently large, the carries do not extend significantly into the positions in I.

The term at digit N, on the left-hand side of (13.2), is $c_e \cdot e!$, which is independent of m. Hence for all large m, its carries are bounded by $1 + \lfloor \log_B(c_e \cdot e!) \rfloor$, which occupies only a small portion of I.

On the other hand, the carries occurring in positions to the right of those in I will never come close to position N. For we have, considering a single term in (13.2),

$$\sum_{r \ge N + 2^m} \frac{a(r, k)}{B^r} \le \sum_{r \ge N + 2^m} \frac{(1 + \log_2 r)^k}{B^r}$$

$$\le \sum_{r \ge N + 2^m} \frac{r}{B^r} \quad \text{(for m sufficiently large)}$$

$$\le \frac{N + 2^m}{B^{N + 2^m - 2}},$$

which gives carries to at most $\lfloor \log_B(N + 2^m) \rfloor + 3$ positions to the left of position $N + 2^m$. Now multiply by H and sum $e + 1$ terms, to get carries at most to position

$$\lfloor \log_B(N + 2^m) \rfloor + 4 + \lfloor \log_B H(e + 1) \rfloor.$$

As $m \to \infty$, these cannot come close to position N.

It follows that the left-hand side of (13.2) looks, in base B, like

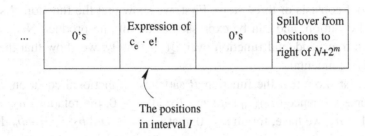

while the right-hand side of (13.2) looks like

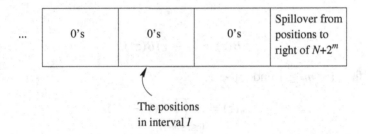

so they cannot be equal. ∎

13.4 Transcendence of the Thue–Morse Number

We prove in this section that the Thue–Morse number is transcendental. Our proof uses the following theorem on analytic functions, which we state without proof.

Theorem 13.4.1 *Suppose f is an analytic function on some nonempty connected open subset Ω of $\mathbb{C}$. Let*

$$Z(f) = \{z \in \Omega : f(z) = 0\}.$$

Then either $Z(f) = \Omega$, or $Z(f)$ has no limit point in Ω.

Theorem 13.4.2 *Let $(a_n)_{n \geq 0}$ be the Thue–Morse sequence with values 0 and 1. Then the Thue–Morse number $\mathcal{T} = \sum_{n \geq 0} a_n 2^{-n}$ is transcendental.*

Proof. First, we replace the sequence $(a_n)_{n \geq 0}$ by the sequence $(b_n)_{n \geq 0}$, where $b_n = 1 - 2a_n$. Now the sequence $(b_n)_{n \geq 0}$ takes the values ± 1, and it suffices to show that the number $\sum_{n \geq 0} b_n 2^{-n} = 2 - 2 \sum_{n \geq 0} a_n 2^{-n}$ is transcendental. We define for $|z| < 1$

$$B(z) = \sum_{n \geq 0} b_n z^n, \tag{13.4}$$

so that the number $\sum_{n \geq 0} b_n 2^{-n}$ is equal to $B(\frac{1}{2})$.

The proof consists of three steps. First, we show that the function B satisfies a functional equation, and can be expressed as an infinite product. Next, we show that B is a transcendental function over $\mathbb{Q}(z)$. Finally, we show that the number $B(\frac{1}{2})$ is transcendental.

Let us first show that the function B satisfies a functional equation. Since the Thue–Morse sequence $(a_n)_{n \geq 0}$ satisfies, for all $n \geq 0$, the relations $a_{2n} = a_n$ and $a_{2n+1} = 1 - a_n$, we have, for all $n \geq 0$, that $b_{2n} = b_n$ and $b_{2n+1} = -b_n$. Hence

$$B(z) = \sum_{n \geq 0} b_n z^n = \sum_{n \geq 0} b_{2n} z^{2n} + z \sum_{n \geq 0} b_{2n+1} z^{2n} = \sum_{n \geq 0} b_n z^{2n} - z \sum_{n \geq 0} b_n z^{2n}.$$

Hence

$$B(z) = (1 - z)B(z^2). \tag{13.5}$$

Let us define, for $m \geq 1$ and $|z| < 1$,

$$W_m(z) = \prod_{0 \leq j \leq m-1} (1 - z^{2^j}). \tag{13.6}$$

Then, iterating Eq. (13.5), we have

$$B(z) = W_m(z)B(z^{2^m}) \tag{13.7}$$

for all $m \geq 1$, and for all z with $|z| < 1$. Since $|z| < 1$, $\lim_{m \to \infty} z^{2^m} = 0$, and since B is continuous we have $\lim_{m \to \infty} B(z^{2^m}) = B(0) = 1$. Thus we find

$$B(z) = \lim_{m \to \infty} W_m(z) = \prod_{j \geq 0} (1 - z^{2^j}) \tag{13.8}$$

for all z with $|z| < 1$. In particular,

$$B(\tfrac{1}{2}) = \lim_{m \to \infty} W_m(\tfrac{1}{2}) = \prod_{j \geq 0} (1 - 2^{-2^j}). \tag{13.9}$$

Now we prove that the function $B(z)$ is transcendental over $\mathbb{Q}(z)$. Although the transcendence of B can be seen as a consequence of more general results (see the Notes for more details), we give a direct elementary proof based upon the functional equation (13.5) satisfied by B.

Suppose that B is algebraic over $\mathbb{Q}(z)$. Then there exist an integer $d \geq 1$ and $d + 1$ polynomials $Q_0, Q_1, \ldots, Q_d$, not all zero, such that

$$\sum_{0 \leq k \leq d} Q_k(z) B^k(z) = 0 \qquad (13.10)$$

for all z with $|z| < 1$. We can suppose that d is minimal, which implies that $Q_0 \neq 0$. Now, replacing z by z^2 in Eq. (13.10) above and using Eq. (13.5) gives

$$\sum_{0 \leq k \leq d} Q_k(z^2) B^k(z^2) = \sum_{0 \leq k \leq d} Q_k(z^2)(1 - z)^{-k} B^k(z),$$

and so, multiplying by $(1 - z)^d$, we get

$$\sum_{0 \leq k \leq d} Q_k(z^2)(1 - z)^{d-k} B^k(z) = 0 \qquad (13.11)$$

for all z with $|z| < 1$. Now, multiplying Eq. (13.10) by $Q_d(z^2)$ and Eq. (13.11) by $Q_d(z)$, and subtracting, we obtain

$$\sum_{0 \leq k \leq d-1} (Q_d(z) Q_k(z^2)(1 - z)^{d-k} - Q_d(z^2) Q_k(z)) B^k(z) = 0 \qquad (13.12)$$

for all z with $|z| < 1$. Since d was chosen to be minimal, this implies that all the coefficients in the sum (13.12) are in fact 0, and in particular, setting $k = 0$, we get

$$Q_d(z) Q_0(z^2)(1 - z)^d = Q_d(z^2) Q_0(z) \qquad (13.13)$$

for all z with $|z| < 1$. If we define the non-negative integers u and v and the polynomials P_0 and P_d by $Q_0(z) = (1 - z)^u P_0(z)$, $Q_d(z) = (1 - z)^v P_d(z)$ and $P_0(1) \neq 0$, $P_d(1) \neq 0$, then Eq. (13.13) implies that

$$(1 - z)^{u+v+d} P_d(z)(1 + z)^u P_0(z^2) = (1 - z)^{u+v}(1 + z)^v P_d(z^2) P_0(z),$$

giving a contradiction when we divide this identity by $(1 - z)^{u+v}$ and set $z = 1$. ∎

We are now ready to show that the number $B(\frac{1}{2})$ is transcendental.

Let us suppose that $B(\frac{1}{2})$ is algebraic of degree g. Let N be a fixed integer such that $N > 2g$. We claim that it is possible to find $N + 1$ polynomials $P_0, P_1, \ldots, P_N$, with integer coefficients and not all 0, such that $\deg P_k \leq N$ for all $k \leq N$ and

$$\sum_{0 \leq k \leq N} P_k(z) B(z)^k = R(z)$$

for all z with $|z| < 1$, where the formal power series R can be written $R(z) = z^{N^2} \sum_{k \geq 0} r_k z^k$, i.e., the first N^2 coefficients of R are zero. This is indeed possible, since the coefficients of the P_k's are $(N + 1)^2$ unknowns, and the condition on R gives rise to N^2 linear homogeneous equations with integer coefficients.

Now we define polynomials $P_{m,k}$, for $m \geq 1$ and $0 \leq k \leq N$, by

$$P_{m,k}(z) = P_k(z^{2^m}).$$

The polynomials $P_{m,k}$ have integer coefficients, and from Eq. (13.7) we have, for all $m \geq 1$ and for all z such that $|z| < 1$,

$$\sum_{0 \leq k \leq N} P_{m,k}(z) W_m(z)^{-k} B(z)^k = \sum_{0 \leq k \leq N} P_{m,k}(z) B(z^{2^m})^k$$

$$= \sum_{0 \leq k \leq N} P_k(z^{2^m}) B(z^{2^m})^k = R(z^{2^m}). \quad (13.14)$$

Now for a polynomial P, define the norm $\|P\|$ as follows

$$\|a_0 + a_1 z + a_2 z^d + \cdots + a^d z^d\| = \max_j |a_j|.$$

Also define the number M by

$$M = \max_{0 \leq k \leq N} \max_{0 \leq x \leq \frac{1}{2}} |P_k(x)|.$$

If for $m \geq 1$ we define the polynomials

$$\widetilde{P}_m(z) = 2^{N2^{m+1}} \sum_{0 \leq k \leq N} P_{m,k}(\tfrac{1}{2}) W_m^{N-k}(\tfrac{1}{2}) z^k, \quad (13.15)$$

then these polynomials $\widetilde{P}_m$ have their coefficients in $\mathbb{Z}$, since $P_{m,k}$ is a polynomial with integer coefficients of degree $\leq N2^{2^m}$, and W_m is a polynomial with integer coefficients of degree $2^m - 1$. Hence the product $P_{m,k} W_m^{N-k}$ is a polynomial with integer coefficients of degree $< N2^{2^{m+1}}$. Furthermore the polynomials $\widetilde{P}_m$ satisfy $\|\widetilde{P}_m\| \leq M 2^{N2^{m+1}}$ for $m \geq 1$.

Then, defining $\beta = B(\tfrac{1}{2})$, and putting $z = \tfrac{1}{2}$ in Eq. (13.14) and $z = B(\tfrac{1}{2}) = \beta$ in Eq. (13.15), we have

$$\widetilde{P}_m(\beta) = 2^{N2^{m+1}} \sum_{0 \leq k \leq N} P_{m,k}(\tfrac{1}{2}) W_m^{N-k}(\tfrac{1}{2}) B(\tfrac{1}{2})^k = 2^{N2^{m+1}} W_m^N(\tfrac{1}{2}) R(2^{-2^m}). \quad (13.16)$$

Since the formal power series $R(z)$ begins with a term in z^{N^2}, we can define

$$M' = \max_{0 \leq x \leq \frac{1}{2}} \frac{|R(x)|}{x^{N^2}}.$$

Then we have

$$\|\widetilde{P}_m\|^{g-1} |\widetilde{P}_m(\beta)| \leq (M2^{N2^{m+1}})^{g-1} 2^{N2^{m+1}} W_m^N(\tfrac{1}{2}) |R(2^{-2^m})|$$

$$\leq M^{g-1} M' 2^{gN2^{m+1}} 2^{-2^m N^2}$$

Hence

$$\|\widetilde{P}_m\|^{g-1} |\widetilde{P}_m(\beta)| \leq M^{g-1} M' 2^{(2g-N)N2^m}. \quad (13.17)$$

Since we chose $N > 2g$, Eq. (13.17) shows that

$$\lim_{m \to \infty} \| \widetilde{P}_m \|^{g-1} |\widetilde{P}_m(\beta)| = 0. \tag{13.18}$$

But $\widetilde{P}_m(\beta) \neq 0$ for m large enough. For if $\widetilde{P}_m(\beta)$ were equal to zero for infinitely many m, then Eq. (13.16) would imply that $R(2^{-2^m})$ was zero for infinitely many values of m. Then, by Theorem 13.4.1, R would be zero, which, in view of the definition of R, would contradict the transcendence of the function B. This fact, the fact that $\deg \widetilde{P}_m = N$, Eq. (13.18), and Lemma 2.3.6 together prove that β cannot be an algebraic number of degree g. ∎

13.5 Transcendence of Morphic Real Numbers

In this section, we prove that any irrational real number whose base-k expansion consists of 0's and 1's, and is a fixed point of a morphism that is either primitive or has constant length ≥ 2, is transcendental. We begin with a generalization of Roth's theorem, due to Ridout, that we state without proof.

Theorem 13.5.1 (Ridout) *Let $\xi \neq 0$ be a real algebraic number. Let ρ, c_1, c_2, c_3 be positive constants, and let λ and μ satisfy $0 \leq \lambda, \mu \leq 1$. Let $r', r'' \geq 0$ be integers, and suppose $\omega_1, \omega_2, \ldots, \omega_{r'+r''}$ are finitely many distinct primes. Assume there exist infinitely many fractions p_n/q_n in lowest terms such that*

$$\left| \frac{p_n}{q_n} - \xi \right| \leq c_1 |q_n|^{-\rho}.$$

Furthermore, suppose that p_n and q_n are not zero and can be written in the form

$$p_n = p_n' \prod_{j=1}^{r'} \omega_j^{e_j}, \qquad q_n = q_n' \prod_{j=r'+1}^{r'+r''} \omega_j^{e_j},$$

where the e_i are non-negative integers that may depend on n, and the p_n''s and q_n''s are positive integers that may depend on n. Finally, suppose that

$$0 < |p_n'| \leq c_2 |p_n|^{\lambda}, \quad 0 < |q_n'| \leq c_3 |q_n|^{\mu}$$

for all $n \geq 0$. Then

$$\rho \leq \lambda + \mu.$$

Note that the same hypotheses without the irreducibility of the fractions p_n/q_n (i.e., there are infinitely many fractions p_n/q_n but they may be not irreducible) yield the same conclusion.

Corollary 13.5.2 *Let ξ be an irrational number. Suppose that, for every integer $n \geq 0$, the base-k expansion of ξ begins with $0.U_n V_n V_n V_n'$, where U_n belongs to $\{0, 1, \ldots, k-1\}^*$, V_n belongs to $\{0, 1, \ldots, k-1\}^+$, and the word V_n'*

is a prefix of V_n. Furthermore suppose that $\lim_{n\to\infty} |V_n| = \infty$, and that there exist real numbers $0 \le \alpha < \infty$ and $\beta > 0$ such that $\limsup_{n\to\infty} |U_n|/|V_n| = \alpha$ and $\liminf_{n\to\infty} |V_n'|/|V_n| = \beta$. Then ξ is a transcendental number.

Proof. Let $r_n = |U_n|$, $s_n = |V_n|$, and $s_n' = |V_n'|$, so $\limsup_{n\to\infty} r_n/s_n = \alpha$ and $\liminf_{n\to\infty} s_n'/s_n = \beta$. Define t_n to be the rational number whose base-k expansion is $t_n = 0.U_n V_n V_n V_n \cdots$. Hence $t_n = p_n/k^{r_n}(k^{s_n} - 1)$ for some integer p_n. Note that

$$|\xi - t_n| < \frac{1}{k^{r_n + 2s_n + s_n'}}.$$

Now,

$$\liminf_{n\to\infty} \frac{s_n}{r_n + s_n} = \left(\limsup_{n\to\infty} \frac{r_n + s_n}{s_n} \right)^{-1} = \frac{1}{1 + \alpha}$$

and

$$\liminf_{n\to\infty} \frac{r_n + 2s_n + s_n'}{r_n + s_n} = 1 + \liminf_{n\to\infty} \frac{s_n}{r_n + s_n} + \liminf_{n\to\infty} \frac{s_n'}{r_n + s_n}$$

$$\ge 1 + \left(1 + \liminf_{n\to\infty} \frac{s_n'}{s_n} \right) \left(\liminf_{n\to\infty} \frac{s_n}{r_n + s_n} \right)$$

$$= 1 + (1 + \beta)\frac{1}{1 + \alpha}.$$

Hence there exist two positive real numbers μ, ρ such that

$$1 + \frac{s_n}{r_n + s_n} < 1 + \mu < \rho < \frac{r_n + 2s_n + s_n'}{r_n + s_n} \tag{13.19}$$

for infinitely many n. With this choice of μ and ρ, let us take $p_n' = p_n$, $\lambda = 1$, $c_2 = 1$, $q_n' = k^{s_n} - 1$. Let us choose the primes $\omega_{r'+1}, \cdots, \omega_{r'+r''}$ to be the prime divisors of k. Finally, defining $e_{r'+1}, \ldots, e_{r'+r''}$ by $k^{r_n} = \prod_{i=r'+1}^{r'+r''} \omega_j^{e_j}$, we can apply Ridout's theorem and deduce that $\rho \le \lambda + \mu$, which contradicts Eq. (13.19). Hence ξ is transcendental. (Note that the t_n's are not necessarily in their irreducible forms, but there are an infinite number of them, since the sequence $(t_n)_n$ converges to ξ, which is irrational from the hypothesis.) $\blacksquare$

Now we are ready for a theorem giving the transcendence of some real numbers whose base-k expansions are fixed points of morphisms.

Theorem 13.5.3 *If the expansion of the real number ξ in some integer base $k \ge 2$ is a non-ultimately-periodic fixed point of a morphism σ that either has constant length or is primitive, and if furthermore this expansion contains an overlap, then the number ξ is transcendental.*

Proof. First, we note that we can suppose without loss of generality that $0 < \xi < 1$. We then write the base-k expansion of ξ as $\xi = 0.UVVa, \cdots$, where U and V

are finite words, and a is the first letter of V. Since the expansion of ξ is a fixed point of the morphism σ, this expansion also begins with $\sigma^n(U)\sigma^n(V)\sigma^n(V)\sigma^n(a)$ for every $n \geq 1$. We can apply the previous corollary with $U_n = \sigma^n(U)$, $V_n = \sigma^n(V)$, and $V'_n = \sigma^n(a)$. What remains to prove is that $|V_n|$ tends to infinity, that $\limsup_{n\to\infty} |U_n|/|V_n| < +\infty$, and that $\liminf_{n\to\infty} |V'_n|/|V_n| > 0$. These results are consequences of the asymptotic behavior $|\sigma^n(W)| \sim c(W)\ell^n$, where $c(W)$ is a positive constant and ℓ is the dominant eigenvalue of the incidence matrix of the morphism σ, when σ is primitive; see Proposition 8.4.1. In the case where σ has constant length $d \geq 2$, then easily $\ell = d$, and $c(W) = |W|$. ∎

We can now state a transcendence result for positive real numbers whose base-k expansions are certain fixed points of morphisms.

Theorem 13.5.4 *Let x be a positive irrational real number whose base-k expansion is a fixed point of a morphism on the alphabet $\{0, 1\}$. If the morphism either has constant length ≥ 2 or is primitive, then the number x is transcendental.*

Proof. We can suppose that the number x satisfies $0 < x < 1$. Let us suppose that x is irrational. If the sequence of base-k digits of x contains an overlap, we can apply Theorem 13.5.3, and the number x is transcendental. If this sequence contains no overlap, we can apply Theorem 1.7.9 to conclude that this sequence is either the Thue–Morse sequence beginning with 0 or the Thue–Morse sequence beginning with 1: the morphism that fixes the sequence of digits of x is either of constant length ≥ 2 or primitive, hence nontrivial. But then the number x is either the Thue–Morse number $t = 0.110100110010110\cdots$ or $2 - t$, and we proved the number t transcendental in Section 13.4 above. ∎

13.6 Transcendence of Characteristic Real Numbers

As we have seen in Section 9.1, with each irrational real number θ, $0 < \theta < 1$, we can associate an infinite word of 0's and 1's, called the characteristic word. We can consider the characteristic word as defining the base-b expansion of a real number. A natural question is, are these numbers transcendental? The following theorem shows that the answer is yes.

Theorem 13.6.1 *Let θ be an irrational real number with $0 < \theta < 1$. Let its associated characteristic word be*

$$\mathbf{f}_\theta = f_1 f_2 f_3 \cdots .$$

Let b be an integer ≥ 2, and define

$$\alpha = \sum_{i \geq 1} f_i b^{-i}.$$

Then α is a transcendental number.

Proof. The idea is to show that α can be well approximated by rationals. The rational numbers we use are those whose base-b expansions are of the form

$$0.X_i \, X_i \, X_i \cdots,$$

where X_i is the ith characteristic block, as defined in Section 9.1.

More precisely, let $\theta = [0, a_1, a_2, \dots]$. Define $c_i = [X_i]_b$ and $d_i = b^{|X_i|} - 1$. Then the base-b expansion of c_i / d_i is easily seen to be $0.X_i \, X_i \, X_i \cdots$. Now we know from Corollary 9.1.12 that for $i \geq 3$ we have $X_i^{e_i}$ is a prefix of X_{i+2}, where $e_i = a_{i+1} + 1 + (q_{i-1} - 2)/q_i$. Hence we have

$$\left| \alpha - \frac{c_i}{d_i} \right| \leq b^{-e_i |X_i|} \leq (b^{|X_i|} - 1)^{-e_i} = d_i^{-e_i}.$$

Now there are two cases to consider: (a) when $a_{i+1} \geq 2$ for infinitely many i, and (b) when $a_{i+1} = 1$ for all i sufficiently large.

In case (a), since $e_i \geq a_{i+1} + 1$, we have $e_i \geq 3$ for infinitely many i. Hence by Roth's theorem (Theorem 2.3.7), α is transcendental.

In case (b), we have $a_{i+1} = 1$ for all i sufficiently large. In this case, it is easy to see that $\lim_{i \to \infty} q_{i-1}/q_i = (\sqrt{5} - 1)/2 \doteq 0.61803$. Hence $(q_{i-1} - 2)/q_i \geq \frac{1}{2}$ for all sufficiently large i, and hence $e_i \geq \frac{5}{2}$ for all sufficiently large i. Again by Roth's theorem, α is transcendental. ∎

13.7 The Thue–Morse Continued Fraction

An old, still-unsolved conjecture asserts that a positive real number whose continued fraction expansion has bounded partial quotients cannot be algebraic of degree ≥ 3, even in the case where the quotients are, say 1 and 2. In the case where these quotients satisfy extra conditions some results are known. We will prove here that the continued fraction $[0, 1, 2, 2, 1, 2, 1, 1, 2, \dots]$, whose partial quotients are given by the Thue–Morse sequence on the alphabet $\{1, 2\}$, is transcendental.

First, we state (without proof) a theorem of W. Schmidt that we will need. We recall the following definition: if ξ is a root of the minimal equation $a\xi^2 + b\xi + c = 0$, with $a, b, c \in \mathbb{Z}$ and $\gcd(|a|, |b|, |c|) = 1$, the *height* $H(\xi)$ of ξ is defined by $H(\xi) = \max(|a|, |b|, |c|)$.

Theorem 13.7.1 (W. Schmidt) *Let x be a real number in $(0, 1)$. We suppose that x is neither rational nor quadratic irrational. If there exist a real number $B > 3$ and infinitely many quadratic irrational numbers ξ_k such that*

$$|x - \xi_k| < H(\xi_k)^{-B},$$

then x is transcendental.

The following proposition will prove useful.

Proposition 13.7.2

(a) *Let $\xi \in (0, 1)$ be a number with periodic continued fraction expansion*

$$\xi = [0, a_1, a_2, \ldots, a_k, a_1, a_2, \ldots, a_k, \ldots].$$

Then the (quadratic irrational) number ξ satisfies $H(\xi) \leq q_k$.
(b) *If $x, y \in (0, 1]$ have the same first k partial quotients $a_1, a_2, \ldots, a_k$, then*

$$|x - y| \leq \frac{1}{q_k^2}.$$

Proof. (a): We have $\xi = [0, a_1, a_2, \ldots, a_k, \xi]$. Hence $\xi = (\xi p_k + p_{k-1})/(\xi q_k + q_{k-1})$, which gives

$$q_k \xi^2 + \xi(q_{k-1} - p_k) - p_{k-1} = 0.$$

Since $\xi \in (0, 1)$, we have $p_n \leq q_n$ for every $n \geq 1$; hence

$$H(\xi) \leq \max(q_k, |q_{k-1} - p_k|, p_{k-1}) \leq q_k.$$

(b): Since $p_k/q_k = [0, a_1, a_2, \cdots, a_k]$, we have $|x - p_k/q_k| \leq 1/q_k q_{k+1}$ and also $|y - p_k/q_k| \leq 1/q_k q_{k+1}$. Furthermore $x - p_k/q_k$ and $y - p_k/q_k$ have the same sign, and this sign depends only on k. Hence

$$|x - y| = \left| \left| x - \frac{p_k}{q_k} \right| - \left| y - \frac{p_k}{q_k} \right| \right| \leq \frac{1}{q_k^2}. \qquad \blacksquare$$

Together with Schmidt's theorem, Proposition 13.7.2 permits us to prove the following theorem.

Theorem 13.7.3 *Let $\xi \in (0, 1)$ be an irrational number with continued fraction expansion $\xi = [0, a_1, a_2, \ldots, a_n, \ldots]$. We suppose that, for an infinite number of k's, the sequence $(a_n)_{n \geq 1}$ begins with the word $U_k V_k$, where $\lim_{k \to \infty} |U_k| = +\infty$, the word V_k is a prefix of U_k, and $\liminf_{k \to \infty} (|U_k| + |V_k|)/|U_k| = \gamma \geq 1$. Let $M = \limsup_{k \to \infty} q_{|U_k|}^{1/|U_k|}$ and $m = \liminf_{k \to \infty} q_{|U_k V_k|}^{1/|U_k V_k|}$. If the inequality $\gamma > (3 \log M)/(2 \log m)$ holds, then the number ξ is transcendental.*

Proof. Define ξ_k by the periodic continued fraction expansion

$$\xi_k = [0, a_1, a_2, \ldots, a_{|U_k|}, a_1, a_2, \ldots, a_{|U_k|}, \ldots].$$

Then by Proposition 13.7.2(a) we have $H(\xi_k) \leq q_{|U_k|}$. But the continued fraction expansion of ξ_k actually begins with $[0, a_1, a_2, \ldots, a_{|U_k|}, a_1, a_2, \ldots, a_{|V_k|}, \ldots]$, as does the expansion of ξ. Hence by Proposition 13.7.2(b), we have $|x - \xi_k| \leq 1/q_{|U_k V_k|^2}$.

To apply Schmidt's theorem, it suffices to show that there exists a $B > 3$ such that $q_{|U_k|}^B < q_{|U_k V_k|}^2$. Namely, we then have

$$|x - \xi_k| \leq \frac{1}{q_{|U_k V_k|}^2} < \frac{1}{q_{|U_k|}^B} \leq H(\xi_k)^{-B}.$$

To prove the existence of such a B, it suffices to show that $3 \log q_{|U_k|} < 2 \log q_{|U_k V_k|}$. Indeed, this inequality is a consequence of the following inequalities:

$$2 \liminf_{k \to \infty} \frac{\log q_{|U_k V_k|}}{\log q_{|U_k|}} = 2 \liminf_{k \to \infty} \frac{\log q_{|U_k V_k|}}{|U_k V_k|} \frac{|U_k|}{\log q_{|U_k|}} \frac{|U_k V_k|}{|U_k|}$$

$$\geq 2 \liminf_{k \to \infty} \frac{\log q_{|U_k V_k|}}{|U_k V_k|} \frac{1}{\limsup_{n \to \infty} \frac{\log q_{|U_k|}}{|U_k|}} \liminf_{n \to \infty} \frac{|U_k V_k|}{|U_k|}$$

$$\geq \frac{2\gamma \log m}{\log M} > 3.$$ ∎

Before applying the result to the Thue–Morse continued fraction, we prove two properties of its denominators.

Proposition 13.7.4 *Let p_n/q_n be the nth convergent to the continued fraction*

$$[0, 1, 2, 2, 1, 2, 1, 1, 2, \dots],$$

where the sequence of 1's and 2's is the Thue–Morse sequence on the alphabet $\{1, 2\}$ that begins with 1. Define the norm $\| \cdot \|$ to be the L^2-norm on the 2×2 matrices. Define the matrices A and B by

$$A = \begin{bmatrix} 1 & 1 \\ 1 & 0 \end{bmatrix}, \qquad B = \begin{bmatrix} 2 & 1 \\ 1 & 0 \end{bmatrix}.$$

Then we have

(a) $\limsup_{n \to \infty} q_n^{1/n} \leq \sqrt{\|AB\|};$

(b) $\liminf_{n \to \infty} q_{5 \cdot 2^k}^{1/5 \cdot 2^k} \geq \sqrt{\rho(AB)}$, *where $\rho(AB)$ is the spectral radius of the matrix AB, as defined in Section 3.3.*

Proof. (a): We have $q_0 = 1$ and $q_{-1} = 0$, and hence

$$\begin{bmatrix} q_n \\ q_{n-1} \end{bmatrix} = \begin{bmatrix} a_n & 1 \\ 1 & 0 \end{bmatrix} \begin{bmatrix} a_{n-1} & 1 \\ 1 & 0 \end{bmatrix} \cdots \begin{bmatrix} a_1 & 1 \\ 1 & 0 \end{bmatrix} \begin{bmatrix} 1 \\ 0 \end{bmatrix}.$$

Since $\left\| \begin{bmatrix} 1 \\ 0 \end{bmatrix} \right\| = 1$, it follows that

$$\left\| \begin{bmatrix} q_{2n} \\ q_{2n-1} \end{bmatrix} \right\| \leq \left\| \begin{bmatrix} a_{2n} & 1 \\ 1 & 0 \end{bmatrix} \begin{bmatrix} a_{2n-1} & 1 \\ 1 & 0 \end{bmatrix} \cdots \begin{bmatrix} a_2 & 1 \\ 1 & 0 \end{bmatrix} \begin{bmatrix} a_1 & 1 \\ 1 & 0 \end{bmatrix} \right\|.$$

Now, since the Thue–Morse sequence on the alphabet $\{A, B\}$ is a fixed point of the morphism $A \longrightarrow AB, B \longrightarrow BA$, each of these products of two matrices is equal either to AB or to BA. Noting that $\|BA\| = \|(AB)^T\| = \|AB\|$, we see that the corresponding norms are all equal to $\|AB\|$. Hence

$$q_{2n}, q_{2n-1} \leq \left\| \begin{bmatrix} q_{2n} \\ q_{2n-1} \end{bmatrix} \right\| \leq \|AB\|^n.$$

We thus have

$$\limsup_{n \to \infty} q_{2n}^{1/(2n)} \leq \sqrt{\|AB\|}$$

and

$$\limsup_{n \to \infty} q_{2n-1}^{1/(2n-1)} \leq \sqrt{\|AB\|}.$$

Hence assertion (a) is proved.

(b): First, notice that the set of 2×2 matrices $\mathcal{M}$ defined by

$$\mathcal{M} = \left\{ M = \begin{bmatrix} x & y \\ z & t \end{bmatrix} : x, y, z, t \in \mathbb{N}, \ x \geq y \geq t, \text{ and } x \geq z \geq t \right\}$$

is stable under multiplication and contains the matrices A and B; hence it contains any product built with these two matrices.

Now, if we define the matrix C_j to be A for $j = 1$ and B for $j = 2$, we see that

$$\begin{bmatrix} q_n \\ q_{n-1} \end{bmatrix} = C_{a_n} C_{a_{n-1}} \cdots C_{a_1} \begin{bmatrix} 1 \\ 0 \end{bmatrix}.$$

Hence, if

$$C_{a_n} C_{a_{n-1}} \cdots C_{a_1} = \begin{bmatrix} \alpha_n & \beta_n \\ \gamma_n & \delta_n \end{bmatrix},$$

then

$$2q_n > q_n + q_{n-1} = \alpha_n + \gamma_n \geq \alpha_n + \delta_n = \mathrm{Tr}(C_{a_n} C_{a_{n-1}} \cdots C_{a_1}).$$

Let us define the matrices Y_n and Z_n by $Y_0 = A$, $Z_0 = B$, and, for $n \geq 0$, $Y_{n+1} = Y_n Z_n$ and $Z_{n+1} = Z_n Y_n$. Hence $Y_1 = AB$, $Z_1 = BA$, $Y_2 = ABBA$, $Z_2 = BAAB \cdots$. In particular $Y_n = C_{a_1} C_{a_2} \cdots C_{a_{2^n}}$, and Z_n is obtained from Y_n by interchanging the A's and B's. We also note that

$$a_1 a_2 \cdots a_{5 \cdot 2^k} = (a_1 \cdots a_{2^{k+1}})(a_{2^{k+1}+1} \cdots a_{2^{k+1}+2^k})$$
$$\times (a_{2^{k+1}+2^k+1} \cdots a_{2^{k+2}})(a_{2^{k+2}} \cdots a_{2^{k+2}+2^{k+1}}),$$

so we have

$$C_{a_1} C_{a_2} \cdots C_{a_{5 \cdot 2^k}} = Y_{k+1} Z_k Y_k Z_k = Y_{k+1} Z_k Y_{k+1}.$$

We thus have

$$2q_{5\cdot 2^k} > \mathrm{Tr}(C_{a_{5\cdot 2^k}} C_{a_{5\cdot 2^{k-1}}} \cdots C_{a_1}) = \mathrm{Tr}((C_{a_1} C_{a_2} \cdots C_{a_{5\cdot 2^k}})^T) = \mathrm{Tr}(Y_{k+1} Z_k Y_{k+1}),$$

and to finish up the proof of assertion (b), it will suffice to prove that

$$\liminf_{n \to \infty} (\mathrm{Tr}(Y_{k+1} Z_k Y_{k+1}))^{1/5\cdot 2^k} \geq \sqrt{\rho(AB)}.$$

This is the purpose of the next lemma.   ∎

Lemma 13.7.5 *With the notation above, we have*

$$\liminf_{n \to \infty} (\mathrm{Tr}(Y_{k+1} Z_k Y_{k+1}))^{1/5\cdot 2^k} \geq \sqrt{\rho(AB)}.$$

Proof. Let $t_k = \mathrm{Tr}(Y_k)$. We easily see that $\mathrm{Tr}(Z_k) = t_k$ for $k \geq 1$. Since Y_k and Z_k are 2×2 matrices of determinant 1 for $k \geq 1$, the Cayley-Hamilton theorem gives, for $k \geq 1$,

$$Y_k^2 = t_k Y_k - I \quad \text{and} \quad Z_k^2 = t_k Z_k - I, \qquad \text{where } I = \begin{bmatrix} 1 & 0 \\ 0 & 1 \end{bmatrix}.$$

Hence

$$
\begin{aligned}
\mathrm{Tr}(Y_{k+1} Z_k Y_{k+1}) = \mathrm{Tr}(Y_{k+1}^2 Z_k) &= \mathrm{Tr}((t_{k+1} Y_{k+1} - I) Z_k) \\
&= t_{k+1} \mathrm{Tr}(Y_{k+1} Z_k) - \mathrm{Tr}(Z_k) \\
&= t_{k+1} \mathrm{Tr}(Y_k Z_k Z_k) - \mathrm{Tr}(Z_k) \\
&= t_{k+1} \mathrm{Tr}(Y_k (t_k Z_k - I)) - \mathrm{Tr}(Z_k) \\
&= t_{k+1} t_k \mathrm{Tr}(Y_k Z_k) - t_{k+1} \mathrm{Tr}(Y_k) - \mathrm{Tr}(Z_k) \\
&= t_{k+1} t_k \mathrm{Tr}(Y_{k+1}) - t_{k+1} t_k - \mathrm{Tr}(Z_k) \\
&= t_{k+1}^2 t_k - t_{k+1} t_k - t_k.
\end{aligned}
$$

It is easily checked that matrices in $\mathcal{M}$ have two real eigenvalues, one of which is non-negative. Since Y_k has determinant $+1$ for $k \geq 1$, we see that Y_k has two positive eigenvalues. Hence $t_k = \mathrm{Tr}(Y_k) \geq \rho(Y_k)$, where $\rho(Y_k)$ is the spectral radius of the matrix Y_k. In order to conclude, it suffices to prove that $\rho(Y_k) \geq \rho(AB)^{2^{k-1}}$, since this will imply first that t_k goes to infinity, second that $\mathrm{Tr}(Y_{k+1} Z_k Y_{k+1})$ is equivalent to $t_{k+1}^2 t_k$ when k goes to infinity, third that this last quantity is larger than $((\rho(AB))^{2^k})^2 (\rho(AB))^{2^{k-1}} = (\sqrt{\rho(AB)})^{5\cdot 2^k}$, and we will be done.

The assertion $\rho(Y_k) \geq \rho(AB)^{2^{k-1}}$ is now proved by induction, using two remarks: first, for any symmetric matrix S we have $\|S\| = \sqrt{\rho(S^T S)} = \sqrt{\rho(S^2)} = \rho(S)$; hence for any matrix M and its transpose M^T, we have $\rho(M) \leq \|M\| = \sqrt{\rho(M^T M)} = \sqrt{\|M^T M\|}$. Second, if k is odd, then $Z_k = Y_k^T$. Suppose we have proved that $\rho(Y_j) \geq \rho(AB)^{2^{j-1}}$ for all $j \leq 2k$. Then, on one hand,

$$
\begin{aligned}
\rho(Y_{2k+1}) = \rho(Y_{2k}Z_{2k}) &= \rho(Y_{2k-1}Z_{2k-1}Z_{2k-1}Y_{2k-1}) \\
&= \rho(Y_{2k-1}^2 Z_{2k-1}^2) \\
&= \rho(Y_{2k-1}^2 (Y_{2k-1}^T)^2) \\
&= \rho((Y_{2k-1}^2)(Y_{2k-1}^2)^T) \\
&\geq (\rho(Y_{2k-1}^2))^2 = (\rho(Y_{2k-1}))^4 \\
&\geq (\rho(AB)^{2^{2k-2}})^4 = (\rho(AB))^{2^{2k}} = (\rho(AB))^{2^{(2k+1)-1}},
\end{aligned}
$$

and on the other hand

$$
\begin{aligned}
\rho(Y_{2k+2}) = \rho(Y_{2k+1}Z_{2k+1}) &= \rho(Y_{2k+1}(Y_{2k+1}^T)) \\
&\geq (\rho(Y_{2k+1}))^2 \\
&\geq ((\rho(AB))^{2^{2k}})^2 \\
&= (\rho(AB))^{2^{(2k+2)-1}}.
\end{aligned}
$$ ∎

We are now ready for the main theorem of this section.

Theorem 13.7.6 *Let* $x = [0, 1, 2, 2, 1, 2, 1, 1, 2, 2, 1, 1, 2, \ldots] \in (0, 1)$ *be the real number whose partial quotients* $a_1, a_2, \ldots$ *are given by the Thue–Morse sequence on* $\{1, 2\}$ *that begins with 1. Then the number* x *is transcendental.*

Proof. The Thue–Morse sequence we are looking at begins with 12212 and is the fixed point beginning with 1 of the morphism σ defined by $\sigma(1) = 12$, $\sigma(2) = 21$. Hence, for any $k \geq 0$, this sequence begins with $\sigma^k(12212) = \sigma^k(122)\sigma^k(12)$. Define the words U_k and V_k by $U_k = \sigma^k(122)$ and $V_k = \sigma^k(12)$. Then V_k is a prefix of U_k, since 12 is a prefix of 122. Furthermore $|U_k| = 3.2^k$ and $|V_k| = 2^{k+1}$. Hence $|U_k|$ tends to infinity when k goes to infinity, and

$$
\gamma = \lim_{k \to \infty} \frac{|U_k| + |V_k|}{|U_k|} = \frac{5}{3}.
$$

Now the partial quotients of x have the property that

$$
M = \limsup_{k \to \infty} q_{|U_k|}^{1/|U_k|} = \limsup_{k \to \infty} q_{3.2^k}^{1/3.2^k} \leq \limsup_{k \to \infty} q_k^{1/k} \leq \sqrt{\|AB\|}
$$

and

$$
m = \liminf_{k \to \infty} q_{|U_k V_k|}^{1/|U_k V_k|} = \liminf_{k \to \infty} q_{5.2^k}^{1/5.2^k} \geq \sqrt{\rho(AB)},
$$

by Proposition 13.7.4.

To apply Theorem 13.7.3 and conclude, we only have to check that $\gamma > (3 \log M)/(2 \log m)$. But $\gamma = \frac{5}{3}$, $M \le \sqrt{\|AB\|}$, and $m \ge \sqrt{\rho(AB)}$. We compute

$$AB = \begin{bmatrix} 3 & 1 \\ 2 & 1 \end{bmatrix}.$$

Hence $\rho(AB) = 2 + \sqrt{3}$ and

$$\|AB\| = \sqrt{\rho((AB)(AB)^T)} = \sqrt{\rho\left(\begin{bmatrix} 10 & 7 \\ 7 & 5 \end{bmatrix}\right)} = \sqrt{\frac{15 + \sqrt{221}}{2}}.$$

Hence

$$\frac{3 \log M}{2 \log m} \le \frac{3 \log \sqrt{\|AB\|}}{2 \log \sqrt{\rho(AB)}} = \frac{3 \log \|AB\|}{2 \log \rho(AB)}$$

$$= \frac{\frac{3}{2} \log \left(\frac{15 + \sqrt{221}}{2}\right)}{2 \log(2 + \sqrt{3})} < 1.5397 < \gamma. \qquad \blacksquare$$

13.8 Exercises

1. Let b, n be integers ≥ 2, and let a be an integer relatively prime to n with $1 \le a < n$. Discuss the mean value and variance of the period of the base-b expansion of a/n.

2. Let $r = \sum_{k \ge 0} 2^{-2^k}$. Show r^2 is a 2-automatic real.

3. Give an example of an irrational 2-automatic real number whose reciprocal is also 2-automatic.

4. Show there are uncountably many Liouville numbers.

5. Find a quadratic equation with coefficients ≤ 30 in absolute value that has a root α within 10^{-4} of
   (a) $\sum_{n \ge 0} 2^{-2^n}$;
   (b) $\sum_{n \ge 0} 10^{-2^n}$.

6. (Plouffe) Define $x_1 = \frac{1}{2}$, and $x_{n+1} = 2x_n/(1 - x_n^2)$ for $n \ge 1$. Define

$$f(x) = \begin{cases} 1 & \text{if } x < 0, \\ 0 & \text{if } x \ge 0. \end{cases}$$

   Show that $\sum_{i \ge 1} f(x_i)/2^i = \arctan(\frac{1}{2})/\pi$.

7. Consider the set T of real numbers x that satisfy (a) $0 < x < 1$ and (b) $1 - x \le \{2^k x\} \le x$ for all $k \ge 0$. Show that the Thue–Morse real number $\mathcal{T}$ is the smallest accumulation point of T, as well as the smallest irrational point.

8. (D. Wilson) Compute the appearance function $R'(n)$, $n = 1, 2, 3$, for the decimal expansions of the fractional parts of π and e. What 1, 2, and 3-digit sequences w maximize $R'(w)$?

9. Let $\mathcal{P}$ be the probability that a randomly chosen language L over $\{0, 1\}$ contains at least one word of length i for all $i \geq 0$. Here by "randomly chosen" we mean that each word w appears in L with probability $\frac{1}{2}$. Show that $\mathcal{P}$ and $\mathcal{T}$ (the Thue–Morse real number) are related.

10. Consider a game in which each of two players takes turns alternately specifying the decimal digit of a number. Suppose the goal of one player is to specify an irrational number. Then this player can win, for example, by specifying any nonperiodic sequence. Suppose the goal is to specify a transcendental number. Is there a winning strategy?

11. Suppose $A_k := \sum_{n \geq 1} 1/(k^n - 1) = \sum_{n \geq 1} \tau(n)/k^n$ where τ is the number-of-divisors function. Show that A_k is irrational for every integer $k \geq 2$.

12. (Vanden Eynden)
   (a) Give an example of a rational number x such that neither x nor $\sqrt{x}$ contains the digit 0 in its decimal expansion.
   (b) Give an example of an irrational number with the same properties.

13. Show that $g(1/b)^2 \notin L(k, b)$, where $g(X) = \sum_{m \geq 1, \ n \geq 0} X^{(k^m - 1)k^n}$.

14. Show that $\int_1^\infty (\sum_{k \geq 1} 1/x^{2^k}) \frac{dx}{x+1} = \gamma$, where $\gamma \doteq 0.57721566$ is Euler's constant.

15. Fix an integer $b \geq 2$. Define a real number to be *primitive morphic real number* if some suffix of its base-b expansion is generated as the fixed point of some primitive morphism.
   (a) Show that the class of primitive morphic real numbers is closed under multiplication by a rational number.
   (b) Show that the class of primitive morphic real numbers is not closed under addition.

16. Give an example of a transcendental function that takes algebraic values at all algebraic arguments.

17. Fix an integer $k \geq 2$, and let α be a real number with $0 \leq \alpha < 1$. Define $L_\alpha = \{w \in \Sigma_k^* : [0.w]_k \leq \alpha\}$. Show that L_α is a regular language if and only if α is rational.

18. Let $k \geq 2$ be an integer, and let $M = (Q, \Sigma_k, \delta, q_0, F)$ be a DFA. Let $D(M)$ be the set of real numbers α, $0 \leq \alpha \leq 1$, such that M accepts every prefix of the base-k representation of α. Show that the measure of $D(M)$ is a rational number.

19. Let k be an integer ≥ 2. A real number α is said to be *normal to base k* if in its base-k representation every finite sequence of digits x occurs with limiting frequency $k^{-|x|}$.
   (a) Suppose k, l are integers ≥ 2 that are multiplicatively dependent. Show that if α is normal to base k, then it is normal to base l.
   (b) Show that the real number $0.123456789101112131415\cdots$ obtained by concatenating the base-10 representations of the integers together in increasing order is normal to base 10.

(c) Show that the real number $0.2357111317192329\cdots$ obtained by concatenating the base-10 representations of the prime numbers together in increasing order is normal to base 10.

(d) Show that the number $\sum_{n\geq 0} 3^{-2^n} 2^{-3^{2^n}}$ is normal to base 2, and give an explicit description of its continued fraction expansion.

(e) Construct an irrational real number that is not normal to any base $k \geq 2$.

20. Prove that $\sum_{\substack{n\geq 1 \\ n \text{ squarefree}}} n2^{-n}$ is irrational.

21. Let $\mathbf{t} = t_0 t_1 t_2 \cdots$ be the Thue–Morse sequence, and let α be the unique real root in the interval $(1, 2)$ of the equation $1 = \sum_{n\geq 1} t_i x^{-i}$. Prove that α is transcendental.

22. Show how to handle the small problem at the end of the proof of Lemma 13.1.4.

23. In analogy with the set of real numbers $L(k, b)$, define $P(k, p)$ to be the set of p-adic numbers $\sum_{i\geq -i_0} a_i p^i$ such that the sequence $(a_i)_{i\geq 0}$ is k-automatic. Prove the analogues of Theorems 13.1.3, 13.1.5, and 13.1.6 for $P(k, p)$.

13.9 Open Problems

1. Prove or disprove: every number whose base-b expansion is a morphic sequence is either rational or transcendental. (Remarks: This conjecture, together with its more restricted form where we consider only k-automatic sequences instead of general morphic sequences, is the most difficult and important open problem in the area. Partial results have been given above in Section 13.5. See the Notes to Section 13.4 below for more information about the k-automatic case.)

2. Provide an explicit example of an automatic real $r \neq 0$ such that $1/r$ is not automatic. (Remark: Applying Corollary 13.2.4 gives three numbers, at least one of which fulfills the desired conditions, but currently we do not know which.)

3. Show that $\log 2$ is not a 2-automatic real. (Remark: Perhaps the identity $\log 2 = \sum_{n\geq 1} 1/(n\cdot 2^n)$ will prove useful.)

4. Show that π is not a 2-automatic real. (Remark: Perhaps the identity

$$\pi = \sum_{n\geq 0} \left(\frac{4}{8n+1} - \frac{2}{8n+4} - \frac{1}{8n+5} - \frac{1}{8n+6} \right) \left(\frac{1}{16} \right)^n,$$

due to Bailey, Borwein, and Plouffe [1997], will prove useful.)

5. In the novel *Microserfs* by Douglas Coupland, there is a scene where the two main characters recite the decimal digits of π in unison. The digits they say are 470183890341. Where is the first occurrence of this block of digits in the decimal expansion of π? (Remark: The web site http://pi.nersc.gov allows you to search for patterns in the base-2 expansion of π.)

6. Consider the class of real numbers whose base-k expansion is a morphic sequence. Is it true that this class is closed under addition? Multiplication?

7. Let $b_1, b_2 \geq 2$ be multiplicatively independent integers. Prove or disprove: $L(k, b_1) \cap L(k, b_2) = \mathbb{Q}$.

8. Consider the real number r whose base-2 expansion is given by $h^\omega(0)$, where h maps $0 \to 010$, and $1 \to 11$. Is r transcendental?

9. Consider the Thue–Morse real number $\mathcal{T}$, with expansion $0.0110100110010110 \cdots$ in base 2. Let the continued fraction expansion of $\mathcal{T}$ be $[a_0, a_1, a_2, \ldots]$. Are the a_i unbounded? (Remark. A calculation reveals the large partial quotient $a_{95} = 867374$.)

10. Consider the number $\sum_{n \geq 0} (-1)^{t_n}/(n+1) \doteq 0.3987610881084188124074305 4 \cdots$, where $(t_n)_{n \geq 0}$ is the Thue–Morse sequence. Find an expression for this number in terms of known constants.

11. (Mahler) Let $(e_i)_{i \geq 1}$ be an infinite sequence over $\{0, 1\}$ that is not ultimately periodic. Prove or disprove: at least one of the two numbers $\sum_{n \geq 1} e_n 2^{-n}$, $\sum_{n \geq 1} e_n 3^{-n}$ is transcendental. (Remarks: See Mendès France [1980]; in conversation he attributes this question to Mahler.)

12. Prove or disprove: the numbers A_k of Exercise 11 are normal in base k. Are these numbers k-automatic?

13. Prove or disprove: the number $\sum_{n \geq 0} t_n/n!$ is transcendental, where $\mathbf{t} = t_0 t_1 t_2 \cdots$ is the Thue–Morse sequence.

14. Find an explicit example of a k-automatic real number x such that $1/x$ is not k-automatic.

13.10 Notes on Chapter 13

Turing's paper is Turing [1936]. For more on Turing computable reals, see, for example, Pour-El and Richards [1989]. Hartmanis and Stearns [1965] suggested that real numbers could be classified according to the computational complexity of their base-k expansions. They asked whether there exist any irrational algebraic numbers that can be computed in real time, a question that is still open.

13.1 Corollary 13.1.7, the fact that the k-automatic reals form a vector space over $\mathbb{Q}$, is due to Lehr [1993]. Also see Lehr, Shallit, and Tromp [1996].

For alternative models of real numbers accepted by finite automata, see Even [1964] and Hartmanis and Stearns [1967].

13.2 For Theorem 13.2.1, see Lehr, Shallit, and Tromp [1996].

13.3 The number $F = \sum_{n \geq 0} 2^{-2^n}$ is sometimes called the "Fredholm number", although Fredholm apparently never studied it. Our proof of the transcendence of F is due to Knight [1991]. For other proofs, see Kempner [1916]; Blumberg [1926]; Mahler [1929]; Loxton and van der Poorten [1978]; Nishioka [1996, Thm. 1.1.2]. Nishioka [2001] proved more general results about algebraic independence.

13.4 For Theorem 13.4.1 see, e.g., Rudin [1966, Thm. 10.18].

Mahler [1929] proved that if $F(x) := \prod_{n \geq 0} \left(1 - x^{2^n}\right)$, then $F(\alpha)$ is transcendental for all algebraic numbers α with $0 < \alpha < 1$. In particular, this gives Theorem 13.4.2. Our proof is essentially that of Dekking [1977], as corrected with a personal communication from Dekking. As mentioned by Dekking in his paper, this proof was inspired by a paper of Cobham [1968a], which was a step towards proving that all numbers in $L(k, b)$ are either rational or transcendental. This last assertion was approached by Cobham [1968b] (who incorrectly claimed a proof) and Loxton and van der Poorten [1982, 1988], but the general result has not been fully proved yet. Unfortunately, some papers (e.g., Morton and Mourant [1991]) and books (e.g., Wolfram [2002]) cite Loxton and van der Poorten's work as if it constituted a complete proof of the assertion about the nonalgebraic character of numbers in $L(k, b)$.

Note that the transcendence over $\mathbb{Q}(X)$ of the Thue–Morse power series $B(X)$ can be deduced from more general results. Carlson [1921] proved that a power series with integer coefficients that converges inside the unit disk either is a rational function or has the unit circle as natural boundary, and hence cannot be algebraic irrational. Another result, due to Szegő [1922], states that a power series with only finitely many different coefficients that converges inside the unit disk either is a rational function or has the unit circle as natural boundary.

Mahler [1987] proved that, for any polynomial P, with $P(0) = 1$ and $P(1) = 0$, and for any integer $g \geq 2$, the infinite product $\prod_{k \geq 0} P(z^{g^k})$ converges, for $|z| < 1$, to a function admitting the unit circle as natural boundary, and hence transcendental.

13.5 For Ridout's theorem, see Ridout [1957] or Mahler [1961, p. 147] The results in this section are based on the work of Ferenczi and Mauduit [1997] and Allouche and Zamboni [1998]. Portions of these results were obtained independently, using a different method, by Nishioka, Tanaka, and Wen [1999].

13.6 Theorem 13.6.1 is due to Böhmer [1926]. It has been rediscovered many times. For more details, see the Notes to Section 9.3. A weaker result was found by Knuth [1964]; compare Corollary 9.3.3.

Using Exercise 9.12, the series in Theorem 13.6.1 is closely related to $\sum_{n \geq 1} \lfloor n\theta \rfloor X^n$, which was studied by Hecke [1921], Hardy and Littlewood [1923a, 1923b], Mahler [1929], Loxton and van der Poorten [1977c, 1977d], and Masser [1999].

13.7 For Schmidt's theorem, see W. Schmidt [1967]. The results in this section are based on work of Queffélec [1998, 2000]. Also see Allouche [2000] and Allouche, Davison, Queffélec, and Zamboni [2001].

14

Multidimensional Automatic Sequences

In Chapter 5 we defined the notion of automatic sequence, and in later chapters we explored the properties of these sequences. By definition, an automatic sequence is a one-sided, one-dimensional sequence. But one-dimensional infinite arrays of items are not the only such objects studied in mathematics; two-dimensional arrays (also called *tables* or *double sequences*; we use these terms interchangeably) are studied, as well as higher-dimensional objects. In this chapter we will examine a generalization of automatic sequences to a multidimensional setting, concentrating on the two-dimensional case. The interested reader will have no problem extending the results to the multidimensional case.

14.1 The Sierpiński Carpet

We start with an example.

Example 14.1.1 Consider the two-dimensional *Sierpiński carpet* array $\mathbf{s} = (s_{i,j})_{i,j \geq 0}$ over $\{0, 1\}$, defined as follows: $s_{i,j} = 0$ if and only if the base-3 expansions of i and j share at least one 1 in an identical position. More precisely, let $0 \leq i, j < 3^n$, and let $x = a_{n-1} \cdots a_0$, $y = b_{n-1} \cdots b_0$ be strings of length n such that $[x]_3 = i$, $[y]_3 = j$. Then $s_{i,j} = 0$ if and only if there exists an index s, $0 \leq s < n$, such that $a_s = b_s = 1$.

Here are the first 9 rows and columns of this array:

$$A = \begin{bmatrix} 1 & 1 & 1 & 1 & 1 & 1 & 1 & 1 & 1 \\ 1 & 0 & 1 & 1 & 0 & 1 & 1 & 0 & 1 \\ 1 & 1 & 1 & 1 & 1 & 1 & 1 & 1 & 1 \\ 1 & 1 & 1 & 0 & 0 & 0 & 1 & 1 & 1 \\ 1 & 0 & 1 & 0 & 0 & 0 & 1 & 0 & 1 \\ 1 & 1 & 1 & 0 & 0 & 0 & 1 & 1 & 1 \\ 1 & 1 & 1 & 1 & 1 & 1 & 1 & 1 & 1 \\ 1 & 0 & 1 & 1 & 0 & 1 & 1 & 0 & 1 \\ 1 & 1 & 1 & 1 & 1 & 1 & 1 & 1 & 1 \end{bmatrix}. \tag{14.1}$$

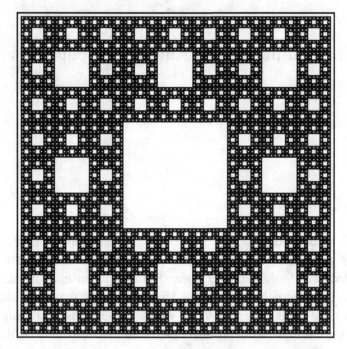

Figure 14.1: The Array **s**, the Sierpiński Carpet.

The Sierpiński carpet has a visually interesting pictorial interpretation, if we let 0 represent a white square and 1 represent a black square. Figure 14.1 displays the first 243 rows and columns of **s**.

The array **s** can be generated by a deterministic finite automaton with output that takes as input the base-3 expansions of i and j, and outputs $s_{i,j}$. Of course, we have to say more precisely how two different numbers can be simultaneously fed into the automaton.

Consider the automaton given in Figure 14.2. It accepts as input a string of *pairs of symbols* chosen from $\{0, 1\}$. The input is intended to represent the base-3 expansion of i (in the first entries) and j (in the second entries). Note that here representations are input starting with the most significant digit, and the shorter of the two representations is padded with leading zeros if necessary.

For example, to determine $s_{4,11}$, we expand 4 and 11 in base 3, obtaining 011 and 102, respectively. Then we input

$$[0, 1][1, 0][1, 2]$$

and get 1 as output.

The array **s** can also be generated by iterating a morphism. Now, however, our morphism must be *two-dimensional*, that is, it must map every symbol to a square array of symbols.

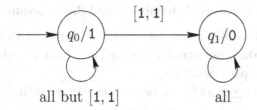

Figure 14.2: Automaton for the Sierpiński Carpet.

Figure 14.3: Iterating Haferman's Morphism.

Consider the morphism h defined as follows:

$$1 \to \begin{bmatrix} 1 & 1 & 1 \\ 1 & 0 & 1 \\ 1 & 1 & 1 \end{bmatrix}, \qquad 0 \to \begin{bmatrix} 0 & 0 & 0 \\ 0 & 0 & 0 \\ 0 & 0 & 0 \end{bmatrix}$$

Then $\mathbf{s}$ can be generated by iterating h, starting with the symbol 1; that is, $\mathbf{s} = h^{\omega}(1)$. For example, $h^2(1)$ gives the array in Eq. (14.1).

Example 14.1.2 Even simple two-dimensional morphisms, when iterated, can produce astounding results. Consider the following example, due to J. Haferman: let a matrix-valued morphism be defined by

$$0 \to \begin{bmatrix} 1 & 1 & 1 \\ 1 & 1 & 1 \\ 1 & 1 & 1 \end{bmatrix}, \qquad 1 \to \begin{bmatrix} 0 & 1 & 0 \\ 1 & 0 & 1 \\ 0 & 1 & 0 \end{bmatrix}.$$

We then obtain, after five iterations, the picture in Figure 14.3.

14.2 Formal Definitions and Basic Results

In this section we formally define our model of multidimensional automatic sequence. First, we extend our notation $(n)_k$ and $[w]_k$, introduced in Section 3.1, to pairs of integers and pairs of strings.

Let k, l be integers ≥ 2, and let w be a string over the alphabet

$$\Sigma_k \times \Sigma_l = \{0, 1, \ldots, k-1\} \times \{0, 1, \ldots, l-1\}.$$

If

$$w = [a_1, b_1][a_2, b_2] \cdots [a_j, b_j],$$

then we define $[w]_{k,l} = ([a_1 \cdots a_j]_k, [b_1 \cdots b_j]_l)$. Similarly, we define $(m, n)_{k,l}$ as follows: if $(m)_k = a_1 \cdots a_i$, and $(n)_l = b_1 \cdots b_j$, then

$$(m, n)_{k,l} = \begin{cases} [0, b_1] \cdots [0, b_{j-i}][a_1, b_{j-i+1}] \cdots [a_i, b_j] & \text{if } j \geq i, \\ [a_1, 0] \cdots [a_{i-j}, 0][a_{i-j+1}, b_1] \cdots [a_i, b_j] & \text{if } i > j. \end{cases}$$

We can also generalize the notion of k-fiber to two dimensions (and higher). If $\mathbf{a} = (a_{m,n})_{m,n \geq 0}$ is a double sequence defined over Δ, then $I_{k,l}(\mathbf{a}, d) = \{(m, n)_{k,l} : a_{m,n} = d\}$.

A *two-dimensional $[k, l]$-DFAO* is a 6-tuple $M = (Q, \Sigma, \delta, q_0, \Delta, \tau)$, where

Q is a finite nonempty set of states;
$\Sigma = \{0, 1, \ldots, k-1\} \times \{0, 1, \ldots, l-1\}$;
$\delta : Q \times \Sigma \to Q$ is the transition function;
q_0 is the initial state;
Δ is the output alphabet; and
$\tau : Q \to \Delta$ is the output mapping.

If $k = l$ then we sometimes write k-DFAO as shorthand for $[k, k]$-DFAO.

We say that such a DFAO generates the two-dimensional array $(u_{i,j})_{i,j \geq 0}$ if for all $m, n \geq 0$, and all $w \in \Sigma_k \times \Sigma_l$ such that $[w]_{k,l} = (m, n)$, we have $u_{m,n} = \tau(\delta(q_0, w))$. We say a two-dimensional array $\mathbf{u}$ is $[k, l]$-automatic if there is a $[k, l]$-DFAO generating it. If $k = l$, then we sometimes say $\mathbf{u}$ is k-automatic.

Just as in Section 5.2, we can show that this definition is robust, in the sense that it does not matter whether the input is read beginning with the least- or most-significant digit. More precisely, we have

Theorem 14.2.1 *The two-dimensional array $(u_{i,j})_{i,j \geq 0}$ is $[k, l]$-automatic if and only if there exists a DFAO $(Q, \Sigma_k \times \Sigma_l, \delta, q_0, \Delta, \tau)$ such that $u_{i,j} = \tau(\delta(q_0, (i, j)^R_{k,l}))$ for all $i, j \geq 0$.*

Proof. Left to the reader. ∎

We can also generalize the notion of k-kernel to two-dimensional arrays. We write

$$K_{k,l}(\mathbf{u}) = \{(u_{k^a \cdot m + r, l^a \cdot n + s})_{m,n \geq 0} : a \geq 0, \ 0 \leq r < k^a, \ 0 \leq s < l^a\}$$

for a two-dimensional array $\mathbf{u} = (u_{n,m})_{m,n \geq 0}$. If $k = l$, we sometimes write K_k instead of $K_{k,k}$.

Theorem 14.2.2 *The two-dimensional array* $\mathbf{u} = (u_{n,m})_{m,n \geq 0}$ *is* $[k, l]$-*automatic if and only if the* $[k, l]$-*kernel* $K_{k,l}(\mathbf{u})$ *is finite.*

Proof. (Sketch.) $\Longrightarrow$: Suppose that $(u_{n,m})_{m,n \geq 0}$ is $[k, l]$-automatic. Then it can be proved (in analogy with Theorem 5.2.3) that there exists a DFAO $(Q, \Sigma, \delta, q_0, \Delta, \tau)$ such that

$$u_{m,n} = \tau \left(\delta \left(q_0, (m, n)_{k,l}^R [0, 0]^t \right) \right)$$

for all $t \geq 0$. Now let w be a word over $(\{0, 1, 2, \ldots, k - 1\} \times \{0, 1, \ldots, l - 1\})^*$ such that

$$|w| = a \quad \text{and} \quad [w]_{k,l} = (i, j).$$

Define $q = \delta(q_0, w^R)$. Since for $(m, n) \neq (0, 0)$ we have

$$((k^a \cdot m + i, l^a \cdot n + j))_{k,l} = (m, n)_{k,l} w,$$

it follows that

$$\delta(q_0, (k^a \cdot m + i, l^a \cdot n + j)_{k,l}^R) = \delta(\delta(q_0, w^R), (m, n)_{k,l}^R)$$
$$= \delta(q, (m, n)_{k,l}^R),$$

except possibly when $(m, n) = (0, 0)$.

Hence, except possibly at $(m, n) = (0, 0)$, the subsequence $(u_{k^a \cdot m + i, l^a \cdot n + j})_{m,n \geq 0}$ is accepted by the same $[k, l]$-DFAO, but with q replacing q_0 as the start state. Hence $|K_{k,l}(\mathbf{u})|$ is finite.

$\Longleftarrow$: Follow Theorem 6.6.2. ∎

Similarly, we can generalize the notion of uniform morphism to two dimensions (and higher). If

$$A = (a_{ij})_{\substack{0 \leq i < m \\ 0 \leq j < n}}$$

is an $m \times n$ matrix with entries in Σ, and $\varphi : \Sigma \to \Delta^{k \times l}$ is a $[k, l]$-uniform matrix-valued morphism, i.e., a map sending each letter in Σ to an $k \times l$ matrix, then $\varphi(A)$

is a $km \times ln$ matrix given by

$$
\begin{bmatrix}
\varphi(a_{00}) & \varphi(a_{01}) & \cdots & \varphi(a_{0,n-1}) \\
\varphi(a_{10}) & \varphi(a_{11}) & \cdots & \varphi(a_{1,n-1}) \\
\vdots & \vdots & \ddots & \vdots \\
\varphi(a_{m-1.0}) & \varphi(a_{m-1,1}) & \cdots & \varphi(a_{m-1,n-1})
\end{bmatrix}.
$$

Theorem 14.2.3 *The sequence* $(u_{m,n})_{m,n \geq 0}$ *is* $[k, l]$-*automatic if and only if it is the image (under a coding) of a fixed point of a* $[k, l]$-*morphism.*

Proof. Imitate the proof of Theorem 6.3.2. ∎

The following theorem gives a method for generating $[k, l]$-automatic two-dimensional arrays:

Theorem 14.2.4 *Let* $(a_m)_{m \geq 0}$ *be a* k-*automatic sequence over an alphabet* Δ_1, *and* $(b_n)_{n \geq 0}$ *be an* l-*automatic sequence over an alphabet* Δ_2. *Let* f *be any function mapping* $\Delta_1 \times \Delta_2 \to \Delta$. *Then* $\mathbf{c} = (f(a_m, b_n))_{m,n \geq 0}$ *is a* $[k, l]$-*automatic two-dimensional array.*

Proof. Since $\mathbf{a} = (a_m)_{m \geq 0}$ is k-automatic, it follows that $K_k(\mathbf{a})$ is finite, say $K_k(\mathbf{a}) = \{(a_{k^i m+r})_{m \geq 0} : (i, r) \in S\}$ for some finite set S. Similarly, since $\mathbf{b} = (b_n)_{n \geq 0}$, it follows that $K_l(\mathbf{b})$ is finite, say $K_l(\mathbf{b}) = \{(b_{l^j n+s})_{n \geq 0} : (j, s) \in T\}$ for some finite set T. It now follows immediately that

$$
K_{k,l}(\mathbf{c}) \subseteq \{(f(a_{k^i m+r}, b_{l^j n+s}))_{m,n \geq 0} : (i, r) \in S, \ (j, s) \in T\},
$$

which is clearly finite. ∎

Example 14.2.5 Consider the sequence $\mathbf{w} := 0101100101100011010 \cdots$ from Exercise 1.49. If we define $f(x, y) = (x + y) \bmod 2$, then we get a two-dimensional 3-automatic sequence $\mathbf{A} = (a_{i,j})_{i,j \geq 0}$, part of which is portrayed in Figure 14.4. Wegner called this "Reverend Back's abbey floor".

We now show that the class of k-automatic two-dimensional arrays is closed under periodic indexing. We also show that taking the generalized diagonal of a k-automatic two-dimensional array yields a k-automatic sequence.

Theorem 14.2.6 *Let* $\mathbf{s} = (s_{m,n})_{m,n \geq 0}$ *be a* k-*automatic two-dimensional array. Then:*

(a) If a, b, c, d, e, f *are integers with* $a, b, d, e \geq 0$, *the two-dimensional array*

$$
(s_{am+bn+c,dm+en+f})_{m,n \geq 0}
$$

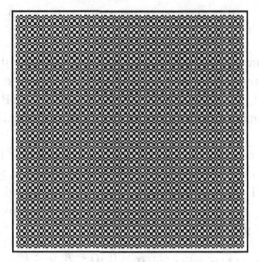

Figure 14.4: Reverend Back's Abbey Floor.

is also k-automatic (where, as usual, entries corresponding to negative indexes are deemed to be 0).

(b) If a, b, c, d are integers with $a, c \geq 0$, the sequence $(s_{an+b,cn+d})_{n\geq 0}$ is k-automatic.

Proof. (a): It is possible to prove this using a method similar to the proof of Theorem 6.8.1, but here we sketch a different proof.

The idea is to use the technique of Theorem 6.8.6. Roughly speaking, on input $(m, n)_k^R$, we transduce the input to $(am + bn + c, dm + en + f)_k^R$ before processing it with the DFAO for **s**. A slight complication comes from the fact that $am + bn + c$, for example, may have more digits than m. We can get around this problem by simulating an input followed by some fixed number of trailing zeros (that may depend on a, b, c, d, e, f), and using a theorem similar to Theorem 5.2.4.

(b): The same idea used for (a) works. On input $(n)_k^R$, we transduce the input to $(an + b, cn + d)_k^R$ before processing it with the DFAO for **s**. ∎

We give a sort of converse of this theorem below.

Theorem 14.2.7 *Let $(s_{m,n})_{m,n\geq 0}$ be a two-dimensional array with values in a finite set such that there exist two integers $a \geq 1$ and $b \geq 1$ for which all the sequences $(s_{am+c,bn+d})_{m,n\geq 0}$ with $c \in \{0, 1, \ldots, a - 1\}$, $d \in \{0, 1, \ldots, b - 1\}$ are k-automatic for some integer $k \geq 2$. Then the two-dimensional array $(s_{m,n})_{m,n\geq 0}$ itself is k-automatic.*

Proof. Our proof will mimic the proof of the analogous claim for the one-dimensional case. We first note that it suffices to prove the following assertions.

(a) If $(w_{am+c,n})_{m,n\geq 0}$ is k-automatic for every $c \in \{0, 1, \ldots, a - 1\}$, then $(w_{m,n})_{m,n\geq 0}$ is k-automatic.

(b) If $(w_{m,bn+d})_{m,n\geq 0}$ is k-automatic for every $d \in \{0, 1, \ldots, b-1\}$, then $(w_{m,n})_{m,n\geq 0}$ is k-automatic.

If (a) and (b) are proved, let $(s_{m,n})_{m,n\geq 0}$ be a sequence that has the property given in Theorem 14.2.7 above. Then for every fixed $d \in [0, b-1]$ the sequence $(s_{am+c,bn+d})_{m,n\geq 0}$ is k-automatic for any $c \in [0, a-1]$. By (a) the sequence $(s_{m,bn+d})_{m,n\geq 0}$ is k-automatic for all $d \in [0, b-1]$. Now, (b) implies that $(s_{m,n})_{m,n\geq 0}$ is k-automatic.

We conclude the proof by showing the validity of (a) and (b). It suffices to prove (a). Suppose that, for some integer $k \geq 2$, for some integer $a \geq 2$, and for every $c \in \{0, 1, \ldots, a-1\}$, the sequence $(w_{am+c,n})_{m,n\geq 0}$ is k-automatic. To prove that the sequence $\mathbf{w} = (w_{m,n})_{m,n\geq 0}$ itself is k-automatic, we have to prove that its k-kernel, i.e., the set of subsequences

$$\{(w_{k^\alpha m+\beta, k^\alpha n+\gamma})_{m,n\geq 0} : \alpha \geq 0, \ 0 \leq \beta, \gamma \leq k^\alpha - 1\}$$

is finite. Therefore it suffices to prove that there are only finitely many sequences of the form

$$(w_{k^\alpha(am+c)+\beta, k^\alpha n+\gamma})_{m,n\geq 0}$$

with $c \in \{0, 1, \ldots, a-1\}$, $\alpha \geq 0$, and $0 \leq \beta, \gamma \leq k^\alpha - 1$. Write $k^\alpha c + \beta = ax + y$, with $0 \leq y \leq a-1$. We have $ax \leq ax + y = k^\alpha c + \beta < k^\alpha (c+1) \leq ak^\alpha$. Hence $x < k^\alpha$, i.e., $x \leq k^\alpha - 1$. Then $w_{k^\alpha(am+c)+\beta, k^\alpha n+\gamma} = w_{ak^\alpha(m+x)+y, k^\alpha n+\gamma}$. The numbers x and y do not depend on (m, n), but only on α, β, and c. Furthermore, $y \leq a-1$ and $x \leq k^\alpha - 1$.

Hence the sequence $(w_{a(k^\alpha m+x)+y, k^\alpha n+\gamma})_{m,n\geq 0}$ is in the k-kernel of the sequence $(w_{am+y,n})_{m,n\geq 0}$, which ensures there are only finitely many possibilities. More precisely, there are finitely many sequences $(w_{am+y,n})_{m,n\geq 0}$ $(0 \leq y \leq a-1)$, and each of them is k-automatic. ∎

Here is another method for generating $[k, l]$-automatic two-dimensional arrays in the case $k = l$:

Theorem 14.2.8 *Let* $\mathbf{s} = (s_i)_{i\geq 0}$ *be a k-automatic sequence. Then* $(s_{am+bn+c})_{m,n\geq 0}$ *is a k-automatic two-dimensional array.*

Proof. As in the proof of Theorem 14.2.6, on input $(m, n)_k^R$, we transduce this input to $(am + bn + c)_k^R$ before feeding it into the DFAO for $\mathbf{s}$. ∎

14.3 Subword Complexity

One can generalize the notion of subword complexity for infinite sequences, as discussed in Chapter 10, to the two-dimensional case. If $\mathbf{x}$ is an infinite two-dimensional array, then for integers $m, n \geq 0$ we define $p_\mathbf{x}(m, n)$ to be the number of distinct $m \times n$ blocks appearing in $\mathbf{x}$.

Theorem 14.3.1 *Let h be a $[k, l]$-uniform matrix-valued morphism, prolongable on a letter $a \in \Sigma$. Let $\tau : \Sigma \to \Delta$ be a coding, and define $\mathbf{x} = \tau(h^\omega(a))$. Given integers $m, n \geq 1$, define $c = 1 + \max(\lfloor \frac{\log m}{\log k} \rfloor, \lfloor \frac{\log n}{\log l} \rfloor)$. Then $p_\mathbf{x}(m, n) \leq d^4 k^c l^c$, where $d = \mathrm{Card}\, \Sigma$.*

Proof. Let r, s be integers such that $k^{r-1} \leq m < k^r$ and $l^{s-1} \leq n < l^s$. Then $c = \max(r, s)$. Let $b = \mathbf{x}[i..i + m - 1, j..j + n - 1]$ be any subblock of dimension $m \times n$. Define $e = \lfloor i/k^c \rfloor$ and $f = \lfloor j/l^c \rfloor$. If M is defined to be the 2×2 matrix $\mathbf{x}[e..e + 1, f..f + 1]$, then $h^c(M) = \mathbf{x}[k^c e..k^c(e + 2) - 1, l^c f..l^c(f + 2) - 1]$, and hence contains b as a subblock. It follows that b is completely determined by the 4 entries of M and the position of b within $h^c(M)$. Hence we find $p_\mathbf{x}(m, n) \leq d^4 k^c l^c$, as desired. ∎

Corollary 14.3.2 *The subword complexity of the two-dimensional k-automatic array $\mathbf{x}$ satisfies $p_\mathbf{x}(m, n) = O(\max(m, n)^2)$.*

Proof. Let $k = l$. Then from Theorem 14.3.1 we get $p_\mathbf{x}(m, n) \leq d^4 k^{2c}$. But

$$k^{2c} = k^{2\max(r,s)} \leq k^{2\max(1 + \frac{\log m}{\log k}, 1 + \frac{\log n}{\log k})} \leq k^2 \max(m, n)^2.$$

It follows that $p_\mathbf{x}(m, n) \leq d^4 k^2 \max(m, n)^2$. ∎

14.4 Formal Power Series

We can also generalize Theorem 12.2.5, the Christol theorem, to the multidimensional case. The proper generalization uses multivariate power series. Let $(u_{m,n})_{m,n \geq 0}$ be a double sequence over a field F; then its *associated formal power series* is

$$G(X, Y) = \sum_{\substack{m \geq 0 \\ n \geq 0}} u_{m,n} X^m Y^n \in F[[X, Y]].$$

We say that G is algebraic if (as before) there exist polynomials $p_0, p_1, \ldots, p_r \in F[X, Y]$ such that

$$\sum_{0 \leq i \leq r} p_i(X, Y) G^i(X, Y) = 0.$$

Theorem 14.4.1 *Let p be a prime number. The sequence $(u_{m,n})_{m,n \geq 0}$, over Δ, is p-automatic if and only if there exists an integer $n \geq 1$ and an injective map*

$b : \Delta \to GF(p^a)$ *such that*

$$\sum_{\substack{m \geq 0 \\ n \geq 0}} b(u_{m,n}) X^m Y^n$$

is algebraic over $GF(p^a)[X, Y]$.

Proof. Mimic the proof of Theorem 12.2.5. ∎

We give an interesting corollary:

Theorem 14.4.2 *Let* $q = p^t$, *p prime*, $t \geq 1$. *Let*

$$G(X, Y) = \sum_{\substack{m \geq 0 \\ n \geq 0}} g_{m,n} X^m Y^n$$

be a formal power series in $GF(q)[[X, Y]]$ *representing an algebraic function (algebraic over* $GF(q)[X, Y]$). *Then the diagonal series*

$$D(Z) = \sum_{m \geq 0} g_{m,m} Z^m$$

is also algebraic (over $GF(q)[Z]$).

Proof. Let G be algebraic. Then the double sequence $\mathbf{g} = (g_{m,n})_{m,n \geq 0}$ is p-automatic. Hence the p-kernel

$$K_p(\mathbf{g}) = \{(g_{p^a \cdot m + r, \, p^a \cdot n + s})_{m,n \geq 0} : a \geq 0, \, 0 \leq r < p^a, \, 0 \leq s < p^a\}$$

is finite. Hence the set

$$S = \{(g_{p^a \cdot m + r, \, p^a \cdot m + r})_{m \geq 0} : a \geq 0, \, 0 \leq r < p^a\}$$

is finite. Define $u_m = g_{m,m}$ and $\mathbf{u} = (u_m)_{m \geq 0}$. Then $K_p(\mathbf{u}) = S$ is finite. Hence $\mathbf{u}$ is automatic, and so, from Theorem 12.2.5,

$$\sum_{m \geq 0} u_m Z^m = \sum_{m \geq 0} g_{m,m} Z^m$$

is algebraic over $GF(q)[Z]$. ∎

14.5 Automatic Sequences in Base $-1 + i$

As we have seen in Theorem 3.10.5, every Gaussian integer $x + yi$ can be represented essentially uniquely in the form

$$x + yi = \sum_{0 \leq j < r} e_j (-1 + i)^j,$$

where $e_j \in \Sigma_2 = \{0, 1\}$. (By "essentially uniquely", we mean that $e_{r-1} \neq 0$ if $(x, y) \neq (0, 0)$.)

Thus for sequences defined over $\mathbb{Z}[i]$ we can define a notion of $(-1 + i)$-automatic sequence, in analogy with k-automatic sequences for integer $k \geq 2$. Such a sequence is accepted by an automaton that takes as input the expansion of $x + yi$ in base $-1 + i$ and outputs an element of Δ. Similarly, for a sequence $\mathbf{a} = (a_z)_{z \in \mathbb{Z}[i]}$ defined over Δ and an element $d \in \Delta$, we can can define the fiber $I_{-1+i}(\mathbf{a}, d)$ to be the set $\{(z)_{-1+i} : a_z = d\}$.

In this section we sketch a proof of the following

Theorem 14.5.1 *The sequence* $(a_z)_{z \in \mathbb{Z}[i]}$ *from the Gaussian integers to a finite set* Δ *is computable by a finite automaton (i.e., is* $(-1 + i)$*-"automatic") if and only if each of the four double sequences*

$\mathbf{a}_1 := (a_{m+ni})_{m,n \geq 0};$
$\mathbf{a}_2 := (a_{-m+ni})_{m,n \geq 0};$
$\mathbf{a}_3 := (a_{m-ni})_{m,n \geq 0};$
$\mathbf{a}_4 := (a_{-m-ni})_{m,n \geq 0}$

is 2-automatic.

Proof. As in the proof of Theorem 5.3.2, it suffices to

(a) create a transducer to convert from two base-2 representations to a single base-$(-1 + i)$ representation; and
(b) show that we can determine, using a finite-state machine, to which quadrant a given $(-1 + i)$-representation belongs.

Here is an outline of the argument assuming (a) and (b) have been completed. If $(a_z)_{z \in \mathbb{Z}[i]}$ is $(-1 + i)$-automatic over an alphabet Δ, then the fibers $I_{-1+i}(\mathbf{a}, d)$ are regular languages for $d \in \Delta$. Further, the partitions of these fibers corresponding to each quadrant, e.g.,

$$I_{-1+i}^{++} = \{(z)_{-1+i} : a_z = d \text{ and } \Re(z) \geq 0, \Im(z) \geq 0\},$$
$$I_{-1+i}^{+-} = \{(z)_{-1+i} : a_z = d \text{ and } \Re(z) \geq 0, \Im(z) \leq 0\},$$
$$I_{-1+i}^{-+} = \{(z)_{-1+i} : a_z = d \text{ and } \Re(z) \leq 0, \Im(z) \geq 0\},$$
$$I_{-1+i}^{--} = \{(z)_{-1+i} : a_z = d \text{ and } \Re(z) \leq 0, \Im(z) \leq 0\}$$

are also regular. Now, the inverse transduction of each of these sets (converting each representation in base $-1 + i$ to a pair of base-2 representations) is regular. Hence each of the double sequences $\mathbf{a}_1$, $\mathbf{a}_2$, $\mathbf{a}_3$, and $\mathbf{a}_4$ is 2-automatic.

On the other hand, if the four double sequences are 2-automatic, then each of the fibers $I_2(\mathbf{a}_1, d)$, $I_2(\mathbf{a}_2, d)$, $I_2(\mathbf{a}_3, d)$, and $I_2(\mathbf{a}_4, d)$ is regular, for all $d \in \Delta$. Hence the transductions of these sets into the corresponding representations in base $-1 + i$

are regular. The unions of the fibers corresponding to the four quadrants are regular, and hence the sequence $(a_z)_{z \in \mathbb{Z}[i]}$ is $(-1 + i)$-automatic. This completes the sketch of the argument.

Now we describe a transducer $M = (Q, \Sigma, \delta, q_0, \Delta, \tau)$ that converts its input from base 2 to base $-1 + i$. (Actually, there are four separate transducers, depending on the quadrant of the input. These transducers differ only in the choice of the initial state. However, it is easier to speak of there being only one transducer.)

The transducer is fed with the the base-2 expansion of the absolute values of x and y, least significant bit first, and it outputs the base-$(-1 + i)$ representation of $z := x + iy$, least significant bit first. We write

$$|x| = [a_{r-1}a_{r-2} \cdots a_1 a_0]_2,$$

$$|y| = [b_{r-1}b_{r-2} \cdots b_1 b_0]_2,$$

and the input to the transducer is a sequence of pairs of bits:

$$[a_0, b_0] \, [a_1, b_1] \, \cdots \, [a_{r-1}, b_{r-1}].$$

The output is $e_0 e_1 \cdots e_{2r-1}$, where

$$z = \sum_{0 \le j < 2r} e_j(-1 + i)^j.$$

This transducer has 32 states. We define

$$Q = \{0, 1\} \times \{-1, 1\} \times \{-1, 1\} \times \{0, 1\} \times \{0, 1\},$$
$$\Sigma = \Sigma_2 \times \Sigma_2,$$
$$\Delta = \Sigma_2.$$

Each state of Q is of the form (d, s_x, s_y, c_x, c_y), where d governs whether the first entry (x) or the second entry (y) is the real part, s_x governs the sign of x, s_y governs the sign of y, c_x is the carry bit for x, and c_y is the carry bit for y. The initial state depends on the sign of x and y, as follows:

$$q_0 = \begin{cases} (0, +1, +1, 0, 0) & \text{if } x \ge 0, y \ge 0, \\ (0, -1, +1, 0, 0) & \text{if } x < 0, y \ge 0, \\ (0, +1, -1, 0, 0) & \text{if } x \ge 0, y < 0, \\ (0, -1, -1, 0, 0) & \text{if } x < 0, y < 0. \end{cases}$$

The idea is for the automaton to keep track of z_j, where $z_0 = z$ and

$$z_{j+1} = \frac{z_j - t_j}{-2i}.$$

for $j \geq 0$. Here t_j is defined as follows, where $z_j = x_j + y_j i$:

$$
t_j = \begin{cases} 0 & \text{if } x_j \equiv 0, \ y_j \equiv 0 \pmod 2, \\ i & \text{if } x_j \equiv 0, \ y_j \equiv 1 \pmod 2, \\ 1 & \text{if } x_j \equiv 1, \ y_j \equiv 0 \pmod 2, \\ -1 + i & \text{if } x_j \equiv 1, \ y_j \equiv 1 \pmod 2. \end{cases}
$$

In fact, the transducer is designed so that the automaton has read

$$[a_0, b_0] \, [a_1, b_1] \, \cdots \, [a_{s-1}, b_{s-1}]$$

(with

$$[a_s, b_s] \, [a_{s+1}, b_{s+1}] \, \cdots \, [a_{r-1}, b_{r-1}]$$

yet to be read), then we are in state (d, s_x, s_y, c_x, c_y), where

$$
z_s = \begin{cases} ([a_{r-1} \cdots a_s]_2 + c_x)s_x + ([b_{r-1} \cdots b_s]_2 + c_y)s_y i & \text{if } d = 0, \\ ([a_{r-1} \cdots a_s]_2 + c_x)s_x i + ([b_{r-1} \cdots b_s]_2 + c_y)s_y & \text{if } d = 1. \end{cases}
$$

Note that it is possible to compute t_s from knowledge of a_s, c_x, b_s, c_y alone. For we have, for $d = 0$,

$$
\begin{aligned}
x_s &\equiv a_s + c_x \pmod 2, \\
y_s &\equiv b_s + c_y \pmod 2,
\end{aligned}
$$

while for $d = 1$ we have

$$
\begin{aligned}
x_s &= b_s + c_y \pmod 2, \\
y_s &= a_s + c_x \pmod 2.
\end{aligned}
$$

We define

$$\delta\big((0, s_x, s_y, c_x, c_y), [a_s, b_s]\big) := \big(1, s_x', s_y', c_x', c_y'\big),$$

where

$$
(s_x', s_y') = \begin{cases} (1, -1) & \text{if } (s_x, s_y) = (1, 1), \\ (1, 1) & \text{if } (s_x, s_y) = (1, -1), \\ (-1, -1) & \text{if } (s_x, s_y) = (-1, 1), \\ (-1, 1) & \text{if } (s_x, s_y) = (-1, -1), \end{cases}
$$

and

$$\frac{(a_s + c_x)s_x + (b_s + c_y)s_y i - t_s}{-2i} = c_x' s_x' i + c_y' s_y'.$$

We also define

$$\delta\big((1, s_x, s_y, c_x, c_y), [a_s, b_s]\big) := \big(0, s_x', s_y', c_x', c_y'\big),$$

where

$$
(s'_x, s'_y) = \begin{cases}
(-1, 1) & \text{if } (s_x, s_y) = (1, 1), \\
(-1, -1) & \text{if } (s_x, s_y) = (1, -1), \\
(1, 1) & \text{if } (s_x, s_y) = (-1, 1), \\
(1, -1) & \text{if } (s_x, s_y) = (-1, -1),
\end{cases}
$$

and

$$
\frac{(a_s + c_x)s_x i + (b_s + c_y)s_y - t_s}{-2i} = c'_x s'_x + c'_y s'_y i.
$$

It can be verified that $0 \le c'_x, c'_y \le 1$. The output map τ is given by

$$
\tau \left((0, s_x, s_y, c_x, c_y), [a_s, b_s] \right) = \begin{cases}
00 & \text{if } a_s + c_x \equiv 0; b_s + c_y \equiv 0 \pmod 2, \\
11 & \text{if } a_s + c_x \equiv 0; b_s + c_y \equiv 1 \pmod 2, \\
10 & \text{if } a_s + c_x \equiv 1; b_s + c_y \equiv 0 \pmod 2, \\
01 & \text{if } a_s + c_x \equiv 1; b_s + c_y \equiv 1 \pmod 2,
\end{cases}
$$

and

$$
\tau \left((1, s_x, s_y, c_x, c_y), [a_s, b_s] \right) = \begin{cases}
00 & \text{if } a_s + c_x \equiv 0; b_s + c_y \equiv 0 \pmod 2, \\
11 & \text{if } a_s + c_x \equiv 1; b_s + c_y \equiv 0 \pmod 2, \\
10 & \text{if } a_s + c_x \equiv 0; b_s + c_y \equiv 1 \pmod 2, \\
01 & \text{if } a_s + c_x \equiv 1; b_s + c_y \equiv 1 \pmod 2.
\end{cases}
$$

Note that the base-$(-1 + i)$ expansion of a Gaussian integer $x + iy$ may be slightly longer than twice the length of the base-2 expansion of $\max(x, y)$. This means that in order to properly convert from base 2 to base $-1 + i$, we need to append leading zeros to the expansion (or, since we work with the least significant digit first, trailing zeros). However, by Exercise 3.41(b), we only need to append a constant number of zeros.

To complete the proof, it suffices to show that we can determine the sign of the real and imaginary parts of z with a finite automaton, given the expansion $(z)_{-1+i}$ in base $-1 + i$.

To do this, we break the representation $(z)^R_{-1+i} = e_0 e_1 e_2 \cdots e_{r-1}$ into groups of size 8, possibly padding with leading zeros at the right (most significant bit), if necessary:

$$
\begin{array}{ccccc}
e_0 e_1 \cdots e_7 & e_8 \cdots e_{15} & e_{16} \cdots e_{23} & \cdots & e_{8k} \cdots e_{8k+7} \\
\| & \| & \| & & \| \\
A_0 & A_1 & A_2 & \cdots & A_k
\end{array}
$$

Next, we create a dictionary mapping each 8-bit pattern $f = f_0 f_1 f_2 \cdots f_7$ to

$$
(\operatorname{sgn} \Re([f]_{-1+i}), \ \operatorname{sgn} \Im([f]_{-1+i}))
$$

as follows:

1	$-1+i$	$-2i$	$2+2i$	-4	$4-4i$	$8i$	$-8-8i$	sgn $\Re$	sgn $\Im$
0	0	0	0	0	0	0	0	0	0
0	0	0	0	0	0	0	1	-1	-1
0	0	0	0	0	0	1	0	0	$+1$
0	0	0	0	0	0	1	1	-1	0
0	0	0	0	0	1	0	0	$+1$	-1
$\vdots$	$\vdots$	$\vdots$	$\vdots$	$\vdots$	$\vdots$	$\vdots$	$\vdots$	$\vdots$	$\vdots$
1	1	1	1	1	1	1	1	-1	-1

Next, we build a finite automaton that computes the above mapping for each 8-bit group $A_0 A_1 \cdots A_k$. Then the correct quadrant is given by the last (i.e., rightmost) nonzero sign in each component. For example,

$$
\begin{array}{cccc}
A_0 & A_1 & A_2 & A_3 \\
(0, +1) & (+1, -1) & (0, -1) & (0, +1) & \to & (+1, +1).
\end{array}
$$

The proof is easy and is left to the reader. ∎

Example 14.5.2 Use the transducer above to convert $43 + 24i$ to base $-1 + i$:

$$
(43)_2 = 101011,
$$
$$
(24)_2 = 011000.
$$

The input is then

$$
[1, 0]\,[1, 0]\,[0, 0]\,[1, 1]\,[0, 1]\,[1, 0],
$$

and the states and outputs are:

State	z_i	Input	Output
$(0, 1, 1, 0, 0)$	$z_0 = 43 + 24i$	$[1, 0]$	10
$(1, 1, -1, 0, 0)$	$z_1 = 21i - 12$	$[1, 0]$	11
$(0, -1, -1, 0, 0)$	$z_2 = -10 - 6i$	$[0, 0]$	00
$(1, -1, 1, 0, 0)$	$z_3 = -5i + 3$	$[1, 1]$	01
$(0, 1, 1, 1, 1)$	$z_4 = 3 + 2i$	$[0, 1]$	10
$(1, 1, -1, 0, 1)$	$z_5 = i - 1$	$[1, 0]$	01
$(0, -1, -1, 0, 0)$	$z_6 = 0$		

Hence the representation of $43 + 24i$ in base $-1 + i$ is 100110001101.

Example 14.5.3 Convert $1 - i$. The input is (note that we need to add leading zeros)

$$
[1, 1]\,[0, 0]\,[0, 0],
$$

and the states and inputs are:

State	z_i	Input	Output
$(0, 1, -1, 0, 0)$	$z_0 = 1 - i$	$[1, 1]$	01
$(1, 1, 1, 1, 1)$	$z_1 = i + 1$	$[0, 0]$	01
$(0, -1, 1, 0, 1)$	$z_2 = i$	$[0, 0]$	11
$(1, -1, -1, 0, 0)$	$z_3 = 0$		

Hence the representation of $1 - i$ in base $-1 + i$ is 111010.

14.6 The Pascal Triangle Modulo d

In this section we study the two-dimensional array defined by taking the Pascal triangle modulo d, where d is an integer ≥ 2, i.e., the two-dimensional sequence $\left(\binom{m}{n}\right)_{m,n \geq 0}$, with as usual $\binom{m}{n} = 0$ if $m < n$. We begin with a general theorem.

Theorem 14.6.1 *Let $R(X)$, $G(X)$ be two polynomials in $\mathcal{R}[X]$, where $\mathcal{R}$ is a finite commutative ring with unit element. Suppose that there exists an integer $k \geq 2$ such that $R(X^k) = R^k(X)$. Then the sequence $(s_{m,n})_{m,n \geq 0}$ defined by $G(X)R(X)^n = \sum_{m \geq 0} s_{m,n} X^m$ is k-automatic.*

Proof. The sequence $\mathbf{s} = (s_{m,n})_{m,n \geq 0}$ is k-automatic if and only if its k-kernel is finite. Clearly, this is equivalent to the existence of a set of sequences S such that:

the set S is finite;
the sequence $\mathbf{s}$ belongs to S;
the set S is invariant under the maps $\varphi_{u,v}$ defined for $0 \leq u, v \leq k - 1$ and any sequence $\mathbf{w} = (w_{m,n})_{m,n \geq 0}$ by

$$\varphi_{u,v}\left((w_{m,n})_{m,n \geq 0}\right) = \left((w_{km+u,kn+v})_{m,n \geq 0}\right).$$

Now, if H is a polynomial in $\mathcal{R}(X)$, say $H(X) = \sum b_n X^n$, define $\Lambda_u(H)$, for $0 \leq u \leq k - 1$, to be the polynomial $\Lambda_u(H)(X) = \sum b_{kn+u} X^n$ (see Definition 12.2.1). Note that $\deg \Lambda_u(H) \leq \frac{\deg H}{k}$, and for two polynomials A and B we have $\Lambda_u(A(X)B(X^k)) = B(X)\Lambda_u(A(X))$ (see Eq. (12.3)).

Now the sequence $\mathbf{s} = (s_{m,n})_{m,n \geq 0}$ is defined by $G(X)R(X)^n = \sum_{m \geq 0} s_{m,n} X^m$. Let $M = \deg G + (k - 1)\deg R$, and let S be the set

$$S := \left\{ \mathbf{a} = (a_{m,n})_{m,n \geq 0}; \ \exists H \in \mathcal{R}[X], \ \deg H \leq M; \right.$$

$$\left. \sum_{m \geq 0} a_{m,n} X^m = H(X)R(X)^n \right\}.$$

As H belongs to a finite set of polynomials (the ring $\mathcal{R}$ is finite), the set S is finite. This set contains the sequence R (take $H = G$). Let us prove that S is stable under the maps $\varphi_{u,v}$. Let $\mathbf{a} = (a_{m,n})_{m,n \geq 0}$ be a sequence in S, and let H be such that $H(X)R(X)^n = \sum_{m \geq 0} a_{m,n} X^m$ for all $n \geq 0$. Then for all $v \leq k - 1$ and for all integers n we have that

$$H(X)R(X)^{kn+v} = \sum_m a_{m,kn+v} X^m = \sum_{0 \leq u < k} X^u \sum_m a_{km+u,kn+v} X^{km}.$$

On the other hand, $H(X)R(X)^{kn+v} = (H(X)R(X)^v)(R(X^k))^n$. Hence

$$\Lambda_u(HR^v)R^n = \sum_m a_{km+u,kn+v} X^m.$$

Since $\deg \Lambda_u(HR^v) \leq \frac{M+(k-1)\deg R}{k} \leq M$, and $k \geq 2$, we deduce that the sequence $(a_{km+u,kn+v})_{m,n \geq 0}$ belongs to S. ∎

We now state the main theorem of this section.

Theorem 14.6.2 *The sequence* $\left(\binom{n}{m} \bmod d\right)_{m,n \geq 0}$ *is k-automatic if and only if the integers d and k are powers of the same prime number p. In this case the sequence is p^j-automatic for any $j \geq 0$.*

Proof. We first prove that, if $d = p^\ell$ for some prime number p and some $\ell \geq 1$, then the sequence $\left(\binom{n}{m} \bmod d\right)_{m,n \geq 0}$ is p-automatic (and thus p^j-automatic for any $j \geq 1$). We know (see Exercise 2.29) that the polynomial $R(X) := (1 + X)^{p^{\ell-1}}$ has the property that $R(X^p) \equiv R^p(X) \pmod{p^\ell}$. Hence, applying Theorem 14.6.1 with $\mathcal{R} := \mathbb{Z}/p^\ell\mathbb{Z}$, $R(X) := (1 + X)^{p^{\ell-1}}$, $G(X) := (1 + X)^t$, and $k := p$, we get

$$\sum_m s_{m,n} X^m \equiv G(X)R(X)^n \equiv (1 + X)^{p^{\ell-1}n+t} \equiv \sum_m \binom{p^{\ell-1}n + t}{m} X^m \pmod{p^\ell},$$

and the sequence

$$\left(\binom{p^{\ell-1}n + t}{m} \bmod p^\ell\right)_{m,n \geq 0}$$

is p-automatic for every $t \geq 0$. This holds in particular for every $t \in [0, p^{\ell-1} - 1]$. Hence, applying Theorem 14.2.7, we have that the sequence $\left(\binom{n}{m} \bmod p^\ell\right)_{m,n \geq 0}$ is p-automatic.

Let us now prove that the sequence $\left(\binom{n}{m} \bmod d\right)_{m,n \geq 0}$ is not k-automatic for any $k \geq 2$ if $d \geq 2$ is not a prime power. We distinguish two cases.

Case 1: The integer d is divisible by two distinct odd primes. We note the following formula, valid on the rational numbers (see Exercise 2.28):

$$\sum_{t \geq 0} \binom{2t}{t} X^t = (1 - 4X)^{-\frac{1}{2}}.$$

Hence, defining the formal power series $F(X) := \sum_{t \geq 0} \binom{2t}{t} X^t$, we have

$$(1 - 4X)F(X)^2 - 1 = 0.$$

As this relation holds in $\mathbb{Z}[[X]]$, it also holds in $\mathbb{Z}/p\mathbb{Z}[[X]]$ for every prime number p. This proves that the series F is algebraic over the field of rational functions $\mathbb{Z}/p\mathbb{Z}(X)$. Furthermore, if $p \neq 2$, this series is not rational: if we had $F = \frac{P}{Q}$ for two polynomials P and Q in $\mathbb{Z}/p\mathbb{Z}[X]$, P and Q coprime, then $(1 - 4X)P^2 = Q^2$, hence Q^2 would divide $1 - 4X$. This would imply that Q is a constant polynomial, and give the desired contradiction.

Hence, from Theorem 12.2.5, the sequence $(\binom{2t}{t} \bmod p)_{t \geq 0}$ is p-automatic, and it is not ultimately periodic if p is an odd prime number.

Now suppose that the sequence $(\binom{n}{m} \bmod d)_{m,n \geq 0}$ is k-automatic for some integer $k \geq 2$. Therefore the one-dimensional sequence $(\binom{2t}{t} \bmod d)_{t \geq 0}$ is k-automatic (Theorem 14.2.6(b)). Let p_1 and p_2 be two different odd prime divisors of d. By "projection", (i.e., using the canonical map from $\mathbb{Z}/d\mathbb{Z}$ to $\mathbb{Z}/p_1\mathbb{Z}$ that consists of "re-reducing" modulo p_1), the sequence $(\binom{2t}{t} \bmod p_1)_{t \geq 0}$ is k-automatic. From what precedes we know that this sequence is p_1-automatic and not ultimately periodic. Hence, from Theorem 11.2.2, k is necessarily a power of p_1.

In the same way k must be a power of p_2, which is a contradiction.

Case 2: The integer d is equal to $2^a p^b$, where p is an odd prime, and $a, b \geq 0$. Here we will study the coefficients $\binom{3t}{t} \bmod 2$. The previous method does not work, since the sequence $(\binom{2t}{t} \bmod 2)_{t \geq 0}$ is ultimately periodic.

Remember that Lucas's theorem asserts that if n and t have binary expansions respectively given by $n = \sum_{q \geq 0} n_q 2^q$ with $n_q \in \{0, 1\}$, and $t = \sum_{q \geq 0} t_q 2^q$ with $t_q \in \{0, 1\}$, then

$$\binom{t}{n} \equiv \prod_{q \geq 0} \binom{t_q}{n_q} \bmod 2.$$

Using this theorem and defining the sequence $\mathbf{u} = (u_t)_{t \geq 0}$ by

$$u_t := \binom{3t}{t} \bmod 2,$$

the reader can check that the following relations hold for all t:

$$u_{2t} = u_t, \qquad u_{4t+1} = u_t, \qquad u_{4t+3} = 0.$$

Hence the sequence is 2-automatic, since its 2-kernel is equal to

$$\{(u_t)_{t\geq 0}, \ (u_{2t+1})_{t\geq 0}, \ 0\}.$$

Furthermore, defining the formal power series G in $\mathbb{Z}/2\mathbb{Z}[[X]]$ by

$$G(X) = \sum_{t\geq 0} u_t X^t,$$

the previous relations imply that

$$XG^3 + G + 1 = 0.$$

This proves that the formal power series G is algebraic over the field of rational functions $\mathbb{Z}/2\mathbb{Z}(X)$, which is not a surprise (Theorem 12.2.5). We can use this relation to prove that G is not a rational function (i.e., the sequence $\mathbf{u}$ is not ultimately periodic). If we had $G = \frac{P}{Q}$ for two polynomials in $\mathbb{Z}/2\mathbb{Z}[X]$, P and Q coprime, then

$$XP^3 + PQ^2 + Q^3 = 0.$$

Hence Q divides X. If Q is constant, we obtain

$$XP^3 + P + 1 = 0,$$

which is not possible (compute the degrees). If $Q = X$, we get

$$XP^3 + X^2P + X^3 = 0;$$

hence

$$P^3 + XP + X^2 = 0.$$

That would imply that X divides P, which is not possible, since P and Q are coprime.

Now suppose that the sequence $(\binom{n}{m} \bmod d)_{m,n\geq 0}$ is k-automatic for some integer $k \geq 2$, and remember that $d = 2^a p^b$. By the same reasoning as in the first case, k must be a power of p. On the other hand, the hypothesis implies that the one-dimensional sequence $(\binom{3t}{t} \bmod d)_{t\geq 0}$ is k-automatic (Theorem 14.2.6(b)). Hence, by projection, the sequence $(\binom{3t}{t} \bmod 2)_{t\geq 0}$ is k-automatic. Since it is 2-automatic and not ultimately periodic, using Cobham's theorem again, we have that k must be a power of 2, which is impossible. ∎

14.7 Exercises

1. A two-dimensional infinite array A

$$
\begin{array}{ccccc}
a_{00} & a_{01} & a_{02} & a_{03} & \cdots \\
a_{10} & a_{11} & a_{12} & a_{13} & \cdots \\
a_{20} & a_{21} & a_{22} & a_{23} & \cdots \\
a_{30} & a_{31} & a_{32} & a_{33} & \cdots \\
\vdots & \vdots & \vdots & \vdots & \ddots
\end{array}
$$

can be "linearized" as follows:

$$A' = a_{00}a_{01}a_{10}a_{02}a_{11}a_{20}a_{03}a_{12}a_{21}a_{30}\cdots$$

Show that if A is two-dimensional k-automatic, then the sequence A' need not necessarily be k-automatic.

2. Give an example of a two-dimensional infinite array such that all its rows and columns are k-automatic, but the array itself is not k-automatic.

3. Suppose $\mathbf{A} = (A_{i,j})_{i,j\geq 0}$ is the table of 0's and 1's such that the ith row is the base-2 representation of i, starting with the least significant digit. The first six rows and columns are as below:

$$
\begin{array}{cccccc}
0 & 0 & 0 & 0 & 0 & 0 \\
1 & 0 & 0 & 0 & 0 & 0 \\
0 & 1 & 0 & 0 & 0 & 0 \\
1 & 1 & 0 & 0 & 0 & 0 \\
0 & 0 & 1 & 0 & 0 & 0 \\
1 & 0 & 1 & 0 & 0 & 0
\end{array}
$$

Show that $\mathbf{A}$ is not k-automatic for any k.

4. Consider the matrix-valued morphism on $\{-1, 1\}$ defined by

$$\varphi(1) = \begin{pmatrix} 1 & 1 \\ 1 & -1 \end{pmatrix}, \qquad \varphi(-1) = \begin{pmatrix} -1 & -1 \\ -1 & 1 \end{pmatrix}.$$

(a) Define $H_n = \varphi^n(1) = (m_{i,j})_{0\leq i,j<2^n}$. Show that $m_{i,j} = (-1)^e$, where $e = \sum_{0\leq t<n} i_t j_t$ and $[i_{n-1}\cdots i_1 i_0]_2 = i$, $[j_{n-1}\cdots j_1 j_0]_2 = j$.

(b) Show that H_n is a *Hadamard matrix*, i.e., that $H_n H_n^T = 2^n I$, where I is the identity matrix. Conclude that $\det H_n = \pm 2^{n\cdot 2^{n-1}}$.

(c) Show that $2^{n/2}$ is an eigenvalue of H_n.

(d) Suppose $v_n = [a_0 a_1 \cdots a_{2^n-1}]$, where $a_i = (-1 + \sqrt{2})^{s_2(i)}$, and s_2 as usual denotes the sum of the base-2 digits. Show that v_n is an eigenvector of H_n.

(e) Define maps T_0 and T_1 that take k-element vectors to $2k$-element vectors as follows: $T_0([v]) = [\text{Concat}(v, v)]$, $T_1([v]) = [\text{Concat}(v, -v)]$. Let $v_{n,i}$ denote the ith row of the matrix H_n (we index starting with row 0). Suppose $[i_{n-1}\cdots i_1 i_0]_2 = i$. Show that $v_{n,i} = T_{i_{n-1}}(T_{i_{n-2}}(\cdots T_{i_1}(T_{i_0}([1]))\cdots))$.

5. Show how to construct an infinite square array over $\{0, 1, 2\}$ such that each of its rows and columns constitutes an infinite squarefree word.

6. Show how to construct an infinite square array over $\{0, 1, 2, 3, 4\}$ such that each of its rows, columns, and diagonals constitutes an squarefree word.

7. Suppose we define the generalized k-kernel of an infinite array $(a_{m,n})_{m,n\geq0}$ to be all subarrays of the form $(a_{k^i m+a, k^j n+b})_{m,n\geq0}$ for $i, j \geq 0$ and $0 \leq a < k^i$, $0 \leq b < k^j$. Give an example of a k-automatic sequence whose generalized k-kernel is infinite.

8. Consider the matrix-valued morphism φ, where

$$0 \to \begin{bmatrix} 0 & 0 \\ 0 & 1 \end{bmatrix}, \quad 1 \to \begin{bmatrix} 0 & 0 \\ 0 & 0 \end{bmatrix}.$$

Let M be the fixed point of φ. Show that for $i \geq 0$, row i of M is purely periodic with minimal period length $2^{v_2(i+1)}$.

9. Give an example of a two-dimensional infinite array over a finite alphabet in which each row and column is purely periodic, but the array is not k-automatic for any k.

10. Prove a two-dimensional analogue of Cobham's theorem (Theorem 11.2.1).

11. Let $\Sigma = \{0, 1\}$ and let φ be the matrix-valued morphism defined by

$$0 \to \begin{bmatrix} 1 & 0 \\ 0 & 1 \end{bmatrix}, \quad 1 \to \begin{bmatrix} 0 & 1 \\ 1 & 0 \end{bmatrix}.$$

Let $\mathbf{w}$ be the infinite array obtained by iterating φ on 0. Let $p_{\mathbf{w}}(n)$, the "subarray complexity", be defined as the number of distinct $n \times n$ subarrays contained in $\mathbf{w}$. Let $\mathbf{t}$ be the Thue–Morse sequence; then determine if $p_{\mathbf{w}}(n) = n p_{\mathbf{t}}(n)$ for all positive integers n.

12. Show that the Sierpiński carpet $\mathbf{s}$ cannot be obtained as the outer product $(f(a_m, b_n))_{m,n\geq0}$ for two 3-automatic sequences $(a_m)_{m\geq0}$ and $(b_n)_{n\geq0}$.

14.8 Open Problems

1. Can an analogue of Theorem 14.5.1 be proved for base-$(-k + i)$, $k \geq 2$?

2. (Currie) Does there exist an infinite two-dimensional array over a finite alphabet, $\mathbf{a} = (a_{i,j})_{-\infty < i, j < \infty}$, such that every line $(a_{ci+d, ei+f})_{i\in\mathbb{Z}}$ (where c, d, e, f are integers with $c, e \neq 0$) is squarefree?

3. A finite or infinite two-dimensional array over $\{0, 1\}$ is *rectilinearly connected* if there is a finite path of 1's from every 1 to every other 1 using only horizontal or vertical steps. Given a two-dimensional morphism h that is prolongable on a letter a, is it decidable whether $h^\omega(a)$ is rectilinearly connected?

14.9 Notes on Chapter 14

14.1 The Sierpiński carpet can be found in Sierpiński [1916] and Mandelbrot [1983, p. 144]. For generalizations, see Reiter [1994].

For applications of two-dimensional automatic sequences to computer graphics, see Shallit [1988b]; Berstel and Nait Abdallah [1989]; Berstel and Morcrette [1989]; Mozes [1989]; Shallit and Stolfi [1989]; Barbé [1993].

For applications to tiling problems, see Salon [1989a]; Allouche and Salon [1990].

14.2 Array-valued morphisms were studied by Siromoney and Subramanian [1985] and Siromoney [1987].

For Reverend Back's abbey floor, see Wegner [1982]; Siromoney and Subramanian [1983].

Theorem 14.2.3, the extension of Cobham's theorem to the two-dimensional case, was proved by Černý and Gruska [1986a, 1986b] (who called two-dimensional automatic sequences "modular trellises") and Salon [1986, 1987, 1989b], independently.

For another generalization of multidimensional automatic sequences, see Tamura [2000].

14.3 Allouche and Berthé [1997] studied the subword complexity of the two-dimensional array $(\binom{m}{n} \bmod d)_{m,n\geq0}$. This was generalized by Berthé [2000a] to the subword complexity of the two-dimensional patterns generated by one-dimensional linear cellular automata. Also see Berthé and Vuillon [2000a].

Nivat has conjectured the following analogue of Theorem 10.2.6: if there exists a pair (m, n) that $p_{\mathbf{w}}(m, n) \leq mn$, then $\mathbf{w}$ has a periodicity vector. This conjecture is still open; for partial results see Epifanio, Koskas, and Mignosi [1999]. Cassaigne [1999b] classified the two-dimensional arrays with subword complexity $mn + 1$. Also see Cassaigne [2000].

Peyrière [1987] studied the frequencies with which particular blocks appear in automatic two-dimensional arrays.

14.4 The material in this section is due to Salon [1986, 1987, 1989b]. Theorem 14.4.2 is originally due to Deligne [1984], who proved the result for *any* field of positive characteristic; our proof is due to Salon. A proof using a generalization of the notion of kernel in the case of an infinite field of positive characteristic was given independently by Sharif and Woodcock [1988] and Harase [1988]; see also the survey of Allouche [1989].

For rational functions defined over $\mathbb{Z}/m\mathbb{Z}$ and two-dimensional automatic sequences, see von Haeseler and Petersen [1998].

14.5 Theorem 14.5.1 is taken nearly verbatim from Allouche, Cateland, Gilbert, Peitgen, Shallit, and Skordev [1997].

Previously, Salon [1989b] had proved a special case of Theorem 14.5.1: namely, that the sum-of-digits function in base $-1 + i$ is two-dimensional 2-automatic, when reduced modulo 2.

14.6 Theorem 14.6.2 was originally proved (with different terminology) by Korec [1990]; a different proof was given by Allouche, von Haeseler, Peitgen, and Skordev [1996]. This was generalized to the two-dimensional patterns generated by one-dimensional linear cellular automata by Allouche, von Haeseler, Lange, Petersen, and Skordev [1997]. The reader can also consult the papers of von Haeseler, Peitgen and Skordev [1993, 1995]; Barbé, von Haeseler, Peitgen, and Skordev [1995]; and Barbé, Peitgen, and Skordev [1999].

15

Automaticity

In this chapter we consider approximation of formal languages by regular languages, and approximation of sequences by automatic sequences.

15.1 Basic Notions

Most functions are not polynomials, but we can approximate a function by polynomials. In the limit, we get a power series expansion for the function.

Similarly, most languages are not regular, but we can approximate languages by regular languages. More precisely, we say a language L' is an *nth-order approximation* to a language L if

$$L \cap \Sigma^{\leq n} = L' \cap \Sigma^{\leq n};$$

we recall that $\Sigma^{\leq n} = \{\epsilon\} \cup \Sigma \cup \cdots \cup \Sigma^n$. We define $A_L(n)$, the *automaticity* of L, to be the number of states in a smallest DFA accepting some nth-order approximation to L.[1] We measure the size of a DFA by counting the number of its states.

Example 15.1.1 Let $L = \{0^n 1^n : n \geq 0\}$. The automaton in Figure 15.1 shows that $A_L(6) \leq 8$. (In fact, one can show that $A_L(6) = 7$.)

Our first theorem proves some basic results about automaticity.

Theorem 15.1.2 *Let $L \subseteq \Sigma^*$ be a language. Then:*

(a) For all $n \geq 0$ we have $A_L(n) \leq A_L(n+1)$.
(b) L is regular if and only if $A_L(n) = O(1)$.
(c) $A_L(n) = A_{\overline{L}}(n)$.

---

[1] We say "a smallest" rather than "the smallest" because there may be many such DFAs.

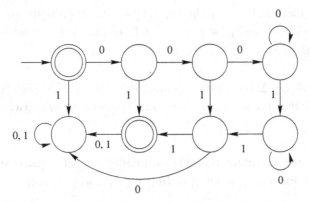

Figure 15.1: Automaton Accepting 6th-Order Approximation to $\{0^n 1^n : n \geq 0\}$.

Proof. (a): If L' is an $(n + 1)$th-order approximation to L, then it is also an nth-order approximation to L.

(b): Suppose L is regular. Then L is accepted by a DFA with r states for some $r \geq 1$. Then $A_L(n) \leq r$ for all $n \geq 0$, so $A_L(n) = O(1)$.

On the other hand, if $A_L(n) = O(1)$, then by part (a) we know that there exist r, n_0 such that $A_L(n) = r$ for all $n \geq n_0$. Hence for all $n \geq n_0$ there exists a DFA M_n with r states such that $L(M_n)$ is an nth-order approximation to L. Since there are only a finite number of distinct DFAs with r states and input alphabet Σ, there must be a DFA M such that $M_n = M$ for infinitely many n. Then $L(M_n) = M$ for infinitely many n, and hence $L(M)$ is an nth-order approximation to L for infinitely many n. Thus $L(M) = L$.

(c): If $M = (Q, \Sigma, \delta, q_0, F)$ is a DFA with r states accepting an nth-order approximation to L, then $M' = (Q, \Sigma, \delta, q_0, Q \setminus F)$ is a DFA with r states accepting an nth-order approximation to $\overline{L}$. ∎

We now introduce the notion of *n-similar* and *n-dissimilar* strings. Let $x, y \in \Sigma^*$. We say x is *n-similar* to y, and write $x \sim_n y$, if for all $z \in \Sigma^*$ with $|xz|, |yz| \leq n$ we have $xz \in L$ if and only if $yz \in L$. If x, y are not *n-similar*, then they are *n-dissimilar*. Note that $\sim_n$ is, in general, not an equivalence relation; see Exercise 1. However, under some conditions $\sim_n$ behaves like an equivalence relation, as the following lemma shows.

Lemma 15.1.3 *If $x \sim_n w$ and $x \sim_n y$ and $|x| \leq \max(|w|, |y|)$, then $w \sim_n y$.*

Proof. Assume, contrary to what we want to prove, that there exists z with $|wz|, |yz| \leq n$, and without loss of generality, that $wz \in L$ and $yz \notin L$. There are two cases to consider, depending on whether $|x| \leq |w|$ or $|x| \leq |y|$.

If $|x| \leq |w|$, then $|xz| \leq |wz|$. But $|wz| \leq n$ by hypothesis, so $|xz| \leq n$. Now $x \sim_n w$ by hypothesis, and $wz \in L$, so $xz \in L$. But then $x \not\sim_n y$, a contradiction.

If $|x| \leq |y|$, then $|xz| \leq |yz|$. Again, $|yz| \leq n$ by hypothesis, so $|xz| \leq n$. But $x \sim_n y$ by hypothesis, and $yz \notin L$, so $xz \notin L$. But $wz \in L$ and $|wz| \leq n$, so $x \not\sim_n w$, a contradiction. ∎

Theorem 15.1.4 *Let Σ be a finite alphabet and let $L \subseteq \Sigma^*$. Then for all integers $n \geq 0$, $A_L(n)$ is the maximum possible cardinality of a set of pairwise n-dissimilar strings.*

Proof. Let r be the maximum possible cardinality of a set of pairwise n-dissimilar strings. Then there exists a set $S = \{w_1, w_2, \dots, w_r\}$ of pairwise n-dissimilar strings.

First we prove that $r \leq A_L(n)$. This inequality is clearly true if $r = 1$, since every DFA has at least one state. Hence assume $r > 1$. Suppose there is a DFA $M = (Q, \Sigma, q_0, \delta, F)$ such that $L(M)$ is an nth-order approximation to L and M has $< r$ states. Then there exist distinct i, j such that $\delta(q_0, w_i) = \delta(q_0, w_j)$. But then for all z we have $\delta(q_0, w_i z) = \delta(q_0, w_j z)$, so for all z we have $w_i z \in L$ if and only if $w_j z \in L$. Hence $w_i \sim_n w_j$, a contradiction.

Now we prove $r \geq A_L(n)$. We show how to construct a DFA M with r states such that $L(M) \cap \Sigma^{\leq n} = L \cap \Sigma^{\leq n}$. For $1 \leq i \leq r$ we define w_i' to be a shortest word such that $w_i' \sim_n w_i$; then $|w_i'| \leq |w_i|$.

We claim there is no word w_i'' such that $|w_i''| < |w_i'|$ and $w_i'' \sim_n w_i'$. For if there were, then by Lemma 15.1.3 we would have $w_i \sim_n w_i''$, contradicting the definition of w_i'. Next, we prove that for all i, j with $i \neq j$ we have $w_i' \not\sim_n w_j'$. Assume $w_i' \sim_n w_j'$. Then by Lemma 15.1.3 $w_i \sim_n w_j'$. But $w_j' \sim_n w_j$ and Lemma 15.1.3 together give $w_i \sim_n w_j$, a contradiction. Thus $S' = \{w_1', w_2', \dots, w_r'\}$ is also a set of pairwise n-dissimilar words.

We observe that $|w_i'| \leq n$ for $1 \leq i \leq r$. For if not, then $|w_i'| > n$ and hence trivially $w_i' \sim_n \epsilon$. But $0 = |\epsilon| < |w_i'|$, a contradiction.

We now define a DFA $M = (Q, \Sigma, \delta, q_0, F)$ as follows: we set $Q = \{q_1, q_2, \dots, q_r\}$. We define δ as follows: for each i, $1 \leq i \leq r$, and each $a \in \Sigma$, we choose a j, $1 \leq j \leq r$, such that $w_i' a \sim_n w_j'$. Such a j always exists because S' is a maximal set of pairwise n-dissimilar words. There may be more than one such j, in which case we choose one arbitrarily. We then define $\delta(q_i, a) := q_j$. We choose a k such that $\epsilon \sim_n w_k$, and define $q_0 := q_k$. Finally, we define $F := \{q_i : w_i' \in L\}$. We claim that $L(M) \cap \Sigma^{\leq n} = L \cap \Sigma^{\leq n}$.

First, we extend the domain of δ to $Q \times \Sigma^*$ in the usual way. We now prove by induction of $|w|$ that if $\delta(q_0, w) = q_j$, then $w \sim_n w_j'$. This is clear for $|w| = 0$. Now assume the result is true for all strings of length $< i$; we prove it for $|w| = i$. Write $w = xa$ with $x \in \Sigma^*, a \in \Sigma$. By induction $\delta(q_0, x) = q_j$, where $x \sim_n w_j'$. We also have $|w_j'| \leq |x|$. Now $xa \sim_n w_j' a$ and $|w_j' a| \leq |xa|$. We have $\delta(q_j, a) = q_k$ for some k, $1 \leq k \leq r$. Then $w_j' a \sim_n w_k'$. Now from Lemma 15.1.3 we have $w_k' \sim_n xa$. But $\delta(q_0, xa) = q_k$, and the result follows.

Now suppose $w \in L(M) \cap \Sigma^{\le n}$. By the definition of acceptance in a DFA, there exists $q_j \in F$ such that $\delta(q_0, w) = q_j$. From the preceding paragraph we have $w \sim_n w'_j$, and so $wz \in L$ if and only if $w'_j z \in L$, for all z with $|wz|, |w'_j z| \le n$. Set $z = \epsilon$, and observe that $q_j \in F$ implies that $w'_j \in L$. Since $|w| \le n$, it follows that $w \in L$.

On the other hand, suppose $w \in L \cap \Sigma^{\le n}$. Then there exists an integer i such that $q_i = \delta(q_0, w)$, so $w \sim_n w'_i$. Hence $wz \in L$ if and only if $w'_i z \in L$, for all z with $|wz|, |w'_i z| \le n$. Take $z = \epsilon$. Then $w'_i \in L$, so $q_i \in F$, and $w \in L(M)$. The proof is now complete.   ∎

As an application of Theorem 15.1.4, consider the following.

Example 15.1.5 Let $L = \{0^n 1^n : n \ge 0\}$. Then $A_L(k) \ge 2n + 1$ for $k = 2n, 2n + 1$, because $\{\epsilon, 0, 00, \ldots, 0^n\}$ forms a pairwise $(2n + 1)$-dissimilar set of strings. To see this, consider 0^j and 0^k for $0 \le j < k \le n$. Then $0^j 1^j \in L$, but $0^j 1^k \notin L$.

The following is a basic result on automaticity.

Theorem 15.1.6 *Let $L \subseteq \Sigma^*$ be a nonregular language. Then $A_L(n) \ge (n + 3)/2$ for infinitely many non-negative integers n.*

Proof. Assume the contrary. Since $A_L(n)$ is an increasing function of n by Theorem 15.1.2 (a), there exists an integer n_0 such that $A_L(n) < (n + 3)/2$ for all $n > n_0$. Since n is an integer, $A_L(n) \le (n + 2)/2$ for all $n > n_0$. By Theorem 15.1.2(b) we know $A_L(n)$ is unbounded, so there exists $r > n_0$ such that $A_L(r + 1) > A_L(r)$. Let M_r and M_{r+1} be DFAs with $A_L(r)$ and $A_L(r + 1)$ states, respectively, such that $L(M_r)$ is an rth-order approximation to L and $L(M_{r+1})$ is an $(r + 1)$th-order approximation to L. Since $A_L(r + 1) > A_L(r)$, there must be a shortest word w accepted by one of M_r, M_{r+1} that is rejected by the other. Clearly $|w| = r + 1$. By Exercise 4.5 we have $r + 1 \le A_L(r) + A_L(r + 1) - 2$. Hence $r + 1 \le (r + 2)/2 + (r + 3)/2 - 2 = r + \frac{1}{2}$, a contradiction.   ∎

15.2 Nondeterministic Automaticity

We may define an analogue of deterministic automaticity for nondeterministic automata. We define $N_L(n)$ to be the number of states in a smallest NFA accepting some nth-order approximation to L.

The following theorem gives the basic properties of nondeterministic automaticity:

Theorem 15.2.1 *Let L be a language. Then*

(a) for all $n \ge 0$ we have $N_L(n) \le N_L(n + 1)$;

(b) L is regular if and only if $N_L(n) = O(1)$;
(c) for all $n \geq 0$ we have $N_L(n) \leq A_L(n) \leq 2^{N_L(n)}$.

Proof. Left to the reader as Exercise 4.   ∎

It is indeed possible for N_L to be exponentially smaller than A_L, as the following example shows:

Example 15.2.2 Let $PAL = \{x \in \{0, 1\}^* : x = x^R\}$, the language of palindromes over $\{0, 1\}$, and let $UNPAL = \overline{PAL}$, the language of nonpalindromes. Then $A_{PAL}(n) = \Omega(2^{n/2})$, as it is easy to see that all strings of length $\lfloor n/2 \rfloor$ form a pairwise n-dissimilar set. Hence $A_{UNPAL}(n) = \Omega(2^{n/2})$ by Theorem 15.1.4. However, we can prove $N_{UNPAL}(n) = \Theta(n)$. The lower bound comes from Theorem 15.2.1 (c). For the upper bound, we can create a DFA with $O(n)$ states that accepts an nth-order approximation to $UNPAL$, as follows: we use a "counter" in the range $[0, n/2]$ to guess a position of a symbol in the first half of the string that is different from the corresponding symbol in the second half. Once a position is guessed, the input symbols are processed until we guess and verify that a mismatch has actually occurred. The construction for $n = 9$ is given in Figure 15.2.

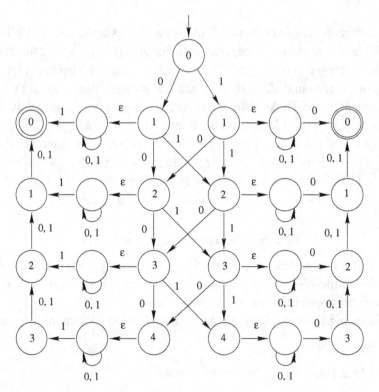

Figure 15.2: Plan for NFA Accepting nth-Order Approximation to *UNPAL*.

Often number theory is useful in bounding $N_L(n)$. Consider the following example.

Example 15.2.3 Let $EQ = \{x \in \{0, 1\}^* : |x|_0 = |x|_1\}$, the language of strings with the same number of 0's as 1's, and let $UNEQ = \overline{EQ}$. It is easy to prove that $A_{EQ}(n) = A_{UNEQ}(n) = n + 1$ for $n \geq 0$. We now prove that $N_{UNEQ}(n) = O((\log n)^2/(\log \log n))$. To nondeterministically accept an nth-order approximation, it suffices to guess a small prime p, and compute $|x|_0 - |x|_1 \pmod{p}$ by using a counter modulo p. The counter is implemented with a cycle of states, increasing the state number when a 0 is encountered, and decreasing it when a 1 is encountered. By Exercise 2.23 if $|x|_0 \neq |x|_1$ then there is a prime $p \leq 4.4 \log n$ with $|x|_0 \not\equiv |x|_1 \pmod{p}$. The total number of states used by this construction is $1 + \sum_{p \leq 4.4 \log n} p = O((\log n)^2/(\log \log n))$, using the estimate given in Theorem 2.13.2.

15.3 Unary Automaticity

In this section we examine properties of automaticity for languages over a 1-letter alphabet.

Definition 15.3.1 Let $\mathbf{w} = a_0 a_1 a_2 \cdots$ be an infinite word. We define $s_{\mathbf{w}}(n)$ to be the length of the longest suffix of $a_0 a_1 \cdots a_n$ that is also a subword of $a_0 a_1 \cdots a_{n-1}$.

Lemma 15.3.2 *Let $\Sigma = \{0\}$, and let $L \subseteq \Sigma^*$. Let $\mathbf{w} = w(L) = a_0 a_1 a_2 \cdots$ be the characteristic word of L, i.e., $a_i = 1$ if $0^i \in L$, and 0 otherwise. Then $A_L(n) = n + 1 - s_{\mathbf{w}}(n)$.*

Proof. Let $t = s_{\mathbf{w}}(n)$, and suppose $a_{n-t+1} \cdots a_n$ is the longest suffix of $a_0 \cdots a_n$ that appears as a subword of $a_0 a_1 \cdots a_{n-1}$. Suppose $a_{n-t+1} \cdots a_n = a_u \cdots a_{u+t-1}$ for some $u \leq n - t$. Then we can create a DFA $M = (Q, \Sigma, \delta, q_0, F)$ accepting an nth-order approximation to L as follows: $Q = \{0, 1, \ldots, n - t\}$, $q_0 = 0$, $F = \{i : 0 \leq i \leq n - t \text{ and } 0^i \in L\}$, and

$$\delta(i, 0) = \begin{cases} i + 1 & \text{if } 0 \leq i < n - t, \\ u & \text{if } i = n - t. \end{cases}$$

It is easy to see that M accepts an nth-order approximation to L, and hence $A_L(n) \leq n - t + 1$.

Now consider the strings $\epsilon, 0, \ldots, 0^{n-t}$. We claim these are pairwise n-dissimilar. For if not, then $0^i \sim_n 0^j$ for some $0 \leq i < j \leq n - t$. But then $0^{i+k} \in L$ if and only if $0^{j+k} \in L$ for $0 \leq k \leq n - j$. It follows that $a_i \cdots a_{i+n-j} = a_j \cdots a_n$, contradicting the maximality of t. ∎

There is a very interesting connection between unary automaticity and the function R' introduced in Section 10.10. More specifically, the functions $s_{\mathbf{w}}(n)$ and $R'_{\mathbf{w}}$ are (weak) inverses of each other.

Theorem 15.3.3 *For $n \geq 0$ we have*

(a) $R'_{\mathbf{w}}(s_{\mathbf{w}}(n) + 1) \geq n + 1$;
(b) $s_{\mathbf{w}}(R'_{\mathbf{w}}(n)) = n$;
(c) $(\liminf_{n\to\infty} s_{\mathbf{w}}(n)/n)^{-1} = \limsup_{n\to\infty} R'_{\mathbf{w}}(n)/n$.

Proof. (a): Let $\mathbf{w} = a_0 a_1 a_2 \cdots$, and let $t = s_{\mathbf{w}}(n)$. Then $a_{n-t} \cdots a_n$ is a suffix of $a_0 \cdots a_n$ of length $t + 1$, and hence is not a subword of $a_0 \cdots a_{n-1}$. It follows that $R'_{\mathbf{w}}(t + 1) > n$.

(b): Let $v = R'_{\mathbf{w}}(n)$. Then $a_0 \cdots a_{v-1}$ contains all subwords of $\mathbf{w}$ of length n, but $a_0 \cdots a_{v-2}$ does not. Hence $a_{v-n} \cdots a_{v-1}$ is not a subword of $a_0 \cdots a_{v-2}$. Thus $a_{v-n} \cdots a_v$ is not a subword of $a_0 \cdots a_{v-1}$. It follows that $s_{\mathbf{w}}(v) < n + 1$.

On the other hand, $a_{v-n+1} \cdots a_v$ is a subword of length n and hence a subword of $a_0 \cdots a_{v-1}$. Thus $s_{\mathbf{w}}(v) \geq n$. Putting this together with the last paragraph, we obtain $s_{\mathbf{w}}(v) = n$.

(c): Let $A = \liminf_{n\to\infty} s_{\mathbf{w}}(n)/n$ and $B = \limsup_{n\to\infty} R'_{\mathbf{w}}(n)/n$. Using (b), we have

$$A \leq \liminf_{n\to\infty} \frac{s_{\mathbf{w}}(R'_{\mathbf{w}}(n))}{R'_{\mathbf{w}}(n)} = \liminf_{n\to\infty} \frac{n}{R'_{\mathbf{w}}(n)} = \frac{1}{B}.$$

On the other hand, let $(n_i)_{i\geq 0}$ be a strictly increasing sequence such that $\lim_{i\to\infty} s_{\mathbf{w}}(n_i)/n_i = A$. Now the sequence $(s_{\mathbf{w}}(n_i))_{i\geq 0}$ is unbounded, for if it were bounded, $(R'_{\mathbf{w}}(s_{\mathbf{w}}(n_i) + 1))_{i\geq 0}$ would be bounded, but by (a) we have $R'_{\mathbf{w}}(s_{\mathbf{w}}(n_i) + 1) \geq n_i + 1$. Hence, replacing the sequence $(n_i)_{i\geq 0}$ by a subsequence if necessary, we can assume that the sequence $(s_{\mathbf{w}}(n_i))_{i\geq 0}$ is strictly increasing. Then by (a) we have

$$B \geq \limsup_{i\to\infty} \frac{R'_{\mathbf{w}}(s_{\mathbf{w}}(n_i) + 1)}{s_{\mathbf{w}}(n_i) + 1} \geq \limsup_{i\to\infty} \frac{n_i + 1}{s_{\mathbf{w}}(n_i) + 1} = \lim_{i\to\infty} \frac{n_i + 1}{s_{\mathbf{w}}(n_i) + 1} = \frac{1}{A}.$$

It follows that $A = 1/B$, as desired. ∎

15.4 Automaticity of Sequences

In the previous section we discussed the automaticity of languages. We can generalize this notion to sequences as follows: let $\mathbf{s} = (s(i))_{i\geq 0}$ be a sequence over a (possibly infinite) alphabet Δ. Let $k \geq 2$ be an integer. We define $A_{\mathbf{s}}^k(n)$ to be the smallest number of states in any k-DFAO M that correctly computes $s(i)$ for $0 \leq i \leq n$. More precisely, we demand that for $M = (Q, \Sigma, \Delta, \delta, q_0, \tau)$, we have

$\tau(\delta(q_0, w^R)) = s(i)$ for all i with $0 \le i \le n$ and all w with $[w]_k = i$. Notice here that we are feeding the automaton with the least significant digit first.

If $\mathbf{s} = (s(i))_{i \ge 0}$ is a sequence with $\mathcal{A}_{\mathbf{s}}^k(n) = O(\log n)$, then we say that $\mathbf{s}$ is *k-quasiautomatic*. In addition to the *k*-automatic sequences, the class of *k*-quasiautomatic sequences contains many sequences that closely resemble automatic sequences, but are not automatic.

We can also generalize automaticity to sets $S \subseteq \mathbb{N}$. By $\mathcal{A}_S^k(n)$ we mean $\mathcal{A}_{\chi_S}^k(n)$, where χ_S is the characteristic sequence associated with the set S.

We now prove

Theorem 15.4.1 *Every paperfolding sequence is 2-quasiautomatic.*

Proof. Let $(f(n))_{n \ge 1}$ be a paperfolding sequence. By Theorem 6.5.2 we know that $f(n) = (-1)^j f(2^k)$ if $n = 2^k(2j + 1)$ for integers $j, k \ge 0$. It follows that we can compute $f(n)$ with a machine that, given an input w with $[w^R] = n$, records the number of leading 0 bits and the two bits that follow. For $n \le N$ this can be done using $O(\log N)$ states. ∎

It is possible for a sequence that is *k*-automatic to have large automaticity relative to other bases. We give an example in a moment; first we prove the following lemma.

Lemma 15.4.2 *Let $\mathbf{s} = (s(i))_{i \ge 0}$ be a sequence, and suppose there exists a constant $d > 0$ such that for all $i \ge 0$ and all a, b with $0 \le a, b < k^i$ and $a \ne b$ there exists $m = O(k^{id})$ such that $s(k^i m + a) \ne s(k^i m + b)$. Then $\mathcal{A}_{\mathbf{s}}^k(n) = \Omega(n^{1/(d+1)}/k)$, where the implied constant in the Ω does not depend on k.*

Proof. If $m = O(k^{id})$ then there exists a constant c such that $m \le ck^{id} - 1$. Let $i = \lfloor (\log_k n - \log_k c)/(d+1) \rfloor$. Then

$$\frac{1}{k}\left(\frac{n}{c}\right)^{1/(d+1)} < k^i \le \left(\frac{n}{c}\right)^{1/(d+1)}.$$

By hypothesis there exists $m \le ck^{id} - 1$ such that $s(k^i m + a) \ne s(k^i m + b)$. However,

$$k^i m + a < (ck^{id} - 1)k^i + k^i = ck^{i(d+1)} \le c(n/c) = n,$$

and a similar bound holds for $k^i m + b$. It follows that if a DFAO M computes $s(j)$ correctly for all $j \le n$, then it must compute different values for $k^i m + a$ and $k^i m + b$. Then, by Exercise 4.17 we have that M must have at least k^i different states. Hence $\mathcal{A}_{\mathbf{s}}^k(n) \ge k^i = \Omega(n^{1/(d+1)}/k)$. ∎

Recall that $\nu_2(n)$ is the exponent of the highest power of 2 that divides n.

Theorem 15.4.3 *Let* $s(i) := v_2(i + 1)$ *mod* 2. *Then* $\mathbf{s} = (s(i))_{i \geq 0}$ *is 2-automatic. However, if* $k \geq 3$ *is odd, then* $A_\mathbf{s}^k(n) = \Omega(n^{1/2}/k)$.

Proof. We leave it to the reader to verify that $(s(i))_{i \geq 0}$ is 2-automatic. Otherwise assume $k \geq 3$ is odd. If we can show that for all a, b with $1 < a < b \leq k^i$ there exists $m < 4k^i$ such that $v_2(mk^i + a) \not\equiv v_2(mk^i + b) \pmod 2$, then the desired result will follow by Lemma 15.4.2.

To see this, let $b - a = 2^c \cdot t$, where t is odd, and let $r = k^i$. Let

$$m \equiv (2^{c+1} - b)r^{-1} \pmod{2^{c+2}};$$

this definition is meaningful, since r is odd. Then $rm + b \equiv 2^{c+1} \pmod{2^{c+2}}$, so $v_2(rm + b) = c + 1$. On the other hand,

$$
\begin{aligned}
rm + a &\equiv 2^{c+1} + a - b &&\pmod{2^{c+2}} \\
&\equiv 2^{c+1} - 2^c \cdot t &&\pmod{2^{c+2}} \\
&\equiv 2^c(2 - t) &&\pmod{2^{c+2}}.
\end{aligned}
$$

Since t is odd, we have $v_2(rm + a) = c$. Now $0 \leq m < 2^{c+2} = 4 \cdot 2^c < 4r$. ∎

15.5 Exercises

1. Show, by giving an example, that $\sim_n$ is not an equivalence relation.
2. Show that the bound of $(n + 3)/2$ in Theorem 15.1.6 is best possible, in the sense that the result is not true if 2 is replaced by any smaller positive real number, or if the 3 is replaced by any larger real number.
3. Show that given any unbounded function $g(n)$, there exists a unary nonregular language L such that $A_L(n) \leq g(n)$ for infinitely many n.
4. Prove Theorem 15.2.1.
5. Show that if L is a unary language, then $A_L(n) \leq n + 1 - \lfloor \log_2 n \rfloor$ infinitely often.
6. (L. Hellerstein) Consider the unary language $L = \{0^{2 \cdot 3 \cdot 5 \cdots p_n} : n \geq 1\}$, where p_n is the nth prime, $p_1 = 2$. Show that $N_{\overline{L}}(n) = O((\log n)^5)$.
7. Give an example of a language L such that $N_L(n) = O((\log n)^2)$ but $A_L(n) \neq O(n^c)$ for any c.
8. Give an example of a nonregular language L with $N_L(n) = O(\log n)$.
9. Using Theorem 15.4.3, show that if $k \geq 3$ and $\mathbf{t} = (t_i)_{i \geq 0}$ is the Thue–Morse sequence, then $A_\mathbf{t}^k(n) = \Omega(n^{1/4}k^{-1/2})$.

15.6 Open Problems

1. Let $L = \{0^i : i \text{ is squarefree}\}$. Find good upper and lower bounds on $N_L(n)$ and $N_{\overline{L}}(n)$.
2. Does there exist a unary language L such that $N_L(n) = O((\log n)^2/(\log \log n))$?

15.7 Notes on Chapter 15

15.1 Automaticity has a long history, with many investigators proving the same results independently. The basic idea of automaticity apparently was first introduced by Trakhtenbrot [1964]. However, he used a slightly different model of computation of finite-state functions, where outputs are associated with transitions rather than states. His results were improved by Grinberg and Korshunov [1966]. Later, Karp [1967] introduced the notion of automaticity as we have defined it here. He also proved Theorem 15.1.6. Breitbart [1971] studied the p-automaticity of the characteristic sequence of kth powers, where p is a prime. Additional results appeared in Breitbart [1973, 1976]. Wolfram [1984] studied the automaticity of cellular automata. Chytil [1986] studied an analogue of automaticity for context-free languages. Dwork and Stockmeyer [1989, 1990] introduced the concept of n-similarity, and proved one direction of Theorem 15.1.4. Kaneps and Freivalds [1990] independently studied n-similarity, and gave a complete proof of Theorem 15.1.4. Condon, Hellerstein, Pottle, and Wigderson [1994, 1998] found a relationship between $A_L(n)$ and deterministic communication complexity. Shallit and Breitbart [1996] wrote a survey in which the above results were summarized and many new results proved.

15.2 The results in this section are from Shallit and Breitbart [1996].

15.3 Pomerance, Robson, and Shallit [1997] studied descriptional complexity of unary languages. Also see Cassaigne [1997b]. Theorem 15.3.3 is due to Allouche and Bousquet-Mélou [1995].

Glaister and Shallit [1998] studied the closure properties of various classes associated with automaticity; also see Shallit [2000].

15.4 Some of the material in this section is taken, more or less verbatim, from Shallit and Breitbart [1996], with permission from Elsevier Science. For Theorem 15.4.3, see Shallit [1996].

16

k-Regular Sequences

Up to now in this book we have dealt almost exclusively with sequences over a finite alphabet. While many interesting sequences, such as the Thue–Morse sequence, are of this form, there are also a large number of well-studied sequences over infinite alphabets such as $\mathbb{Z}$. This naturally suggests the question of how the class of k-automatic sequences can be fruitfully generalized to the case of infinite alphabets.

In this chapter we examine one such possible generalization, called the class of k-regular sequences.

16.1 Basics

Recall that a sequence is k-automatic if and only if its k-kernel is finite. We say a sequence is k-regular if the $\mathbb{Z}$-module generated by its k-kernel is finitely generated. (More general definitions are possible, but we do not cover them here.)

Example 16.1.1 Consider the function $s_2(n)$ introduced in Chapter 3, which counts the sum of the bits in the binary expansion of n. If $i \geq 0$ and $0 \leq b < 2^i$, then we have $s_2(2^i n + b) = s_2(n) + s_2(b)$. It follows that every element of the 2-kernel of the sequence $(s_2(n))_{n \geq 0}$ can be written as a $\mathbb{Z}$-linear combination of the sequence $(s_2(n))_{n \geq 0}$ and the constant sequence 1. This is a prototypical example of a k-regular sequence.

Let R be a $\mathbb{Z}$-module, i.e., an abelian group, written additively, and let k be an integer ≥ 2. We say that a sequence $(a(n))_{n \geq 0}$ taking values in R is a k-regular sequence if there exist a finite number of sequences over R, $\{(a_1(n))_{n \geq 0}, \ldots, (a_s(n))_{n \geq 0}\}$, such that every sequence in $K_k(a)$ is a $\mathbb{Z}$-linear combination of the a_i. More precisely:

Definition 16.1.2 We say a sequence $(a(n))_{n \geq 0}$ is k-regular if for every integer $i \geq 0$ and $0 \leq b < k^i$ there exist $c_1, c_2, \ldots, c_s \in \mathbb{Z}$ such that for all integers $n \geq 0$ we have

$$a(k^i n + b) = \sum_{1 \leq j \leq s} c_j a_j(n).$$

438

We then have the following theorem, which gives several alternative character-izations of the class of k-regular sequences.

Theorem 16.1.3 *Let* $\Sigma_k = \{0, 1, \ldots, k - 1\}$. *The following are equivalent:*

(a) $\mathbf{a} = (a(n))_{n \geq 0}$ *is k-regular.*

(b) *The $\mathbb{Z}$-module generated by the k-kernel of $(a(n))_{n \geq 0}$ is generated by a finite number of its subsequences of the form $(a(k^f n + b))_{n \geq 0}$, where $f \geq 0$ and $0 \leq b < k^f$.*

(c) *There exists an integer E such that for all integers $e > E$, each subsequence of the form $(a(k^e n + b))_{n \geq 0}$ with $0 \leq b < k^e$ can be expressed as a $\mathbb{Z}$-linear combination of subsequences of the form $(a(k^f n + c))_{n \geq 0}$ with $0 \leq f \leq E$ and $0 \leq c < k^f$.*

(d) *There exist an integer r and r sequences $\mathbf{a}_i = (a_i(n))_{n \geq 0}$, $1 \leq i \leq r$, with $\mathbf{a} = \mathbf{a}_1$, such that for each i, $1 \leq i \leq r$, the k subsequences $(a_i(kn + b))_{n \geq 0}$ are $\mathbb{Z}$-linear combina-tions of the $\mathbf{a}_i$.*

(e) *There exist an integer r, r sequences $\mathbf{a}_i = (a_i(n))_{n \geq 0}$, and a matrix-valued morphism $\mu : \Sigma_k \to \mathbb{Z}^{r \times r}$ such that if*

$$V(n) := \begin{bmatrix} a_1(n) \\ a_2(n) \\ \vdots \\ a_r(n) \end{bmatrix},$$

then $V(kn + b) = \mu(b)V(n)$ *for* $0 \leq b < k$.

Proof. (a) $\Longrightarrow$ (b): Let $\mathcal{K}$ denote the k-kernel of $(a(n))_{n \geq 0}$. Then $\langle \mathcal{K} \rangle$, the $\mathbb{Z}$-module generated by $\mathcal{K}$, is finitely generated. Hence there exist sequences $a_1, a_2, \ldots, a_j$ such that $\langle \mathcal{K} \rangle = \langle a_1, a_2, \ldots, a_j \rangle$. But then each a_i is a $\mathbb{Z}$-linear combination of elements from $\mathcal{K}$, and there are only finitely many a_i. Thus $\langle \mathcal{K} \rangle$ is generated by finitely many members of $\mathcal{K}$.

(b) $\Longrightarrow$ (c): Let $\langle \mathcal{K} \rangle$ be the $\mathbb{Z}$-module generated by $\mathcal{K}$. Suppose $\langle \mathcal{K} \rangle = \langle a_1, a_2, \ldots, a_s \rangle$ where $a_i(n) = a(k^{f_i} n + b_i)$ and $f_i \geq 0$ for $n \geq 0$, $1 \leq i \leq s$, and $0 \leq b_i < k^{f_i}$ for $1 \leq i \leq s$. Let $E = \max_{1 \leq i \leq s} f_i$. Then for all $e > E$ and $0 \leq b < k^e$, there exist $c_1, c_2, \ldots, c_i$ such that $a(k^e n + b) = \sum_{1 \leq i \leq s} c_i a_i(n)$, as desired.

(c) $\Longrightarrow$ (d): Take as the r sequences the set $\mathcal{T}$ of subsequences $a_i(n) = a(k^{f_i} n + b_i)$ with $0 \leq f_i \leq E$ and $0 \leq b_i < k^{f_i}$. Then

$$a_i(kn + a) = a(k^{f_i}(kn + a) + b_i) = a(k^{f_i+1}n + ak^{f_i} + b_i),$$

which, if $f_i + 1 \leq E$, is an element of $\mathcal{T}$, and if $f_i + 1 > E$, is a linear combination of elements of $\mathcal{T}$.

(d) $\Longrightarrow$ (e): Follows trivially.

(e) $\Longrightarrow$ (a): We show that $a(k^e n + b)$ can be expressed as a linear combination of the a_i. Express b in base k, possibly with leading zeros, as $\sum_{0 \leq i < e} b_i k^i$. Then an easy induction shows that $V(k^e n + b) = \mu(b_0 b_1 \cdots b_{e-1})V(n)$, and this expresses $a(k^e n + b)$ as a linear combination of the a_i. ∎

Example 16.1.4 Recall from Section 3.7 that $s_{-2}(n)$ denotes the sum of the digits of n when expressed in base -2. We show that $(s_{-2}(n))_{n \geq 0}$ is a 2-regular sequence. It is readily verified that for all strings $x \in \{0, 1\}^*$ we have

$$
\begin{aligned}
[x1]_{-2} = 2n + 1 & \quad \Leftrightarrow \quad & [x0]_{-2} = 2n, \\
[x00]_{-2} = 4n & \quad \Leftrightarrow \quad & [x]_{-2} = n, \\
[x110]_{-2} = 8n + 2 & \quad \Leftrightarrow \quad & [x1]_{-2} = 2n + 1, \\
[x0010]_{-2} = 16n + 14 & \quad \Leftrightarrow \quad & [x1010]_{-2} = 16n + 6, \\
[x11010]_{-2} = 32n + 6 & \quad \Leftrightarrow \quad & [x1]_{-2} = 2n + 1, \\
[x01010]_{-2} = 32n + 22 & \quad \Leftrightarrow \quad & [x010]_{-2} = 8n + 6.
\end{aligned}
$$

It follows that

$$
\begin{aligned}
s_{-2}(2n + 1) &= s_{-2}(2n) + 1, \\
s_{-2}(4n) &= s_{-2}(n), \\
s_{-2}(8n + 2) &= s_{-2}(2n + 1) + 1, \\
s_{-2}(16n + 14) &= s_{-2}(16n + 6) - 1, \\
s_{-2}(32n + 6) &= s_{-2}(2n + 1) + 2, \\
s_{-2}(32n + 22) &= s_{-2}(8n + 6) + 1.
\end{aligned}
$$

Hence the $\mathbb{Z}$-module generated by the 2-kernel of $(s_{-2}(n))_{n \geq 0}$ is generated by

$$\{(s_{-2}(n))_{n \geq 0}, (s_{-2}(2n))_{n \geq 0}, (s_{-2}(4n+2))_{n \geq 0}, (s_{-2}(8n+6))_{n \geq 0}, (s_{-2}(16n+6))_{n \geq 0}\},$$

together with the constant sequence 1.

If we now let

$$
V(n) = \begin{bmatrix} s_{-2}(n) \\ s_{-2}(2n) \\ s_{-2}(4n + 2) \\ s_{-2}(8n + 6) \\ s_{-2}(16n + 6) \\ 1 \end{bmatrix},
$$

then $V(2n) = \mu(0)V(n)$ and $V(2n + 1) = \mu(1)V(n)$, where

$$
\mu(0) = \begin{bmatrix} 0 & 1 & 0 & 0 & 0 & 0 \\ 1 & 0 & 0 & 0 & 0 & 0 \\ 0 & 1 & 0 & 0 & 0 & 2 \\ 0 & 0 & 0 & 0 & 1 & 0 \\ 0 & 1 & 0 & 0 & 0 & 3 \\ 0 & 0 & 0 & 0 & 0 & 1 \end{bmatrix}, \quad
\mu(1) = \begin{bmatrix} 0 & 1 & 0 & 0 & 0 & 1 \\ 0 & 0 & 1 & 0 & 0 & 0 \\ 0 & 0 & 0 & 1 & 0 & 0 \\ 0 & 0 & 0 & 0 & 1 & -1 \\ 0 & 0 & 0 & 1 & 0 & 1 \\ 0 & 0 & 0 & 0 & 0 & 1 \end{bmatrix}.
$$

In a similar fashion, it can be shown that $(s_{-2}(-n))_{n \geq 0}$ is a 2-regular sequence.

The next theorem gives a connection between k-regular and k-automatic sequences.

Theorem 16.1.5 *A sequence is k-regular and takes on only finitely many values if and only if it is k-automatic.*

Proof. Suppose $(a(n))_{n \geq 0}$ is k-automatic. Then by definition it takes finitely many values. By Theorem 6.6.2 its k-kernel is finite, and so the $\mathbb{Z}$-module generated by its k-kernel is finitely generated.

Now suppose $(a(n))_{n \geq 0}$ is k-regular and takes on finitely many values. By Theorem 16.1.3 there exist r sequences $a_1 = a, a_2, \ldots, a_r$, which can be taken to be in the k-kernel of $(a(n))_{n \geq 0}$, and a matrix-valued morphism $\mu : \Sigma_k \to \mathbb{Z}^{r \times r}$ such that

$$V(n) = \begin{bmatrix} a_1(n) \\ a_2(n) \\ \vdots \\ a_r(n) \end{bmatrix}$$

satisfies $V(kn + b) = \mu(b)V(n)$ for $0 \leq b < k$. Let $\mathcal{V}$ be the finite set of values of $(V(n))_{n \geq 0}$, and define the k-uniform morphism σ by $\sigma(v) = c_0 c_1 \cdots c_{k-1}$, where $v \in \mathcal{V}$ and $c_b = \mu(b)v$ for $0 \leq b < k$. Then the infinite word

$$V(0)V(1)V(2) \cdots$$

is a fixed point of σ, and $a = a_1$ is an image of this fixed point. By Theorem 6.3.2, it follows that a is k-automatic. ∎

Corollary 16.1.6 *If the integer sequence $(a(n))_{n \geq 0}$ is k-regular, then for all integers $m \geq 1$, the sequence $(a(n) \bmod m)_{n \geq 0}$ is k-automatic.*

The converse, however, does not hold; see Exercise 2.

16.2 Robustness of the *k*-Regularity Concept

In this theorem we show that several variations on the definition of k-regular sequences still result in the same class of sequences.

Theorem 16.2.1 *Let $\mathbf{s} = (s(n))_{n \geq 0}$ and $\mathbf{t} = (t(n))_{n \geq 0}$ be k-regular sequences, and let λ be an integer. Then the sequences $\mathbf{s} + \mathbf{t} = (s(n) + t(n))_{n \geq 0}$, $\lambda \mathbf{s} = (\lambda s(n))_{n \geq 0}$, and $\mathbf{st} = (s(n)t(n))_{n \geq 0}$ are all k-regular.*

Proof. Let $s_1 = s, s_2, \ldots, s_m$ be sequences generating the $\mathbb{Z}$-module generated by the k-kernel of $\mathbf{s}$, and let $t_1 = t, t_2, \ldots, t_n$ be sequences generating the $\mathbb{Z}$-module generated by the k-kernel of $\mathbf{t}$. Then the $m + n$ sequences $s_1, s_2, \ldots, s_m$, $t_1, t_2, \ldots, t_n$ generate the $\mathbb{Z}$-module generated by the k-kernel of $\mathbf{s} + \mathbf{t}$. Similarly, the mn sequences $s_i t_j$ with $1 \leq i \leq m, 1 \leq j \leq n$ generate the $\mathbb{Z}$-module generated by the k-kernel of **st**. Finally, the sequences λs_i, $1 \le i \le m$, generate the $\mathbb{Z}$-module generated by the k-kernel of λ**s**. ∎

Theorem 16.2.2 *Let* $(s(n))_{n \ge 0}$ *be a k-regular sequence. Then for $a \ge 1$, $b \ge 0$, the sequence* $(s(an + b))_{n \ge 0}$ *is k-regular.*

Proof. The proof is analogous to the proof of Theorem 6.8.1, and is left to the reader. ∎

Theorem 16.2.3 *Let* $\Sigma_k = \{0, 1, \dots, k-1\}$.

(a) *If a sequence* $(s(n))_{n \ge 0}$ *taking values in a $\mathbb{Z}$-module R is k-regular, then there exist a matrix-valued morphism* $\mu : \Sigma_k \to \mathbb{Z}^{r \times r}$ *and vectors λ, κ with entries in R such that* $s(n) = \lambda \mu(a_e \cdots a_2 a_1) \kappa$ *for all integers $n \ge 0$ and all strings $w = a_1 a_2 \cdots a_e \in \Sigma_k^*$ such that* $[w]_k = n$.

(b) *If there exist a matrix-valued morphism* $\mu : \Sigma^k \to \mathbb{Z}^{r \times r}$ *and vectors λ, κ with entries in R such that* $s(n) = \lambda \mu(a_e \cdots a_2 a_1) \kappa$ *for all integers $n \ge 0$, where $(n)_k = a_1 a_2 \cdots a_e$ is the canonical base-k representation of n, then* $(s(n))_{n \ge 0}$ *is k-regular.*

Proof. (a): By Theorem 16.1.3(e), there exists μ such that if

$$V(m) = \begin{bmatrix} s_1(m) \\ \vdots \\ s_r(m) \end{bmatrix}$$

with $s_1 = s$, then $V(km + a) = \mu(a)V(m)$ for all $m \ge 0$ and $0 \le a < k$. Suppose $w = a_1 a_2 \cdots a_e$ and $[w]_k = n$. Then a simple induction gives $V(k^e m + n) = \mu(a_e \cdots a_1)V(m)$ for all $m \ge 0$. Now set $m = 0$, and let $\lambda = [1\ 0\ 0 \cdots 0]$ and $\kappa = V(0)$. The result follows.

(b): Now suppose that $s(n) = \lambda \mu(a_e \cdots a_1) \kappa$ where $(n)_k = a_1 a_2 \cdots a_e$ is the canonical base-k representation of n.

Define $V(n) = \mu(a_e \cdots a_1) \kappa$, and let $v_1, v_2, \dots, v_r$ be such that

$$V(n) = \begin{bmatrix} v_1(n) \\ \vdots \\ v_r(n) \end{bmatrix}.$$

Then from our construction of V, it follows that $V(kn + a) = \mu(aa_e a_{e-1} \cdots a_1)\kappa = \mu(a)V(n)$, except possibly when $n = 0$ and $a = 0$. (This special case arises because the canonical representation of kn is $a_1 a_2 \cdots a_e 0$ for $n \ge 1$, but not for $n = 0$.) Set $v' = V(0) - \mu(0)V(0)$; then $V(kn) = \mu(0)V(n) + v'u(n)$, where $(u(n))_{n \ge 0}$ is the sequence defined by $u(0) = 1$ and $u(i) = 0$ for all $i \ge 1$. By Theorem 16.1.3(d), each of the sequences $v_i(n)$, $1 \le i \le n$, is k-regular, and hence $\lambda V(n)$ is k-regular, by Theorem 16.2.1. ∎

The following corollary shows that our characterization of k-regular sequences is robust, in the sense that it does not depend on the order in which base-k expansion is processed (left to right or right to left):

Corollary 16.2.4 *Let $\Sigma_k = \{0, 1, \ldots, k-1\}$ and R be a $\mathbb{Z}$-module. A sequence $(s(n))_{n \ge 0}$ taking values in R is k-regular if and only if there exist a matrix-valued morphism $\mu' : \Sigma_k \to \mathbb{Z}^{r \times r}$ and vectors λ', κ' with values in R such that $s(n) = \lambda' \mu'(a_1 \cdots a_e) \kappa'$ for all $w = a_1 a_2 \cdots a_e \in \Sigma_k^*$ such that $[w]_k = n$.*

Proof. Take the transpose of the identity $s(n) = \lambda \mu(a_e \cdots a_1) \kappa$, and set $\lambda' = \kappa^T$, $\kappa' = \lambda^T$, and $\mu'(i) = \mu(i)^T$ for $0 \le i < k$. ∎

Finally, we show that, similar to the situation for automatic sequences in Theorem 6.8.6, a sequence remains k-regular if we transduce an input's digits before applying the matrix product.

Theorem 16.2.5 *Let $T = (Q, \Sigma_k, \delta, q_0, \Sigma_k, \tau)$ be a finite-state transducer. Suppose there exist a matrix-valued morphism $\mu : \Sigma^k \to \mathbb{Z}^{r \times r}$ and two vectors λ, κ such that $s_n := \lambda \mu(b_1 \cdots b_i) \kappa$, where $b_1 b_2 \cdots b_i = T((n)_k)$. Then the sequence $(s_n)_{n \ge 0}$ is k-regular.*

Proof. The basic idea is to show how to simulate the computations of T while accumulating the matrix product at the same time. To do so, for $a \in \Sigma_k$ we define $\mu'(a)$ to be a square matrix of dimension $r|Q|$ as follows: if $\tau(q_i, a) = c_1 \cdots c_s$, and $\delta(q_i, a) = q_j$ then the entries in rows $(i-1)r, \ldots, ir - 1$ and columns $(j-1)r, \ldots, jr - 1$ equal $\mu(c_1 \cdots c_s)$. A simple induction now shows that if on input $w = d_1 \cdots d_u$ the machine T enters state $\delta(q_0, w) = q_l$ and outputs $T(w) = e_1 \cdots e_v$, then the entries in rows $0, \ldots, r - 1$ and columns $(l-1)r, \ldots, lr - 1$ of $\mu'(d_1 \cdots d_u)$ equal $\mu(e_1 \cdots e_v)$.

Now define $\lambda' = [\lambda \, 0 \, 0 \, \cdots \, 0]$ and

$$\kappa' = \begin{bmatrix} \kappa \\ \kappa \\ \vdots \\ \kappa \end{bmatrix}.$$

It now follows that if $b_1 b_2 \cdots b_i = T((n)_k)$, then $\lambda \mu(b_1 \cdots b_i) \kappa = \lambda' \mu(e_1 \cdots e_h) \kappa'$, where $e_1 e_2 \cdots e_h = (n)_k$. ∎

16.3 Further Results

Theorem 16.3.1 *Let $(s(n))_{n \geq 0}$ be a k-regular sequence over $\mathbb{Z}$. Then there exists a constant c such that $s(n) = O(n^c)$.*

Proof. Let the base-k expansion of n be $n = \sum_{0 \leq i < t} a_i k^i$. Then $t \leq 1 + \log_k n$. We use the characterization of Theorem 16.1.3(e). Then $V(n) = \mu(a_0 a_1 \cdots a_{t-1}) V(0)$. If v is a d-dimensional vector, define the norm

$$\|v\| = \sum_{1 \leq i \leq r} |v_i|,$$

and if M is a $r \times r$ matrix, define

$$\|M\| = \max_{1 \leq i \leq r} \sum_{1 \leq j \leq r} |M_{ij}|.$$

Then it is easy to see that $\|Mv\| \leq \|M\| \|v\|$. Thus

$$s(n) \leq \|V(n)\| \leq \|\mu(a_0)\| \|\mu(a_1)\| \cdots \|\mu(a_{t-1})\| \|V(0)\|.$$

Now let $e = \max_{0 \leq i < t} \|\mu(i)\|$ and $d = \|V(0)\|$. Then we have $s(n) \leq e^{1 + \log_k n} d \leq edn^{\log_k e} = O(n^c)$, where $c = \log_k e$. ∎

Theorem 16.3.2 *Let F be a field with $a \in F$. Then the sequence of powers $(a^n)_{n \geq 0}$ is k-regular if and only if $a = 0$ or a is a root of unity.*

Proof. One direction is simple, since if $a = 0$ or a root of unity, the sequence of powers is periodic, and hence k-regular.

For the other direction, assume $(a^n)_{n \geq 0}$ is k-regular. Then there exist an integer $r < \infty$ and integers $\lambda_0, \lambda_1, \ldots, \lambda_{r-1}$, not all zero, such that $\sum_{0 \leq j < r} \lambda_j a^{k^j \cdot n} = 0$ for all $n \geq 0$.

Now we use the Vandermonde determinant identity (see Exercise 1), which states that

$$\det \begin{bmatrix} 1 & b_0 & b_0^2 & \cdots & b_0^m \\ 1 & b_1 & b_1^2 & \cdots & b_1^m \\ \vdots & \vdots & \vdots & \ddots & \vdots \\ 1 & b_m & b_m^2 & \cdots & b_m^m \end{bmatrix} = \prod_{0 \leq i < j \leq m} (b_j - b_i). \tag{16.1}$$

It follows that the sequences $(b_j^n)_{n \geq 0}$ are linearly independent if and only if the numbers $b_0, b_1, \ldots, b_m$ are distinct. Hence the numbers $1, a^k, a^{k^2}, \ldots, a^{k^r}$ are not all distinct, and we must have $a^{k^j} = a^{k^l}$ for some $j \neq l$. Thus either $a = 0$ or a is a root of unity. ∎

We now prove a weak analogue of Cobham's theorem (Theorem 11.2.1) for k-regular sequences.

Proposition 16.3.3 *Let k and l be integers ≥ 2. Let $\mathbf{x}$ be a sequence that is both k-regular and l-regular. Then $\mathbf{x}$ is kl-regular.*

Proof. Suppose $\mathbf{x} = (x_n)_{n\geq 0}$ is both k-regular and l-regular, for $k, l \geq 2$. Since $\mathbf{x}$ is k-regular, we know there exist sequences $(x_n^{(1)})_{n\geq 0}, \ldots, (x_n^{(d)})_{n\geq 0}$, each of the form $(x_{k^\alpha n+\beta})_{n\geq 0}$ where $\alpha \geq 0$ and $0 \leq \beta < k^\alpha$, such that $(x_n^{(1)})_{n\geq 0} = (x_n)_{n\geq 0}$, and any sequence $(x_{k^\gamma n+\delta})_{n\geq 0}$ with $\gamma \geq 0$ and $0 \leq \delta < k^\gamma$ is a $\mathbb{Z}$-linear combination of the sequences $(x_n^{(i)})_{n\geq 0}$ for $i = 1, 2, \ldots, d$.

Since the sequence $(x_n)_{n\geq 0}$ is l-regular, it follows from Theorem 16.2.2 that the sequences $(x_n^{(i)})_{n\geq 0}$ are also l-regular, for each of them is of the form $(x_{k^\alpha n+\beta})_{n\geq 0}$. Hence for each $i = 1, 2, \ldots, d$ there exist sequences $(x_n^{(i,1)})_{n\geq 0}, (x_n^{(i,2)})_{n\geq 0}, \ldots,$ $(x_n^{(i,e_i)})_{n\geq 0}$, each of the form $(x_{l^\alpha n+\beta}^{(i)})_{n\geq 0}$, where $\alpha \geq 0$ and $0 \leq \beta < l^\alpha$, such that $(x_n^{(i,1)})_{n\geq 0} = (x_n^{(i)})_{n\geq 0}$. Further, any sequence $(x_{l^\gamma n+\delta}^{(i)})_{n\geq 0}$ with $\gamma \geq 0$ and $0 \leq \delta < l^\gamma$ is a linear combination of the sequences $(x_n^{(i,j)})_{n\geq 0}$, for $j = 1, 2, \ldots, e_i$.

Now let $\alpha \geq 0$ and $0 \leq \beta < (kl)^\alpha$, and consider the sequence $(x_{(kl)^\alpha n+\beta})_{n\geq 0}$. Let $\beta = k^\alpha q + r$, with $q \geq 0$ and $0 \leq r < k^\alpha$. Then $k^\alpha q \leq k^\alpha q + r = \beta < (kl)^\alpha$. Hence $q < l^\alpha$.

We then have $x_{(kl)^\alpha n+\beta} = x_{k^\alpha(l^\alpha n+q)+r}$. The sequence $(x_{k^\alpha n+r})_{n\geq 0}$ is a $\mathbb{Z}$-linear combination of the sequences $(x_n^{(i)})_{n\geq 0}$, with $i = 1, 2, \ldots, d$. Hence the sequence $(x_{(kl)^\alpha n+\beta})_{n\geq 0}$ is the same linear combination of the sequences $(x_{l^\alpha n+q}^{(i)})$, and hence, since $q < l^\alpha$, a linear combination of the sequences $(x_n^{(i,j)})_{n\geq 0}$, with $i = 1, 2, \ldots, d$ and $j = 1, 2, \ldots, e_i$. ∎

16.4 *k*-Regular Power Series

In this section we consider the properties of power series whose coefficients form a k-regular sequence.

First we discuss convolution. If $\mathbf{a} = (a(n))_{n\geq 0}$ and $\mathbf{b} = (b(n))_{n\geq 0}$ are two sequences taking values in a $\mathbb{Z}$-module R, we define their convolution $\mathbf{c} = \mathbf{a} \star \mathbf{b}$ as follows: if $\mathbf{c} = (c(n))_{n\geq 0}$, then

$$c(n) = \sum_{i+j=n} a(i)b(j).$$

Theorem 16.4.1 *Let $\mathbf{a} = (a(n))_{n\geq 0}$ and $\mathbf{b} = (b(n))_{n\geq 0}$ be k-regular sequences. Then $\mathbf{c} = \mathbf{a} \star \mathbf{b}$ is k-regular.*

Proof. Since $\mathbf{a}$ and $\mathbf{b}$ are k-regular, there exist sequences $\mathbf{a}_1, \mathbf{a}_2, \ldots, \mathbf{a}_p$ that generate the $\mathbb{Z}$-module generated by the k-kernel of $\mathbf{a}$, and sequences $\mathbf{b}_1, \mathbf{b}_2, \ldots, \mathbf{b}_q$ that generate the $\mathbb{Z}$-module generated by the k-kernel of $\mathbf{b}$. We want to find a basis for $\mathcal{C}$, the $\mathbb{Z}$-module generated by the k-kernel of $\mathbf{c}$. We write $\mathbf{u}_{i,j} = \mathbf{a}_i \star \mathbf{b}_j$ for $1 \leq i \leq p, 1 \leq j \leq q$, and $\mathbf{u}_{i,j} = (u_{ij}(n))_{n\geq 0}$.

We claim that the set $\mathcal{M}$ of the $2pq$ sequences $(u_{ij}(n))_{n\geq 0}$ and $(u_{ij}(n-1))_{n\geq 0}$, $1 \leq i \leq p$, $1 \leq j \leq q$, generates $\mathcal{C}$. (We define $u_{ij}(-1) = 0$.)

Let us write $\mathbf{A}_{e,s} = (a(k^e n + s))_{n\geq 0}$ and $\mathbf{B}_{f,t} = (b(k^f n + t))_{n\geq 0}$. Now every sequence of the form $\mathbf{A}_{e,s}$ for $e \geq 0, 0 \leq s < k^e$ is a linear combination of the $\mathbf{a}_i$, and every subsequence of the form $\mathbf{B}_{f,t}$ for $f \geq 0, 0 \leq t < k^f$ is a linear combination of the $\mathbf{b}_i$. It follows that $\langle \mathcal{M} \rangle$ contains all sequences of the form

$$\mathbf{A}_{e,s} \star \mathbf{B}_{f,t}, \tag{16.2}$$

where $e, f \geq 0$ and $0 \leq s < k^e, 0 \leq t < k^f$. Similarly, $\langle \mathcal{M} \rangle$ contains all sequences of the form

$$\mathcal{S}^{-1}(\mathbf{A}_{e,s} \star \mathbf{B}_{f,t}), \tag{16.3}$$

where $\mathcal{S}$ is the shift operator introduced in Section 1.1. Thus to prove the claim it suffices to show how to write all sequences of the form $(c(k^g n + d))_{n\geq 0}$ as linear combinations of the sequences in Eqs. (16.2) and (16.3).

We do this as follows:

$$c(k^g n + d) = \sum_{0 \leq i \leq d} (\mathbf{A}_{g,i} \star \mathbf{B}_{g,d-i})[n] + \sum_{d < j < k^g} (\mathbf{A}_{g,j} \star \mathbf{B}_{g,k^g+d-j})[n-1].$$

Verification is left to the reader. ∎

Let R be a ring. We say a power series in $R[[X]]$ is k-regular if its sequence of coefficients forms a k-regular sequence.

Corollary 16.4.2 *The set of k-regular power series forms a ring.*

Proof. If $A(X)$ and $B(X)$ are two k-regular formal power series, then $A + B$ is k-regular by Theorem 16.2.1 and AB is k-regular by Theorem 16.4.1. ∎

Theorem 16.4.3 *Let F be an algebraically closed field (e.g., $\mathbb{C}$). Let $(s(n))_{n\geq 0}$ be a sequence with values in F. Let $f(X) = \sum_{n\geq 0} s(n)X^n$ be a formal power series in $F[[X]]$. Assume that f represents a rational function of X, i.e., there exist polynomials p, q such that $f(X) = p(X)/q(X)$. Then $(s(n))_{n\geq 0}$ is k-regular if and only if the poles of f are roots of unity.*

Proof. Note that by assumption, 0 is not a pole of f.

Suppose the poles of f are roots of unity. Then, using expansion by partial fractions, we can write

$$f(X) = \sum_i \frac{c_i}{(1 - \zeta_i X)^{e_i}},$$

where $c_i \in F$, the e_i are non-negative integers, and each ζ_i is a root of unity. To prove that $(s(n))_{n \geq 0}$ is k-regular, it suffices to show that the power series $(1 - \zeta_i X)^{-1}$ is k-regular. But this power series has periodic coefficients and hence is k-regular.

For the converse, suppose $f(X) = p(X)/q(X)$ for polynomials p, q, and f is k-regular. Let $1/\zeta$ be one of the poles of f; we may assume $\zeta \neq 0$. We can then write

$$f(X) = \frac{p(X)}{q(X)} = \frac{r(X)}{s(X)(1 - \zeta X)^e},$$

where r, s are polynomials, and $r(X)$ and $1 - \zeta X$ are relatively prime. Then there exist two polynomials u, v such that $u(X)r(X) + v(X)(1 - \zeta X)^e = 1$. Now

$$u(X)f(X)s(X) + v(X) = (1 - \zeta X)^{-e} \tag{16.4}$$

is also a k-regular power series. But $(1 - \zeta X)^{e-1}$ is a polynomial, and hence a k-regular power series, so its product with (16.4) is k-regular, and thus $(1 - \zeta X)^{-1}$ is k-regular. But the coefficients of this power series are ζ^n, which, by Theorem 16.3.2, is k-regular if and only if ζ is a root of unity. ∎

16.5 Additional Examples

In this section we give some additional examples of k-regular sequences.

Let us start with some of the sequences defined in Section 3.2.

Example 16.5.1 Recall that $S_k(n) = \sum_{0 \leq i < n} s_k(i)$, where $s_k(n)$ is the sum of the base-k digits of n. It follows from Theorem 16.4.1 that $(S_k(n))_{n \geq 0}$ is k-regular, but this can also be established directly. From Exercise 3.5, it follows that

$$S_k(kn + a) = kS_k(n) + as_k(n) + k(k-1)n/2 + a(a-1)/2.$$

Hence the k-kernel of $(S_k(n))_{n \geq 0}$ is generated by the sequences $(S_k(n))_{n \geq 0}$, $(s_k(n))_{n \geq 0}$, $(n)_{n \geq 0}$, and the constant sequence 1.

Example 16.5.2 Gauss proved that an integer $n \geq 0$ is the sum of three integer squares if and only if n is not of the form $4^a(8k + 7)$ for integers $a, k \geq 0$. It is easy to prove that the sequence $(t(n))_{n \geq 0}$ defined by

$$\bullet \quad t(n) = \begin{cases} 0 & \text{if there exist } a, k \text{ such that } n = 4^a(8k + 7), \\ 1 & \text{otherwise} \end{cases}$$

is 2-automatic; see Exercise 6.34. Hence, the sequence $Q(n) := \sum_{1 \leq i \leq n} t(i)$, which counts the number of positive integers $\leq n$ that are the sum of the three squares, is 2-regular.

Example 16.5.3 The Danish composer Per Nørgård (1932–) used a particular mathematical sequence $(c_n)_{n\geq0}$, called by some commentators the "infinity series", in many of his musical compositions. Here $(c_n)_{n\geq0}$ is defined by $c_0 = 0$, and for $n \geq 0$ we have $c_{2n} = -c_n$ and $c_{2n+1} = c_n + 1$. The first few values of this sequence are given in the following table:

n	0	1	2	3	4	5	6	7	8	9	10	11	12	13	14	15	16
c_n	0	1	−1	2	1	0	−2	3	−1	2	0	1	2	−1	−3	4	1
m_n	G	A♭	F♯	A	A♭	G	F	B♭	F♯	A	G	A♭	A	F♯	E	B	A♭

For example, the first 1024 notes of the second movement of his symphony *Voyage into the Golden Screen* (1968) are defined as follows: the nth note of the composition m_n is the note offset by c_n halftones of the chromatic scale from G (sol). The sequence $(c_n)_{n\geq0}$ is 2-regular. Note that $t_n = c_n \bmod 2$, where $\mathbf{t} = (t_n)_{n\geq0}$ is the Thue–Morse sequence.

Example 16.5.4 Consider Kimberling's sequence $(c_n)_{n\geq1}$ defined by $c_n = \frac{1}{2}\left(n/2^{\nu_2(n)} + 1\right)$. The first few terms are

$$1, 1, 2, 1, 3, 2, 4, 1, 5, 3, 6, 2, 7, 4, 8, \ldots.$$

This sequence has the pleasant property that deleting the first occurrence of each positive integer in it leaves the sequence unchanged. It is easily verified that $c(2n) = c(n)$ and $c(2n - 1) = n$ for $n \geq 1$, and hence $(c(n))_{n\geq1}$ is 2-regular.

It is also possible to define the notion of k-regular two-dimensional array (in analogy with Chapter 14). Roughly speaking, a two-dimensional array $(a(m, n))_{m,n\geq0}$ is k-regular if there exist a finite number of two-dimensional arrays $(a_i(m, n))_{m,n\geq0}$ such that each subarray of the form $(a(k^e m + a, k^e n + b))_{m,n\geq0}$ with $e \geq 0$ and $0 \leq a, b < k^e$ can be written as a $\mathbb{Z}$-linear combination of the a_i.

Example 16.5.5 Let r, s be non-negative integers with base-2 representation given by $\sum_{0\leq i<t} c_i 2^i$ and $\sum_{0\leq i<t} d_i 2^i$, respectively. Define the *nim sum* of two integers, $r \oplus s$, to be the integer given by $\sum_{0\leq i<t}((c_i + d_i)\bmod 2)2^i$. Consider the two-dimensional array $\mathbf{N} = (m \oplus n)_{m,n\geq0}$. The first few rows and columns of this array are given in Table 16.1.

It is easily seen that $\mathbf{N}$ is 2-regular, as we find

$$\mathbf{N}[2i, 2j] = \mathbf{N}[2i + 1, 2j + 1] = 2\mathbf{N}[i, j],$$
$$\mathbf{N}[2i + 1, 2j] = \mathbf{N}[2i, 2j + 1] = 2\mathbf{N}[i, j] + 1.$$

Table 16.1.

$\oplus$	0	1	2	3	4	5	6	7	8	9
0	0	1	2	3	4	5	6	7	8	9
1	1	0	3	2	5	4	7	6	9	8
2	2	3	0	1	6	7	4	5	10	11
3	3	2	1	0	7	6	5	4	11	10
4	4	5	6	7	0	1	2	3	12	13
5	5	4	7	6	1	0	3	2	13	12
6	6	7	4	5	2	3	0	1	14	15
7	7	6	5	4	3	2	1	0	15	14
8	8	9	10	11	12	13	14	15	0	1
9	9	8	11	10	13	12	15	14	1	0

16.6 Exercises

1. Prove the Vandermonde identity, Eq. (16.1).
2. Give an example of a sequence $(a_n)_{n\geq 0}$ over $\mathbb{Z}$ such that $(a_n \bmod m)_{n\geq 0}$ is 2-automatic for all $m \geq 1$, but $(a_n)_{n\geq 0}$ is not 2-regular.
3. Let α be a real number. Show that $(\lfloor n\alpha \rfloor)_{n\geq 0}$ is k-regular if and only if α is rational.
4. (a) Show there is a unique monotone infinite sequence $\mathbf{d} = (d(n))_{n\geq 0}$ of non-negative integers such that $d(d(n)) = 2n$ for $n \neq 1$.
   (b) Show that d is 2-regular.
5. (a) Show there is unique monotone infinite sequence $\mathbf{e} = (e(n))_{n\geq 0}$ of non-negative integers such that $e(e(n)) = 3n$ for $n \geq 0$.
   (b) Show that e is 3-regular.
6. Give an example of a 2-regular power series $f(X)$ with zero constant term such that $f(f(X))$ is not 2-regular.
7. Let $U(X) = \sum_{i\geq 0} u_i X^i$ be a 2-regular power series with $u_0 = 0$. Show that $V(X) := U(X) + U(X^2) + U(X^4) + \cdots$ is also 2-regular.
8. Give an example of a 2-regular power series $f(X)$ such that $f(1/2) \notin L(2, 2)$. (For the definition of $L(k, b)$, see Section 13.1.)
9. Recall that $e_{2;P}(n)$ counts the number of occurrences of the pattern P in the binary expansion of n. Show that $e_{2;P}(n)$ is 2-regular for all nonzero patterns P.
10. Show that a sequence $(S(n))_{n\geq 0}$ is 2-regular if and only if its pattern transform (cf. Section 3.3) $(\hat{S}(n))_{n\geq 0}$ is 2-regular.
11. Recall that $s_k(n)$ counts the sum of the digits in the base-k expansion of n. Show that $(s_k(s_k(n)))_{n\geq 0}$ is not a k-regular sequence. Hint: Use Exercise 3.8.
12. Let $f(n)$ be the number of representations of n as a sum of 3 triangular numbers, where order matters. (A *triangular* number is an integer of the form $n(n+1)/2$.) Thus, for example, $t(15) = 6$, since we may take $15 + 0 + 0$ and its rearrangements and $6 + 6 + 3$ and its rearrangements.

(a) Show that

$$f(27n + 12) = 3f(n + 1),$$
$$f(27n + 21) = 5f(3n + 2),$$
$$f(81n + 3) = 4f(9n),$$
$$f(81n + 57) = 4f(9n + 6)$$

for $n \geq 0$.

(b) Show that, despite the identities in part (a), the sequence $(f(n))_{n \geq 0}$ is not 3-regular.

13. As in Example 3.3.8, define $B_k(n)$ to be the number of blocks of adjacent identical digits in the binary expansion of n. Thus, for example, $B_2(118687) = 5$, since $(118687)_2 = 11100111110011111$. Prove that $(B_k(n))_{n \geq 0}$ is k-regular. Also find its expansion as a sum of pattern sequences, as in Section 3.3.

14. Define $h(n)$ to be the length of the longest block of contiguous 1's in the binary expansion of n.

(a) Show that $h(2n) = h(n)$ and $h(2n + 1) = \max(h(n), v_2(n + 1) + 1)$ for $n \geq 0$.

(b) Show that h is not 2-regular.

15. Define $a_0 = 0$, $a_1 = 1$, and, for $n \geq 1$, $a_{2n} = a_n$ and $a_{2n+1} = a_n + a_{n+1}$. Set $A(X) = \sum_{n \geq 0} a_n X^n$. Find a functional equation for $A(X)$.

16. Suppose $(n)_k = a_r a_{r-1} \cdots a_0$. As usual, let $s_k(n) = \sum_{0 \leq i \leq r} a_i$. Define $h_k(n) = \sum_{0 \leq i \leq r} i a_i$. Show that

$$\sum_{n \geq 0} (-1)^{s_2(n)} X^{h_2(2n+1)} = 1 + \sum_{n \geq 1} (-1)^n (X^{n(3n-1)/2} + X^{n(3n+1)/2}).$$

17. Show that $(a_n)_{n \geq 0}$ is a 2-regular sequence if and only if the transformed sequence

$$a_0, a_0, a_0, a_1, a_0, a_1, a_2, a_3, a_0, a_1, \ldots, a_{2^3-1}, a_0, a_1, \ldots, a_{2^4-1}, \ldots$$

is 2-regular.

18. (M. LeBrun) Consider the sequence $(m_n)_{n \geq 0}$ defined as follows: if $(n)_k = a_r a_{r-1} \cdots a_0$, then $m_n = [a_0 a_r a_{r-1} \cdots a_1]_k$. Show that $(m_n)_{n \geq 0}$ is k-regular.

19. (N. J. A. Sloane) Consider the sequence $(d_n)_{n \geq 0}$ defined as follows: if $(n)_k = a_r a_{r-1} \cdots a_0$, then $d_n = [b_{r-1} \cdots b_0]_k$, where $b_i = (a_{i+1} + a_i) \bmod k$. Show that $(d_n)_{n \geq 0}$ is k-regular.

20. If $\mathbf{s} = (s_i)_{i \geq 0}$ is a sequence of real numbers, let $\Delta \mathbf{s}$ denote the first difference sequence $(s_{i+1} - s_i)_{i \geq 0}$. Similarly, let $\Delta^n \mathbf{s}$ be the nth iterated difference sequence. Give an example of a k-regular sequence $\mathbf{s}$ such that $\Delta^n \mathbf{s}$ is unbounded for all $n \geq 0$.

21. Nim is a two-player game, with each player moving alternately. The initial configuration consists of three piles of counters, with no two piles containing the same number. A player moves by selecting a pile and removing any number of counters from it. A player wins when, at the conclusion of his/her turn, no

counters are left. Show that the second player has a forced win if and only if the number of counters is initially $(i, j, i \oplus j)$, where $\oplus$ is the nim sum defined in Example 16.5.5.

22. Suppose the sequence $(a_n)_{n\geq 0}$ is 2-regular. Prove that the two-dimensional array given by $(a_{i\oplus j})_{i,j\geq 0}$ is 2-regular. Here $\oplus$ is the nim sum defined in Example 16.5.5.

23. Let $M(n)$ be the array formed by the first 2^n rows and columns of the nim-sum array $\mathbf{N}$ introduced in Example 16.5.5. Let $M = \lim_{n\to\infty} M(n)$.

    (a) Show that $M(n) = g^n([0])$, where g is the map that sends a $k \times k$ array $[A]$ to the $2k \times 2k$ array given by

    $$\begin{bmatrix} [A] & [A] + k \\ [A] + k & [A] \end{bmatrix},$$

    where by $[A] + k$ we mean the array obtained by adding k to each element of $[A]$.

    (b) Show that M is a fixed point of the matrix-valued morphism h that sends n to the 2×2 array

    $$\begin{bmatrix} 2n & 2n+1 \\ 2n+1 & 2n \end{bmatrix}.$$

    (c) Let L be the infinite array generated by the following greedy algorithm: set $L_{0,0} = 0$, and fill in subsequent entries $L_{i,j}$ in order of increasing sum $i + j$. Once all elements above and to the left of row i and column j have been filled in, define $L_{i,j}$ to be the least non-negative integer that does not appear previously in either row i or column j. Then $L = M$.

    (d) Let $H(n)$ be the Hadamard matrix introduced in Exercise 14.4. Show that the columns of $H(n)$ span the eigenspace of the matrix $M(n)$.

    (e) Show that $M(n)$ is of rank $n + 1$ and its characteristic polynomial is $X^{2^n-n-1}(X - 2^{n-1}(2^n - 1))\prod_{0\leq r<n}(X + 2^{n+r-1})$.

24. Show that $(s_2(m + n))_{m,n\geq 0}$ is a 2-regular array. How about $(s_2(mn))_{m,n\geq 0}$?

25. For $n \geq 1$ define $u(n) = \sum_{0\leq i<n} \binom{2i}{i}$.

    (a) Show that $v_3(\binom{2n}{n}) = s_3(n) - \frac{1}{2}s_3(2n)$.

    (b) Show that $v_3(u(n)) = v_3(\binom{2n}{n}) + 2v_3(n)$.

    (c) Prove that $(v_3(u(n)))_{n\geq 1}$ is a 3-regular sequence.

26. Give an example of a 2-regular sequence $(a_n)_{n\geq 0}$ such that $(\text{sgn}(a_n))_{n\geq 0}$ is not 2-automatic.

27. Suppose one is given a k-regular sequence $(s(n))_{n\geq 0}$, say, by being provided with λ, μ, κ in its linear representation. Show that the following problems are unsolvable. Hint: Use Hilbert's tenth problem (Theorem 4.6.3).

    (a) There exists i such that $s(i) = 0$.

    (b) For all i we have $s(i) \geq 0$.

28. Define $a(0) = 2$, $a(1) = 3$, and let $a(n)$ be the least integer $> a(n - 1)$ such that $a(n) \neq a(k) + a(k - 1)$ for all k with $1 \leq k < n$.

    (a) Show that $a(2n) = 3n + 1 + (\lfloor \log_2 n \rfloor \bmod 2)$ for $n \geq 1$ and $a(2n + 1) = 3n + 3$ for $n \geq 0$. Conclude that $(a(n))_{n\geq 0}$ is 2-regular.

(b) Let $b(n) = a(n) + a(n-1)$ for $n \geq 1$. Show that $b(n) = 3n + 2 - (\lfloor \log_2 n \rfloor \bmod 2)$ for $n \geq 1$.

(c) Show that $\{a(n) : n \geq 0\} \cup \{b(n) : n \geq 1\} = \{2, 3, 4, \ldots\}$.

29. Let $(a(n))_{n \geq 0}$ be a k-regular sequence with $0 < a(0) < a(1) < a(2) \cdots$, and let b be an integer ≥ 2. Show that the number

$$.(a(0))_b(a(1))_b(a(2))_b \cdots$$

is irrational.

30. (D. Fux) Define $b(n) = \sum_{0 \leq k \leq n} \left(\binom{n}{k} \bmod 3 \right)$.

(a) Show that $b(n) = 2^{a-1}(3^{b+1} - 1)$, where a is the number of occurrences of the digit 1 and b is the number of occurrences of the digit 2 in the base-3 expansion of n.

(b) Conclude that $(b(n))_{n \geq 0}$ is a 3-regular sequence.

31. (Morton and Mourant) Let G be an abelian group, written additively, and let $\mathbf{a} = (a(n))_{n \geq 0}$ be a sequence of elements of G. For $n, q \geq 0$ we define a subword of $\mathbf{a}$ of length k^q as follows:

$$X_n^q = (a(nk^q), a(nk^q + 1), \ldots, a(nk^q + k^q - 1)).$$

For a subword $v = b_1 b_2 \cdots b_t$, we define $v - c = d_1 d_2 \cdots d_t$, where $d_i = b_i - c$ for $1 \leq i \leq t$. Define the group $\Gamma_k(G)$ to be the set of all sequences $\mathbf{a}$ for which the sequence of blocks $(X_n^q - a(n))_{n \geq 0}$ is purely periodic, for all $q \geq 0$. In other words, a sequence $\mathbf{a}$ is in $\Gamma_k(G)$ if there exists an integer $M \geq 1$ such that $X_m^q - a(m) = X_n^q - a(n)$ if $m \equiv n \pmod{M}$.

(a) Show that $\mathbf{a} \in \Gamma_k(\mathbb{Z})$ if and only if the sequence $(a(n) - a(\lfloor n/k \rfloor))_{n \geq 0}$ is purely periodic, and conclude that every sequence in $\Gamma_k(\mathbb{Z})$ is k-regular.

(b) Let $\mathbf{a}$ be a sequence in $\Gamma_k(G)$, i.e., if $m \equiv n \pmod{M}$, then $X_m^q - a(m) = X_n^q - a(n)$ for all $q \geq 0$. Then there exist a morphism $\varphi : (G \times \mathbb{Z}/M\mathbb{Z}) \to (G \times \mathbb{Z}/M\mathbb{Z})^k$ and a coding $\tau : G \times \mathbb{Z}/M\mathbb{Z} \to G$ such that $\mathbf{a} = \tau(\varphi^\omega([a(0), 0]))$. Conclude that if G is finite, the sequence $\mathbf{a}$ is k-automatic.

32. Define $x_0 = 0$ and $x_{n+1} = 1/(1 + 2\lfloor x_n \rfloor - x_n)$. Define the sequences $(p_n)_{n \geq 0}$, $(q_n)_{n \geq 0}$ by $p_n/q_n = x_n$, where $q_n \geq 1$ and $\gcd(p_n, q_n) = 1$.

(a) Show that $q_n = p_{n+1}$ for $n \geq 0$.

(b) Show that $(p_n)_{n \geq 0}$ and $(q_n)_{n \geq 0}$ are 2-regular.

(c) Show that the sequence $(x_n)_{n \geq 0}$ enumerates every non-negative rational number exactly once.

33. Let $(b_n)_{n \geq 0}$ be a k-regular sequence with $b_0 = 0$. Show that $(c_n)_{n \geq 0}$ is a k-regular sequence, where $c_n := \sum_{j \geq 0} b_{\lfloor n/k^j \rfloor}$.

34. Let $s_{-k}(n)$ denote the sum of the digits of n when expressed in base $-k$ (see Theorem 3.7.2). Show that the sequences $(s_{-k}(n))_{n \geq 0}$ and $(s_{-k}(n) - s_{-k}(-n))_{n \geq 0}$ are both k-regular sequences.

35. Show that the sequence $(k_n)_{n \geq 0}$, which counts the number of partitions of n into Fibonacci number parts, can be computed as follows: there exist a

matrix-valued morphism $\mu : \Sigma_k \to \mathbb{Z}^{r \times r}$ and vectors λ, κ with integer entries such that $k(n) = \lambda \mu((n)_F) \kappa$, where $(n)_F$ denotes Fibonacci representation.

36. Show that the following problem is unsolvable: given the matrix representation of a k-regular sequence $(a_n)_{n \geq 0}$, decide if there exists an index n such that $a_n = 0$. Hint: Use Theorem 4.6.3.

16.7 Open Problems

1. (F. Beukers) Define the Apéry numbers $u_n = \sum_{0 \leq k \leq n} \binom{n}{k}^2 \binom{n+k}{k}^2$.
   (a) Prove or disprove: $5^{e_{5;1}(n)+e_{5;3}(n)} \mid u_n$.
   (b) Define $b_n = v_5(u_n)$. Is $(b_n)_{n \geq 0}$ a 5-regular sequence?
   (Remark: See, for example, Coster [1988, p. 42].)

2. Let

$$a(n) = \sum_{0 \leq k \leq n} \binom{n}{k} \binom{n+k}{k}.$$

Let $b(n) = v_3(a(n))$. Prove or disprove that

$$b(n) = \begin{cases} b(\lfloor n/3 \rfloor) + (\lfloor n/3 \rfloor \bmod 2) & \text{if } n \equiv 0, 2 \pmod 3, \\ b(\lfloor n/9 \rfloor) + 1 & \text{if } n \equiv 1 \pmod 3. \end{cases}$$

3. Let f be a polynomial with rational coefficients. Prove or disprove: $(v_p(f(n)))_{n \geq 0}$ is a p-regular sequence.

4. Define $f(n) = \min_{k \geq n+1}(k - v_2(k))$. Is $(f(n))_{n \geq 0}$ a 2-regular sequence? (Remark: See Lengyel [1994].)

5. Suppose $(a_n)_{n \geq 0}$ and $(b_n)_{n \geq 0}$ are 2-regular sequences taking non-negative integer values. Must $(a_n \oplus b_n)_{n \geq 0}$ be 2-regular? Here $\oplus$ is the nim sum defined in Example 16.5.5. How about the two-dimensional array $(a_i \oplus b_j)_{i, j \geq 0}$?

6. The nim product $m \otimes n$ is defined as follows:
   (a) If $m = 2^{2^a}$ for some integer $a \geq 0$ and $n < m$, then $m \otimes n = mn$.
   (b) If $m = 2^{2^a}$ for some integers $a \geq 0$, then $m \otimes m = 3m/2$.
   (c) For other products, use the fact that $\otimes$ is associative and distributes over $\oplus$.
   Prove or disprove: the sequence $(n \otimes n)_{n \geq 0}$ is 2-regular. How about the two-dimensional array $(m \otimes n)_{m, n \geq 0}$?

7. Determine all the units (invertible elements) of the ring of k-regular power series.

8. Suppose $(S(n))_{n \geq 0}$ and $(T(n))_{n \geq 0}$ are k-regular sequences over $\mathbb{Z}$ and $T(n) \neq 0$ for all n. Prove or disprove: if $S(n)/T(n)$ is always an integer, then $(S(n)/T(n))_{n \geq 0}$ is k-regular.

9. Prove or disprove: if $S(n)$ is a k-regular sequence over $\mathbb{Z}$ and unbounded, then it takes on infinitely many composite values. (A integer ≥ 2 is *composite* if it is not a prime.)

10. Prove or disprove: $(\lfloor \frac{1}{2} + \log_2 n \rfloor)_{n \geq 1}$ is not a 2-regular sequence.

11. Consider expanding n in Fibonacci representation instead of base k. Generalizing Exercise 35, develop an analogous theory of "Fibonacci-regular" sequences, and prove analogues of some of the theorems in this chapter. (Remark: See, for example, Allouche, Scheicher, and Tichy [2000].)

12. Let $(a(n))_{n \geq 0}$ be a k-regular sequence with $0 < a(0) < a(1) < a(2) \cdots$, and let b be an integer ≥ 2. Prove that the number

$$.(a(0))_b (a(1))_b (a(2))_b \cdots$$

is transcendental.

16.8 Notes on Chapter 16

16.1 Allouche and Shallit [1992] coined the term k-regular. Their paper contains thirty examples of k-regular sequences from the literature, some of which are reproduced in this chapter. The sequences we study in this chapter are $(\mathbb{Z}, k)$-regular in the notation of the original paper.

   The theory of k-regular sequences is closely linked to the theory of rational series, for which see Berstel and Reutenauer [1988].

16.2 Theorems 16.2.1–16.2.3 are from Allouche and Shallit [1992].

16.3 Theorem 16.3.3 is a weak analogue of Cobham's theorem (Theorem 11.2.1). It seems reasonable to conjecture that if a sequence $(s_n)_{n \geq 0}$ is both k_1-regular and k_2-regular, with k_1 and k_2 multiplicatively independent, then the associated power series $\sum_{n \geq 0} s_n X^n \in \mathbb{Z}[[X]]$ must be a rational function. See Randé [1993].

16.4 Dumas [1993b] studied the algebraic properties of k-regular sequences and the asymptotic properties of the corresponding power series on $\mathbb{C}$.

   Randé [1993] also studied the properties of k-regular power series. He obtained partial results on their hypertranscendence and a Cobham-like theorem.

   It is known that k-regular power series satisfy Mahler functional equations; see Randé [1992], Dumas [1993a], and Becker [1994]. Becker [1994] also obtained transcendence results on the values of k-regular power series at algebraic points.

   Nishioka [1996, Chapter 5] discussed transcendence and k-regular sequences.

16.5 For the "infinity series" of Per Nørgård, see Kullberg [1996].

   Example 16.5.4 is due to Kimberling [1995, 1997].

   For the game of nim (Example 16.5.5), see, for example, Bouton [1902]; E. H. Moore [1909]; Ball [1939]; Uspensky and Heaslet [1939]; Pedoe [1958]; Yaglom and Yaglom [1967, p. 19]; Conway [1976, Chapter 6]; and Berlekamp, Conway, and Guy [1982].

17

Physics

Quasicrystals, discovered by the materials scientist Dan Shechtman in April 1982, are materials that are intermediate between crystalline and random structures (*glasses*). They are formed, for example, in certain alloys of aluminum with other metals, such as copper or manganese.

After their experimental discovery, it was proposed that a certain tiling of the plane, known as the Penrose tiling, could serve as a theoretical model of the structure of these alloys. The Penrose tiling is not periodic, but possesses fivefold symmetry and looks "regular"; see Figures 17.1 and 17.2. Figure 17.1 illustrates a partial Penrose tiling of the plane by two kinds of pieces: *kites* and *darts*. A local fivefold symmetry can be seen (local invariance under a rotation of $2\pi/5$). In Figure 17.2, some pieces have been glued together to form *bow ties* that are either long or short. If such an infinite line, or *worm*, of short and long bow ties can be found in a Penrose tiling, the bow ties are arranged according to the infinite Fibonacci word $\mathbf{f} = 010010100100101001010\cdots$ introduced in Section 7.1, where a short bow tie is replaced by 0 and a long one by 1.

Because the Fibonacci word is associated with the Penrose tiling, theoretical physicists began to study the properties of $\mathbf{f}$. This word being both morphic and Sturmian, physicists became more generally interested in morphic sequences – including automatic sequences – and Sturmian sequences. Such sequences, when they are not ultimately periodic, are somewhere between periodicity (order) and chaos (disorder). They might correspond to (one-dimensional) materials having physical properties between crystals and glasses, and might be a good theoretical model of one-dimensional quasicrystals.

In this chapter we present a few examples of the use of automatic and morphic sequences in physics. Many more occurrences of these sequences can be found in the literature.

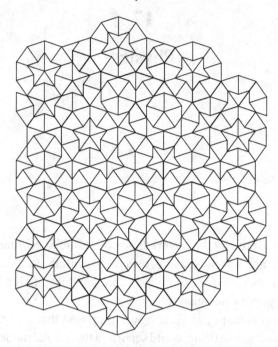

Figure 17.1: A Partial Penrose Tiling of the Plane.

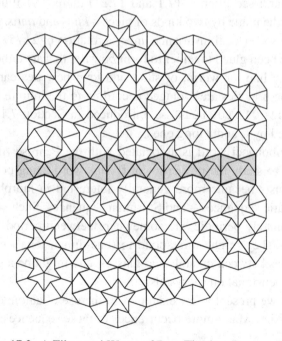

Figure 17.2: A Fibonacci Worm of Bow Ties in a Penrose Tiling.

17.1 The One-Dimensional Ising Model

The Ising model is a relatively simple model that describes the magnetic properties of matter. We restrict ourselves to the one-dimensional case; this corresponds to one-dimensional samples.

Definition 17.1.1 An *Ising chain* of N *sites* consists of the set $\{0, 1, \ldots, N - 1\}$ and a finite sequence $\eta = (\eta_0, \ldots, \eta_{N-1})$ of ± 1's of length N. A *configuration* of the Ising chain is a finite sequence of N *spins*, i.e., of ± 1's, say $\sigma = (\sigma_0, \sigma_1, \ldots, \sigma_{N-1})$.

To describe the magnetic properties of matter, we imagine that neighboring sites are interacting in that σ_q and σ_{q+1} create an interaction field proportional to the product $-\sigma_q \sigma_{q+1}$. We also suppose there is an external field whose action at each site q is proportional to σ_q. The following definition is more precise.

Definition 17.1.2 The *Hamiltonian* or *energy* of a configuration σ is defined by

$$\mathcal{H}_\eta(\sigma) = -J \sum_{0 \leq q < N} \eta_q \sigma_q \sigma_{q+1} - H \sum_{0 \leq q < N} \sigma_q,$$

where the chain is *cyclic*, i.e., $\sigma_N = \sigma_0$, where $J > 0$ is a given parameter called the *coupling constant* or *binding energy*, and where H is a real parameter called the *external field*. Let $T > 0$ be a parameter called the *temperature*, and let $\beta := 1/kT$, where k is a universal constant called the Boltzmann constant. The *partition function* $Z_N(\eta, J, H, \beta)$ is defined by

$$Z_N(\eta, J, H, \beta) := \sum_{\sigma \in \{-1, +1\}^N} \exp(-\beta \mathcal{H}_\eta(\sigma)).$$

Finally, the *transfer matrix* of the Ising chain at site q is the matrix M_q defined by

$$M_q := \begin{pmatrix} e^{\beta H + \beta J \eta_q} & e^{\beta H - \beta J \eta_q} \\ e^{-\beta H - \beta J \eta_q} & e^{-\beta H + \beta J \eta_q} \end{pmatrix}.$$

Recall that the trace of a matrix M, denoted by $\mathrm{tr}(M)$, is the sum of the diagonal entries of the matrix M. We now give a theorem showing how to compute the partition function using transfer matrices.

Theorem 17.1.3 *We have*

$$Z_N(\eta, J, H, \beta) = \mathrm{tr} \left(\prod_{0 \leq q \leq N-1} M_q \right).$$

Proof. We write

$$Z_N(\eta, J, H, \beta) = \sum_{\sigma \in \{-1,+1\}^N} \exp(-\beta \mathcal{H}_\eta(\sigma))$$

$$= \sum_{\sigma \in \{-1,+1\}^N} \prod_{0 \le q < N} \exp(\beta J \eta_q \sigma_q \sigma_{q+1} + \beta H \sigma_q)$$

$$= \sum_{\sigma_0 = \pm 1} \sum_{\sigma_1 = \pm 1, \dots, \sigma_{N-1} = \pm 1} \prod_{0 \le q \le N-1} \exp(\beta J \eta_q \sigma_q \sigma_{q+1} + \beta H \sigma_q)$$

Denote the entries of M_q by $(M_q)_{u,v}$, with $1 \le u, v \le 2$. We then have

$$e^{\beta J \eta_q \sigma_q \sigma_{q+1} + \beta H \sigma_q} = \begin{cases} e^{\beta J \eta_q + \beta H} = (M_q)_{1,1} & \text{if } \sigma_q = +1, \ \sigma_{q+1} = +1, \\ e^{-\beta J \eta_q + \beta H} = (M_q)_{1,2} & \text{if } \sigma_q = +1, \ \sigma_{q+1} = -1, \\ e^{-\beta J \eta_q - \beta H} = (M_q)_{2,1} & \text{if } \sigma_q = -1, \ \sigma_{q+1} = +1, \\ e^{\beta J \eta_q - \beta H} = (M_q)_{2,2} & \text{if } \sigma_q = -1, \ \sigma_{q+1} = -1. \end{cases}$$

In other words,

$$e^{\beta J \eta_q \sigma_q \sigma_{q+1} + \beta H \sigma_q} = (M_q)_{\frac{3-\sigma_q}{2}, \frac{3-\sigma_{q+1}}{2}}.$$

Hence

$$Z_N(\eta, J, H, \beta) = \sum_{\sigma_0 = \pm 1} \sum_{\sigma_1 = \pm 1, \dots, \sigma_{N-1} = \pm 1} \prod_{0 \le q \le N-1} (M_q)_{\frac{3-\sigma_q}{2}, \frac{3-\sigma_{q+1}}{2}}$$

$$= \sum_{1 \le \tau_0 \le 2} \sum_{1 \le \tau_1, \tau_2, \dots, \tau_{N-1} \le 2} \prod_{0 \le q \le N-1} (M_q)_{\tau_q, \tau_{q+1}}$$

$$= \sum_{1 \le \tau_0 \le 2} \sum_{1 \le \tau_1, \tau_2, \dots, \tau_{N-1} \le 2} (M_0)_{\tau_0, \tau_1} (M_1)_{\tau_1, \tau_2} \cdots (M_{N-1})_{\tau_{N-1}, \tau_0},$$

where we have used the fact that that $\sigma_n = \sigma_0$, and hence $\tau_n = \tau_0$. Note that the inner sum is the (τ_0, τ_0) entry of the matrix $M_0 M_1 \cdots M_{N-1}$, and hence $Z_N(\eta, J, H, \beta)$ is the trace of this matrix product. ∎

The Ising model is called *homogeneous* if all the η_q are equal; that is, if $\eta_q = \eta_0 = \pm 1$ for all q. In this case all the matrices M_q are equal, and the partition function is easily computed.

Theorem 17.1.4 *The partition function of the N-site homogeneous Ising chain, with coupling constant J and external field H, and where $\eta_q = \eta$ for all q, is given by*

$$Z_N(\eta, J, H, \beta) = \lambda_1^N + \lambda_2^N,$$

where λ_1 and λ_2 are the (not necessarily distinct) complex eigenvalues of the matrix

$$M := \begin{pmatrix} z^{\alpha+\eta} & z^{\alpha-\eta} \\ z^{-\alpha-\eta} & z^{-\alpha+\eta} \end{pmatrix},$$

the complex numbers z and α being defined by $z := e^{\beta J}$ and $\alpha := H/J$.

Proof. Using Theorem 17.1.3, we have $Z_N(\eta, J, H, \beta) = \mathrm{tr}(M^N)$, where

$$M := \begin{pmatrix} z^{\alpha+\eta} & z^{\alpha-\eta} \\ z^{-\alpha-\eta} & z^{-\alpha+\eta} \end{pmatrix}.$$

Let λ_1 and λ_2 be the (not necessarily distinct) complex eigenvalues of the matrix M. We know there exists an invertible matrix P such that $M = PDP^{-1}$, where

$$D = \begin{pmatrix} \lambda_1 & a \\ 0 & \lambda_2 \end{pmatrix} \qquad \text{for some complex number } a.$$

We thus have $M^N = PD^N P^{-1}$. Hence $\mathrm{tr}(M^N) = \mathrm{tr}(D^N) = \lambda_1^N + \lambda_2^N$, and the theorem follows. ∎

17.2 The Rudin–Shapiro Sequence and the One-Dimensional Ising Model

In this section we prove that the one-dimensional Ising model and the Rudin–Shapiro sequence are closely linked. More precisely, define the *complete* Rudin–Shapiro sequence $(u(n))_{n \geq 0}$ by

$$u(n) := e_{2;11}(n) = \sum_{q \geq 0} \varepsilon_q^{(2)}(n)\varepsilon_{q+1}^{(2)}(n),$$

where $n = \sum_{q \geq 0} \varepsilon_q^{(2)}(n)2^q$, with $\varepsilon_q^{(2)}(n) \in \{0, 1\}$. Note that the ordinary Rudin–Shapiro sequence $(r_n)_{n \geq 0}$ (defined in Example 5.1.5) is actually the sequence $((-1)^{u(n)})_{n \geq 0}$. Also note the following useful relation:

$$u(2m + 1) = \frac{1}{2}(1 - (-1)^m) + u(m).$$

Let us change the definition of $(u(n))$ slightly. For a fixed $N \geq 1$ and any integer $n \in [0, 2^N - 1]$, we define

$$\tilde{\varepsilon}_q(n) := \begin{cases} \varepsilon_q^{(2)}(n) & \text{if } q \leq N - 1, \\ \varepsilon_0^{(2)}(n) & \text{if } q = N. \end{cases}$$

We then define the quantity $u(n, N)$ by

$$u(n, N) := \sum_{0 \leq q < N} \tilde{\varepsilon}_q(n)\tilde{\varepsilon}_{q+1}(n).$$

Note that $u(n, N)$ counts the number of occurrences of the word 11 in the "cyclic" binary expansion of n, and that, easily,

$$u(n, N) = \begin{cases} u(n) & \text{if } 0 \leq n < 2^{N-1}, \\ u(n) + \frac{1}{2}(1 - (-1)^n) & \text{if } 2^{N-1} \leq n < 2^N. \end{cases}$$

Let $\mu_q(n)$ be defined by $\mu_q(n) := 1 - 2\tilde{\varepsilon}_q(n)$. Note that $\mu_q(n) = \pm 1$. The proof of the following lemma is straightforward.

Lemma 17.2.1 *We have*

$$u(n, N) = \frac{N}{4} - \frac{1}{2}\left(\sum_{0 \leq q < N} \mu_q(n)\right) + \frac{1}{4}\left(\sum_{0 \leq q < N} \mu_q(n)\mu_{q+1}(n)\right).$$

We are now ready for the main result of this section. Define $R(N, x)$ and $S(N, x)$ by

$$R(N, x) := \sum_{0 \leq n < 2^N} \exp(2i\pi x u(n)) \quad \text{and} \quad S(N, x) := \sum_{0 \leq n < 2^N} \exp(2i\pi x u(n, N)).$$

We have

Theorem 17.2.2 *Let x be a real number. Then*

(a) $S(N, x) - R(N, x) = (\exp(2i\pi x) - 1)(S(N - 1, x) - R(N - 2, x));$

(b) $S(N, x) = \exp(\frac{i\pi N x}{2}) Z_N(1, \frac{1}{2}, -1, i\pi x)$, *where $Z_N(1, \frac{1}{2}, -1, i\pi x)$ is the partition function of the homogeneous Ising model with all η_i equal to 1, with $J = \frac{1}{2}$, $H = -1$, and where the temperature T is a purely imaginary number such that $1/kT = \beta = i\pi x$.*

Proof. We prove assertion (a) as follows:

$$S(N, x) - R(N, x)$$

$$= \sum_{0 \leq n < 2^N} \exp(2i\pi x u(n, N)) - \sum_{0 \leq n < 2^N} \exp(2i\pi x u(n))$$

$$= \sum_{2^{N-1} \leq n < 2^N} \exp(2i\pi x u(n))\left(\exp\left(2i\pi x\left(\frac{1}{2}(1 - (-1)^n)\right)\right) - 1\right)$$

$$= (\exp(2i\pi x) - 1)\sum_{2^{N-2} \leq m < 2^{N-1}} \exp(2i\pi x u(2m + 1))$$

$$= (\exp(2i\pi x) - 1)\sum_{2^{N-2} \leq m < 2^{N-1}} \exp\left(2i\pi x\left(\frac{1}{2}(1 - (-1)^m) + u(m)\right)\right)$$

$$= (\exp(2i\pi x) - 1) \sum_{2^{N-2} \leq m < 2^{N-1}} \exp(2i\pi x u(m, N-1))$$

$$= (\exp(2i\pi x) - 1)(S(N-1, x) - R(N-2, x)).$$

Assertion (b) is a consequence of the observation that there is a bijection between the set $\{-1, +1\}^N$ of all possible configurations of an N-dimensional Ising chain and all the integers in $[0, 2^N - 1]$. Hence

$$S(N, x)$$
$$= \sum_{0 \leq n < 2^N} \exp(2i\pi x u(n, N))$$

$$= \exp\left(\left(\frac{i\pi N x}{2}\right) \sum_{0 \leq n < 2^N} \left(\exp i\pi x \left(\frac{1}{2} \sum_{0 \leq q < N} \mu_q(n)\mu_{q+1}(n) - \sum_{0 \leq q < N} \mu_q(n)\right)\right)\right.$$

$$= \exp\left(\frac{i\pi N x}{2}\right) Z_N(1, \tfrac{1}{2}, -1, i\pi x)$$

with the notation used in Definition 17.1.2.   ∎

Remark. The reader should note that although the temperature in Theorem 17.2.2 is a purely imaginary complex number, the transfer matrix method of Theorem 17.1.3 and Theorem 17.1.4 is still valid.

Corollary 17.2.3 *Let x be a real number. Define the complex numbers $\lambda_1(x)$, $\lambda_2(x)$ and $\gamma = \gamma(x)$ by*

$$\lambda_1(x) := \left(\cos(\pi x) + (-\sin^2(\pi x) + \exp(-2i\pi x))^{1/2}\right) \exp(i\pi x/2),$$
$$\lambda_2(x) := \left(\cos(\pi x) - (-\sin^2(\pi x) + \exp(-2i\pi x))^{1/2}\right) \exp(i\pi x/2).$$

$$\gamma(x) := \begin{cases} \lambda_1(x) & \text{if } |\lambda_1(x)| \geq \lambda_2(x), \\ \lambda_2(x) & \text{if } |\lambda_1(x)| < \lambda_2(x). \end{cases}$$

(Here $U^{1/2}$ is any complex square root of U.)

(a) *If $x \equiv \frac{1}{2}$ (mod 1), then $S(N, x) = (1 + (-1)^N)2^{N/2}$.*
(b) *If $x \not\equiv \frac{1}{2}$ (mod 1), then $S(N, x) = \exp(i\pi N x/2)(\lambda_1(x)^N + \lambda_2(x)^N) \sim (e^{i\pi x}\gamma)^N$ (when N goes to ∞).*
(c) *We have $\sqrt{2} \leq |\gamma| \leq 2$. Furthermore $|\gamma| = \sqrt{2}$ if and only if $x \equiv \frac{1}{2}$ (mod 1), and $|\gamma| = 2$ if and only if $x \equiv 0$ (mod 1).*

Proof. Parts (a) and (b) follow directly from Theorem 17.2.2 and Theorem 17.1.4 above. The corresponding transfer matrix is

$$M := \begin{pmatrix} e^{-i\pi x/2} & e^{-3i\pi x/2} \\ e^{i\pi x/2} & e^{3i\pi x/2} \end{pmatrix},$$

and the eigenvalues of M are $\left(\cos(\pi x) \pm (-\sin^2(\pi x) + \exp(-2i\pi x))^{1/2}\right)$ $\exp(i\pi x/2)$. Note for (b) that if $x \not\equiv \frac{1}{2}$ (mod 1), then $|\lambda_1(x)| \neq |\lambda_2(x)|$.

Part (c) is a consequence of the relation

$$|\gamma|^2 = \frac{1 + (1 + 4T^4)^{1/2} + \sqrt{2}(1 - 2T^2 + (1 + 4T^4)^{1/2})^{1/2}}{1 + T^2},$$

where $T := \tan(\pi x)$. ∎

17.3 Distribution Results for the Rudin–Shapiro Sequence

We now apply the connection between the Rudin–Shapiro sequence and the quantity $u(n, N)$ defined at the beginning of Section 17.2 to prove distribution results for the complete Rudin–Shapiro sequence.

Theorem 17.3.1 *Let $(u_n)_{n \geq 0}$ be the complete Rudin–Shapiro sequence. Let x be a real number that is not an integer. Then there exists $\alpha = \alpha(x) \in (0, 1)$ such that $\sum_{n < N} \exp(2i\pi x u(n)) = O(N^\alpha)$. In particular, the sequence $(xu(n))_{n \geq 0}$ is uniformly distributed modulo 1 for all irrational numbers x.*

Proof. In order to prove the uniform distribution property, it suffices, using Weyl's theorem (Theorem 2.5.6), to prove that for all integers $h \neq 0$ we have $\lim_{N \to \infty} \frac{1}{N} \sum_{n < N} e^{2\pi i h x u(n)} = 0$. Hence it suffices to prove the $O(N^\alpha)$ estimate of the theorem. This is done in two steps: first, we assume that N is a power of 2, and second, we deduce the result for any N.

We first prove that, for $x \not\equiv \frac{1}{2}$ (mod 1), we have $R(N, x) = \sum_{0 \leq n < 2^N} \exp(2i\pi x u(n)) \sim c(x)(\gamma(x)e^{i\pi x})^N$, where $c(x)$ is a nonzero constant depending only on x, and $\gamma = \gamma(x)$ was defined in Corollary 17.2.3.

Define for $j = 1, 2$ the numbers y_j by $y_j := \lambda_j \exp(i\pi x/2)$, where λ_1 and λ_2 were defined in Corollary 17.2.3. Let $E := \exp(2i\pi x) - 1$. Using Theorem 17.2.2 and Corollary 17.2.3, we have

$$R(N, x) - ER(N - 2, x) = y_1^{N-1}(y_1 - E) + y_2^{N-1}(y_2 - E).$$

Hence

$$\sum_{0 \leq k < \lfloor N/2 \rfloor} E^k(R(N - 2k, x) - ER(N - 2k - 2, x))$$

$$= (y_1 - E)\left(\sum_{0 \leq k < \lfloor N/2 \rfloor} y_1^{N-2k-1}E^k\right) + (y_2 - E)\left(\sum_{0 \leq k < \lfloor N/2 \rfloor} y_2^{N-2k-1}E^k\right).$$

This gives, using that $E \neq y_1^2$ and $E \neq y_2^2$ (since x is not an integer and $E = y_1 y_2$),

$$R(N, x) - E^{\lfloor N/2 \rfloor} R(N - 2\lfloor N/2 \rfloor, x)$$
$$= \frac{y_1 - E}{y_1^2 - E} y_1^{N+1} + \frac{y_2 - E}{y_2^2 - E} y_2^{N+1} - W E^{\lfloor N/2 \rfloor},$$

where

$$W := \frac{y_1 - E}{y_1^2 - E} y_1^{N+1-2\lfloor N/2 \rfloor} + \frac{y_2 - E}{y_2^2 - E} y_2^{N+1-2\lfloor N/2 \rfloor}.$$

To compute W, it suffices to use that $y_1 y_2 = E$ and $y_1 + y_2 = \exp(i\pi x/2)(\lambda_1 + \lambda_2) = E + 2$. This gives easily $W = 1$ if N is even, and $W = 2$ if N is odd. Noting that $R(0, x) = 1$ and $R(1, x) = 2$, we thus have

$$R(N, x) = \frac{y_1 - E}{y_1^2 - E} y_1^{N+1} + \frac{y_2 - E}{y_2^2 - E} y_2^{N+1}.$$

Using the number γ defined in Corollary 17.2.3, we see that, for $x \not\equiv \frac{1}{2} \pmod 1$ we have

$$R(N, x) \sim c(\gamma \exp(i\pi x))^N$$

when N goes to ∞. Here $c = c(x)$ is a constant that depends only on x. Furthermore, for every x we have

$$R(N, x) = O(\gamma(x)^N).$$

To extend this bound to the general sum $T(N, x) := \sum_{0 \leq n < N} \exp(2i\pi xu(n))$, we proceed by splitting this sum according to the binary expansion of N. More precisely, let $N = 2^{M_0} + 2^{M_1} + \cdots$, where $M_0 > M_1 > M_2 > \cdots$. We write $T(N, x) = \sum_1 + \sum_2 + \cdots$, where

$$\Sigma_1 := \sum_{0 \leq n < 2^{M_0}} \exp(2i\pi xu(n)),$$

$$\Sigma_2 := \sum_{2^{M_0} \leq n < 2^{M_0} + 2^{M_1}} \exp(2i\pi xu(n)),$$

$$\Sigma_3 := \sum_{2^{M_0} + 2^{M_1} \leq n < 2^{M_0} + 2^{M_1} + 2^{M_2}} \exp(2i\pi xu(n)),$$

$$\vdots$$

Each of the sums Σ_ℓ has the form $\sum_{K \leq n < K + 2^M} \exp(2i\pi xu(n))$, where $2^{M+1} \mid K$. But, if $0 \leq n < 2^M$ and $2^{M+1} \mid K$, then $u(n + K) = u(n) + u(K)$. Hence

$$\sum_{K \leq n < K + 2^M} \exp(2i\pi xu(n)) = \exp(2i\pi xu(K)) R(M, x).$$

Thus, using the first part of this proof, we obtain

$$
\begin{aligned}
|T(N, x)| &\leq |R(M_0, x)| + |R(M_1, x)| + |R(M_2, x)| + \cdots \\
&\leq c'(|\gamma(x)|^{M_0} + |\gamma(x)|^{M_1} + |\gamma(x)|^{M_2} + \cdots) \\
&\leq c'(|\gamma(x)|^{M_0} + |\gamma(x)|^{M_0-1} + |\gamma(x)|^{M_0-2} + \cdots) \\
&\leq c''|\gamma(x)|^{M_0} \leq c''|\gamma(x)|^{\log N/\log 2}.
\end{aligned}
$$

This finally gives

$$
|T(N, x)| \leq c'' N^{\alpha(x)}, \qquad \text{where} \ \ \alpha(x) := \frac{\log|\gamma(x)|}{\log 2}. \qquad \blacksquare
$$

Remark. Note that the exponent $\alpha(x)$ is < 1 if x is not an integer. Also note that this exponent is optimal. Of course, we have $\gamma(\frac{1}{2}) = \sqrt{2}$, and hence $\alpha(\frac{1}{2}) = \frac{1}{2}$. Hence the case $x = \frac{1}{2}$ gives the classical bound in $\sqrt{N}$ for the sum $\sum_{0 \leq k < N} (-1)^{u(n)}$.

17.4 The One-Dimensional Schrödinger Operator

Another example of the occurrence of automatic and morphic sequences in physics is given by the study of one-dimensional Schrödinger operators. Schrödinger operators were introduced to describe vibrations of atoms, considered roughly as masses separated by springs.

Let $(u_n)_{n \geq 0}$ be a uniformly recurrent sequence. We associate with this sequence the set Ω (also called the *hull* or *induced subshift*) that consists of all infinite words (or sometimes two-sided infinite words) that have the same (finite) subwords as the sequence $(u_n)_{n \geq 0}$.

With such a structure is associated a family of discrete one-dimensional Schrödinger operators $(H_\omega)_{\omega \in \Omega}$ as follows: if $\omega = (\omega_n)_{n \geq 0}$ belongs to Ω, then

$$
(H_\omega \phi)(n) := \phi(n + 1) + \phi(n - 1) + \omega_n \phi(n).
$$

Studying the spectrum of the operator H_ω leads one to look at the *tight-binding Schrödinger equation* in one dimension,

$$
E\phi(n) = \phi(n + 1) + \phi(n - 1) + \omega_n \phi(n),
$$

where E is a (usually) real number, the *energy*. Note that this equation can also be written as:

$$
G(n) = T_n G(n - 1),
$$

$$
\text{where} \ G(n) := \begin{pmatrix} \phi(n + 1) \\ \phi(n) \end{pmatrix} \ \text{and} \ T_n := \begin{pmatrix} E - \omega_n & -1 \\ 1 & 0 \end{pmatrix}.
$$

By iterating,

$$
G(n) = (T_n T_{n-1} \cdots T_0) G(0).
$$

Since the determinants of all matrices T_n are equal to 1, all these matrices as well as their products satisfy Cayley-Hamilton equations of the form

$$M^2 - \text{tr}(M)M + I = 0, \qquad \text{where } I = \begin{pmatrix} 1 & 0 \\ 0 & 1 \end{pmatrix}.$$

Hence it is of interest to find easy ways to compute quantities such as $\text{tr}(T_n T_{n-1} \cdots T_0)$. In particular, if the sequence $(\omega_n)_{n\geq 0}$ is a fixed point of a morphism, then the same property holds for the sequence of matrices $(T_n)_{n\geq 0}$. If the sequence $(T_n)_{n\geq 0}$ is a fixed point of a morphism σ, the computation of the trace of the product $T_n T_{n-1} \cdots T_0$, when $n + 1 = |\sigma^k(T_0)|$ for some integer k, leads to recurrence relations and to *trace maps*, an example of which is given below.

Example 17.4.1 Let $\mathbf{f} = (f_n)_{n\geq 0} = 0100101001001010\cdots$ be the Fibonacci word of Section 7.1. Suppose that the sequence of matrices $(T_n)_{n\geq 0}$ takes two values A and B, and that there exists a map θ from $\{0, 1\}$ to $\{A, B\}$ such that $T_n = \theta(f_n)$ for each $n \geq 0$. Define the matrix M_k by $M_k = T_{F_k-1}T_{F_k-2}\cdots T_{F_0}$, where F_k is the kth Fibonacci number. An easy consequence of Theorem 7.1.1 is that, for $k \geq 3$,

$$M_{k+1} = M_{k-1}M_k.$$

Hence

$$
\begin{aligned}
\text{tr}(M_{k+2}) &= \text{tr}(M_k M_{k+1}) = \text{tr}(M_k M_{k-1} M_k) \\
&= \text{tr}((M_k)^2 M_{k-1}) = \text{tr}((\text{tr}(M_k)M_k - I)M_{k-1}) \\
&= \text{tr}(M_k)\text{tr}(M_k M_{k-1}) - \text{tr}(M_{k-1}) = \text{tr}(M_k)\text{tr}(M_{k-1}M_k) - \text{tr}(M_{k-1}) \\
&= \text{tr}(M_k)\text{tr}(M_{k+1}) - \text{tr}(M_{k-1}).
\end{aligned}
$$

In other words, defining $\Theta(k) := \text{tr}(M_k)$, the following recurrence relation holds for $k \geq 3$:

$$\Theta(k + 2) = \Theta(k)\Theta(k + 1) - \Theta(k - 1).$$

Let us return to the family of discrete one-dimensional Schrödinger operators $(H_\omega)_{\omega\in\Omega}$ defined above by

$$(H_\omega\phi)(n) = \phi(n + 1) + \phi(n - 1) + \omega_n\phi(n).$$

With this operator is associated a measure called its *spectral measure*. The Lebesgue decomposition allows to write this measure as the sum of three measures: one is absolutely continuous (i.e., proportional to the Lebesgue measure), and one is singular continuous (i.e., every point has measure 0, and the support of the measure has zero Lebesgue measure), and one is pure point (i.e., it is a sum of Dirac measures). The spectral properties of H_ω then determine the "conductivity properties" of the given structure. Informally, a structure corresponding to an absolutely continuous spectrum behaves like a conductor, while a structure corresponding to a pure point spectrum behaves like an insulator. It is generally expected that the interme-diate spectral case – singular continuous spectrum – corresponds to intermediate transport properties.

Although singular continuous spectra do not occur for periodic structures, it seems that they do typically occur for one-dimensional quasicrystals. One important result in this respect is Theorem 17.4.2 below. Note that this theorem deduces a sin-gular continuous spectrum from combinatorial properties of the sequence $(u_n)_{n \geq 0}$. Suppose that the subwords of u occur with well-defined, positive frequencies. (This is the case for a large class of sequences, including Sturmian sequences and se-quences generated by primitive morphisms.) Then the following theorem holds, which we state without proof.

Theorem 17.4.2 *Let $(u_n)_{n \geq 0}$ be a non-ultimately-periodic sequence on a finite alphabet. We suppose that $(u_n)_{n \geq 0}$ contains arbitrarily long palindromic subwords. Furthermore we suppose that the frequency of occurrence of every subword of $(u_n)_{n \geq 0}$ exists and is positive. Then, for uncountably many $\omega \in \Omega$, the operator H_ω has purely singular continuous spectrum.*

Remarks. Note that, since the sequence $(u_n)_{n \geq 0}$ is uniformly recurrent, it suffices to assume that $(u_n)_{n \geq 0}$ is not periodic.

Theorem 17.4.2 applies to any sequence that is generated by a primitive morphism and contains arbitrarily long palindromic subwords. Furthermore, it applies to all Sturmian sequences, and, more generally, to all sequences defined by circle maps.

Finally, there is a similar, purely combinatorial sufficient condition for singular continuous spectrum in terms of the *powers* occurring in the sequence $(u_n)_{n \geq 0}$.

In view of Theorem 17.4.2 above, it is interesting to know whether a given sequence contains arbitrarily long palindromic subwords. More precisely, we can define the *palindrome complexity* of a sequence $\mathbf{u} = (u_n)_{n \geq 0}$ on a finite alphabet as follows: $\mathrm{pal}_{\mathbf{u}}(k)$ is defined to be the number of different palindromic subwords of length k that occur in the sequence $\mathbf{u}$. Several examples are given in the exercises at the end of this chapter.

17.5 Exercises

1. (a) Prove that the palindrome complexity $\mathbf{v}$ of the Fibonacci sequence $\mathbf{f}$ is given by $\mathrm{pal}_{\mathbf{v}}(k) = 2$ if k is odd, and $\mathrm{pal}_{\mathbf{v}}(k) = 1$ if k is even.
   (b) Prove that the same result as above holds for *any* Sturmian sequence.
   (c) Prove the converse. Hence a sequence $\mathbf{z}$ is Sturmian if and only if its palindrome complexity satisfies $\mathrm{pal}_{\mathbf{z}}(k) = 2$ if k is odd, and $\mathrm{pal}_{\mathbf{z}}(k) = 1$ if k is even.
2. Show that no paperfolding sequence on two symbols contains palindromes of length 14 or larger.

3. Define the class of *generalized Rudin–Shapiro sequences* as follows: let $(u_n)_{n \geq 1}$
   be any paperfolding sequence. Then the associated generalized Rudin–Shapiro
   sequence is the sequence $(v_n)_{n \geq 1}$ defined by

   $$v_n := \left(\sum_{1 \leq k \leq n} u_k \right) \bmod 2.$$

   Show that no generalized Rudin–Shapiro sequence has palindromes of length
   15 or larger.
4. Let **d** be the period-doubling sequence of Example 6.3.4.
   (a) Prove that **d** contains no palindromic subword of even length ≥ 4.
   (b) Prove that, for every odd $k \geq 5$, we have $\mathrm{pal}_{\mathbf{d}}(k) = \mathrm{pal}_{\mathbf{d}}(2k - 1) = \mathrm{pal}_{\mathbf{d}}(2k + 1)$.
   (c) Prove that the sequence $(\mathrm{pal}_{\mathbf{d}}(k))_{k \geq 0}$ is 2-automatic.
5. Prove that the palindrome complexity of a fixed point of a primitive morphism
   is bounded.

17.6 Notes on Chapter 17

In 1974, Penrose [1974] introduced a nonperiodic tiling of the plane with fivefold
symmetry, now called the Penrose tiling. It was popularized by Gardner [1977].
Another early reference on this tiling is Penrose [1979].

De Bruijn [1981b, 1981c] studied the algebraic properties of the Penrose tiling.
(Three other papers of de Bruijn [1986, 1987, 1997] were devoted later to the study
of Penrose or related tilings.)

In 1984 a paper of Shechtman, Blech, Gratias, and Cahn [1984] appeared, relating
how they built an alloy of aluminum and manganese that has fivefold symmetry.
Since this is not possible for a crystal, and since this alloy is not a glass (an X-ray
diffraction picture shows Bragg peaks), the term *quasicrystal* was coined for such
materials (by Levine and Steinhardt [1984]). This discovery led several physicists
to begin studying one-dimensional structures based on the Fibonacci sequence **f**.
The papers of Kohmoto, Kadanoff, and Tang [1983] and Ostlund, Pandit, Randit,
Schellnhuber, and Siggia [1983] on Schrödinger equations with an almost periodic
potential appeared in the same issue of *Physical Review Letters*. Interesting early
papers on quasicrystallography include Pleasants [1985] and Lunnon and Pleasants
[1987]. The book of Senechal [1995] is a worthwhile survey.

17.1 The Ising model was first described in a paper by E. Ising [1925]. A general
     presentation is given, for example, in the book of Huang [1987].
17.2 It seems that the first time that "physical" properties of automatic sequences
     (i.e., pointwise images of fixed points of uniform morphisms) were studied
     was in Allouche and Mendès France [1985a]. This section is essentially taken
     verbatim from that paper. Schrödinger equations with an automatic potential

were first studied (in the case where the potential is given by the Thue–Morse sequence) by Axel, Allouche, Kleman, Mendès France, and Peyrière [1986]. This was done in more detail in Axel and Peyrière [1989] and Bellissard [1990]. Many results for automatic sequences, morphic sequences, and other sequences were published at the end of the 1980s and the beginning of the 1990s. See, in particular, Bellissard, Iochum, Scoppola, and Testard [1989]; Bellissard, Bovier, and Ghez [1991, 1992, 1993]; Bellissard [1992]; and Bovier and Ghez [1993a, 1993b, 1995].

17.3   Theorem 17.3.1 is due to Allouche and Mendès France [1985a]. It can also be proved without the Ising formalism, but a feature of this presentation is to emphasize the relations between two very different objects, the one-dimensional Ising model and the Rudin–Shapiro sequence. Related results for the Rudin–Shapiro sequence are given in Allouche and Mendès France [1985b].

A small sample of other papers on the one-dimensional Ising model in relation with morphic sequences is Allouche and Mendès France [1986a; 1986b]; Mendès France [1986, 1990a, 1991]; Hermisson, Grimm, and Baake [1997]; and Kamae and Mendès France [1996].

17.4   Theorem 17.4.2 is due to Hof, Knill, and Simon [1995]. They conjectured in particular that the Rudin–Shapiro sequence does not have arbitrarily long subwords that are palindromes. This was later proved by Allouche [1997b] and by Baake [1999].

Palindrome complexity was introduced by Allouche, Baake, Cassaigne, and Damanik [2001].

For a neat mathematical treatment of Schrödinger operators, see Guille-Biel [1997].

The trace map was studied by Allouche and Peyrière [1986], Kolář and Nori [1990]; Peyrière [1991]; Baake, Grimm, and Joseph [1993]; Grimm and Baake [1994]; Roberts and Baake [1994a, 1994b]; and Avishai, Berend, and Glaubman [1994]. Among other references we mention the survey of Peyrière [1995].

For Schrödinger equations or related one-dimensional structures we mention only a few papers. Aubry, Godrèche, and Luck [1987, 1988] studied structures intermediate between quasiperiodicity and randomness. Quasiperiodicity and one- and two-dimensional tilings were studied in particular by Godrèche and Luck [1989a, 1989b]; Godrèche [1990]; and Godrèche, Luck, Janner, and Janssen [1993].

The Fibonacci sequence **f** and generalizations occur, for example, in Gumbs and Ali [1998a, 1988b, 1989], Kolář and Ali [1990], Salejda [1995a], and Baake, Hermisson, and Pleasants [1997].

The Thue–Morse sequence and generalizations occur, for example, in van Enter and Miękisz [1990]; Lin and Tao [1990]; Delyon and Peyrière [1991]; Axel and Terauchi [1991] (see comment in Kolář [1994] and reply to comment in Axel and Terauchi [1994]); Zhong, Yan, and You [1991]; Huang, Gumbs, and Kolář [1992] (see comment in Liviotti and Erdős [1995], and reply to comment in Huang, Gumbs, and Kolar [1995]); Lin and Tao [1992]; van Enter and Miękisz [1992]; Zhong, You, and Yan [1992]; Roy and Khan [1994b, 1994c]; Tao [1994] (where the Fibonacci sequence is also discussed); Turban, Berche, and Berche [1994]; Roy, Khan, and Basu [1995]; Roy, Basu, and Khan [1995]; Chakrabarti, Karmakar, and Moitra [1995] (see comment in Fan and Lin [1995]); Gumbs, Dubey, Salman, Mahmoud, and Huang [1995] (where the Fibonacci sequence is also discussed); de Brito, da Silva, and Nazareno [1995] (where the Fibonacci sequence is also discussed); Gasparian, Ruiz, Ortuño, and Cuevas [1996] (where the Fibonacci sequence is also discussed); Liviotti [1996]; Deych, Zaslavsky, and Lisyansky [1997]; Liu [1997]; Musikhin, Il'in, Rabizo, and Bakueva [1997]; Tong [1997]; Gazeau and Miękisz [1998]; Ghosh and Karmakar [1998]; and Pan, Jiao, Jin, Hu, and Jiang [1998].

Riklund, Severin, and Liu [1987], Qin, Ma, and Tsai [1990], and Ryu, Oh, and Lee [1993] are other papers on the Thue–Morse sequence in physics. For an appraisal of these and other papers, see Bovier and Ghez [1995], where it is pointed out that some results are in error.

The Rudin–Shapiro sequence occurs, for example, in Dulea, Johansson, and Riklund [1992a, 1992b]; Axel, Allouche, and Wen [1992]; de Oliveira [1995] (where the Fibonacci and the Thue–Morse sequences are also discussed); Lin and Goda [1997] (where the Thue–Morse sequence is also discussed); Hörnquist and Ouchterlony [1998] (where the Fibonacci and the Thue–Morse sequences are also discussed); Lindquist and Riklund [1998]; Pinho, Haddad, and Salinas [1998]; and Lennholm and Hörnquist [1999] (where the Fibonacci and the Thue–Morse sequences are also discussed).

Other examples of morphic sequences are given in Karevski and Turban [1996].

More results and references can be found in Axel and Gratias [1995].

For quasicrystals in botany, see Rivier [1986].

Appendix

Hints, References, and Solutions
for Selected Exercises

In this appendix we provide some hints, references, and solutions for certain exercises in the main text.

A.1 Chapter 1

5. See Shallit and Wang [1999].
14. Cummings [1996] showed there are exactly 117 distinct abelian squarefree strings over a 3-letter alphabet.
15. See I. Stewart [1995].
16. See Jacobs [1992, p. 104].
17. See Berstel [1989a]. For other papers on Langford strings, see Langford [1958]; Priday [1959]; Davies [1959]; Marcus and Păun [1989]; Păun [1992].
18. See Euwe [1929]; Morse [1938]; MacMurray [1938]; Morse and Hedlund [1944]; I. Stewart [1995].
19. Here are some squares in English: atlatl, murmur, testes, and tartar. The uncommon word tratratratra is the name of an extinct lemur from Madagascar – a fourth power!

    A square in German is nennen, which is the verb "to call".

    A square in Italian is restereste, which means "you would remain". An overlap in Italian is intinti, which is, e.g., bread dipped into soup. Palindromes in Italian include avallava (agree) and onorarono (to give honor).

    An overlap in French is entente.

    Some squares in Spanish include arar (to plow); enarenar (to run aground); and adorador (worshipper). The overlap adoradora means a female worshipper. The Spanish verb reconocer (to recognize) is a palindrome.

    The Danish words farfar, mormor, and purpur are squares. The Danish words sneppens, snerrens, and regninger are palindromes.

    The Dutch words enen, kerker, and tenten are squares, while koekoek is an overlap.

    The Swedish word rattar is a palindrome that means "steering wheels".
21. Zech [1958] observed that the decimal expansion of e is not squarefree.
24. See Kfoury [1988b].
26. See Fine and Wilf [1965].
31. See Allouche and Cosnard [1983].
32. See Shyr [1977].
43. For more information about borders, see Silberger [1971], Harborth [1974], Nielsen [1973], and Blom [1994].

45. Given a word w, use any linear-time pattern-matching algorithm to find the first occurrence of w in $w'w$, where w' is w with the first character removed. The position j where w first matches $w'w$ is the length of the shortest x such that $w = x^k$. If this position is $|w|$, then w is primitive.

49. See Wegner [1982]; Istrail [1983]; Keranen [1983]; Arnold [1983]; Boasson [1983]; Crochemore [1983c]; Ehrenfeucht and Rozenberg [1983e]; Harju [1983]; Klop [1983]; Skyum [1983]; and Verraedt, De Bra, and Gyssens [1983].

51. See Shyr and Thierrin [1977].

52. See Lyndon and Schützenberger [1962] and D. Chu and Town [1978].

54. See Loftus, Shallit, and Wang [2000].

56. Yes, for if w is a subword, it appears as a subword of $\mu^{2t}(0)$ for some t. But this is a palindrome.

57. See Fraenkel and Simpson [1998].

58. See Friedman [2001].

59. See Shallit and Wang [2001].

61. It must contain 0102010, which is squarefree. Then it cannot continue with any symbol and still be squarefree.

62. See Xie [1996, p. 6].

A.2 Chapter 2

5. (b): We prove the claim by induction on k. Clearly the result is true for $k = 1$. Now assume the result is true for $k < r$; we prove it for $k = r$. Assume there was an infinite antichain $a_1, a_2, \ldots$. Since $a_2, a_3, \ldots$ are incomparable with a_1, each a_i has a coordinate where it is less than the corresponding coordinate of a_1. Since there are only a finite number of possible coordinates, some coordinate has the property that infinitely many of the a_i are less than a_1 in that coordinate. Without loss of generality, assume the first coordinate has that property, and number the infinitely many elements as $b_1, b_2, b_3, \ldots$. Now there are only a finite number of non-negative integers less than $a_1(1)$, so there exists some non-negative integer such that infinitely many of the b_i have $b_i(1) = d$ for some integer $d < a_1(1)$. Number these infinitely many elements as $c_1, c_2, c_3, \ldots$. All of these are incomparable, and all have the first coordinate the same. By removing the first coordinate we get an infinite number of incomparable elements in $\mathbb{N}^{k-1}$, a contradiction.

6. This is a version of Fatou's lemma. See, for example, Fatou [1904]; Salem [1963, Chapter 1, §3, Lemma 2]; van der Waerden [1986].

7. See Bloom [1995].

10. See, for example, Parent [1984].

11. See Selfridge [1960].

15. See, e.g., Woodcock and Sharif [1990, Theorem 4.1].

19. See Erdős [1948].

26. Let x, y be generators. Then $xy = xy(yx)^2 = x(yy)xyx = xxyx = yx$, since the exponent is 2.

27. See Morse and Hedlund [1944]; Restivo and Reutenauer [1985].

30. See Brubaker [1971].

31. See, for example, Artin [1924, §12]; de Mathan [1970]; Baum and Sweet [1976]; Mendès France and van der Poorten [1991]; W. Schmidt [2000].

A.3 Chapter 3

2. See Berndt and Bhargava [1993].

4. See Fine [1965].

5. For a similar formula for $k = 10$, see d'Ocagne [1886].

8. Suppose $(s_k(n))_{n \geq 0}$ did satisfy a linear recurrence. Then there would be a j and integers $a_0, a_1, \ldots, a_j$, $a_0 \neq 0$, such that $a_0 s_k(n) = a_1 s_k(n + 1) + \cdots + a_j s_k(n + j)$ for all n sufficiently large. Now consider n of the form $k^t - 1$ as $t \to \infty$. Then the absolute value of the left-hand side grows without bound, but the right-hand side is eventually constant.

9. (a) Suppose $(s_k(n) \bmod m)_{n \geq 0}$ were ultimately periodic. But
   $s_k(n - 1) - s_k(n) + 1 = (k - 1)v_k(n)$, so $((k - 1)v_k(n) \bmod m)_{n \geq 0}$ would also be ultimately periodic. Assume it is, with period d, for all $n \geq c$. Write $d = k^a \cdot r$, with $k \nmid r$. Choose $n = k^s + k^{a+1}$ with s such that $k^s > c$, $s > a + 1$. Then $n + d = k^s + k^{a+1} + d$, so $v_k(n) = a + 1$. But $v_k(n + d) = a$. Now $(k - 1)(a + 1) \equiv (k - 1)a \pmod{m}$ implies $k - 1 \equiv 0 \pmod{m}$.
   For the other direction, use Exercise 3.6. See Morton and Mourant [1991].

   (b) See Allouche and Shallit [2000]. For an interesting generalization, see Frid [2001].

15. See Clements and Lindström [1965]; Lindström and Zetterström [1967]; Wilf [1968]; Kano and Shiokawa [1988]; and Kano [1991].

16. See Granville [1995].

22. (a)–(c) See Brillhart and Carlitz [1970].
   (d) See Rudin [1959].
   (e)–(f) See Brillhart [1973].
   (g)–(j) See Brillhart, Lomont, and Morton [1976].

23. See Brillhart and Morton [1978].

24. See Shallit [1999].

26. See Barbier [1887a, 1887b] and Thompson [1959].

27. See Shallit [1991a].

29. See Lenard [1991].

32. See Trigg [1949].

33. See Samborski [1977].

35. See Knuth [1988]; Arnoux [1989].

36. See Glaisher [1899a, 1899b]; Fine [1947]; Stolarsky [1977].

39. See Bateman and Bradley [1997].

40. See Klosinski, Alexanderson, and Larson [1985, Prob. B-5].

42. See S. Golomb [1975].

43. See Levine [1988] and Bowman and White [1989].

44. See Segal and Lepp [1969].

45. See Graham and Pollak [1970] and Rabinowitz and Gilbert [1991].

46. See Knuth [1989].

47. See M. Golomb [1993].

51. See Olivier [1975, 1976]. Also see Cooper and Kennedy [1997] for the s_{10} case.

52. See Klosinski, Alexanderson, and Hillman [1982]; Shallit [1984]; Allouche and Shallit [1990].

53. See Lindström [1997].

55. See Erdős [1994]; Dombi and Valko [1997]; Blecksmith, McCallum, and Selfridge [1998].

56. See Reitwiesner [1960]; Güntzer and Paul [1987]; Jedwab and Mitchell [1989]. For applications, see, e.g., Morain and Olivos [1990] and Koblitz [1992]. For further analysis see Thuswaldner [1999b]; O'Connor [1999]; Prodinger [2000]; Bosma [2001]. For generalizations to other bases, see Clark and Liang [1973]; Arno and Wheeler [1993]; Heuberger [1999].

57. See Allouche [1997b] and Baake [1999].

58. See Chang and Tsai [2000].

59. See Hickerson [1974].

A.4 Chapter 4

1. (d): Choose $z = a^p b^{(p-1)!}$ where p is a prime $> n$, the pumping-lemma constant.
7. See, for example, Moore [1971].
10. See Main [1985].
11. Take L to be the language of all finite subwords of **t**, the Thue–Morse word. Then L is clearly factorial. Assume L has an infinite regular subset. Then by the pumping lemma there would be a word in L of the form uv^3w with $v \neq \epsilon$. But **t** contains no cubes, a contradiction.
12. See Birget [1992] or Glaister and Shallit [1996].
13. One regular expression for the set of strings over $\{0, 1\}$ having an even number of occurrences of the subword 11 is $0^* + 0^* 1(1(00^* 1)^* 1 + 00^* 1)^* 0^*$.
15. See Mootha [1993] and Holzer and Rossmanith [1996].
16. This question was raised by Marcus [1964] and answered by Friant [1969].

A.5 Chapter 5

1. See, for example, Lehmer [1947].
3. The following DFAO generates the sequence $(s_{-2}(n) \bmod 2)_{n \geq 0}$. Let $Q = \{0, 1, \ldots, 11\}$, $\Delta = \Sigma = \{0, 1\}$, $q_0 = 0$, and δ, τ be as defined in Table A.1.
6. Let $(a(n))_{n \geq 0}$ be the sequence generated by the 2-DFAO given in Figure A.1.
8. See de Weger [1991].
10. Consider $S = T = \{2, 3, \ldots, \}$.
11. See Shallit [1982b].

A.6 Chapter 6

3. See Allouche, Astoorian, Randall, and Shallit [1994]; Hinz [1996].
5. See Niederreiter and Vielhaber [1996].
6. See Allouche and Cohen [1985].
9. (a) See, for example, Kakutani [1967]; Dekking [1979a].
13. (a): See Davis and Knuth [1970, p. 74].
14. See Davis and Knuth [1970, p. 136].

Table A.1. *DFAO for Exercise 3*

q	$\delta(q, 0)$	$\delta(q, 1)$	$\tau(q)$
0	0	1	0
1	2	3	1
2	4	5	0
3	6	7	1
4	2	6	1
5	0	8	0
6	9	10	1
7	5	3	0
8	11	8	1
9	9	11	1
10	1	10	0
11	4	7	0

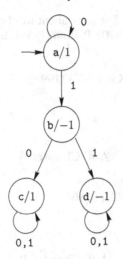

Figure A.1: A 2-DFAO.

16. This property was observed by Davis and Knuth [1970] for the regular paperfolding sequence, and by Dekking, Mendès France, and van der Poorten [1982] for the general case.
17. A counterexample can be deduced using the work of Boyd [1997, 2001]. Here is one solution: let one partition be

$$[0,1,6,7,9,12,14,15,16,17,18,21,24,27,30,31,32,33,36,39,42,45,46,47,48,49,51,54,56,57,62,63]$$

and the other be

$$[2,3,4,5,8,10,11,13,19,20,22,23,25,26,28,29,34,35,37,38,40,41,43,44,50,52,53,55,58,59,60,61].$$

21. (a): See Main [1985]; Berstel [1986a]; Main, Bucher, and Haussler [1987].
    (b): See Grazon [1987].
22. See Wilson and Shallit [1992].
24. See Séébold [1986].
26. See Shallit [1980]; Lunnon, Pleasants, and Stephens [1979].
29. Define $a_k(n) := e_{2;(k)_2}(n)$. We have

$$R_n = a_1(n) - 3a_3(n) - \sum_{n \geq 2} a_{2^n+1}(n) + 2 \sum_{n \geq 2} a_{2^n+2^{n-1}+1}(n).$$

30. See Arshon [1937]; Berstel [1979]; Kitaev [2000a]; Currie [2001]; Séébold [2001].
31. This is true for $n \leq 4$, but for $n = 5$ there is the following counterexample:

$$[1,-1,-1,1,1,-1,1,-1,1,1,1,-1,-1,-1,1,1,-1,1,1,-1,1,-1,-1,1,-1,1,1,1,-1,-1,-1,-1,-1].$$

35. See Yazdani [2001].
36. See Compton [1999].
40. See Doche and Mendès France [2000].

A.7 Chapter 7

3. See Allouche, Bétréma, and Shallit [1989].
5. The characteristic sequence of $(2^n + n)_{n \geq 0}$ has this property.
6. See Wang and Shallit [1999].

8. See Berstel and Brlek [1987]. For more about word chains, see Diwan [1986]; Arnold and Brlek [1989]; Althöfer [1990]; P. Roth [1989]; Bousquet-Mélou [1992]; Merekin [1998].
12. See de Luca [1981].
13. See Cummings, Moore, and Karhumäki [1996].
14. See Allouche and Dress [1990].
24. See Pansiot [1981b].

A.8 Chapter 8

4. See Minc [1988, p. 16, Theorem 4.4].
7. See, for example, Cobham [1972, p. 186].
9. See Wang and Shallit [1999].

A.9 Chapter 9

4. See Szüsz [1985] for $\beta = 0$ and Bowman [1995] for the general case.
8. This exercise is about Beatty sequences, and there is a large literature. The result was apparently first stated by Rayleigh [1877], without proof; see Schoenberg [1982]. The sequences were named after Beatty [1926]; also see Bricard [1926] and Faucheux [1926]. For additional papers, see Sprague [1938]; Lambek and Moser [1954]; Skolem [1957]; Bang [1957]; Komatsu [1995a].
9. See Uspensky [1927] and Graham [1963].
10. See, for example, Wythoff [1907]; Ahrens [1910]; Coxeter [1953]. For additional results on and generalization of Wythoff's game, see Connell [1959]; Fraenkel and Borosh [1973]; Silber [1976]; Horadam [1978]; Fraenkel [1984]; Turner [1989]; Stolarsky [1991]; Porta and Stolarsky [1991].
11. See Downey and Griswold [1984]; Granville and Rasson [1988]; Gault and Clint [1988]; Dilcher [1993]; Grytczuk [1996b].
12. See Anderson, Brown, and Shiue [1995].
14. See Anderson, Brown, and Shiue [1995]; Anderson [1997].
15. See Kimberling [1998a].
16. See Diamond [1989].
17. See Boshernitzan and Fraenkel [1984].
18. For parts (a) and (b), see, for example, Knuth, Morris, and Pratt [1977]; Berstel and Séébold [2002]. For part (f) see Gambaudo, Lanford, and Tresser [1984].
19. See T. Brown and Shiue [1995]; T. Brown [2002].

A.10 Chapter 10

6. See Grillenberger [1973].
7. Let $L \subseteq \{0, 1\}^*$ be the language of all words not containing two consecutive 1's, and write $L = \{w_1, w_2, \dots\}$. Then consider $\mathbf{w} = w_1 0 w_2 0 w_3 0 \cdots$.
10. See Brlek [1989]; de Luca and Varricchio [1989a].
12. See Tromp and Shallit [1995].
18. See Gumbs and Ali [1988a, 1988b, 1989].
22. For more on the Kolakoski word, see Kolakoski [1965]; Kimberling [1979]; Dekking [1979a, 1997]; Weakley [1989]; Keane [1991]; Culik and Karhumäki [1992]; Culik, Karhumäki, and Lepistö [1992]; Păun [1993]; Carpi [1993c, 1994]; Lepistö [1994]; Chvátal [1994]; Shen and Huang [1996]; Steacy [1996].

23. See Justin and Pirillo [1991]; de Luca and Varricchio [1991a, 1991b]; Pirillo and Varricchio [1996].
28. Suppose $p_{\mathbf{x}}(n) = k^n$. Then the string 0 cannot occur with bounded gaps, since the string 1^n must appear somewhere.
30. This construction is mentioned in Cobham [1972, pp. 169–170]; he attributes it to Keane [1968]. Also see Allouche and Shallit [1993].
33. See Ehrenfeucht and Rozenberg [1982a].
36. See Ehrenfeucht and Rozenberg [1982a].
37. See Durand [1998b, p. 91].
40. See Benczur [1986].
42. See Rauzy [1983]; Rote [1994].
43. See Cassaigne and Karhumäki [1995a, 1995b].
45. See Wang and Shallit [1998].
46. See Shallit [1993].
48. See Allouche and Bousquet-Mélou [1994a, p. 264].
52. See Avgustinovich, Fon-Der-Flaass, and Frid [2001].

A.11 Chapter 11

2. Suppose j and k are multiplicatively dependent. Then by Theorem 2.5.7 we know there exist integers $l \geq 2$, $a, b \geq 1$ such that $j = l^a$, $k = l^b$. Then it is easy to construct an l-DFAO accepting the powers of k. Hence by Theorem 6.6.4 the characteristic sequence of the powers of k is j-automatic.

   On the other hand, the characteristic sequence of the powers of k is clearly k-automatic. If it were also j-automatic for j, k multiplicatively independent, then by Theorem 11.2.1 it would be ultimately periodic, which it isn't.

A.12 Chapter 12

2. The equation is $X^3(1 + X)^4 A^4 + (1 + X)^4 A + X = 0$.
8. See M. Newman [1960]. For generalizations, see Mordell [1961, 1965a, 1965b]; Meijer [1963]; Popken [1963].
20. See Mahler [1980].
23. This is a classical result due to Jacobi; see, for example, Hardy and Wright [1985, Chapter XVII].
24. See Woodcock and Sharif [1989]; Allouche, Gouyou-Beauchamps, and Skordev [1998]; Allouche [1999].
30. We have $C(X) = [1, X^2, \overline{2X^2 + 1, X^2 + 2}]$.
31. See Rees [1946].
32. See Baum and Sweet [1976].

A.13 Chapter 13

1. See Girstmair [1997].
6. See Borwein and Girgensohn [1995].
11. See Erdős [1948]; Borwein [1991]; Duverney [1996].
12. See Vanden Eynden [1995].
15. See Ketkar and Zamboni [1998].
17. See Even [1964].
18. See Hartmanis and Stearns [1967].

19. (a) See W. Schmidt [1960].
    (b) See Champernowne [1933]. This number was proved transcendental by Mahler [1937b]. Amou [1991] proved that the irrationality measure of this number is 10.
    (c) See Copeland and Erdős [1946].
    (d) See Korobov [1990].
    (e) See G. Martin [2001].
20. See Chen and Ruzsa [1999].
21. See Allouche and Cosnard [2000].
22. See Eilenberg [1974, Proposition 3.1, p. 106]; Wilson and Shallit [1992].

A.14 Chapter 14

2. Define $a(m, n) := s_k(mn)$. Then every row and column of $\mathbf{a} := (a(m, n))_{m,n \geq 0}$ is k-automatic, but $\mathbf{a}$ is not k-automatic. If it were, then its diagonal would be k-automatic, by Theorem 14.2.6. But the diagonal is $s_k(n^2)$, which is not k-automatic, by Theorem 6.10.1.
4. See Yarlagadda and Hershey [1982].
5. See Bean, Ehrenfeucht, and McNulty [1979]; Currie and Simpson [2002].
6. See Currie and Simpson [2002].
10. See Bruyère, Hansel, Michaux, and Villemaire [1994].

A.15 Chapter 15

2. See, for example, Shallit and Breitbart [1996].
3. See Dwork and Stockmeyer [1990].
7. See Shallit [2000].
8. See Shallit and Breitbart [1996].
9. See Shallit [1996].

A.16 Chapter 16

5. See Propp [1979].
6. Let $f(X) = X + X^2 + X^4 + X^8 + \cdots$; this is 2-regular. Assume $f(f(X))$ is 2-regular. Then the coefficient of X^n in $f(f(X))$ is just the number of ways to write n as a sum of r powers of 2, where r itself is some power of 2. Now consider $n = 2^{2^t}$; then there are at least $(2^t - 1)!$ such representations. But this contradicts Theorem 16.3.1.
10. See Allouche and Shallit [1992].
12. (a): See Hirschhorn and Sellers [1996, 1999].
15. See Mullhaupt [1986]; Becker [1994].
23. (c): See, for example, Yaglom and Yaglom [1967].
25. See Strauss and Shallit [1990].
27. (a) See Allouche and Shallit [1992].
29. This follows from Martinez [2001].
31. See Morton and Mourant [1989, 1991].
32. Here $(p_n)_{n \geq 0}$ is the *Stern-Brocot sequence*. See Allouche and Shallit [1992]; Calkin and Wilf [2000].
35. This result was found by Shallit in 1996 and published in Shallit [1999]. Also see Berstel [2001]. For related material, see Carlitz [1968]; Klarner [1968]; Robbins [1996]; Ardila [2002].

A.17 Chapter 17

1. See Droubay [1995] for the palindrome complexity of the Fibonacci sequence, and Droubay and Pirillo [1999] for the palindrome complexity of general Sturmian sequences.
2. See Allouche [1997b] and Baake [1999].
3. For generalized Rudin–Shapiro sequences, see Mendès France and Tenenbaum [1981]. For the solution to the exercise, see Allouche [1997b].
4. See Damanik [2000].
5. See Damanik and Zare [2000].

Bibliography

[Aberkane 2001] A. Aberkane. Exemples de suites de complexité inférieure à $2n$. *Bull. Belg. Math. Soc.* **8** (2001), 161–180.

[Adams 1985] W. W. Adams. The algebraic independence of certain Liouville continued fractions. *Proc. Amer. Math. Soc.* **95** (1985), 512–516.

[Adams and Davison 1977] W. W. Adams and J. L. Davison. A remarkable class of continued fractions. *Proc. Amer. Math. Soc.* **65** (1977), 194–198.

[Adian 1979] S. I. Adian. *The Burnside Problem and Identities in Groups.* Springer-Verlag, 1979.

[Adian 1980] S. I. Adian. Classification of periodic words and their application in group theory. In J. L. Mennicke, editor, *Burnside Groups (Proceedings, Bielefeld, Germany, 1977)*, Vol. 806 of *Lecture Notes in Mathematics*, pp. 1–40. Springer-Verlag, 1980.

[Adler and Li 1977] A. Adler and S.-Y. R. Li. Magic cubes and Prouhet sequences. *Amer. Math. Monthly* **84** (1977), 618–627.

[Agrawal 1974a] D. P. Agrawal. Negabinary carry-look-ahead adder and fast multiplier. *Electronics Letters* **10** (1974), 312–313. Corrigendum, **10** (1974), 468.

[Agrawal 1974b] D. P. Agrawal. Negabinary complex-number multiplier. *Electronics Letters* **10** (1974), 502–503.

[Agrawal 1975a] D. P. Agrawal. Negabinary parallel counters. *Digital Processes* **1** (1975), 75–85.

[Agrawal 1975b] D. P. Agrawal. On negabinary division and square-rooting. *Digital Processes* **1** (1975), 267–275.

[Agrawal 1975c] D. P. Agrawal. Arithmetic algorithms in a negative base. *IEEE Trans. Comput.* **C-24** (1975), 998–1000.

[Agrawal 1977] D. P. Agrawal. Comments on "A note on base −2 arithmetic logic". *IEEE Trans. Comput.* **C-26** (1977), 511.

[Agrawal 1978] D. P. Agrawal. On arithmetic inter-relationships and hardware interchangeability of negabinary and binary systems. In *Proc. 4th IEEE Symp. Computer Arithmetic*, pp. 88–96. IEEE Press, 1978.

[Agronomof 1926] N. Agronomof. Sobre una función numérica. *Revista Matemática Hispano-Americana* **1** (1926), 267–269.

[Ahrens 1910] W. Ahrens. *Mathematische Unterhaltungen und Spiele*, Vol. 1. Teubner, Leipzig, 1910.

[Akiyama 1999] S. Akiyama. Self affine tiling and Pisot numeration system. In S. Kanemitsu and K. Györy, editors, *Number Theory and Its Applications*, pp. 7–17. Kluwer, 1999.

[Akiyama 2000] S. Akiyama. Cubic Pisot units with finite beta expansions. In F. Halter-Koch and R. F. Tichy, editors, *Algebraic Number Theory and Diophantine Analysis*, pp. 11–26. Walter de Gruyter, 2000.

481

[Akiyama and Pethö 2002]   S. Akiyama and A. Pethö. On canonical number systems. *Theoret. Comput. Sci.* **270** (2002), 921–933.

[Akiyama and Thuswaldner 2000]   S. Akiyama and J. M. Thuswaldner. Topological properties of two-dimensional number systems. *J. Théorie Nombres Bordeaux* **12** (2000), 69–79.

[Alessandri 1993]   P. Alessandri. Codage des rotations. Technical report, Mémoire de D. E. A., Université Claude Bernard, Lyon I, 1993.

[Alessandri and Berthé 1998]   P. Alessandri and V. Berthé. Three distance theorems and combinatorics on words. *Enseign. Math.* **44** (1998), 103–132.

[Allauzen 1998]   C. Allauzen. Une caractérisation simple des nombres de Sturm. *J. Théorie Nombres Bordeaux* **10** (1998), 237–241.

[Allen 1968]   D. Allen, Jr. On a characterization of the nonregular set of primes. *J. Comput. System Sci.* **2** (1968), 464–467.

[Allouche 1982]   J.-P. Allouche. Somme des chiffres et transcendance. *Bull. Soc. Math. France* **110** (1982), 279–285.

[Allouche 1983]   J.-P. Allouche. Théorie des nombres et automates, Thèse d'État, Université de Bordeaux, 1983.

[Allouche 1984]   J.-P. Allouche. Suites infinies à répétitions bornées. In *Séminaire de Théorie des Nombres de Bordeaux*, 1983–1984. Exposé no. 20.

[Allouche 1987]   J.-P. Allouche. Automates finis en théorie des nombres. *Exposition. Math.* **5** (1987), 239–266.

[Allouche 1989]   J.-P. Allouche. Note sur un article de Sharif et Woodcock. *Séminaire de Théorie des Nombres de Bordeaux* **1** (1989), 163–187.

[Allouche 1990]   J.-P. Allouche. Sur la transcendance de la série formelle π. *Séminaire de Théorie des Nombres de Bordeaux* **2** (1990), 103–117.

[Allouche 1992]   J.-P. Allouche. The number of factors in a paperfolding sequence. *Bull. Austral. Math. Soc.* **46** (1992), 23–32.

[Allouche 1994a]   J.-P. Allouche. Sur la complexité des suites infinies. *Bull. Belg. Math. Soc.* **1** (1994), 133–143.

[Allouche 1994b]   J.-P. Allouche. Note on the cyclic towers of Hanoi. *Theoret. Comput. Sci.* **123** (1994), 3–7.

[Allouche 1996]   J.-P. Allouche. Transcendence of the Carlitz–Goss Gamma function at rational arguments. *J. Number Theory* **60** (1996), 318–328.

[Allouche 1997a]   J.-P. Allouche. Note on the transcendence of a generating function. In A. Laurincikas and E. Manstavicius, editors, *Proceedings of the Palanga Conference for the 75th birthday of Prof. Kubilius*, Vol. 4 of *New Trends in Probability and Statistics*, pp. 461–465. VSP, 1997.

[Allouche 1997b]   J.-P. Allouche. Schrödinger operators with Rudin–Shapiro potentials are not palindromic. *J. Math. Phys.* **38** (1997), 1843–1848.

[Allouche 1999]   J.-P. Allouche. Transcendence of formal power series with rational coefficients. *Theoret. Comput. Sci.* **218** (1999), 143–160.

[Allouche 2000]   J.-P. Allouche. Nouveaux résultats de transcendance de réels à développement non aléatoire. *Gazette des Mathématiciens* **84** (2000), 19–34.

[Allouche, Astoorian, Randall, and Shallit 1994]   J.-P. Allouche, D. Astoorian, J. Randall, and J. Shallit. Morphisms, squarefree strings, and the Tower of Hanoi puzzle. *Amer. Math. Monthly* **101** (1994), 651–658.

[Allouche, Baake, Cassaigne, and Damanik 2001]   J.-P. Allouche, M. Baake, J. Cassaigne, and D. Damanik. Palindrome complexity. Unpublished manuscript, 2001.

[Allouche and Berthé 1997]   J.-P. Allouche and V. Berthé. Triangle de Pascal, complexité et automates. *Bull. Belg. Math. Soc.* **4** (1997), 1–23.

[Allouche, Bétréma, and Shallit 1989]   J.-P. Allouche, J. Bétréma, and J. Shallit. Sur des points fixes de morphismes du monoïde libre. *RAIRO Inform. Théor. App.* **23** (1989), 235–249.

[Allouche and Bousquet-Melou 1994a]   J.-P. Allouche and M. Bousquet-Mélou. Canonical positions for the factors in the paperfolding sequences. *Theoret. Comput. Sci.* **129** (1994), 263–278.

[Allouche and Bousquet-Melou 1994b]   J.-P. Allouche and M. Bousquet-Mélou. Facteurs des suites de Rudin-Shapiro généralisées. *Bull. Belg. Math. Soc.* **1** (1994), 145–164.

[Allouche and Bousquet-Mélou 1995]   J.-P. Allouche and M. Bousquet-Mélou. On the conjectures of Rauzy and Shallit for infinite words. *Comment. Math. Univ. Carolinae* **36** (1995), 705–711.

[Allouche, Cateland, Gilbert, Peitgen, Shallit, and Skordev 1997]   J.-P. Allouche, E. Cateland, W. J. Gilbert, H.-O. Peitgen, J. Shallit, and G. Skordev. Automatic maps in exotic numeration systems. *Theory Comput. Systems* **30** (1997), 285–331.

[Allouche and Cohen 1985]   J.-P. Allouche and H. Cohen. Dirichlet series and curious infinite products. *Bull. Lond. Math. Soc.* **17** (1985), 531–538.

[Allouche, Cohen, Mendès France, and Shallit 1987]   J.-P. Allouche, H. Cohen, M. Mendès France, and J. O. Shallit. De nouveaux curieux produits infinis. *Acta Arith.* **49** (1987), 141–153.

[Allouche and Cosnard 1983]   J.-P. Allouche and M. Cosnard. Itérations de fonctions unimodales et suites engendrées par automates. *C. R. Acad. Sci. Paris* **296** (1983), 159–162.

[Allouche and Cosnard 2000]   J.-P. Allouche and M. Cosnard. The Komornik–Loreti constant is transcendental. *Amer. Math. Monthly* **107** (2000), 448–449.

[Allouche, Davison, Queffélec, and Zamboni 2001]   J.-P. Allouche, L. J. Davison, M. Queffélec, and L. Q. Zamboni. Transcendence of Sturmian or morphic continued fractions. *J. Number Theory* **91** (2001), 39–66.

[Allouche and Dress 1990]   J.-P. Allouche and F. Dress. Tours de Hanoï et automates. *RAIRO Inform. Théor. App.* **24** (1990), 1–15.

[Allouche, Gouyou-Beauchamps, and Skordev 1998]   J.-P. Allouche, D. Gouyou-Beauchamps, and G. Skordev. Transcendence of binomial and Lucas' formal power series. *J. Algebra* **210** (1998), 577–592.

[Allouche, von Haeseler, Lange, Petersen, and Skordev 1997]   J.-P. Allouche, F. von Haeseler, E. Lange, A. Petersen, and G. Skordev. Linear cellular automata and automatic sequences. *Parallel Computing* **23**(11) (1997), 1577–1592.

[Allouche and Johnson 1995]   J.-P. Allouche and T. Johnson. Finite automata and morphisms in assisted musical composition. *J. New Music Research* **24** (1995), 97–108.

[Allouche and Johnson 1996]   J.-P. Allouche and T. Johnson. Narayana's cows and delayed morphisms. In *Cahiers du GREYC, Troisièmes Journées d'Informatique Musicale (JIM 96)*, Vol. 4, pp. 2–7, 1996.

[Allouche and Liardet 1991]   J.-P. Allouche and P. Liardet. Generalized Rudin–Shapiro sequences. *Acta Arith.* **60** (1991), 1–27.

[Allouche, Lubiw, Mendès France, van der Poorten and Shallit 1996]   J.-P. Allouche, A. Lubiw, M. Mendès France, A. J. van der Poorten, and J. Shallit. Convergents of folded continued fractions. *Acta Arith.* **77** (1996), 77–96.

[Allouche and Mendès France 1985a]   J.-P. Allouche and M. Mendès France. Suite de Rudin–Shapiro et modèle d'Ising. *Bull. Soc. Math. France* **113** (1985), 273–283.

[Allouche and Mendès France 1985b]   J.-P. Allouche and M. Mendès France. On an extremal property of the Rudin–Shapiro sequence. *Mathematika* **32** (1985), 33–38.

[Allouche and Mendès France 1986a]   J.-P. Allouche and M. Mendès France. Finite automata and zero temperature quasicrystal Ising chain. *J. Physique* **47** (1986), C3.63–C3.73. Colloque C3, Supplement to No. 7.

[Allouche and Mendès France 1986b]   J.-P. Allouche and M. Mendès France. Quasicrystal Ising chain and automata theory. *J. Statist. Phys.* **42** (1986), 809–821.

[Allouche, Mendès France, and Peyrière 2000]   J.-P. Allouche, M. Mendès France, and
    J. Peyrière. Automatic Dirichlet series. *J. Number Theory* **81** (2000), 359–373.
[Allouche, Mendès France, and Tenenbaum 1988]   J.-P. Allouche, M. Mendès
    France, and G. Tenenbaum. Entropy: an inequality. *Tokyo J. Math.* **11** (1988),
    323–328.
[Allouche, Morton, and Shallit 1992]   J.-P. Allouche, P. Morton, and J. Shallit. Pattern
    spectra, substring enumeration, and automatic sequences. *Theoret. Comput. Sci.* **94**
    (1992), 161–174.
[Allouche and Peyrière 1986]   J.-P. Allouche and J. Peyrière. Sur une formule de
    récurrence sur les traces de produits de matrices associés à certaines substitutions.
    C. R. Acad. Sci. Paris Sér. II **302** (1986), 1135–1136.
[Allouche, Peyrière, Wen, and Wen 1998]   J.-P. Allouche, J. Peyrière, Z. X. Wen, and
    Z. Y. Wen. Hankel determinants of the Thue-Morse sequence. *Ann. Inst. Fourier
    (Grenoble)* **48** (1998), 1–27.
[Allouche and Salon 1990]   J.-P. Allouche and O. Salon. Finite automata, quasicrystals,
    and Robinson tilings. In I. Hargittai, editor, *Quasicrystals, Networks, and Molecules
    of Fivefold Symmetry*, pp. 97–105. VCH Publishers, New York, 1990.
[Allouche and Salon 1993]   J.-P. Allouche and O. Salon. Sous-suites polynomiales de
    certaines suites automatiques. *J. Théorie Nombres Bordeaux* **5** (1993), 111–
    121.
[Allouche, Scheicher, and Tichy 2000]   J.-P. Allouche, K. Scheicher, and R. F. Tichy.
    Regular maps in generalized number systems. *Math. Slovaca* **50** (2000), 41–58.
[Allouche and Shallit 1990]   J.-P. Allouche and J. O. Shallit. Sums of digits and the
    Hurwitz zeta function. In K. Nagasaka and E. Fouvry, editors, *Proc. Japanese-French
    Symposium Held in Tokyo, Japan, October 10–13, 1988*, Vol. 1434 of *Lecture Notes
    in Computer Science*, pp. 19–30. Springer-Verlag, 1990.
[Allouche and Shallit 1992]   J.-P. Allouche and J. O. Shallit. The ring of k-regular
    sequences. *Theoret. Comput. Sci.* **98** (1992), 163–197.
[Allouche and Shallit 1993]   J.-P. Allouche and J. O. Shallit. Complexité des suites de
    Rudin-Shapiro généralisées. *J. Théorie Nombres Bordeaux* **5** (1993), 283–302.
[Allouche and Shallit 1999]   J.-P. Allouche and J. O. Shallit. The ubiquitous
    Prouhet–Thue–Morse sequence. In C. Ding, T. Helleseth, and H. Niederreiter,
    editors, *Sequences and Their Applications, Proceedings of SETA '98*, pp. 1–16.
    Springer-Verlag, 1999.
[Allouche and Shallit 2000]   J.-P. Allouche and J. O. Shallit. Sums of digits, overlaps,
    and palindromes. *Discrete Math. & Theoret. Comput. Sci.* **4** (2000), 1–10.
[Allouche and Thakur 1999]   J.-P. Allouche and D. S. Thakur. Automata and
    transcendence of the Tate period in finite characteristic. *Proc. Amer. Math. Soc.* **127**
    (1999), 1309–1312.
[Allouche, von Haeseler, Peitgen, and Skordev 1996]   J.-P. Allouche, F. von Haeseler,
    H.-O. Peitgen, and G. Skordev. Linear cellular automata, finite automata and Pascal's
    triangle. *Discrete Appl. Math.* **66** (1996), 1–22.
[Allouche and Zamboni 1998]   J.-P. Allouche and L. Q. Zamboni. Algebraic irrational
    binary numbers cannot be fixed points of non-trivial constant-length or primitive
    morphisms. *J. Number Theory* **69** (1998), 119–124.
[Althöfer 1990]   I. Althöfer. Tight lower bounds for the length of word chains. *Inform.
    Process. Lett.* **34** (1990), 275–276.
[Altman, Gaujal, and Hordijk 2000]   E. Altman, B. Gaujal, and A. Hordijk. Balanced
    sequences and optimal routing. *J. Assoc. Comput. Mach.* **47** (2000), 752–775.
[Alzer 1995]   H. Alzer. Note on an extremal property of the Rudin–Shapiro sequence.
    Abh. Math. Sem. Univ. Hamburg **65** (1995), 243–248.
[Amou 1991]   M. Amou. Approximation to certain transcendental decimal fractions by
    algebraic numbers. *J. Number Theory* **37** (1991), 231–241.

[Anderson 1997]   P. G. Anderson. Elementary problem B-838. *Fibonacci Quart.* **35** (1997), 371–372. Solution, **36** (1998), 186.

[Anderson, Brown, and Shiue 1995]   P. G. Anderson, T. C. Brown, and P. J.-S. Shiue. A simple proof of a remarkable continued fraction identity. *Proc. Amer. Math. Soc.* **123** (1995), 2005–2009.

[Angel and Morrison 1991]   E. S. Angel and D. R. Morrison. Speeding up Bresenham's algorithm. *IEEE Comput. Graphics Appl.* **11** (1991), 16–17.

[Apostolico 1984]   A. Apostolico. On context constrained squares and repetitions in a string. *RAIRO Inform. Théor. Appl.* **18** (1984), 147–159.

[Apostolico 1992]   A. Apostolico. Optimal parallel detection of squares in strings. *Algorithmica* **8** (1992), 282–319.

[Apostolico and Breslauer 1996]   A. Apostolico and D. Breslauer. An optimal $O(\log \log N)$-time parallel algorithm for detecting all squares in a string. *SIAM J. Comput.* **25** (1996), 1318–1331.

[Apostolico, Breslauer, and Galil 1992]   A. Apostolico, D. Breslauer, and Z. Galil. Optimal parallel algorithms for periods, palindromes, and squares. In W. Kuich, editor, *Proc. 19th Int'l Conf. on Automata, Languages, and Programming (ICALP)*, Vol. 623 of *Lecture Notes in Computer Science*, pp. 296–307. Springer-Verlag, 1992.

[Apostolico, Breslauer, and Galil 1995]   A. Apostolico, D. Breslauer, and Z. Galil. Parallel detection of all palindromes in a string. *Theoret. Comput. Sci.* **141** (1995), 163–173.

[Apostolico and Brimkov 2000]   A. Apostolico and V. E. Brimkov. Fibonacci arrays and their two-dimensional repetitions. *Theoret. Comput. Sci.* **237** (2000), 263–273.

[Apostolico and Fraenkel 1987]   A. Apostolico and A. S. Fraenkel. Robust transmission of unbounded strings using Fibonacci representations. *IEEE Trans. Inform. Theory* **33** (1987), 238–245.

[Apostolico and Preparata 1983]   A. Apostolico and F. P. Preparata. Optimal off-line detection of repetitions in a string. *Theoret. Comput. Sci.* **22** (1983), 297–315.

[Arbib 1968]   M. A. Arbib, editor. *Algebraic Theory of Machines, Languages, and Semigroups*. Academic Press, 1968.

[Arcelli and Massarotti 1978]   C. Arcelli and A. Massarotti. On the parallel generation of digital straight lines. *Computer Graphics and Image Processing* **7** (1978), 67–83.

[Ardila 2002]   F. M. Ardila. On the coefficients of a Fibonacci power series. *Fibonacci Quart.*, to appear. Electronic preprint at `http://arxiv.org/abs/math.CO/0110160`, 2002.

[Arno and Wheeler 1993]   S. Arno and F. S. Wheeler. Signed digit representations of minimal Hamming weight. *IEEE Trans. Comput.* **42** (1993), 1007–1010.

[Arnold 1983]   A. Arnold. Solution to problem 12. *Bull. European Assoc. Theor. Comput. Sci.*, No. 19 (February 1983), 12–13.

[Arnold and Brlek 1989]   A. Arnold and S. Brlek. Optimal word chains for the Thue–Morse word. *Inform. Comput.* **83** (1989), 140–151.

[Arnoux 1989]   P. Arnoux. Some remarks about Fibonacci multiplication. *Appl. Math. Lett.* **2** (1989), 319–320.

[Arnoux, Ferenczi, and Hubert 1999]   P. Arnoux, S. Ferenczi, and P. Hubert. Trajectories of rotations. *Acta Arith.* **87** (1999), 209–217.

[Arnoux and Mauduit 1996]   P. Arnoux and C. Mauduit. Complexité de suites engendrées par des récurrences unipotentes. *Acta Arith.* **76** (1996), 85–97.

[Arnoux, Mauduit, Shiokawa, and Tamura 1994a]   P. Arnoux, C. Mauduit, I. Shiokawa, and J. I. Tamura. Complexity of sequences defined by billiards in the cube. *Bull. Soc. Math. France* **122** (1994), 1–12.

[Arnoux, Mauduit, Shiokawa, and Tamura 1994b]   P. Arnoux, C. Mauduit, I. Shiokawa, and J. I. Tamura. Rauzy's conjecture on the cubic billiards. *Tokyo J. Math.* **17** (1994), 211–218.

[Arnoux and Rauzy 1991]   P. Arnoux and G. Rauzy. Représentation géométrique de
   suites de complexité $2n + 1$. *Bull. Soc. Math. France* **119** (1991), 199–215.
[Arshon 1937]   S. E. Arshon. Proof of the existence of asymmetric infinite sequences.
   Mat. Sbornik **2** (1937), 769–779. In Russian, with French abstract.
[Artin 1924]   E. Artin. Quadratische Körper im Gebiet der höheren Kongruenzen I, II.
   Math. Z. **19** (1924), 153–246. Reprinted in *Collected Papers*, pp. 1–104.
[Atkinson 1978]   K. E. Atkinson. *An Introduction to Numerical Analysis*. Wiley, 1978.
[Atkinson 1981]   M. D. Atkinson. The cyclic towers of Hanoi. *Inform. Process. Lett.* **13**
   (1981), 118–119.
[Aubry, Godrèche, and Luck 1987]   S. Aubry, C. Godrèche, and J.-M. Luck. A structure
   intermediate between quasiperiodic and random. *Europhys. Lett.* **4** (1987), 639–643.
[Aubry, Godrèche, and Luck 1988]   S. Aubry, C. Godrèche, and J.-M. Luck. Scaling
   properties of a structure intermediate between quasiperiodic and random. *J. Statist.
   Phys.* **51** (1988), 1033–1075.
[Autebert, Beauqier, Boasson, and Nivat 1979]   J. M. Autebert, J. Beauquier,
   L. Boasson, and M. Nivat. Quelques problèmes ouverts en théorie des langages
   algébriques. *RAIRO Inform. Théor. Appl.* **13** (1979), 363–379.
[Avgustinovich 1994]   S. V. Avgustinovich. The number of different subwords of given
   length in the Morse–Hedlund sequence. *Sibirsk. Zh. Issled. Oper.* **1** (1994), 3–7, 103.
   In Russian. English translation in A. D. Korshunov, editor, *Discrete Analysis and
   Operations Research*, Kluwer, 1996, pp. 1–5.
[Avgustinovich, Fon-Der-Flaass, and Frid 2001]   S. V. Avgustinovich, D. G.
   Fon-Der-Flaass, and A. E. Frid. Arithmetical complexity of infinite words. To appear,
   Proc. 3rd Int. Conf. on Words, Languages, and Combinatorics, 2001.
   http://www.math.nsc.ru/LBRT/k4/Frid_arithm.ps
[Avishai, Berend, and Glaubman 1994]   Y. Avishai, D. Berend, and D. Glaubman.
   Minimum-dimension trace maps for substitution sequences. *Phys. Rev. Lett.* **72**
   (1994), 1842–1845.
[Axel, Allouche, Kleman, Mendès France, and Peyrière 1986]   F. Axel, J.-P. Allouche,
   M. Kleman, M. Mendès France, and J. Peyrière. Vibrational modes in a one
   dimensional "quasi-alloy": the Morse case. *J. Physique* **47** (1986), C3.181–C3.186.
   Colloque C3, Supplement to No. 7.
[Axel, Allouche, and Wen 1992]   F. Axel, J.-P. Allouche, and Z.-Y. Wen. On certain
   properties of high-resolution X-ray diffraction spectra of finite-size generalized
   Rudin-Shapiro multilayer heterostructures. *J. Phys.: Condens. Matter* **4** (1992),
   8713–8728.
[Axel and Gratias 1995]   F. Axel and D. Gratias, editors. *Beyond Quasicrystals*. Les
   Éditions de Physique; Springer, 1995.
[Axel and Peyrière 1989]   F. Axel and J. Peyrière. Spectrum and extended states in a
   harmonic chain with controlled disorder: effects of the Thue-Morse symmetry.
   J. Statist. Phys. **57** (1989), 1013–1047.
[Axel and Terauchi 1991]   F. Axel and H. Terauchi. High-resolution X-ray diffraction
   spectra of Thue–Morse GaAs–AlAs heterostructures: towards a novel description of
   disorder. *Phys. Rev. Lett.* **66** (1991), 2223–2226.
[Axel and Terauchi 1994]   F. Axel and H. Terauchi. Reply to comment: High-resolution
   X-ray-diffraction spectra of Thue-Morse GaAs–AlAs heterostructures. *Phys. Rev.
   Lett.* **73** (1994), 1308.
[Baake 1999]   M. Baake. A note on palindromicity. *Lett. Math. Phys.* **49** (1999),
   217–227.
[Baake, Elser, and Grimm 1997]   M. Baake, V. Elser, and U. Grimm. The entropy of
   square-free words. *Mathematical and Computer Modelling* **26** (1997), 13–26.
[Baake, Grimm, and Joseph 1993]   M. Baake, U. Grimm, and D. Joseph. Trace maps,
   invariants, and some of their applications. *Int. J. Mod. Phys. B* **7** (1993), 1527–1550.

[Baake, Hermisson, and Pleasants 1997]   M. Baake, J. Hermisson, and P. A. B. Pleasants. The torus parametrization of quasiperiodic LI-classes. *J. Phys. A: Math. Gen.* **30** (1997), 3029–3056.

[Bailey, Borwein, and Plouffe 1997]   D. Bailey, P. Borwein, and S. Plouffe. On the rapid computation of various polylogarithmic constants. *Math. Comp.* **66** (1997), 903–913.

[Baker 1990]   A. Baker. *Transcendental Number Theory*. Cambridge University Press, 1990.

[Baker, McNulty, and Taylor 1989]   K. A. Baker, G. F. McNulty, and W. Taylor. Growth problems for avoidable words. *Theoret. Comput. Sci.* **69** (1989), 319–345.

[Balasiński and Mrówka 1957]   W. Balasiński and S. Mrówka. On algorithms of arithmetical operations. *Bull. Acad. Polon. Sci. Cl. III* **5** (1957), 803–804.

[Ball 1939]   W. W. R. Ball. *Mathematical Recreations and Essays*. MacMillan, New York, 1939. Revised by H. S. M. Coxeter.

[Bang 1957]   T. Bang. On the sequence $[n\alpha]$, $n = 1, 2, \cdots$. Supplementary note to the preceding paper by Th. Skolem. *Math. Scand.* **5** (1957), 69–76.

[Barbé 1993]   A. M. Barbé. Artistic design with fractal matrices. *Visual Computer* **9** (1993), 233–238.

[Barbé, Peitgen, and Skordev 1999]   A. Barbé, H.-O. Peitgen, and G. Skordev. Automaticity of coarse-graining invariant orbits of one-dimensional linear cellular automata. *Int. J. Bifur. Chaos Appl. Sci. Engrg.* **9** (1999), 67–95.

[Barbé, von Haeseler, Peitgen, and Skordev 1995]   A. Barbé, F. von Haeseler, H.-O. Peitgen, and G. Skordev. Coarse-graining invariant patterns of one-dimensional two-state linear cellular automata. *Int. J. Bifur. Chaos Appl. Sci. Engrg.* **5** (1995), 1611–1631.

[Barbier 1887a]   E. Barbier. On suppose écrite la suite naturelle des nombres; quel est le $(10^{1000})^{\text{ième}}$ chiffre écrit? *C. R. Acad. Sci. Paris* **105** (1887), 795–798.

[Barbier 1887b]   E. Barbier. On suppose écrite la suite naturelle des nombres; quel est le $(10^{10000})^{\text{ième}}$ chiffre écrit? *C. R. Acad. Sci. Paris* **105** (1887), 1238–1239.

[Bar-Hillel, Perles, and Shamir 1961]   Y. Bar-Hillel, M. Perles, and E. Shamir. On formal properties of simple phrase structure grammars. *Z. Phonetik. Sprachwiss. Kommuniationsforsch.* **14** (1961), 143–172.

[Baron and Urbanek 1989]   G. Baron and F. Urbanek. Factorial languages with quadratically upper bounded growth functions and nonlinearly upper bounded subword complexities. *Inform. Process. Lett.* **32** (1989), 267–269.

[Baryshnikov 1995]   Yu. Baryshnikov. Complexity of trajectories in rectangular billiards. *Commun. Math. Phys.* **174** (1995), 43–56.

[Bassily and Kátai 1995]   N. L. Bassily and I. Kátai. Distribution of the values of q-additive functions on polynomial sequences. *Acta Math. Hung.* **68** (1995), 353–361.

[Bateman and Bradley 1997]   P. Bateman and D. Bradley. Problem 10589 [*sic*, should be 10596]. *Amer. Math. Monthly* **104** (1997), 456. Solution by D. Callan and R. Stong, **106** (1999), 366–367.

[Baum and Sweet 1976]   L. E. Baum and M. M. Sweet. Continued fractions of algebraic power series in characteristic 2. *Ann. Math.* **103** (1976), 593–610.

[Bazinet and Siddiqi 1972]   J. Bazinet and J. A. Siddiqi. On nonstrongly regular matrices. *Proc. Amer. Math. Soc.* **34** (1972), 428–432.

[Beals and Thakur 1998]   R. M. Beals and D. S. Thakur. Computational classification of numbers and algebraic properties. *Int. Math. Research Notices*, No. 15 (1998), 799–818.

[Bean, Ehrenfeucht, and McNulty 1979]   D. A. Bean, A. Ehrenfeucht, and G. McNulty. Avoidable patterns in strings of symbols. *Pacific J. Math.* **85** (1979), 261–294.

[Beatty 1926]   S. Beatty. Problem 3173. *Amer. Math. Monthly* **33** (1926), 159. Solution, **34** (1927), 159–160.

[Beauquier 1985]   D. Beauquier. Ensembles reconnaissables de mots biinfinis. In M. Nivat and D. Perrin, editors, *Automata on Infinite Words*, Vol. 192 of *Lecture Notes in Computer Science*, pp. 28–46. Springer-Verlag, 1985.

[Beck 1991]   J. Beck. Flat polynomials on the unit circle – note on a problem of Littlewood. *Bull. Lond. Math. Soc.* **23** (1991), 269–277.

[Becker 1993]   P.-G. Becker. Automatische Folgen und Transzendenz in positiver Charakteristik. *Arch. Math. (Basel)* **61** (1993), 68–74.

[Becker 1994]   P.-G. Becker. k-regular power series and Mahler-type functional equations. *J. Number Theory* **49** (1994), 269–286.

[Bellissard 1990]   J. Bellissard. Spectral properties of Schrödinger's operator with a Thue–Morse potential. In J.-M. Luck, P. Moussa, and M. Waldschmidt, editors, *Number Theory and Physics*, Vol. 47 of *Springer Proceedings in Physics*, pp. 140–150. Springer-Verlag, 1990.

[Bellissard 1992]   J. Bellissard. Gap labelling theorems for Schrödinger's operators. In M. Waldschmidt, P. Moussa, J.-M. Luck, and C. Itzykson, editors, *From Number Theory to Physics*, pp. 538–630. Springer-Verlag, 1992.

[Bellissard, Bovier, and Ghez 1991]   J. Bellissard, A. Bovier, and J.-M. Ghez. Spectral properties of a tight binding Hamiltonian with period doubling potential. *Commun. Math. Phys.* **135** (1991), 379–399.

[Bellissard, Bovier, and Ghez 1992]   J. Bellissard, A. Bovier, and J.-M. Ghez. Gap labelling theorems for one-dimensional discrete Schrödinger operators. *Rev. Math. Phys.* **4** (1992), 1–37.

[Bellissard, Bovier, and Ghez 1993]   J. Bellissard, A. Bovier, and J.-M. Ghez. Discrete Schrödinger operators with potentials generated by substitutions. In W. F. Ames, E. M. Harrell, and J. V. Herod, editors, *Differential Equations with Applications to Mathematical Physics*, pp. 13–23. Academic Press, Boston, 1993.

[Bellissard, Iochum, Scoppola, and Testard 1989]   J. Bellissard, B. Iochum, E. Scoppola, and D. Testard. Spectral properties of one-dimensional quasi-crystals. *Commun. Math. Phys.* **125** (1989), 527–543.

[Bellman and Shapiro 1948]   R. Bellman and H. N. Shapiro. On a problem in additive number theory. *Ann. Math.* **49** (1948), 333–340.

[Benczur 1986]   A. Benczur. On the expected time of the first occurrence of every k bit long patterns in the symmetric Bernoulli process. *Acta Math. Hung.* **47** (1986), 233–238.

[Bender and Kochman 1993]   E. A. Bender and F. Kochman. The distribution of subword counts is usually normal. *Eur. J. Combinatorics* **14** (1993), 265–275.

[Bendisch 1985]   J. Bendisch. Generalized sequencing problem "Towers of Hanoi". *Z. Oper. Res.* **29** (1985), 31–45.

[Bercoff 1995]   C. Bercoff. A family of tag systems for paperfolding sequences. In E. W. Mayr and C. Puech, editors, *STACS 95, Proc. 12th Symp. Theoretical Aspects of Comp. Sci.*, Vol. 900 of *Lecture Notes in Computer Science*, pp. 303–312. Springer-Verlag, 1995.

[Bercoff 1996]   C. Bercoff. A family of p-uniform tag systems for p-paperfolding sequences. In J. Dassow, G. Rozenberg, and A. Salomaa, editors, *Developments in Language Theory II*, pp. 3–12. World Scientific, 1996.

[Bercoff 1997]   C. Bercoff. Uniform tag systems for paperfolding sequences. *Discrete Appl. Math.* **77** (1997), 119–138.

[Berend and Frougny 1994]   D. Berend and C. Frougny. Computability by finite automata and Pisot bases. *Math. Systems Theory* **27** (1994), 275–282.

[Berenstein, Kanal, Lavine, and Olson 1987]   C. A. Berenstein, L. N. Kanal, D. Lavine, and E. C. Olson. A geometric approach to subpixel registration accuracy. *Computer Vision, Graphics, and Image Processing* **40** (1987), 334–360.

[Berenstein and Lavine 1988]   C. A. Berenstein and D. Lavine. On the number of digital straight line segments. *IEEE Trans. Pattern Anal. Machine Intell.* **10** (1988), 880–887.

[Berlekamp, Conway, and Guy 1982]   E. R. Berlekamp, J. H. Conway, and R. K. Guy. *Winning Ways for Your Mathematical Plays*, Vol. 1. Academic Press, Toronto, 1982.

[Berndt and Bhargava 1993]   B. C. Berndt and S. Bhargava. Ramanujan – for lowbrows. *Amer. Math. Monthly* **100** (1993), 644–656.

[Bernoulli 1772]   J. Bernoulli. Sur une nouvelle espece de calcul. In *Recueil pour les Astronomes*, Vol. I, pp. 255–284. Berlin, 1772.

[Berstel 1978]   J. Berstel. Sur la construction de mots sans carré. *Séminaire de Théorie des Nombres* (1978–1979), 18.01–18.15.

[Berstel 1979]   J. Berstel. Sur les mots sans carré définis par un morphisme. In H. A. Maurer, editor, *Proc. 6th Int. Conf. on Automata, Languages, and Programming (ICALP)*, Vol. 71 of *Lecture Notes in Computer Science*, pp. 16–25. Springer-Verlag, 1979.

[Berstel 1980a]   J. Berstel. Mots sans carré et morphismes itérés. *Discrete Math.* **29** (1980), 235–244.

[Berstel 1980b]   J. Berstel. Mots de Fibonacci. *Séminaire d'Informatique Théorique, LITP* **6–7** (1980–1981), 57–78.

[Berstel 1982]   J. Berstel. Three questions on square-free words. *Bull. Eur. Assoc. Theor. Comput. Sci.*, No. 17 (June 1982), 178.

[Berstel 1984a]   J. Berstel. Some recent results on squarefree words. In M. Fontet and K. Mehlhorn, editors, *STACS 84, Proc. 1st Symp. Theoretical Aspects of Comp. Sci.*, Vol. 166 of *Lecture Notes in Computer Science*, pp. 14–25. Springer-Verlag, 1984.

[Berstel 1986a]   J. Berstel. Every iterated morphism yields a co-CFL. *Inform. Process. Lett.* **22** (1986), 7–9.

[Berstel 1986b]   J. Berstel. Fibonacci words – a survey. In G. Rozenberg and A. Salomaa, editors, *The Book of L*, pp. 13–27. Springer-Verlag, 1986.

[Berstel 1989a]   J. Berstel. Langford strings are squarefree. *Bull. Eur. Assoc. Theor. Comput. Sci.*, No. 37 (February 1989), 127–129.

[Berstel 1990]   J. Berstel. Tracé de droites, fractions continues et morphismes itérés. In M. Lothaire, editor, *Mots: Mélanges Offerts à M.-P. Schützenberger*, pp. 298–309. Hermès, 1990.

[Berstel 1992]   J. Berstel. Axel Thue's work on repetitions in words. In P. Leroux and C. Reutenauer, editors, *Séries Formelles et Combinatoire Algébrique*, Vol. 11 of *Publications du LaCim*, pp. 65–80. Université du Québec à Montréal, 1992.

[Berstel 1994]   J. Berstel. A rewriting of Fife's theorem about overlap-free words. In J. Karhumäki, H. Maurer, and G. Rozenberg, editors, *Results and Trends in Theoretical Computer Science*, Vol. 812 of *Lecture Notes in Computer Science*, pp. 19–29. Springer-Verlag, 1994.

[Berstel 1995]   J. Berstel. *Axel Thue's Papers on Repetitions in Words: a Translation*. Number 20 in Publications du Laboratoire de Combinatoire et d'Informatique Mathématique. Université du Québec à Montréal, February 1995.

[Berstel 1999]   J. Berstel. On the index of Sturmian words. In J. Karhumäki, H. Maurer, G. Păun, and G. Rozenberg, editors, *Jewels are Forever*, pp. 287–294. Springer-Verlag, 1999.

[Berstel 2001]   J. Berstel. An exercise on Fibonacci representations. *Theoret. Informatics Appl.* **35** (2001), 491–498.

[Berstel and Boasson 1999]   J. Berstel and L. Boasson. Partial words and a theorem of
    Fine and Wilf. *Theoret. Comput. Sci.* **218** (1999), 135–141.
[Berstel and Brlek 1987]   J. Berstel and S. Brlek. On the length of word chains. *Inform.
    Process. Lett.* **26** (1987/88), 23–28.
[Berstel and Morcrette 1989]   J. Berstel and M. Morcrette. Compact representation of
    patterns by finite automata. In A. Gagalowicz, editor, *Pixim 89: Proc. 2nd Int. Conf.*,
    pp. 387–395. Hermes – ACM SIGGRAPH France, 1989.
[Berstel and Nait Abdallah 1989]   J. Berstel and A. Nait Abdallah. Tétrarbres engendrés
    par des automates finis. Technical Report 89-7, Rapp. Tec. Lab. Info. Theor. Prog.,
    January 1989.
[Berstel and Pocchiola 1993]   J. Berstel and M. Pocchiola. A geometric proof of the
    enumeration formula for Sturmian words. *Int. J. Algebra Comput.* **3** (1993), 349–355.
[Berstel and Pocchiola 1996]   J. Berstel and M. Pocchiola. Random generation of finite
    Sturmian words. *Discrete Math.* **153** (1996), 29–39.
[Berstel and Reutenauer 1988]   J. Berstel and C. Reutenauer. *Rational Series and Their
    Languages*, Vol. 12 of *EATCS Monographs on Theoretical Computer Science.*
    Springer-Verlag, 1988.
[Berstel and Séébold 1993a]   J. Berstel and P. Séébold. A characterization of overlapfree
    morphisms. *Discrete Appl. Math.* **46** (1993), 275–281.
[Berstel and Séébold 1993b]   J. Berstel and P. Séébold. A characterization of Sturmian
    morphisms. In A. Borzyszkowski and S. Sokolowski, editors, *Proc. 18th Symposium,
    Mathematical Foundations of Computer Science 1993*, Vol. 711 of *Lecture Notes in
    Computer Science*, pp. 281–290. Springer-Verlag, 1993.
[Berstel and Séébold 1994a]   J. Berstel and P. Séébold. Morphismes de Sturm. *Bull.
    Belg. Math. Soc.* **1** (1994), 175–189.
[Berstel and Séébold 1994b]   J. Berstel and P. Séébold. A remark on morphic Sturmian
    words. *RAIRO Inform. Théor. App.* **28** (1994), 255–263.
[Berstel and Séébold 2002]   J. Berstel and P. Séébold. Sturmian words. In M. Lothaire,
    editor, *Algebraic Combinatorics on Words*, Vol. 90 of *Encyclopedia of Mathematics
    and Its Applications*, pp. 45–110. Cambridge University Press, 2002.
[Berthé 1992]   V. Berthé. De nouvelles preuves "automatiques" de transcendance pour la
    fonction zêta de Carlitz. In D. F. Coray and Y.-F. S. Pétermann, editors, *Journées
    Arithmétiques de Genève*, Vol. 209 of *Astérisque*, pp. 159–168, 1992.
[Berthé 1993]   V. Berthé. Fonction ζ de Carlitz et automates. *J. Théorie Nombres
    Bordeaux* **5** (1993), 53–77.
[Berthé 1994]   V. Berthé. Automates et valeurs de transcendance du logarithme de
    Carlitz. *Acta Arith.* **66** (1994), 369–390.
[Berthé 1995]   V. Berthé. Combinaisons linéaires de $\zeta(s)/\pi^s$ sur $\mathbb{F}_q(x)$, pour
    $1 \le s \le q - 2$. *J. Number Theory* **53** (1995), 272–299.
[Berthé 1996]   V. Berthé. Fréquences des facteurs des suites sturmiennes. *Theoret.
    Comput. Sci.* **165** (1996), 295–309.
[Berthé 2000a]   V. Berthé. Complexité et automates cellulaires linéaires. *Theoret.
    Informatics Appl.* **34** (2000), 403–423.
[Berthé 2000b]   V. Berthé. Sequences of low complexity: automatic and Sturmian
    sequences. In *Topics in Symbolic Dynamics and Applications (Temuco, 1997)*,
    Vol. 279 of *London Math. Soc. Lecture Note Ser.*, pp. 1–34. Cambridge University
    Press, 2000.
[Berthé 2001]   V. Berthé. Autour du système de numération d'Ostrowski. *Bull. Belg.
    Math. Soc.* **8** (2001), 209–239.
[Berthé and Tijdeman 2002]   V. Berthé and R. Tijdeman. Balance properties of
    multi-dimensional words. *Theoret. Comput. Sci.* **273** (2002), 197–224.
[Berthé and Vuillon 2000a]   V. Berthé and L. Vuillon. Suite doubles de basse
    complexité. *J. Théorie Nombres Bordeaux* **12** (2000), 179–208.

[Berthé and Vuillon 2000b]   V. Berthé and L. Vuillon. Tilings and rotations on the torus: a two-dimensional generalization of Sturmian sequences. *Discrete Math.* **223** (2000), 27–53.

[Berthé and Vuillon 2001]   V. Berthé and L. Vuillon. Palindromes and two-dimensional Sturmian sequences. *J. Automata, Languages, and Combinatorics* **6** (2001), 121–138.

[Bertrand-Mathis 1989]   A. Bertrand-Mathis. Comment écrire les nombres entiers dans une base que n'est pas entière. *Acta Math. Hung.* **54** (1989), 237–241.

[Bès 1997]   A. Bès. Undecidable extensions of Büchi arithmetic and Cobham–Semënov theorem. *J. Symbolic Logic* **62** (1997), 1280–1296.

[Bésineau 1972]   J. Bésineau. Indépendance statistique d'ensembles liés à la fonction "somme des chiffres". *Acta Arith.* **20** (1972), 401–416.

[Birget 1992]   J.-C. Birget. Intersection and union of regular languages and state complexity. *Inform. Process. Lett.* **43** (1992), 185–190.

[Blanchard and Mendès France 1982]   A. Blanchard and M. Mendès France. Symétrie et transcendance. *Bull. Sci. Math.* **106** (1982), 325–335.

[Blanchard 1989]   F. Blanchard. β-expansions and symbolic dynamics. *Theoret. Comput. Sci.* **65** (1989), 131–141.

[Blanchard and Kurka 1998]   F. Blanchard and P. Kurka. Language complexity of rotations and Sturmian sequences. *Theoret. Comput. Sci.* **209** (1998), 179–193.

[Blanchet-Sadri and Hegstrom 2002]   F. Blanchet-Sadri and R. A. Hegstrom. Partial words and a theorem of Fine and Wilf revisited. *Theoret. Comput. Sci.* **270** (2002), 401–419.

[Blecksmith, Filaseta, and Nicol 1993]   R. Blecksmith, M. Filaseta, and C. Nicol. A result on the digits of a^n. *Acta Arith.* **64** (1993), 331–339.

[Blecksmith and Laud 1995]   R. Blecksmith and P. W. Laud. Some exact number theory computations via probability mechanisms. *Amer. Math. Monthly* **102** (1995), 893–903.

[Blecksmith, McCallum, and Selfridge 1998]   R. Blecksmith, M. McCallum, and J. L. Selfridge. 3-smooth representations of integers. *Amer. Math. Monthly* **105** (1998), 529–543.

[Bleuzen-Guernalec 1985]   N. Bleuzen-Guernalec. Suites points fixes de transductions uniformes. *C. R. Acad. Sci. Paris* **300** (1985), 85–88.

[Bleuzen-Guernalec and Blanc 1989]   N. Bleuzen-Guernalec and G. Blanc. Production en temps réel et complexité de structure de suites infinies. *RAIRO Inform. Théor. App.* **23** (1989), 195–216.

[Blom 1994]   G. Blom. Problem 94-20. *SIAM Review* **36** (1994), 657. Solution by O. P. Lossers, **37** (1995), 619–620.

[Bloom 1971]   D. M. Bloom. Elementary problem proposal E 2307. *Amer. Math. Monthly* **78** (1971), 792. Solution by D. E. Knuth, **79** (1972), 773–774.

[Bloom 1995]   D. M. Bloom. A one-sentence proof that $\sqrt{2}$ is irrational. *Math. Mag.* **68** (1995), 286.

[Blumberg 1926]   H. Blumberg. Note on a theorem of Kempner concerning transcendental numbers. *Bull. Amer. Math. Soc.* **32** (1926), 351–356.

[Boardman, Garrett, and Robson 1986]   J. T. Boardman, C. Garrett, and G. C. A. Robson. A recursive algorithm for the optimal solution of a complex allocation problem using a dynamic programming formulation. *Computer J.* **29** (1986), 182–186.

[Boasson 1983]   L. Boasson. An answer to P12: www^R is cube-free. *Bull. European Assoc. Theor. Comput. Sci.*, No. 19, (February 1983), 13–14.

[Boffa and Point 1991]   M. Boffa and F. Point. Identités de Thue-Morse dans les groupes. *C. R. Acad. Sci. Paris* **312** (1991), 667–670.

[Boffa and Point 1992]   M. Boffa and F. Point. m-identities. *C. R. Acad. Sci. Paris* **314** (1992), 879–880.

[Böhmer 1926]   P. E. Böhmer. Über die Transzendenz gewisser dyadischer Brüche. *Math. Annalen* **96** (1926), 367–377. Erratum, **96** (1926), 735.

[Bongiovanni, Luccio, and Zorat 1975]   G. Bongiovanni, F. Luccio, and A. Zorat. The discrete equation of the straight line. *IEEE Trans. Comput.* **C-24** (1975), 310–313.

[Borel 1997]   J.-P. Borel. Opérations sur les mots de Christoffel. *C. R. Acad. Sci. Paris* **325** (1997), 239–242.

[Borel 2001]   J.-P. Borel. Image par homographie de mots de Christoffel. *Bull. Belg. Math. Soc.* **8** (2001), 241–255.

[Borel and Laubie 1991]   J.-P. Borel and F. Laubie. Construction de mots de Christoffel. *C. R. Acad. Sci. Paris* **313** (1991), 483–485.

[Borel and Laubie 1993]   J.-P. Borel and F. Laubie. Quelques mots sur la droite projective réelle. *J. Théorie Nombres Bordeaux* **5** (1993), 23–51.

[Borwein and Girgensohn 1995]   J. M. Borwein and R. Girgensohn. Addition theorems and binary expansions. *Canad. J. Math.* **47** (1995), 262–273.

[Borwein 1991]   P. Borwein. On the irrationality of $\sum(1/(q^n + r))$. *J. Number Theory* **37** (1991), 253–259.

[Borwein and Ingalls 1994]   P. Borwein and C. Ingalls. The Prouhet–Tarry–Escott problem revisited. *Enseign. Math.* **40** (1994), 3–27.

[Borwein and Mossinghoff 2000]   P. Borwein and M. Mossinghoff. Rudin-Shapiro-like polynomials in L_4. *Math. Comp.* **69** (2000), 1157–1166.

[Boshernitzan and Fraenkel 1984]   M. Boshernitzan and A. S. Fraenkel. A linear algorithm for nonhomogeneous spectra of numbers. *J. Algorithms* **5** (1984), 187–198.

[Bosma 2001]   W. Bosma. Signed bits and fast exponentiation. *J. Théorie Nombres Bordeaux* **13** (2001), 27–41.

[Bousquet-Mélou 1992]   M. Bousquet-Mélou. The number of minimal word chains computing the Thue-Morse word. *Inform. Process. Lett.* **44** (1992), 57–64.

[Boute 2000]   R. T. Boute. Zeroless positional number representation and string ordering. *Amer. Math. Monthly* **107** (2000), 437–444.

[Bouton 1902]   C. L. Bouton. Nim, a game with a complete mathematical theory. *Ann. Math.* **3** (1902), 35–39.

[Bovier and Ghez 1993a]   A. Bovier and J.-M. Ghez. Schrödinger operators with substitution potentials. In N. Boccara, E. Goles, S. Martinez, and P. Picco, editors, *Cellular Automata and Cooperative Systems*, Vol. 396 of *NATO ASI Series, Series C: Mathematical and Physical Sciences*, pp. 67–83. Kluwer, 1993.

[Bovier and Ghez 1993b]   A. Bovier and J.-M. Ghez. Spectral properties of one-dimensional Schrödinger operators with potentials generated by substitutions. *Commun. Math. Phys.* **158** (1993), 45–66. Erratum, **166** (1994), 431–432.

[Bovier and Ghez 1995]   A. Bovier and J.-M. Ghez. Remarks on the spectral properties of tight binding and Kronig–Penney models with substitution sequences. *J. Phys. A: Math. Gen.* **28** (1995), 2313–2324.

[Bowman 1988]   D. Bowman. A new generalization of Davison's theorem. *Fibonacci Quart.* **26** (1988), 40–45.

[Bowman 1995]   D. Bowman. Approximation of $\lfloor n\alpha + s \rfloor$ and the zero of $\{n\alpha + s\}$. *J. Number Theory* **50** (1995), 128–144.

[Bowman and White 1989]   D. Bowman and T. White. Advanced problem 6609. *Amer. Math. Monthly* **96** (1989), 743. Solution, **98** (1991), 279–281.

[Boyd 1997]   D. W. Boyd. On a problem of Byrnes concerning polynomials with restricted coefficients. *Math. Comp.* **66** (1997), 1697–1703.

[Boyd 2001]   D. W. Boyd. On a problem of Byrnes concerning polynomials with restricted coefficients, II. *Math. Comp.* **71** (2001), 1205–1217.

[Brandenburg 1983]   F.-J. Brandenburg. Uniformly growing k-th power-free homomorphisms. *Theoret. Comput. Sci.* **23** (1983), 69–82.

[Braunholtz 1963]   C. H. Braunholtz. An infinite sequence of three symbols with no adjacent repeats. *Amer. Math. Monthly* **70** (1963), 675–676.

[Breitbart 1971]   Y. Breitbart. On automaton and "zone" complexity of the predicate "to be a *k*th power of an integer". *Dokl. Akad. Nauk SSSR* **196** (1971), 16–19. In Russian. English translation in *Soviet Math. Dokl.* **12** (1971), 10–14.

[Breitbart 1973]   Y. Breitbart. Complexity of the calculation of predicates by finite automata. PhD thesis, Technion, Haifa, Israel, June 1973.

[Breitbart 1976]   Y. Breitbart. Some bounds on the complexity of predicate recognition by finite automata. *J. Comput. System Sci.* **12** (1976), 336–349.

[Bresenham 1965]   J. E. Bresenham. Algorithm for computer control of a digital plotter. *IBM Systems J.* **4** (1965), 25–30.

[Bresenham 1982]   J. E. Bresenham. Incremental line compaction. *Computer J.* **25** (1982), 116–120.

[Bresenham 1985]   J. E. Bresenham. Run length slice algorithm for incremental lines. In R. A. Earnshaw, editor, *Fundamental Algorithms for Computer Graphics*, Vol. F17 of *NATO ASI Series*, pp. 59–104. Springer-Verlag, 1985.

[Breusch 1954]   R. Breusch. A proof of the irrationality of π. *Amer. Math. Monthly* **61** (1954), 631–632.

[Bricard 1926]   R. Bricard. Sur un problème relatif aux nombres incommensurables. *Nouvelles Annales de Mathématiques* **1** (1926), 100–103.

[Brillhart 1973]   J. Brillhart. On the Rudin-Shapiro polynomials. *Duke Math. J.* **40** (1973), 335–353.

[Brillhart and Carlitz 1970]   J. Brillhart and L. Carlitz. Note on the Shapiro polynomials. *Proc. Amer. Math. Soc.* **25** (1970), 114–118.

[Brillhart, Erdős, and Morton 1983]   J. Brillhart, P. Erdős, and P. Morton. On sums of Rudin-Shapiro coefficients II. *Pacific J. Math.* **107** (1983), 39–69.

[Brillhart, Lomont, and Morton 1976]   J. Brillhart, J. S. Lomont, and P. Morton. Cyclotomic properties of the Rudin-Shapiro polynomials. *J. Reine Angew. Math.* **288** (1976), 37–65.

[Brillhart and Morton 1978]   J. Brillhart and P. Morton. Über Summen von Rudin–Shapiroschen Koeffizienten. *Illinois J. Math.* **22** (1978), 126–148.

[Brillhart and Morton 1996]   J. Brillhart and P. Morton. A case study in mathematical research: the Golay–Rudin–Shapiro sequence. *Amer. Math. Monthly* **103** (1996), 854–869.

[Brinkhuis 1983]   J. Brinkhuis. Non-repetitive sequences on three symbols. *Quart. J. Math. Oxford* **34** (1983), 145–149.

[de Brito, da Silva, and Nazareno 1995]   P. E. de Brito, C. A. A. da Silva, and H. N. Nazareno. Field-induced localization in Fibonacci and Thue–Morse lattices. *Phys. Rev. B* **51** (1995), 6096–6099.

[Britton 1973]   J. L. Britton. The existence of infinite Burnside groups. In W. W. Boone, F. B. Cannonito, and R. C. Lyndon, editors, *Word Problems: Decision Problems and the Burnside Problem in Group Theory*, Vol. 71 of *Studies in Logic and the Foundations of Mathematics*, pp. 67–348. North-Holland, 1973.

[Britton 1980]   J. L. Britton. Erratum: The existence of infinite Burnside groups. In W. W. Boone and G. Higman, editors, *Word Problems II*, Vol. 95 of *Studies in Logic and the Foundations of Mathematics*, p. 71. North-Holland, 1980.

[Brlek 1989]   S. Brlek. Enumeration of factors in the Thue-Morse word. *Discrete Appl. Math.* **24** (1989), 83–96.

[Brons 1974]   R. Brons. Linguistic methods for the description of a straight line on a grid. *Computer Graphics and Image Processing* **3** (1974), 48–62.

[Brons 1985]   R. Brons. Theoretical and linguistic methods for describing straight lines. In R. A. Earnshaw, editor, *Fundamental Algorithms for Computer Graphics*, Vol. F17 of *NATO ASI Series*, pp. 19–57. Springer-Verlag, 1985.

[Brousseau 1976]   Br. A. Brousseau. Towers of Hanoi with more pegs. *J. Recreational Math.* **8** (1976), 165–176.

[J. Brown 1961]   J. L. Brown, Jr. Note on complete sequences of integers. *Amer. Math. Monthly* **68** (1961), 557–560.

[J. Brown 1963]   J. L. Brown, Jr. A generalization of semi-completeness for integer sequences. *Fibonacci Quart.* **1** (1963), 3–15.

[J. Brown 1964]   J. L. Brown, Jr. Zeckendorf's theorem and some applications. *Fibonacci Quart.* **2** (1964), 163–168.

[J. Brown 1965]   J. L. Brown, Jr. A new characterization of the Fibonacci numbers. *Fibonacci Quart.* **3** (1965), 1–8.

[T. Brown 1971]   T. C. Brown. Is there a sequence on four symbols in which no two adjacent segments are permutations of one another? *Amer. Math. Monthly* **78** (1971), 886–888.

[T. Brown 1991]   T. C. Brown. A characterisation of the quadratic irrationals. *Canad. Math. Bull.* **34** (1991), 36–41.

[T. Brown 1993]   T. C. Brown. Descriptions of the characteristic sequence of an irrational. *Canad. Math. Bull.* **36** (1993), 15–21.

[T. Brown 1994]   T. C. Brown. Powers of digital sums. *Fibonacci Quart.* **32** (1994), 207–210.

[T. Brown 2002]   T. C. Brown. Applications of standard Sturmian words to elementary number theory. *Theoret. Comput. Sci.* **273** (2002), 5–9.

[T. Brown and Shiue 1995]   T. C. Brown and P. J.-S. Shiue. Sums of fractional parts of integer multiples of an irrational. *J. Number Theory* **50** (1995), 181–192.

[Brubaker 1971]   D. A. Brubaker. A proof that not both πe and $\pi + e$ are algebraic. *Math. Mag.* **44** (1971), 267.

[Bruckstein 1990]   A. M. Bruckstein. The self-similarity of digital straight lines. In *Proc. 10th Intl. Conf. on Pattern Recognition*, Vol. I, pp. 485–490. IEEE Press, 1990.

[Bruckstein 1991]   A. M. Bruckstein. Self-similarity properties of digitized straight lines. In R. A. Melter, A. Rosenfeld, and P. Bhattacharya, editors, *Vision Geometry*, Vol. 119 of *Contemporary Mathematics*, pp. 1–20. Amer. Math. Soc., 1991.

[de Bruijn 1946]   N. G. de Bruijn. A combinatorial problem. *Proc. Konin. Neder. Akad. Wet.* **49** (1946), 758–764.

[de Bruijn 1975]   N. G. de Bruijn. Acknowledgement of priority to C. Flye Sainte-Marie on the counting of circular arrangements of $2n$ zeros and ones that show each n-letter word exactly once. Technical Report 75-WSK-06, Department of Mathematics and Computing Science, Eindhoven University of Technology, The Netherlands, June 1975.

[de Bruijn 1981b]   N. G. de Bruijn. Algebraic theory of Penrose's nonperiodic tilings of the plane. I. *Proc. Konin. Neder. Akad. Wet.* **84** (1981), 39–52.

[de Bruijn 1981c]   N. G. de Bruijn. Algebraic theory of Penrose's nonperiodic tilings of the plane. II. *Proc. Konin. Neder. Akad. Wet.* **84** (1981), 53–66.

[de Bruijn 1986]   N. G. de Bruijn. Quasicrystals and their Fourier transform. *Indag. Math.* **89** (1986), 123–152.

[de Bruijn 1987]   N. G. de Bruijn. Modulated quasicrystals. *Indag. Math.* **90** (1987), 121–132.

[de Bruijn 1997]   N. G. de Bruijn. Remarks on Penrose tilings. In R. L. Graham and J. Nešetřil, editors, *The Mathematics of Paul Erdős II*, pp. 264–283. Springer-Verlag, 1997.

[Bruyère, Hansel, Michaux, and Villemaire 1994]   V. Bruyère, G. Hansel, C. Michaux, and R. Villemaire. Logic and p-recognizable sets of integers. *Bull. Belg. Math. Soc.* **1** (1994), 191–238. Corrigendum, *Bull. Belg. Math. Soc.* **1** (1994), 577.

[Brzozowski 1962]   J. Brzozowski. A survey of regular expressions and their applications. *IEEE Trans. Electrol. Comput.* **11** (1962), 324–335.

[Brzozowski, Culik, and Gabrielian 1971]   J. A. Brzozowski, K. Culik II, and
A. Gabrielian. Classification of noncounting events. *J. Comput. System Sci.* **5** (1971),
41–53.

[Büchi 1960]   J. R. Büchi. Weak secord-order arithmetic and finite automata. *Z. Math.
Logik Grundlagen Math.* **6** (1960), 66–92. Reprinted in S. Mac Lane and D. Siefkes,
editors, *The Collected Works of J. Richard Büchi*, Springer-Verlag, 1990,
pp. 398–424.

[Buck and Robbins 1995]   M. W. Buck and D. P. Robbins. The continued fraction
expansion of an algebraic power series satisfying a quartic equation. *J. Number
Theory* **50** (1995), 335–344.

[Bullett and Sentenac 1994]   S. Bullett and P. Sentenac. Ordered orbits of the shift,
square roots, and the devil's staircase. *Proc. Cambridge Phil. Soc.* **115** (1994),
451–481.

[Bundschuh 1980]   P. Bundschuh. Über eine Klasse reeler transzendenter Zahlen mit
explizit angebbarer g-adischer und Kettenbruch-Entwicklung. *J. Reine Angew. Math.*
318 (1980), 110–119.

[Buneman and Levy 1980]   P. Buneman and L. Levy. The towers of Hanoi problem.
Inform. Process. Lett. **10** (1980), 243–244.

[Burdík, Frougny, Gazeau, and Krejcar 2000]   C. Burdík, C. Frougny, J.-P. Gazeau, and
R. Krejcar. Beta-integers as a group. In J.-M. Gambaudo, P. Hubert, P. Tisseur, and
S. Vaienti, editors, *Dynamical Systems: From Crystal to Chaos*, pp. 125–136. World
Scientific, 2000.

[Burnside 1902]   W. Burnside. On an unsettled question in the theory of discontinuous
groups. *Quart. J. Pure Appl. Math.* **33** (1902), 230–238.

[Burris and Nelson 1971]   S. Burris and E. Nelson. Embedding the dual of Π_∞ in the
lattice of equational classes of semigroups. *Algebra Universalis* **1** (1971), 248–
253.

[Bush 1940]   L. E. Bush. An asymptotic formula for the average sum of the digits of
integers. *Amer. Math. Monthly* **47** (1940), 154–156.

[Bush 1955]   L. E. Bush. The William Lowell Putnam mathematical competition. *Amer.
Math. Monthly* **62** (1955), 558–564. See problem 5, p. 561, and solution on p. 564.

[Byrnes 1977]   J. S. Byrnes. On polynomials with coefficients of modulus one. *Bull.
Lond. Math. Soc.* **9** (1977), 171–176.

[Cai 1996a]   T. Cai. On 2-Niven and 3-Niven numbers. *Fibonacci Quart.* **34** (1996),
118–120.

[Cai 1996b]   T. Cai. On k-self-numbers and universal generated numbers. *Fibonacci
Quart.* **34** (1996), 144–146.

[Calkin and Wilf 2000]   N. Calkin and H. S. Wilf. Recounting the rationals. *Amer. Math.
Monthly* **107** (2000), 360–363.

[Capocelli 1989]   R. M. Capocelli. Comments and additions to "Robust transmission of
unbounded strings using Fibonacci representations". *IEEE Trans. Inform. Theory* **35**
(1989), 191–193.

[Carlitz 1968]   L. Carlitz. Fibonacci representations. *Fibonacci Quart.* **6** (1968),
193–220.

[Carlitz, Scoville, and Hoggatt 1972a]   L. Carlitz, R. Scoville, and V. E. Hoggatt, Jr.
Fibonacci representations. *Fibonacci Quart.* **10** (1972), 1–28. Addendum, **10** (1972),
527–530.

[Carlitz, Scoville, and Hoggatt 1972b]   L. Carlitz, R. Scoville, and V. E. Hoggatt, Jr.
Lucas representations. *Fibonacci Quart.* **10** (1972), 29–42.

[Carlson 1921]   F. Carlson. Über Potenzreihen mit ganzzahligen Koeffizienten. *Math.
Zeitschrift* **9** (1921), 1–13.

[Carpi 1983a]   A. Carpi. On the size of a square-free morphism on a three letter alphabet.
Inform. Process. Lett. **16** (1983), 231–235.

[Carpi 1983b]   A. Carpi. A solution to Berstel's problem no. P1. *Bull. European Assoc. Theor. Comput. Sci.*, No. 19 (February 1983), 2–4.

[Carpi 1984]   A. Carpi. On the centers of the set of weakly squarefree words on a two-letter alphabet. *Inform. Process. Lett.* **19** (1984), 187–190.

[Carpi 1993a]   A. Carpi. Overlap-free words and finite automata. *Theoret. Comput. Sci.* **115** (1993), 243–260.

[Carpi 1993b]   A. Carpi. On abelian power-free morphisms. *Int. J. Algebra Comput.* **3** (1993), 151–167.

[Carpi 1993c]   A. Carpi. Repetitions in the Kolakovski [*sic*] sequence. *Bull. European Assoc. Theor. Comput. Sci.*, No. 50 (June 1993), 194–196.

[Carpi 1994]   A. Carpi. On repeated factors in C^∞-words. *Inform. Process. Lett.* **52** (1994), 289–294.

[Carpi 1998]   A. Carpi. On the number of Abelian square-free words on four letters. *Discrete Appl. Math.* **81** (1998), 155–167.

[Carpi 1999]   A. Carpi. On abelian squares and substitutions. *Theoret. Comput. Sci.* **218** (1999), 61–81.

[Carpi and de Luca 1986]   A. Carpi and A. de Luca. Square-free words on partially commutative free monoids. *Inform. Process. Lett.* **22** (1986), 125–131.

[Carpi and de Luca 1990]   A. Carpi and A. de Luca. Non-repetitive words relative to a rewriting system. *Theoret. Comput. Sci.* **72** (1990), 39–53.

[Cassaigne 1993a]   J. Cassaigne. Unavoidable binary patterns. *Acta Informatica* **30** (1993), 385–395.

[Cassaigne 1993b]   J. Cassaigne. Counting overlap-free binary words. In P. Enjalbert, A. Finkel, and K. W. Wagner, editors, *STACS 93, Proc. 10th Symp. Theoretical Aspects of Comp. Sci.*, Vol. 665 of *Lecture Notes in Computer Science*, pp. 216–225. Springer-Verlag, 1993.

[Cassaigne 1996]   J. Cassaigne. Special factors of sequences with linear subword complexity. In J. Dassow, G. Rozenberg, and A. Salomaa, editors, *Developments in Language Theory II*, pp. 25–34. World Scientific, 1996.

[Cassaigne 1997a]   J. Cassaigne. Complexité et facteurs spéciaux. *Bull. Belg. Math. Soc.* **4** (1997), 67–88.

[Cassaigne 1997b]   J. Cassaigne. On a conjecture of J. Shallit. In P. Degano, R. Gorrieri, and A. Marchetti-Spaccamela, editors, *Proc. 24th Int. Conf. on Automata, Languages, and Programming (ICALP)*, Vol. 1256 of *Lecture Notes in Computer Science*, pp. 693–704. Springer-Verlag, 1997.

[Cassaigne 1998]   J. Cassaigne. Sequences with grouped factors. In *DLT '97: Developments in Language Theory III*, pp. 211–222. Aristotle University of Thessaloniki, 1998.

[Cassaigne 1999a]   J. Cassaigne. Limit values of the recurrence quotient of sturmian sequences. *Theoret. Comput. Sci.* **218** (1999), 3–12.

[Cassaigne 1999b]   J. Cassaigne. Double sequences with complexity $mn + 1$. *J. Automata, Languages, and Combinatorics* **4** (1999), 153–170.

[Cassaigne 2000]   J. Cassaigne. Subword complexity and periodicity in two or more dimensions. In G. Rozenberg and W. Thomas, editors, *Developments in Language Theory 1999*, pp. 14–21. World Scientific, 2000.

[Cassaigne, Ferenczi, and Zamboni 2000]   J. Cassaigne, S. Ferenczi, and L. Q. Zamboni. Imbalances in Arnoux–Rauzy sequences. *Ann. Inst. Fourier (Grenoble)* **50** (2000), 1265–1276.

[Cassaigne and Karhumäki 1995a]   J. Cassaigne and J. Karhumäki. Toeplitz words, generalized periodicity and periodically iterated morphisms. Technical Report 36, LITP, Institut Blaise Pascal, 1995.

[Cassaigne and Karhumäki 1995b]   J. Cassaigne and J. Karhumäki. Toeplitz words, generalized periodicity and periodically iterated morphisms. In D.-Z. Du and M. Li, editors, *Computing and Combinatorics, First Annual Conference (COCOON '95)*, Vol. 959 of *Lecture Notes in Computer Science*, pp. 244–253. Springer-Verlag, 1995.

[Cassels 1957]   J. W. S. Cassels. *An Introduction to Diophantine Approximation.* Cambridge University Press, 1957.

[Castelli, Mignosi, and Restivo 1999]   M. G. Castelli, F. Mignosi, and A. Restivo. Fine and Wilf's theorem for three periods and a generalization of Sturmian words. *Theoret. Comput. Sci.* **218** (1999), 83–94.

[Castle and Pitteway 1985]   C. M. A. Castle and M. L. V. Pitteway. An application of Euclid's algorithm to drawing straight lines. In R. A. Earnshaw, editor, *Fundamental Algorithms for Computer Graphics*, Vol. F17 of *NATO ASI Series*, pp. 135–139. Springer-Verlag, 1985.

[Cateland 1992]   E. Cateland. Suites digitales et suites k-régulières. PhD thesis, Université Bordeaux I, 1992.

[Cauchy 1840]   A. Cauchy. Sur les moyens d'éviter les erreurs dans les calculs numériques. *C. R. Acad. Sci. Paris* **11** (1840), 789–798.

[Černý 1984]   A. Černý. On generalized words of Thue-Morse. In M. P. Chytil and V. Koubek, editors, *Proc. 11th Symposium, Mathematical Foundations of Computer Science 1984*, Vol. 176 of *Lecture Notes in Computer Science*, pp. 232–239. Springer-Verlag, 1984.

[Černý 1986]   A. Černý. On generalized words of Thue-Morse. *Acta Math. Univ. Comenianae* **48–49** (1986), 299–309.

[Černý and Gruska 1986a]   A. Černý and J. Gruska. Modular trellises. In G. Rozenberg and A. Salomaa, editors, *The Book of L*, pp. 45–61. Springer-Verlag, 1986.

[Černý and Gruska 1986b]   A. Černý and J. Gruska. Modular real-time trellis automata. *Fund. Inform.* **9** (1986), 253–282.

[Chakrabarti, Karmakar, and Moitra 1995]   A. Chakrabarti, S. N. Karmakar, and R. K. Moitra. Role of a new type of correlated disorder in extended electronic states in the Thue-Morse lattice. *Phys. Rev. Lett.* **74** (1995), 1403–1406.

[Champernowne 1933]   D. G. Champernowne. The construction of decimals normal in the scale of ten. *J. London Math. Soc.* (1933), 254–260.

[Chan 1989]   T.-H. Chan. A statistical analysis of the Towers of Hanoi problem. *Internat. J. Comput. Math.* **28** (1989), 57–65.

[Chang and Tsai 2000]   K.-N. Chang and S.-C. Tsai. Exact solution of a minimal recurrence. *Inform. Process. Lett.* **75** (2000), 61–64.

[Chang and Shyr 1995]   R. K. Chang and H. J. Shyr. Global and coglobal languages. *Algebra Colloq.* **2** (1995), 11–22.

[Chekhova 2000]   N. Chekhova. Covering numbers of rotations. *Theoret. Comput. Sci.* **230** (2000), 97–116.

[Chekhova, Hubert, and Messaoudi 2001]   N. Chekhova, P. Hubert, and A. Messaoudi. Propriétés combinatoires, ergodiques et arithmétiques de la substitution de Tribonacci. *J. Théorie Nombres Bordeaux* **13** (2001), 371–394.

[Chen and Ruzsa 1999]   Y.-G. Chen and I. Z. Ruzsa. On the irrationality of certain series. *Period. Math. Hung.* **38** (1999), 31–37.

[Cheo and Yien 1955]   P.-H. Cheo and S.-C. Yien. A problem on the k-adic representation of positive integers. *Acta Math. Sinica* **5** (1955), 433–438.

[Chérif and de Mathan 1993]   H. Chérif and B. de Mathan. Irrationality measures of Carlitz zeta values in characteristic p. *J. Number Theory* **44** (1993), 260–272.

[Choffrut 1992]   C. Choffrut. Iterated substitutions and locally catenative systems: a decidability result in the binary case. In G. Rozenberg and A. Salomaa, editors, *Lindenmayer Systems: Impacts on Theoretical Computer Science, Computer Graphics, and Developmental Biology*, pp. 49–92. Springer-Verlag, 1992.

[Choffrut and Karhumäki 1997]   C. Choffrut and J. Karhumäki. Combinatorics of words. In G. Rozenberg and A. Salomaa, editors, *Handbook of Formal Languages*, Vol. 1, pp. 329–438. Springer-Verlag, 1997.

[Chomsky 1956]   N. Chomsky. Three models for the description of language. *IRE Trans. Inform. Theory* **2** (1956), 113–124.

[Chomsky 1959]   N. Chomsky. On certain formal properties of grammars. *Inform. Comput.* **2** (1959), 137–167.

[Chomsky 1962]   N. Chomsky. Context-free grammars and pushdown storage. In Quarterly Progress Report No. 65, pp. 187–194. MIT Research Lab. Electron, Cambridge, Mass., 1962.

[Chomsky and Schützenberger 1963]   N. Chomsky and M. P. Schützenberger. The algebraic theory of context-free languages. In P. Braffort and D. Hirschberg, editors, *Computer Programming and Formal Systems*, pp. 118–161. North Holland, Amsterdam, 1963.

[Choquet 1980]   G. Choquet. Répartition des nombres $k(3/2)^n$; et ensembles associés. *C. R. Acad. Sci. Paris* **290** (1980), A575–A580.

[Christoffel 1875]   E. B. Christoffel. Observatio arithmetica. *Annali di Matematica* **6** (1875), 145–152.

[Christoffel 1888]   E. B. Christoffel. Lehrsätze über arithmetische Eigenschaften der Irrationalzahlen. *Annali di Matematica Pura ed Applicata, Series II* **15** (1888), 253–276.

[Christol 1970]   G. Christol. Sur une opération analogue à l'opération de Cartier en caractéristique nulle. *C. R. Acad. Sci. Paris* **271** (1970), 1–3.

[Christol 1972]   G. Christol. Opération de Cartier et vecteurs de Witt. *Séminaire Delange-Pisot-Poitou* **12** (1972), 13.1–13.7.

[Christol 1979]   G. Christol. Ensembles presque périodiques k-reconnaissables. *Theoret. Comput. Sci.* **9** (1979), 141–145.

[Christol 1986]   G. Christol. Fonctions et éléments algébriques. *Pacific J. Math.* **125** (1986), 1–37.

[Christol, Kamae, Mendès France, and Rauzy 1980]   G. Christol, T. Kamae, M. Mendès France, and G. Rauzy. Suites algébriques, automates et substitutions. *Bull. Soc. Math. France* **108** (1980), 401–419.

[D. Chu and Town 1978]   D. D. Chu and H.-S. Town. Another proof on a theorem of Lyndon and Schützenberger in a free monoid. *Soochow J. Math.* **4** (1978), 143–146.

[I. Chu and Johnsonbaugh 1991]   I.-P. Chu and R. Johnsonbaugh. The four-peg tower of Hanoi puzzle. *SIGCSE Bull.* **23**(3) (1991), 2–4.

[Chuan 1992]   W.-F. Chuan. Fibonacci words. *Fibonacci Quart.* **30** (1992), 68–76.

[Chuan 1993a]   W.-F. Chuan. Embedding Fibonacci words into Fibonacci word patterns. In G. E. Bergum, A. N. Philippou, and A. F. Horadam, editors, *Applications of Fibonacci Numbers, Vol. 5*, pp. 113–122. Kluwer, 1993.

[Chuan 1993b]   W.-F. Chuan. Symmetric Fibonacci words. *Fibonacci Quart.* **31** (1993), 251–255.

[Chuan 1995a]   W.-F. Chuan. Extraction property of the golden sequence. *Fibonacci Quart.* **33** (1995), 113–122.

[Chuan 1995b]   W.-F. Chuan. Generating Fibonacci words. *Fibonacci Quart.* **33** (1995), 104–112.

[Chuan 1999]   W.-F. Chuan. Sturmian morphisms and α-words. *Theoret. Comput. Sci.* **225** (1999), 129–148.

[Chung and Graham 1976]   F. R. K. Chung and R. L. Graham. On the set of distances determined by the union of arithmetic progressions. *Ars Combin.* **1** (1976), 57–76.

[Chvátal 1994]   V. Chvátal. Notes on the Kolakoski sequence. Technical Report 93-84, DIMACS, March 1994. Revised.

[Chytil 1986]   M. P. Chytil. Almost context-free languages. *Fund. Inform.* **9** (1986), 283–322.

[Clark and Liang 1973]   W. E. Clark and J. J. Liang. On arithmetic weight for a general radix representation of integers. *IEEE Trans. Inform. Theory* **19** (1973), 823–826.

[Clements and Lindström 1965]   G. F. Clements and B. Lindström. A sequence of (±1)-determinants with large values. *Proc. Amer. Math. Soc.* **16** (1965), 548–550.

[Cobham 1968a]   A. Cobham. A proof of transcendence based on functional equations. Technical Report RC-2041, IBM Yorktown Heights, March 25 1968.

[Cobham 1968b]   A. Cobham. On the Hartmanis-Stearns problem for a class of tag machines. In *IEEE Conference Record of 1968 Ninth Annual Symposium on Switching and Automata Theory*, pp. 51–60, 1968. Also appeared as IBM Research Technical Report RC-2178, August 23 1968.

[Cobham 1969]   A. Cobham. On the base-dependence of sets of numbers recognizable by finite automata. *Math. Systems Theory* **3** (1969), 186–192.

[Cobham 1972]   A. Cobham. Uniform tag sequences. *Math. Systems Theory* **6** (1972), 164–192.

[Cohn 1974]   H. Cohn. Some direct limits of primitive homotopy words and of Markoff geodesics. In L. Greenberg, editor, *Discontinuous Groups and Riemann Surfaces*, Vol. 79 of *Annals of Mathematics Studies*, pp. 81–98. Princeton University Press, 1974.

[Coifman, Geshwind, and Meyer 2001]   R. Coifman, F. Geshwind, and Y. Meyer. Noiselets. *Appl. Comput. Harmon. Anal.* **10** (2001), 27–44.

[Collet and Eckmann 1980]   P. Collet and J.-P. Eckmann. *Iterated Maps on the Interval as Dynamical Systems*, Vol. 1 of *Progress in Physics*. Birkhäuser, 1980.

[Colson 1726]   J. Colson. A short account of negativo-affirmative arithmetick. *Phil. Trans.* **34**(396) (1726), 161–173.

[Compton 1999]   K. J. Compton. A van der Waerden variant. *Electronic J. Combinatorics* **6** (1999), #R22 (electronic), http://www.combinatorics.org/Volume_6/Abstracts/v6i1r22.html

[Condon, Hellerstein, Pottle, and Wigderson 1994]   A. Condon, L. Hellerstein, S. Pottle, and A. Wigderson. On the power of finite automata with both nondeterministic and probabilistic states. In *Proc. 26th Annual ACM Symp. Theor. Comput. (STOC)*, pp. 676–685, 1994.

[Condon, Hellerstein, Pottle, and Wigderson 1998]   A. Condon, L. Hellerstein, S. Pottle, and A. Wigderson. On the power of finite automata with both nondeterministic and probabilistic states. *SIAM J. Comput.* **27** (1998), 739–762.

[Connell 1959]   I. G. Connell. A generalization of Wythoff's game. *Canad. Math. Bull.* **2** (1959), 181–190.

[Conway 1976]   J. H. Conway. *On Numbers and Games*. Academic Press, 1976.

[Cooper and Kennedy 1985]   C. N. Cooper and R. E. Kennedy. On an asymptotic formula for the Niven numbers. *Internat. J. Math. & Math. Sci.* **8** (1985) 537–543.

[Cooper and Kennedy 1988]   C. N. Cooper and R. E. Kennedy. A partial asymptotic formula for the Niven numbers. *Fibonacci Quart.* **26** (1988), 163–168.

[Cooper and Kennedy 1989]   C. N. Cooper and R. E. Kennedy. Chebyshev's inequality and natural density. *Amer. Math. Monthly* **96** (1989), 118–124.

[Cooper and Kennedy 1993]   C. N. Cooper and R. E. Kennedy. On consecutive Niven numbers. *Fibonacci Quart.* **31** (1993), 146–151.

[Cooper and Kennedy 1997]   C. N. Cooper and R. E. Kennedy. On the set of positive integers which are relatively prime to their digital sum and its complement. *J. Inst. Math & Comp. Sci. (Math. Ser.)* **10** (1997), 173–180.

[Copeland and Erdős 1946]   A. H. Copeland and P. Erdős. Note on normal numbers. *Bull. Amer. Math. Soc.* **52** (1946), 857–860.

[Coquet 1979]   J. Coquet. Sur la mesure spectrale des suites q-multiplicatives. *Ann. Inst. Fourier (Grenoble)* **29** (1979), 163–170.

[Coquet 1980]   J. Coquet. Sur certaines suites uniformément équiréparties modulo 1. *Acta Arith.* **36** (1980), 157–162.

[Coquet 1983]   J. Coquet. A summation formula related to the binary digits. *Inventiones Math.* **73** (1983), 107–115.

[Coquet 1986]   J. Coquet. Power sums of digital sums. *J. Number Theory* **22** (1986), 161–176.

[Coquet and van den Bosch 1986]   J. Coquet and P. van den Bosch. A summation formula involving Fibonacci digits. *J. Number Theory* **22** (1986), 139–146.

[Coquet, Kamae, and Mendès France 1977]   J. Coquet, T. Kamae, and M. Mendès France. Sur la mesure spectrale de certaines suites arithmétiques. *Bull. Soc. Math. France* **105** (1977), 369–384.

[Coquet and Mendès France 1977]   J. Coquet and M. Mendès France. Suites à spectre vide et suites pseudo-aléatoires. *Acta Arith.* **32** (1977), 99–106.

[Coquet and Toffin 1981]   J. Coquet and P. Toffin. Représentations des entiers naturels et indépendance statistique. *Bull. Sci. Math.* **105** (1981), 289–298.

[Cori and Formisano 1990]   R. Cori and M. R. Formisano. Partially abelian squarefree words. *RAIRO Inform. Théor. Appl.* **24** (1990), 509–520.

[Cori and Formisano 1991]   R. Cori and M. R. Formisano. On the number of partially abelian square-free words on a three-letter alphabet. *Theoret. Comput. Sci.* **81** (1991), 147–153.

[Coster 1988]   M. Coster. Congruence properties of coefficients of certain algebraic power series. *Compositio Math.* **68** (1988), 41–57.

[Coven 1975]   E. M. Coven. Sequences with minimal block growth II. *Math. Systems Theory* **8** (1975), 376–382.

[Coven and Hedlund 1973]   E. M. Coven and G. A. Hedlund. Sequences with minimal block growth. *Math. Systems Theory* **7** (1973), 138–153.

[Coxeter 1953]   H. S. M. Coxeter. The golden section, phyllotaxis, and Wythoff's game. *Scripta Math.* **19** (1953), 135–143.

[Crisp, Moran, Pollington, and Shiue 1993]   D. Crisp, W. Moran, A. Pollington, and P. Shiue. Substitution invariant cutting sequences. *J. Théorie Nombres Bordeaux* **5** (1993), 123–137.

[Crochemore 1981]   M. Crochemore. An optimal algorithm for computing the repetitions in a word. *Inform. Process. Lett.* **12** (1981), 244–250.

[Crochemore 1982a]   M. Crochemore. Sharp characterizations of squarefree morphisms. *Theoret. Comput. Sci.* **18** (1982), 221–226.

[Crochemore 1982b]   M. Crochemore. A solution to Berstel's problem no. P3. *Bull. European Assoc. Theor. Comput. Sci.*, No. 18 (October 1982), 9–11.

[Crochemore 1983a]   M. Crochemore. Recherche linéaire d'un carré dans un mot. *C. R. Acad. Sci. Paris* **296** (1983), 781–784.

[Crochemore 1983b]   M. Crochemore. Tests sur les morphismes faiblement sans carré. In L. J. Cummings, editor, *Combinatorics on Words: Progress and Perspectives*, pp. 63–89. Academic Press, 1983.

[Crochemore 1983c]   M. Crochemore. A solution to P12, number 18, October 1982. *Bull. European Assoc. Theor. Comput. Sci.*, No. 19 (February 1983), 15–16.

[Crochemore 1984]   M. Crochemore. Linear searching for a square in a word. *Bull. European Assoc. Theor. Comput. Sci.*, No. 24 (1984), 66–72.

[Crochemore 1986]   M. Crochemore. Transducers and repetitions. *Theoret. Comput. Sci.* **45** (1986), 63–86.

[Crochemore and Goralcik 1991]   M. Crochemore and P. Goralcik. Mutually avoiding ternary words of small exponent. *Int. J. Algebra Comput.* **1** (1991), 407–410.

[Crochemore, Le Rest, and Wender 1983]   M. Crochemore, M. Le Rest, and P. Wender. An optimal test on finite unavoidable sets of words. *Inform. Process. Lett.* **16** (1983), 179–180.

[Crochemore and Rytter 1991]   M. Crochemore and W. Rytter. Efficient parallel algorithms to test square-freeness and factorize strings. *Inform. Process. Lett.* **38** (1991), 57–60.

[Crowe 1956]   D. W. Crowe. The *n*-dimensional cube and the towers of Hanoi. *Amer. Math. Monthly* **63** (1956), 29–30.

[Cucker and Gabarró 1989]   F. Cucker and J. Gabarró. Nonrecursive functions have transcendental generating series. *RAIRO Inform. Théor. App.* **23** (1989), 445–448.

[Culik and Harju 1984]   K. Culik II and T. Harju. The ω-sequence equivalence problem for D0L systems is decidable. *J. Assoc. Comput. Mach.* **31** (1984), 282–298.

[Culik and Karhumäki 1992]   K. Culik II and J. Karhumäki. Iterative devices generating infinite words. In *STACS 92, Proc. 9th Symp. Theoretical Aspects of Comp. Sci.*, Vol. 577 of *Lecture Notes in Computer Science*, pp. 531–543. Springer-Verlag, 1992.

[Culik and Karhumäki 1994b]   K. Culik II and J. Karhumäki. Iterative devices generating infinite words. *Internat. J. Found. Comp. Sci.* **5** (1994), 69–97.

[Culik, Karhumäki, and Lepistö 1992]   K. Culik II, J. Karhumäki, and A. Lepistö. Alternating iteration of morphisms and the Kolakovski [*sic*] sequence. In G. Rozenberg and A. Salomaa, editors, *Lindenmayer Systems*, pp. 93–103. Springer-Verlag, 1992.

[Cull and Ecklund 1982]   P. Cull and E. F. Ecklund, Jr. On the Towers of Hanoi and generalized Towers of Hanoi problems. *Congr. Numer.* **35** (1982), 229–238.

[Cull and Ecklund 1985]   P. Cull and E. F. Ecklund, Jr. Towers of Hanoi and analysis of algorithms. *Amer. Math. Monthly* **92** (1985), 407–420.

[Cull and Gerety 1985]   P. Cull and C. Gerety. Is Towers of Hanoi really hard? *Congr. Numer.* **47** (1985), 237–242.

[Cummings 1978]   L. J. Cummings. On the construction of Thue sequences. In *Proc. 9th Southeast Conf. Combinat. Graph Theory Comput.*, pp. 235–242, 1978. (= *Congr. Numer.* **21**).

[Cummings 1981]   L. J. Cummings. Overlapping substrings and Thue's problem. In *Proc. 3rd Caribbean Conf. Combinat. and Comput.*, pp. 99–109, 1981.

[Cummings 1983]   L. J. Cummings. Strongly *q*th power-free strings. *Ann. Discrete Math.* **17** (1983), 247–252.

[Cummings 1993]   L. J. Cummings. Gray codes and strongly square-free strings. In R. Capocelli, A. de Santis, and U. Vaccaro, editors, *Sequences II: Methods in Communication, Security, and Computer Science*, pp. 439–446. Springer-Verlag, 1993.

[Cummings 1996]   L. J. Cummings. Strongly square-free strings on three letters. *Australasian J. Combin.* **14** (1996), 259–266.

[Cummings and Mays 2001]   L. J. Cummings and M. Mays. A one-sided Zimin construction. *European J. Combinatorics* **8**(1) (2001), R27 (electronic), http://www.combinatorics.org/Volume_8/Abstracts/v8i1r27.html

[Cummings, Moore, and Karhumäki 1996]   L. J. Cummings, D. Moore, and J. Karhumäki. Borders of Fibonacci strings. *J. Combin. Math. Combin. Comput.* **20** (1996), 81–87.

[Cummings and Smyth 1997]   L. J. Cummings and W. F. Smyth. Weak repetitions in strings. *J. Combin. Math. Combin. Comput.* **24** (1997), 33–48.

[Currie 1984]   J. D. Currie. A direct proof of a result of Thue. *Utilitas Math.* **25** (1984), 299–302.

[Currie 1993]   J. D. Currie. Open problems in pattern avoidance. *Amer. Math. Monthly* **100** (1993), 790–793.

[Currie 1995a]   J. D. Currie. On the structure and extendibility of k-power free words. *Eur. J. Combin.* **16** (1995), 111–124.

[Currie 1995b]   J. D. Currie. A note on antichains of words. *Electronic J. Combin.* **2** (1995), #R21 (electronic), http://www.combinatorics.org/Volume_2/volume2.html#R21

[Currie 1996]   J. D. Currie. Non-repetitive words: ages and essences. *Combinatorica* **16** (1996), 19–40.

[Currie 2001]   J. D. Currie. No iterated morphism generates any Arshon sequence of odd order. Unpublished manuscript, 2001.

[Currie and Bendor-Samuel 1992]   J. Currie and A. Bendor-Samuel. Words without near-repetitions. *Canad. Math. Bull.* **35** (1992), 161–166.

[Currie and Shelton 1996a]   J. D. Currie and R. O. Shelton. Cantor sets and Dejean's conjecture. *J. Automata Languages Combin.* **1** (1996), 113–127.

[Currie and Shelton 1996b]   J. D. Currie and R. O. Shelton. Cantor sets and Dejean's conjecture. In J. Dassow, G. Rozenberg, and A. Salomaa, editors, *Developments in Language Theory II*, pp. 35–43. World Scientific, 1996.

[Currie and Simpson 2002]   J. D. Currie and J. Simpson. Non-repetitive tilings. *Elect. J. Combinatorics* **9** (2002), #R28, (electronic), http://www.combinatorics.org/Volume_9/Abstracts/v9i1n10.html

[Damamme and Hellegouarch 1988]   G. Damamme and Y. Hellegouarch. Propriétés de transcendance des valeurs de la fonction zéta de Carlitz. *C. R. Acad. Sci. Paris* **307** (1988), 635–637.

[Damamme and Hellegouarch 1991]   G. Damamme and Y. Hellegouarch. Transcendence of the values of the Carlitz zeta function by Wade's method. *J. Number Theory* **39** (1991), 257–278.

[Damanik 2000]   D. Damanik. Local symmetries in the period-doubling sequence. *Discrete Appl. Math.* **100** (2000), 115–121.

[Damanik and Lenz 2002]   D. Damanik and D. Lenz. The index of Sturmian sequences. *Eur. J. Combin.* **23** (2002), 23–29.

[Damanik and Zare 2000]   D. Damanik and D. Zare. Palindrome complexity bounds for primitive substitution sequences. *Discrete Math.* **222** (2000), 259–267.

[Danilov 1972]   L. V. Danilov. Some classes of transcendental numbers. *Mat. Zametki* **12** (1972), 149–154. In Russian. English translation in *Math. Notes Acad. Sci. USSR* **12** (1972), 524–527.

[Dartyge and Mauduit 2000]   C. Dartyge and C. Mauduit. Nombres presque premiers dont l'écriture en base r ne comporte pas certains chiffres. *J. Number Theory* **81** (2000), 270–291.

[Davies 1959]   R. O. Davies. On Langford's problem (II). *Math. Gazette* **43** (1959), 253–255.

[Davio, Deschamps, and Gossart 1978]   M. Davio, J. P. Deschamps, and C. Gossart. Complex arithmetic. Technical Report R369, MBLE Research Laboratory, Brussels, Belgium, May 1978.

[Davis and Knuth 1970]   C. Davis and D. E. Knuth. Number representations and dragon curves – I, II. *J. Recreational Math.* **3** (1970), 66–81, 133–149.

[Davis and Weyuker 1983]   M. D. Davis and E. J. Weyuker. *Computability, Complexity, and Languages: Fundamentals of Theoretical Computer Science*. Academic Press, 1983.

[Davison 1977]   J. L. Davison. A series and its associated continued fraction. *Proc. Amer. Math. Soc.* **63** (1977), 29–32.

[Daykin 1960]   D. E. Daykin. Representation of natural numbers as sums of generalised Fibonacci numbers. *J. London Math. Soc.* **35** (1960), 143–161.

[De Felice 1983]   C. De Felice. An answer to Berstel's problem n. P2. *Bull. European Assoc. Theor. Comput. Sci.*, No. 19, (February 1983), 5–7.

[Dean 1965]   R. A. Dean. A sequence without repeats on x, x^{-1}, y, y^{-1}. *Amer. Math. Monthly* **72** (1965), 383–385. checked.

[Dejean 1972]   F. Dejean. Sur un théorème de Thue. *J. Combin. Theory Ser. A* **13** (1972), 90–99.

[Dekking 1976]   F. M. Dekking. On repetitions of blocks in binary sequences. *J. Combin. Theory Ser. A* **20** (1976), 292–299.

[Dekking 1977]   F. M. Dekking. Transcendance du nombre de Thue–Morse. *C. R. Acad. Sci. Paris* **285** (1977), 157–160.

[Dekking 1979a]   F. M. Dekking. Regularity and irregularity of sequences generated by automata. In *Séminaire de Théorie des Nombres de Bordeaux*, pp. 9.01–9.10, 1979–1980.

[Dekking 1979b]   F. M. Dekking. Strongly non-repetitive sequences and progression-free sets. *J. Combin. Theory Ser. A* **27** (1979), 181–185.

[Dekking 1992]   F. M. Dekking. On the Thue-Morse measure. *Acta Univ. Carolinae – Math. Phys.* **33** (1992), 35–40.

[Dekking 1994]   F. M. Dekking. Iteration of maps by an automaton. *Discrete Math.* **126** (1994), 81–86.

[Dekking 1997]   F. M. Dekking. What is the long range order in the Kolakoski sequence? In R. V. Moody, editor, *The Mathematics of Long-Range Aperiodic Order*, Vol. 489 of *NATO ASI Ser., Ser. C., Math. Phys. Sci.*, pp. 115–125. Kluwer, 1997.

[Dekking, Mendès France, and van der Poorten 1982]   F. M. Dekking, M. Mendès France, and A. J. van der Poorten. Folds! *Math. Intelligencer* **4** (1982), 130–138, 173–181, 190–195. Erratum, **5** (1983), 5.

[Delange 1972]   H. Delange. Sur les fonctions q-additives ou q-multiplicatives. *Acta Arith.* **21** (1972), 285–298.

[Delange 1975]   H. Delange. Sur la fonction sommatoire de la fonction "somme des chiffres". *Enseign. Math.* **21** (1975), 31–47.

[Deligne 1984]   P. Deligne. Intégration sur un cycle évanescent. *Inventiones Math.* **76** (1984), 129–143.

[Delyon and Peyrière 1991]   F. Delyon and J. Peyrière. Recurrence of the eigenstates of a Schrödinger operator with automatic potential. *J. Statist. Phys.* **64** (1991), 363–368.

[Denef and Lipshitz 1987]   J. Denef and L. Lipshitz. Algebraic power series and diagonals. *J. Number Theory* **26** (1987), 46–67.

[Deshouillers 1979]   J.-M. Deshouillers. La répartition modulo 1 des puissances de rationnels dans l'anneau des séries formelles sur un corps fini. In *Séminaire de Théorie des Nombres de Bordeaux*, pp. 5.01–5.22, 1979–1980.

[Devroye and Goudjil 1998]   L. Devroye and A. Goudjil. A study of random Weyl trees. *Random Structures and Algorithms* **12** (1998), 271–295.

[Deych, Zaslavsky, and Lisyansky 1997]   L. I. Deych, D. Zaslavsky, and A. A. Lisyansky. Wave localization in generalized Thue-Morse superlattices with disorder. *Phys. Rev. E* **56** (1997), 4780–4790.

[Diamond 1989]   H. G. Diamond. Elementary problem E 3353. *Amer. Math. Monthly* **96** (1989), 838. Solution, *Amer. Math. Monthly* **98** (1991), 271–272.

[Didier 1997]   G. Didier. Échanges de trois d'intervalles et suites sturmiennes. *J. Théorie Nombres Bordeaux* **9** (1997), 463–478.

[Didier 1998a]   G. Didier. Combinatoire des codages des rotations. *Acta Arith.* **85** (1998), 157–177.

[Didier 1998b]   G. Didier. Codage de rotations et fractions continues. *J. Number Theory* **71** (1998), 275–306.

[Didier 1999]   G. Didier. Caractérisation des N-écritures et application á l'étude des suites de complexité ultimement $n + c^{\text{ste}}$. *Theoret. Comput. Sci.* **215** (1999), 31–49.

[Dietmeyer 1963]   D. L. Dietmeyer. Conversion from positive to negative and imaginary radix. *IEEE Trans. Elect. Comput.* **12** (1963), 20–22.

[Dilcher 1993]   K. Dilcher. On a class of iterative recurrence relations. In G. E. Bergum, A. N. Philippou, and A. F. Horadam, editors, *Applications of Fibonacci Numbers, Vol. 5*, pp. 143–158. Kluwer, 1993.

[Diwan 1986]   A. A. Diwan. A new combinatorial complexity measure for languages. Technical report, Computer Science Group, Tata Institute, Bombay, 1986.

[Doche and Mendès France 2000]   C. Doche and M. Mendès France. Integral geometry and real zeros of Thue-Morse polynomials. *Experimental Math.* **9** (2000), 339–350.

[Dombi and Valkó 1997]   G. Dombi and B. Valkó. On a problem of Erdős. *Acta Math. Hung.* **77** (1997), 47–56.

[Dömösi, Hauschildt, Horváth, and Kudlek 1999]   P. Dömösi, D. Hauschildt, G. Horváth, and M. Kudlek. Some results on small context-free grammars generating primitive words. *Publ. Math. (Debrecen)* **54** (1999), 667–686. Supplement.

[Dömösi, Horváth, and Ito 1993]   P. Dömösi, S. Horváth, and M. Ito. Formal languages and primitive words. *Publ. Math. (Debrecen)* **42** (1993), 315–321.

[Dömösi, Horváth, Ito, Kászonyi, and Katsura 1993]   P. Dömösi, S. Horváth, M. Ito, L. Kászonyi, and M. Katsura. Formal languages consisting of primitive words. In Z. Ésik, editor, *Fundamentals of Computation Theory: FCT '93*, Vol. 710 of *Lecture Notes in Computer Science*, pp. 194–203. Springer-Verlag, 1993.

[Dömösi, Horváth, Ito, Kászonyi, and Katsura 1994a]   P. Dömösi, S. Horváth, M. Ito, L. Kászonyi, and M. Katsura. Some results on primitive words. In *Proc. Conf. on Semigroups, Automata, and Languages*, pp. 33–35. 1994.

[Dömösi, Horváth, Ito, Kászonyi, and Katsura 1994b]   P. Dömösi, S. Horváth, M. Ito, L. Kászonyi, and M. Katsura. Some combinatorial properties of words, and the Chomsky-hierarchy. In M. Ito and H. Jürgensen, editors, *Words, Languages, and Combinatorics II*, pp. 105–123. World Scientific, 1994.

[Dorst and Smeulders 1984]   L. Dorst and A. W. M. Smeulders. Discrete representation of straight lines. *IEEE Trans. Pattern Analysis & Machine Intell.* **PAMI-6** (1984), 450–463.

[Downey and Griswold 1984]   P. J. Downey and R. E. Griswold. On a family of nested recurrences. *Fibonacci Quart.* **22** (1984), 310–317.

[Drazin and Griffith 1952]   M. P. Drazin and J. S. Griffith. On the decimal representation of integers. *Proc. Cambridge Phil. Soc.* **48** (1952), 555–565.

[Dringó and Kátai 1981]   L. Dringó and I. Kátai. Some remarks concerning the sum of digits of integers. *Acta Math. Acad. Sci. Hung.* **37** (1981), 165–172.

[Drmota and Larcher 2001]   M. Drmota and G. Larcher. The sum-of-digits function and uniform distribution modulo 1. *J. Number Theory* **89** (2001), 65–96.

[Drmota and Skalba 1995]   M. Drmota and M. Skalba. Sign-changes of the Thue-Morse fractal function and Dirichlet L-series. *Manuscripta Math.* **86** (1995), 519–541.

[Drmota and Skalba 2000a]   M. Drmota and M. Skalba. The parity of the Zeckendorf sum-of-digits function. *Manuscripta Math.* **101** (2000), 361–383.

[Drmota and Skalba 2000b]   M. Drmota and M. Skalba. Rarified sums of the Thue–Morse sequence. *Trans. Amer. Math. Soc.* **352** (2000), 609–642.

[Droubay 1995]   X. Droubay. Palindromes in the Fibonacci word. *Inform. Process. Lett.* **55** (1995), 217–221.

[Droubay and Pirillo 1999]   X. Droubay and G. Pirillo. Palindromes and Sturmian words. *Theoret. Comput. Sci.* **223** (1999), 73–85.

[Dubuc and Elqortobi 1990]   S. Dubuc and A. Elqortobi. Le maximum de la fonction de Knopp. *INFOR (Information Systems and Operational Research)* **28** (1990), 311–323.

[Dudeney 1908]   H. E. Dudeney. *The Canterbury Puzzles, and Other Curious Problems.* E. P. Dutton, 1908.

[Dudley 1994]   U. Dudley. Smith numbers. *Math. Mag.* **67** (1994), 62–65.

[Dulea, Johansson, and Riklund 1992a]   M. Dulea, M. Johansson, and R. Riklund. Localization of electrons and electromagnetic waves in a deterministic aperiodic system. *Phys. Rev. B* **45** (1992), 105–114.

[Dulea, Johansson, and Riklund 1992b]   M. Dulea, M. Johansson, and R. Riklund. Trace-map invariant and zero-energy states of the tight-binding Rudin-Shapiro model. *Phys. Rev. B* **46** (1992), 3296–3304.

[Dulucq and Gouyou-Beauchamps 1990]   S. Dulucq and D. Gouyou-Beauchamps. Sur les facteurs des suites de Sturm. *Theoret. Comput. Sci.* **71** (1990), 381–400.

[Dumas 1993a]   P. Dumas. Récurrences Mahlériennes, suites automatiques, études asymptotiques. PhD thesis, Université de Bordeaux I, 1993.

[Dumas 1993b]   P. Dumas. Algebraic aspects of *b*-regular series. In A. Lingas, R. Karlsson, and A. Carlsson, editors, *Proc. 20th Int. Conf. on Automata, Languages, and Programming (ICALP)*, Vol. 700 of *Lecture Notes in Computer Science*, pp. 457–468. Springer-Verlag, 1993.

[Dumont 1983]   J.-M. Dumont. Discrépance des progressions arithmétiques dans la suite de Morse. *C. R. Acad. Sci. Paris* **297** (1983), 145–148.

[Dumont and Thomas 1993]   J.-M. Dumont and A. Thomas. Digital sum moments and substitutions. *Acta Arith.* **64** (1993), 205–225.

[Dumont and Thomas 1997]   J. M. Dumont and A. Thomas. Gaussian asymptotic properties of the sum-of-digits function. *J. Number Theory* **62** (1997), 19–38.

[Dupain 1979]   Y. Dupain. Discrépance de la suite $(\{n\alpha\})$, $\alpha = (1 + \sqrt{5})/2$. *Ann. Inst. Fourier (Grenoble)* **29** (1979), 81–106.

[Dupain and Sós 1980]   Y. Dupain and V. T. Sós. On the one-sided boundedness of discrepancy-function of the sequence $\{n\alpha\}$. *Acta Arith.* **37** (1980), 363–374.

[Durand 1998a]   F. Durand. A generalization of Cobham's theorem. *Theory Comput. Systems* **31** (1998), 169–185.

[Durand 1998b]   F. Durand. A characterization of substitutive sequences using return words. *Discrete Math.* **179** (1998), 89–101.

[Durand 1998c]   F. Durand. Sur les ensembles d'entiers reconnaissables. *J. Théorie Nombres Bordeaux* **10** (1998), 65–84.

[Duverney 1996]   D. Duverney. À propos de la série $\sum_{n=1}^{+\infty} x^n/(q^n - 1)$. *J. Théorie Nombres Bordeaux* **8** (1996), 173–181.

[Dwork and Stockmeyer 1989]   C. Dwork and L. Stockmeyer. On the power of 2-way probabilistic finite state automata. In *Proc. 30th Ann. Symp. Found. Comput. Sci.*, pp. 480–485. IEEE Press, 1989.

[Dwork and Stockmeyer 1990]   C. Dwork and L. Stockmeyer. A time complexity gap for two-way probabilistic finite-state automata. *SIAM J. Comput.* **19** (1990), 1011–1023.

[Edlin 1999]   A. E. Edlin. The number of binary cube-free words of length up to 47 and their numerical analysis. *J. Diff. Equations Appl.* **5** (1999), 353–354.

[Edwards and Price 1970]   R. E. Edwards and J. F. Price. A naively constructive approach to boundedness principles, with applications to harmonic analysis. *Enseign. Math.* **16** (1970), 255–296.

[Ehrenfeucht, Lee, and Rozenberg 1975]   A. Ehrenfeucht, K. P. Lee, and G. Rozenberg. Subword complexities of various classes of deterministic developmental languages without interaction. *Theoret. Comput. Sci.* **1** (1975), 59–75.

[Ehrenfeucht, Lee, and Rozenberg 1976]   A. Ehrenfeucht, K. P. Lee, and G. Rozenberg. On the number of subwords of everywhere growing DTOL-languages. *Discrete Math.* **15** (1976), 223–234.

[Ehrenfeucht and Rozenberg 1973]   A. Ehrenfeucht and G. Rozenberg. A limit theorem for sets of subwords in deterministic T0L languages. *Inform. Process. Lett.* **2** (1973), 70–73.

[Ehrenfeucht and Rozenberg 1978]   A. Ehrenfeucht and G. Rozenberg. Simplifications of homomorphisms. *Inform. Control* **38** (1978), 298–309.

[Ehrenfeucht and Rozenberg 1981a]   A. Ehrenfeucht and G. Rozenberg. On the subword complexity and square-freeness of formal languages. In P. Deussen, editor, *Theoretical Computer Science: 5th GI Conference*, Vol. 104 of *Lecture Notes in Computer Science*, pp. 1–4. Springer-Verlag, Berlin, 1981.

[Ehrenfeucht and Rozenberg 1981b]   A. Ehrenfeucht and G. Rozenberg. On the subword complexity of D0L languages with a constant distribution. *Inform. Process. Lett.* **13** (1981), 108–113.

[Ehrenfeucht and Rozenberg 1981c]   A. Ehrenfeucht and G. Rozenberg. On the subword complexity of square-free D0L languages. *Theoret. Comput. Sci.* **16** (1981), 25–32.

[Ehrenfeucht and Rozenberg 1982a]   A. Ehrenfeucht and G. Rozenberg. On the subword complexity of homomorphic images of languages. *RAIRO Inform. Théor. Appl.* **16** (1982), 303–316.

[Ehrenfeucht and Rozenberg 1982b]   A. Ehrenfeucht and G. Rozenberg. Repetitions in homomorphisms and languages. In M. Nielsen and E. M. Schmidt, editors, *Proc. 9th Int. Conf. on Automata, Languages, and Programming (ICALP)*, Vol. 140 of *Lecture Notes in Computer Science*, pp. 192–196. Springer-Verlag, 1982.

[Ehrenfeucht and Rozenberg 1983a]   A. Ehrenfeucht and G. Rozenberg. On the subword complexity of locally catenative D0L-languages. *Inform. Process. Lett.* **16** (1983), 7–9.

[Ehrenfeucht and Rozenberg 1983b]   A. Ehrenfeucht and G. Rozenberg. On the subword complexity of m-free D0L-languages. *Inform. Process. Lett.* **17** (1983), 121–124.

[Ehrenfeucht and Rozenberg 1983c]   A. Ehrenfeucht and G. Rozenberg. On the size of the alphabet and the subword complexity of square-free D0L languages. *Semigroup Forum* **26** (1983), 215–223.

[Ehrenfeucht and Rozenberg 1983d]   A. Ehrenfeucht and G. Rozenberg. On the separating power of EOL systems. *RAIRO Inform. Théor. Appl.* **17** (1983), 13–22.

[Ehrenfeucht and Rozenberg 1983e]   A. Ehrenfeucht and G. Rozenberg. A solution to P12. *Bull. European Assoc. Theor. Comput. Sci.*, No. 19 (February 1983), 16–18.

[Eilenberg 1974]   S. Eilenberg. *Automata, Languages, and Machines*, Vol. A. Academic Press, 1974.

[Ekhad and Zeilberger 1998]   S. B. Ekhad and D. Zeilberger. There are more than $2^{(n/17)}$ n-letter ternary square-free words. *J. Integer Sequences* **1** (1998), 98.1.9 (electronic), http://www.math.uwaterloo.ca/JIS/zeil.html

[Ellis and Steele 1981]   M. H. Ellis and J. M. Steele. Fast sorting of Weyl sequences using comparisons. *SIAM J. Comput.* **10** (1981), 88–95.

[Elser 1983]   V. Elser. Repeat-free sequences. Technical Report LBL-16632, Lawrence Berkeley Laboratories, September 1983.

[Èminyan 1991]   K. M. Èminyan. On the Dirichlet divisor problem in some sequences of natural numbers. *Izv. Akad. Nauk SSSR Ser. Mat* **55** (1991), 680–686. In Russian. English translation in *Math. USSR Izvestiya* **38** (1992), 669–675.

[Èminyan 1994]   K. M. Èminyan. On the representation of numbers with given properties of the binary expansion by sums of two squares. *Trudy. Math. Inst. Steklova* **207** (1994), 377–382. In Russian. English translation in *Proc. Steklov Inst. Math.* **207** (1995), 347–351.

[van Enter and Miękisz 1990]   A. C. D. van Enter and J. Miękisz. Breaking of periodicity at positive temperatures. *Commun. Math. Phys.* **134** (1990), 647–651.

[van Enter and Miękisz 1992]   A. C. D. van Enter and J. Miękisz. How should one define a (weak) crystal? *J. Statist. Phys.* **66** (1992), 1147–1153.

[Entringer, Jackson, and Schatz 1974]   R. C. Entringer, D. E. Jackson, and J. A. Schatz. On nonrepetitive sequences. *J. Combin. Theory Ser. A* **16** (1974), 159–164.

[Epifanio, Koskas, and Mignosi 1999]   C. Epifanio, M. Koskas, and F. Mignosi. On a conjecture on bidimensional words. Technical Report 78, Dipartimento di Matematica e Appl., Università di Palermo, 1999.

[Er 1982]   M. C. Er. A representation approach to the Tower of Hanoi problem. *Computer J.* **25** (1982), 442–447.

[Er 1983a]   M. C. Er. An iterative solution to the generalized Towers of Hanoi problem. *BIT* **23** (1983), 295–302.

[Er 1983b]   M. C. Er. An analysis of the generalized Towers of Hanoi problem. *BIT* **23** (1983), 429–435.

[Er 1984a]   M. C. Er. The generalized colour Towers of Hanoi: an iterative algorithm. *Computer J.* **27** (1984), 278–282.

[Er 1984b]   M. C. Er. An iterative algorithm for the cyclic Towers of Hanoi problem. *Internat. J. Comput. Inform. Sci.* **13** (1984), 123–129.

[Er 1984c]   M. C. Er. A generalization of the cyclic Towers of Hanoi: an iterative solution. *Internat. J. Comput. Math.* **15** (1984), 129–140.

[Er 1984d]   M. C. Er. The cyclic Towers of Hanoi: a representation approach. *Computer J.* **27** (1984), 171–175.

[Er 1985a]   M. C. Er. The complexity of the generalised cyclic Towers of Hanoi problem. *J. Algorithms* **6** (1985), 351–358.

[Er 1985b]   M. C. Er. The Towers of Hanoi and binary numerals. *J. Info. Optim. Sci.* **6** (1985), 147–152.

[Er 1985c]   M. C. Er. Towers of Hanoi with black and white discs. *J. Inform. Optim. Sci.* **6** (1985), 87–94.

[Er 1986a]   M. C. Er. A time and space efficient algorithm for the cyclic Towers of Hanoi problem. *J. Inform. Processing* **9** (1986), 163–165.

[Er 1986b]   M. C. Er. Performance evaluations of recursive and iterative algorithms for the Towers of Hanoi problem. *Computing. Arch. Informatik und Numerik* **37** (1986), 93–102.

[Er 1986c]   M. C. Er. The cyclic Towers of Hanoi and pseudoternary codes. *J. Inform. Optim. Sci.* **7** (1986), 271–277.

[Er 1987a]   M. C. Er. An algorithmic solution to the multi-Tower Hanoi problem. *J. Inform. Optim. Sci.* **8** (1987), 91–100.

[Er 1987b]   M. C. Er. A general algorithm for finding a shortest path between two *n*-configurations. *Inform. Sci.* **42** (1987), 137–141.

[Er 1987c]   M. C. Er. Counter examples to adjudicating a Towers of Hanoi contest. *Int. J. Comput. Math.* **21** (1987), 123–131.

[Er 1989]   M. C. Er. A linear space algorithm for solving the Towers of Hanoi problem by using a virtual disc. *Inform. Sci.* **47** (1989), 47–52.

[Erdős 1948]   P. Erdős. Advanced problem 4321. *Amer. Math. Monthly* **55** (1948), 642. Solution by R. Steinberg, **57** (1950), 347.

[Erdős 1961]   P. Erdős. Some unsolved problems. *Magyar Tud. Akad. Mat. Kutató Int. Közl.* **6** (1961), 221–254.

[Erdős 1962]   P. Erdős. On the representation of large integers as sums of distinct summands taken from a fixed set. *Acta Arith.* **7** (1962), 345–354.

[Erdős 1994]   P. Erdős. Quickie Q814. *Math. Mag.* **67** (1994), 67, 74.

[Euwe 1929]   M. Euwe. Mengentheoretische Betrachtungen über das Schachspiel. *Proc. Konin. Akad. Wetenschappen Amsterdam* **32** (1929), 633–642.

[Evdokimov 1968]   A. A. Evdokimov. Strongly asymmetric sequences generated by a finite number of symbols. *Dokl. Akad. Nauk SSSR* **179** (1968), 1268–1271. In Russian. English translation in *Soviet Math. Dokl.* **9** (1968), 536–539.

[Even 1964]   S. Even. Rational numbers and regular events. *IEEE Trans. Electr. Comput.* **13** (1964), 740–741.

[Evey 1963]   J. Evey. Application of pushdown store machines. In *Proc. 1963 Fall Joint Computer Conference*, pp. 215–227. AFIPS Press, 1963.

[Fabre 1994]   S. Fabre. Une généralisation du théorème de Cobham. *Acta Arith.* **67** (1994), 197–208.

[Fagnot 1996]   I. Fagnot. Langage de Lukasiewicz et diagonales de séries formelles. *J. Théorie Nombres Bordeaux* **8** (1996), 31–45.

[Fagnot 1997]   I. Fagnot. On the subword equivalence problem for morphic words. *Discrete Appl. Math.* **75** (1997), 231–253.

[Fagnot and Vuillon 2002]   I. Fagnot and L. Vuillon. Generalized balances in Sturmian words. *Discrete Appl. Math.* **121** (2002), 83–101.

[Fan and Lin 1995]   L. Fan and Z. Lin. Comment on "Role of a new type of correlated disorder in extended electronic states in the Thue-Morse lattice". *Phys. Rev. Lett.* **75** (1995), 2903.

[Fatou 1904]   P. Fatou. Sur les séries entières à coefficients entiers. *C. R. Acad. Sci. Paris* **138** (1904), 342–344.

[Faucheux 1926]   Faucheux. Sur une question concernant des suites de nombres incommensurables. *Nouvelles Ann. Math.* **1** (1926), 237–239.

[Feigenbaum 1978]   M. J. Feigenbaum. Quantitative universality for a class of nonlinear transformations. *J. Statist. Phys.* **19** (1978), 25–52.

[Ferenczi 1992]   S. Ferenczi. Tiling the Morse sequence. *Theoret. Comput. Sci.* **94** (1992), 215–221.

[Ferenczi 1995]   S. Ferenczi. Les transformations de Chacon: combinatoire, structure géométrique, lien avec les systèmes de complexité $2n + 1$. *Bull. Soc. Math. France* **123** (1995), 271–292.

[Ferenczi 1996]   S. Ferenczi. Rank and symbolic complexity. *Ergod. Theory & Dynam. Syst.* **16** (1996), 663–682.

[Ferenczi 1999]   S. Ferenczi. Complexity of sequences and dynamical systems. *Discrete Math.* **206** (1999), 145–154.

[Ferenczi, Holton, and Zamboni 2001]   S. Ferenczi, C. Holton, and L. Q. Zamboni. Structure of three interval exchange transformations I. An arithmetic study. *Ann. Inst. Fourier (Grenoble)* **51** (2001), 861–901.

[Ferenczi and Kása 1999]   S. Ferenczi and Z. Kása. Complexity for finite factors of infinite sequences. *Theoret. Comput. Sci.* **218** (1999), 177–195.

[Ferenczi and Mauduit 1997]   S. Ferenczi and C. Mauduit. Transcendence of numbers with a low complexity expansion. *J. Number Theory* **67** (1997), 146–161.

[Ferenczi, Mauduit, and Nogueira 1996]   S. Ferenczi, C. Mauduit, and A. Nogueira. Substitution dynamical systems: algebraic characterization of eigenvalues. *Ann. Sci. École Norm. Sup.* **29** (1996), 519–533.

[Ferns 1965]   H. H. Ferns. On the representation of integers as sums of distinct Fibonacci numbers. *Fibonacci Quart.* **3** (1965), 21–30. Acknowledgment and errata, **3** (1965), 160.

[Fife 1980]   E. D. Fife. Binary sequences which contain no BBb. *Trans. Amer. Math. Soc.* **261** (1980), 115–136.

[Fife 1983]   E. D. Fife. Irreducible binary sequences. In L. J. Cummings, editor, *Combinatorics on Words: Progress and Perspectives*, pp. 91–100. Academic Press, 1983.

[Figà-Talamanca and Price 1973]   A. Figà-Talamanca and J. F. Price. Rudin-Shapiro sequences on compact groups. *Bull. Austral. Math. Soc.* **8** (1973), 241–245.

[Filipponi 1986]   P. Filipponi. A note on the representation of integers as a sum of distinct Fibonacci numbers. *Fibonacci Quart.* **24** (1986), 336–343.

[Fine 1947]   N. J. Fine. Binomial coefficients modulo a prime. *Amer. Math. Monthly* **54** (1947), 589–592.

[Fine 1965]   N. J. Fine. The distribution of the sum of digits (mod p). *Bull. Amer. Math. Soc.* **71** (1965), 651–652.

[Fine and Wilf 1965]   N. J. Fine and H. S. Wilf. Uniqueness theorems for periodic functions. *Proc. Amer. Math. Soc.* **16** (1965), 109–114.

[Flajolet 1985]   P. Flajolet. Ambiguity and transcendence. In W. Brauer, editor, *Proc. 12th Int. Conf. on Automata, Languages, and Programming (ICALP)*, Vol. 194 of *Lecture Notes in Computer Science*, pp. 179–188. Springer-Verlag, 1985.

[Flajolet 1987]   P. Flajolet. Analytic models and ambiguity of context-free languages. *Theoret. Comput. Sci.* **49** (1987), 283–309.

[Flajolet, Grabner, Kirschenhofer, Prodinger, and Tichy 1994]   P. Flajolet, P. Grabner, P. Kirschenhofer, H. Prodinger, and R. F. Tichy. Mellin transforms and asymptotics: digital sums. *Theoret. Comput. Sci.* **123** (1994), 291–314.

[Fliess 1969]   M. Fliess. Du produit de Hurwitz de deux séries formelles. *C. R. Acad. Sci. Paris Sér. A* **268** (1969), 535–537.

[Fliess 1974a]   M. Fliess. Sur divers produits de séries formelles. *Bull. Soc. Math. France* **102** (1974), 181–191.

[Florek 1951]   K. Florek. Une remarque sur la répartition des nombres $n\xi$ (mod 1). *Colloq. Math.* **2** (1951), 323–324.

[Flye Sainte-Marie 1894]   C. Flye Sainte-Marie. Question 48. *L'Intermédiaire Math.* **1** (1894), 107–110.

[Forslund 1995]   R. R. Forslund. A logical alternative to the existing positional number system. *Southwest J. Pure Appl. Math.* **1** (1995), 27–29.

[Foster 1987]   D. M. E. Foster. Estimates for a remainder term associated with the sum of digits function. *Glasgow Math. J.* **29** (1987), 109–129.

[Foster 1991]   D. M. E. Foster. A lower bound for a remainder term associated with the sum of digits function. *Proc. Edinburgh Math. Soc.* **34** (1991), 121–142.

[Foster 1992]   D. M. E. Foster. Averaging the sum of digits function to an even base. *Proc. Edinburgh Math. Soc.* **35** (1992), 449–455.

[Fournier 1990]   J.-C. Fournier. Pour en finir avec la dérécursivation du problème des Tours de Hanoï. *RAIRO Inform. Théor. App.* **24** (1990), 17–35.

[Fraenkel 1984]   A. S. Fraenkel. Wythoff games, continued fractions, cedar trees and Fibonacci searches. *Theoret. Comput. Sci.* **29** (1984), 49–73.

[Fraenkel 1985]   A. S. Fraenkel. Systems of numeration. *Amer. Math. Monthly* **92** (1985), 105–114.

[Fraenkel 1989]   A. S. Fraenkel. The use and usefulness of numeration systems. *Inform. Comput.* **81** (1989), 46–61.

[Fraenkel and Borosh 1973]   A. S. Fraenkel and I. Borosh. A generalization of Wythoff's game. *J. Combin. Theory Ser. A* **15** (1973), 175–191.

[Fraenkel and Holzman 1995]   A. S. Fraenkel and R. Holzman. Gap problems for integer part and fractional part sequences. *J. Number Theory* **50** (1995), 66–86.

[Fraenkel, Levitt, and Shimshoni 1972]   A. S. Fraenkel, J. Levitt, and M. Shimshoni. Characterization of the set of values $f(n) = [n\alpha]$, $n = 1, 2, \ldots$. *Discrete Math.* **2** (1972), 335–345.

[Fraenkel, Mushkin, and Tassa 1978]   A. S. Fraenkel, M. Mushkin, and U. Tassa. Determination of $[n\theta]$ by its sequences of differences. *Canad. Math. Bull.* **21** (1978), 441–446.

[Fraenkel, Seeman, and Simpson 2001]   A. S. Fraenkel, T. Seeman, and J. Simpson. The subword complexity of a two-parameter family of sequences. *Electronic J. Combin.* **8**(2) (2001), #R10 (electronic), http://www.combinatorics.org/ Volume_8/Abstracts/v8i2r10.html

[Fraenkel and Simpson 1995]   A. S. Fraenkel and J. Simpson. How many squares must a binary sequence contain? *Electronic J. Combin.* **2** (1995), #R2.

[Fraenkel and Simpson 1998]   A. S. Fraenkel and J. Simpson. How many squares can a string contain? *J. Combin. Theory. Ser. A* **82** (1998), 112–120.

[Frame 1949]   J. S. Frame. Continued fractions and matrices. *Amer. Math. Monthly* **56** (1949), 98–103.

[Franĕk, Karaman, and Smyth 2000]   F. Franĕk, A. Karaman, and W. F. Smyth. Repetitions in Sturmian strings. *Theoret. Comput. Sci.* **249** (2000), 289–303.

[Fredricksen 1982]   H. Fredricksen. A survey of full length nonlinear shift register cycle algorithms. *SIAM Rev.* **24** (1982), 195–221.

[Fredricksen 1992]   H. Fredricksen. Gray codes and the Thue–Morse–Hedlund sequence. *J. Combin. Math. Combin. Comput.* **11** (1992), 3–11.

[Freeman 1961a]   H. Freeman. On the encoding of arbitrary geometric configurations. *IRE Trans. Electron. Comput.* **EC-10** (1961), 260–268.

[Freeman 1961b]   H. Freeman. Techniques for the digital computer analysis of chain-encoded arbitrary plane curves. In *Proc. 17th National Electronics Conference*, pp. 421–432, 1961.

[Freeman 1970]   H. Freeman. Boundary encoding and processing. In B. S. Lipkin and A. Rosenfeld, editors, *Picture Processing and Psychopictorics*, pp. 241–266. Academic Press, 1970.

[Freitag and Phillips 1996]   H. T. Freitag and G. M. Phillips. On the Zeckendorf form of F_{kn}/F_n. *Fibonacci Quart.* **34** (1996), 444–446.

[Fresnel, Koskas, and de Mathan 2000]   J. Fresnel, M. Koskas, and B. de Mathan. Automata and transcendence in positive characteristic. *J. Number Theory* **80** (2000), 1–24.

[Friant 1969]   J. Friant. Grammaires génératives pour l'ensemble des nombres premiers. *Rev. Roumaine Math. Pures Appl.* **14** (1969), 473–488.

[Frid 1997a]   A. E. Frid. The subword complexity of fixed points of binary uniform morphisms. In B. S. Chlebus and L. Czaja, editors, *Fundamentals of Computation Theory: FCT '97*, Vol. 1279 of *Lecture Notes in Computer Science*, pp. 179–187. Springer-Verlag, 1997.

[Frid 1997b]   A. E. Frid. On the subword complexity of iteratively generated infinite words. *Diskretn. Anal. Issled. Oper. Ser. 1* **4**(1) (1997), 53–59. In Russian. English translation in *Discrete Appl. Math.* **114** (2001), 115–120.

[Frid 1998]   A. Frid. On uniform D0L words. In M. Morvan, C. Meinel, and D. Krob, editors, *STACS 98, Proc. 15th Symp. Theoretical Aspects of Comp. Sci.*, Vol. 1373 of *Lecture Notes in Computer Science*, pp. 544–554. Springer-Verlag, 1998.

[Frid 1999a]   A. E. Frid. On factor graphs of D0L words. *Diskretn. Anal. Issled. Oper. Ser. 1* **6**(4) (1999), 92–103. In Russian. English translation in *Discrete Appl. Math.* **114** (2001), 121–130.

[Frid 1999b]   A. Frid. Applying a uniform marked morphism to a word. *Discrete Math. & Theoret. Comput. Sci.* **3** (1999), 125–140.

[Frid 2001]   A. E. Frid. Overlap-free symmetric D0L words. *Discrete Math. & Theoret. Comput. Sci.* **4** (2001), 357–362.

[Frid and Avgustinovich 1999]   A. E. Frid and S. V. Avgustinovich. On bispecial words and subword complexity of D0L sequences. In C. Ding, T. Helleseth, and H. Niederreiter, editors, *Sequences and Their Applications, Proceedings of SETA '98*, pp. 191–204. Springer-Verlag, 1999.

[Fried and Sós 1992]   E. Fried and V. T. Sós. A generalization of the three-distance theorem for groups. *Algebra Universalis* **29** (1992), 136–149.

[Friedman 2001]   H. M. Friedman. Long finite sequences. *J. Combin. Theory. Ser. A* **95** (2001), 102–144.

[Frobenius 1908]   G. Frobenius. Über Matrizen aus positiven Elementen. *Sitzungsber. Preuss. Akad. Wiss. Berlin* (1908–9), 471–476, 514–518.

[Frobenius 1912]   G. Frobenius. Über Matrizen aus nicht negativen Elementen. *Sitzungsber. Preuss. Akad. Wiss. Berlin* (1912), 456–477.

[Frougny 1986]   C. Frougny. Fibonacci numeration systems and rational functions. In J. Gruska, B. Rovan, and J. Wiedermann, editors, *Proc. 12th Symposium, Mathematical Foundations of Computer Science 1986*, Vol. 233 of *Lecture Notes in Computer Science*, pp. 350–359. Springer-Verlag, 1986.

[Frougny 1988]   C. Frougny. Linear numeration systems of order two. *Inform. Comput.* **77** (1988), 233–259.

[Frougny 1989a]   C. Frougny. Linear numeration systems, θ-developments and finite automata. In B. Monien and R. Cori, editors, *STACS 89, Proc. 6th Symp. Theoretical Aspects of Comp. Sci.*, Vol. 349 of *Lecture Notes in Computer Science*, pp. 144–155. Springer-Verlag, 1989.

[Frougny 1989b]   C. Frougny. Systèmes de numération lineaires et automates finis (Thèse d'État). Technical Report 89-69, Laboratoire Informatique Théorique et Programmation, Université P. et M. Curie, Université Paris VII, September 1989.

[Frougny 1991]   C. Frougny. Fibonacci representations and finite automata. *IEEE Trans. Inform. Theory* **37** (1991), 393–399.

[Frougny 1992a]   C. Frougny. Representations of numbers and finite automata. *Math. Systems Theory* **25** (1992), 37–60.

[Frougny 1992b]   C. Frougny. Systèmes de numération linéaires et θ-représentations. *Theoret. Comput. Sci.* **94** (1992), 223–236.

[Frougny 1992d]   C. Frougny. How to write integers in non-integer base. In I. Simon, editor, *1st Latin American Symposium on Theoretical Informatics (LATIN '92)*, Lecture Notes in Computer Science, pp. 154–164. Springer-Verlag, 1992.

[Frougny 2000]   C. Frougny. Number representation and finite automata. In F. Blanchard, A. Maass, and A. Nogueira, editors, *Topics in Symbolic Dynamics and Applications*, Vol. 279 of *London Math. Soc. Lect. Note Ser.*, pp. 207–228. Cambridge University Press, 2000.

[Frougny and Sakarovitch 1999]   C. Frougny and J. Sakarovitch. Automatic conversion from Fibonacci representation to representation in base ϕ, and a generalization. *Int. J. Algebra Comput.* **9** (1999), 351–384.

[Frougny and Solomyak 1992]   C. Frougny and B. Solomyak. Finite beta-expansions. *Ergod. Theory & Dynam. Sys.* **12** (1992), 713–723.

[Furstenberg 1967]   H. Furstenberg. Algebraic functions over finite fields. *J. Algebra* **7** (1967), 271–277.

[Furstenberg 1981]   H. Furstenberg. *Recurrence in Ergodic Theory and Combinatorial Number Theory*. Princeton University Press, 1981.

[Gaafar 1977]   M. Gaafar. Convexity verification, block-chords, and digital straight lines. *Computer Graphics and Image Processing* **6** (1977), 361–370.

[Gabarró 1985]   J. Gabarró. Some applications of the interchange lemma. *Bull. European Assoc. Theor. Comput. Sci.*, No. 25 (February 1985), 19–21.

[Galambos 1973]   J. Galambos. Probabilistic theorems concerning expansions of real numbers. *Period. Math. Hung.* **3** (1973), 101–113.

[Gambaudo, Lanford, and Tresser 1984]   J.-M. Gambaudo, O. Lanford III, and C. Tresser. Dynamic symbolique des rotations. *C. R. Acad. Sci. Paris* **299** (1984), 823–826.

[Gantmacher 1960]   F. R. Gantmacher. *The Theory of Matrices*. Chelsea, 1960.

[Gardner 1961a]   M. Gardner. Mathematical games. *Sci. Amer.* **204**(1) (1961), 164–172.

[Gardner 1961b]   M. Gardner. Mathematical games. *Sci. Amer.* **204**(2) (1961), 146–154.

[Gardner 1967a]   M. Gardner. *The Numerology of Dr. Matrix*. Simon & Schuster, New York, 1967.

[Gardner 1967b]   M. Gardner. Mathematical games. *Sci. Amer.* **216**(3) (1967), 124–129. Additional material in **216** (4) (April 1967), 116–123; **217** (1) (July 1967), 112–116.

[Gardner 1973]   M. Gardner. How to turn a chessboard into a computer and to calculate with negabinary numbers. *Sci. Amer.* **228**(4) (1973), 106–111.

[Gardner 1975a]   M. Gardner. From rubber ropes to rolling cubes, a miscellany of refreshing problems. *Sci. Amer.* **232**(3) (1975), 112–116.

[Gardner 1975b]   M. Gardner. Six sensational discoveries that somehow or another have escaped public attention. *Sci. Amer.* **232**(4) (1975), 126–133.

[Gardner 1977]   M. Gardner. Extraordinary non-periodic tiling that enriches the theory of tiles. *Sci. Amer.* **236**(1) (1977), 110–121.

[Gasparian, Ruiz, Ortuño, and Cuevas 1996]   V. Gasparian, J. Ruiz, M. Ortuño, and E. Cuevas. Brewster anomaly in Fibonacci and Thue-Morse dielectric multilayers. *Electromagnetics* **16** (1996), 313–322.

[Gault and Clint 1987]   D. Gault and M. Clint. A fast algorithm for the Towers of Hanoi problem. *Computer J.* **30** (1987), 376–378.

[Gault and Clint 1988]   D. Gault and M. Clint. "Curiouser and curiouser" said Alice. Further reflections on an interesting recursive function. *Internat. J. Comput. Math.* **26** (1988), 35–43.

[Gauss 1832]   C. F. Gauss. Theoria residuorum biquadraticorum. Commentatio secunda. *Comm. Soc. Reg. Sci. Gottingen* **7** (1832), 1–34. Reprinted in *Werke*, Georg Olms Verlag, Hildesheim, 1973, pp. 93–148.

[Gazeau and Miękisz 1998]   J.-P. Gazeau and J. Miękisz. A symmetry group of a Thue-Morse quasicrystal. *J. Phys. A: Math. Gen.* **31** (1998), L435–L440.

[Gedeon 1992]   T. D. Gedeon. The Reve's puzzle: an interactive solution produced by transformation. *Computer J.* **35** (1992), 186–187.

[Geelen and Simpson 1993]   J. F. Geelen and R. J. Simpson. A two dimensional Steinhaus theorem. *Australasian J. Combin.* **8** (1993), 169–197.

[Gelfond 1968]   A. O. Gelfond. Sur les nombres qui ont des propriétés additives et multiplicatives données. *Acta Arith.* **13** (1968), 259–265.

[Gessel 1981]   I. M. Gessel. Two theorems on rational power series. *Utilitas Math.* **19** (1981), 247–254.

[Ghosh and Karmakar 1998]   A. Ghosh and S. N. Karmakar. Trace map of a general aperiodic Thue-Morse chain: electronic properties. *Phys. Rev. B* **58** (1998), 2586–2590.

[Giancarlo and Mignosi 1994]   R. Giancarlo and F. Mignosi. Generalizations of the periodicity theorem of Fine and Wilf. In S. Tison, editor, *Trees in Algebra and Programming – CAAP '94*, Vol. 787 of *Lecture Notes in Computer Science*, pp. 130–141. Springer-Verlag, 1994.

[Gilbert 1981a]   W. J. Gilbert. Radix representations of quadratic fields. *J. Math. Anal. Appl.* **83** (1981), 264–274.

[Gilbert 1981b]   W. J. Gilbert. Geometry of radix representations. In C. Davis, B. Grünbaum, and F. A. Sherk, editors, *The Geometric Vein: The Coxeter Festschrift*, pp. 129–139. Springer-Verlag, 1981.

[Gilbert 1982a]   W. J. Gilbert. Fractal geometry derived from complex bases. *Math. Intelligencer* **4** (1982), 78–86.

[Gilbert 1982b]   W. J. Gilbert. Complex numbers with three radix expansions. *Canad. J. Math.* **34** (1982), 1335–1348.

[Gilbert 1986]   W. J. Gilbert. The fractal dimension of sets derived from complex bases. *Canad. Math. Bull.* **29** (1986), 495–500.

[Gilbert 1987]   W. J. Gilbert. Complex bases and fractal similarity. *Ann. Sci. Math. Québec* **11** (1987), 65–77.

[Gilbert and Green 1979]   W. J. Gilbert and R. J. Green. Negative based number systems. *Math. Mag.* **52** (1979), 240–244.

[Ginsburg and Spanier 1963]   S. Ginsburg and E. Spanier. Quotients of context free languages. *J. Assoc. Comput. Mach.* **10** (1963), 487–492.

[Girstmair 1997]   K. Girstmair. Digit variance and Dedekind sums. *J. Number Theory* **65** (1997), 197–205.

[Glaisher 1899a]   J. W. L. Glaisher. On the residue of a binomial-theorem coefficient with respect to a prime modulus. *Quart. J. Pure Appl. Math.* **30** (1899), 150–156.

[Glaisher 1899b]   J. W. L. Glaisher. A congruence theorem relating to sums of binomial-theorem coefficients. *Quart. J. Pure Appl. Math.* **30** (1899), 361–383.

[Glaister and Shallit 1996]   I. Glaister and J. O. Shallit. A lower bound technique for the size of nondeterministic finite automata. *Inform. Process. Lett.* **59** (1996), 75–77.

[Glaister and Shallit 1998]   I. Glaister and J. O. Shallit. Automaticity III: Polynomial automaticity and context-free languages. *Comput. Complexity* **7** (1998), 371–387.

[Godrèche 1990]   C. Godrèche. Types of order and diffraction spectra for tilings of the line. In J.-M. Luck, P. Moussa, and M. Waldschmidt, editors, *Number Theory and Physics*, Vol. 47 of *Springer Proceedings in Physics*, pp. 86–99. Springer-Verlag, 1990.

[Godrèche and Luck 1989a]   C. Godrèche and J.-M. Luck. Tilings of the plane and their diffraction spectra. In M. V. Jaric and S. Lundqvist, editors, *Proceedings of the Anniversary Adriatico Research Conference on Quasicrystals (Trieste, Italy, July 4–7 1989)*, pp. 144–160. World Scientific, 1989.

[Godrèche and Luck 1989b]   C. Godrèche and J.-M. Luck. Quasiperiodicity and randomness in tilings of the plane. *J. Statist. Phys.* **55** (1989), 1–28.

[Godrèche, Luck, Janner, and Janssen 1993]   C. Godrèche, J.-M. Luck, A. Janner, and T. Janssen. Fractal atomic surfaces of self-similar quasiperiodic tilings of the plane. *J. Physique I* **3** (1993), 1921–1939.

[Golay 1949]   M. J. E. Golay. Multi-slit spectrometry. *J. Optical Soc. Amer.* **39** (1949), 437–444.

[Golay 1951]   M. J. E. Golay. Static multislit spectrometry and its application to the panoramic display of infrared spectra. *J. Optical Soc. Amer.* **41** (1951), 468–472.

[Goldstein, Kelly, and Speer 1992]   S. Goldstein, K. Kelly, and E. R. Speer. The fractal structure of rarefied sums of the Thue-Morse sequence. *J. Number Theory* **42** (1992), 1–19.

[Goldstine 1976]   J. Goldstine. Bounded AFLs. *J. Comput. System Sci.* **12** (1976), 399–419.

[M. Golomb 1993]   M. Golomb. Problem 10333. *Amer. Math. Monthly* **100** (1993), 797. Solution by A. N. 't Woord, **103** (1996), 703–704.

[S. Golomb 1975]   S. W. Golomb. Elementary problem E 2529. *Amer. Math. Monthly* **82** (1975), 400. Solution by R. E. Shafer, **83** (1976), 487–488.

[Good 1946]   I. J. Good. Normal recurring decimals. *J. London Math. Soc.* **21** (1946), 167–169.

[Good 1993]   J. Good. Enigma and Fish. In F. H. Hinsley and A. Stripp, editors, *The Codebreakers: The Inside Story of Bletchley Park*, pp. 149–166. Oxford University Press, 1993.

[Goormaghtigh 1949]   R. Goormaghtigh. A bit of numerology. *Scripta Math.* **15** (1949), 91.

[Goralcik and Vanicek 1991]   P. Goralcik and T. Vanicek. Binary patterns in binary words. *Int. J. Algebra Comput.* **1** (1991), 387–391.

[Gottlieb 1969]   A. J. Gottlieb. Puzzle corner. *Technology Review* **71**(6) (1969), 98–100. Also see **72** (2) (December 1969), 72–75; **72** (6) (April 1970), 88–89.

[Gottschalk 1963]   W. H. Gottschalk. Substitution minimal sets. *Trans. Amer. Math. Soc.* **109** (1963), 467–491.

[Gottschalk and Hedlund 1955]   W. H. Gottschalk and G. A. Hedlund. *Topological Dynamics*, Vol. 36 of *AMS Colloquium Publications*. Amer. Math. Soc., 1955.

[Gottschalk and Hedlund 1964]   W. H. Gottschalk and G. A. Hedlund. A characterization of the Morse minimal set. *Proc. Amer. Math. Soc.* **15** (1964), 70–74.

[Gould, Kim, and Hoggatt 1977]   H. W. Gould, J. B. Kim, and V. E. Hoggatt, Jr. Sequences associated with t-ary coding of Fibonacci's rabbits. *Fibonacci Quart.* **15** (1977), 311–318.

[Gouvêa 1993]   F. Q. Gouvêa. *p-adic Numbers*. Springer-Verlag, 1993.

[Grabner 1993a]   P. J. Grabner. Completely q-multiplicative functions: the Mellin transform approach. *Acta Arith.* **65** (1993), 85–96.

[Grabner 1993b]   P. J. Grabner. A note on the parity of the sum-of-digits function. In *Séminaire Lotharingien de Combinatoire (Gerolfingen, 1993)*, Prépubl. Inst. Rech. Math. Av., Univ. Louis Pasteur, pp. 35–42, 1993–1994.

[Grabner, Kirschenhofer, and Prodinger 1998]   P. J. Grabner, P. Kirschenhofer, and H. Prodinger. The sum-of-digits function for complex bases. *J. London Math. Soc.* **57** (1998), 20–40.

[Grabner, Kirschenhofer, Prodinger, and Tichy 1993]   P. J. Grabner, P. Kirschenhofer, H. Prodinger, and R. F. Tichy. On the moments of the sum-of-digits function. In G. E. Bergum, A. N. Philippou, and A. F. Horadam, editors, *Applications of Fibonacci Numbers, Vol. 5*, pp. 263–271. Kluwer, 1993.

[Grabner and Thuswaldner 2000]   P. J. Grabner and J. M. Thuswaldner. On the sum of digits function for number systems with negative bases. *Ramanujan J.* **4** (2000), 201–220.

[Graham 1963]   R. L. Graham. On a theorem of Uspensky. *Amer. Math. Monthly* **70** (1963), 407–409.

[Graham, Knuth, and Patashnik 1989]   R. L. Graham, D. E. Knuth, and O. Patashnik. *Concrete Mathematics*. Addison-Wesley, 1989.

[Graham and van Lint 1968]   R. L. Graham and J. H. van Lint. On the distribution of $n\theta$ modulo 1. *Canad. J. Math.* **20** (1968), 1020–1024.

[Graham and Pollak 1970]   R. L. Graham and H. O. Pollak. Note on a nonlinear recurrence related to $\sqrt{2}$. *Math. Mag.* **43** (1970), 143–145.

[Granville 1995]   A. Granville. Binomial coefficients (mod p^q). Unpublished manuscript, 1995.

[Granville and Rasson 1988]   V. Granville and J. P. Rasson. A strange recursive relation. *J. Number Theory* **30** (1988), 238–241.

[Grassl and Mullhaupt 1989]   R. M. Grassl and A. P. Mullhaupt. Hook and shifted hook numbers. *Discrete Math.* **79** (1989/90), 153–167.

[Grazon 1987]   A. Grazon. An infinite word language which is not co-CFL. *Inform. Process. Lett.* **24** (1987), 81–85.

[Green and Rees 1952]   J. A. Green and D. Rees. On semi-groups in which $x^r = x$. *Proc. Cambridge Phil. Soc.* **48** (1952), 35–40.

[Grillenberger 1973]   C. Grillenberger. Construction of strictly ergodic systems. *Z. Wahrscheinlichkeitstheorie Verw. Gebiete* **25** (1973), 323–334.

[Grimm 2001]   U. Grimm. Improved bounds on the number of ternary square-free words. *J. Integer Sequences* **4** (2001), 01.2.7 (electronic), http://www.math.uwaterloo.ca/JIS/VOL4/GRIMM/words.html

[Grimm and Baake 1994]   U. Grimm and M. Baake. Non-periodic Ising quantum chains and conformal invariance. *J. Statist. Phys.* **74** (1994), 1233–1245.

[Grinberg and Korshunov 1966]   V. S. Grinberg and A. D. Korshunov. Asymptotic behavior of the maximum of the weight of a finite tree. *Problemy Peredachi Informatsii* **2** (1966), 96–99. In Russian. English translation in *Problems of Information Transmission* **2** (1966), 75–78.

[Grossman 1985]   E. H. Grossman. Number bases in quadratic fields. *Stud. Sci. Math. Hung.* **20** (1985), 55–58.

[Grounds and Silberger 1993]   W. V. Grounds and D. M. Silberger. How many primitive words? *Ann. Soc. Math. Pol. Ser. I: Comment. Math.* **33** (1993), 57–59.

[Grundman 1994]   H. G. Grundman. Sequences of consecutive n-Niven numbers. *Fibonacci Quart.* **32** (1994), 174–175.

[Grünwald 1885]   V. Grünwald. Intorno all'aritmetica dei sistemi numerici a base negativa con particolare riguardo al sistema numerico a base negativo-decimale per lo studio delle sue analogie coll'aritmetica (decimale). *Giorn. Mat. Battaglini* **23** (1885), 203–221. Errata, p. 367.

[Grytczuk 1996a]   J. Grytczuk. Infinite self-similar words. *Discrete Math.* **161** (1996), 133–141.

[Grytczuk 1996b]   J. Grytczuk. Intertwined infinite binary words. *Discrete Appl. Math.* **66** (1996), 95–99.

[Guille-Biel 1997]   C. Guille-Biel. Sparse Schrödinger operators. *Rev. Math. Phys.* **9** (1997), 315–341.

[Gumbs and Ali 1988a]   G. Gumbs and M. K. Ali. Dynamical maps, Cantor spectra, and localization for Fibonacci and related quasiperiodic lattices. *Phys. Rev. Lett.* **60** (1988), 1081–1084.

[Gumbs and Ali 1988b]   G. Gumbs and M. K. Ali. Scaling and eigenstates for a class of one-dimensional quasiperiodic lattices. *J. Phys. A: Math. Gen.* **21** (1988), L517–L521.

[Gumbs and Ali 1989]   G. Gumbs and M. K. Ali. Electronic properties of the tight-binding Fibonacci Hamiltonian. *J. Phys. A: Math. Gen.* **22** (1989), 951–970.

[Gumbs, Dubey, Salman, Mahmoud, and Huang 1995]   G. Gumbs, G. S. Dubey, A. Salman, B. S. Mahmoud, and D. Huang. Statistical and transport properties of quasiperiodic layered structures: Thue-Morse and Fibonacci. *Phys. Rev. B* **52** (1995), 210–219.

[Güntzer and Paul 1987]   U. Güntzer and M. Paul. Jump interpolation search trees and symmetric binary numbers. *Inform. Process. Lett.* **26** (1987–1988), 193–204.

[Gupta 1989]   N. Gupta. On groups in which every element has finite order. *Amer. Math. Monthly* **96** (1989), 297–308.

[Gupta, Madan, and Tewari 1994]   S. K. Gupta, S. Madan, and U. B. Tewari. The conjugation operator on $A_q(G)$. *Proc. Amer. Math. Soc.* **121** (1994), 163–166.

[Guy 1994]   R. K. Guy. *Unsolved Problems in Number Theory*. Springer-Verlag, 2nd edition, 1994.

[Hadamard 1899]   J. Hadamard. Théorème sur les séries entières. *Acta Math.* **22** (1899), 55–63.

[von Haeseler, Peitgen, and Skordev 1993]   F. von Haeseler, H.-O. Peitgen, and G. Skordev. Cellular automata, matrix substitutions and fractals. *Ann. Math. Artificial Intell.* **8** (1993), 345–362.

[von Haeseler, Peitgen, and Skordev 1995]   F. von Haeseler, H.-O. Peitgen, and G. Skordev. Global analysis of self-similarity features of cellular automata: selected examples. *Physica D* **86** (1995), 64–80.

[von Haeseler and Petersen 1998]   F. von Haeseler and A. Petersen. Automaticity of rational functions. *Beiträge zur Algebra und Geometrie* **39** (1998), 219–229.

[Halava, Harju, and Ilie 2000]   V. Halava, T. Harju, and L. Ilie. Periods and binary words. *J. Combin. Theory. Ser. A* **89** (2000), 298–303.

[Hall 1957]   M. Hall, Jr. Solution of the Burnside problem for exponent 6. *Proc. Nat. Acad. Sci. U.S.A.* **43** (1957), 751–753.

[Hall 1958]   M. Hall, Jr. Solution of the Burnside problem for exponent six. *Illinois J. Math.* **2** (1958), 764–786.

[Hall 1964]   M. Hall, Jr. Generators and relations in groups – the Burnside problem. In T. L. Saaty, editor, *Lectures on Modern Mathematics*, Vol. II, pp. 42–92. Wiley, 1964.

[Halmos 1950]   P. R. Halmos. *Measure Theory*. D. Van Nostrand, 1950.

[Halton 1965]   J. H. Halton. The distribution of the sequence $\{n\xi\}$ ($n = 0, 1, 2, \dots$). *Proc. Cambridge Phil. Soc.* **61** (1965), 665–670.

[Hamm and Shallit 1999]   D. Hamm and J. Shallit. Characterization of finite and one-sided infinite fixed points of morphisms on free monoids. Technical Report CS-99-17, University of Waterloo, Department of Computer Science, July 1999.

[Hamming 1950]   R. W. Hamming. Error detecting and error correcting codes. *Bell System Tech. J.* **29** (1950), 147–160.

[Hansel 1982]   G. Hansel. A propos d'un théorème de Cobham. In D. Perrin, editor, *Actes de la Fête des Mots*, pp. 55–59. Greco de Programmation, CNRS, Rouen, 1982.

[Hansel 1998]   G. Hansel. Systèmes de numération indépendants et syndéticité. *Theoret. Comput. Sci.* **204** (1998), 119–130.

[Harase 1988]   T. Harase. Algebraic elements in formal power series rings. *Israel J. Math.* **63** (1988), 281–288.

[Harase 1989]   T. Harase. Algebraic elements in formal power series rings II. *Israel J. Math.* **67** (1989), 62–66.

[Harborth 1974]   H. Harborth. Endliche 0-1-Folgen mit gleichen Teilblöcken. *J. Reine Angew. Math.* **271** (1974), 139–154.

[Hardy 1967]   G. H. Hardy. *A Mathematician's Apology*. Cambridge University Press, 1967.

[Hardy and Littlewood 1923a]   G. H. Hardy and J. E. Littlewood. Some problems of diophantine approximation: the analytic character of the sum of Dirichlet's series considered by Hecke. *Abh. Math. Sem. Univ. Hamburg* **3** (1923), 57–68.

[Hardy and Littlewood 1923b]   G. H. Hardy and J. E. Littlewood. Some problems of diophantine approximation: the analytic properties of certain Dirichlet's series associated with the distribution of numbers to modulus unity. *Trans. Cambridge Phil. Soc.* **22** (1923), 519–533.

[Hardy and Wright 1985]   G. H. Hardy and E. M. Wright. *An Introduction to the Theory of Numbers*. Oxford University Press, 5th edition, 1985.

[Harju 1983]   T. Harju. On repetition free morphisms. *Bull. Eur. Assoc. Theor. Comput. Sci.*, No. 19 (February 1983), 18–20.

[Harju 1986]   T. Harju. On cyclically overlap-free words in binary alphabets. In G. Rozenberg and A. Salomaa, editors, *The Book of L*, pp. 125–130. Springer-Verlag, 1986.

[Harju and Karhumäki 1997]   T. Harju and J. Karhumäki. Morphisms. In G. Rozenberg and A. Salomaa, editors, *Handbook of Formal Languages*, Vol. 1, pp. 439–510. Springer-Verlag, 1997.

[Harju and Linna 1986]   T. Harju and M. Linna. On the periodicity of morphisms on free monoids. *RAIRO Inform. Théor. Appl.* **20** (1986), 47–54.

[Harrison 1978]   M. A. Harrison. *Introduction to Formal Language Theory*. Addison-Wesley, 1978.

[Hartmanis and Shank 1968]   J. Hartmanis and H. Shank. On the recognition of primes by automata. *J. Assoc. Comput. Mach.* **15** (1968), 382–389.

[Hartmanis and Shank 1969]   J. Hartmanis and H. Shank. Two memory bounds for the recognition of primes by automata. *Math. Systems Theory* **3** (1969), 125–129.

[Hartmanis and Stearns 1965]   J. Hartmanis and R. E. Stearns. On the computational complexity of algorithms. *Trans. Amer. Math. Soc.* **117** (1965), 285–306.

[Hartmanis and Stearns 1967]   J. Hartmanis and R. E. Stearns. Sets of numbers defined by finite automata. *Amer. Math. Monthly* **74** (1967), 539–542.

[Hautus and Klarner 1971]   M. L. J. Hautus and D. A. Klarner. The diagonal of a double power series. *Duke Math. J.* **38** (1971), 229–235.

[Hawkins and Mientka 1956]   D. Hawkins and W. E. Mientka. On sequences which contain no repetitions. *Math. Student* **24** (1956), 185–187.

[Hayes 1977]   P. J. Hayes. A note on the Towers of Hanoi problem. *Computer J.* **20** (1977), 282–285.

[Head 1981]   T. Head. Fixed languages and the adult languages of 0L schemes. *Int. J. Comput. Math.* **10** (1981), 103–107.

[Head and Lando 1986]   T. Head and B. Lando. Fixed and stationary ω-words and ω-languages. In G. Rozenberg and A. Salomaa, editors, *The Book of L*, pp. 147–156. Springer-Verlag, 1986.

[Heath 1972]   F. G. Heath. Origins of the binary code. *Sci. Amer.* **227**(2) (1972), 76–83.

[Heath-Brown 1992]   D. R. Heath-Brown. Zero-free regions for Dirichlet *L*-functions, and the least prime in an arithmetic progression. *Proc. Lond. Math. Soc.* **64** (1992), 265–338.

[Hecke 1921]   E. Hecke. Über analytische Funktionen und die Verteilung von Zahlen mod. eins. *Abh. Math. Sem. Univ. Hamburg* **1** (1921), 54–76.

[Hedlund 1944]   G. A. Hedlund. Sturmian minimal sets. *Amer. J. Math.* **66** (1944), 605–620.

[Hedlund 1967]   G. A. Hedlund. Remarks on the work of Axel Thue on sequences. *Nordisk Mat. Tidskr.* **15** (1967), 148–150.

[Heinis 2002]   A. Heinis. The $P(n)/n$ function for bi-infinite words. *Theoret. Comput. Sci.* **273** (2002), 35–46.

[Hendel and Monteferrante 1994]   R. J. Hendel and S. A. Monteferrante. Hofstadter's extraction conjecture. *Fibonacci Quart.* **32** (1994), 98–107.

[Heppner 1976]   E. Heppner. Über die Summe der Ziffern natürlicher Zahlen. *Ann. Univ. Sci. Budapest Eötvös Sect. Math.* **19** (1976), 41–43.

[Herman, Lindenmayer, and Rozenberg 1975]   G. T. Herman, A. Lindenmayer, and G. Rozenberg. Description of developmental languages using recurrence systems. *Math. Systems Theory* **8** (1975), 316–341.

[Hermisson, Grimm, and Baake]   J. Hermisson, U. Grimm, and M. Baake. Aperiodic Ising quantum chains. *J. Phys. A: Math. Gen.* **30** (1997), 7315–7335.

[Hermite 1873]   C. Hermite. Sur la fonction exponentielle. *C. R. Acad. Sci. Paris* **77** (1873), 18–24, 74–79, 226–233, 285–293. Reprinted in *Oeuvres*, Vol. III, 150–181.

[Herstein 1975]   I. N. Herstein. *Topics in Algebra*. Wiley, New York, 2nd edition, 1975.

[Heuberger 1999]   C. Heuberger. Minimal expansions in redundant number systems and shortest paths in graphs. *Computing* **63** (1999), 341–349.

[Hickerson 1974]   D. R. Hickerson. A relationship between an integer and the one with the reversed order of digits. *Math. Mag.* **47** (1974), 36–39.

[Higgins 1987]   P. M. Higgins. The naming of popes and a Fibonacci sequence in two noncommuting indeterminates. *Fibonacci Quart.* **25** (1987), 57–61.

[Hinz 1989a]   A. M. Hinz. The tower of Hanoi. *Enseign. Math.* **35** (1989), 289–321.

[Hinz 1989b]   A. M. Hinz. An iterative algorithm for the Tower of Hanoi with four pegs. *Computing* **42** (1989), 133–140.

[Hinz 1992a]   A. M. Hinz. Pascal's triangle and the tower of Hanoi. *Amer. Math. Monthly* **99** (1992), 538–544.

[Hinz 1992b]   A. M. Hinz. Shortest paths between regular states of the towers of Hanoi. *Info. Sci.* **63** (1992), 173–181.

[Hinz 1996]   A. M. Hinz. Square-free tower of Hanoi sequences. *Enseign. Math.* **42** (1996), 257–264.

[Hinz 1999]   A. M. Hinz. The Tower of Hanoi. In K.-P. Shum, E. J. Taft, and Z.-X. Wan, editors, *Algebras and Combinatorics (Proc. of ICAC '97)*, pp. 277–289. Springer-Verlag, 1999.

[Hirschhorn and Sellers 1996]   M. D. Hirschhorn and J. A. Sellers. On representations of a number as a sum of three triangles. *Acta Arith.* **77** (1996), 289–301.

[Hirschhorn and Sellers 1999]   M. D. Hirschhorn and J. A. Sellers. On representations of a number as a sum of three squares. *Discrete Math.* **199** (1999), 85–101.

[Hodgson 1983]   B. Hodgson. Décidabilité par automate fini. *Ann. Sci. Math. Québec* **7** (1983), 39–57.

[Hof, Knill, and Simon 1995]   A. Hof, O. Knill, and B. Simon. Singular continuous spectrum for palindromic Schrödinger operators. *Commun. Math. Phys.* **174** (1995), 149–159.

[Hoggatt 1972]   V. E. Hoggatt, Jr. Generalized Zeckendorf theorem. *Fibonacci Quart.* **10** (1972), 89–93.

[Hoit 1999] A. Hoit. The distribution of generalized sum-of-digits functions in residue classes. *J. Number Theory* **79** (1999), 194–216.

[Hollander 1998] M. Hollander. Greedy numeration systems and regularity. *Theory Comput. Systems* **31** (1998), 111–133.

[Holton and Zamboni 1999] C. Holton and L. Q. Zamboni. Descendants of primitive substitutions. *Theory Comput. Systems* **32** (1999), 133–157.

[Holton and Zamboni 2000] C. Holton and L. Q. Zamboni. Iteration of maps by primitive substitutive sequences. In J.-M. Gambaudo, P. Hubert, P. Tisseur, and S. Vaienti, editors, *Dynamical Systems: From Crystal to Chaos*, pp. 137–143. World Scientific, 2000.

[Holzer and Rossmanith 1996] M. Holzer and P. Rossmanith. A simpler grammar for Fibonacci numbers. *Fibonacci Quart.* **34** (1996), 465–466.

[Honkala 1986] J. Honkala. A decision method for the recognizability of sets defined by number systems. *RAIRO Inform. Théor. Appl.* **20** (1986), 395–403.

[Hopcroft and Ullman 1979] J. E. Hopcroft and J. D. Ullman. *Introduction to Automata Theory, Languages, and Computation*. Addison-Wesley, 1979.

[Horadam 1978] A. F. Horadam. Wythoff pairs. *Fibonacci Quart.* **16** (1978), 147–151.

[Hörnquist and Ouchterlony] M. Hörnquist and T. Ouchterlony. Quantum dots in aperiodic order. *Physica E* **3** (1998), 213–223.

[Horváth 1995] S. Horváth. Strong interchangeability and nonlinearity of primitive words. In A. Nijholt, G. Scollo, and R. Steetskamp, editors, *Proc. Worskhop AMiLP '95*, pp. 173–178, 1995.

[Horváth and Ito 1999] S. Horváth and M. Ito. Decidable and undecidable problems of primitive words, regular and context-free languages. *J. Universal Comput. Sci.* **5** (1999), 532–541.

[Houselander 1974] L. S. Houselander. Cellular-array negabinary multiplier. *Electronics Letters* **10** (1974), 168–169.

[Huang, Gumbs, and Kolář 1992] D. Huang, G. Gumbs, and M. Kolář. Localization in a one-dimensional Thue-Morse chain. *Phys. Rev. B* **46** (1992), 11479–11486.

[Huang, Gumbs, and Kolář 1995] D. Huang, G. Gumbs, and M. Kolář. Reply to "Comment on 'Localization in a one-dimensional Thue-Morse chain' ". *Phys. Rev. B* **51** (1995), 3276.

[Huang 1987] K. Huang. *Statistical Mechanics*. Wiley, 2nd edition, 1987.

[Hubert 1995a] P. Hubert. Dynamique symbolique des billards polygonaux rationnels. PhD thesis, Université d'Aix-Marseille II, 1995.

[Hubert 1995b] P. Hubert. Complexité des suites définies par des billards rationnels. *Bull. Soc. Math. France* **123** (1995), 257–270.

[Hubert 2000] P. Hubert. Suites équilibrées. *Theoret. Comput. Sci.* **242** (2000), 91–108.

[Huffman 1954] D. A. Huffman. The synthesis of sequential switching circuits. *J. Franklin Inst.* **257** (1954), 161–190; 275–303.

[Hungerford 1974] T. W. Hungerford. *Algebra*. Springer-Verlag, 1974.

[Hurwitz and Kritikos 1986] A. Hurwitz and N. Kritikos. *Lectures on Number Theory*. Springer-Verlag, 1986.

[Iliopoulos, Moore, and Smyth 1996] C. S. Iliopoulos, D. Moore, and W. F. Smyth. A linear algorithm for computing all the squares of a Fibonacci string. In M. E. Houle and P. Eades, editors, *Proceedings of CATS '96 (Computing: The Australasian Theory Symposium)*, Vol. 18 of *Austral. Comput. Sci. Commun.*, pp. 57–63, 1996.

[Iliopoulos, Moore, and Smyth 1997] C. S. Iliopoulos, D. Moore, and W. F. Smyth. A characterization of the squares in a Fibonacci string. *Theoret. Comput. Sci.* **172** (1997), 281–291.

[Ising 1925] E. Ising. Beitrag zur Theorie der Ferromagnetismus. *Z. Physik* **31** (1925), 253–258.

[Istrail 1977]   S. Istrail. On irreductible languages and nonrational numbers. *Bull. Math. Soc. Sci. Math. R. S. Roumanie* **21** (1977), 301–308.

[Istrail 1983]   S. Istrail. A solution to Wegner's problem P12. *Bull. Eur. Assoc. Theor. Comput. Sci.*, No. 19 (February 1983), 20–24.

[M. Ito and Katsura 1991]   M. Ito and M. Katsura. Context-free languages consisting of non-primitive words. *Int. J. Comput. Math.* **40** (1991), 157–167.

[M. Ito, Katsura, Shyr, and Yu 1988]   M. Ito, M. Katsura, H. J. Shyr, and S. S. Yu. Automata accepting primitive words. *Semigroup Forum* **37** (1988), 45–52.

[S. Ito 1989]   S. Ito. On the fractal curves induced from the complex radix expansion. *Tokyo J. Math.* **12** (1989), 299–320.

[S. Ito 1991]   S. Ito. On a dynamical system related to sequences $[nx + y] - [(n - 1)x + y]n = 1, 2, \ldots$. In *Dynamical Systems and Related Topics (Nagoya, 1990)*, Vol. 9 of *Adv. Ser. Dyn. Syst.*, pp. 192–197. World Scientific, 1991.

[S. Ito and Sano 2001]   S. Ito and Y. Sano. On periodic β-expansions of Pisot numbers and Rauzy fractals. *Osaka J. Math.* **38** (2001), 349–368.

[S. Ito and Yasutomi 1990]   S. Ito and S. Yasutomi. On continued fractions, substitutions and characteristic sequences $[nx + y] - [(n - 1)x + y]$. *Japanese J. Math.* **16** (1990), 287–306.

[Ivanov 1992]   S. V. Ivanov. On the Burnside problem on periodic groups. *Bull. Amer. Math. Soc.* **27** (1992), 257–260.

[Ivanov 1994]   S. V. Ivanov. The free Burnside groups of sufficiently large exponents. *Int. J. Algebra Comput.* **4** (1994), 1–308.

[Jacobs 1992]   K. Jacobs. *Invitation to Mathematics*. Princeton University Press, 1992.

[Jacobs and Keane 1969]   K. Jacobs and M. Keane. 0–1-sequences of Toeplitz type. *Z. Wahrscheinlichkeitstheorie und Verw. Gebiete* **13** (1969), 123–131.

[Jacobson 1974]   N. Jacobson. *Basic Algebra I*. W. H. Freeman, San Francisco, 1974.

[Jedwab and Mitchell 1989]   J. Jedwab and C. J. Mitchell. Minimum weight modified signed-digit representations and fast exponentiation. *Electron. Lett.* **25** (1989), 1171–1172.

[Jeffreys 1973]   H. Jeffreys. *Scientific Inference*. Cambridge University Press, 1973.

[Ježek 1976]   J. Ježek. Intervals in the lattice of varieties. *Algebra Universalis* **6** (1976), 147–158.

[Joshi 1971]   V. S. Joshi. A note on self-numbers. *Math. Student* **39** (1971), 327–328.

[Jungen 1931]   R. Jungen. Sur les séries de Taylor n'ayant que des singularités algébrico-logarithmiques sur leur cercle de convergence. *Comment. Math. Helvetici* **3** (1931), 266–306.

[Justin 1972]   J. Justin. Characterization of the repetitive commutative semigroups. *J. Algebra* **21** (1972), 87–90.

[Justin 2000]   J. Justin. On a paper by Castelli, Mignosi, Restivo. *Theoret. Inform. Appl.* **34** (2000), 373–377.

[Justin and Pirillo 1991]   J. Justin and G. Pirillo. Shirshov's theorem and ω-permutability of semigroups. *Adv. Math.* **87** (1991), 151–159.

[Justin and Pirillo 2001]   J. Justin and G. Pirillo. Fractional powers in Sturmian words. *Theoret. Comput. Sci.* **255** (2001), 363–376.

[Kahane 1980]   J.-P. Kahane. Sur les polynômes à coefficients unimodulaires. *Bull. Lond. Math. Soc.* **12** (1980), 321–342.

[Kakutani 1967]   S. Kakutani. Ergodic theory of shift transformations. In *Proc. 5th Berkeley Symposium on Mathematical Statistics and Probability*, Vol. II, pp. 405–414. University of California Press, 1967.

[Kamae 1990]   T. Kamae. Number theoretic problems involving two independent bases. In J. H. Loxton, editor, *Number Theory and Cryptography*, Vol. 154 of *London Math. Soc. Lect. Note Ser.*, pp. 196–203. Cambridge University Press, 1990.

[Kamae 2001]   T. Kamae. Sequence entropy and the maximal pattern complexity of infinite words. Unpublished manuscript, 2001.

[Kamae and Mendès France 1996]   T. Kamae and M. Mendès France. A continuous family of automata: the Ising automata. *Ann. Inst. H. Poincaré Phys. Théor.* **64** (1996), 349–372.

[Kamae, Tamura, and Wen 1999]   T. Kamae, J.-I. Tamura, and Z.-Y. Wen. Hankel determinants for the Fibonacci word and Padé approximation. *Acta Arith.* **89** (1999), 123–161.

[Kamae and Zamboni 2001]   T. Kamae and L. Zamboni. Maximal pattern complexity for discrete systems. Unpublished manuscript, 2001.

[Kanani and O'Keefe 1973]   D. V. Kanani and K. H. O'Keefe. A note on conditional-sum addition for base -2 systems. *IEEE Trans. Comput.* **C-22** (1973), 626.

[Kaneps and Freivalds 1990]   J. Kaneps and R. Freivalds. Minimal nontrivial space complexity of probabilistic one-way Turing machines. In B. Rovan, editor, *Proc. 15th Symposium, Mathematical Foundations of Computer Science 1990*, Vol. 452 of *Lecture Notes in Computer Science*, pp. 355–361. Springer-Verlag, 1990.

[Kano 1991]   H. Kano. On the sums of digits in integers. *Proc. Japan Acad. Ser. A* **67** (1991), 148–150.

[Kano and Shiokawa 1988]   H. Kano and I. Shiokawa. On sums of digits in integer sequences. *Sem. Math. Sci.* **12** (1988), 43–48. In Japanese.

[Kaprekar 1956]   D. R. Kaprekar. Self-numbers. *Scripta Math.* **22** (1956), 80–81.

[Karevski and Turban 1996]   D. Karevski and L. Turban. Log-periodic corrections to scaling: exact results for aperiodic Ising quantum chains. *J. Phys. A: Math. Gen.* **29** (1996), 3461–3470.

[Karhumäki 1981]   J. Karhumäki. On strongly cube-free ω-words generated by binary morphisms. In F. Gécseg, editor, *Fundamentals of Computation Theory: Proceedings of the 1981 International FCT-Conference*, Vol. 117 of *Lecture Notes in Computer Science*, pp. 182–189. Springer-Verlag, 1981.

[Karhumäki 1983]   J. Karhumäki. On cube-free ω-words generated by binary morphisms. *Discrete Appl. Math.* **5** (1983), 279–297.

[Karhumäki, Plandowski, and Rytter 2000]   J. Karhumäki, W. Plandowski, and W. Rytter. On the complexity of computing the order of repetition of a string. In G. Rozenberg and W. Thomas, editors, *Developments in Language Theory 1999*, pp. 178–184. World Scientific, 2000.

[Kari and Thierrin 1998]   L. Kari and G. Thierrin. Word insertions and primitivity. *Utilitas Math.* **53** (1998), 49–61.

[Karp 1967]   R. M. Karp. Some bounds on the storage requirements of sequential machines and Turing machines. *J. Assoc. Comput. Mach.* **14** (1967), 478–489.

[Kászonyi and Katsura 1997]   L. Kászonyi and M. Katsura. On the context-freeness of a class of primitive words. *Publ. Math. (Debrecen)* **51** (1997), 1–11.

[Kátai 1967]   I. Kátai. On the sum of digits of prime numbers. *Ann. Univ. Sci. Budapest Eötvös Sect. Math.* **10** (1967), 89–93.

[Kátai 1977a]   I. Kátai. On the sum of digits of primes. *Acta Math. Acad. Sci. Hung.* **30** (1977), 169–173.

[Kátai 1977b]   I. Kátai. Change of the sum of digits by multiplication. *Acta Sci. Math. (Szeged)* **39** (1977), 319–328.

[Kátai and Kovács 1980]   I. Kátai and B. Kovács. Kanonische Zahlensysteme in der Theorie der quadratischen algebraischen Zahlen. *Acta Sci. Math. (Szeged)* **42** (1980), 99–107.

[Kátai and Kovács 1981]   I. Kátai and B. Kovács. Canonical number systems in imaginary quadratic fields. *Acta Math. Acad. Sci. Hung.* **37** (1981), 159–164.

[Kátai and Mogyoródi 1968]   I. Kátai and J. Mogyoródi. On the distribution of digits. *Publ. Math. (Debrecen)* **15** (1968), 57–68.

[Kátai and Szabó 1975]   I. Kátai and J. Szabó. Canonical number systems for complex integers. *Acta Sci. Math. (Szeged)* **37** (1975), 255–260.

[Katok and Stepin 1967]   A. B. Katok and A. M. Stepin. Approximations in ergodic theory. *Uspekhi Mat. Nauk* **22** (1967), 81–106. In Russian. English translation in *Russian Math. Surveys* **22** (1967), 76–102.

[Kautz 1965]   W. H. Kautz. Fibonacci codes for synchronization control. *IEEE Trans. Inform. Theory* **11** (1965), 284–292.

[Kawai 1984]   H. Kawai. α-additive functions and uniform distribution modulo one. *Proc. Japan Acad. Ser. A* **60** (1984), 299–301.

[Keane 1968]   M. Keane. Generalized Morse sequences. *Z. Wahrscheinlichkeitstheorie Verw. Gebiete* **10** (1968), 335–353.

[Keane 1970]   M. Keane. Irrational rotations and quasi-ergodic measures. In *Probabilités, 1970–1971*, Vol. 1 of *Publications des Séminaires de Mathématiques*, pp. 17–26. Dép. Math. et Informat., Univ. Rennes, 1970.

[Keane 1975]   M. S. Keane. Interval exchange transformations. *Math. Z.* **141** (1975), 25–31.

[Keane 1991]   M. S. Keane. Ergodic theory and subshifts of finite type. In T. Bedford, M. Keane, and C. Series, editors, *Ergodic Theory, Symbolic Dynamics, and Hyperbolic Spaces*, pp. 35–70. Oxford University Press, 1991.

[Keller 1972]   T. J. Keller. Generalizations of Zeckendorf's theorem. *Fibonacci Quart.* **10** (1972), 95–102; 111–112.

[Kelley 1955]   J. L. Kelley. *General Topology*. Springer-Verlag, 1955.

[Kempner 1916]   A. J. Kempner. On transcendental numbers. *Trans. Amer. Math. Soc.* **17** (1916), 476–482.

[Kempner 1936]   A. J. Kempner. Anormal systems of numeration. *Amer. Math. Monthly* **43** (1936), 610–617.

[Kennedy 1982]   R. E. Kennedy. Digital sums, Niven numbers, and natural density. *Crux Math.* **8** (1982), 129–133.

[Kennedy and Cooper 1984]   R. E. Kennedy and C. N. Cooper. On the natural density of the Niven numbers. *College Math. J.* **15** (1984), 309–312.

[Kennedy and Cooper 1989a]   R. E. Kennedy and C. N. Cooper. Niven repunits and $10^n \equiv 1 \pmod n$. *Fibonacci Quart.* **27** (1989), 139–143.

[Kennedy and Cooper 1991]   R. E. Kennedy and C. N. Cooper. An extension of a theorem by Cheo and Yien concerning digital sums. *Fibonacci Quart.* **29** (1991), 145–149.

[Kennedy and Cooper 1993]   R. E. Kennedy and C. N. Cooper. Sums of powers of digital sums. *Fibonacci Quart.* **31** (1993), 341–345.

[Kennedy, Goodman, and Best 1980]   R. E. Kennedy, T. A. Goodman, and C. A. Best. Mathematical discovery and Niven numbers. *MATYC J.* **14**(3) (Winter 1980), 20–25.

[Keränen 1983]   V. Keränen. On L. Wegner's problem in the general case. *Bull. Eur. Assoc. Theor. Comput. Sci.*, No. 19 (February 1983), 24–31.

[Keränen 1985]   V. Keränen. On k-repetition free words generated by length uniform morphisms over a binary alphabet. In W. Brauer, editor, *Proc. 12th Int. Conf. on Automata, Languages, and Programming (ICALP)*, Vol. 194 of *Lecture Notes in Computer Science*, pp. 338–347. Springer-Verlag, 1985.

[Keränen 1986]   V. Keränen. On the k-freeness of morphisms on free monoids. *Ann. Acad. Sci. Fenn.* **61** (1986), 1–55.

[Keränen 1987]   V. Keränen. On the k-freeness of morphisms on free monoids. In F. J. Brandenburg, G. Vidal-Naquet, and M. Wirsing, editors, *STACS 87, Proc. 4th Symp. Theoretical Aspects of Comp. Sci.*, Vol. 247 of *Lecture Notes in Computer Science*, pp. 180–188. Springer-Verlag, 1987.

[Keränen 1992]   V. Keränen. Abelian squares are avoidable on 4 letters. In W. Kuich, editor, *Proc. 19th Int. Conf. on Automata, Languages, and Programming (ICALP)*,

Vol. 623 of *Lecture Notes in Computer Science*, pp. 41–52. Springer-Verlag, 1992.

[Kervaire, Saffari, and Vaillancourt 1986]   M. Kervaire, B. Saffari, and R. Vaillancourt. Une méthode de détection de nouveaux polynômes vérifiant l'identité de Rudin-Shapiro. *C. R. Acad. Sci. Paris* **302** (1986), 95–98.

[Ketkar and Zamboni 1998]   P. Ketkar and L. Q. Zamboni. Primitive substitutive numbers are closed under rational multiplication. *J. Théorie Nombres Bordeaux* **10** (1998), 315–320.

[Kfoury 1985]   A. J. Kfoury. Definability by deterministic and nondeterministic programs (with applications to first-order dynamic logic). *Inform. Control* **65** (1985), 98–121.

[Kfoury 1988a]   A.-J. Kfoury. A linear-time algorithm to decide whether a binary word contains an overlap. *RAIRO Inform. Théor. App.* **22** (1988), 135–145.

[Kfoury 1988b]   A.-J. Kfoury. Square-free and overlap-free words. In G. Mirkowska and H. Rasiowa, editors, *Mathematical Problems in Computation Theory*, Vol. 21 of *Banach Center Publications*, pp. 285–297. PWN – Polish Scientific Publishers, Warsaw, 1988.

[Kieffer 1988]   J. C. Kieffer. Sturmian minimal systems associated with the iterates of certain functions on an interval. In J. C. Alexander, editor, *Dynamical Systems. Proceedings, University of Maryland, 1986–87*, Vol. 1342 of *Lecture Notes in Mathematics*, pp. 354–360. Springer-Verlag, 1988.

[Kimberling 1979]   C. Kimberling. Advanced problem 6281. *Amer. Math. Monthly* **86** (1979), 793.

[Kimberling 1995]   C. Kimberling. Numeration systems and fractal sequences. *Acta Arith.* **73** (1995), 103–117.

[Kimberling 1997]   C. Kimberling. Fractal sequences and interspersions. *Ars Combin.* **45** (1997), 157–168.

[Kimberling 1998a]   C. Kimberling. Palindromic sequences from irrational numbers. *Fibonacci Quart.* **36** (1998), 171–173.

[Kimberling 1998b]   C. Kimberling. Edouard Zeckendorf. *Fibonacci Quart.* **36** (1998), 416–418.

[Kirschenhofer 1983]   P. Kirschenhofer. Subblock occurrences in the q-ary representation of n. *SIAM J. Algebraic Discrete Methods* **4** (1983), 231–236.

[Kirschenhofer 1990]   P. Kirschenhofer. On the variance of the sum of digits function. In H. Hlawka and R. F. Tichy, editors, *Number-Theoretic Analysis: Seminar, Vienna 1988–89*, Vol. 1452 of *Lecture Notes in Computer Science*, pp. 112–116. Springer-Verlag, Berlin, 1990.

[Kitaev 2000a]   S. Kitaev. There are no iterative morphisms that define the Arshon sequence and the σ-sequence. Unpublished manuscript, 2000. http://www.cs. chalmers.se/~kitaev

[Klarner 1968]   D. A. Klarner. Partitions of N into distinct Fibonacci numbers. *Fibonacci Quart.* **6** (1968), 235–244.

[Klavžar and Milutinović 1997]   S. Klavžar and U. Milutinović. Graphs $S(n, k)$ and a variant of the Tower of Hanoi problem. *Czech. Math. J.* **47** (1997), 95–104.

[Kleene 1956]   S. C. Kleene. Representation of events in nerve nets and finite automata. In *Automata Studies*, pp. 3–42. Princeton University Press, 1956.

[B. Klein 1972]   B. G. Klein. Homomorphisms of symbolic dynamical systems. *Math. Systems Theory* **6** (1972), 107–122.

[C. Klein and Minsker 1993]   C. S. Klein and S. Minsker. The super towers of Hanoi problem: large rings on small rings. *Discrete Math.* **114** (1993), 283–295.

[Klepinin and Sukhanov 1999]   A. V. Klepinin and E. V. Sukhanov. On combinatorial properties of the Arshon sequence. *Diskretn. Anal. Issled. Oper.* **6**(6) (1999),

23–40. In Russian. English translation in *Discrete Appl. Math.* **114** (2001), 155–169.

[Klop 1983]   J. W. Klop. A solution to problem P12 (number 18, October 1982). *Bull. European Assoc. Theor. Comput. Sci.*, No. 19 (February 1983), 31–33.

[Klosinski, Alexanderson, and Hillman 1982]   L. F. Klosinski, G. L. Alexanderson, and A. P. Hillman. The William Lowell Putnam mathematical competition. *Amer. Math. Monthly* **89** (1982), 679–686.

[Klosinski, Alexanderson, and Larson 1985]   L. F. Klosinski, G. L. Alexanderson, and L. C. Larson. The William Lowell Putnam mathematical competition. *Amer. Math. Monthly* **92** (1985), 560–567.

[Klosinski, Alexanderson, and Larson 1993]   L. F. Klosinski, G. L. Alexanderson, and L. C. Larson. The fifty-third William Lowell Putnam mathematical competition. *Amer. Math. Monthly* **100** (1993), 755–767.

[Kmošek 1979]   M. Kmošek. Rozwiniecie niektórych liczb niewymiernych na ulamki lańcuchowe. Master's thesis, University of Warsaw, 1979.

[Knight 1991]   M. J. Knight. An "ocean of zeros" proof that a certain non-Liouville number is transcendental. *Amer. Math. Monthly* **98** (1991), 947–949.

[Knuth 1960]   D. E. Knuth. An imaginary number system. *Comm. ACM* **3** (1960), 245–247. Errata, **4** (1961), 355.

[Knuth 1964]   D. E. Knuth. Transcendental numbers based on the Fibonacci sequence. *Fibonacci Quart.* **2** (1964), 43–44, 52.

[Knuth 1968]   D. E. Knuth. *The Art of Computer Programming. Volume 1: Fundamental Algorithms*. Addison-Wesley, 1968.

[Knuth 1969]   D. E. Knuth. *The Art of Computer Programming. Volume 2: Seminumerical Algorithms*. Addison-Wesley, 1st edition, 1969.

[Knuth 1973]   D. E. Knuth. *The Art of Computer Programming. Volume 3: Sorting and Searching*. Addison-Wesley, 1973.

[Knuth 1981]   D. E. Knuth. *The Art of Computer Programming. Volume 2: Seminumerical Algorithms*. Addison-Wesley, 2nd edition, 1981.

[Knuth 1988]   D. E. Knuth. Fibonacci multiplication. *Appl. Math. Lett.* **1** (1988), 57–60.

[Knuth 1989]   D. E. Knuth. Elementary problem E 3303. *Amer. Math. Monthly* **96** (1989), 54. Solution by J. H. Nieto, **97** (1990), 348–349.

[Knuth, Morris, and Pratt 1977]   D. E. Knuth, J. Morris, and V. Pratt. Fast pattern matching in strings. *SIAM J. Comput.* **6** (1977), 323–350.

[Kobayashi 1986]   Y. Kobayashi. Repetition-free words. *Theoret. Comput. Sci.* **44** (1986), 175–197.

[Kobayashi 1988]   Y. Kobayashi. Enumeration of irreducible binary words. *Discrete Appl. Math.* **20** (1988), 221–232.

[Koblitz 1984]   N. Koblitz. *p-adic Numbers, p-adic Analysis, and Zeta-Functions*. Springer-Verlag, 2nd edition, 1984.

[Koblitz 1992]   N. Koblitz. CM-curves with good cryptographic properties. In J. Feigenbaum, editor, *Advances in Cryptology – CRYPTO '91 Proceedings*, Vol. 576 of *Lecture Notes in Computer Science*, pp. 279–287. Springer-Verlag, 1992.

[Kobuchi 1977]   Y. Kobuchi. Two characterization theorems of locally catenative developmental systems. *Inform. Process. Lett.* **6** (1977), 120–124.

[Kobuchi and Wood 1981]   Y. Kobuchi and D. Wood. On the complete simulation of D0L schemes and locally catenative schemes. *Inform. Comput.* **49** (1981), 64–80.

[Kohmoto, Kadanoff, and Tang 1983]   M. Kohmoto, L. P. Kadanoff, and C. Tang. Localization problem in one dimension: mapping and escape. *Phys. Rev. Lett.* **50** (1983), 1870–1872.

[Koksma 1936]   J. F. Koksma. *Diophantische Approximationen*. J. Springer, 1936.

[Kolakoski 1965]   W. Kolakoski. Elementary problem 5304. *Amer. Math. Monthly* **72**
   (1965), 674. Solution, **73** (1966), 681–682.
[Kolář 1994]   M. Kolář. Comment: High-resolution X-ray-diffraction spectra of
   Thue–Morse GaAs–AlAs heterostructures. *Phys. Rev. Lett.* **73** (1994), 1307.
[Kolář and Ali 1990]   M. Kolář and M. K. Ali. One-dimensional generalized Fibonacci
   tilings. *Phys. Rev. B* **41** (1990), 7108–7112.
[Kolář and Nori 1990]   M. Kolář and F. Nori. Trace maps of general substitutional
   sequences. *Phys. Rev. B* **42** (1990), 1062–1065.
[Kolden 1949]   K. Kolden. Continued fractions and linear substitutions. *Arch. Math.
   Naturvidenskab* **50** (1949), 141–196.
[Kolpakov and Kucherov 1999b]   R. Kolpakov and G. Kucherov. Finding maximal
   repetitions in a word in linear time. In *Proc. 40th Ann. Symp. Found. Comput. Sci.*,
   pp. 596–604. IEEE Press, 1999.
[Komatsu 1995a]   T. Komatsu. The fractional part of $n\theta + \phi$ and Beatty sequences. *J.
   Théorie Nombres Bordeaux* **7** (1995), 387–406.
[Komatsu and van der Poorten 1996]   T. Komatsu and A. J. van der Poorten. Substitution
   invariant Beatty sequences. *Japanese J. Math.* **22** (1996), 349–354.
[Koplowitz, Lindenbaum, and Bruckstein 1990]   J. Koplowitz, M. Lindenbaum, and
   A. Bruckstein. The number of digital straight lines on an $N \times N$ grid. *IEEE Trans.
   Inform. Theory* **36** (1990), 192–197.
[Korec 1990]   I. Korec. Pascal triangles modulo n and modular trellises. *Comput. and
   Artificial Intelligence* **9** (1990), 105–113.
[Körmendi 1986b]   S. Körmendi. Canonical number systems in $\mathbb{Q}(\sqrt[3]{2})$. *Acta Sci. Math.
   (Szeged)* **50** (1986), 351–357.
[Körner 1979]   T. W. Körner. A Rudin-Shapiro type theorem. *Illinois J. Math.* **23** (1979),
   217–240.
[Körner 1980]   T. W. Körner. On a polynomial of Byrnes. *Bull. Lond. Math. Soc.* **12**
   (1980), 219–224.
[Korobov 1990]   A. N. Korobov. Continued fractions of certain normal numbers. *Mat.
   Zametki* **47**(2) (1990), 28–33. In Russian. English translation in *Math. Notes Acad.
   Sci. USSR* **47** (1990), 128–132.
[Kosaraju 1994]   S. R. Kosaraju. Computation of squares in a string. In M. Crochemore
   and D. Gusfield, editors, *Combinatorial Pattern Matching, 1994*, Vol. 807 of *Lecture
   Notes in Computer Science*, pp. 146–150. Springer-Verlag, 1994.
[Koskas 1996]   M. Koskas. About the p-paperfolding words. *Theoret. Comput. Sci.* **158**
   (1996), 35–51.
[Koskas 1998]   M. Koskas. Complexité de suites de Toeplitz. *Discrete Math.* **183** (1998),
   161–183.
[Kovács 1981a]   B. Kovács. Canonical number systems in algebraic number fields. *Acta
   Math. Acad. Sci. Hung.* **37** (1981), 405–407.
[Kovács and Pethö 1991]   B. Kovács and A. Pethö. Number systems in integral domains,
   especially in orders of algebraic number fields. *Acta Sci. Math. (Szeged)* **55** (1991),
   287–299.
[Kuich 1997]   W. Kuich. Semirings and formal power series. In G. Rozenberg and
   A. Salomaa, editors, *Handbook of Formal Languages*, Vol. 1, pp. 609–677.
   Springer-Verlag, 1997.
[Kuich and Salomaa 1986]   W. Kuich and A. Salomaa. *Semirings, Automata, Languages*.
   Springer-Verlag, 1986.
[Kuipers and Niederreiter 1974]   L. Kuipers and H. Niederreiter. *Uniform Distribution of
   Sequences*. Wiley, 1974.
[Kullberg 1996]   E. Kullberg. Beyond infinity: On the infinity series – the DNA of
   hierarchical music. In A. Beyer, editor, *The Music of Per Nørgård: Fourteen
   Interpretive Essays*, pp. 71–93. Scolar Press, 1996.

[Laakso 1996]   T. Laakso. Musical rendering of an infinite repetition-free string. In C. Gefwert, P. Orponen, and J. Seppänen, editors, *Proc. Logic, Mathematics and the Computer*, Vol. 14 of *Publications of the Finnish Artificial Intelligence Society, Symposiosarja*, pp. 292–297. 1996.

[Lalanne 1840]   L. Lalanne. Note sur quelques propositions d'arithmologie élémentaire. *C. R. Acad. Sci. Paris* **11** (1840), 903–905.

[Lallement 1979]   G. Lallement. *Semigroups and Combinatorial Applications*. John Wiley & Sons, 1979.

[Lambek and Moser 1954]   J. Lambek and L. Moser. Inverse and complementary sequences of natural numbers. *Amer. Math. Monthly* **61** (1954), 454–458.

[Lambert 1761]   J. H. Lambert. Mémoire sur quelques propriétés remarquables des quantités transcendantes circulaires et logarithmiques. *Histoire Acad. Roy. Sci. et Belles Lettr., Berlin* (1761), 265–322. Reprinted in A. Speiser, editor, *Opera Mathematica*, Vol. II, pp. 112–159. Orell Füssli Verlag, 1948.

[Landau 1903]   E. Landau. Über die Maximalordnung der Permutationen gegebenen Grades. *Arkiv Math. und Physik* **5** (1903), 92–103.

[Lang 1995]   S. Lang. *Introduction to Diophantine Approximations*. Springer-Verlag, 1995.

[Langevin 1991]   M. Langevin. Stimulateur cardiaque et suites de Farey. *Period. Math. Hung.* **23** (1991), 75–86.

[Langford 1958]   C. D. Langford. Problem. *Math. Gazette* **42** (1958), 228.

[Lasjaunias 1997]   A. Lasjaunias. Diophantine approximation and continued fraction expansions of algebraic power series in positive characteristic. *J. Number Theory* **65** (1997), 206–225.

[Lasjaunias 1999]   A. Lasjaunias. Continued fractions for algebraic formal power series over a finite base field. *Finite Fields Appl.* **5** (1999), 46–56.

[Lasjaunias 2000a]   A. Lasjaunias. Quartic power series in $\mathbb{F}_3((T^{-1}))$ with bounded partial quotients. *Acta Arith.* **95** (2000), 49–59.

[Lasjaunias 2000b]   A. Lasjaunias. A survey of Diophantine approximation in fields of power series. *Monatsh. Math.* **130** (2000), 211–229.

[Laubie 1991]   F. Laubie. Prolongements homographiques des substitutions de mots de Christoffel. *C. R. Acad. Sci. Paris* **313** (1991), 565–567.

[Laubie and Laurier 1995]   F. Laubie and E. Laurier. Calcul de multiples de mots de Christoffel. *C. R. Acad. Sci. Paris* **320** (1995), 765–768.

[Lavallée 1985]   I. Lavallée. Note sur le problème des tours de Hanoï. *Rev. Roumaine Math. Pures Appl.* **30** (1985), 433–438.

[Lazarkiewicz and Balasiński 1961]   A. Lazarkiewicz and W. Balasiński. A simple experimental computer with negative basis. *Math. Comp.* **15** (1961), 275–285.

[Lecomte and Rigo 2001]   P. B. A. Lecomte and M. Rigo. Numeration systems on a regular language. *Theory Comput. Systems* **34** (2001), 27–44.

[Leconte 1985]   M. Leconte. *K*th power-free codes. In M. Nivat and D. Perrin, editors, *Automata on Infinite Words*, Vol. 192 of *Lecture Notes in Computer Science*, pp. 172–187. Springer-Verlag, 1985.

[Leech 1957]   J. Leech. A problem on strings of beads. *Math. Gazette* **41** (1957), 277–278.

[Legendre 1830]   A.-M. Legendre. *Théorie des Nombres*. Firmin Didot Frères, Paris, 1830.

[Lehmer 1947]   D. H. Lehmer. The Tarry–Escott problem. *Scripta Math.* **13** (1947), 37–41.

[Lehr 1993]   S. Lehr. Sums and rational multiples of q-automatic sequences are q-automatic. *Theoret. Comput. Sci.* **108** (1993), 385–391.

[Lehr, Shallit, and Tromp 1996]   S. Lehr, J. Shallit, and J. Tromp. On the vector space of the automatic reals. *Theoret. Comput. Sci.* **163** (1996), 193–210.

[Leighton and Scott 1939]   W. Leighton and W. T. Scott. A general continued fraction expansion. *Bull. Amer. Math. Soc.* **45** (1939), 596–605.

[Leinfellner 1999]   H. Leinfellner. New results on rarefied sums of the Thue-Morse sequence. In *Beiträge zur zahlentheoretischen Analysis*, Vol. 338 of *Grazer Math. Ber.*, pp. 9–30. Karl-Franzens-Univ., 1999.

[Leiss 1983a]   E. L. Leiss. Solving the "Towers of Hanoi" on graphs. *J. Combin. Inform. System Sci.* **8** (1983), 81–89.

[Leiss 1983b]   E. L. Leiss. On restricted Hanoi problems. *J. Combin. Inform. System Sci.* **8** (1983), 277–285.

[Leiss 1984]   E. L. Leiss. Finite Hanoi problems: how many discs can be handled? *Congr. Numer.* **44** (1984), 221–229.

[Lekkerkerker 1952]   C. G. Lekkerkerker. Voorstelling van natuurlijke getallen door een som van getallen van Fibonacci. *Simon Stevin* **29** (1952), 190–195.

[Lenard 1991]   A. Lenard. Problem E 3474. *Amer. Math. Monthly* **98** (1991), 956. Solution, **101** (1994), 177–179.

[Lengyel 1994]   T. Lengyel. Characterizing the 2-adic order of the logarithm. *Fibonacci Quart.* **32** (1994), 397–401.

[Lennholm and Hörnquist 1999]   E. Lennholm and M. Hörnquist. Role of aperiodic order for fluxon dynamics in Josephson junction arrays. *Phys. Rev. E* **59** (1999), 381–389.

[Lepistö 1994]   A. Lepistö. Repetitions in Kolakoski sequence. In G. Rozenberg and A. Salomaa, editors, *Developments in Language Theory*, pp. 130–143. World Scientific, 1994.

[Lepistö 1996]   A. Lepistö. On the computational complexity of infinite words. In J. Dassow, G. Rozenberg, and A. Salomaa, editors, *Developments in Language Theory II*, pp. 350–359. World Scientific, 1996.

[LeVeque 1956]   W. J. LeVeque. *Topics in Number Theory*, Vol. II. Addison-Wesley, 1956.

[Levi 1994]   F. W. Levi. On semigroups. *Bull. Calcutta Math. Soc.* **36** (1944), 141–146.

[Levine and Steinhardt 1984]   D. Levine and P. J. Steinhardt. Quasicrystals: a new class of ordered solids. *Phys. Rev. Lett.* **53** (1984), 2477–2480.

[Levine 1988]   E. Levine. Problem 386. *College Math. J.* **19** (1988), 448. Solution, **21** (1990), 151–152.

[Li and Reingold 1989]   Z. Li and E. M. Reingold. Solution of a divide-and-conquer maximin recurrence. *SIAM J. Comput.* **18** (1989), 1188–1200.

[Liang 1979]   F. M. Liang. A short proof of the $3d$ distance theorem. *Discrete Math.* **28** (1979), 325–326.

[van de Liefvoort 1992]   A. van de Liefvoort. An iterative algorithm for the Reve's puzzle. *Computer J.* **35** (1992), 91–92.

[Lin and Goda 1997]   Z. Lin and M. Goda. Long-range correlations in quantum systems with aperiodic Hamiltonians. *Physica E* **55** (1997), 2632–2639.

[Lin and Tao 1990]   Z. Lin and R. Tao. Quantum Ising model on Thue–Morse aperiodic chain. *Phys. Lett. A* **150** (1990), 11–13.

[Lin and Tao 1992]   Z. Lin and R. Tao. Phase transition of quantum Ising spin models on g-letter generalized Thue-Morse aperiodic chains. *J. Phys. A: Math. Gen.* **25** (1992), 2483–2488.

[Lindemann 1882]   F. Lindemann. Ueber die Zahl π. *Math. Annalen* **20** (1882), 213–225.

[Lindenbaum and Bruckstein 1993]   M. Lindenbaum and A. Bruckstein. On recursive, $O(N)$ partitioning of a digitized curve into digital straight segments. *IEEE Trans. Pattern Anal. Machine Intell.* **15** (1993), 949–953.

[Lindenbaum and Koplowitz 1991]   M. Lindenbaum and J. Koplowitz. A new parameterization of digital straight lines. *IEEE Trans. Pattern Anal. Machine Intell.* **13** (1991), 847–852.

[Lindquist and Riklund 1998]   B. Lindquist and R. Riklund. Deterministic aperiodic one-dimensional systems with all states extended, one of which is periodic. *J. Phys. Soc. Japan* **67** (1998), 1672–1676.

[Lindström 1997]   B. Lindström. On the binary digits of a power. *J. Number Theory* **65** (1997), 321–324.

[Lindström and Zetterström 1967]   B. Lindström and H.-O. Zetterström. A combinatorial problem in the k-adic number system. *Proc. Amer. Math. Soc.* **18** (1967), 166–170.

[Liouville 1844]   J. Liouville. Sur des classes très étendues de quantités dont la valeur n'est ni algébrique, ni même reductible à des irrationelles algébriques. *C. R. Acad. Sci. Paris* **18** (1844), 883–885, 910–911.

[Littlewood 1966]   J. E. Littlewood. On polynomials $\sum \pm z^m$, $\sum e^{\alpha m i} z^m$, $z = e^{\theta i}$. *J. London Math. Soc.* **41** (1966), 367–376.

[Littlewood 1968]   J. E. Littlewood. *Some Problems in Real and Complex Analysis.* Heath, 1968. See Problem 19.

[Liu 1997]   N.-h. Liu. Propagation of light waves in Thue-Morse dielectric multilayers. *Phys. Rev. B* **55** (1997), 3543–3547.

[Liviotti 1996]   E. Liviotti. A study of the structure factor of Thue–Morse and period-doubling chains by wavelet analysis. *J. Phys.: Condens. Matter* **8** (1996), 5007–5015.

[Liviotti and Erdős 1995]   E. Liviotti and P. Erdős. Comment on "Localization in a one-dimensional Thue-Morse chain". *Phys. Rev. B* **51** (1995), 3273–3275.

[Loftus, Shallit, and Wang 2000]   J. Loftus, J. O. Shallit, and M.-w. Wang. New problems of pattern avoidance. In G. Rozenberg and W. Thomas, editors, *Developments in Language Theory 1999*, pp. 185–199. World Scientific, 2000.

[Lopez and Narbel 2001]   L.-M. Lopez and P. Narbel. Substitutions and interval exchange transformations of rotation class. *Theoret. Comput. Sci.* **255** (2001), 323–344.

[Loraud 1995]   N. Loraud. β-shift, systèmes de numération et automates. *J. Théorie Nombres Bordeaux* **7** (1995), 473–498.

[Lothaire 1983]   M. Lothaire. *Combinatorics on Words*, Vol. 17 of *Encyclopedia of Mathematics and Its Applications*. Addison-Wesley, 1983.

[Lothaire 2002]   M. Lothaire. *Algebraic Combinatorics on Words*, Vol. 90 of *Encyclopedia of Mathematics and Its Applications*. Cambridge University Press, 2002.

[Loxton and van der Poorten 1977c]   J. H. Loxton and A. J. van der Poorten. Arithmetic properties of certain functions in several variables II. *J. Austral. Math. Soc. Ser. A* **24** (1977), 393–408.

[Loxton and van der Poorten 1977d]   J. H. Loxton and A. J. van der Poorten. Arithmetic properties of certain functions in several variables III. *Bull. Austral. Math. Soc.* **16** (1977), 15–47.

[Loxton and van der Poorten 1978]   J. H. Loxton and A. J. van der Poorten. Algebraic independence properties of the Fredholm series. *J. Austral. Math. Soc. Ser. A* **26** (1978), 31–45.

[Loxton and van der Poorten 1982]   J. H. Loxton and A. J. van der Poorten. Arithmetic properties of the solutions of a class of functional equations. *J. Reine Angew. Math.* **330** (1982), 159–172.

[Loxton and van der Poorten 1988]   J. H. Loxton and A. J. van der Poorten. Arithmetic properties of automata: regular sequences. *J. Reine Angew. Math.* **392** (1988), 57–69.

[Lu 1988]   X.-M. Lu. Towers of Hanoi problem with arbitrary $k \geq 3$ pegs. *Int. J. Comput. Math.* **24** (1988), 39–54.

[Lu 1989]   X.-M. Lu. An iterative solution for the 4-peg Towers of Hanoi. *Computer J.* **32** (1989), 187–189.

[Lu and Dillon 1994]   X.-M. Lu and T. S. Dillon. A note on parallelism for the Towers of Hanoi. *Math. and Computer Modelling* **20**(3) (1994), 1–6.

[Lu and Dillon 1995]   X.-M. Lu and T. S. Dillon. Parallelism for multipeg Towers of Hanoi. *Math. and Computer Modelling* **21**(3) (1995), 3–17.

[Lu and Dillon 1996]   X.-M. Lu and T. S. Dillon. Nonrecursive solution to parallel multipeg Towers of Hanoi: a decomposition approach. *Math. and Computer Modelling* **24**(3) (1996), 29–35.

[de Luca 1981]   A. de Luca. A combinatorial property of the Fibonacci words. *Inform. Process. Lett.* **12** (1981), 193–195.

[de Luca 1983]   A. de Luca. Sul problema di Burnside per i semigruppi e i linguaggi. *Rend. Sem. Mat. Fis. Milano* **53** (1983), 207–220. In Italian.

[de Luca 1984]   A. de Luca. On the product of square-free words. *Discrete Math.* **52** (1984), 143–157.

[de Luca 1990]   A. de Luca. On the Burnside problem for semigroups. (1990), 185–200.

[de Luca 1995]   A. de Luca. A division property of the Fibonacci word. *Inform. Process. Lett.* **54** (1995), 307–312.

[de Luca 1996]   A. de Luca. On standard Sturmian morphisms. In F. Meyer auf der Heide and B. Monien, editors, *Proc. 23rd Int. Conf. on Automata, Languages, and Programming (ICALP)*, Vol. 1099 of *Lecture Notes in Computer Science*, pp. 403–415. Springer-Verlag, 1996.

[de Luca 1997]   A. de Luca. Standard Sturmian morphisms. *Theoret. Comput. Sci.* **178** (1997), 205–224.

[de Luca and Mignosi 1994]   A. de Luca and F. Mignosi. Some combinatorial properties of Sturmian words. *Theoret. Comput. Sci.* **136** (1994), 361–385.

[de Luca and Mione 1994]   A. de Luca and L. Mione. On bispecial factors of the Thue-Morse word. *Inform. Process. Lett.* **49** (1994), 179–183.

[de Luca and Varricchio 1989a]   A. de Luca and S. Varricchio. Some combinatorial properties of the Thue-Morse sequence and a problem in semigroups. *Theoret. Comput. Sci.* **63** (1989), 333–348.

[de Luca and Varricchio 1989b]   A. de Luca and S. Varricchio. Factorial languages whose growth function is quadratically upper bounded. *Inform. Process. Lett.* **30** (1989), 283–288.

[de Luca and Varricchio 1990]   A. de Luca and S. Varricchio. A combinatorial theorem on p-power-free words and an application to semigroups. *RAIRO Inform. Théor. Appl.* **24** (1990), 205–228.

[de Luca and Varricchio 1991a]   A. de Luca and S. Varricchio. Finiteness and iteration conditions for semigroups. *Theoret. Comput. Sci.* **87** (1991), 315–327.

[de Luca and Varricchio 1991b]   A. de Luca and S. Varricchio. Combinatorial properties of uniformly recurrent words and an application to semigroups. *Int. J. Algebra Comput.* **1** (1991), 227–245.

[Lucas 1884]   E. Lucas. Le calcul et les machines à calculer. *Assoc. Française Avance. Sci. C. R.* **13** (1884), 111–141.

[Lunnon 1986]   W. F. Lunnon. The Reve's puzzle. *Computer J.* **29** (1986), 478.

[Lunnon and Pleasants 1987]   W. F. Lunnon and P. A. B. Pleasants. Quasicrystallographic tilings. *J. Math. Pures Appl.* **66** (1987), 217–263.

[Lunnon and Pleasants 1992]   W. F. Lunnon and P. A. B. Pleasants. Characterization of two-distance sequences. *J. Austral. Math. Soc. Ser. A* **43** (1992), 198–218.

[Lunnon, Pleasants, and Stephens 1979]   W. F. Lunnon, P. A. B. Pleasants, and N. M. Stephens. Arithmetic properties of Bell numbers to a composite modulus I. *Acta Arith.* **35** (1979), 1–16.

[Ly 2000]   O. Ly. Automatic graphs and graph D0L-systems. In M. Nielsen and B. Rovan, editors, *Proc. 25th Symposium, Mathematical Foundations of Computer*

Science 2000, Vol. 1893 of *Lecture Notes in Computer Science*, pp. 539–548.
Springer-Verlag, 2000.

[Lyndon 1951]   R. C. Lyndon. Advanced problem 4454. *Amer. Math. Monthly* **58** (1951),
569. Solution by W. Gustin, **60** (1953), 51.

[Lyndon and Schützenberger 1962]   R. C. Lyndon and M. P. Schützenberger. The
equation $a^M = b^N c^P$ in a free group. *Michigan Math. J.* **9** (1962), 289–298.

[Lysënok 1992]   I. G. Lysënok. Infinity of Burnside groups of period 2^k for $k \geq 13$.
Uspekhi. Mat. Nauk **47** (1992), 201–202. In Russian. English translation in *Russian
Math. Surveys* **47** (1992), 229–230.

[Lysënok 1996]   I. G. Lysënok. Infinite Burnside groups of even period. *Izv. Ross. Akad.
Nauk Ser. Mat.* **60** (1996), 3–224. In Russian.

[MacCluer 2000]   C. R. MacCluer. The many proofs and applications of Perron's
theorem. *SIAM Rev.* **42** (2000), 487–498.

[MacMurray 1938]   D. MacMurray. A mathematician gives an hour to chess. *Chess
Review* (October 1938). Reprinted in Bruce Pandolfini, editor, *The Best of Chess Life
& Review – Vol. I 1933–1960*, p. 84. Simon and Schuster, 1988.

[Maes 1999]   A. Maes. An automata-theoretic decidability proof for first-order theory of
$\langle \mathbb{N}, <, P \rangle$ with morphic predicate P. *J. Automata, Languages, and Combin.* **4**
(1999), 229–245.

[Magnus, Karrass, and Solitar 1976]   W. Magnus, A. Karrass, and D. Solitar.
Combinatorial Group Theory. Dover, 2nd revised edition, 1976.

[Mahler 1927]   K. Mahler. On the translation properties of a simple class of arithmetical
functions. *J. Math. and Phys.* **6** (1927), 158–163.

[Mahler 1929]   K. Mahler. Arithmetische Eigenschaften der Lösungen einer Klasse von
Funktionalgleichungen. *Math. Annalen* **101** (1929), 342–366. Corrigendum, **103**
(1930), 532.

[Mahler 1937a]   K. Mahler. Ueber die Dezimalbruchentwicklung gewisser
Irrationalzahlen. *Mathematica B (Zutphen, Holland)* **6** (1937–1938), 2–16.

[Mahler 1937b]   K. Mahler. Arithmetische Eigenschaften einer Klasse von
Dezimalbrüchen. *Proc. Konin. Neder. Akad. Wet.* **40** (1937), 421–428.

[Mahler 1961]   K. Mahler. *Lectures on Diophantine Approximations, Part I: g-adic
Numbers and Roth's Theorem*. University of Notre Dame, 1961.

[Mahler 1968]   K. Mahler. An unsolved problem on the powers of 3/2. *J. Austral. Math.
Soc.* **8** (1968), 313–321.

[Mahler 1976a]   K. Mahler. *Lectures on Transcendental Numbers*, Vol. 546 of *Lecture
Notes in Mathematics*. Springer-Verlag, 1976.

[Mahler 1976b]   K. Mahler. On a class of transcendental numbers. *Comm. Pure Appl.
Math.* **29** (1976), 716–725.

[Mahler 1980]   K. Mahler. On a special function. *J. Number Theory* **12** (1980),
20–26.

[Mahler 1987]   K. Mahler. On two analytic functions. *Acta Arith.* **49** (1987), 15–20.

[Main 1985]   M. G. Main. An infinite square-free co-CFL. *Inform. Process. Lett.* **20**
(1985), 105–107.

[Main 1989]   M. G. Main. Detecting leftmost maximal periodicities. *Discrete Appl.
Math.* **25** (1989), 145–153.

[Main, Bucher, and Haussler 1987]   M. G. Main, W. Bucher, and D. Haussler.
Applications of an infinite square-free co-CFL. *Theoret. Comput. Sci.* **49** (1987),
113–119.

[Main and Lorentz 1984]   M. G. Main and R. J. Lorentz. An $O(n \log n)$ algorithm for
finding all repetitions in a string. *J. Algorithms* **5** (1984), 422–432.

[Main and Lorentz 1985]   M. G. Main and R. J. Lorentz. Linear time recognition of
squarefree strings. In A. Apostolico and Z. Galil, editors, *Combinatorial Algorithms
on Words*, pp. 271–278. Springer-Verlag, Berlin, 1985.

[Majumdar 1994a]   A. A. K. Majumdar. The generalized four-peg Tower of Hanoi problem. *Optimization* **29** (1994), 349–360.

[Majumdar 1994b]   A. A. K. Majumdar. A note on the iterative algorithm for the Reve's puzzle. *Computer J.* **37** (1994), 463–464.

[Majumdar 1995]   A. A. K. Majumdar. The generalized *p*-peg Tower of Hanoi problem. *Optimization* **32** (1995), 175–183.

[Majumdar 1996]   A. A. K. Majumdar. Generalized multi-peg Tower of Hanoi problem. *J. Austral. Math. Soc. Ser. B* **38** (1996), 201–208.

[Makowski 1966]   A. Makowski. On Kaprekar's "junction numbers". *Math. Student* **34** (1966), 77.

[Mallol, López, and Serrato 1996]   C. Mallol, J. López, and D. Serrato. Hanoi Towers: study of a class of algorithms. *Nova J. Math. Game Theory Algebra* **5** (1996), 383–388.

[Mandelbrot 1983]   B. B. Mandelbrot. *The Fractal Geometry of Nature*. W.H. Freeman and Company, San Francisco, 1983.

[Marcus 1964]   S. Marcus. Sur un modèle de H. B. Curry pour le langage mathématique. *C. R. Acad. Sci. Paris* **258** (1964), 1954–1956.

[Marcus and Păun 1989]   S. Marcus and G. Păun. Langford strings, formal languages, and contextual ambiguity. *Int. J. Comput. Math.* **26** (1989), 179–191.

[Markoff 1882]   A. A. Markoff. Sur une question de Jean Bernoulli. *Math. Ann.* **19** (1882), 27–36.

[G. Martin 2001]   G. Martin. Absolutely abnormal numbers. *Amer. Math. Monthly* **108** (2001), 746–754.

[J. Martin 1976]   J. C. Martin. Generalized Morse sequences on *n* symbols. *Proc. Amer. Math. Soc.* **54** (1976), 379–383.

[J. Martin 1977]   J. C. Martin. The structure of generalized Morse minimal sets on *n* symbols. *Trans. Amer. Math. Soc.* **232** (1977), 343–355.

[M. Martin 1934]   M. H. Martin. A problem in arrangements. *Bull. Amer. Math. Soc.* **40** (1934), 859–864.

[Martinez 2001]   P. Martinez. Some new irrational decimal fractions. *Amer. Math. Monthly* **108** (2001), 250–253.

[Masser 1999]   D. W. Masser. Algebraic independence properties of the Hecke–Mahler series. *Quart. J. Math. Oxford* **50** (1999), 207–230.

[Massias 1984]   J.-P. Massias. Majoration explicite de l'ordre maximum d'un élément du groupe symétrique. *Ann. Fac. Sci. Toulouse* **6** (1984), 269–281.

[de Mathan 1970]   B. de Mathan. Approximations diophantiennes dans un corps local. *Bull. Soc. Math. France Suppl. Mém.* **21** (1970), 1–93.

[Matiyasevich 1993]   Yu. V. Matiyasevich. *Hilbert's Tenth Problem*. The MIT Press, 1993.

[Matula 1976]   D. W. Matula. Radix arithmetic: digital algorithms for computer architecture. In R. T. Yeh, editor, *Applied Computation Theory: Analysis, Design, Modeling*, pp. 374–448. Prentice-Hall, 1976.

[Matula 1978]   D. W. Matula. Basic digit sets for radix representation of the integers. In *Proc. 4th IEEE Symp. Computer Arithmetic*, pp. 1–9. IEEE Press, 1978.

[Matula 1982]   D. W. Matula. Basic digit sets for radix representation. *J. Assoc. Comput. Mach.* **29** (1982), 1131–1143.

[Mauclaire 1987]   J.-L. Mauclaire. Sur certaines fonctions définies par les chiffres des entiers. *Proc. Japan Acad. Ser. A* **63** (1987), 201–204.

[Mauclaire 1994]   J.-L. Mauclaire. A characterization of generalized Rudin-Shapiro sequences with values in a locally compact abelian group. *Acta Arith.* **68** (1994), 213–217.

[Mauclaire 1997]   J.-L. Mauclaire. Some consequences of a result of Jean Coquet. *J. Number Theory* **62** (1997), 1–18.

[Mauduit 2001]   C. Mauduit. Multiplicative properties of the Thue–Morse sequence. *Period. Math. Hung.* **43** (2001), 137–153.

[Mauduit and Sárközy 1996]   C. Mauduit and A. Sárközy. On the arithmetic structure of sets characterized by sum of digits properties. *J. Number Theory* **61** (1996), 25–38.

[McCulloch and Pitts 1943]   W. S. McCulloch and W. Pitts. A logical calculus of the ideas immanent in nervous activity. *Bull. Math. Biophy.* **5** (1943), 115–133.

[McDaniel and Yates 1989]   W. L. McDaniel and S. Yates. The sum of digits function and its application to a generalization of the Smith number problem. *Nieuw Arch. Wiskunde* **7** (1989), 39–51.

[McIlroy 1974]   M. D. McIlroy. The number of 1's in binary integers: bounds and extremal properties. *SIAM J. Comput.* **3** (1974), 255–261.

[McIlroy 1984]   M. D. McIlroy. A note on discrete representation of lines. *AT&T Tech. J.* **64** (1984), 481–490.

[McMullen and Price 1976]   J. R. McMullen and J. F. Price. Rudin-Shapiro sequences for arbitrary compact groups. *J. Austral. Math. Soc. Ser. A* **22** (1976), 421–430.

[McNaughton 1963]   R. McNaughton. Review of two papers by Büchi. *J. Symbolic Logic* **28** (1963), 100–102.

[McNaughton and Zalcstein 1975]   R. McNaughton and Y. Zalcstein. The Burnside problem for semigroups. *J. Algebra* **34** (1975), 292–299.

[Mealy 1955]   G. H. Mealy. A method for synthesizing sequential circuits. *Bell System Tech. J.* **34** (1955), 1045–1079.

[Meijer 1963]   H. G. Meijer. Irrational power series. *Proc. Konin. Neder. Akad. Wet. Ser. A* **66** (1963), 682–690. (= *Indag. Math.* **25**).

[Mendès France 1967]   M. Mendès France. Nombres normaux, applications aux fonctions pseudoaléatoires. *J. d'Analyse Math.* **20** (1967), 1–56.

[Mendès France 1970]   M. Mendès France. Fonctions *g*-additives et les suites à spectre vide. In *Séminaire Delange-Pisot-Poitou*, pp. 10.01–10.06, 1970–1971.

[Mendès France 1973a]   M. Mendès France. Sur les fractions continues limitées. *Acta Arith.* **23** (1973), 207–215.

[Mendès France 1973b]   M. Mendès France. Les suites à spectre vide et la répartition modulo 1. *J. Number Theory* **5** (1973), 1–15.

[Mendès France 1973c]   M. Mendès France. Les suites additives et leur répartition (mod. 1). In *Séminaire de Théorie des Nombres de Bordeaux*, pp. 8.01–8.06, 1973–1974.

[Mendès France 1980]   M. Mendès France. Sur les décimales des nombres algébriques réels. In *Séminaire de Théorie des Nombres de Bordeaux*, pp. 28.01–28.07, 1979–1980.

[Mendès France 1981]   M. Mendès France. Principe de la symétrie perturbée. In M.-J. Bertin, editor, *Séminaire de Théorie des Nombres, Paris 1979–80*, pp. 77–98. Birkhäuser, 1981.

[Mendès France 1984b]   M. Mendès France. Automates et nombres transcendants. In *Séminare de Théorie des Nombres 1982–83*, pp. 173–183. Birkhäuser, 1984.

[Mendès France 1984c]   M. Mendès France. Folding paper and thermodynamics. *Phys. Reports* **103** (1984), 161–172.

[Mendès France 1986]   M. Mendès France. The Ising chain with nonconstant external field. *J. Statist. Phys.* **45** (1986), 89–97.

[Mendès France 1989]   M. Mendès France. Chaos implies confusion. In M. M. Dodson and J. A. G. Vickers, editors, *Number Theory and Dynamical Systems*, Vol. 134 of *London Math. Soc. Lect. Note Ser.*, pp. 137–152. Cambridge University Press, 1989.

[Mendès France 1990a]   M. Mendès France. The inhomogeneous Ising chain and paperfolding. In J.-M. Luck, P. Moussa, and M. Waldschmidt, editors, *Number Theory and Physics*, Vol. 47 of *Springer Proceedings in Physics*, pp. 195–202. Springer-Verlag, 1990.

[Mendès France 1990b]   M. Mendès France. The Rudin-Shapiro sequence, Ising chain, and paperfolding. In B. C. Berndt, H. G. Diamond, H. Halberstam, and A. Hildebrand, editors, *Analytic Number Theory. Proceedings of a Conference in Honor of Paul T. Bateman*, Vol. 85 of *Portugal. Math.*, pp. 367–382. Birkhäuser, Boston, 1990.

[Mendès France 1991]   M. Mendès France. Opacity of an automaton. Application to the inhomogeneous Ising chain. *Commun. Math. Phys.* **139** (1991), 341–352.

[Mendès France and van der Poorten 1981]   M. Mendès France and A. J. van der Poorten. Arithmetic and analytic properties of paper folding sequences. *Bull. Austral. Math. Soc.* **24** (1981), 123–131.

[Mendès France and van der Poorten 1991]   M. Mendès France and A. J. van der Poorten. Some explicit continued fraction expansions. *Mathematika* **38** (1991), 1–9.

[Mendès France, van der Poorten, and Shallit 1999]   M. Mendès France, A. J. van der Poorten, and J. O. Shallit. On lacunary formal power series and their continued fraction expansion. In K. Györy, H. Iwaniec, and J. Urbanowicz, editors, *Number Theory in Progress*, Vol. 1, pp. 321–326. Walter de Gruyter, 1999.

[Mendès France and Shallit 1989]   M. Mendès France and J. Shallit. Wire bending. *J. Combin. Theory. Ser. A* **50** (1989), 1–23.

[Mendès France and Tenenbaum 1981]   M. Mendès France and G. Tenenbaum. Dimension des courbes planes, papiers pliés et suites de Rudin–Shapiro. *Bull. Soc. Math. France* **109** (1981), 207–215.

[Mendès France and Yao 1997]   M. Mendès France and J.-Y. Yao. Transcendence and the Carlitz–Goss gamma function. *J. Number Theory* **63** (1997), 396–402.

[Merekin 1998]   Yu. V. Merekin. Upper bounds for the complexity of sequences generated by symmetric boolean functions. *Diskretn. Anal. Issled. Oper. Ser. 1* **5**(3) (1998), 38–43, 96. In Russian. English translation in *Discrete Appl. Math.* **114** (2001), 227–231.

[Michaux and Point 1986]   C. Michaux and F. Point. Les ensembles k-reconnaissables sont définissables dans $\langle \mathbf{N}, +, V_k \rangle$. *C. R. Acad. Sci. Paris* **303** (1986), 939–942.

[Michaux and Villemaire 1993]   C. Michaux and R. Villemaire. Cobham's theorem seen through Büchi's theorem. In *Proc. 20th Int. Conf. on Automata, Languages, and Programming (ICALP)*, Vol. 700 of *Lecture Notes in Computer Science*, pp. 325–334. Springer-Verlag, 1993.

[Michaux and Villemaire 1996a]   C. Michaux and R. Villemaire. Presburger arithmetic and recognizability of sets of natural numbers by automata: new proofs of Cobham's and Semenov's theorems. *Ann. Pure Appl. Logic* **77** (1996), 251–277.

[Michaux and Villemaire 1996b]   C. Michaux and R. Villemaire. Open questions around Büchi and Presburger arithmetics. In W. Hodges, M. Hyland, C. Steinhorn, and J. Truss, editors, *Logic: From Foundations to Applications, European Logic Colloquium*, pp. 355–385. Clarendon, Oxford, 1996.

[Michel 1975]   P. Michel. Sur les ensembles minimaux engendrés par les substitutions de longueur non constante. PhD thesis, Université de Rennes, 1975.

[Michel 1976a]   P. Michel. Stricte ergodicité d'ensembles minimaux de substitution. In J.-P. Conze and M. S. Keane, editors, *Théorie Ergodique: Actes des Journées Ergodiques, Rennes 1973/1974*, Vol. 532 of *Lecture Notes in Mathematics*, pp. 189–201. Springer-Verlag, 1976.

[Mignosi 1989]   F. Mignosi. Infinite words with linear subword complexity. *Theoret. Comput. Sci.* **65** (1989), 221–242.

[Mignosi 1991]   F. Mignosi. On the number of factors of Sturmian words. *Theoret. Comput. Sci.* **82** (1991), 71–84.

[Mignosi and Pirillo 1992]   F. Mignosi and G. Pirillo. Repetitions in the Fibonacci infinite word. *RAIRO Inform. Théor. App.* **26** (1992), 199–204.

[Mignosi and Séébold 1993b]   F. Mignosi and P. Séébold. Morphismes sturmiens et règles de Rauzy. *J. Théorie Nombres Bordeaux* **5** (1993), 221–233.

[Mignosi, Shallit, and Wang 2001]   F. Mignosi, J. Shallit, and M.-w. Wang. Variations on a theorem of Fine and Wilf. In J. Sgall, A. Pultr, and P. Kolman, editors, *Proc. 26th Symposium, Mathematical Foundations of Computer Science 2001*, Vol. 2136 of *Lecture Notes in Computer Science*, pp. 512–523. Springer-Verlag, 2001.

[Mignosi and Zamboni 2002]   F. Mignosi and L. Q. Zamboni. On the number of Arnoux–Rauzy words. *Acta Arith.* **101** (2002), 121–129.

[Miller 1987]   W. Miller. The maximum order of an element of a finite symmetric group. *Amer. Math. Monthly* **94** (1987), 497–506.

[Mills and Robbins 1986]   W. H. Mills and D. P. Robbins. Continued fractions for certain algebraic power series. *J. Number Theory* **23** (1986), 388–404.

[Minc 1988]   H. Minc. *Nonnegative Matrices*. Wiley, 1988.

[Minsker 1991]   S. Minsker. The Towers of Antwerpen problem. *Inform. Process. Lett.* **38** (1991), 107–111.

[Minsky and Papert 1966]   M. Minsky and S. Papert. Unrecognizable sets of numbers. *J. Assoc. Comput. Mach.* **13** (1966), 281–286.

[Mirsky 1949]   L. Mirsky. A theorem on representations of integers in the scale of r. *Scripta Math.* **15** (1949), 11–12.

[Mitrana 1997a]   V. Mitrana. Some remarks on morphisms and primitivity. *Bull. European Assoc. Theor. Comput. Sci.*, No. 62 (June 1997), 213–216.

[Mitrana 1997b]   V. Mitrana. Primitive morphisms. *Inform. Process. Lett.* **64** (1997), 277–281.

[Mkaouar 1995]   M. Mkaouar. Sur le développement en fraction continue de la série de Baum et Sweet. *Bull. Soc. Math. France* **123** (1995), 361–374.

[Montgomery 1994]   H. L. Montgomery. *Ten Lectures on the Interface between Analytic Number Theory and Harmonic Analysis*, Vol. 84 of *CBMS Regional Conference Series in Mathematics*. Amer. Math. Soc., 1994.

[E. Moore 1956]   E. F. Moore. Gedanken experiments on sequential machines. In C. E. Shannon and J. McCarthy, editors, *Automata Studies*, pp. 129–153. Princeton University Press, 1956.

[E. H. Moore 1909]   E. H. Moore. A generalization of the game called Nim. *Ann. Math.* **11** (1909–1910), 93–94.

[Moore 1971]   F. R. Moore. On the bounds for state-set size in the proofs of equivalence between deterministic, nondeterministic, and two-way finite automata. *IEEE Trans. Comput.* **20** (1971), 1211–1214.

[Mootha 1993]   V. K. Mootha. Unary Fibonacci numbers are context-sensitive. *Fibonacci Quart.* **31** (1993), 41–43.

[Morain and Olivos 1990]   F. Morain and J. Olivos. Speeding up the computations on an elliptic curve using addition–subtraction chains. *RAIRO Inform. Théor. Appl.* **24** (1990), 531–544.

[Mordell 1961]   L. J. Mordell. Irrational power series. *Proc. Amer. Math. Soc.* **12** (1961), 522–526.

[Mordell 1965a]   L. J. Mordell. Irrational power series II. *Acta Arith.* **11** (1965), 181–188.

[Mordell 1965b]   L. J. Mordell. Irrational power series. III. *Proc. Amer. Math. Soc.* **16** (1965), 819–821.

[De Morgan 1840]   A. De Morgan. Description of a calculating machine invented by Mr. Thomas Fowler, of Torrington in Devonshire. *Abstracts of the Papers Printed in the Phil. Trans. Roy. Soc. London* **4** (1837–1843), 243–244.

[Morse 1921]   M. Morse. Recurrent geodesics on a surface of negative curvature. *Trans. Amer. Math. Soc.* **22** (1921), 84–100.

[Morse 1938]   M. Morse. Abstract 360: a solution of the problem of infinite play in chess. *Bull. Amer. Math. Soc.* **44** (1938), 632.

[Morse and Hedlund 1938]   M. Morse and G. A. Hedlund. Symbolic dynamics. *Amer. J. Math.* **60** (1938), 815–866.

[Morse and Hedlund 1940]   M. Morse and G. A. Hedlund. Symbolic dynamics II. Sturmian trajectories. *Amer. J. Math.* **62** (1940), 1–42.

[Morse and Hedlund 1944]   M. Morse and G. A. Hedlund. Unending chess, symbolic dynamics, and a problem in semigroups. *Duke Math. J.* **11** (1944), 1–7.

[Morton 1981]   P. Morton. A note on recurrences satisfied by special classes of polynomials. *Houston J. Math.* **7** (1981), 97–101.

[Morton and Mourant 1989]   P. Morton and W. Mourant. Paper folding, digit patterns, and groups of arithmetic fractals. *Proc. Lond. Math. Soc.* **59** (1989), 253–293.

[Morton and Mourant 1991]   P. Morton and W. J. Mourant. Digit patterns and transcendental numbers. *J. Austral. Math. Soc. Ser. A* **51** (1991), 216–236.

[Mossé 1992]   B. Mossé. Puissances de mots et reconnaissabilité des points fixes d'une substitution. *Theoret. Comput. Sci.* **99** (1992), 327–334.

[Mossé 1996a]   B. Mossé. Reconnaissabilité des substitutions et complexité des suites automatiques. *Bull. Soc. Math. France* **124** (1996), 329–346.

[Mouline 1990]   J. Mouline. Contribution à l'étude de la complexité des suites substitutives. PhD thesis, Université de Provence, April 1990.

[Moulin-Ollagnier 1992]   J. Moulin-Ollagnier. Proof of Dejean's conjecture for alphabets with 5, 6, 7, 8, 9, 10 and 11 letters. *Theoret. Comput. Sci.* **95** (1992), 187–205.

[Mozes 1989]   S. Mozes. Tilings, substitutions systems and dynamical systems generated by them. *J. d'Analyse Math.* **53** (1989), 139–186.

[Mukherjee and Karner 1998]   M. Mukherjee and G. Karner. Irrational numbers of constant type – a new characterization. *New York J. Math.* **4** (1998), 31–34.

[Mullhaupt 1986]   A. Mullhaupt. Discrete self-similarity. Unpublished manuscript, dated June 27, 1986.

[Murata and Mauclaire 1988]   L. Murata and J.-L. Mauclaire. An explicit formula for the average of some q-additive functions. In T. Mitsui, K. Nagasaka, and T. Kano, editors, *Prospects of Mathematical Science*, pp. 141–156. World Scientific, 1988.

[Murugesan 1977]   S. Murugesan. Negabinary arithmetic circuits using binary arithmetic. *IEE J. Electronic Circuits and Systems* **1**(2) (1977), 77–78.

[Musikhin, Il'in, Rabizo, and Bakueva 1997]   S. F. Musikhin, V. I. Il'in, O. V. Rabizo, and L. G. Bakueva. Optical and electrical properties of the CdS/PbS Thue-Morse and Fibonacci superlattices at visible and UV regions. In *Proc. 23rd Int. Symp. Compound Semiconductors*, Vol. 155 of *Inst. Phys. Conf. Ser.*, pp. 141–143, 1997.

[Myhill 1957]   J. Myhill. Finite automata and the representation of events. Technical Report WADD TR-57-624, Wright Patterson Air Force Base, Ohio, 1957.

[Nadler 1961]   M. Nadler. Division and square root in the quater-imaginary number system. *Comm. ACM* **4** (1961), 192–193.

[Nathanson 1972]   M. Nathanson. On the greatest order of an element of the symmetric group. *Amer. Math. Monthly* **79** (1972), 500–501.

[Nelson 1967]   A. H. Nelson. Investigation to discovery with a negative base. *Math. Teacher* **60** (1967), 723–726.

[Nerode 1958]   A. Nerode. Linear automaton transformations. *Proc. Amer. Math. Soc.* **9** (1958), 541–544.

[D. Newman 1969]   D. J. Newman. On the number of binary digits in a multiple of three. *Proc. Amer. Math. Soc.* **21** (1969), 719–721.

[D. Newman and Byrnes 1990]   D. J. Newman and J. S. Byrnes. The L^4 norm of a polynomial with coefficients ±1. *Amer. Math. Monthly* **97** (1990), 42–45.

[D. Newman and Slater 1975]   D. J. Newman and M. Slater. Binary digit distribution over naturally defined sequences. *Trans. Amer. Math. Soc.* **213** (1974), 71–78.

[M. Newman 1960]   M. Newman. Irrational power series. *Proc. Amer. Math. Soc.* **11** (1960), 699–702.

[Nicolas 1969a]   J.-L. Nicolas. Calcul de l'ordre maximum d'un élément du groupe symétrique S_n. *RAIRO Inform. Théor. App.* **3** (1969), 43–50.

[Niederreiter and Vielhaber 1996]   H. Niederreiter and M. Vielhaber. Tree complexity and a doubly exponential gap between structured and random sequences. *J. Complexity* **12** (1996), 187–198.

[Nielsen 1973]   P. T. Nielsen. A note on bifix-free sequences. *IEEE Trans. Inform. Theory* **IT-19** (1973), 704–706.

[Nishioka 1996]   K. Nishioka. *Mahler Functions and Transcendence*, Vol. 1631 of *Lecture Notes in Mathematics*. Springer-Verlag, 1996.

[Nishioka 2001]   K. Nishioka. Algebraic independence of Fredholm series. *Acta Arith.* **100** (2001), 315–327.

[Nishioka, Shiokawa, and Tamura 1992]   K. Nishioka, I. Shiokawa, and J. Tamura. Arithmetical properties of a certain power series. *J. Number Theory* **42** (1992), 61–87.

[Nishioka, Tanaka, and Wen 1999]   K. Nishioka, T.-A. Tanaka, and Z.-Y. Wen. Substitution in two symbols and transcendence. *Tokyo J. Math.* **22** (1999), 127–136.

[Nivat and Perrin 1982]   M. Nivat and D. Perrin. Ensembles reconnaissables de mots biinfinis. In *Proc. Fourteenth Ann. ACM Symp. Theor. Comput.*, pp. 47–59. ACM, 1982.

[Nivat and Perrin 1986]   M. Nivat and D. Perrin. Ensembles reconnaissables de mots biinfinis. *Canad. J. Math.* **38** (1986), 513–537.

[Niven 1947]   I. Niven. A simple proof that π is irrational. *Bull. Amer. Math. Soc.* **53** (1947), 509.

[Niven 1963]   I. Niven. *Irrational Numbers*. MAA, 1963.

[Niven 1969]   I. Niven. Formal power series. *Amer. Math. Monthly* **76** (1969), 871–889.

[Noland 1962]   H. Noland. Advanced problem 5030: an infinite sequence of 3 symbols with no adjacent repeats. *Amer. Math. Monthly* **69** (1962), 439. Solution by C. H. Braunholtz, **70** (1963), 675–676.

[Noland 1990]   H. Noland. Problem proposal 1350. *Math. Mag.* **63** (1990), 189. Solution by N. F. Lindquist, D. G. Poole, and A. J. Schwenk, **64** (1991), 199–203.

[Noonan and Zeilberger 1999]   J. Noonan and D. Zeilberger. The Goulden–Jackson cluster method: extensions, applications and implementations. *J. Diff. Equations. Appl.* **5** (1999), 355–377.

[Nørgård 1999]   P. Nørgård. *Trommebogen/Drumbook*. Edition Wilhelm Hansen, Copenhagen, 1999.

[Novikov and Adian 1968]   P. S. Novikov and S. I. Adian. Infinite periodic groups I, II, III. *Izv. Akad. Nauk. SSSR Ser. Mat.* **32** (1968), 212–244, 251–524, 709–731.

[Nürnberg 1983]   R. Nürnberg. All generalized Morse-sequences are loosely Bernoulli. *Math. Z.* **182** (1983), 403–407.

[d'Ocagne 1886]   M. d'Ocagne. Sur certaines sommations arithmétiques. *Jornal de sciencias mathematicas e astronomicas* **7** (1886), 117–128.

[O'Connor 1999]   L. O'Connor. An analysis of exponentiation based on formal languages. In J. Stern, editor, *Advances in Cryptology – EUROCRYPT '99 Proceedings*, Vol. 1592 of *Lecture Notes in Computer Science*, pp. 375–388. Springer-Verlag, 1999.

[Odlyzko 1978]   A. M. Odlyzko. Non-negative digit sets in positional number systems. *Proc. Lond. Math. Soc.* **37** (1978), 213–229.

[Oettinger 1961]   A. G. Oettinger. Automatic syntactic analysis and the pushdown store. In *Structure of Language and Its Mathematical Aspects*, Vol. 12 of *Proc. Symp. Appl. Math.*, pp. 104–129, 1961.

[Okada, Sekiguchi, and Shiota 1995]   T. Okada, T. Sekiguchi, and Y. Shiota.
    Applications of binomial measures to power sums of digital sums. *J. Number Theory*
    52 (1995), 256–266.

[de Oliveira 1995]   C. R. de Oliveira. Dynamical localization for nonperiodic systems.
    Phys. Lett. A **201** (1995), 419–424.

[Olivier 1971a]   M. Olivier. Sur le développement en base g des nombres premiers. *C. R.
    Acad. Sci. Paris* **272** (1971), 937–939.

[Olivier 1975]   M. Olivier. Sur la probabilité que n soit premier à la somme de ses
    chiffres. *C. R. Acad. Sci. Paris* **280** (1975), 543–545.

[Olivier 1976]   M. Olivier. Fonctions g-additives et formule asymptotique pour la
    propriété $(n, f(n)) = q$. *Acta Arith.* **31** (1976), 361–384.

[Ol'shanskii 1982]   A. Yu. Ol'shanskii. On the Novikov–Adian theorem. *Math. USSR
    Sbornik* **46** (1982), 203–236. In Russian.

[Oltikar and Wayland 1983]   S. Oltikar and K. Wayland. Construction of Smith numbers.
    Math. Mag. **56** (1983), 36–37.

[Osbaldestin 1991]   A. H. Osbaldestin. Digital sum problems. In H.-O. Peitgen, J. M.
    Henriques, and L. F. Penedo, editors, *Fractals in the Fundamental and Applied
    Science*, pp. 307–328. Elsevier, 1991.

[Ostlund, Pandit, Randit, Schellnhuber, and Siggia 1983]   S. Ostlund, R. Pandit,
    D. Rand, H. J. Schellnhuber, and E. D. Siggia. Schrödinger equation with an almost
    periodic potential. *Phys. Rev. Lett.* **50** (1983), 1873–1876.

[Ostrowski 1922]   A. Ostrowski. Bemerkungen zur Theorie der Diophantischen
    Approximationen. *Abh. Math. Sem. Hamburg* **1** (1922), 77–98, 250–251. Reprinted
    in *Collected Mathematical Papers*, Vol. 3, pp. 57–80.

[Pan, Jiao, Jin, Hu, and Jiang 1998]   F. M. Pan, Z. K. Jiao, G. J. Jin, A. Hu, and S. S.
    Jiang. Phonon properties of W/Ti Thue-Morse superlattices. *Phys. Lett. A* **245**
    (1998), 483–488.

[Pansiot 1981a]   J. J. Pansiot. The Morse sequence and iterated morphisms. *Inform.
    Process. Lett.* **12** (1981), 68–70.

[Pansiot 1981b]   J. J. Pansiot. A decidable property of iterated morphisms. In P. Deussen,
    editor, *Theoretical Computer Science, Proc. 5th GI-Conference*, Vol. 104 of *Lecture
    Notes in Computer Science*, pp. 152–158. Springer-Verlag, 1981.

[Pansiot 1983]   J.-J. Pansiot. Hiérarchie et fermeture de certaines classes de
    tag-systèmes. *Acta Inform.* **20** (1983), 179–196.

[Pansiot 1983b]   J.-J. Pansiot. Mots infinis de Fibonacci et morphismes itérés. *RAIRO
    Inform. Théor. App.* **17** (1983), 131–135.

[Pansiot 1984a]   J.-J. Pansiot. Complexité des facteurs des mots infinis engendrés par
    morphismes itérés. In J. Paredaens, editor, *Proc. 11th Int. Conf. on Automata,
    Languages, and Programming (ICALP)*, Vol. 172 of *Lecture Notes in Computer
    Science*, pp. 380–389. Springer-Verlag, 1984.

[Pansiot 1984b]   J.-J. Pansiot. Bornes inferieures sur la complexité des facteurs des mots
    infinis engendrés par morphismes itérés. In M. Fontet and K. Mehlhorn, editors,
    STACS 84, Proc. 1st Symp. Theoretical Aspects of Comp. Sci., Vol. 166 of *Lecture
    Notes in Computer Science*, pp. 230–240. Springer-Verlag, 1984.

[Pansiot 1984c]   J.-J. Pansiot. A propos d'une conjecture de F. Dejean sur les répétitions
    dans les mots. *Discrete Appl. Math.* **7** (1984), 297–311.

[Pansiot 1985]   J.-J. Pansiot. On various classes of infinite words obtained by iterated
    mappings. In M. Nivat and D. Perrin, editors, *Automata on Infinite Words*, Vol. 192 of
    Lecture Notes in Computer Science, pp. 188–197. Springer-Verlag, 1985.

[Pansiot 1986]   J.-J. Pansiot. Decidability of periodicity for infinite words. *RAIRO
    Inform. Théor. Appl.* **20** (1986), 43–46.

[Parent 1984]   D. P. Parent. *Exercises in Number Theory*. Springer-Verlag, 1984.

[Parry 1960]   W. Parry. On the β-expansions of real numbers. *Acta Math. Acad. Sci.
    Hung.* **11** (1960), 401–416.

[Parry 1981]   W. Parry. Self-generation of self-replicating maps of an interval. *Ergod. Theory & Dynam. Syst.* **1** (1981), 197–208.

[Parvaix 1998]   B. Parvaix. Contribution à l'étude des mots sturmiens. PhD thesis, Université de Limoges, 1998. Also appeared as Publications du Laboratoire de Combinatoire et d'Informatique Mathématique, Université du Queébec à Montréal, **25**, 1998.

[de Parville 1884]   H. de Parville. Récreations mathématiques: la tour d'Hanoï et la question du Tonkin. *La Nature* **12** (1884), 285–286.

[Patel 1990]   R. B. Patel. The first non-trivial self-twin in an even base k. *Indian J. Pure Appl. Math.* **21** (1990), 320–325.

[Patel 1991a]   R. B. Patel. A note on some particular forms of a self-number. *Bull. Calcutta Math. Soc.* **83** (1991), 325–329.

[Patel 1991b]   R. B. Patel. Some tests for k-self-numbers. *Math. Student* **56** (1991), 206–210.

[Paul 1975]   M. E. Paul. Minimal symbolic flows having minimal block growth. *Math. Systems Theory* **8** (1975), 309–315.

[Păun 1992]   G. Păun. Anti-Langford sequences. In M. Ito, editor, *Words, Languages and Combinatorics*, pp. 410–421. World Scientific, 1992.

[Păun 1993]   G. Păun. How much Thue is Kolakovski? [*sic*]. *Bull. European Assoc. Theor. Comput. Sci.*, No. 49 (February 1993), 183–185.

[Păun, Santean, Thierin, and Yu 2002]   G. Păun, N. Santean, G. Thierrin, and S. Yu. On the robustness of primitive words. *Discrete Appl. Math.* **117** (2002), 239–252.

[Păun and Thierrin 1997]   G. Păun and G. Thierrin. Morphisms and primitivity. *Bull. European Assoc. Theor. Comput. Sci.*, No. 61 (February 1997), 85–88.

[Pawlak 1959]   Z. Pawlak. An electronic digital computer based on the "−2" system. *Bull. Acad. Polon. Sci. Sér. Sci. Tech.* **7** (1959), 713–721.

[Pawlak 1960]   Z. Pawlak. The organization of a digital digital computer based on the "−2" system. *Bull. Acad. Polon. Sci. Sér. Sci. Tech.* **8** (1960), 253–258.

[Pawlak and Wakulicz 1957]   Z. Pawlak and A. Wakulicz. Use of expansions with a negative basis in the arithmometer of a digital computer. *Bull. Acad. Polon. Sci. Cl. III* **5** (1957), 233–235.

[Pedoe 1958]   D. Pedoe. *The Gentle Art of Mathematics*. Dover, New York, 1958.

[Peitgen, Jürgens, and Saupe 1992]   H.-O. Peitgen, H. Jürgens, and D. Saupe. *Chaos and Fractals: New Frontiers of Science*. Springer-Verlag, 1992.

[Penney 1964]   W. Penney. A numeral system with a negative base. *Math. Student Journal* **11**(4) (1964), 1–2.

[Penney 1965]   W. Penney. A "binary" system for complex numbers. *J. Assoc. Comput. Mach.* **12** (1965), 247–248.

[Pennington 1956]   J. V. Pennington. Elementary problem E 1226. *Amer. Math. Monthly* **63** (1956), 491. Solution by T. F. Mulcrone, **64** (1957), 197–198.

[Penrose 1974]   R. Penrose. The rôle of aesthetics in pure and applied mathematical research. *Bull. Inst. Math. Appl.* **10** (1974), 266–271.

[Penrose 1979]   R. Penrose. Pentaplexity: a class of non-periodic tilings of the plane. *Math. Intelligencer* **2** (1979), 32–37.

[Perrin 1990]   D. Perrin. Finite automata. In J. van Leeuwen, editor, *Handbook of Theoretical Computer Science, Volume B: Formal Models and Semantics*, pp. 1–57. Elsevier–MIT Press, 1990.

[Perrin 1995a]   D. Perrin. Les débuts de la théorie des automates. *Technique et Sci. Inform.* **14** (1995), 409–443.

[Perrin and Pin 1993]   D. Perrin and J.-E. Pin. Mots infinis. Technical Report 93.40, Laboratoire Informatique Théorique et Programmation, Institut Blaise Pascal, July 1993.

[Perron 1907a]   O. Perron. Grundlagen für eine Theorie des Jacobischen Kettenbruchalgorithmus. *Math. Annalen* **64** (1907), 11–76.

[Perron 1907b]   O. Perron. Zur Theorie der Matrices. *Math. Annalen* **64** (1907), 248–263.

[Perron 1960]   O. Perron. *Irrationalzahlen*. Walter de Gruyter, 1960.

[Petersen 1994]   H. Petersen. The ambiguity of primitive words. In P. Enjalbert, E. W. Mayr, and K. W. Wagner, editors, *STACS 94, Proc. 11th Symp. Theoretical Aspects of Comp. Sci.*, Vol. 775 of *Lecture Notes in Computer Science*, pp. 679–690. Springer-Verlag, 1994.

[Petersen 1996]   H. Petersen. On the language of primitive words. *Theoret. Comput. Sci.* **161** (1996), 141–156.

[Pettorossi 1985]   A. Pettorossi. Towers of Hanoi problems: deriving iterative solutions by program transformations. *BIT* **25** (1985), 327–334.

[Peyrière 1978]   J. Peyrière. Mandelbrot random beadsets and birth processes with interaction. Technical Report RC 7417 (#31952), IBM Thomas J. Watson Research Center, December 1978.

[Peyrière 1987]   J. Peyrière. Fréquence des motifs dans les suites doubles invariantes par une substitution. *Ann. Sci. Math. Québec* **11** (1987), 133–138.

[Peyrière 1991]   J. Peyrière. On the trace map for products of matrices associated with substitutive sequences. *J. Statist. Phys.* **62** (1991), 411–414.

[Peyrière 1995]   J. Peyrière. Trace maps. In F. Axel and D. Gratias, editors, *Beyond Quasicrystals*, pp. 465–480. Les Éditions de Physique; Springer, 1995.

[Pinho, Haddad, and Salinas 1998]   S. T. R. Pinho, T. A. S. Haddad, and S. R. Salinas. Critical behavior of the Ising model on a hierarchical lattice with aperiodic interactions. *Physica A* **257** (1998), 515–520.

[Pirillo 1991]   G. Pirillo. On some properties of the Thue infinite word. In A. Barlotti, G. Lunardon, F. Mazzocca, N. Melone, D. Olanda, A. Pasini, and G. Tallini, editors, *Combinatorics '88: Proc. Int. Conf. on Incidence Geometries and Combinatorial Structures*, Vol. 2, pp. 325–329. Mediterranean Press, Commenda di Rende, Italy, 1991.

[Pirillo 1993]   G. Pirillo. Fibonacci numbers and words. In *Séminaire Lotharingien de Combinatoire (Gerolfingen, 1993)*, number 34 in Prépubl. Inst. Rech. Math. Av., Univ. Louis Pasteur, Strasbourg, 1993, pp. 77–85, 1993.

[Pirillo 1999]   G. Pirillo. Some factorizations of the Fibonacci word. *Algebra Colloq.* **6** (1999), 361–368.

[Pirillo and Varricchio 1996]   G. Pirillo and S. Varricchio. Some combinatorial properties of infinite words and applications to semigroup theory. *Discrete Math.* **153** (1996), 239–251.

[Pitteway 1985]   M. L. V. Pitteway. The relationship between Euclid's algorithms and run-length encoding. In R. A. Earnshaw, editor, *Fundamental Algorithms for Computer Graphics*, Vol. F17 of *NATO ASI Series*, pp. 105–112. Springer-Verlag, 1985.

[Pitteway and Green 1982]   M. L. V. Pitteway and A. J. R. Green. Bresenham's algorithm with run line coding shortcut. *The Computer Journal* **25** (1982), 114–115.

[Pleasants 1970]   P. A. B. Pleasants. Non-repetitive sequences. *Proc. Cambridge Phil. Soc.* **68** (1970), 267–274.

[Pleasants 1985]   P. A. B. Pleasants. Quasicrystallography: some interesting new patterns. In H. Iwaniec, editor, *Elementary and Analytic Theory of Numbers*, Vol. 17 of *Banach Center Publications*, pp. 439–461. PWN – Polish Scientific Publishers, 1985.

[Point and Bruyère 1997]   F. Point and V. Bruyère. On the Cobham–Semenov theorem. *Theory Comput. Systems* **30** (1997), 197–220.

[Pomerance, Robson, and Shallit 1997]   C. Pomerance, J. M. Robson, and J. Shallit. Automaticity II: Descriptional complexity in the unary case. *Theoret. Comput. Sci.* **180** (1997), 181–201.

[Poole 1992]   D. Poole. The bottleneck Towers of Hanoi problem. *J. Recreational Math.* **24** (1992), 203–207.

[Poole 1994]   D. G. Poole. The towers and triangles of Professor Claus (or, Pascal knows Hanoi). *Math. Mag.* **67** (1994), 323–344.

[van der Poorten 1990]   A. J. van der Poorten. Notes on continued fractions and recurrence sequences. In J. H. Loxton, editor, *Number Theory and Cryptography*, Vol. 154 of *London Math. Soc. Lect. Note Ser.*, pp. 86–97. Cambridge University Press, 1990.

[van der Poorten and Shallit 1992]   A. J. van der Poorten and J. O. Shallit. Folded continued fractions. *J. Number Theory* **40** (1992), 237–250.

[Popken 1963]   J. Popken. Irrational power series. *Proc. Konin. Neder. Akad. Wet. Ser. A* **66** (1963), 691–694. (= *Indag. Math.* **25**).

[Porges 1945]   A. Porges. A set of eight numbers. *Amer. Math. Monthly* **52** (1945), 379–382.

[Porta and Stolarsky 1990]   H. Porta and K. B. Stolarsky. Half-silvered mirrors and Wythoff's game. *Canad. Math. Bull.* **33** (1990), 119–125.

[Porta and Stolarsky 1991]   H. A. Porta and K. B. Stolarsky. Wythoff pairs as semigroup invariants. *Adv. Math.* **85** (1991), 69–82.

[Post 1965]   E. Post. Absolutely unsolvable problems and relatively undecidable propositions: account of an anticipation. In M. Davis, editor, *The Undecidable*, pp. 338–433. Raven Press, 1965.

[Pour-El and Richards 1989]   M. B. Pour-El and J. I. Richards. *Computability in Analysis and Physics*. Springer-Verlag, 1989.

[Priday 1959]   C. J. Priday. On Langford's problem (I). *Math. Gazette* **43** (1959), 250–253.

[Prodinger 1982]   H. Prodinger. Generalizing the sum of digits function. *SIAM J. Algebraic Discrete Methods* **3** (1982), 35–42.

[Prodinger 1983]   H. Prodinger. Non-repetitive sequences and Gray code. *Discrete Math.* **43** (1983), 113–116.

[Prodinger 2000]   H. Prodinger. On binary representations of integers with digits $-1, 0, 1$. *Integers* **0** (2000), #A08 (electronic), http://www.integers-ejcnt.org/\break vol0.html

[Prodinger and Urbanek 1979]   H. Prodinger and F. J. Urbanek. Infinite 0–1-sequences without long adjacent identical blocks. *Discrete Math.* **28** (1979), 277–289.

[Propp 1979]   J. Propp. Problem proposal 474. *Crux Math.* **5** (1979), 229. Solution by G. Patruno, **6** (1980), 198.

[Prouhet 1851]   E. Prouhet. Mémoire sur quelques relations entre les puissances des nombres. *C. R. Acad. Sci. Paris* **33** (1851), 225.

[Qin, Ma, and Tsai 1990]   M.-G. Qin, H.-R. Ma, and C.-H. Tsai. A renormalisation analysis of the one-dimensional Thue-Morse aperiodic chain. *J. Phys.: Condens. Matter* **2** (1990), 1059–1072.

[Queffélec 1987a]   M. Queffelec. Une nouvelle propriété des suites de Rudin-Shapiro. *Ann. Inst. Fourier (Grenoble)* **37** (1987), 115–138.

[Queffélec 1987b]   M. Queffélec. *Substitution Dynamical Systems – Spectral Analysis*, Vol. 1294 of *Lecture Notes in Mathematics*. Springer-Verlag, 1987.

[Queffélec 1998]   M. Queffélec. Transcendance des fractions continue de Thue-Morse. *J. Number Theory* **73** (1998), 201–211.

[Queffélec 2000]   M. Queffélec. Irrational numbers with automaton-generated continued fraction expansion. In J.-M. Gambaudo, P. Hubert, P. Tisseur, and S. Vaienti, editors, *Dynamical Systems: From Crystal to Chaos*, pp. 190–198. World Scientific, 2000.

[Rabin 1985]   M. O. Rabin. Discovering repetitions in strings. In A. Apostolico and Z. Galil, editors, *Combinatorial Algorithms on Words*, pp. 279–288. Springer-Verlag, Berlin, 1985.

[Rabin and Scott 1959]   M. O. Rabin and D. Scott. Finite automata and their decision
  problems. *IBM J. Res. Develop.* **3** (1959), 115–125.
[Rabinowitz and Gilbert 1991]   S. Rabinowitz and P. Gilbert. A nonlinear recurrence
  yielding binary digits. *Math. Mag.* **64** (1991), 168–171.
[Raimi 1976]   R. A. Raimi. The first digit problem. *Amer. Math. Monthly* **83** (1976),
  521–538.
[Ramshaw 1981]   L. Ramshaw. On the discrepancy of the sequence formed by the
  multiples of an irrational numbers. *J. Number Theory* **13** (1981), 138–175.
[Randé 1992]   B. Randé. Équations fonctionnelles de Mahler et applications aux suites
  p-régulières. PhD thesis, Université Bordeaux I, September 1992.
[Randé 1993]   B. Randé. Récurrences 2- et 3-mahlériennes. *J. Théorie Nombres
  Bordeaux* **5** (1993), 101–109.
[A. Rao 1966]   A. N. Rao. On a technique for obtaining numbers with a multiplicity of
  generators. *Math. Student* **34** (1966), 79–84.
[G. Rao, Rao, and Krishnamurthy 1974]   G. S. Rao, M. N. Rao, and E. V. Krishnamurthy.
  Logical design of a negative binary adder-subtracter. *Int. J. Electronics* **36** (1974),
  537–542.
[Rauzy 1979]   G. Rauzy. Échanges d'intervalles et transformations induites. *Acta Arith.*
  34 (1979), 315–328.
[Rauzy 1983]   G. Rauzy. Suites à termes dans un alphabet fini. In *Séminaire de Théorie
  des Nombres de Bordeaux*, pp. 25.01–25.16, 1982–1983.
[van Ravenstein 1985]   T. van Ravenstein. On the discrepancy of the sequence formed
  from multiples of an irrational number. *Bull. Austral. Math. Soc.* **31** (1985),
  329–338.
[van Ravenstein 1988]   T. van Ravenstein. The three gap theorem (Steinhaus conjecture).
  J. Austral. Math. Soc. Ser. A **45** (1988), 360–370.
[van Ravenstein 1989]   T. van Ravenstein. Optimal spacing of points on a circle.
  Fibonacci Quart. **27** (1989), 18–24.
[van Ravenstein, Winley, and Tognetti 1990]   T. van Ravenstein, G. Winley, and
  K. Tognetti. Characteristics and the three gap theorem. *Fibonacci Quart.* **28** (1990),
  204–214.
[Ray and Majumdar 1995a]   G. C. Ray and A. A. K. Majumdar. A note on the bottleneck
  Tower of Hanoi problem. *Chittagong Univ. Stud. Part II: Science* **19** (1995), 27–34.
[Ray and Majumdar 1995b]   G. C. Ray and A. A. K. Majumdar. A note on the cyclic
  Tower of Hanoi problem. *Chittagong Univ. Stud. Part II: Science* **19** (1995), 75–81.
[Rayleigh 1877]   J. W. S. Rayleigh. *The Theory of Sound*, Vol. 1. Macmillan, London,
  1877–1878. Reprinted by Dover, New York, 1945.
[Razafy Andriamampianina 1989]   D. Razafy Andriamampianina. Le p-pliage de papier.
  Ann. Fac. Sci. Toulouse **10** (1989), 401–414.
[Razafy Andriamampianina 1992]   D. Razafy Andriamampianina. Le p-pliage de papier
  et les polynômes. *C. R. Acad. Sci. Paris* **314** (1992), 875–878.
[Razafy Andriamampianina 1996]   D. Razafy Andriamampianina. Suites de Toeplitz,
  p-pliage, suites automatiques et polynômes. *Acta Math. Hung.* **73** (1996), 179–190.
[Read 1979]   R. C. Read. Problem no. 32: Sequences avoiding repeated subsequences.
  Ann. New York Acad. Sci. **319** (1979), 586–587. Proc. 2nd Int. Conf. Combin. Math.
[Recamán 1973]   B. Recamán. Elementary problem E 2408. *Amer. Math. Monthly* **80**
  (1973), 434. Solution by D. W. Bange, **81** (1974), 407.
[Recher 1992]   F. Recher. Propriétés de transcendance de séries formelles provenant de
  l'exponentielle de Carlitz. *C. R. Acad. Sci. Paris* **315** (1992), 245–250.
[Recht and Rosenman 1947]   L. Recht and M. Rosenman. Advanced problem 4247.
  Amer. Math. Monthly **54** (1947), 232. Solution by E. P. Starke, **55** (1948), 588–592.
[Rees 1946]   D. Rees. Note on a paper by I. J. Good. *J. London Math. Soc.* **21** (1946),
  169–172.

[de Regt 1967]   M. P. de Regt. Negative radix arithmetic. *Computer Design* **6**(5) (May 1967), 52–63. Series continues in **6** (6) (June 1967), 56–65; **6** (7) (July 1967), 36–43; **6** (8) (August 1967), 36–44; **6** (9) (September 1967), 44–50; **6** (10) (October 1967), 68–73; **6** (12) (December 1967), 70–77; **7** (1) (January 1968), 62–66.

[Reiter 1994]   C. A. Reiter. Sierpiński fractals and GCDs. *Computers and Graphics* **18** (1994), 885–891.

[Reitwiesner 1960]   G. W. Reitwiesner. *Binary arithmetic*, Vol. 1 of *Advances in Computers*, pp. 231–308. Academic Press, 1960.

[Rényi 1957]   A. Rényi. Representations for real numbers and their ergodic properties. *Acta Math. Acad. Sci. Hung.* **8** (1957), 477–493.

[Restivo 1977]   A. Restivo. Mots sans répétitions et langages rationnels bornés. *RAIRO Inform. Théor. App.* **11** (1977), 197–202.

[Restivo and Reutenauer 1985]   A. Restivo and C. Reutenauer. Rational languages and the Burnside problem. *Theoret. Comput. Sci.* **40** (1985), 13–30.

[Restivo and Salemi 1983]   A. Restivo and S. Salemi. On weakly square free words. *Bull. European Assoc. Theor. Comput. Sci.*, No. 21, (October 1983), 49–56.

[Restivo and Salemi 1985a]   A. Restivo and S. Salemi. Overlap free words on two symbols. In M. Nivat and D. Perrin, editors, *Automata on Infinite Words*, Vol. 192 of *Lecture Notes in Computer Science*, pp. 198–206. Springer-Verlag, 1985.

[Restivo and Salemi 1985b]   A. Restivo and S. Salemi. Some decision results on nonrepetitive words. In A. Apostolico and Z. Galil, editors, *Combinatorial Algorithms on Words*, pp. 289–295. Springer-Verlag, 1985.

[Restivo and Salemi 2002]   A. Restivo and S. Salemi. Binary patterns in infinite binary words. In W. Brauer, H. Ehrig, J. Karhumäki, and A. Salomaa, editors, *Formal and Natural Computing*, Vol. 2300 of *Lecture Notes in Computer Science*, pp. 107–116. Springer-Verlag, 2002.

[Reutenauer 1981]   C. Reutenauer. A new characterization of the regular languages. In S. Even and O. Kariv, editors, *Proc. 8th Int. Conf. on Automata, Languages, and Programming (ICALP)*, Vol. 115 of *Lecture Notes in Computer Science*, pp. 177–183. Springer-Verlag, 1981.

[Ribenboim 1985]   P. Ribenboim. Representation of real numbers by means of Fibonacci numbers. *Enseign. Math.* **31** (1985), 249–259.

[Richomme 1999]   G. Richomme. Another characterization of Sturmian words (one more). *Bull. European Assoc. Theor. Comput. Sci.*, No. 67 (February 1999), 173–175.

[Richomme and Séébold 1999]   G. Richomme and P. Séébold. Characterization of test-sets for overlap-free morphisms. *Discrete Appl. Math.* **98** (1999), 151–157.

[Richomme and Wlazinski 2000]   G. Richomme and F. Wlazinski. About cube-free morphisms. In H. Reichel and S. Tison, editors, *STACS 2000, Proc. 17th Symp. Theoretical Aspects of Comp. Sci.*, Vol. 1770 of *Lecture Notes in Computer Science*, pp. 99–109. Springer-Verlag, 2000.

[Richomme and Wlazinski 2002]   G. Richomme and F. Wlazinski. Some results on k-power-free morphisms. *Theoret. Comput. Sci.* **273** (2002), 119–142.

[Rider 1966]   D. Rider. Transformations of Fourier coefficients. *Pacific J. Math.* **19** (1966), 347–355.

[Rider 1969]   D. Rider. Closed subalgebras of $L^1(T)$. *Duke Math. J.* **36** (1969), 105–115.

[Ridout 1957]   D. Ridout. Rational approximations to algebraic numbers. *Mathematika* **4** (1957), 125–131.

[Rigo 2000]   M. Rigo. Generalization of automatic sequences for numeration systems on a regular language. *Theoret. Comput. Sci.* **244** (2000), 271–281.

[Rigo 2001]   M. Rigo. Numeration systems on a regular language: arithmetic operations, recognizability and formal power series. *Theoret. Comput. Sci.* **269** (2001), 469–498.

[Rigo 2002]   M. Rigo. Construction of regular languages and recognizability of polynomials. *Discrete Math.* **254** (2002), 485–496.

[Riklund, Severin, and Liu 1987]   R. Riklund, M. Severin, and Y. Liu. The Thue–Morse aperiodic crystal, a link between the Fibonacci quasicrystal and the periodic crystal. *Int. J. Modern Phys. B* **1** (1987), 121–132.

[Risley and Zamboni 2000]   R. N. Risley and L. Q. Zamboni. A generalization of Sturmian sequences: combinatorial structure and transcendence. *Acta Arith.* **95** (2000), 167–184.

[Ritchie 1963]   R. W. Ritchie. Finite automata and the set of squares. *J. Assoc. Comput. Mach.* **10** (1963), 528–531.

[Rivier 1986]   N. Rivier. A botanical quasicrystal. *J. de Phys. Colloq.* **47 C3** (1986), 299–309.

[Robbins 1996]   N. Robbins. Fibonacci partitions. *Fibonacci Quart.* **34** (1996), 306–313.

[Robert 1994]   A. Robert. A good basis for computing with complex numbers. *Elem. Math.* **49** (1994), 111–117.

[Roberts and Baake 1994a]   J. A. G. Roberts and M. Baake. The dynamics of trace maps. In J. Seimenis, editor, *Hamiltonian Mechanics: Integrability and Chaotic Behaviour*, NATO ASI Series, Series B: Physics, pp. 275–285. Plenum Press, 1994.

[Roberts and Baake 1994b]   J. A. G. Roberts and M. Baake. Trace maps as 3D reversible dynamical systems with an invariant. *J. Statist. Phys.* **74** (1994), 829–888.

[Roberts 1958]   J. B. Roberts. A new proof of a theorem of Lehmer. *Canad. J. Math.* **10** (1958), 191–194.

[Robson 1989]   J. M. Robson. Separating strings with small automata. *Inform. Process. Lett.* **30** (1989), 209–214.

[Rockett and Szüsz 1992]   A. M. Rockett and P. Szüsz. *Continued Fractions*. World Scientific, 1992.

[Rohl and Gedeon 1986]   J. S. Rohl and T. D. Gedeon. The Reve's puzzle. *Computer J.* **29** (1986), 187–188.

[Rokicki and Knuth 1987]   T. G. Rokicki and D. E. Knuth. A programming and problem-solving seminar. Technical Report STAN-CS-87-1154, Department of Computer Science, Stanford University, April 1987.

[Ronse 1985]   C. Ronse. A simple proof of Rosenfeld's characterization of digital straight line segments. *Pattern Recog. Lett.* **3** (1985), 323–326.

[Rosaz 1995]   L. Rosaz. Unavoidable languages, cuts and innocent sets of words. *RAIRO Inform. Théor. Appl.* **29** (1995), 339–382.

[Rosaz 1998]   L. Rosaz. Inventories of unavoidable languages and the word-extension conjecture. *Theoret. Comput. Sci.* **201** (1998), 151–170.

[Rosenfeld 1974]   A. Rosenfeld. Digital straight line segments. *IEEE Trans. Comput.* **23** (1974), 1264–1269.

[Rosenfeld and Kim 1982]   A. Rosenfeld and C. E. Kim. How a digital computer can tell whether a line is straight. *Amer. Math. Monthly* **89** (1982), 230–235.

[Ross and Winklmann 1982]   R. Ross and K. Winklmann. Repetitive strings are not context-free. *RAIRO Inform. Théor. Appl.* **16** (1982), 191–199.

[Rosser and Schoenfeld 1962]   J. B. Rosser and L. Schoenfeld. Approximate formulas for some functions of prime numbers. Illinois *J. Math.* **6** (1962), 64–94.

[Rote 1994]   G. Rote. Sequences with subword complexity 2n. *J. Number Theory* **46** (1994), 196–213.

[K. Roth 1955]   K. F. Roth. Rational approximations to algebraic numbers. *Mathematika* **2** (1955), 1–20. Corrigendum, p. 168.

[P. Roth 1989]   P. Roth. A note on word chains and regular languages. *Inform. Process. Lett.* **30** (1989), 15–18.

[P. Roth 1991]   P. Roth. l-occurrences of avoidable patterns. In C. Choffrut and M. Jantzen, editors, *STACS 91, Proc. 8th Symp. Theoretical Aspects of Comp. Sci.*, Vol. 480 of *Lecture Notes in Computer Science*, pp. 42–49. Springer-Verlag, 1991.

[P. Roth 1992]   P. Roth. Every binary pattern of length six is avoidable on the two-letter alphabet. *Acta Inform.* **29** (1992), 95–107.

[T. Roth 1974]   T. Roth. The Tower of Brahma revisited. *J. Recreational Math.* **7** (1974), 116–119.

[Rothstein and Weiman 1976]   J. Rothstein and C. Weiman. Parallel and sequential specification of a context sensitive language for straight lines on grids. *Computer Graphics and Image Processing* **5** (1976), 106–124.

[Roy, Basu, and Khan 1995]   C. L. Roy, C. Basu, and A. Khan. Density of states of generalised Thue-Morse lattice and related issues. *Phys. Lett. A* **198** (1995), 424–432.

[Roy and Khan 1994b]   C. L. Roy and A. Khan. Relativistic impact on the Landauer resistance of Thue-Morse lattices. *J. Phys.: Condens. Matter* **6** (1994), 4493–4504.

[Roy and Khan 1994c]   C. L. Roy and A. Khan. Landauer resistance of Thue–Morse and Fibonacci lattices and some related issues. *Phys. Rev. B* **49** (1994), 14979–14983.

[Roy, Khan, and Basu 1995]   C. L. Roy, A. Khan, and C. Basu. A study of Landauer resistance and related issues of the generalized Thue–Morse lattice. *J. Phys.: Condens. Matter* **7** (1995), 1843–1853.

[Rozenberg 1981]   G. Rozenberg. On subwords of formal languages. In F. Gécseg, editor, *Fundamentals of Computation Theory: Proceedings of the 1981 International FCT Conference*, Vol. 117 of *Lecture Notes in Computer Science*, pp. 328–333. Springer-Verlag, 1981.

[Rozenberg and Lindenmayer 1973]   G. Rozenberg and A. Lindenmayer. Developmental systems with locally catenative formula. *Acta Inform.* **2** (1973), 214–248.

[Rozenberg and Salomaa 1980]   G. Rozenberg and A. Salomaa. *The Mathematical Theory of L Systems*, Vol. 90 of *Pure and Applied Mathematics*. Academic Press, 1980.

[Rudin 1959]   W. Rudin. Some theorems on Fourier coefficients. *Proc. Amer. Math. Soc.* **10** (1959), 855–859.

[Rudin 1966]   W. Rudin. *Real and Complex Analysis*. McGraw-Hill, 1966.

[Ryu, Oh, and Lee 1993]   C. S. Ryu, G. Y. Oh, and M. H. Lee. Electronic properties of a tight-binding and a Kronig–Penney model of the Thue-Morse chain. *Phys. Rev. B* **48** (1993), 132–141.

[Saffari 1986]   B. Saffari. Une fonction extrémale liée à la suite de Rudin–Shapiro. *C. R. Acad. Sci. Paris* **303** (1986), 97–100.

[Saffari 1987]   B. Saffari. Structure algébrique sur les couples de Rudin–Shapiro. *C. R. Acad. Sci. Paris* **304** (1987), 127–130.

[Sajo 1984]   A. Sajo. On subword complexity functions. *Discrete Appl. Math.* **8** (1984), 209–212.

[Salejda 1995a]   W. Salejda. Lattice dynamics of the binary aperiodic chains of atoms I. Fractal dimension of phonon spectra. *Int. J. Modern Phys. B* **9** (1995), 1429–1451.

[Salem 1963]   R. Salem. *Algebraic numbers and Fourier analysis*. D. C. Heath, 1963.

[Salem and Zygmund 1954]   R. Salem and A. Zygmund. Some properties of trigonometric series whose terms have random signs. *Acta Math.* **91** (1954), 245–301.

[Salomaa 1973]   A. Salomaa. *Formal Languages*. Academic Press, 1973.

[Salon 1986]   O. Salon. Suites automatiques à multi-indices. In *Séminaire de Théorie des Nombres de Bordeaux*, pp. 4.01–4.27, 1986–1987.

[Salon 1987]   O. Salon. Suites automatiques à multi-indices et algébricité. *C. R. Acad. Sci. Paris* **305** (1987), 501–504.

[Salon 1989a]   O. Salon. Quelles tuiles! (pavages apériodiques du plan et automates bidimensionnels). *Sém. Théorie Nombres Bordeaux* **1** (1989), 1–25.

[Salon 1989b]   O. Salon. Propriétés arithmétiques des automates multidimensionnels. PhD thesis, Université Bordeaux I, 1989.

[Samborski 1977]   J. R. Samborski. Elementary problem E 2667. *Amer. Math. Monthly* **84** (1977), 567. Solution by O. P. Lossers, **85** (1978), 825.

[Sankar, Chakrabarti, and Krishnamurthy 1973a]   P. V. Sankar, S. Chakrabarti, and E. V. Krishnamurthy. Arithmetic algorithms in a negative base. *IEEE Trans. Comput.* **C-22** (1973), 120–125.

[Sankar, Chakrabarti, and Krishnamurthy 1973b]   P. V. Sankar, S. Chakrabarti, and E. V. Krishnamurthy. Deterministic division algorithm in a negative base. *IEEE Trans. Comput.* **C-22** (1973), 125–128.

[Sanov 1940]   I. N. Sanov. Solution of the Burnside problem for exponent 4. *Učen. Zap. Leningrad Univ.* **10** (1940), 166–170.

[Santini-Bouchard 1997]   M.-L. Santini-Bouchard. Échanges de trois intervalles et suite minimales. *Theoret. Comput. Sci.* **174** (1997), 171–191.

[Sarkar 2000]   U. K. Sarkar. On the design of a constructive algorithm to solve the multi-peg towers of Hanoi problem. *Theoret. Comput. Sci.* **237** (2000), 407–421.

[U. Schmidt 1987a]   U. Schmidt. Avoidable patterns on 2 letters. In F. J. Brandenburg, G. Vidal-Naquet, and M. Wirsing, editors, *STACS 87, Proc. 4th Symp. Theoretical Aspects of Comp. Sci.*, Vol. 247 of *Lecture Notes in Computer Science*, pp. 189–197. Springer-Verlag, 1987.

[U. Schmidt 1987b]   U. Schmidt. Long unavoidable patterns. *Acta Inform.* **24** (1987), 433–445.

[U. Schmidt 1989]   U. Schmidt. Avoidable patterns on two letters. *Theoret. Comput. Sci.* **63** (1989), 1–17.

[W. Schmidt 1960]   W. Schmidt. On normal numbers. *Pacific J. Math.* **10** (1960), 661–672.

[W. Schmidt 1967]   W. Schmidt. On simultaneous approximations of two algebraic numbers by rationals. *Acta Math.* **119** (1967), 27–50.

[W. Schmidt 1980]   W. M. Schmidt. *Diophantine Approximation*, Vol. 785 of *Lecture Notes in Mathematics*. Springer-Verlag, 1980.

[W. Schmidt 2000]   W. M. Schmidt. On continued fractions and Diophantine approximation in power series fields. *Acta Arith.* **95** (2000), 139–166.

[Schoenberg 1982]   I. J. Schoenberg. *Mathematical Time Exposures*. MAA, Washington, D.C., 1982.

[Schoenfeld 1976]   L. Schoenfeld. Sharper bounds for the Chebyshev functions $\theta(x)$ and $\psi(x)$. II. *Math. Comp.* **30** (1976), 337–360. Corrigenda, **30** (1976), 900.

[Schützenberger 1964]   M. P. Schützenberger. On the synchronizing properties of certain prefix codes. *Inform. Control* **7** (1964), 23–36.

[Schützenberger 1968]   M. P. Schützenberger. A remark on acceptable sets of numbers. *J. Assoc. Comput. Mach.* **15** (1968), 300–303.

[Scorer, Grundy, and Smith 1944]   R. S. Scorer, P. M. Grundy, and C. A. B. Smith. Some binary games. *Math. Gazette* **28** (1944), 96–103.

[Scott and Wall 1940]   W. T. Scott and H. S. Wall. Continued fraction expansions for arbitrary power series. *Ann. Math.* **41** (1940), 328–349.

[Séébold 1982]   P. Séébold. Morphismes itérés, mot de Morse, et mot de Fibonacci. *C. R. Acad. Sci. Paris* **295** (1982), 439–441.

[Séébold 1985a]   P. Séébold. Generalized Thue–Morse sequences. In G. Goos and J. Hartmanis, editors, *Fundamentals of Computation Theory: FCT '85*, Vol. 199 of *Lecture Notes in Computer Science*, pp. 402–411. Springer-Verlag, 1985.

[Séébold 1985c]   P. Séébold. Overlap-free sequences. In M. Nivat and D. Perrin, editors, *Automata on Infinite Words*, Vol. 192 of *Lecture Notes in Computer Science*, pp. 207–215. Springer-Verlag, 1985.

[Séébold 1985d]   P. Séébold. Sequences generated by infinitely iterated morphisms. *Discrete Appl. Math.* **11** (1985), 255–264.

[Séébold 1986]   P. Séébold. Complément à l'étude des suites de Thue–Morse généralisées. *RAIRO Inform. Théor. Appl.* **20** (1986), 157–181.

[Séébold 1991]   P. Séébold. Fibonacci morphisms and Sturmian words. *Theoret. Comput. Sci.* **88** (1991), 365–384.

[Séébold 1998]   P. Séébold. On the conjugation of standard morphisms. *Theoret. Comput. Sci.* **195** (1998), 91–109.

[Séébold 2000]   P. Séébold. On some generalizations of the Thue–Morse morphism. Technical Report 2000-14, Laboratoire de Recherche en Informatique, Amiens, France, November 2000.

[Séébold 2001]   P. Séébold. Some properties of the Prouhet morphisms and words. Technical Report 2001-04, Laboratoire de Recherche en Informatique, Amiens, France, June 2001.

[Segal and Lepp 1969]   A. C. Segal and B. Lepp. Elementary problem E 2204. *Amer. Math. Monthly* **76** (1969), 1138. Solution by N. J. Fine, **77** (1970), 1009–1010.

[Seki and Kobuchi 1991]   S. Seki and Y. Kobuchi. On standard locally catenative L schemes. *Theoret. Comput. Sci.* **83** (1991), 237–248.

[Selfridge 1960]   J. L. Selfridge. Elementary problem proposal E 1408. *Amer. Math. Monthly* **67** (1960), 290. Solution, **67** (1960), 924–925.

[Semenov 1977]   A. L. Semenov. Presburgerness of predicates regular in two number systems. *Sibirskii Mat. Zh.* **18** (1977), 403–418. In Russian. English translation in *Siberian J. Math.* **18** (1977), 289–300.

[Senechal 1995]   M. Senechal. *Quasicrystals and Geometry*. Cambridge University Press, 1995.

[Senge and Straus 1973]   H. G. Senge and E. G. Straus. PV-numbers and sets of multiplicity. *Period. Math. Hung.* **3** (1973), 93–100.

[Series 1985]   C. Series. The geometry of Markoff numbers. *Math. Intelligencer* **7**(3) (1985), 20–29.

[Shallit 1979]   J. O. Shallit. Simple continued fractions for some irrational numbers. *J. Number Theory* **11** (1979), 209–217.

[Shallit 1980]   J. O. Shallit. A triangle for the Bell numbers. In V. E. Hoggatt, Jr. and M. Bicknell-Johnson, editors, *A Collection of Manuscripts Related to the Fibonacci Sequence, 18th Anniversary Volume*, pp. 69–71. Fibonacci Association, 1980.

[Shallit 1982a]   J. O. Shallit. Simple continued fractions for some irrational numbers, II. *J. Number Theory* **14** (1982), 228–231.

[Shallit 1982b]   J. O. Shallit. Explicit descriptions of some continued fractions. *Fibonacci Quart.* **20** (1982), 77–81.

[Shallit 1984]   J. O. Shallit. Advanced problem 6450. *Amer. Math. Monthly* **91** (1984), 59–60. Solution, **92** (1985), 513–514.

[Shallit 1985]   J. O. Shallit. On infinite products associated with sums of digits. *J. Number Theory* **21** (1985), 128–134.

[Shallit 1988a]   J. O. Shallit. A generalization of automatic sequences. *Theoret. Comput. Sci.* **61** (1988), 1–16.

[Shallit 1988b]   J. O. Shallit. Fractals, bitmaps, and APL. *APL Quote-Quad* **18**(3) (March 1988), 24–32.

[Shallit 1991a]   J. O. Shallit. Quickie problem proposal Q785. *Math. Mag.* **64** (1991), 351, 357.

[Shallit 1991b]   J. O. Shallit. Characteristic words as fixed points of homomorphisms. Technical Report CS-91-72, Department of Computer Science, University of Waterloo, December 1991.

[Shallit 1992]   J. O. Shallit. Real numbers with bounded partial quotients. *Enseign. Math.* **38** (1992), 151–187.

[Shallit 1993]   J. O. Shallit. On the maximum number of distinct factors of a binary string. *Graphs and Combinatorics* **9** (1993), 197–200.

[Shallit 1994]   J. O. Shallit. Numeration systems, linear recurrences, and regular sets. *Inform. Comput.* **113** (1994), 331–347.

[Shallit 1996]   J. O. Shallit. Automaticity IV: Sequences, sets, and diversity. *J. Théorie Nombres Bordeaux* **8** (1996), 347–367.

[Shallit 1999]   J. O. Shallit. Number theory and formal languages. In D. A. Hejhal, J. Friedman, M. C. Gutzwiller, and A. M. Odlyzko, editors, *Emerging Applications of Number Theory*, Vol. 109 of *IMA Volumes in Mathematics and Its Applications*, pp. 547–570. Springer-Verlag, 1999.

[Shallit 2000]   J. O. Shallit. Automaticity and rationality. *J. Automata, Languages, and Combinatorics* **5** (2000), 255–268.

[Shallit and Breitbart 1996]   J. O. Shallit and Y. Breitbart. Automaticity I: Properties of a measure of descriptional complexity. *J. Comput. System Sci.* **53** (1996), 10–25.

[Shallit and Stolfi 1989]   J. O. Shallit and J. Stolfi. Two methods for generating fractals. *Computers and Graphics* **13** (1989), 185–191.

[Shallit and Swart 1999]   J. O. Shallit and D. Swart. An efficient algorithm for computing the i'th letter of $\varphi^n(a)$. In *Proc. 10th ACM–SIAM Symp. Discrete Algorithms (SODA)*, pp. 768–775, 1999.

[Shallit and Wang 1999]   J. O. Shallit and M.-w. Wang. On two-sided infinite fixed points of morphisms. In G. Ciobanu and G. Păun, editors, *Fundamentals of Computation Theory: FCT '99*, Vol. 1684 of *Lecture Notes in Computer Science*, pp. 488–499. Springer-Verlag, 1999.

[Shallit and Wang 2001]   J. O. Shallit and M.-w. Wang. Weakly self-avoiding words and a construction of Friedman. *Eur. J. Combin.* **8**(1) (2001), N2 (electronic), http://www.combinatorics.org/Volume_8/Abstracts/v8i1n2.abs.tex

[Shallit and Wang 2002]   J. O. Shallit and M.-w. Wang. On two-sided infinite fixed points of morphisms. *Theoret. Comput. Sci.* **270** (2002), 659–675.

[Shannon 1950]   C. E. Shannon. A symmetrical notation for numbers. *Amer. Math. Monthly* **57** (1950), 90–93.

[Shapiro 1952]   H. S. Shapiro. Extremal problems for polynomials and power series. Master's thesis, MIT, 1952.

[Sharif 1993]   H. Sharif. Children products of formal power series. *Math. Japonica* **38** (1993), 319–324.

[Sharif and Woodcock 1988]   H. Sharif and C. F. Woodcock. Algebraic functions over a field of positive characteristic and Hadamard products. *J. London Math. Soc.* **37** (1988), 395–403.

[Shechtman, Blech, Gratias, and Cahn 1984]   D. Shechtman, I. Blech, D. Gratias, and J. W. Cahn. Metallic phase with long-range orientational order and no translational symmetry. *Phys. Rev. Lett.* **53** (1984), 1951–1953.

[Shelton 1981a]   R. Shelton. Aperiodic words on three symbols. *J. Reine Angew. Math.* **321** (1981), 195–209.

[Shelton 1981b]   R. Shelton. Aperiodic words on three symbols. II. *J. Reine Angew. Math.* **327** (1981), 1–11.

[Shelton 1983]   R. O. Shelton. On the structure and extendibility of square-free words. In L. J. Cummings, editor, *Combinatorics on Words: Progress and Perspectives*, pp. 101–118. Academic Press, 1983.

[Shelton and Soni 1982]   R. O. Shelton and R. P. Soni. Aperiodic words on three symbols. III. *J. Reine Angew. Math.* **330** (1982), 44–52.

[Shelton and Soni 1985]   R. O. Shelton and R. P. Soni. Chains and fixing blocks in irreducible binary sequences. *Discrete Math.* **54** (1985), 93–99.

[Shen and Huang 1996]   C. Shen and Y. Huang. Some properties of C^∞-words with applications. *Southeast Asian Bull. Math.* **20** (1996), 19–30.

[Shepherd, Van Eetvelt, Wyatt-Millington, and Barton 1995]   S. J. Shepherd, P. W. J. Van Eetvelt, C. W. Wyatt-Millington, and S. K. Barton. Simple coding scheme to reduce peak factor in QPSK multicarrier modulation. *Electron. Lett.* **31** (1995), 1131–1132.

[Shiokawa 1974]   I. Shiokawa. g-adical analogues of some arithmetical functions. *Math. J. Okayama Univ.* **17** (1974), 75–94.

[Shiokawa 1974a]   I. Shiokawa. On the sum of digits of prime numbers. *Proc. Japan Acad.* **50** (1974), 551–554.

[Shiokawa 1974b]   I. Shiokawa. On a problem in additive number theory. *Math. J. Okayama Univ.* **16** (1974), 167–176.

[Shiu 1999]   P. Shiu. A function from Diophantine approximations. *Publ. Inst. Math. (Beograd)* **65** (1999), 52–62.

[Shur 1996a]   A. M. Shur. Binary words avoided by the Thue-Morse sequence. *Semigroup Forum* **53** (1996), 212–219.

[Shur 1996b]   A. M. Shur. Binary avoidability and Thue-Morse words. *Dokl. Akad. Nauk* **348** (1996), 598–599. In Russian. English translation in *Doklady Math.* **53** (1996), 405–406.

[Shyr 1977]   H. Shyr. A strongly primitive word of arbitrary length and its application. *Int. J. Comput. Math.* **6** (1977), 165–170.

[Shyr 1991]   H. J. Shyr. *Lecture Notes: Free Monoids and Languages*. Hon Min Book Company, Taiwan, 1991.

[Shyr 1996]   H. J. Shyr. Completely $\mu(n)$-reducible languages and n-annihilators of languages. *Soochow J. Math.* **22** (1996), 339–356.

[Shyr and Thierrin 1977]   H. J. Shyr and G. Thierrin. Disjunctive languages and codes. In M. Karpiński, editor, *Fundamentals of Computation Theory: Proceedings of the 1977 International FCT-Conference*, Vol. 56 of *Lecture Notes in Computer Science*, pp. 171–176. Springer-Verlag, 1977.

[Shyr and Tu 1991]   H. J. Shyr and F. K. Tu. Local distribution of non-primitive words. In *Proc. SEAMS Conf. on Ordered Structures and Algebra of Computer Languages*, pp. 202–217. 1991.

[Shyr and Yu 1994a]   H. J. Shyr and S. S. Yu. Languages defined by two functions. *Soochow J. Math.* **20** (1994), 279–296.

[Shyr and Yu 1994b]   H. J. Shyr and S. S. Yu. Non-primitive words in the language p^+q^+. *Soochow J. Math.* **20** (1994), 535–546.

[Shyr and Yu 1994c]   H. J. Shyr and S. S. Yu. Annihilators of languages. In M. Ito and H. Jürgensen, editors, *Words, Languages, and Combinatorics II*, pp. 406–414. World Scientific, 1994.

[Sierpiński 1916]   W. Sierpiński. On curves which contains the image of any given curve. *Mat. Sbornik* **30** (1916), 267–287. In Russian. French translation in *Oeuvres Choisies*, II, 107–119.

[Silber 1976]   R. Silber. A Fibonacci property of Wythoff pairs. *Fibonacci Quart.* **14** (1976), 380–384.

[Silberger 1971]   D. M. Silberger. Borders and roots of a word. *Portugal. Math.* **30** (1971), 191–199.

[Siromoney 1987]   R. Siromoney. Advances in array languages. In H. Ehrig, M. Nagl, G. Rozenberg, and A. Rosenfeld, editors, *Graph-Grammars and Their Application to Computer Science*, Vol. 291 of *Lecture Notes in Computer Science*, pp. 549–563. Springer-Verlag, 1987.

[Siromoney and Subramanian 1983]   R. Siromoney and K. G. Subramanian. Generative grammar for the cube-free abbey floor. *Bull. Eur. Assoc. Theor. Comput. Sci.*, No. 20 (June 1983), 160–162.

[Siromoney and Subramanian 1985]   R. Siromoney and K. G. Subramanian. Square-free and cube-free arrays. Technical Report MATH 15/85, Department of Mathematics, Madras Christian College, July 1985.

[Skolem 1957]   T. Skolem. On certain distributions of integers in pairs with given differences. *Math. Scand.* **5** (1957), 57–68.

[Skyum 1983]   S. Skyum. A solution to P12, number 18, October 1982. *Bull. Eur. Assoc. Theor. Comput. Sci.*, No. 19 (February 1983), 34–37.

[Slater 1950]   N. B. Slater. The distribution of the integers N for which $\{\theta N\} < \phi$. *Proc. Cambridge Phil. Soc.* **46** (1950), 525–534.

[Slater 1964]   N. B. Slater. Distribution problems and physical applications. *Compositio Math.* **16** (1964), 176–183.

[Slater 1967]   N. B. Slater. Gaps and steps for the sequence $n\theta$ mod 1. *Proc. Cambridge Phil. Soc.* **63** (1967), 1115–1123.

[Slekys and Avizienis 1978]   A. G. Slekys and A. Avizienis. A modified bi-imaginary number system. In *Proc. 4th IEEE Symp. Computer Arithmetic*, pp. 48–55. IEEE Press, 1978.

[Sloane 1973]   N. J. A. Sloane. The persistence of a number. *J. Recreational Math.* **6** (1973), 97–98.

[Smith 1876]   H. J. S. Smith. Note on continued fractions. *Messenger Math.* **6** (1876), 1–14.

[Smyth 2000]   W. F. Smyth. Repetitive perhaps, but certainly not boring. *Theoret. Comput. Sci.* **249** (2000), 343–355.

[Solinas 1989]   J. A. Solinas. On the joint distribution of digital sums. *J. Number Theory* **33** (1989), 131–151.

[Solomyak 1992]   B. Solomyak. Substitutions, adic transformations, and beta-expansions. In P. Walters, editor, *Symbolic Dynamics and Its Applications*, Vol. 135 of *Contemporary Math.*, pp. 361–372. Amer. Math. Soc., 1992.

[Songster 1956]   G. F. Songster. Investigation of a number system using -2 as a base for use in automatic digital computers. Master's thesis, University of Pennsylvania, Moore School of Electrical Engineering, Philadelphia, June 1956.

[Songster 1962]   G. F. Songster. Some representation systems for the real numbers. PhD thesis, University of Pennsylvania, Philadelphia, May 1962.

[Songster 1963]   G. F. Songster. Negative-base number-representation systems. *IEEE Trans. Electr. Comput.* **12** (1963), 274–277.

[Sós 1957]   V. T. Sós. On the theory of diophantine approximations. I. *Acta Math. Acad. Sci. Hung.* **8** (1957), 461–471.

[Sós 1958]   V. T. Sós. On the distribution mod 1 of the sequence $n\alpha$. *Ann. Univ. Sci. Budapest Eötvös Sect. Math.* **1** (1958), 127–134.

[Spencer 1952]   S. M. Spencer, Jr. Transcendental numbers over certain function fields. *Duke Math. J.* **19** (1952), 93–105.

[Sprague 1938]   R. Sprague. Ein Satz über Teilfolgen der Reihe der natürlichen Zahlen. *Math. Annalen* **115** (1938), 153–156.

[Sproull 1982]   R. F. Sproull. Using program transformations to derive line-drawing algorithms. *ACM Trans. on Graphics* **1** (1982), 259–273.

[Steacy 1996]   R. Steacy. Structure in the Kolakoski sequence. *Bull. Eur. Assoc. Theor. Comput. Sci.*, No. 59 (June 1996), 173–182.

[Stein 1982]   A. H. Stein. Exponential sums of an iterate of the binary sum-of-digit function. *Indiana Univ. Math. J.* **31** (1982), 309–315.

[Stein and Stux 1978]   A. H. Stein and I. E. Stux. A mean value theorem for binary digits. *Pacific J. Math.* **75** (1978), 565–577.

[B. Stewart 1939]   B. M. Stewart. Problem 3918. *Amer. Math. Monthly* **46** (1939), 363. Solution by J. S. Frame and B. M. Stewart, **48** (1941), 216–219.

[B. Stewart 1960]   B. M. Stewart. Sums of functions of digits. *Canad. J. Math.* **12** (1960), 374–389.

[C. Stewart 1980]   C. Stewart. On the representation of an integer in two different bases. *J. Reine Angew. Math.* **319** (1980), 63–72.

[I. Stewart 1995]   I. Stewart. Mathematical recreations: the never-ending chess game. *Scientific American* **273**(4) (1995), 182–183. Addenda, **274** (3) (March 1996), 109.

[Stockmeyer 1994]   P. K. Stockmeyer. Variations on the four-post Tower of Hanoi puzzle. *Congr. Numer.* **102** (1994), 3–12.

[Stolarsky 1976]   K. B. Stolarsky. Beatty sequences, continued fractions, and certain shift operators. *Canad. Math. Bull.* **19** (1976), 473–482.

[Stolarsky 1977]   K. B. Stolarsky. Power and exponential sums of digital sums related to binomial coefficient parity. *SIAM J. Appl. Math.* **32** (1977), 717–730.

[Stolarsky 1978]   K. B. Stolarsky. The binary digits of a power. *Proc. Amer. Math. Soc.* **71** (1978), 1–5.

[Stolarsky 1979]   K. B. Stolarsky. The number of bits in a product of odd integers. In M. B. Nathanson, editor, *Number Theory: Carbondale, 1979*, Vol. 751 of *Lecture Notes in Mathematics*, pp. 283–293. Springer-Verlag, 1979.

[Stolarsky 1980]   K. B. Stolarsky. Integers whose multiples have anomalous digit frequencies. *Acta Arith.* **38** (1980), 117–128.

[Stolarsky 1991]   K. B. Stolarsky. From Wythoff's nim to Chebyshev's inequality. *Amer. Math. Monthly* **98** (1991), 889–900.

[Stolboushkin 1983]   A. P. Stolboushkin. Regular dynamic logic is not interpretable in deterministic context-free dynamic logic. *Inform. Control* **58** (1983), 94–107.

[Stolboushkin and Taitslin 1983]   A. P. Stolboushkin and M. A. Taitslin. Deterministic dynamic logic is strictly weaker than dynamic logic. *Inform. Control* **57** (1983), 48–55.

[Strauss and Shallit 1990]   N. Strauss and J. Shallit. Advanced problem 6625. *Amer. Math. Monthly* **97** (1990), 252. Solution by D. Zagier, **99** 1992, 66–69.

[Sundar Raj and Koplowitz 1986]   P. A. Sundar Raj and J. Koplowitz. On bit reduction of chain coded line drawings. *Pattern Recog. Lett.* **4** (1986), 99–102.

[Surányi 1958]   J. Surányi. On the distribution mod 1 of the sequence $n\alpha$. *Ann. Univ. Sci. Budapest Eötvös Sect. Math.* **1** (1958), 107–111.

[Świerczkowski 1958]   S. Świerczkowski. On successive settings of an arc on the circumference of a circle. *Fundamenta Math.* **46** (1958), 187–189.

[Szegö 1922]   G. Szegö. Über Potenzreihen mit endlich vielen verschiedenen Koeffizienten. *Sitzungsber. Preuss. Akad. Wiss. Phys.-Math. Klasse* (1922), 88–91. Reprinted in *Collected Papers*, Vol. 1, 557–561.

[Szüsz 1985]   P. Szüsz. On a theorem of Fraenkel, Levitt, and Shimshoni. *Discrete Math.* **56** (1985), 75–77.

[Tamura 1992]   J.-I. Tamura. Transcendental numbers having explicit g-adic and Jacobi–Perron expansions. *Sém. Théorie Nombres Bordeaux* **4** (1992), 75–95.

[Tamura 1995]   J.-I. Tamura. A class of transcendental numbers having explicit g-adic and Jacobi–Perron expansions of arbitrary dimension. *Acta Arith.* **71** (1995), 301–329.

[Tamura 1999]   J.-I. Tamura. Padé approximation for words generated by certain substitutions, and Hankel determinants. In S. Kanemitsu and K. Györy, editors, *Number Theory and Its Applications*, pp. 309–346. Kluwer, 1999.

[Tamura 2000]   J.-I. Tamura. Certain partitions of a lattice. In J.-M. Gambaudo, P. Hubert, P. Tisseur, and S. Vaienti, editors, *Dynamical Systems: From Crystal to Chaos*, pp. 199–219. World Scientific, 2000.

[Tang 1963]   S. C. Tang. An improvement and generalization of Bellman–Shapiro's theorem on a problem in additive number theory. *Proc. Amer. Math. Soc.* **14** (1963), 199–204.

[Tao 1994]   R. Tao. Extended states in aperiodic systems. *J. Phys. A: Math. Gen.* **27** (1994), 5069–5077.

[Tapsoba 1987]   T. Tapsoba. Complexité de suites automatiques. PhD thesis, Université Aix-Marseille II, 1987. Thèse de troisième cycle.

[Tapsoba 1994]   T. Tapsoba. Automates calculant la complexité de suites automatiques. *J. Théorie Nombres Bordeaux* **6** (1994), 127–134.

[Tapsoba 1995]   T. Tapsoba. Minimum complexity of automatic non-Sturmian sequences. *RAIRO Inform. Théor. Appl.* **29** (1995), 285–291.

[Tenenbaum 1995]   G. Tenenbaum. *Introduction to Analytic and Probabilistic Number Theory*. Cambridge University Press, 1995.

[Tenenbaum 1997]   G. Tenenbaum. Sur la non-dérivabilité de fonctions périodiques associées à certaines formules sommatoires. In R. L. Graham and J. Nešetřil, editors, *The Mathematics of Paul Erdős*, pp. 117–128. Springer-Verlag, 1997.

[Terr 1996]   D. C. Terr. On the sums of digits of Fibonacci numbers. *Fibonacci Quart.* **34** (1996), 349–355.

[Thakur 1996b]   D. S. Thakur. Automata-style proof of Voloch's result on transcendence. *J. Number Theory* **58** (1996), 60–63.

[Thakur 1996c]   D. S. Thakur. Transcendence of gamma values for $\mathbb{F}_q(T)$. *Ann. Math.* **144** (1996), 181–188.

[Thakur 1998]   D. S. Thakur. Automata and transcendence. In V. Kumar Murty and M. Waldschmidt, editors, *Number Theory. Ramanujan Mathematical Society, January 3–6, 1996*, Vol. 210 of *Contemporary Math.*, pp. 387–399. Amer. Math. Soc., 1998.

[Thakur 1999]   D. S. Thakur. Diophantine approximation exponents and continued fractions for algebraic power series. *J. Number Theory* **79** (1999), 284–291.

[Thompson 1959]   G. C. Thompson. Elementary problem proposal E 1371. *Amer. Math. Monthly* **66** (1959), 512. Solution, **67** (1960), 183.

[Thue 1906]   A. Thue. Über unendliche Zeichenreihen. *Norske vid. Selsk. Skr. Mat. Nat. Kl.* **7** (1906), 1–22. Reprinted in *Selected Mathematical Papers of Axel Thue*, T. Nagell, editor, Universitetsforlaget, Oslo, 1977, pp. 139–158.

[Thue 1912]   A. Thue. Über die gegenseitige Lage gleicher Teile gewisser Zeichenreihen. *Norske vid. Selsk. Skr. Mat. Nat. Kl.* **1** (1912), 1–67. Reprinted in *Selected Mathematical Papers of Axel Thue*, T. Nagell, editor, Universitetsforlaget, Oslo, 1977, pp. 413–478.

[Thuswaldner 1998a]   J. M. Thuswaldner. The sum of digits function in number fields. *Bull. Lond. Math. Soc.* **30** (1998), 37–45.

[Thuswaldner 1998b]   J. M. Thuswaldner. Fractal dimension of sets induced by bases of imaginary quadratic fields. *Math. Slovaca* **48** (1998), 365–371.

[Thuswaldner 1999a]   J. M. Thuswaldner. The sum of digits function in number fields: distribution in residue classes. *J. Number Theory* **74** (1999), 111–125.

[Thuswaldner 1999b]   J. M. Thuswaldner. Summatory functions of digital sums occurring in cryptography. *Period. Math. Hung.* **38** (1999), 111–130.

[Thuswaldner 2000]   J. M. Thuswaldner. The complex sum of digits function and primes. *J. Théorie Nombres Bordeaux* **12** (2000), 133–146.

[Tichy and Turnwald 1986]   R. F. Tichy and G. Turnwald. On the discrepancy of some special sequences. *J. Number Theory* **26** (1986), 351–366.

[Tijdeman 1999]   R. Tijdeman. On the minimal complexity of infinite words. *Indag. Math.* **10** (1999), 123–129.

[Toeplitz 1928]   O. Toeplitz. Beispiele zur Theorie der fastperiodischen Funktionen. *Math. Annalen* **98** (1928), 281–295.

[Tompkins and Wakelin 1950]   C. B. Tompkins and J. H. Wakelin. *High-Speed Computing Devices*. McGraw-Hill, 1950. Reprinted by Tomash Publishers, 1983.

[Tong 1997]   P. Tong. Structure of the electronic energy spectrum of the one-dimensional nondiagonal Thue-Morse lattice. *Phys. Lett. A* **228** (1997), 195–201.

[Toshimitsu 1997]   T. Toshimitsu. q-additive functions and algebraic independence. *Arch. Math. (Basel)* **69** (1997), 112–119.

[Toshimitsu 1998]   T. Toshimitsu. Strongly q-additive functions and algebraic independence. *Tokyo J. Math.* **21** (1998), 107–113.

[Trakhtenbrot 1964]   B. A. Trakhtenbrot. On an estimate for the weight of a finite tree. *Sibirskii Mat. Zh.* **5** (1964), 186–191. In Russian.

[Trigg 1949]   C. W. Trigg. Elementary problem proposal E 887. *Amer. Math. Monthly* **56** (1949), 632. Solution by N. G. Gunderson, **57** (1950), 338.

[Troi and Zannier 1995]   G. Troi and U. Zannier. Note on the density constant in the distribution of self-numbers. *Boll. Unione Mat. Italiana, Ser. A* **9-A** (1995), 143–148.

[Trollope 1967]   J. R. Trollope. Generalized bases and digital sums. *Amer. Math. Monthly* **74** (1967), 690–694.

[Trollope 1968]   J. R. Trollope. An explicit expression for binary digital sums. *Math. Mag.* **41** (1968), 21–25.

[Tromp and Shallit 1995]   J. Tromp and J. Shallit. Subword complexity of a generalized Thue–Morse word. *Inform. Process. Lett.* **54** (1995), 313–316.

[Trotter and Winkler 1987]   W. T. Trotter and P. Winkler. Arithmetic progressions in partially ordered sets. *Order* **4** (1987), 37–42.

[Turban, Berche, and Berche 1994]   L. Turban, P.-E. Berche, and B. Berche. Surface magnetization of aperiodic Ising systems: a comparative study of the bond and site problems. *J. Phys. A: Math. Gen.* **27** (1994), 6349–6366.

[Turing 1936]   A. M. Turing. On computable numbers, with an application to the Entscheidungsproblem. *Proc. Lond. Math. Soc.* **42** (1936), 230–265.

[Turner 1988]   J. C. Turner. Fibonacci word patterns and binary sequences. *Fibonacci Quart.* **26** (1988), 233–246.

[Turner 1989]   J. C. Turner. The alpha and the omega of the Wythoff pairs. *Fibonacci Quart.* **27** (1989), 76–86.

[Twaddle 1963]   R. D. Twaddle. A look at the base negative ten. *Math. Teacher* **56** (1963), 88–90.

[Uchida 1999]   Y. Uchida. On p and q-additive functions. *Tokyo J. Math.* **22** (1999), 83–97.

[Urzyczyn 1983]   P. Urzyczyn. Nontrivial definability by flow-chart programs. *Inform. Control* **58** (1983), 59–87.

[Uspensky 1927]   J. V. Uspensky. On a problem arising out of the theory of a certain game. *Amer. Math. Monthly* **34** (1927), 516–521.

[Uspensky 1946a]   J. V. Uspensky. Sobre un problema de Juan Bernoulli. *Rev. Un. Mat. Argentina* **11** (1946), 141–154, 165–183, 239–255. In Spanish.

[Uspensky 1946b]   J. V. Uspensky. Sobre un problema de Juan Bernoulli. *Rev. Un. Mat. Argentina* **12** (1946), 10–19. In Spanish.

[Uspensky and Heaslet 1939]   J. V. Uspensky and M. A. Heaslet. *Elementary Number Theory*. McGraw-Hill, New York, 1939.

[Vaidya 1969]   A. M. Vaidya. On Kaprekar's tests for self numbers. *Math. Student* **37** (1969), 212–214.

[Vanden Eynden 1995]   C. Vanden Eynden. Problem 10439. *Amer. Math. Monthly* **102** (1995), 273. Solution by L. L. Foster and J. W. Grossman, **104** (1995), 873.

[Vandeth 2000]   D. Vandeth. Sturmian words and words with a critical exponent. *Theoret. Comput. Sci.* **242** (2000), 283–300.

[Veech 1984a]   W. A. Veech. The metric theory of interval exchange transformations I. Generic spectral properties. *Amer. J. Math.* **106** (1984), 1331–1359.

[Veech 1984b]   W. A. Veech. The metric theory of interval exchange transformations II. Approximation by primitive interval exchanges. *Amer. J. Math.* **106** (1984), 1361–1387.

[Veech 1984c]   W. A. Veech. The metric theory of interval exchange transformations III. The Sah–Arnoux–Fathi invariant. *Amer. J. Math.* **106** (1984), 1389–1422.

[Venkov 1970]   B. A. Venkov. *Elementary Number Theory*. Wolters-Noordhoff, Groningen, 1970.

[Verraedt, De Bra, and Gyssens 1983]   R. Verraedt, P. De Bra, and M. Gyssens. A solution to P12. *Bull. Euro. Assoc. Theor. Comput. Sci.*, No. 19 (February 1983), 37–39.

[Villemaire 1992a]   R. Villemaire. Joining k- and l-recognizable sets of natural numbers. In *STACS 92, Proc. 9th Symp. Theoretical Aspects of Comp. Sci.*, Vol. 577 of *Lecture Notes in Computer Science*, pp. 83–94. Springer-Verlag, 1992.

[Villemaire 1992b]   R. Villemaire. The theory of $\langle \mathbf{N}, +, V_k, V_l \rangle$ is undecidable. *Theoret. Comput. Sci.* **106** (1992), 337–349.

[Villemaire 1992c]   R. Villemaire. $\langle \mathbb{N}, +, V_2, V_3 \rangle$ est indécidable. *C. R. Acad. Sci. Paris* **314** (1992), 775–777.

[Voss 1991]   K. Voss. Coding of digital straight lines by continued fractions. *Computers and Artificial Intelligence* **10** (1991), 75–80.

[Vuillon 1998]   L. Vuillon. Combinatoire des motifs d'une suite sturmienne bidimensionnelle. *Theoret. Comput. Sci.* **209** (1998), 261–285.

[Wade 1941]   L. I. Wade. Certain quantities transcendental over $GF(p^n, x)$. *Duke Math. J.* **8** (1941), 701–720.

[Wade 1943]   L. I. Wade. Certain quantities transcendental over $GF(p^n, x)$ (2). *Duke Math. J.* **10** (1943), 587–594.

[Wade 1944]   L. I. Wade. Two types of function field transcendental numbers. *Duke Math. J.* **11** (1944), 755–758.

[Wade 1946]   L. I. Wade. Transcendence properties of the Carlitz ψ-functions. *Duke Math. J.* **13** (1946), 79–85.

[Wadel 1957]   L. B. Wadel. Negative base number systems. *IRE Trans. Electron. Comput.* **6** (1957), 123.

[Wadel 1961]   L. B. Wadel. Conversion from conventional to negative-base number representation. *IRE Trans. Electron. Comput.* **10** (1961), 779.

[van der Waerden 1986]   B. L. van der Waerden. Verallgemeinerung eines Satzes von Fatou. *Österreich. Akad. Wiss. Math.-Natur. Kl. Sitzungsber. II* **195** (1986), 191–194.

[Walsh 1982]   T. R. Walsh. The towers of Hanoi revisited: moving the rings by counting the moves. *Inform. Process. Lett.* **15** (1982), 64–67.

[Walsh 1983a]   T. R. Walsh. Iteration strikes back – at the cyclic Towers of Hanoi. *Inform. Process. Lett.* **16** (1983), 91–93.

[Walsh 1983b]   T. R. Walsh. A case for iteration. *Congr. Numer.* **40** (1983), 409–417.

[Walsh 1998]   T. R. Walsh. The generalized towers of Hanoi for space-deficient computers and forgetful humans. *Math. Intelligencer* **20** (1998), 32–38.

[Wang and Shallit 1998]   M.-w. Wang and J. Shallit. On minimal words with given subword complexity. *Electronic J. Combinatorics* **5** (1998), #R35 (electronic), `http://www.combinatorics.org/Volume_5/Abstracts/v5i1r35.html`

[Wang and Shallit 1999]   M.-w. Wang and J. Shallit. An inequality for non-negative matrices. *Linear Algebra Appl.* **290** (1999), 135–144.

[Wantiez 1994]   P. Wantiez. Identités dans les groupes d'exposant quatre. *Commun. Algebra* **22** (1994), 4033–4050.

[Wantiez 1995]   P. Wantiez. Quelques resultats sur les m-identités. *Commun. Algebra* **23** (1995), 4375–4392.

[Weakley 1989]   W. D. Weakley. On the number of C^∞-words of each length. *J. Combin. Theory. Ser. A* **51** (1989), 55–62.

[de Weger 1991]   B. M. M. de Weger. Elementary problem E 3470. *Amer. Math. Monthly* **98** (1991), 955. Solution by M. Vowe and O. P. Lossers, **101** (1994), 83–84. Also see **102** (1995), 936.

[Wegner 1982]   L. Wegner. Problem P12: is www^R cube-free? *Bull. Eur. Assoc. Theor. Comput. Sci.*, No. 18, (October 1982), 120.

[Wells 1963]   C. H. Wells, Jr. Using a negative base for number notation. *Math. Teacher* **56** (1963), 91–93.

[Wen and Wen 1994a]   Z.-X. Wen and Z.-Y. Wen. Some properties of the singular words of the Fibonacci word. *European J. Combinatorics* **15** (1994), 587–598.

[Wen and Yao 2002]   Z.-Y. Wen and J.-Y. Yao. Transcendence, automata theory and gamma functions for polynomial rings. *Acta Arith.* **101** (2002), 39–51.

[Wilansky 1982]   A. Wilansky. Smith numbers. *Two-Year College Math. J.* **13** (1982), 21.

[Wilf 1968]   H. S. Wilf. Hadamard determinants, Möbius functions, and the chromatic number of a graph. *Bull. Amer. Math. Soc.* **74** (1968), 960–964.

[Wilson and Shallit 1992]   D. Wilson and J. O. Shallit. The "$3x + 1$" problem and finite automata. *Bull. European Assoc. Theor. Comput. Sci.*, No. 46, (February 1992), 182–185.

[Wlazinski 2001]   F. Wlazinski. A test-set for k-power-free binary morphisms. *RAIRO Inform. Théor. App.* **35** (2001), 437–452.

[Wolfram 1984]   S. Wolfram. Computation theory of cellular automata. *Commun. Math. Phys.* **96** (1984), 15–57.

[Wolfram 2002]   S. Wolfram. *A New Kind of Science.* Wolfram Media, 2002.

[Wood 1981]   D. Wood. The towers of Brahma and Hanoi revisited. *J. Recreational Math.* **14** (1981), 17–24.

[Wood 1983]   D. Wood. Adjudicating a Towers of Hanoi contest. *Internat. J. Comput. Math.* **14** (1983), 199–207.

[Woodcock and Sharif 1989]   C. F. Woodcock and H. Sharif. On the transcendence of certain series. *J. Algebra* **121** (1989), 364–369.

[Woodcock and Sharif 1990]   C. F. Woodcock and H. Sharif. Hadamard products of rational formal power series. *J. Algebra* **128** (1990), 517–527.

[Woods 1978]   D. R. Woods. Elementary problem proposal E 2692. *Amer. Math. Monthly* **85** (1978), 48. Solution by D. Robbins, **86** (1979), 394–395.

[Wozny and Zamboni 2001]   N. Wozny and L. Q. Zamboni. Frequencies of factors in Arnoux-Rauzy sequences. *Acta Arith.* **96** (2001), 261–278.

[Wright 1948]   E. M. Wright. The Prouhet–Lehmer problem. *J. London Math. Soc.* **23** (1948), 279–285.

[Wright 1959]   E. M. Wright. Prouhet's 1851 solution of the Tarry–Escott problem of 1910. *Amer. Math. Monthly* **66** (1959), 199–201.

[J. Wu and Chen 1992]   J.-S. Wu and R.-J. Chen. The towers of Hanoi problem with parallel moves. *Inform. Process. Lett.* **44** (1992), 241–243.

[J. Wu and Chen 1993]   J.-S. Wu and R.-J. Chen. The towers of Hanoi problem with cyclic parallel moves. *Inform. Process. Lett.* **46** (1993), 1–6.

[L. Wu 1980]   L.-D. Wu. On the Freeman's conjecture about the chain code of a line. In *Proc. 5th Int. Conf. on Pattern Recognition*, Vol. 1, pp. 32–34. IEEE Press, 1980.

[L. Wu 1982]   L.-D. Wu. On the chain code of a line. *IEEE Trans. Pattern Anal. Machine Intell.* **4** (1982), 347–353.

[Wythoff 1907]   W. A. Wythoff. A modification of the game of nim. *Nieuw Archief voor Wiskunde* **7** (1907), 199–202.

[Xie 1996]   H. Xie. *Grammatical Complexity and One-Dimensional Dynamical Systems.* World Scientific, 1996.

[Yaglom and Yaglom 1967]   A. M. Yaglom and I. M. Yaglom. *Challenging Mathematical Problems with Elementary Solutions*, Vol. II. Holden-Day, 1967. Translated by J. McCawley.

[Yao 1997a]   J.-Y. Yao. Critères de non-automaticité et leurs applications. *Acta Arith.* **80** (1997), 237–248.

[Yao 1997b]   J.-Y. Yao. Généralisations de la suite de Thue–Morse. *Ann. Sci. Math. Québec* **21** (1997), 177–189.

[Yarlagadda and Hershey 1982]   R. Yarlagadda and J. Hershey. A note on the eigenvectors of Hadamard matrices of order 2^n. *Linear Algebra Appl.* **45** (1982), 43–53.

[Yasutomi 1996]   S.-I. Yasutomi. The complexity of generalized Sturmian sequences. *Tokyo J. Math.* **19** (1996), 155–168.

[Yazdani 2001]   S. Yazdani. Multiplicative functions and k-automatic sequences. *J. Théorie Nombres Bordeaux* **13** (2001), 651–658.

[J. Yu 1991]   J. Yu. Transcendence and special zeta values in characteristic *p*. *Ann. Math.* **134** (1991), 1–23.

[J. Yu 1992]   J. Yu. Transcendence in finite characteristic. In D. Goss, D. R. Hayes, and M. I. Rosen, editors, *The Arithmetic of Function Fields*, pp. 253–264. Walter de Gruyter, Berlin, 1992.

[S. Yu 1997]   S. Yu. Regular languages. In G. Rozenberg and A. Salomaa, editors, *Handbook of Formal Languages*, Vol. 1, pp. 41–110. Springer-Verlag, 1997.

[X. Yu 1995]   X. Yu. A new solution for Thue's problem. *Inform. Process. Lett.* **54** (1995), 187–191.

[Yuen 1975]   C. K. Yuen. A note on base −2 arithmetic logic. *IEEE Trans. Comput.* **C-24** (1975), 325–329.

[Zannier 1982]   U. Zannier. On the distribution of self-numbers. *Proc. Amer. Math. Soc.* **85** (1982), 10–14.

[Zariski and Samuel 1960]   O. Zariski and P. Samuel. *Commutative Algebra*, Vol. II. D. Van Nostrand, 1960.

[Zech 1958]   T. Zech. Wiederholungsfreie Folgen. *Z. Angew. Math. Mech.* **38** (1958), 206–209.

[Zeckendorf 1972]   E. Zeckendorf. Représentation des nombres naturels par une somme de nombres de Fibonacci ou de nombres Lucas. *Bull. Soc. Roy. Liége* **41** (1972), 179–182.

[Zhong, Yan, and You 1991]   J. X. Zhong, J. R. Yan, and J. Q. You. Electronic properties of the generalized Thue–Morse lattices: a dynamical-map approach. *J. Phys.: Condens. Matter* **3** (1991), 6293–6298.

[Zhong, You, and Yan 1992]   J. X. Zhong, J. Q. You, and J. R. Yan. The exact Green function of a one-dimensional Thue-Morse lattice. *J. Phys.: Condens. Matter* **4** (1992), 5959–5965.

[Zimin 1982]   A. I. Zimin. Blocking sets of terms. *Mat. Sbornik* **119** (1982), 363–375, 447. In Russian. English translation in *Math. USSR Sbornik*, **47** (1984), 353–364.

[Zohar 1970]   S. Zohar. Negative radix conversion. *IEEE Trans. Comput.* **C-19** (1970), 222–226. See comments by Wadel, Pawlak, and Zohar in **C-20** (1971), 587.

Index